Ethnic Fermented Foods and Beverages of India: Science History and Culture

Jyoti Prakash Tamang
Editor

Ethnic Fermented Foods and Beverages of India: Science History and Culture

Springer

Editor
Jyoti Prakash Tamang
School of Sciences, Department of Microbiology
Sikkim University
Gangtok, Sikkim, India

ISBN 978-981-15-1485-2 ISBN 978-981-15-1486-9 (eBook)
https://doi.org/10.1007/978-981-15-1486-9

This Springer imprint is published by the registered company Springer Nature Singapore Pte Ltd.
The registered company address is: 152 Beach Road, #21-01/04 Gateway East, Singapore 189721, Singapore

Preface

India has one of the oldest civilizations in the world with vast history, food culture, bio-resources, and diverse ethnicity. India is administratively divided into 28 states and 9 union territories and is the world's second largest populated country with >120 billion people speaking more than 1652 languages, 216 dialects, and 23 official languages. Ethnic food has a cultural connotation in India and linked to diverse ethnicity which represents the traits of wisdom and knowledge of ancient Indian people on culinary and also on the right choice of "bio-actively enriched balance diets" since *Rig Vedic* period around 1500 BCE to 1000 BCE. Gastronomy of Indian ethic foods is based on *Vedic* food habits with adoption of some Mughal and English cuisines, the present-day Indian foods of more than 2000 diverse ethnic groups of people are unparalleled and unique to each other. It is remarkable to note that unlike many countries India has more than 350 diverse types of major or common and minor or region-specific ethnic fermented foods and beverages (234 fermented foods, 84 alcoholic beverages, and 32 traditionally prepared amylolytic starters) which are traditionally prepared and consumed by ethnic people of India for 5000 years. Indian gastronomy and culinary with huge diversity of ethnic foods are always pride of our rich dietary culture and heritage.

This book has 23 chapters covering different types of ethnic fermented foods and alcoholic beverages of all states of India. I am grateful to all contributing authors and co-authors who accepted our invitation to write this book. I could bring all of them in the same platform who have been working on various aspects of Indian fermented foods and beverages. I am also grateful to Springer Nature for publishing this comprehensive book on Indian ethnic foods. We hope this book will be read by researchers, students, teachers, nutritionists, dieticians, food entrepreneurs, agriculturalist, government policy makers, ethnologists, sociologists, and electronic media persons who keep interest on the health benefits of fermented foods and beverages. This book *Ethnic Fermented Foods and Beverages of India: Science History and Culture* is the first of this kind on compilation of various ethnic fermented foods and alcoholic beverages of India.

I dedicate this book to our ancestors who invented the technology, knowledge, wisdom, and skill of food fermentation and created the platform of "ethno-microbiology" for research to study in depth of metagenomics and determination of bioactive molecules in ethnic fermented foods and beverages contributing health benefits to the consumers.

Gangtok, India

Jyoti Prakash Tamang

Contents

Editor and Contributors

About the Editor

Jyoti Prakash Tamang is one of the pioneer microbiologists working on the interpretation of "ethno-microbiology" to molecular food microbiology associated with age-old ethnic fermented foods and beverages of India, Nepal, and Bhutan, a field he has investigated for the past 33 years using a range of metagenomics/omics and bioinformatics tools. He has also sparked significant interest in ethnic fermented foods by establishing knowledge-based resources and promoting academic interest in India. He has more than 154 publications to his credit, including several books with international publishers—*Himalayan Fermented Foods: Microbiology, Nutrition, and Ethnic Values* (CRC Press, New York 2010), *Fermented Foods and Beverages of the World* (CRC Press, 2010), *Health Benefits of Fermented Foods and Beverages* (CRC Press, 2015), and *Ethnic Fermented Foods and Alcoholic Beverages of Asia* (Springer Nature, New Delhi 2016). He has received several national and international awards and is currently the senior-most Professor in Microbiology and Dean of Sikkim Central University, Gangtok, India. Prof. Tamang is the International Centre for Integrated Mountain Development (ICIMOD) Mountain Chair for the year 2019–21 and also Visiting Professor of National Institute of Industrial Science and Technology, Japan.

Contributors

T. Ajungla Department of Botany, Nagaland University, Zunheboto, Nagaland, India

Rehana Akhter Department of Food Science and Technology, University of Kashmir, Srinagar, Jammu and Kashmir, India

Uday Shriramrao Annapure Department of Food Engineering and Technology, Institute of Chemical Technology (ICT), Mumbai, Maharashtra, India

Usha Antony College of Fish Nutrition and Food Technology, Tamilnadu Dr J Jayalalithaa Fisheries University, Madhavaram, Chennai, Tamilnadu, India

K. A. Anu Appaiah Microbiology and Fermentation Technology Department, CSIR-Central Food Technological Research Institute, Mysuru, India

Ibemhal D. Asem Department of Biochemistry, Manipur University, Imphal, Manipur, India

Madhumita Barooah Department of Agricultural Biotechnology, Assam Agricultural University, Jorhat, Assam, India

Rwivoo Baruah Microbiology and Fermentation Technology Department, CSIR-Central Food Technological Research Institute, Mysuru, India

Esmil Beliya Pt. Ravishankar Shukla University, Raipur, Chhattisgarh, India

Arun Beniwal Chemical Biology Lab, Department of Biotechnology, Indian Institute of Technology Roorkee, Roorkee, Uttarakhand, India

Kailash N. Bhardwaj Uttarakhand State Council for Science and Technology (UCOST), Vigyan Dham, Jhajra, Dehradun, Uttarakhand, India

Keshani Bhushan Department of Microbiology, Punjab Agricultural University, Ludhiana, Punjab, India

Koel Biswas Microbiology Laboratory, Department of Biotechnology and Bioinformatics, North-Eastern Hill University, Shillong, Meghalaya, India

Sudipta Sankar Bora DBT-North East Centre for Agricultural Biotechnology, Jorhat, Assam, India

Ramachandran Chelliah Department of Food Science and Biotechnology, Kangwon National University, Chuncheon, Republic of Korea

Irene Furtado Department of Microbiology, Goa University, Taleigao, Goa, India

Aarti Suryakant Ghanate Department of Food Engineering and Technology, Institute of Chemical Technology (ICT), Mumbai, Maharashtra, India

Kuntal Ghosh Department of Biological Science, Midnapore City College, Paschim Medinipur, West Bengal, India

Tamoghna Ghosh Chemical Biology Lab, Department of Biotechnology, Indian Institute of Technology Roorkee, Roorkee, Uttarakhand, India

Gunajit Goswami DBT-North East Centre for Agricultural Biotechnology, Jorhat, Assam, India

A. G. Greeshma Dublin Health Services Management, Abu Dhabi, UAE

Prakash M. Halami Microbiology and Fermentation Technology Department, CSIR-Central Food Technological Research Institute, Mysuru, India

Prabodh Shirish Halde Marico Limited, Mumbai, Maharashtra, India

Shankar Ilango Department of Biotechnology and Centre for Food Technology, Anna University, Chennai, Tamilnadu, India

S. K. Jadhav Pt. Ravishankar Shukla University, Raipur, Chhattisgarh, India

Kumaraswamy Jeyaram Institute of Bioresources and Sustainable Development (IBSD), Aizawl, Mizoram, India

Santa Ram Joshi Microbiology Laboratory, Department of Biotechnology and Bioinformatics, North-Eastern Hill University, Shillong, Meghalaya, India

Sunita Kumari Kamal KVK, Palamu, Birsa Agricultural University, Ranchi, Jharkhand, India

Naveen Kango Department of Microbiology, Dr. Harisingh Gour Vishwavidyalaya, Sagar, Madhya Pradesh, India

S. S. Kanwar Department of Microbiology, CSK Himachal Pradesh Agriculture University, Palampur, Himachal Pradesh, India

Rupesh Kapale Department of Botany, Government PG College, Chhindwara, Madhya Pradesh, India

Santosh Keisam Advanced Level Institutional Biotech Hub, Modern College, Imphal, Manipur, India

Welfareson Khongriah Microbiology Laboratory, Department of Biotechnology and Bioinformatics, North-Eastern Hill University, Shillong, Meghalaya, India

Asangla Kichu Department of Botany, Nagaland University, Zunheboto, Nagaland, India

Bharathi Kulkarni Clinical Division, National Institute of Nutrition, Indian Council of Medical Research, Jamai Osmania, Hyderabad, Telangana, India

H. Lalhlenmawia Department of Pharmacy, Regional Institute of Paramedical and Nursing Sciences (RIPANS), Aizawl, Mizoram, India

Ranendra Kumar Majumdar College of Fisheries, Central Agricultural University (Imphal), Agartala, Tripura, India

F. A. Masoodi Department of Food Science and Technology, University of Kashmir, Srinagar, Jammu and Kashmir, India

Keshab Chandra Mondal Department of Microbiology, Vidyasagar University, Midnapore, West Bengal, India

Saswati Parua Mondal Department of Physiology, Bajkul Milani Mahavidyalaya, Purba Medinipur, West Bengal, India

Nimisha R. Nair Department of Integrated BSc-MSc Biotechnology (IBC), Kerala Agriculture University Vellayani, Thiruvananthapuram, Kerala, India

K. Madhavan Nampoothiri Microbial Processes and Technology Division, CSIR—National Institute for Interdisciplinary Science and Technology (NIIST), Trivandrum, Kerala, India

Naveen Kumar Navani Chemical Biology Lab, Department of Biotechnology, Indian Institute of Technology Roorkee, Roorkee, Uttarakhand, India

Debananda S. Ningthoujam Department of Biochemistry, Manipur University, Imphal, Manipur, India

Gloria Nyenthang Department of Botany, Nagaland University, Zunheboto, Nagaland, India

Ravindranadh Palika Biochemistry Division, National Institute of Nutrition, Indian Council of Medical Research, Jamai Osmania, Hyderabad, Telangana, India

Jai Shankar Paul Pt. Ravishankar Shukla University, Raipur, Chhattisgarh, India

Jashbhai B. Prajapati SMC College of Dairy Science, Anand Agricultural University, Anand, Gujarat, India

Biswajit Pramanik Laboratory of Biotechnology and Plant-Microbe Interaction, Department of Forestry, North Eastern Regional Institute of Science and Technology, Nirjuli, Arunachal Pradesh, India

Raghu Pullakhandam Biochemistry Division, National Institute of Nutrition, Indian Council of Medical Research, Jamai Osmania, Hyderabad, Telangana, India

Jeelani Raja Department of Food Science and Technology, University of Kashmir, Srinagar, Jammu and Kashmir, India

Teena Rajiv Clinical Division, National Institute of Nutrition, Indian Council of Medical Research, Jamai Osmania, Hyderabad, Telangana, India

Sudha Rani Ramakrishnan School of Food Science and Biotechnology, Kyungpook National University, Daegu, Republic of Korea

Sajad Ahmad Rather Department of Food Science and Technology, University of Kashmir, Srinagar, Jammu and Kashmir, India

Kavitha Ravichandran Department of Biotechnology and Centre for Food Technology, Anna University, Chennai, Tamilnadu, India

Bhaskar Jyoti Sharma Laboratory of Biotechnology and Plant-Microbe Interaction, Department of Forestry, North Eastern Regional Institute of Science and Technology, Nirjuli, Arunachal Pradesh, India

G. D. Sharma Atal Bihari Vajpayee University, Bilaspur, Chhattisgarh, India

Isha Sharma Department of Microbiology, Dr. Harisingh Gour Vishwavidyalaya, Sagar, Madhya Pradesh, India

Karuna Shrivastava Laboratory of Biotechnology and Plant-Microbe Interaction, Department of Forestry, North Eastern Regional Institute of Science and Technology, Nirjuli, Arunachal Pradesh, India

Seema Singh KVK, Dumka, Birsa Agricultural University, Ranchi, Jharkhand, India

Usha Singh Department of Food and Nutrition, Dr. Rajendra Prasad Central Agricultural University, Pusa, Bihar, India

M. P. Soumya Microbial Processes and Technology Division, CSIR—National Institute for Interdisciplinary Science and Technology (NIIST), Trivandrum, Kerala, India

V. Sreeja SMC College of Dairy Science, Anand Agricultural University, Anand, Gujarat, India

Jyoti Prakash Tamang Department of Microbiology, DAICENTRE (DBT-AIST International Centre for Translational and Environmental Research) and Bioinformatics Centre, School of Life Sciences, Sikkim University (Central University), Gangtok, Sikkim, India

Anand Singh Thangjam AICRP on Post-Harvest Engineering Technology, Directorate of Research, Central Agricultural University, Imphal, Manipur, India

K. Thanzami Department of Pharmacy, Regional Institute of Paramedical and Nursing Sciences (RIPANS), Aizawl, Mizoram, India

Namrata Thapa Biotech Hub, Department of Zoology, Nar Bahadur Bhandari Degree College, Gangtok, Sikkim, India

Shubhra Tiwari Pt. Ravishankar Shukla University, Raipur, Chhattisgarh, India

Sheryanne Velho-Pereira Department of Microbiology, St. Xavier's College, Mapusa, Goa, India

Romi Wahengbam Life Sciences Division, Institute of Advanced Study in Science and Technology (IASST), Guwahati, Assam, India

Touseef Ahmed Wani Department of Food Science and Technology, University of Kashmir, Srinagar, Jammu and Kashmir, India

Lydia Yeptho Department of Botany, Nagaland University, Zunheboto, Nagaland, India

History and Culture of Indian Ethnic Fermented Foods and Beverages

1

Jyoti Prakash Tamang

Abstract

"भारत" (Bharat) has one of the oldest civilizations in the world with vast history, food culture, bio-resources, ethnicity, and customs. Ethnic food has a cultural connotation in India and is linked to diverse ethnicity which represents the traits of wisdom and knowledge of ancient Indian people on culinary and also on the right choice of "bioactively enriched balance diets" since *Rig Vedic* period. Diversity of Indian fermented foods and beverages is related to diversity of ethnicity with unparalleled food culture of each community. More than 350 types of major and region-specific ethnic fermented foods and alcoholic beverages are produced either naturally or by adding mixed starter cultures using indigenous knowledge of food fermentation in India. Diversity of microorganisms ranges from mycelia fungi to enzyme-producing to alcohol-producing yeasts and Gram-positive and few Gram-negative bacteria with several functional properties. Functional microorganisms play important roles in the traditional fermentation processes by their functional properties enhancing several health-promoting benefits to the consumers such as bio-preservation of perishable foods, bio-enrichment of nutritional value, protective properties, bioavailability of minerals, production of antioxidants, antimicrobial activities, non-production of biogenic amines, and probiotics properties. Microbial diversity in ethnic fermented foods contributes significant genetic resources due to diverse food cultures of the multiethnic groups of people in India. It has been noticed that consumption of few uncommon ethnic foods is declining in many states of India due to change in life style, shifting from cultural food habit to commercial foods and fast foods affecting drastically the traditional culinary practices, and also due to climate change in some places.

J. P. Tamang (✉)
Department of Microbiology, School of Life Sciences, DAICENTER (DBT-AIST International Centre for Translational and Environmental Research) and Bioinformatics Centre, Sikkim University (Central University), Gangtok, India

J. P. Tamang (ed.), *Ethnic Fermented Foods and Beverages of India: Science History and Culture*, https://doi.org/10.1007/978-981-15-1486-9_1

Keywords
Food culture · Indian ethnic foods · Fermented foods · History of foods

1.1 Introduction

"भारत" (Bharat) in Sanskrit, an ancient name for modern India, has one of the oldest civilizations in the world with vast history, bio-resources, ethnicity, rich cultural heritage, culinary, cuisine, and customs. India with a total area of 3,287,263 km^2 is administratively divided into 28 states and 9 union territories and is the world's second largest populated country with >120 billion people speaking more than 1652 languages, 216 dialects, and 23 official languages (www.censusindia.gov.in). More than 2000 diverse ethnic groups of people are living in India (Majumder 2001). The dominant racial communities in India are Indo-Aryan (72%), Dravidian (25%), and Mongoloid and other minority groups (3%) (Sousa 2019). The main racial groups of Indian ethnic people belong to six groups, viz., Negrito, proto-Australoids or Austrics, Mongoloids, Mediterranean or Dravidian, Western Brachycephals, and Nordic Aryans according to Guha (1937). India is divided into six physiographic regions: (1) the Himalayas in north and northeast; (2) Peninsular Plateau with Aravalli, Vindhyachal and Satpura ranges, Eastern Ghats and Western Ghats and plateaus Malwa Plateau, Chhota Nagpur Plateau, Southern Granulite Terrain, Deccan and Kutch Kathiawar; (3) Indo-Gangetic Plain; (4) Thar Desert in West; (5) Coastal Plains: Eastern Ghats folds and Western Ghats folds; and (6) Islands—the Andaman and Nicobar Islands and the Lakshadweep Islands. Agricultural farming of locally available agro-resources, pastoral and animal husbandry, fishery, and agrarian system of traditional agriculture have been traced back to Indus Valley Civilization in India (Fuller 2015). Since then mixed-agriculture farming in all geomorphological areas under different agroclimatic conditions is being practiced by farmers in the field of animal husbandry, forestry, fishery, etc. to supplement daily food resources in local diets.

1.2 History of Indian Food Culture

Ethnic food has a cultural connotation in India and is linked to diverse ethnicity which represents the traits of wisdom and knowledge of ancient Indian people on culinary and also on the right choice of "bioactively enriched balance diets" since Rig Vedic period around 1500 BCE to 1000 BCE (Jamison and Brereton 2014) to present-day edibles. The genesis of Indian food culture goes back to Indus Valley Civilization also known as Harappan Civilization around 8000 years ago (Sarkar et al. 2016), where cultivation of cereal crops such as wheat and barley as staple diets as well as domestication of cattle for milk and meat is represented by pastoral and farming communities (Possehl 2002). Archaeological evidence suggests that the Harappan people might have invented baking of roasted wheat seeds on griddle to prepare *chapati* (baked wheat-based bread) as staple diet (Achaya 1994). Ancient Indian dish called *pulao/pallao* which is prepared by cooking together rice and meat (Achaya 1994) has been mentioned in *Yajnavalkya Smriti*, one of the texts of Hinduism composed in Sanskrit during 300–500 BCE (Olivelle 2006). The origin of Indian agriculture systems was mentioned in *Rig Veda* (Sanskrit hymns) with cultivations of vegetables, fruits, spice, etc. (Vishnu-Mittren 1978) which supplemented the local diets in ancient India. One of the oldest Hindu sacred texts in Sanskrit is the *Bhagavad Gita* which was written 5000 years ago. In Chap. 17, text (*sloka*) 8–10 of the *Bhagavad Gita* says that "every human being is born with innate faith," which can be of three kinds—*sāttvic* "सात्वकि" (mode of goodness), *rājasic* "राजसकि" (mode of passion), or *tāmasic* "तामसकि" (mode of ignorance). Sāttvic food denotes food for longevity, prosperity, intelligence, strength, health, and happiness, which includes fruits, vegetables, legumes, cereals, and sweets. *Rājasic* food signifies activity, passion, and restlessness, which include hot, sour, spicy, alliaceous plants *including onions* and *garlic* and salty foods. *Tāmasic* food is intoxicating and unhealthy, which generally causes dullness and passivity (Tamang and Samuel 2010). The influence of religions effected the Indian dietary culture toward vegetarianism due to the advent of Buddhism during 600 BCE (Harvey 2013) and also Jainism during 700 BCE (Dundas 2002) and also due to favorable agroclimatic conditions where a variety of fruits, vegetables, and grains could easily be grown throughout the year. Influence of *ayurdeva aahar* (Kulkarni 2002) comprising mostly milk and milk products, cereals as *chapati* or cooked rice "चावल", lentils/legumes "दाल", fruits and vegetables has been observed in Indian diets.

India has a long history of invasion from foreigners which has not only changed the traditional demography and culture but also drastically influenced the ancient Indian food cultures too, sometimes by accepting or sharing the foods of different communities due to acceptability or preferences of foods. Mughal Empire which ruled India from 1526 to 1857 Richards (1995) has introduced foods including many nonvegetarian dishes such as *biriyani*, *mughlai paratha*, *murgh musallam*, *kebabs*, *malai kofta*, *rezala*, etc. into Indian cuisine. East India Company ruled India for 100 years (1757–1858), and the British Crown ruled India for 89 years (1858–1947). British influenced the ethnic Indian dietary cultures and supplemented many English foods known as Anglo-Indian foods and culinary into Indian cuisine during the 189 years of British rule such as breads/loaves, soup, salad, fried/poached eggs, butter, potatoes, tomato, sweet potatoes, peanuts, squash, cabbage, cheese, porridge, puddings, sausages, hams, tea, and alcoholic beverages such as whisky, brandy, rum, and wine. The 189 years of British rule in India also influenced and changed the feeding practices by using cutlery, which was initially adopted by the affluent and educated Indians and now extensively and commonly adopted by new generations in modern India, ignoring the age-old practice of feeding by hands. The Portuguese ruled Goa for 451 years from 1510 to 1961 (Mitragotri 1992) and influenced the food habits of original Goans. Today the non-fermented Indian cuisine is the blend of ancient *Ayurveda aadhar* and Mughlai and Anglo-Indian foods. However ethnic fermented foods and beverages have not been influenced by Mughal or British cuisine, which has maintained its originality as ancient and heritage foods of India prepared and consumed by diverse ethnic communities.

Preparations and consumption of some ancient foods have been restricted due to religions and customary beliefs through food laws and taboos (Kwon and Tamang 2015). Hinduism practices the concept of purity and pollution, which determines interpersonal and intercaste relationships (Kilara and Iya 1992). Kitchens of the Brahmin Hindu produce two types of meals, *kaccha* "कच्चा," which means raw or uncooked, and *pakka* "पक्का" meaning ripe and cooked. *Kaccha* foods are regarded as inferior foods and are likely to be contaminated; hence to make it edible, there are strict codes of cooking, serving, and eating at kitchen. *Pakka* food is fried in traditionally prepared clarified butter called *ghee* so it is high-quality and nutritious foods. Hindus are traditionally vegetarians, but many non-Brahmins are nonvegetarians. Cow is a sacred animal for Hindus. Brahmin Hindus do not eat onion and garlic since these alliaceous plants are religiously classified as *tāmasic* or intoxicants. In the culinary history of India, *thali*, traditional platter used for serving all ingredients of foods, has been recorded in Indus Valley Civilization around 3500 BCE to 2500 BCE (INTACH 2016 www.intach.org). Historically, during Vedic period, *thali* was a ritual cooking in a pot to boil rice as *pakka* food (Achaya 2009). Later on, use of *thali* system with varieties of dishes serve in a round metal platter has become a part of Indian food culture. *Ahara* is a food that has been considered as a balanced nutritious diet in Ayurveda and is also defined as a substance which is swallowed through the throat after eating (Sawhney and Versha 2018). After the advent of Islam in India, consumption of foods is governed by Islamic dietary laws for Muslim as per Quran, the holy book of Muslim (Hussaini 1993). Islamic dietary laws forbid the consumption of meats of pig and dead animals, blood in any form, food previously offered to Gods, and alcohol and any intoxicant substances (Achaya 1998). As per the Islamic dietary laws, slaughtering of animals involves killing through a cut to the jugular vein in the neck which is known as *halal*, meaning lawful as defined in Quran (Twaigery and Spillman 1989). Traditionally Buddhists avoid eating meat and fish out of respect for life (Hinnells 1997). Varieties of spices are used in India along with the addition of desirable amount of salt. Diverse Indian communities in modern India from Vedic period have developed different methods of culinary practices to make presentable ethnic foods and beverages with ancient touch in history of dietary culture.

Traditional feeding practice by hands is an ancient dietary practice since Vedic periods for the last 6000 years ago in India (Prakash 1987; Hegde et al. 2018). Sitting on the floor cross-legged while eating by hands has been practiced since *Vedic* India and remains a unique food culture of Indian communities. Sitting on the floor cross-legged while having food as practiced traditionally in India is a yogic posture called *sukhasan* which is said to massage the abdominal muscles, to activate blood circulation in the lower part of the body, and to increase pliability (Sinha 1996).

Gastronomy of Indian ethnic foods is based on *Vedic* food habits with adoption of some Mughal and English cuisines; the present-day Indian foods of more than 2000 diverse ethnic groups of people (Kulkarni 2002) are unparalleled and unique to each other. Due to varied agroclimatic conditions, geographical locations, and availability of agro-produces, cooked rice "चावल" is a staple diet followed by lentil thick soup called dal "दाल" in East India and Northeast India; rice-legume-based fermented food (*idli*, *dosa*) is a staple diet along with cooked rice in South India, whereas wheat-based baked flat *chapati* is a staple diet along with *dal* in North, West, and Central India (Fig. 1.1). These diets are supplemented with many fermented and non-fermented food items including vegetable, milk products, pickle/*acchar*, *papad*, etc. in thali system.

1.3 History of Indian Fermented Foods

Dietary culture of ethnicity predicts the cultural history of ethnic communities (Tamang and Samuel 2010). Delicacy of foods invokes innovation in culinary and diversification of cuisine to supplement taste, flavor, aroma, and palatability of the final edible product. The ancestors of Indian people might have innovated the traditional food fermentation technology to get fermented and alcoholic beverages and for the preservation of perishable plant and animal resources and enhancement of new preferred flavor and taste to increase delicacy in diets. Innovative food fermentation technology, based on wisdom and native skill of the ancestors of Indian people, includes natural fermentation, "back-slopping" fermentation (using previous cultures or inoculum), smoking, drying, salting processes, and alcohol production, which were developed to preserve the foods for consumption and to impart health benefits. The wisdom and native culinary skills of ancestral people of India for the production of fermented and non-fermented foods have been passed from generation to generation through mothers to daughters, fathers to sons, with few modifications and adjustments based on local prevailing conditions and also involvement of self-practice, family tradition, and community knowledge. This symbolizes a remarkable step in the dietary history of Indian societies. Evidence-based and archaeological findings such as fossils, food remnants of fermented foods except

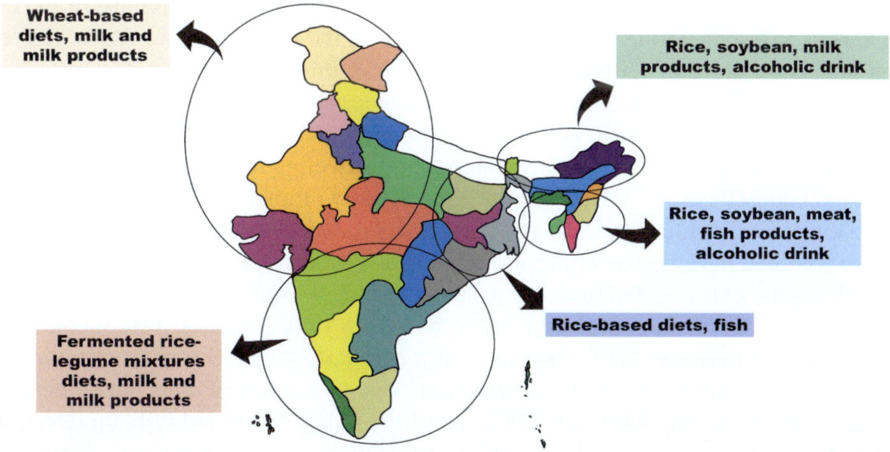

Fig. 1.1 Gastronomical map of India showing geographical distribution of food habits

some in Indus Valley Survey (Sarkar et al. 2016), and detailed historical records and monuments on origins of Indian fermented foods and beverages have not been documented yet. However, some ancient fermented foods and beverages are mentioned in *Rig Veda*, and literatures of Jains, and Buddhism in Sanskrit (Achaya 1994, 2009). We believe that most of the Indian fermented foods belong to *sattvic* category of Hindu philosophy, due to their high nutritive value, functionality, and health-promoting benefits, and some acidic foods such as fermented vegetable and bamboo products are *rajasic* category foods due to the sour taste. Indian fermented foods may not fall under *Vedic* classification of *tamasic* foods since ethnic foods may not cause dullness and passivity to consumers.

Historical evidences on preparation and consumption of some culturally adopted Indian fermented and alcoholic beverages are recorded here. *Dosa* or *dosai*, a traditional fermented mixture of rice and black gram cooked flat, thin, layered pancake-like batter of South India, has been prepared and consumed in Tamil Nadu around the first century CE as mentioned in the Tamil Sangam literature (Achaya 2003). Another popular ethnic food of South India known as *idli* or *idly* or *iddalige*, fermented rice and black lentils consumed as round steamed spongy savory staple food, was first mentioned in Kannada literature known as *Vaddaradhane*, the earliest extant prose work in Kannada written by great Kannada writer Shivakotiacharya in 920 CE (Krishna Jois 1969). *Idli* and *dosa* are the heritage foods of Tamil, Kannada, Telegu, and Malayalam people of Dravidian origins in India, who use these foods in every religious and social occasion. *Jalebi*, a fermented, cereal-based coiled, fried, crispy, sweet food, has been known in India since 1450 CE and is probably of Arabic or Persian origin (Gode 1943). *Dhokla*, a fermented mixture of wheat and Bengal gram of Gujarat, was mentioned in Gujarati literature known as *Varanaka Samuchaya* in 1520 CE; however, it was recorded that *dukkia*, considered as precursor to *dhokla*, was mentioned in Gujarati Jain literature in 1066 CE indicating its origin since 1520 CE (Achaya 1998). *Papad*, a popular crispy fermented pulse dried wafer-like product, first appeared in Buddhist Jain literature in 500 BCE as *parpata* (Prakash 1961), which may be considered as precursor to *papad*. Oral history on the origin of *gundruk* and *sinki*, fermented vegetable products of the Himalayas, was documented in details by Tamang (2010a). The word *gundruk* might have originated from the Newar (one of the major ethnic groups of the Nepali) word "gunnu" meaning dried taro stalk (Tamang 2010a). *Kinema*, a fermented sticky soybean of Kirat Nepali, might have originated in east Nepal bordering Darjeeling hills and Sikkim in India around 600 BCE to 100 CE during the Kirat dynasty (Tamang 2010a).

Domestication of animals for milk and milk products as well as dairy system has been recorded in Harappan civilization for the last 8000 years (Sarkar et al. 2016). In *Rig Veda*, the oldest sacred book of the Hindus which is a collection of Vedic Sanskrit hymns, sanctity and importance of cow as well as milk products were mentioned (Jamison and Brereton 2014). Periods of history of *dahi* production and consumption in India are variable. *Veda* and *Upanisad* mentioned the origin of *dahi* and fermented milk products during 6000–4000 BC, one of the oldest fermented milk products of the Hindus (Yegna Narayan Aiyar 1953). Preparation and consumption of *dahi* have been recorded since 2000 BCE in India (Prakash 1961). *Dahi* plays an important part in the socioreligious habits in Indian subcontinent and is considered as sacred item in many of their festivals and religious ceremony both by Hindus and Buddhists. *Dahi Handi* is celebrated every year the day after "Krishna Janmashtami" where communities hang an earthen pot filled with *dahi* at a height difficult to reach. There is a legend in Hindu that Lord Krishan as a child used to steal *dahi* and *ghee* (butter) kept at difficult place, which may be correlated to history of *dahi*, *lassi*, and *ghee* (butter), the widely consumed milk products during Lord Krishna's time about 3000 BCE (Bryant 2007). *Dahi* was mentioned in *Rig Veda* as *karambha*, a blended *dah*i rice dish (Achaya 1998).

Indus Valley Civilization traced back to 8000 years (Sarkar et al. 2016) appeared to have alcohol drinking culture mostly through fermentation (Singh et al. 2010) and distillation (Achaya 1991), based on findings of clay pot items from excavation sites (Mahdihassan 1979). During the *Vedic* period of Indian history, liquor God called *Soma* was worshipped by *Soma rus*, a refreshing sweet juice prepared from leafless vine (*Sarcostemma acidum*) (Sarma 1939) by the Vedic Aryans which had medicinal values (Bose 1922). Highly alcoholic distilled liquor prepared by fermenting millets called *sura* and the alcoholic product fermented from flowers called *parisrut* were the earliest alcoholic beverages of India around 1500 BCE during the Vedic period (Mahdihassan 1981). No historical records on vinification (fermentation of grapes into wine), malting, and brewing (such as beer) processes are available in Indian food culture. It indicates that alcoholic beverages are fermented either by natural fermentation of plant or cereals or by using traditionally prepared dry amylolytic starters with different vernacular names such as *marcha*, *hamei*, *dabai*, *phub*, etc. (Tamang 2010a). Wine, beer, and whisky are not traditional drink or culture of India though these alcoholic drinks are becoming popular in modern Indian food culture. The popular Himalayan mild alcoholic beverages called *kodo ko jaanr* prepared from fermented finger millets by traditionally prepared dry amylolytic starter culture was mentioned in the history of Nepal during the Kirat dynasty in 625 BCE to 100 CE (Adhikari and Ghimirey 2000). During Malla dynasty in Nepal, the Newar, one of the ethnic communities of Gorkha, used to ferment alcoholic beverages from rice in 880 CE (Khatri 1987). The first record of *chyang*, fermented finger millet alcoholic beverage similar to *kodo ko jaanr* in Sikkim, was first mentioned in Gazetteer of Sikkim by Risley (1928) which quoted "*marwa* or *chyang*, is a kind of beer brewed by everyone in Sikkim, and might be called their staple food and drink." In historical documents on Darjeeling and Sikkim, brief descriptions of fermented millet beverages as alcoholic drink consumed by the local ethnic people were mentioned by Hooker (1854) and Gorer (1938).

1.4 Diversity of Ethnic Fermented Foods of India

Indian fermented foods are classified into ten major groups on the basis of substrates (raw materials) used from plant/animal sources:

Plant-based fermented foods: (1) fermented cereal foods; (2) fermented non-soybean legume foods; (3) fermented soybean foods; and (4) fermented vegetable foods
Animal-based fermented foods: (1) fermented dairy foods; (2) fermented/sun-dried/smoked fish products; and (3) fermented/sun-dried/smoked meat products
Alcoholic beverages: (1) amylolytic starters for production of alcoholic beverages and (2) alcoholic beverages
Miscellaneous fermented products: fermented tea, crabs, fruits, etc.

Very few outside the country have realized that India is the center of the huge diversity food culture comprising ethnic fermented and non-fermented ethnic foods and alcoholic beverages. Ethnic fermented foods and beverages are defined as foods produced by the ethnic people using their native knowledge from locally available agro-/bioresources of plant or animal origins either naturally or by adding starter culture(s) containing functional microorganisms/"back-slopping," which convert the substrates (raw/cooked) biochemically and sensorially into edible products with enhanced nutritive value, flavor, and health-promoting benefits that are culturally and socially acceptable to the consumers (Tamang 2010b, c). Due to varied ethnicity along with unparalleled food habits based on stable cereal diets, the ethnic people have adopted distinct food habits and preferences for specific fermented foods and beverages as state/region-wise in India (Table 1.1). Daily per capita consumption of ethnic fermented foods and alcoholic beverages is not available for any other state in India except Sikkim which shows 163.8 g/capita daily consumption of fermented foods and alcoholic beverages representing 12.6% of total daily diet (Tamang et al. 2007). Indian fermented foods are considered as essential for food and nutritional security of the region and are produced using the native skill of ethnic people practicing the crude knowledge of "microbiology," what I termed as "ethno-microbiology." Microbial community ranging from lactic acid bacteria, non-lactic acid bacteria, acetic acid bacteria, bacilli, micrococci, Gram-negative bacteria, filamentous mold, to enzyme- and alcohol-producing yeasts is associated with Indian fermented foods and alcoholic beverages (Chettri and Tamang 2015; Tamang et al. 2007, 2016a; Shangpliang et al. 2018; Sha et al. 2017, 2018, 2019), and some of them have functionality and health benefits (Tamang et al. 2009, 2016b). Most of Indian ethnic fermented foods are naturally fermented, except the alcoholic beverages, which are produced by using consortia of microbiota consisting of filamentous mold, yeasts, and bacteria in the form of dry, cereal-based amylolytic starter (Tamang 2010c; Tamang et al. 2016a). Some common as well as lesser-known or region-specific ethnic fermented foods and alcoholic beverages have been studied extensively focusing on microbiology, biochemistry, nutritional value, health benefits, upgradation of traditional technology, etc. for the last 50 years.

It is remarkable to note that unlike many countries, India has more than 350 diverse types of major or common and minor or region-specific ethnic fermented foods and beverages (234 fermented foods, 84 alcoholic beverages, and 32 traditionally prepared amylolytic starters) which are traditionally prepared and consumed by ethnic people of India for 5000 years (Tables 1.2, 1.3, 1.4, 1.5, 1.6, 1.7, 1.8, 1.9, 1.10 and 1.11). This indicates that India has rich and huge fermented food and alcoholic beverage diversity comparable to the rest of the world. However, only 17% of ethnic fermented foods and beverages have been studied scientifically and reported till date, and the rest 83% of ethnic fermented foods and beverages are yet to be documented or studied in details. Some of the common and region-specific ethnic fermented foods and alcoholic beverages of India have been extensively studied focusing on microbiology, biochemistry, nutrition, health benefits, optimization of traditional processing methods, value addition, etc. for several years. References of various researchers working on some common and region-specific ethnic fermented foods and beverages are documented:

Table 1.1 Geographical distribution of substrate-based ethnic fermented foods in India

Category	Sensory character (based on pH)	Microbiota groups	State-wise distribution of traditional consumers
Fermented cereal foods	Mild acidic	LAB, yeasts	All except Nagaland, Mizoram, Manipur, Meghalaya, West Bengal, Tripura, Odisha, Jharkhand, Bihar, Chhattisgarh
Fermented milk foods	Acidic	LAB, yeasts, molds	All, except Nagaland, Mizoram, Manipur, and Meghalaya
Fermented non-soybean legume foods	Mostly alkaline	LAB, *Bacillus* spp., yeasts	Punjab, Gujarat, Rajasthan, UP, Haryana, Madhya Pradesh
Fermented rice-legume mixture foods	Acidic	LAB, *Bacillus* spp., yeasts	Tamil Nadu, Karnataka, Kerala, Andhra Pradesh, Telangana
Fermented soybean foods	Mostly alkaline	LAB, *Bacillus* spp., yeasts	Sikkim and GTA, Nagaland, Mizoram, Manipur, Meghalaya, Arunachal Pradesh
Fermented vegetable foods	Acidic	LAB	Sikkim and GTA, Nagaland, Mizoram, Manipur, Meghalaya, Assam, Arunachal Pradesh
Fermented/sun-dried/smoked fish products	Mild acidic	LAB, micrococci	Sikkim and GTA, Nagaland, Mizoram, Manipur, Meghalaya, Assam, Arunachal Pradesh, Odisha, West Bengal, Goa, Coastal regions of Tamil Nadu, Kerala
Fermented/sun-dried/smoked meat products	Mild acidic	LAB, micrococci	Sikkim and GTA, Nagaland, Mizoram, Manipur, Meghalaya, Assam, Arunachal Pradesh, Jammu and Kashmir, Himachal Pradesh, Uttarakhand, Goa
Amylolytic starters for production of alcoholic beverages	Mild acidic, alcoholic	Filamentous molds, amylolytic and alcohol-producing yeasts and bacteria	Sikkim and GTA, Nagaland, Mizoram, Manipur, Meghalaya, Assam, Arunachal Pradesh, Tripura, Jammu and Kashmir, Himachal Pradesh, Uttarakhand, West Bengal, Odisha

(continued)

Table 1.1 (continued)

Category	Sensory character (based on pH)	Microbiota groups	State-wise distribution of traditional consumers
Fermented beverages	Alcoholic	LAB, yeasts, and molds	Cereal-based alcoholic beverages by amylolytic starters: Sikkim and GTA, Nagaland, Mizoram, Manipur, Meghalaya, Assam, Arunachal Pradesh, Tripura, Jammu and Kashmir, Himachal Pradesh, Uttarakhand, West Bengal, Odisha, Madhya Pradesh, Jharkhand, Chhattisgarh
			Toddy and other natural alcoholic fermentation: Goa, Kerala, Tamil Nadu, Orrisa, West Bengal, Karnataka, Andhra Pradesh
Miscellaneous fermented products (fermented tea, crabs, fruits, etc.)	Acidic and alkaline	AAB, LAB, yeasts, molds	Fermented tea, fruits, crabs: Ladakh, Sikkim, Himachal Pradesh, Uttarakhand, Nagaland, Mizoram, Tripura

LAB lactic acid bacteria, *GTA* Gorkhaland Territorial Administration

1. Fermented milk (cow's/buffalo's/yak's) products of India such as *dahi* (Mohanan et al. 1984; Ghosh and Rajorhia 1990; Rathi et al. 1990; Sharma et al. 1993; Gupta et al. 2000; Rajpal and Kansal 2009; Behare et al. 2009; Agarwal and Bhasin 2002; Jain et al. 2010; Shruti and Kansal 2011; Jadhav et al. 2013; Balamurugan et al. 2014; Mohania et al. 2014), *shrikhand* (Boghra and Mathur 2000; Devshete et al. 2013; Dandile et al. 2014; Singh and Singh 2014; Singh et al. 2014b; Karche et al. 2015;), *lassi* (Patidar and Prajapati 1988; Padghan et al. 2015; Patel et al. 2015; Sudheendra Ch et al. 2018), *chhurpi*, *churkam*, *chhu*, *dahi*, *philu*, and *somar* (Katiyar et al. 1991; Tamang et al. 2000; Dewan and Tamang 2006, 2007; Rai et al. 2016; Shangpliang et al. 2018).
2. Fermented legume products such as *dhokla* (Joshi et al. 1989; Kanekar and Joshi 1993; Lohekar and Arya 2014; Shobha and Joshi 2016; Suman and Khetarpaul 2018), *khaman* (Shrestha et al. 2017), *wari* (Batra 1986; Sandhu and Soni 1989; Soni and Sandhu 1990), and *maseura* (Chettri and Tamang 2008).
3. Fermented soybean foods such as *kinema* (Sarkar et al. 1994, 1996, 1997, 1998; Sarkar and Tamang 1994, 1995; Tamang and Nikkuni 1996, 1998; Omizu et al. 2011; Rai et al. 2014; Tamang 1999, 2003, 2015; Tamang et al. 2002; Chettri et al. 2016), *tungrymbai* (Sohliya et al. 2009; Chettri and Tamang 2014, 2015; Thokchom and Joshi 2015), *hawaijar* (Jeyaram et al. 2008a, 2009; Keishing and Banu 2013; Singh et al. 2014, 2018; Thokchom and Joshi 2015; Keisam et al.

Table 1.2 Ethnic fermented cereal foods of India

S.no.	Products	Agro-produce and ingredient	Sensory character and edibles	State/region
1.	*Adai dosa*	Rice, Bengal gram, red gram, black gram, green gram	Slight acidic, like *dosa*; breakfast	Tamil Nadu
2.	*Ambil*	Rice	Cooked slurry; staple	Maharashtra
3.	*Aeersa*	Rice and jaggary	Fried, rounded sweet dish; dessert	Chhattisgarh
4.	*Ambali* or *amboli*	Millet flour and rice	Thick fermented batter; pancake like *uttapam*; staple	Maharashtra, Karnataka, Tamil Nadu, Andhra Pradesh, Kerala, Telangana
5.	*Appam*	Rice, black gram	White color, Concave shaped, bulged at the center with crisp edges; staple	Kerala
6.	*Appam/ kallappam* or *vellayppam*	Rice, black gram	Concave shaped with crisp edges; like *idli*; staple	Kerala
7.	*Aska*	Rice flour	Side dish	Himachal Pradesh
8.	*Babru*	Wheat or rice flour	Pancake; breakfast, snack	Himachal Pradesh
9.	*Bagpinni*	Roasted barley flour *sattu* with *chhyang*	Ball-like; snack	Himachal Pradesh
10.	*Bhatabaru*	Wheat flour, milk, sugar	Sweet, deep fried, oval shaped; snack	Himachal Pradesh
11.	*Bhatooru* or *Bhaturu*	Wheat or barley flour	Baked breads; staple	Himachal Pradesh
12.	*Bedvin roti*	Wheat flour, black gram and walnut	Baked; staple	Himachal Pradesh
13.	*Bobra*	Rice, wheat, jiggery	Sweet dish; dessert	Chhattisgarh
14.	*Boree Basi*	Rice	Baked; breakfast	Chhattisgarh
15.	*Chakuli*	Rice and black gram	Fermented batter, fried; breakfast	Odisha, West Bengal
16.	*Chhuchipatra pitha*	Rice and black gram	Pancake, snack	Odisha, West Bengal
17.	*Chitou*	Rice, black gram, sugar and grated coconut	Fermented batter, baked; snack	Odisha, West Bengal
18.	*Chilra*	Wheat and buckwheat flour	Similar to *dosa*; staple	Himachal Pradesh
19.	*Chzot* or *Girda*	Wheat	Baked, staple	Jammu and Kashmir
20.	*Czochwor*	Wheat, baker's yeast	Baked; staple	Jammu and Kashmir

(continued)

Table 1.2 (continued)

S.no.	Products	Agro-produce and ingredient	Sensory character and edibles	State/region
21.	*Dahi rice* or *Thayir sadham*	Rice—*dahi*	Mildly sour taste; staple	Tamil Nadu
22.	*Dehori*	Rice—*dahi*	Round, fried dough into jaggery syrup	Chhattisgarh
23.	*Dosa* or *dosai* or *dose*	Rice and black gram	Thin, crisp pancake; shallow fried; staple	Tami Nadu, Kerala, Karnataka, Andhra Pradesh, Telangana
24.	*Dhuska*	Rice-lentil	Ball like, fried; snacks; breakfast	Jharkhand
25.	*Enduri pitha*	Rice and black gram	Fermented batter; steamed; snack	Odisha, West Bengal
26.	*Gulgule*	Wheat flour	Cooked as flat pancake; snack	Himachal Pradesh
27.	*Hakua*	Rice	Strong off-flavor; therapeutic uses	Darjeeling hills, Sikkim
28.	*Handwa*	Rice, red gram, and Bengal gram flour	Fermented, baked, spongy; snack	Gujarat
29.	*Idli*	Rice and black gram	Mild acidic, soft, moist, spongy; breakfast	Tamil Nadu, Kerala, Karnataka, Andhra Pradesh, Telangana, Pondicherry
30.	*Jalebi*	Wheat flour	Crispy sweet, deep fried pretzels; snacks	All
31.	*Khadi-badi*	Wheat flour—*dahi* mixture	Boiled, fried; snack	Bihar, Jharkhand
32.	*Khandvi*	Cereal	Soft, slippery texture, mild sour; snack	Gujarat
33.	*Koozh* or *koozhu*	Sorghum, pearl millet, little millet, and foxtail millet flour	Slightly sour, cooked; breakfast	Tamil Nadu
34.	*Lawaas*	Wheat	Baked breads; staple	Jammu and Kashmir
35.	*Marchu*	Wheat flour	Baked breads; staple	Himachal Pradesh
36.	*Munha pitha* or *podo pitha*	Rice and black gram	Spongy like *idli*; staple	Odisha
37.	*Naan*	Wheat flour	Leaved bread; baked; staple	Punjab, Haryana, Delhi, UP, Himachal Pradesh
38.	*Neyyappam*	Rice and jaggary	Solid, spherical, oily; snack	Kerala
39.	*Olyafenya*	Rice	Thin, round shape, fried flat batter like *dosa*; breakfast	Maharashtra
40.	*Pakk*	Barley, butter (*ghee*), *lassi*	Cooked; staple	Himachal Pradesh

(continued)

Table 1.2 (continued)

S.no.	Products	Agro-produce and ingredient	Sensory character and edibles	State/region
41.	*Pazhaya sadham*	Rice	Cooked; breakfast	Tamil Nadu
42.	*Pej* or *Peja*	Rice, *dahi*, salt	Cooked; staple	Madhya Pradesh
43.	*Podo pitha*	Rice and black gram	Fermented batter, fried; breakfast	Odisha, West Bengal
44.	*Poita bhat*	Rice	Cooked, acidic, sour in taste; staple	Assam
45.	*Pua*	Wheat flour	Fermented batter fried, served hot as sweet confectionary	Bihar, Jharkhand
46.	*Puda/Pudla*	Maize, Bengal gram	Solid food, pancake; snack	Tamil Nadu
47.	*Rabadi* or *rabri*	Wheat, barley, pearl millet or maize, milk	Mild-acidic, thick slurry-like product; drink	Punjab, Haryana, Rajasthan
48.	*Salpapdya*	Rice	Dried, circular shape, like *papad*	Maharashtra
49.	*Sannas*	Rice, black gram	Soft round steamed rice cakes; staple	Karnataka
50.	*Selroti*	Rice wheat flour	Pretzel-like; deep fried; bread; staple	Darjeeling hills, Sikkim, Northeast India, Uttarakhand, Himachal Pradesh
51.	*Seera*	Wheat grains	Dried, sweet dish	Himachal Pradesh
52.	*Siddu*	Wheat flour, walnut/opium seeds/black gram	Steamed bread, oval-shaped; staple	Himachal Pradesh
53.	*Thatte idli*	Rice, tapioca, black gram	Spongy, *idli*-like; breakfast	Karnataka
54.	*Tchog*	Roasted barley flour, *chhang*	Solid ball-like dough; staple	Himachal Pradesh
55.	*Unniyappam*	Rice flour and jaggary	Soft, spongy; snack	Kerala
56.	*Uttappam* or *Uthappam*	Rice, black gram	Thick, soft, non-crispy; staple or side dish	Kerala, Tamil Nadu, Karnataka

2019), *bekang* (Chettri and Tamang 2014, 2015), and *akhonii* (Tamang et al. 2012; Singh et al. 2014a).

4. Fermented cereal products of India such as *rabadi* (Khetarpaul and Chauhan 1989, 1990a, b, 1991; Gupta et al. 1992a, b; Modha and Pal 2011) and *sel roti* (Yonzan and Tamang 2009, 2010, 2013).
5. Fermented rice black gram food of South India such as *dosa* (Soni et al. 1985, 1986; Soni and Sandhu 1989a; Palanisamy et al. 2012; Srinivasan et al. 2013; Gupta and Tiwari 2014), *idli* (Mukherjee et al. 1965; Steinkraus et al. 1967; Soni and Sandhu 1989b, 1991; Sridevi et al. 2010; Vijayendra et al. 2010; Ghosh and

Table 1.3 Ethnic fermented non-soybean legume foods of India

S.no.	Fermented foods	Agro-produce and ingredient	Sensory character and edibles	State/region
1.	*Amriti*	Black gram	Rings, deep fried; snack	West Bengal
2.	*Bhallae*	Black gram	Deep-fried patties; snack	Punjab, Haryana
3.	*Bijori*	Black lentils	Flat round, fried; snack	Chhattisgarh
4.	*Borhe*	Black gram	Staple	Himachal Pradesh
5.	*Dangalbari*	Black gram, spices	Staple	Himachal Pradesh
6.	*Dhokla*	Bengal gram and rice	Mild acidic, slightly salty-sour taste, spongy; snack	Gujarat
7.	*Dhuska bara*	Chickpea, black gram, rice *dahi*	Deep fried; snacks	Chhattisgarh
8.	*Handvo*	Lentil-rice	Spicy with cake-like texture; savory	Gujarat
9.	*Khaman*	Bengal gram	Mild acidic, spongy; breakfast	Gujarat
10.	*Masauyra*	Black gram	Cone-shaped hollow, brittle, similar to *wari*, side dish	Sikkim and Darjeeling hills
11.	*Mashbari*	Black gram, spices	Staple food	Himachal Pradesh
12.	*Pao/undoh*	Wheat; baker yeasts	Bread; staple	Goa
13.	*Papad*	Black gram	Circular wafers; snack	All
14.	*Rakhiya Bari*	Black lentils, flesh of ash gourd	Fermented, dried round nuggets	Chhattisgarh
15.	*Sepubari*	Black gram	Fermented paste, side dish	Himachal Pradesh
16.	*Vadai*	Black gram	Paste, fried patties; snack	South India
17.	*Wadi* or *wari* or *bari*	Black gram	Ball-like, brittle; condiment	Punjab, Haryana, UP, Himachal Pradesh, Delhi

Chattopadhyay 2011; Agaliya and Jeevaratnam 2013; Kalki and Shetty 2015; Shrivastava and Ananthanarayan 2015; Chelliah et al. 2016; Chandrasekar Rajendran et al. 2017; Regubalan and Ananthanarayan 2018, 2019), and *uttapam* (Dubey and Jeevaratnam 2015; Saraniya and Jeevaratnam 2015).

6. Fermented vegetable products such as *gundruk* (Tamang et al. 2005, 2009; Tamang and Tamang 2010) and *sinki* (Tamang and Sarkar 1993; Tamang et al. 2005, 2009).
7. Fermented bamboo shoot products such as *mesu* (Tamang and Sarkar 1996; Tamang et al. 2008), *soibum* (Giri and Janmejay 1994; Sarangthem and Singh 2003; Tamang et al. 2008; Jeyaram et al. 2010; Romi et al. 2014, 2015), *soidon* (Tamang et al. 2008; Sonar and Halami 2014; Romi et al. 2015; Sonar et al. 2015), and *khorisa* (Sharma and Barooah 2017).

Table 1.4 Ethnic fermented soybean foods of India

S.no.	Fermented foods	Agro-produce and ingredient	Sensory character and edibles	State/region
1.	*Aagya*	Soybean	Alkaline, sticky; side dish	Arunachal Pradesh
2.	*Aakhone* or *Akhonii* or *Axone*	Soybean	Alkaline, sticky, paste; side dish	Nagaland
3.	*Bari*	Soybean	Alkaline, sticky; soup	Sikkim
4.	*Bethu*	Soybean	Alkaline, sticky; soup; side dish	Manipur, Nagaland
5.	*Bekang* or *Bekang-um*	Soybean	Alkaline, sticky, paste; curry	Mizoram
6.	*Hakhu mata*	Soybeans	Alkaline, sticky, paste; curry	Manipur
7.	*Hawaijar*	Soybean	Alkaline, sticky; curry	Manipur
8.	*Kinema*	Soybean	Alkaline, sticky; curry	Sikkim, Darjeeling hills, Assam
9.	*Khuichang*	Soybean	Alkaline, sticky; side dish	Manipur
10.	*Libi chhurpi*	Soybean	Alkaline, sticky, crushed, dried; curry	Arunachal Pradesh
11.	*Peron naming*	Soybeans	Alkaline, paste, ball; side dish	Arunachal Pradesh
12.	*Pehak*	Soybean	Alkaline, after fermentation; side dish	Arunachal Pradesh
13.	*Peruyaan*	Soybean	Alkaline, sticky; side dish	Arunachal Pradesh
14.	*Theisui*	Soybean	Alkaline, sticky; side dish	Manipur, Nagaland
15.	*Tungrymbai*	Soybean	Alkaline, sticky; curry, soup	Meghalaya
16.	*Yanni perung*	Soybean	Alkaline, paste, ball; side dish	Arunachal Pradesh

8. Fermented and traditionally preserved fish products such as *ngari* (Thapa et al. 2004; Abdhul et al. 2014; Devi et al. 2015; Majumdar et al. 2015a; Thapa 2016; Singh et al. 2018), *hentek* (Thapa et al. 2004; Aarti et al. 2016, 2017; Singh et al. 2018;), *namsing* (Chowdhury et al. 2018), *sukuti* (Thapa et al. 2006; Thapa and Pal 2007), *sidra* (Thapa et al. 2006; Thapa and Pal 2007), *tungtap* (Thapa et al. 2004; Rapsang et al. 2011; Rapsang and Joshi 2012), *sheedal* (Majumdar et al. 2009, 2015b; Kakati and Goswami 2013a, b; Muzaddadi and Basu 2003, 2012; Muzaddadi 2015), *bordia*, *karati*, and *lashim* (Thapa et al. 2007; Thapa 2016).
9. Fermented meat products such as *chartayshya*, *jamma*, *arjia* (Oki et al. 2011), *kargyong*, *satchu*, *suka ko masu* (Rai et al. 2009, 2010a, b), and *sa-um* (De Mandal et al. 2018).
10. Traditionally prepared dry amylolytic starters such as *marcha* (Tamang and Sarkar 1995; Tsuyoshi et al. 2005; Tamang et al. 1988, 2007; Jeyaram et al.

Table 1.5 Ethnic fermented milk foods of India

S.no.	Fermented foods	Agro-produce and ingredient	Sensory character and edibles	State/region
1.	*Chhash*	Cow/buffalo milk	Non-alcoholic beverage; mildly sour taste	Gujarat, Rajasthan
2.	*Chhu* or *sheden*	Cow or yak milk	Soft, strongly flavored; curry	Darjeeling hills, Sikkim
3.	*Chhur chirpen*	Yak milk and crab apple	Pressed, light yellowish brown; side dish	Arunachal Pradesh
4.	*Chhur singba* or *chhur mingba*	Yak milk	Pressed, light yellowish brown; side dish	Arunachal Pradesh
5.	*Chhura*	Dzomo (crossed breed of cow and yak), yak milk	Hard mass; masticator	Ladakh
6.	*Chhurpi* (soft variety)	Cow or yak milk	Soft, cheese-like; curry, pickle	Sikkim, Darjeeling hills, Arunachal Pradesh
7.	*Chhurpi* (hard variety)	Cow or yak milk	Hard mass; masticator	Sikkim, Darjeeling hills, Arunachal Pradesh, Ladakh, Himachal Pradesh, Uttarakhand
8.	*Chhurapi*	Yak milk	Soft, cheese-like; curry, pickle	Arunachal Pradesh
9.	*Chhurpupu/churtang*	Yak/cow milk	4–5-year-old *chhurpi*; side dish	Arunachal Pradesh
10.	*Churkham*	Fresh and old *chhurpi*	Soft cheese packed in yak skin and sun dried, eaten as masticator, mouth freshener	Arunachal Pradesh
11.	*Dahi*	Cow/buffalo/yak milk	Curd; savoury	All
12.	*Dahi Vada*	*Dahi* and legumes mixtures	Soft semiliquid, side dish	Bihar, Jharkhand, UP
13.	*Dudh chhurpi*	Cow milk	Hard mass; masticator	Darjeeling hills, Sikkim
14.	*Ghee/Gheu*	Cow/buffalo milk	Butter	All
15.	*Kadhi*	Cow/buffalo milk/mixed milk	Prepared from sour *dahi* or *chhash*, sour; savory	Gujarat, Rajasthan
16.	*Khalari* or *Maesh-kraej*	Cow/buffalo/goat milk	Semisoft, cottage cheese; side dish	Jammu and Kashmir
17.	*Lassi*	Cow/buffalo milk	Buttermilk; refreshing beverage	All
18.	*Maa/Maar*	Yak milk	Butter	Sikkim, Ladakh

(continued)

Table 1.5 (continued)

S.no.	Fermented foods	Agro-produce and ingredient	Sensory character and edibles	State/region
19.	*Marchang*	Yak ghee and barley flour "kongpu"	Side dish	Arunachal Pradesh
20.	*Mattha*	Milk	Acidic, spicy, like buttermilk	Maharashtra
21.	*Mohi*	Yak milk	Butter milk; refreshment	Sikkim, Darjeeling hills
22.	*Mishti dahi*	Cow/buffalo milk and sugar	Mild acidic, thick gel, sweet; dessert	West Bengal, Odisha, Assam
23.	*Paneer*	Whey of cow milk	Soft, cheese-like product; curry	North India
24.	*Pheuja/Suja*	Tea-yak butter	Fermented butter tea	Sikkim, Ladakh
25.	*Philu*	Yak milk	Cream, fried curry with butter; side dish	Sikkim
26.	*Phrung*	Yak milk	Hard mass; masticator	Arunachal Pradesh
27.	*Phuh*	Milk	Soft cheese; side dish	Jammu and Kashmir
28.	*Raita*	Cow/buffalo milk	Viscous custard like; side dish	Gujarat, Rajasthan
29.	*Shrikhand*	Cow, buffalo milk	Acidic, sweet, viscous; dessert	Gujarat, Rajasthan, Haryana
30.	*Shyow*	Yak milk	Curd; savoury	Sikkim, Ladakh
31.	*Somar*	Milk	Paste, flavored; condiment	Darjeeling hills, Sikkim
32.	*Tara*	Dzomo (crossed breed of cow and yak), yak milk	Butter milk; refreshment	Ladakh
33.	*Zamuthdod*	Milk	Like *dahi*; savory	Jammu and Kashmir

2011; Sha et al. 2016, 2017, 2018), *dawdim* (Anupma et al. 2018; Sha et al. 2018, 2019), *xaj-pittha* (Bora et al. 2016), *hamei* (Tamang et al. 2007; Jeyaram et al. 2008b; Mangang et al. 2017; Sha et al. 2018, 2019), *chowan* (Anupma et al. 2018; Sha et al. 2018, 2019), *humao* (Anupma et al. 2018; Sha et al. 2018, 2019), *khekhrii* (Anupma et al. 2018; Sha et al. 2018, 2019), *thiat* (Anupma et al. 2018; Sha et al. 2017, 2018), and *phut* (Anupma et al. 2018; Sha et al. 2018, 2019).

11. Alcoholic beverages and distilled liquor such as *kodo ko jaanr* or *chyang* (Bhatia et al. 1977; Basappa 2002; Basappa et al. 1997; Thapa and Tamang 2004, 2006; Angmo and Bhalla 2014; Ray et al. 2016), *bhaati jaanr* (Tamang and Thapa 2006), *toddy* (Shamala and Sreekantiah 1988; Kunjithapatham and Balasubramaniam 2012), *kanji* (Sura et al. 2001; Goel et al. 2005; Kingston et al. 2010), *haria* (Ghosh et al. 2014, 2015a, b), *judima* (Chakrabarty et al. 2014), *zutho* (Teramoto et al. 2002), *sura* (Thakur et al. 2004; Thakur 2013), and *feni* (Khorjuvenkar 2016).

Table 1.6 Ethnic fermented vegetable and bamboo shoot products of India

S.no.	Fermented foods	Agro-produce and ingredient	Sensory character and edibles	State/region
1.	*Anishi*	Leaves of *Colocasia* sp.	Fermented; sour; curry	Nagaland
2.	*Ankamthu*	Mustard leaf	Liquid, sour; condiment	Mizoram, Manipur, Nagaland
3.	*Antramthu*	Mustard leaves	Dried/semiliquid; pickle	Assam
4.	*Bastanga*	Shoots of *Dendrocalamus hamiltonii*, *Bambusa tulda*	Sour, acidic; curry	Nagaland
5.	*Cutocie*	Cucumber fruits and leaves	Liquid, flavored; curry	Nagaland
6.	*Ekung*	Young bamboo tender shoots	Sour-acidic; curry, soup	Arunachal Pradesh
7.	*Eup*	Young bamboo tender shoots	Dry, acidic; curry, soup	Arunachal Pradesh
8.	*Ganang tamdui*	Mustard leaf	Sour liquid; condiment	Manipur, Nagaland
9.	*Goyang*	Leafy vegetables	Freshly fermented; juice as condiment, soup	Sikkim, Darjeeling hills
10.	*Gundruk*	Fresh leaves of local vegetables	Dried, sour-acidic; soup, pickle	Sikkim, Darjeeling hills, Assam, Nagaland, Uttarakhand, Himachal Pradesh
11.	*Hirring*	Topmost part of tender bamboo shoots	Sour-acidic; curry, soup	Arunachal Pradesh
12.	*Hungrii*	Leaves of *Brassica* sp.	Sour, acidic; by pit fermentation; side dish	Nagaland
13.	*Inziangsang*	Leafy vegetables	Dried, sour; soup, curry	Nagaland, Manipur
14.	*Inziang-dui*	Mustard leaves	Liquid, sour; condiment	Nagaland, Manipur
15.	*Kahudi*	Rapeseed	Sour, acidic; pickle	Assam
16.	*Khalpi*	Cucumber	Sour; pickle	Sikkim
17.	*Khorisa*	Young bamboo shoot	Acidic, sour; curry and pickle	Assam
18.	*Kharoli*	Black or whole mustard seeds	Pickle	Assam
19.	*Lung-siej*	Young bamboo shoots	Sour-acidic; curry	Meghalaya
20.	*Mesu*	Young bamboo tender shoots	Sour; pickle	Sikkim, Darjeeling hills
21.	*Melye Amiley*	Bamboo shoot	Sour, acidic; curry	Tripura

(continued)

Table 1.6 (continued)

S.no.	Fermented foods	Agro-produce and ingredient	Sensory character and edibles	State/region
22.	*Miya mikhri*	Young bamboo shoots	Sour, acidic, solid; curry	Assam
23.	*Moiya pangsung* or *moiya koshak*	Bamboo shoot	Sour, acidic; curry	Tripura
24.	*Panitenga*	Rapeseeds	Sour, acidic, pickle	Assam
25.	*Sinki*	Radish taproot	Dried, sour-acidic; soup, pickle	Sikkim
26.	*Soibum*	Succulent bamboo shoots	Sour-acidic; curry	Manipur
27.	*Soidon*	Bamboo shoot tips	Sour-acidic; curry	Manipur
28.	*Soijim* or *soijin* or *soidon mahi*	Bamboo shoot	Liquid, sour; condiment	Manipur
29.	*Tuai-um*	Bamboo shoot	Slightly acidic; side dish	Mizoram
30.	*Thunkhiang* or *Thunkheng*	Bamboo shoot	Solid, sour; side dish	Assam, Manipur
31.	*Thunbin*	Bamboo shoot	Sour, acidic, side dish	Manipur
32.	*Tsutocie*	Cucumber	Thick sluggish green paste; condiment	Nagaland
33.	*Tuaithur*	Bamboo shoot	Solid, wet, sour; curry, pickle	Assam
34.	*Tuaihroi*	Bamboo shoot	Solid, dry, sour; curry	Assam
35.	*Ziangsang*	Mustard leaf	Liquid, sour, acidic; side dish	Nagaland, Manipur

Few fermented foods and beverages have been commercialized and produced in large scale for Indian domestic markets, and some of them are exported too. These include *dahi*, *lassi*, *dosa*, *idli*, *shrikhand*, *misti doi*, *dhokla*, *papad*, *wari*, *rabadi*, *utaapam*, *ngari*, and *toddy*. However, 97% of ethnic fermented foods and beverages in India are still prepared by traditional methods for home consumption at region-specific level, and some of these food products are sold at local markets as unorganised sector business to supply local consumers. Some people are economically dependent upon products of ethnic foods which generate the marginal revenue for their livelihood. Such unsector home-based revenue generation from traditional food processing has not been recorded or documented by any policy-making bodies and financial institutions in India.

Table 1.7 Ethnic preserved and fermented fish products of India

S.no.	Fermented foods	Agro-produce and ingredient	Sensory character and edibles	*State/region*
1.	*Ayaiba*	Fish	Smoked fish; pickle, curry	Manipur, Mizoram
2.	*Bordia*	*Pseudeutropius atherinoides*	Dried, salted; side dish	Assam, Manipur, Mizoram, Arunachal Pradesh
3.	*Chucha*	Small fish species (*Gudusia chapra*, *Amblypharyngodon mola*, *Aspidoparia morar*, *Chela laubuca*, *Puntius ticto*, etc.)	Sun-dried fish product; side dish	Assam
4.	*Gnuchi*	*Schizothorax* spp., *Labeo* sp.	Smoked; side dish	Darjeeling hills, Sikkim
5.	*Godak*	*Puntius* spp.	Paste; curry, pickle	Tripura
6.	*Hentak*	*Esomus danricus*, petioles of *Alocasia macrorrhiza*	Paste; side dish	Manipur
7.	*Hidal*	Small fish	Fermented; side dish	Assam
8.	*Hukoti*	Small fish	Dried; side dish	Assam
9.	*Ithiitongba*	Small freshwater fishes	Fermented; side dish	Manipur
10.	*Jawla*	Shrimp	Sun-dried; side dish, pickle	Maharashtra
11.	*Khara masa*	Fish	Sun-dried, salted; side dish, pickle	Maharashtra
12.	*Karndi*	Fish	Sun-dried; side dish, pickle	Maharashtra
13.	*Karati*	*Gudusia chapra*	Dried, salted; side dish	Assam, Meghalaya
14.	*Lashim*	*Cirrhinus reba*	Dried, salted; side dish	Assam, Meghalaya
15.	*Lona Illis*	Indian shad (*Tenualosa ilisha*)	Salted, fermented; side dish	Tripura
16.	*Mio*	Fish	Dried; curry	Arunachal Pradesh
17.	*Naakangba*	Fish	Sun-dried; pickle, curry	Manipur, Nagaland
18.	*Naduba siyan*	Fish	Sun-dried fish product	Assam

(continued)

Table 1.7 (continued)

S.no.	Fermented foods	Agro-produce and ingredient	Sensory character and edibles	*State/region*
19.	*Nah-grain*	Small fish	Dry, solid; pickle, curry	Assam
20.	*Namsing*	*Puntius* spp., *Amblypharyngodon mola*, *Channa* spp., Colocasia (*Alocasia macrorrhiza*), or Shizu leaves (*Euphorbia neriifolia*)	Sun-dried fish products; curry	Assam
21.	*Ngiiyi-yaan*	Fish	Fermented; curry	Arunachal Pradesh
22.	*Ngari*	*Puntius sophore*	Fermented; side dish	Manipur
23.	*Sepaa*	Small fish	Dried; side dish	Tripura, Nagaland, Arunachal Pradesh
24.	*Shidal* or *sheedal*	*Puntis*	Semi-fermented; curry and pickle	Tripura, Assam
25.	*Sidra*	*Puntius sarana*	Dried; curry	Darjeeling hills, Sikkim
26.	*Sode*	Shrimp	Sun-dried; side dish	Maharashtra
27.	*Sukhi bombil* or *Sukhi Mandeli* or *Sukhi Makul*	Fish	Sun-dried; side dish	Maharashtra
28.	*Sukako maacha*	*Schizothorax* spp.	Dried or smoked; curry	Darjeeling hills, Sikkim
29.	*Sukuti*	*Harpadon nehereus*	Dried, salted; curry	Darjeeling hills, Sikkim
30.	*Tengli*	Fish	Sun-dried; side dish	Maharashtra, Goa
31.	*Tungtap*	*Danio* sp.	Fermented; side dish, pickle	Meghalaya
32.	*Vaktya*	Fish	Sun-dried; side dish	Maharashtra

1.5 Conclusion

Environmental ecosystems have food bioresources, and the ancient Indian people, by their native skills, based on trial and error, have successfully selected and preserved the essential microbiota from the environment to produce fermented foods and alcoholic beverages from the available agro-, floral, and fauna resources. Ethnic fermented foods and beverages are an important source of microbial diversity in

Table 1.8 Ethnic preserved and fermented meat products of India

S.no.	Fermented foods	Agro-produce and ingredient	Sensory character and edibles	State/region
1.	*Arjia*	Large intestine of chevon	Sausage; curry	Uttarakhand, Himachal Pradesh
2.	*Ashikumna* or *Thevochie*	Pork fats	Semisolid, flavored; curry	Nagaland
3.	*Bagjinam*	Pork	Fermented pork; curry	Nagaland
4.	*Bongkarot*	Beef	Liquid, sticky; curry	Manipur
5.	*Bongthu*	Cow or buffalo fat	Thick paste, oily; curry	Manipur
6.	*Chartayshya*	Chevon	Dried, smoked meat; curry	Uttarakhand, Himachal Pradesh
7.	*Dingkyo*	Mithun (*Bos frontalis*) meat	Smoked meat; side dish	Arunachal Pradesh
8.	*Faak kargyong*	Pork	Sausage, soft or hard, brownish; curry	Darjeeling hills, Sikkim
9.	*Gwag ruum*	Pork fat	Thick paste, oily; curry	Manipur
10.	*Guaighi kang*	Cow or buffalo skin	Solid, dried, hard; side dish	Manipur
11.	*Honoheingrain*	Pig/boar meat	Hard, dried meat; curry	Assam
12.	*Jamma*	Intestine of chevon, finger millet	Sausage, soft; curry	Uttarakhand, Himachal Pradesh
13.	*Jang kap*	Buffalo skin	Fermented; cooked; side dish	Nagaland
14.	*Khyopeh*	Yak meat inside the abdomen of dead yak	Fermented; cooked; side dish	Sikkim
15.	*Lang kargyong*	Beef	Sausage-soft; curry	Darjeeling hills, Sikkim, Ladakh
16.	*Lang satchu*	Beef	Dried, smoked meat; curry	Darjeeling hills, Sikkim
17.	*Lang chilu*	Beef fat	Hard, used as an edible oil; condiment	Darjeeling hills, Sikkim
18.	*Luk chilu*	Sheep fat	Hard, solid, used as edible oil; condiment	Darjeeling hills, Sikkim
19.	*Lang kheuri*	Beef	Chopped intestine of beef; curry	Darjeeling hills, Sikkim
20.	*Lukter*	Pork/beef	Dried meat; pickle	Arunachal Pradesh
21.	*Pikey Pila*	Pork meat	Smoked meat; pickle	Arunachal Pradesh
22.	*Saayung*	Pork meat	Solid, hard; side dish	Manipur
23.	*Sahro*	Any meat, usually pork	Solid, dried, hard; side dish	Manipur

(continued)

Table 1.8 (continued)

S.no.	Fermented foods	Agro-produce and ingredient	Sensory character and edibles	State/region
24.	*Sathu*	Pork fat	Thick paste, oily; curry	Manipur
25.	*Sa-um*	Pork fat	Creamy to pale yellow; semisolid; side dish	Mizoram
26.	*Suka ko masu*	Buffalo meat	Dried, smoked; curry	Darjeeling hills, Sikkim
27.	*Yak chilu*	Yak fat	Hard, edible oil; condiment	Darjeeling hills, Sikkim
28.	*Yak kargyong*	Yak	Sausage, soft; curry	Darjeeling hills, Sikkim, Arunachal Pradesh
29.	*Yak kheuri*	Yak	Chopped intestine of yak; curry	Darjeeling hills, Sikkim
30.	*Yak satchu*	Yak meat	Dried, smoked meat; curry	Darjeeling hills, Sikkim

India. Due to geographical location, diverse ethnicity, varied agroclimatic conditions, and cultural practices, Indian fermented foods and beverages have maximum microbial diversity representing different taxa including Gram-positive (several species of lactic acid bacteria, non-lactic bacteria, bacilli, micrococci) and few Gram-negative bacteria, filamentous mold, and various enzyme- and alcoholic-producing yeasts with functionality. This shows that Indian fermented foods and beverages contribute a huge microbial genetic resource locked inside the ethnic foods. Some of these strains possess protective and functional properties, rendering them potential candidates for use as starter culture(s) for controlled and optimized production of ethnic fermented foods with improved quality. Information on the characteristics of the microorganisms isolated from the ethnic fermented foods significantly enriches the database of microbial diversity from food ecosystems of India. Some strains can be exploited for production of enzymes and bioactive compounds and for other industrial uses and also to impart health-promoting benefits to consumers. Presently many region-specific ethnic fermented foods and beverages are rarely seen in local markets. Native microorganisms with vast biological importance and potential genetic resources which are associated with ethnic fermented foods are forced to disappear. Survey on consumption and production of ethnic fermented food and beverage in all states in India and calculation of *per capita* consumption is an urgent need to be addressed by the food policy makers mostly by Ministry of Food Processing Industries, Government of India. Introduction of ethnic fermented foods in the syllabus at the master's level in microbiology and food sciences of all universities may be initiated. Sikkim University has already incorporated the separate paper on ethnic fermented foods and beverages of the world in microbiology at postgraduate and PhD levels.

Table 1.9 Ethnic dry amylolytic starters for alcohol production of India

S.no.	Fermented foods	Agro-produce and ingredient	Sensory character and edibles	State/region
1.	*Angkur*	Rice, herbs	Starter to prepare alcoholic beverage—*jou bishi*	Assam
2.	*Apop pitha*	Rice, herbs	Starter to make *aapong*-alcoholic beverages	Assam
3.	*Bakhar*	Rice flour, ginger	Starter to ferment alcoholic beverage—*pachwai*	Uttarakhand, Himachal Pradesh
4.	*Balam*	Roasted wheat flour and spices	Starter to ferment alcoholic beverage—*chhang, jaan*	Uttarakhand
5.	*Bhekur-pitha*	Rice, herbs	Dry starter to ferment alcoholic beverage—*laopani*	Assam
6.	*Chamri*	Rice, bark of *Albizia myriophylla*	Dry starter to ferment alcoholic beverage—*acham, patso*	Manipur
7.	*Chowan*	Rice, herbs	Dry starter to ferment alcoholic beverage—*langi*	Tripura
8.	*Damdim*	Rice, herbs	Dry starter to ferment alcoholic beverage—*rakzu*	Mizoram
9.	*Dhehli*	Herbal mixture of 36 herbs and roasted barley flour	Starter to ferment alcoholic beverage—*sura*	Himachal Pradesh
10.	*Emao*	Rice, herbs	Starter to ferment alcoholic beverages	Assam
11.	*Hamei*	Rice, wild herbs	Dry, mixed starter to ferment alcoholic beverages—*atingba, waiyu*	Manipur
12.	*Humao*	Rice, barks of wild plants	Dry, flat, cake-like starter for *judima* production	Assam
13.	*Ipoh/Siye*	Rice and powder of seeds and bark of locally available plants	Starter to ferment alcoholic beverages—*apong* and *ennog*	Arunachal Pradesh
14.	*Khai*	Rice, bark of wild tree *khaipuroi*	Dry starter, solid, to ferment alcoholic beverage—*timpui, pheijou, zouju*	Manipur, Nagaland
15.	*Keem*	Wheat; plants	Starter to ferment alcoholic beverages—*soor*	Himachal Pradesh, Uttarakhand
16.	*Khekhrii* or *Khrei*	Germinated rice	Starter to ferment alcoholic beverages—*zutho/zhuchu*	Nagaland
17.	*Malera/treh*	Wheat flour	Starter to ferment *bhatooru/chilra*	Himachal Pradesh
18.	*Modor pitha*	Rice and 31 plant materials	Starter to ferment alcoholic beverages—*sujen*	Assam and Arunachal Pradesh

(continued)

Table 1.9 (continued)

S.no.	Fermented foods	Agro-produce and ingredient	Sensory character and edibles	State/region
19.	*Marcha*	Rice, wild herbs, spices	Dry, mixed starter to ferment alcoholic beverages—*kodo ko jaanr or chyang, makai ko jaanr, bhaati jaanr*	Darjeeling hills, Sikkim
20.	*Opop*	Rice, wild herbs, spices	Dry, mixed starter to ferment alcoholic beverages—*nyongin*	Arunachal Pradesh
21.	*Paa*	Rice, wild herbs, spices	Dry, mixed starter to ferment alcoholic beverages—*sira-oa*	Arunachal Pradesh
22.	*Pee*	Rice, wild herbs, spices	Dry, mixed starter to ferment alcoholic beverages—*sira-oa*	Arunachal Pradesh
23.	*Perok kushi*	Rice, herbs	Starter to make *sujen*-alcoholic beverage	Assam
24.	*Phab*	Wheat, wild herbs	Dry, mixed starter to ferment alcoholic beverages—*chyang*	Sikkim, Ladakh, Arunachal Pradesh
25.	*Pham/ phab*	Rice and leaves of *Solanum khasianum*	Starter to ferment alcoholic beverages—*themsing, chhang, arrak, kinnauri*	Arunachal Pradesh, Jammu and Kashmir, Himachal Pradesh
26.	*Phut*	Rice, wild herbs, spices	Dry, mixed starter to ferment alcoholic beverages	Arunachal Pradesh
27.	*Ranu dabai*	Rice, herbs	Starter to ferment alcoholic beverages—*jhara* or *haria*	West Bengal, Odisha, Jharkhand
28.	*Ranu goti*	Rice, herbs	Starter to ferment alcoholic beverages—*handia* and *mahua*	Bihar, Jharkhand, Madhya Pradesh
29.	*Thap*	Rice, herbs	Starter to ferment alcoholic beverages—*arak/hor alank ua*	Assam
30.	*Thiat*	Rice powder, powder of *Amomum aromaticum* Roxb. leaves	Starter to ferment alcoholic beverage—*kiad*	Meghalaya
31.	*Vekur pitha*	Rice, leaves of some local plants	Starter to ferment alcoholic beverages—*ahom*	Assam
32.	*Xaaz* or *Xaj pitha*	Rice, leaves of some local plants	Starter to ferment alcoholic beverages—*xaj pani* or *koloh pani*	Assam

Table 1.10 Ethnic fermented beverages and alcoholic drinks of India

S.no.	Alcoholic products	Raw materials and starter	Nature and use	State/region
1.	*Acham*	Rice, *chamri*	Distilled liquor	Manipur
2.	*Ahom*	Rice and *vekur pitha*	Aromatic, alcoholic, and sweet; refreshing drink	Assam
3.	*Angoori* or *kinnauri*	Grapes	Distillate; drinks	Himachal Pradesh, Jammu and Kashmir
4.	*Apong* or *ennog*	Rice, *Apop pitha*	Mild-alcoholic, beverage	Assam, Arunachal Pradesh
5.	*Ark* or *ara*	Barley, apple, wild apricot	Light brown drinks	Himachal Pradesh
6.	*Arak* or *Hor alank*	Rice, *Thap*	Alcoholic beverage	Assam
7.	*Aara*	Cereals, *paa*, *pee*	Clear distilled liquor; alcoholic drink	Arunachal Pradesh
8.	*Atingba*	Rice, *hamei*	Mild-alcoholic, sweet-sour, food beverage	Manipur
9.	*Basi*	Rice	Alcoholic beverage	Madhya Pradesh
10.	*Bhaati jaanr*	Rice, *marcha*	Mild-alcoholic, sweet-sour, food beverage	Darjeeling hills, Sikkim
11.	*Bhang-chyang*	Maize-rice/barley and *pham*	Extract of *mingri*; alcoholic beverages	Arunachal Pradesh
12.	*Buza*	Barley	Thick liquor	Ladakh
13.	*Chakti*	Jaggery	Filtrate; drinks	Himachal Pradesh
14.	*Chhang/ Chyang/ Chee*	Finger millet/barley and *phab*	Mild-alcoholic, slightly sweet-acidic; alcoholic beverage	Sikkim, Ladakh, Himachal Pradesh
15.	*Chhind*	Extracted sap of chhind (*Phoenix sylvestris*)	Mild alcoholic drink	Chhattisgarh
16.	*Chulli*	Apricot	Alcoholic beverages; filtrate; alcoholic drink	Himachal Pradesh
17.	*Daru*	Cereal	Alcoholic beverages; filtrate; jiggery	Himachal Pradesh
18.	*Dekuijao*	Rice	Alcoholic beverage, drunk directly	Nagaland
19.	*Duizou*	Red rice, *khekhrii*	Alcoholic drink	Nagaland
20.	*Ennog*	Black rice, paddy husk, *phut*	Alcoholic beverage	Arunachal Pradesh
21.	*Faapar ko jaanr*	Buck wheat, *marcha*	Mild-alcoholic, slightly acidic; alcoholic beverage	Darjeeling hills, Sikkim
22.	*Feni*	Cashew apple/ coconut	Distilled liquor	Goa

(continued)

Table 1.10 (continued)

S.no.	Alcoholic products	Raw materials and starter	Nature and use	State/region
23.	*Gahoon ko jaanr*	Wheat, *marcha*	Mild-alcoholic, slightly acidic; alcoholic beverage	Darjeeling hills, Sikkim
24.	*Ghanti*	Apple and apricot	Distillate; drinks	Himachal Pradesh
25.	*Handia*	Rice, *ranu*	Alcoholic beverage	Chhattisgarh, Madhya Pradesh
26.	*Haria*	Rice and *bhakar* or *dabai* or *mod pitha*	Alcoholic beverage	Assam, Bihar, Jharkhand, West Bengal, Odisha, Madhya Pradesh
27.	*Jann or jaan*	Rice, wheat, and other cereals and *balam*	Mild-alcoholic, slightly acidic; alcoholic beverage	Uttarakhand, Himachal Pradesh
28.	*Jao ko jaanr*	Barley and *marcha*	Mild-alcoholic, slightly acidic; alcoholic beverage	Darjeeling hills, Sikkim
29.	*Jhara*	Rice, *ranu dabai*	Sweetness, bitter taste, alcoholic beverage	West Bengal, Odisha
30.	*Jou*	Rice, *khekhrii*	Alcoholic beverage	Nagaland
31.	*Jou Bishi*	Rice, *angkur*	Alcoholic beverage	Assam
32.	*Judima*	Rice, *humao*	Alcoholic beverage, drunk directly	Assam
33.	*Juharo*	Rice, *humao*	Distilled liquor, drunk directly/with water	Assam
34.	*Juhning*	Rice	Alcoholic beverage; drunk directly	Assam
35.	*Kanji*	Beet root and carrot and *torami*	Thick filtrate, drunk directly	Karnataka, Tamil Nadu
36.	*Kiad*	Rice, *thiat*	Distillate; drunk directly	Meghalaya
37.	*Kodo ko jaanr*	Finger millet and *marcha*	Mild-alcoholic, slightly sweet-acidic; alcoholic beverage	Sikkim, Darjeeling hills
38.	*Khor*	Germinated rice	Non-distilled, mild alcoholic liquor	Manipur
39.	*Langi*	Rice, *chowan*	Alcoholic beverage	Tripura
40.	*Laopani*	Rice, *bhekur pitha*	Alcoholic beverage	Assam
41.	*Lugri*	Barley	Alcoholic beverage	Himachal Pradesh, Ladakh
42.	*Madda feni*	Cash	Distillate part of *sur*	Goa
43.	*Madhu*	Rice, *khekhrii*	Distilled liquor	Nagaland
44.	*Madua Apong*	Finger millet, *phab*	Distilled liquor	Arunachal Pradesh

(continued)

Table 1.10 (continued)

S.no.	Alcoholic products	Raw materials and starter	Nature and use	State/region
45.	*Mahua*	Dried corollas of *Madhuca longifolia* and *ranu*	Distilled liquor	Chhattisgarh, Madhya Pradesh
46.	*Mingari/Lohpani*	Maize, rice, or barley and *pham*	Alcoholic beverages	Arunachal Pradesh
47.	*Makai ko jaanr*	Maize, *marcha*	Mild-alcoholic, sweet-sour, food beverage	Darjeeling hills, Sikkim
48.	*Neera*	Saps of species of palm tree	Nonalcoholic beverage	Karnataka, Goa
49.	*Nchiangne*	Red rice	Distilled liquor	Nagaland
50.	*Nduijao*	Rice	Alcoholic beverage, drunk directly	Nagaland
51.	*Nyongin*	Rice, *opop*	Filtrate alcoholic liquor	Arunachal Pradesh
52.	*Oa*	Rice-fillet mixture, *paa*	Filtrate alcoholic liquor	Arunachal Pradesh
53.	*Opo*	Rice-millet and	Soft, alcoholic beverage	Arunachal Pradesh
54.	*Pachwai*	Rice, *bhakar*	Alcoholic beverage	West Bengal, Himachal Pradesh, Uttarakhand
55.	*Patso*	Rice, *chamri*	Non-distilled, mild alcoholic beverage	Manipur
56.	*Pendum*	Rice, herbs, *ranu* or *bhakar*	Alcoholic beverage	Madhya Pradesh
57.	*Pheijou*	Rice, husk, *khai*	Non-distilled, mild alcoholic beverage	Manipur, Nagaland
58.	*Pona/Pone*	Rice	Mild-alcoholic, sweet-sour, food beverage; paste	Arunachal Pradesh
59.	*Rak*	Apple, wild almond, apricot and jiggery, and *phab*	Filtrate; drunk directly	Himachal Pradesh
60.	*Raksi*	Cereals and *marcha*	Clear distilled liquor; alcoholic drink	Sikkim, Darjeeling hills
61.	*Rakzu*	Rice and *dawdim*	High alcoholic beverage, clear distilled liquor	Mizoram
62.	*Ruhi*	Rice	Distilled liquor	Nagaland
63.	*Salfi*	Trunk sap of Salfi (*Caryota urens*)	Milky alcoholic juice	Chhattisgarh
64.	*Sez*	Rice	Fermented cooked rice; mild alcoholic; snack	Uttaranchal
65.	*Sira-Oa*	Rice-fillet mixture, *paa*	Distilled liquor	Arunachal Pradesh

(continued)

Table 1.10 (continued)

S.no.	Alcoholic products	Raw materials and starter	Nature and use	State/region
66.	*Soor*	Fruits and *keem*	Distillate; drinks	Himalayan region
67.	*Sujen*	Rice and *mod pitha* or *perok kushi*	Sweet taste	West Bengal, Assam
68.	*Sur*	Coconut spadix or inflorescence	Alcoholic drink	Goa
69.	*Sura*	Finger millet	Food beverage; staple	Himachal Pradesh
70.	*Simal tarul ko jaanr*	Cassava tuber and *marcha*	Mild-alcoholic, sweet-sour; food beverage	Darjeeling hills, Sikkim
71.	*Themsing*	Finger millet/barley	Alcoholic beverages	Arunachal Pradesh
72.	*Timpui*	Rice, *khai*	Non-distilled, mild alcoholic beverage	Manipur
73.	*Tin-zu*	Rice and *dawdim*	High alcoholic beverage	Mizoram
74.	*Toddy* or *tari* or *kallu*	Sap extracted from palm, palmyra tree (*Borassus* spp.)	Fermented juice, thick alcoholic drinks	South India, Bihar, West Bengal, Jharkhand, Orissa
75.	*Xaj pani/ koloh pani*	Rice, *xaj pittha*	Alcoholic beverage	Assam
76.	*Waiyu*	Rice, husk, *hamei*	Non-distilled, mild alcoholic beverage	Manipur
77.	*Yu angouba*	Germinated rice	Non-distilled, mild alcoholic liquor	Manipur
78.	*Yu*	Rice, *hamei*	Distilled liquor	Manipur
79.	*Zoungao*	Germinated rice	Non-distilled, mild alcoholic liquor	Manipur, Nagaland
80.	*Zu*	Rice	Alcoholic beverages	Assam
81.	*Zouju/ zouzu*	Rice, *khai*	Distilled liquor	Manipur, Nagaland
82.	*Zutho/ Zhuchu*	Rice, *khekhrii*	Milky white, alcoholic beverage	Nagaland
83.	*Zupui*	Rice and *dawdim*	High alcoholic beverage	Mizoram
84.	*Zufang*	Sticky rice and *dawdim*	Mild alcoholic, sweet beverage	Mizoram

Table 1.11 Miscellaneous fermented products of India

S.no.	Fermented foods	Agro-produce and ingredient	Sensory character and edibles	State/region
1.	*Achar* or *chutney*	Domesticated or wild fruits, vegetables, plant parts mixed with oil, salt	Acidic, hot, and sour; pickles	All
2.	*Aithu*	Crab	Alkaline, strong flavor; side dish	Manipur
3.	*Aite-um*	Juvenile fresh water crab	Alkaline with strong ammonical flavor; side dish	Mizoram
4.	*Ai-um*	Fresh water crab and sesame seeds	Alkaline with ammonical flavor; side dish	Mizoram
5.	*Batema*	Elephant foot yam (*Amorphophallus paeoniifolius*, Nicolson)	Solid, round-cake like; side dish	Tripura
6.	*Bell dzu/dzii*	Passion fruit	Mild alcoholic, sweet beverage	Nagaland
7.	Black tea	Tea	Nonalcoholic drink	All
8.	*Bochu-mba*	Flower of *Bombax ceiba* (L.), commonly known as "cotton tree"	Dried; side dish	Tripura
9.	*Chhi-um*	Sesame seeds	Coarse paste with ammonical flavor; side dish	Mizoram
10.	*Chuk*	Fruits	Dark-brown paste, sour bitter taste; therapeutic uses	Sikkim, Darjeeling hills
11.	*Japangangnagtsu*	Crab (*Scylla* sp.) with sesame	Fermented; side dish	Nagaland
12.	*Kombucha* or tea fungus	Tea liquor	Nonalcoholic drink	Ladakh, Himachal Pradesh
13.	*Kothal dzu*	Jackfruit	Mild alcoholic, sweet beverage	Nagaland
14.	*Khulushi dzii*	Gooseberry	Mild alcoholic, sweet beverage	Nagaland
15.	*Tsugu* or *Tsiingen ngashi*	Crab and sesame seeds	Semisolid, flavored; curry	Nagaland

References

Aarti C, Khusro A, Arasu MV, Agastian P, Al-Dhabi NA (2016) Biological potency and characterization of antibacterial substances produced by Lactobacillus pentosus isolated from Hentak, a fermented fish product of North-East India. Springerplus 5:1743. https://doi.org/10.1186/s40064-016-3452-2.

Aarti C, Khusro A, Varghese R, Arasu MV, Agastian P, Al-Dhabi NA, Ilavenil S, Choi KC (2017) In vitro studies on probiotic and antioxidant properties of Lactobacillus brevis strain LAP2 isolated from Hentak, a fermented fish product of North-East India. LWT—Food Sci Technol 86:438–446

Abdhul K, Ganesh M, Shanmughapriya S, Kanagavel M, Anbarasu K, Natarajaseenivasan K (2014) Antioxidant activity of exopolysaccharide from probiotic strain *Enterococcus faecium* (BDU7) from Ngari. Int J Biol Macromol 70:450–454

Achaya KT (1991) Alcoholic fermentation and its products in ancient India. Indian J Hist Sci 26(2):123–129

Achaya KT (1994) Indian food: a historical companion. Oxford University Press, Delhi

Achaya KT (1998) A historical dictionary of Indian food. Oxford University Press, Delhi

Achaya KT (2003) The story of our food. Oxford University Press, Delhi

Achaya KT (2009) The illustrated foods of India. Oxford University Press, Delhi

Adhikari RR, Ghimirey H (2000) Nepalese society and culture. Vidharthi Pushtak Bhandar, Kathmandu (Nepali)

Agaliya JP, Jeevaratnam K (2013) Molecular characterization of lactobacilli isolated from fermented idli batter. Braz J Microbiol 44:1199–1206

Agarwal KN, Bhasin SK (2002) Feasibility studies to control acute diarrhea in children by feeding fermented milk preparations actimel and Indian Dahi. Eur J Clin Nutr 56:S56–S59

Angmo K, Bhalla TC (2014) Preparation of Phabs—an indigenous starter culture for production of traditional alcoholic beverage, Chhang, in Ladakh. Indian J Tradit Knowl 13:347–351

Anupma A, Pradhan P, Sha SP, Tamang JP (2018) Traditional skill of ethnic people of the Eastern Himalayas and North East India in preserving microbiota as dry amylolytic starters. Indian J Tradit Knowl 17(1):184–190

Balamurugan R, Chandragunasekaran AS, Chellappan G, Rajaram K, Ramamoorthi G, Ramakrishna BS (2014) Probiotic potential of lactic acid bacteria present in home made curd in southern India. Indian J Med Res 140(3):345–355

Basappa SC (2002) Investigations on Chhang form finger millet (Eleucine coracana Gaertn.) and its commercial prospects. Indian Food Ind 21(1):46–51

Basappa SC, Somashekar D, Agrawal R, Suma K, Bharathi K (1997) Nutritional composition of fermented ragi (chhang) by phab and defined starter cultures as compared to unfermented ragi (Eleucine coracana G.). Int J Food Sci 48:313–319

Batra LR (1986) Microbiology of some fermented cereals and grains legumes of India and vicinity. In: Hesseltine CW, Wang HL (eds) Indigenous fermented food of non-western origin. J. Cramer, Berlin, pp 85–104

Behare PV, Singh R, Singh RP (2009) Exopolysaccharides-producing mesophilic lactic cultures for preparation of fat-free dahi- an Indian fermented milk. J Dairy Res 76:90–97

Bhatia AK, Singh RP, Atal CK (1977) Chhang—the fermented beverage of Himalayan folk. Indian Food Packer 4:1–8

Boghra VR, Mathur ON (2000) Physico-chemical status of major milk constituents and minerals at various stages of shrikhand preparation. J Food Sci Technol 37:111–115

Bora SS, Keot J, Das S, Sarma K, Barooah M (2016) Metagenomics analysis of microbial communities associated with a traditional rice wine starter culture (Xaj-pitha) of Assam, India. 3 Biotech 6(2):153. https://doi.org/10.1007/s13205-016-0471-1

Bose DK (1922) Wine in ancient India. Connor, Kolkata

Bryant EF (2007) Krishna: a sourcebook. Oxford University Press, New Delhi

Chakrabarty J, Sharma GD, Tamang JP (2014) Traditional technology and product characterization of some lesser-known ethnic fermented foods and beverages of North Cachar Hills District of Assam. Indian J Tradit Knowl 13(4):706–715

Chandrasekar Rajendran SC, Chamlagain B, Kariluoto S, Piironen V, Saris PEJ (2017) Biofortification of riboflavin and folate in idli batter, based on fermented cereal and pulse, by Lactococcus lactis N8 and Saccharomyces boulardii SAA 655. J Appl Microbiol 122(6):1663–1671

Chelliah R, Ramakrishnan SR, Prabhu PR, Antony U (2016) Evaluation of antimicrobial activity and probiotic properties of wild strain Pichia kudriavzevii isolated from frozen idli batter. Yeast 33:385–401

Chettri R, Tamang JP (2008) Microbiological evaluation of maseura, an ethnic fermented legume-based condiment of Sikkim. J Hill Res 21(1):1–7

Chettri R, Tamang JP (2014) Functional properties of tungrymbai and bekang, naturally fermented soybean foods of India. Int J Fermented Foods 3:87–103

Chettri R, Tamang JP (2015) Bacillus species isolated from tungrymbai and bekang, naturally fermented soybean foods of India. Int J Food Microbiol 197:72–76

Chettri R, Bhutia M, Tamang JP (2016) Poly-γ-glutamic acid (PGA)-producing Bacillus species isolated from Kinema, Indian fermented soybean food. Front Microbiol 7:971. https://doi.org/10.3389/fmicb.2016.00971

Chowdhury N, Goswami G, Hazarika S, Sharma-Pathak S, Barooah M (2018) Microbial dynamics and nutritional status of namsing: a traditional fermented fish product of Mishing community of Assam. Proc Natl Acad Sci India Sect B Biol Sci 89:1–12. https://doi.org/10.1007/s40011-018-1022-9

Dandile UM, Pawar BK, Choudhari DM (2014) Sensory quality of Shrikhand prepared by using cardamom and saffron. Res J Animal Hus Dairy Sci 5:1–5

De Mandal S, Singh SS, Bose Muthukumaran R, Thanzami K, Kumar V, Senthil Kumar N (2018) Metagenomic analysis and the functional profiles of traditional fermented pork fat 'sa-um' of Northeast India. AMB Express 8:163

Devi KR, Deka M, Jeyaram K (2015) Bacterial dynamics during yearlong spontaneous fermentation for production of ngari, a dry fermented fish product of Northeast India. Int J Food Microbiol 199:62–71

Devshete NG, Hembade VR, Mandwade KG, Burbade R (2013) Preparation of omega-3 enriched probiotic Shrikhand using walnut powder. Int J Proc Post Harv Technol 4:74–78

Dewan S, Tamang JP (2006) Microbial and analytical characterization of Chhu—a traditional fermented milk product of Sikkim Himalayas. J Sci Ind Res 65:747–752

Dewan S, Tamang JP (2007) Dominant lactic acid bacteria and their technological properties isolated from the Himalayan ethnic fermented milk products. Antonie Van Leeuwenhoek 92:343–352

Dubey A, Jeevaratnam K (2015) Probiotic screening of lactobacilli isolates from uttapam batter fermented supplementing with *piper betel* leaves. Adv Microbiol 5:858–870

Dundas P (2002) The jains, 2nd edn. Routledge, London

Fuller DQ (2015) Harappan seeds and agriculture: some considerations. Antiquity 75:410–414. https://doi.org/10.1017/S0003598X00061068

Ghosh D, Chattopadhyay P (2011) Preparation of idli batter, its properties and nutritional improvement during fermentation. J Food Sci Technol 48(5):610–615

Ghosh J, Rajorhia GS (1990) Selection of starter culture for production of indigenous fermented milk product (Misti dahi). Lait 70:147–154

Ghosh K, Maity C, Adak A, Halder SK, Jana A, Das A, Parua S, Das Mohapatra PK, Pati BR, Mondal KC (2014) Ethnic preparation of Haria, a rice-based fermented beverage, in the province of Lateritic West Bengal, India. Ethnobot Res Appl 12:39–49

Ghosh K, Ray M, Adak A, Dey P, Halder SK, Das A, Jana A, Parua S, Das Mohapatra PK, Pati BR, Mondal KC (2015a) Microbial, saccharifying and antioxidant properties of an Indian rice based fermented beverage. Food Chem 168:196–202

Ghosh K, Ray M, Adak A, Halder SK, Das A, Jana A, Parua S, Vágvölgyi C, Das Mohapatra PK, Pati BR, Mondal KC (2015b) Role of probiotic Lactobacillus fermentum KKL1 in the preparation of a rice based fermented beverage. Bioresour Technol 188:161–168

Giri SS, Janmejay LS (1994) Changes in soluble sugars and other constituents of bamboo shoots in soibum fermentation. J Food Sci Technol 31(6):500–502

Gode PK (1943) Some notes on the history of Indian dietetics with special reference to the history of jalebi. New Indian Antiq 6:169–181

Goel G, Kakkar M, Puniya A, Teotia U, Singh K (2005) Microbiological and chemical profiles of Kanji: a naturally fermented product of carrot. Asian J Microbiol Biotechnol Environ Sci 7(4):815–817

Gorer G (1938) The Lepchas of Sikkim. Gian Publishing House, Delhi

Guha BS (1937) An outline of racial ethnology of India. In: Hora SL (ed) An outline of the field sciences in India. Indian Science Congress Association, Calcutta, pp 125–139

Gupta A, Tiwari SK (2014) Probiotic potential of Lactobacillus plantarum LD1 isolated from batter of Dosa, a South Indian fermented food. Probiotics Antimicrob Proteins 6(2):73–81

Gupta M, Khetarpaul N, Chauhan BM (1992a) Preparation, nutritional value and acceptability of barley rabadi—an indigenous fermented food of India. Plant Foods Hum Nutr 42:351–358

Gupta M, Khetarpaul N, Chauhan BM (1992b) Rabadi fermentation of wheat: changes in phytic acid content and in vitro digestibility. Plant Foods Hum Nutr 42:109–116

Gupta RC, Mann B, Joshi VK, Prasad DN (2000) Microbiological, chemical and ultrastructural characteristics of misti doi (sweetened dahi). J Food Sci Technol 37:54–57

Harvey P (2013) An introduction to Buddhism: teachings, history and practices, 2nd edn. Cambridge University Press, Cambridge, pp 88–90

Hegde S, Nair LP, Chandran H, Irshad H (2018) Traditional Indian way of eating—an overview. J Ethnic Foods 5(1):20–23. https://doi.org/10.1016/j.jef.2018.02.001

Hinnells JR (1997) A new handbook of living religions. Pengiun, London

Hooker JD (1854) Himalayan journals: notes of a naturalist in Bengal, the Sikkim and Nepal Himalayas, the Khasia mountains. John Murray, London

Hussaini MM (1993) Islamic dietary concepts and practices. Islamic Food and nutrition Council of America, Bedford Park, IL

INTACH (The Indian National Trust for Arts and Cultural Heritage) (2016) Haryana Newsletter. INTACH, New Delhi, p 36

Jadhav SR, Shandilya UK, Kansal VK (2013) Exploring the ameliorative potential of probiotic dahi containing Lactobacillus acidophilus and bifidobacterium bifidum on dextran sodium sulphate induced colitis in mice. J Dairy Res 80:21–27

Jain S, Yadav H, Sinha PR, Kapila S, Naito Y, Marotta F (2010) Anti-allergic effects of probiotic dahi through modulation of the gut immune system. Turk J Gastroenterol 21:244–250

Jamison S, Brereton J (2014) The Rigveda: the earliest religious poetry of India. Oxford University Press, New Delhi

Jeyaram K, Mohendro Singh W, Premarani T, Ranjita Devi A, Selina Chanu K, Talukdar NC, Rohinikumar Singh M (2008a) Molecular identification of dominant microflora associated with 'Hawaijar'—a traditional fermented soybean (*Glycine max* L.) food of Manipur, India. Int J Food Microbiol 122:259–268

Jeyaram K, Singh M, Capece A, Romano P (2008b) Molecular identification of yeast species associated with "Hamei"—a traditional starter used for rice wine production in Manipur, India. Int J Food Microbiol 124:115–125

Jeyaram J, Anand Singh T, Romi W, Ranjita Devi A, Mohendro Singh W, Dayanidhi H, Rajmuhon Singh N, Tamang JP (2009) Traditional fermented foods of Manipur. Indian J Tradit Knowl 8(1):115–121

Jeyaram K, Romi W, Singh TA, Devi AR, Devi SS (2010) Bacterial species associated with traditional starter cultures used for fermented bamboo shoot production in Manipur state of India. Int J Food Microbiol 143:1–8

Jeyaram K, Tamang JP, Capece A, Patrizia Romano P (2011) Geographical markers for *Saccharomyces cerevisiae* strains with similar technological origins domesticated for rice-based ethnic fermented beverages production in North East India. Antonie Van Leeuwenhoek 100(4):569–578

Joshi N, Godbole SH, Kanekar P (1989) Microbial and biochemical changes during dhokla fermentation with special reference to flavour compounds. J Food Sci Technol 26(2):113–115

Kakati BK, Goswami UC (2013a) Microorganisms and the nutritive value of traditional fermented fish products of Northeast India. Global J Biosci Biotechnol 2(1):124–127

Kakati BK, Goswami UC (2013b) Characterization of the traditional fermented fish product Shidol of Northeast India prepared from Puntius sophore and *Setipinna phasa*. Indian J Tradit Knowl 12:85–90

Kalki CS, Shetty PK (2015) Isolation and characterization of exopolysaccharide from Leuconostoc lactis KC117496 isolated from idli batter. Int J Biol Macromol 90:100–106. https://doi.org/10.1016/j.ijbiomac.2015.02.007

Kanekar P, Joshi N (1993) Lactobacillus fermentum, Leuconostoc mesenteroides and Hansenula silvicola contributing to acetoin and folic acid during dhokla fermentation. Indian J Microbiol 33:111–117

Karche RV, Thakare VM, Bhagat AV, Shirsath SA (2015) Microbiological quality of cow milk shrikhand blended with sapota pulp. Int J Food Agric Vet Sci 5:18–22

Katiyar SK, Bhasin AK, Bhatia AK (1991) Traditionally processed and preserved milk products of Sikkimese Tribes. Sci Cult 57(10–11):256–258

Keisam S, Tuikhar N, Ahmed G, Jeyaram K (2019) Toxigenic and pathogenic potential of enteric bacterial pathogens prevalent in the traditional fermented foods marketed in the Northeast region of India. Int J Food Microbiol 296:21–30

Keishing S, Banu AT (2013) Hawaijar—a fermented soya of Manipur, India. IOSR J Environ Sci Toxicol Food Technol 4(2):29–33

Khatri PK (1987) Nepali Samaj ra Sanskriti (Prachin-Madhyakal). Shaja Prakashan, Kathmandu (Nepali)

Khetarpaul N, Chauhan BM (1989) Effect of fermentation by pure cultures of yeasts and lactobacilli on phytic acid and polyphenol content of pearl millet. J Food Sci 54:78–781

Khetarpaul N, Chauhan BM (1990a) Fermentation of pearl millet flour with yeasts and lactobacilli: in vitro digestibility and utilization of fermented flour for weaning mixtures. Plant Foods Hum Nutr 40:167–173

Khetarpaul N, Chauhan BM (1990b) Effect of fermentation by pure cultures of yeasts and lactobacilli on the available carbohydrate content of pearl millet. Food Chem 36:287–293

Khetarpaul N, Chauhan BM (1991) Effect of pure sequential culture fermentation by yeasts and lactobacilli on HCl-extractability of minerals from pearl millet. Food Chem 39:347–355

Khorjuvenkar SNP (2016) Diversity among yeast isolated from naturally fermented cashew apple juice. PhD thesis, Goa University

Kilara A, Iya KK (1992) Food practices of the Hindu. Food Technol 46:94–104

Kingston JJ, Radhika M, Roshini PT, Raksha MA, Murali HS, Batra HV (2010) Molecular characterization of lactic acid bacteria recovered from natural fermentation of beet root and carrot Kanji. Indian J Microbiol 50:292–298

Krishna Jois SN (1969) Supa sastra of mangarasa (AD 1516). University of Mysore, Karnataka, pp 3–252

Kulkarni PH (2002) Ayurvedic aahar: the scientific diet. Sri Satguru, Delhi

Kunjithapatham S, Balasubramaniam H (2012) Physicochemical properties of fresh and stored coconut palm toddy. J Psychol Psychother 1:397. https://doi.org/10.4172/scientificreports.397

Kwon DY, Tamang JP (2015) Religious ethnic foods. J Ethnic Foods 2:45–46

Lohekar AS, Arya AB (2014) Development of value added instant 'dhokla' mix. In J Food Nutr Sci 3(4):78–83

Mahdihassan S (1979) Distillation assembly of pottery in ancient India with a single item of special construction. Visvesvaran Indol J 17:264

Mahdihassan S (1981) Parisrut the earliest distilled liquor of Vedic times or of about 1500 BC. Indian J Hist Sci 16(2):223–229

Majumdar RK, Basu S, Nayak BB (2009) Assessment of nutritional quality of 'Shidal' a fermented fish product of Northeast India. Indian J. Fisher Assoc 36:25–34

Majumdar RK, Bejjanki SK, Roy D, Shitole S, Saha A, Narayan B (2015a) Biochemical and microbial characterization of Ngari and Hentaak-traditional fermented fish products of India. J Food Sci Technol 52:8284–8291

Majumdar RK, Roy D, Bejjanki S, Bhaskar N (2015b) Chemical and microbial properties of shidal, a traditional fermented fish of Northeast India. J Food Sci Technol 53(1):401–410

Majumder PP (2001) Ethnic populations of India as seen from an evolutionary perspective. J Biosci Indian Acad Sci 26(4):533–545

Mangang KCS, Das AJ, Deka SC (2017) Comparative shelf life study of two different rice beers prepared using wild-type and established microbial starters. J Inst Brew 123:579–586
Mitragotri VR (1992) A socio-cultural history of Goa from the bhoja to Vijayanagar rulers. PhD thesis, Goa University, Goa
Modha H, Pal D (2011) Optimization of Rabadi-like fermented milk beverage using pearl millet. J Food Sci Technol 48(2):190–196
Mohanan KR, Shankar PA, Laxminarayana H (1984) Microflora of dahi prepared under household conditions of Bangalore. J Food Sci Technol 21:45–46
Mohania D, Kansal VK, Kruzliak P, Kumari A (2014) Probiotic dahi containing Lactobacillus acidophilus and bifidobacterium bifidum modulates the formation of aberrant crypt foci, mucin-depleted foci, and cell proliferation on 1,2-dimethylhydrazine-induced colorectal carcinogenesis in wistar rats. Rejuvenation Res 17:325–333
Mukherjee SK, Albury CS, Pederson AG, Steinkraus KH (1965) Role of Leuconostoc mesenteroides in leavening the batter of idli, a fermented food of India. App Microbiol 13(2):227–231
Muzaddadi AU (2015) Minimisation of fermentation period of shidal from barbs (Puntius spp.). Fish Technol 52:34–41
Muzaddadi AU, Basu S (2003) Seedal-an indigenous fermented fishery product of North-east India. Fisher Chim 23:30–32
Muzaddadi AU, Basu S (2012) An accelerated process for fermented fish (seedal) production in North-east region of India. Indian J Anim Sci 82:98–106
Oki K, Rai AK, Sato S, Watanabe K, Tamang JP (2011) Lactic acid bacteria isolated from ethnic preserved meat products of the Western Himalayas. Food Microbiol 28:1308–1315
Olivelle P (2006) Between the empires: society in India 300 BCE to 400 CE. Oxford University Press, Oxford. ISBN 978-0-19-977507-1
Omizu Y, Tsukamoto C, Chettri R, Tamang JP (2011) Determination of saponin contents in raw soybean and fermented soybean foods of India. J Sci Ind Res 70:533–538
Padghan PV, Mann B, Rajeshkumar, Sharma R, Kumar A (2015) Studies on bio-functional activity of traditional lassi. Indian J Tradit Know 1:124–131
Palanisamy BD, Rajendran V, Sathyaseelan S, Bhat R, Venkatesan BP (2012) Enhancement of nutritional value of finger millet-based food (Indian dosa) by co-fermentation with horse gram flour. Int J Food Sci Nutr 63(1):5–15
Patel P, Patel V, Subhash R (2015) Probiotic lassi preparation and sensory evaluation using *L. paracasei* immobilized dry fruit pieces. Int J Fermented Foods 4(1–2):23–33
Patidar SK, Prajapati JB (1988) Standardization and evaluation of lassi prepared using Lb. acidophilus and S. thermophiles. J Food Sci Technol 35:428–431
Possehl GL (2002) The Indus civilization: a contemporary perspective. Altamira Press, Lanham, MD
Prakash O (1961) Food and drinks in ancient India. Munshi Ram Monoharlal Publishers, Delhi
Prakash O (1987) Economy and food in ancient India, part II food. Bharatiya Vidya Prakashan, Delhi
Rai AK, Palni U, Tamang JP (2009) Traditional knowledge of the Himalayan people on the production of indigenous meat products. Indian J Tradit Knowl 8(1):104–109
Rai AK, Tamang JP, Palni U (2010a) Microbiological studies of ethnic meat products of the Eastern Himalayas. Meat Sci 85:560–567
Rai AK, Tamang JP, Palni U (2010b) Nutritional value of lesser-known ethnic meat products of the Himalayas. J Hill Res 23(1&2):22–25
Rai R, Kharel N, Tamang JP (2014) HACCP model of kinema, a fermented soybean food. J Sci Ind Res 73:588–592
Rai R, Shangpliang HNJ, Tamang JP (2016) Naturally fermented milk products of the Eastern Himalayas. J Ethnic Foods 3:270–275
Rajpal S, Kansal VK (2009) Probiotic dahi containing Lactobacillus acidophilus, Bifidobacterium bifidum stimulates immune system in mice. Milchwissenschaft 64:147–150
Rapsang GF, Joshi SR (2012) Bacterial diversity associated with tungtap, an ethnic traditionally fermented fish product of Meghalaya. Indian J Tradit Knowl 11:134–138

Rapsang GF, Kumar R, Joshi SR (2011) Identification of Lactobacillus puhozihii from tungtap: A traditionally fermented fish food, and analysis of its bacteriocinogenic potential. Afr J Biotechnol 10:12237–12243

Rathi SD, Deshmukh DK, Ingle UM, Syed HM (1990) Studies on the physico-chemical properties of freeze-dried dahi. Indian J Dairy Sci 43:249–251

Ray S, Bagyaraj DJ, Thilagar G, Tamang JP (2016) Preparation of Chyang, an ethnic fermented beverage of the Himalayas using different raw cereals. J Ethnic Foods 3:297–299

Regubalan B, Ananthanarayan L (2018) Shelf life improvement of idli batter by addition of mustard essential oil as bio-preservative. J Food Sci Technol 55(9):3417–3426

Regubalan B, Ananthanarayan L (2019) Investigation of biogenic amines content in fermented idli batter during storage. J Food Sci Technol 56(4):1775–1784

Richards JF (1995) The Mughal Empire. Cambridge University Press, London

Risley HH (1928) The gazetteer of Sikkim. D. K. Publishing Distributors (P) Ltd, New Delhi

Romi W, Keisam S, Ahmed G, Jeyaram K (2014) Reliable differentiation of Meyerozyma guilliermondii from Meyerozyma caribbica by internal transcribed spacer restriction fingerprinting. BMC Microbiol 14:52–62

Romi W, Ahmed G, Jeyaram K (2015) Three-phase succession of autochthonous lactic acid bacteria to reach a stable ecosystem within 7 days of natural bamboo shoot fermentation as revealed by different molecular approaches. Mol Ecol 24(13):3372–3389. https://doi.org/10.1111/mec.13237

Sandhu DK, Soni SK (1989) Microflora associated with Indian Punjabi warri fermentation. J Food Sci Technol 26:21–25

Sarangthem K, Singh TN (2003) Microbial bioconversion of metabolites from fermented succulent bamboo shoots into phytosterols. Curr Sci 84(12):1544–1547

Saraniya A, Jeevaratnam K (2015) In vitro probiotics evaluation of phytase producing Lactobacillus species isolated from Uttapam batter and their application in soy milk fermentation. J Food Sci Technol 52(9):5631–5640

Sarkar PK, Tamang JP (1994) The influence of process variables and inoculum composition on the sensory quality of kinema. Food Microbiol 11:317–325

Sarkar PK, Tamang JP (1995) Changes in the microbial profile and proximate composition during natural and controlled fermentation of soybeans to produce kinema. Food Microbiol 12:317–325

Sarkar PK, Tamang JP, Cook PE, Owens JD (1994) Kinema—a traditional soybean fermented food: proximate composition and microflora. Food Microbiol 11:47–55

Sarkar PK, Jones LJ, Gore W, Craven GS (1996) Changes in soya bean lipid profiles during kinema production. J Sci Food Agric 71:321–328

Sarkar PK, Jones LJ, Craven GS, Somerset SM, Palmer C (1997) Amino acid profiles of kinema, a soybean-fermented food. Food Chem 59(1):69–75

Sarkar PK, Morrison E, Tingii U, Somerset SM, Craven GS (1998) B-group vitamin and mineral contents of soybeans during kinema production. J Sci Food Agric 78:498–502

Sarkar A, Mukherjee AD, Bera MK, Das A, Juyal N, Morthekai R, Deshpande RD, Shinde VS, Rao KS (2016) Oxygen isotope in archaeological bioapatites from India: Implications to climate change and decline of Bronze Age Harappan civilization. Sci Rep 6:26555. https://doi.org/10.1038/srep26555

Sarma PJ (1939) The art of healing. In: Rigveda annals of medical history, 3rd series, vol 1. P.B. Hoeber, New York, p 538

Sawhney S, Versha P (2018) Concept of ahara in ayurveda. Int J Ayurvedic Med 6(5):1124–1130

Sha SP, Anupma A, Pradhan P, Prasad GS, Tamang JP (2016) Identification of yeasts by PCR-mediated DGGE in marcha, an ethnic amylolytic starter of India. J Ethnic Foods 3:292–296

Sha SP, Jani K, Sharma A, Anupma A, Pradhan P, Shouche Y, Tamang JP (2017) Analysis of bacterial and fungal communities in Marcha and Thiat, traditionally prepared amylolytic starters of India. Sci Rep 7:10967. https://doi.org/10.1038/s41598-017-11609-y.

Sha SP, Suryavanshi MS, Jani K, Sharma A, Shouche Y, Tamang JP (2018) Diversity of yeasts and molds by culture-dependent and culture-independent methods for mycobiome surveillance of traditionally prepared dried starters for the production of Indian alcoholic beverages. Front Microbiol 9:2237. https://doi.org/10.3389/fmicb.2018.02237

Sha SP, Suryavanshi MS, Tamang JP (2019) Mycobiome diversity in traditionally prepared starters for alcoholic beverages in India by high-throughput sequencing method. Front Microbiol 10:348. https://doi.org/10.3389/fmicb.2019.003482237

Shamala TR, Sreekantiah KR (1988) Microbiological and biochemical studies on traditional Indian palm wine fermentation. Food Microbiol 5:157–162

Shangpliang HNK, Rai R, Keisam S, Jeyaram K, Tamang JP (2018) Bacterial community in naturally fermented milk products of Arunachal Pradesh and Sikkim of India analysed by high-throughput amplicon sequencing. Sci Rep 8:1532. https://doi.org/10.1038/s41598-018-19524-6

Sharma N, Barooah M (2017) Microbiology of khorisa, its proximate composition and probiotic potential of lactic acid bacteria present in Khorisa, a traditional fermented bamboo shoot product of Assam. Indian J Nat Prod Resour 8:78–88

Sharma NK, Gill JPS, Joshi DV, Kwatra MS (1993) Microflora of Indian fermented milk products (dahi). Indian J Dairy Sci 46:85–87

Shobha D, Joshi N (2016) Evaluation of maize dhokla for physical, sensory and functional parameters. Int J Agric Sci 8(36):1755–1758

Shrestha R, Mehta M, Shah G (2017) Isolation and characterization of khaman fermenting microorganisms. Trends Biosci 10(8):1574–1576

Shrivastava N, Ananthanarayan L (2015) Use of the backslopping method for accelerated and nutritionally enriched idli fermentation. J Sci Food Agric 95(10):2081–2087

Shruti S, Kansal VK (2011) Effect of feeding probiotic dahi containing Lactobacillus acidophilus and Bifidobacterium bifidum on enzymes that catalyze carcinogen activation and detoxification in rats. Milchwissenschaft 66:244–247

Singh D, Singh J (2014) Shrikhand: a delicious and healthful traditional Indian fermented dairy dessert. Trends Biosci 7:153–155

Singh NL, Ramprasad, Mishra PK, Shukla SK, Kumar J, Singh R (2010) Alcoholic fermentation techniques in early Indian tradition. Indian J Hist Sci 45(2):163–173

Singh TA, Devi KR, Ahmed G, Jeyaram K (2014) Microbial and endogenous origin of fibrinolytic activity in traditional fermented foods of Northeast India. Food Res Int 55:356–362

Singh BR, Karan R, Singh V (2014a) Microbial quality and safety of Axone—Akhuni, a fermented soybean food of Nagaland. Noto Med 2014:13

Singh D, Singh J, Kumar S, Verma T (2014b) Microbiological evaluation of soy fortified shrikhand by using response surface methodology. Int J Appl Biol Pharm Technol 5:1–7

Singh SS, Mandal SE, Lalnunmawii E, Kumar NS (2018) Antimicrobial, antioxidant and probiotics characterization of dominant bacterial isolates from traditional fermented fish of Manipur, North-East India. J Food Sci Technol 55:1870–1879

Sinha SC (1996) Dictionary of philosophy. Anmol Publications, New Delhi

Sohliya I, Joshi SR, Bhagobaty RK, Kumar R (2009) Tungrymbai—a traditional fermented soybean food of the ethnic tribes of Meghalaya. Indian J Tradit Knowl 8(4):559–561

Sonar NR, Halami PM (2014) Phenotypic identification and technological attributes of native lactic acid bacteria present in fermented bamboo shoot products from North-East India. J Food Sci Technol 51(12):4143–4148. https://doi.org/10.1007/s13197-014-1456-x

Sonar NR, Vijayendra SVN, Prakash M, Saikia M, Tamang JP, Halami PM (2015) Nutritional and functional profile of traditional fermented bamboo shoot based products from Arunachal Pradesh and Manipur states of India. Int Food Res J 22(2):788–797

Soni SK, Sandhu DK (1989a) Nutritional improvement of Indian dosa batter by yeast enrichment and black gram replacement. J Ferment Bioeng 68(1):1–4

Soni SK, Sandhu DK (1989b) Fermentation of idli: effects of changes in raw materials and physico-chemical conditions. J Cereal Sci 10:227–238

Soni SK, Sandhu DK (1990) Biochemical and nutritional changes associated with Indian Punjab Wari fermentation. J Food Sci Technol 27:82–85

Soni SK, Sandhu DK (1991) Role of yeast domination in Indian idli batter fermentation. World J Microbiol Biotechnol 7:505–507

Soni SK, Sandhu DK, Vilkhu KS (1985) Studies on dosa—an indigenous Indian fermented food: some biochemical changes accompanying fermentation. Food Microbiol 2:175–181
Soni SK, Sandhu DK, Vilkhu KS, Kamra N (1986) Microbiological studies on dosa fermentation. Food Microbiol 3:45–53
Sousa G (2019) Biggest ethnic groups in India. WorldAtlas. worldatlas.com/articles/biggest-ethnic-groups-in-india.html. Accessed 18 July 2019
Sridevi J, Halami PM, Vijayendra SVN (2010) Selection of starter cultures for idli batter fermentation and their effect on quality of idli. J Food Sci Technol 47:557–563
Srinivasan K, Kunjithapatham S, Manickam L (2013) Traditional Indian breakfast (idli and dosa) with enhanced nutritional content using millets. Nutr Diet 70:241–246. https://doi.org/10.1111/1747-0080.12020
Steinkraus KH, van Veer AG, Thiebeau DB (1967) Studies on idli—an Indian fermented black gram-rice food. Food Technol 21(6):110–113
Sudheendra Ch VK, Madhavi TV, Gopikrishna G, Shah RK (2018) Study of microbial changes in probiotic and synbiotic lassi during storage. Pharm Innov 7(2):94–97
Suman, Khetarpaul N (2018) Sensory, microbial, texture and nutritional evaluation of okara supplemented probiotic dhokla. Int J Curr Microbiol Appl Sci 7(4):1274–1283
Sura K, Garg S, Garg FC (2001) Microbiological and biochemical changes during fermentation of Kanji. J Food Sci Technol 38:165–167
Tamang JP (1999) Development of pulverised starter for kinema production. J Food Sci Technol 36:475–478
Tamang JP (2003) Native microorganisms in fermentation of kinema. Indian J Microbiol 43(2):127–130
Tamang JP (2010a) Himalayan fermented foods: microbiology, nutrition, and ethnic values. CRC Press, Taylor & Francis Group, New York
Tamang JP (2010b) Diversity of fermented foods. In: Tamang JP, Kailasapathy K (eds) Fermented foods and beverages of the world. CRC Press, Taylor & Francis Group, New York, pp 41–84
Tamang JP (2010c) Diversity of fermented beverages. In: Tamang JP, Kailasapathy K (eds) Fermented foods and beverages of the world. CRC Press, Taylor & Francis Group, New York, pp 85–125
Tamang JP (2015) Naturally fermented ethnic soybean foods of India. J Ethnic Foods 2:8–17
Tamang JP, Nikkuni S (1996) Selection of starter culture for production of kinema, a fermented soybean food of the Himalaya. World J Microbiol Biotechnol 12:629–635
Tamang JP, Nikkuni S (1998) Effect of temperatures during pure culture fermentation of Kinema. World J Microbiol Biotechnol 14(6):847–850
Tamang JP, Samuel D (2010) Dietary culture and antiquity of fermented foods and beverages. In: Tamang JP, Kailasapathy K (eds) Fermented foods and beverages of the world. CRC Press, Taylor & Francis Group, New York, pp 1–40
Tamang JP, Sarkar PK (1993) Sinki—a traditional lactic acid fermented radish tap root product. J Gen Appl Microbiol 39:395–408
Tamang JP, Sarkar PK (1995) Microflora of marcha: an amylolytic fermentation starter. Microbios 81:115–122
Tamang JP, Sarkar PK (1996) Microbiology of mesu, a traditionally fermented bamboo shoot product. Int J Food Microbiol 29(1):49–58
Tamang B, Tamang JP (2010) In situ fermentation dynamics during production of gundruk and khalpi, ethnic fermented vegetables products of the Himalayas. Indian J Microbiol 50(Suppl 1):S93–S98
Tamang JP, Thapa S (2006) Fermentation dynamics during production of bhaati jaanr, a traditional fermented rice beverage of the Eastern Himalayas. Food Biotechnol 20(3):251–261
Tamang JP, Sarkar PK, Hesseltine CW (1988) Traditional fermented foods and beverages of Darjeeling and Sikkim—a review. J Sci Food Agric 44:375–385
Tamang JP, Dewan S, Olasupo NA, Schillinger V, Holzapfel WH (2000) Identification and enzymatic profiles of predominant lactic acid bacteria isolated from soft variety chhurpi, a traditional cheese typical of the Sikkim Himalayas. Food Biotechnol 14(1 & 2):99–112

Tamang JP, Thapa S, Dewan S, Yasuka J, Fudou R, Yamanaka S (2002) Phylogenetic analysis of Bacillus strains isolated from fermented soybean foods of Asia: Kinema, Chungkokjang and Natto. J Hill Res 15(2):56–62
Tamang JP, Tamang B, Schillinger U, Franz CMAP, Gores M, Holzapfel WH (2005) Identification of predominant lactic acid bacteria isolated from traditionally fermented vegetable products of the Eastern Himalayas. Int J Food Microbiol 105:347–356
Tamang JP, Dewan S, Tamang B, Rai A, Schillinger U, Holzapfel WH (2007) Lactic acid bacteria in Hamei and Marcha of North East India. Indian J Microbiol 47:119–125
Tamang B, Tamang JP, Schillinger U, Franz CMAP, Gores M, Holzapfel WH (2008) Phenotypic and genotypic identification of lactic acid bacteria isolated from ethnic fermented tender bamboo shoots of North East India. Int J Food Microbiol 121:35–40
Tamang JP, Tamang B, Schillinger U, Guigas C, Holzapfel WH (2009) Functional properties of lactic acid bacteria isolated from ethnic fermented vegetables of the Himalayas. Int J Food Microbiol 135:28–33
Tamang JP, Tamang N, Thapa S, Dewan S, Tamang BM, Yonzan H, Rai AK, Chettri R, Chakrabarty J, Kharel N (2012) Microorganisms and nutritional value of ethnic fermented foods and alcoholic beverages of North East India. Indian J Tradit Knowl 11:7–25
Tamang JP, Holzapfel WH, Watanabe K (2016a) Diversity of microorganisms in global fermented foods and beverages. Front Microbiol 7:377. https://doi.org/10.3389/fmicb.2016.00377
Tamang JP, Shin DH, Jung SJ, Chae SW (2016b) Functional properties of microorganisms in fermented foods. Front Microbiol 7:578. https://doi.org/10.3389/fmicb.2016.00578
Teramoto Y, Yoshida S, Ueda S (2002) Characteristics of a rice beer (zutho) and a yeast isolated from the fermented product in Nagaland, India. Int J Food Microbiol 18(9):813–816
Thakur N (2013) Characterization of traditional fermentation processes used for the production of some alcoholic beverages (Chhang, Sura and Jau Chhang) in Himachal Pradesh. PhD thesis, Himachal Pradesh University, Summer Hill, Shimla, India
Thakur N, Savitri, Bhalla TC (2004) Characterization of traditional fermented foods and beverages of Himachal Pradesh. Indian J Tradit Knowl 3:325–335
Thapa N (2016) Ethnic fermented and preserved fish products of India and Nepal. J Ethnic Foods 3:69–77
Thapa N, Pal J (2007) Proximate composition of traditionally processed fish products of the Eastern Himalayas. J Hill Res 20(2):75–77
Thapa S, Tamang JP (2004) Product characterization of kodo ko jaanr: fermented finger millet beverage of the Himalayas. Food Microbiol 21:617–622
Thapa S, Tamang JP (2006) Microbiological and physico-chemical changes during fermentation of kodo ko jaanr, a traditional alcoholic beverage of the Darjeeling hills and Sikkim. Indian J Microbiol 46(4):333–341
Thapa N, Pal J, Tamang JP (2004) Microbial diversity in ngari, hentak and tungtap, fermented fish products of Northeast India. World J Microbiol Biotechnol 20(6):599–607
Thapa N, Pal J, Tamang JP (2006) Phenotypic identification and technological properties of lactic acid bacteria isolated from traditionally processed fish products of the Eastern Himalayas. Int J Food Microbiol 107(1):33–38
Thapa N, Pal J, Tamang JP (2007) Microbiological profile of dried fish products of Assam. Indian J. Fish 54(1):121–125
Thokchom S, Joshi SR (2015) Physicochemical analysis of ethnically fermented soybean products of North-East India and molecular characterization of associated lactic acid bacteria. Proceedings of the National Academy of Sciences, India Section B. Biol Sci 85:527–533
Tsuyoshi N, Fudou R, Yamanaka S, Kozaki M, Tamang N, Thapa S, Tamang JP (2005) Identification of yeast strains isolated from marcha in Sikkim, a microbial starter for amylolytic fermentation. Int J Food Microbiol 99:135–146
Twaigery S, Spillman D (1989) An introduction to Muslim dietary laws. Food Technol 43(2):88–90
Vijayendra SVN, Rajashree K, Halami PM (2010) Characterization of a heat stable anti-listerial bacteriocin produced by vancomycin sensitive *Enterococcus faecium* isolated from idli batter. Indian J Microbiol 50(2):243–246

Vishnu-Mittren (1978) Origins and history of agriculture in the Indian sub-continent. J Hum Evol 7(1):31–36. https://doi.org/10.1016/S0047-2484(78)80033-5
Yegna Narayan Aiyar AK (1953) Dairying in ancient India. Indian Dairyman 5:77–83
Yonzan H, Tamang JP (2009) Traditional processing of Selroti: a cereal based ethnic fermented food of Nepalis. Indian J Tradit Knowl 8(1):110–114
Yonzan H, Tamang JP (2010) Microbiology and nutritional value of selroti, an ethnic fermented cereal food of the Himalayas. Food Biotechnol 24(3):227–247
Yonzan H, Tamang JP (2013) Optimization of traditional processing of Selroti, a popular cereal-based fermented food. J Sci Ind Res 72:43–47

Ethnic Fermented Foods and Beverages of Arunachal Pradesh

2

Karuna Shrivastava, Biswajit Pramanik, Bhaskar Jyoti Sharma, and Greeshma A.G

Abstract

This book chapter deals with different types of traditionally prepared ethnic fermented foods and beverages by various ethnic groups of people in Arunachal Pradesh. Some of these foods have been used by specific local tribal community. The report describes the ingredients, methods of preparation, consumption details and importance of different traditionally prepared fermented food products. Common fermented foods from local clips in Arunachal Pradesh mainly include plant-based products such as fermented food made from bamboo, fermented food made from leafy vegetables and fermented food items made from soybean, while fermented foods of animal origin include food made from milk and fermented food made from meat/fish, etc. A few traditionally fermented products of specific tribal groups being documented for the first time include *iting* of *Adi* tribe; *ngiiyi-yaan* and *sira-o* of *Apatani* tribe; *kupe*, *agya*, *nyongin* and *pone* of *Galo* tribe; fermented soybean and *poling* of *Nyishi* tribe; *anpo*, *anpo-shi-anjita*, *aara* and *lip chhuro* of Monpa tribe; *aara*, *chhuk chhoro*, *phaak* and *moh* of *Sherdukpen* tribe; and *yu* of *Idu-Mishmi* tribes. A few new products like *lukter*, *pehak* and *churasabji* are also new fermented products to this list used by different tribes.

Keywords

Traditional · Fermented food products · Alcoholic beverages · Arunachal Pradesh

K. Shrivastava (✉) · B. Pramanik · B. J. Sharma
Laboratory of Biotechnology and Plant-Microbe Interaction, Department of Forestry, North Eastern Regional Institute of Science and Technology, Nirjuli, Arunachal Pradesh, India
e-mail: ks@nerist.ac.in

Greeshma A.G
Dublin Health Services Management, Abu Dhabi, United Arab Emirates

J. P. Tamang (ed.), *Ethnic Fermented Foods and Beverages of India: Science History and Culture*, https://doi.org/10.1007/978-981-15-1486-9_2

2.1 Introduction

Arunachal Pradesh, the Land of Rising Sun, is the North Eastern most satae of India occupying the area of about 83.743 km^2 which is segmented in 25 districts. The state is known for the world's largest variety of ethnic tribal groups and subgroups which include more than 25 major tribes and 125 subtribes (Tiwari and Mahanta 2007). The people of Arunachal Pradesh are peace loving and enjoy the festivals and food together. They are identified by their own language, clothes, headgears, foods and their use patterns, customs and cultures. The major tribes include *Adi*, *Apatani*, *Nyishi*, *Monpa*, *Mishmi*, *Sherdukpen*, *Noktey*, etc. The state's diverse and unique rich culture and traditional heritage are still remaining fresh and well-preserved. The state is a part of Eastern Himalayan ranges located between 26°28′ and 29°30′ N latitudes and 91°30′ and 97°30′ E longitudes. The state consists of mountainous ranges with the topography rapidly rising to 7000 m at its highest peak and sloping to the plains of Assam (http://www.arunachalpradesh.gov.in/666-2/). Numerous river valleys dissect the precipitous terrain of Arunachal. The climate of Arunachal Pradesh varies with low- to high-altitude (100–1500 m) areas having humid subtropical climate, and very high-altitude areas (3500–5500 m) experience temperate and alpine climate. Arunachal Pradesh receives 2000–5000 m (79–197 in.) of rainfall annually of which 70–80% obtained between May and October (Dhar and Nandargi 2004). The population of Arunachal Pradesh state is 1,383,727 (https://www.census2011.co.in/census/state/districtlist/arunachal+pradesh.html). Arunachal Pradesh is also the richest reservoir of plant diversity in India and is one of the 'biodiversity hotspots' of the world, supporting about 50% of India's biodiversity (Mao and Hynniewta 2000; Chatterjee et al. 2006). Food plays a very important role in defining the identity of one ethnic group from the other (Das et al. 2016). The diverse food culture of Arunachalee people mainly comprises of non-fermented and fermented ethnic foods and alcoholic beverages (Sathe and Mandal 2016). Even ethnic people residing in urban areas also have high demand of ethnic fermented foods due to their diverse traditional food culture and lifestyle (Asati and Yadav 2003; Medhi et al. 2013).

To sustain their livelihood, people in rural areas chiefly practise diverse trade and profession including jhum and wet rice cultivation, horticulture, fish farming, carpet making, wood carving and breeding of mithun, yak, sheep and other livestocks. The people in urban areas of Arunachal are mostly educated and semi-educated, either in government jobs or working as social workers, contractors, town dwellers and traders. The blend of diverse culture and religion provides them ample scope to prosper in many respects (http://districts.nic.in/districts.php?sid=AR).

Arunachalee people rely on their age-old methods of food material fermentation to enhance the taste, flavour, texture and shelf life. They *practise* distinct cuisine with lot of rice varieties and greatly depended on nonvegetarian cuisine, rice beer, etc. Ethnic fermented foods and alcoholic beverages are part and parcel of their daily lifestyle. Many of these fermented foods are area tribe specific with unique ingredients and method of preparation. Locally available materials such as milk, vegetables, bamboo, soybean, meat, fish and cereals are commonly used as substrates (Das and Deka 2012; Tamang et al. 2012). Their ethnic and nutritious food is simple to cook and delicious to consume. Their food items also include use of lots of wild herbs with medicinal properties and indigenous fresh spices which are probably contributing towards taste as tribal cookery avoids use of any kind of oil or dry or packed Indian spices. The staple food of Arunachal Pradesh is rice, along with fish, meat and green vegetables. Lettuce is the most common vegetable, usually prepared by boiling with ginger, coriander and green chillies. Boiled rice cakes wrapped in leaves are a popular snack. The names of some common ethnic foods include *etting*, *sa pila*, *atto pila*, *empi*, *rokpi takeng*, *thukpa*, *pasa*, *makat*, *pee chim khim*, *phoi hom*, *momo*, etc.

Ethnic fermented foods are an intrinsic part of diet of the ethnic tribes in Arunachal Pradesh. Various types of indigenous fermented foods and beverages are commonly prepared and consumed by different ethnic groups of Arunachal Pradesh for centuries; however, daily per capita consumption of ethnic fermented foods and alcoholic beverages in this state needs to be quantified. Also, systematic studies are required on the nutritive and medicinal value of these products to evaluate their potential for use on a large scale. Food in this part of country mostly includes natural ingredients of forest origin or those available in their surroundings. The age-old production technologies however have been passed through oral dialects from elders to youngsters and an integral part of their traditional culture. Fermented foods and drinks are so important for them that they are part of daily cuisines as well as required on all occasions, ceremonies and other important events, not restricted to even women and children. A few common fermented foods are known as *eup*, *ikhing*, *bamboo tenga*, *pikey pila*, *tapyo*, *churapi*, *anpo*, *ziang-sang*, etc., and ethnic alcoholic beverages are known by various names, i.e. *apong*, *o*, *sira-o*, *opo*, *madua apong*, etc. by various tribes of this region (Shrivastava et al. 2012).

Though documentations of a number of fermented foods, beverages and other products of Arunachal Pradesh have been carried out by earlier workers (Singh et al. 2007; Das et al. 2016; Rawat et al. 2018; Tamang et al. 2012; Sha et al. 2018; Rai et al. 2016; Anupama et al. 2018; Roy et al. 2017), many are still to be included in the list. They also need to be verified and validated. In this book chapter, the detailed characteristics of ethnic fermented foods and beverages from Arunachal Pradesh are documented that were studied earlier through literature survey, personal interviews and discussions with local people. The ethnic food and beverages have also been categorised and listed as tribe-wise. The method of preparation and mode of consumption of these fermented products with their nature, importance and future prospects are also discussed where information has been sparse. Table 2.1 provides an overview of different fermented foods, and Table 2.2 contains the details of fermented beverages prepared and utilised by some of the major tribal groups of Arunachal Pradesh (Table 2.3).

Table 2.1 Fermented foods, beverages and their main substrates prepared by different tribal groups of Arunachal Pradesh

Food	Substrate	Sensory property of product	Major ethnic consumers
Ngiiyi-yaan	Fish and meat	Greyish, solid, strong flavour, soft	*Apatani*
Pikey pila		Sticky, semisolid, dark in colour, strong flavour	*Apatani*
Lukter		Dry and spicy	Different tribes
Dingkyo		Acidic, sour and dry meat	*Nyishi*
Churapi/chhurpi/ durkha	Yak milk	Cottage cheese-like, milky in colour, masticator	*Monpa*, *Khambas*
Chhurpupu		Chewable and dry	*Monpa*
Chhur chirpen		Pressed, light yellowish brown, hard, side dish	*Monpa*
Churkham		Chewable and cheesy	*Khambas*
Marchang		Fried	*Monpa*
Churasabji		Strong and spicy	All
Eup, *eyup* and *hiyup*	Bamboo	Acidic, sour, dry, curry	Adi, Nyishi
Iting, *bamboo tenga* or *hitak* and *kupe*, *hirring* and *heccha*		Acidic, soft, sour	Adi, Nyishi, *Apatani*, *Galo*
Ikhung or *ekung*, *hikhung*, *eeku*, *hikhu/hikku and hihi/hi*		Acidic, sour, soft, curry	*Adi*, *Nishi*, *Galo*, *Apatani*
Eepe and eep		Acidic and sour	*Galo*
Anpo and anpo-shi-anjita		Solid, dark brown, sour and slightly pungent	*Idu-Mishmi*
Gundruk	Vegetables	Acidic, sour, dry, soup, side dish	*Adi* adopted from Gorkha
Pila		Dark brown coloured, liquid, salty	*Apatani*
Tapyo		Salty, hard solid or powder	*Apatani*
Peru-yaan, *agya*	Soybean	Alkaline, sticky, emit strong ammoniacal flavour	*Apatani*, *Galo*, *Nyishi*
Pehak		Light brown, sticky, spicy and strong flavour	All
Libi chhurpi		Dry, emit strong aroma	*Monpa*
Apong	Rice and wild herbs	Alcoholic and sweet when fresh, becomes sour with age	All

(continued)

Table 2.1 (continued)

Food	Substrate	Sensory property of product	Major ethnic consumers
Opo	Rice and rice husk	Clear, blackish in colour, strong with sweet alcoholic aroma	*Adi*, *Nyishi*, *Galo* and *Mishmi*
Pone		Clear, dark-coloured drink with strong aroma	*Galo*
Pona	Rice	Milky with sweet alcoholic aroma	*Nyishi*
Nyongin		Alcoholic and sweet in taste	*Galo*
Yu		Alcoholic, sweet when fresh, becomes slightly sour with strong alcoholic aroma	*Idu-Mishmi*
Madua apong	Millet	Golden yellow in colour, transparent with pleasant alcoholic aroma	All
Poling		Alcoholic and sweet with fruity aroma	*Nyishi*
Rakshi	Finger millets/ barley/rice	Distilled liquor	Adopted from Gorkha community by *Monpa*, *Sherdukpen* and *Idu-Mishmi*
Themsing	Finger millets/ barley with ash or charcoal	Golden in colour with good alcoholic aroma	*Monpa*
Mingri, *lohpani* and *bhangchang*	Finger millet/ barley/rice/ maize	Whitish in colour, with week aroma	*Monpa*
O	Rice and millet	Alcoholic and sweet when fresh, becomes sour with age	*Apatani*
Sira-o		Clear, with strong and good alcoholic aroma	*Apatani*

2.2 Fermented Food Products of Plants and Plant Parts

Locally grown species of plants and their parts are used in fermentation in order to enhance digestibility, nutrition, taste and shelf life of cooked food as well as to make the foods safer for consumption. The use of wild edible plants (WEPs) as food is not only an integral part of the culture and tradition of many indigenous communities throughout the world (Konsam et al. 2016) but also the source of their food security. North East India is blessed with rich plant and animal resources. Arunachal Pradesh is considered one of the richest botanical regions in India, hence regarded as one of the biodiversity hotspots of the world (Singh et al. 2010), where people largely depend on shifting cultivation and forest-based food products for their food and

Table 2.2 Fermented food products which are specific to certain ethnic communities in Arunachal Pradesh

Tribe	Name of fermented food	Main ingredients used	Use pattern/ storage	Use/cultural importance	Source
Adi	*Eup* or *eyup*	Dry bamboo shoots	As an ingredient of different dishes such as boiled vegetables, cooked meat, chutney, etc. or as pickle	Regular	Tamang et al. (2008, (2012), Tamang and Tamang (2009), Rawat et al. 2018, Das et al. (2016), Sonar and Halami (2014), Nongdam (2015), Roy et al. (2017)
	Iting	Fresh bamboo shoots	As an ingredient of different dishes such as boiled vegetables, cooked meat, chutney, etc. or as pickle	Regular	Local knowledge holders
	Ikhung or ekung	Small pieces of tender bamboo shoots	As an ingredient of different dishes such as boiled vegetables, cooked meat, chutney, etc. or as pickle	Regular	Tamang et al. (2012), Tamang and Tamang (2009), Das et al. (2016), Rawat et al. (2018), Roy et al. (2017)
Apatani	*Pikey pila*	Pork fat, pila and king chilli	As chutney	Regular	Tiwari and Mahanta (2007), Rinya (2016)
	Pila	Leaves, stem or shoot portion of the banana	Mixed with fermented pork for preparing pikey pila	Regular	Tiwari and Mahanta (2007)
	Tapyo	Leaves of *Clerodendrum colebrookianum*	As salt; instead of pila in preparing pikey pila	Served during Myoko festival along with Ó′	Tiwari and Mahanta (2007), Rawat et al. (2018), Rinya (2016)
	Peru-yaan	Fermented soybean	As chutney by crushing fermented soybean with chilly	Regular	Tamang et al. (2009), Tamang et al. (2012); Das and Deka (2012), Tamang (2015)

(continued)

Table 2.2 (continued)

Tribe	Name of fermented food	Main ingredients used	Use pattern/ storage	Use/cultural importance	Source
	Hikhu/ hikku	Small pieces of fresh bamboo shoots	As an ingredient of different dishes such as boiled vegetables, cooked meat, chutney, etc. or as pickle	Regular	Rawat et al. (2018), Tamang et al. (2012), Roy et al. (2017)
	Hihi/hi	Hikhu	As soup and flavouring agent	Regular	Tamang et al. (2012)
	Ngiiyi-yaan	Fresh fish	As chutney or pickle to enhance flavour in food	Regular	Local knowledge holders
	Hirring	Fresh bamboo shoots	Use for making of pikey and as pickle	Regular	Rawat et al. (2018), Tamang et al. (2008, 2012), Tamang and Tamang (2009), Das and Deka (2012), Das et al. (2016), Sonar and Halami (2014), Nongdam (2015), Roy et al. (2017)
	Heccha	Fresh and long bamboo shoots	As an ingredient of boiled vegetable or meat	Regular	Sonar and Halami (2014), Nongdam (2015), Roy et al. (2017)
Galo	*Kupe*	Fresh and long bamboo shoots	Consumed raw or delicious curry is made with vegetables, meat or fish	Regular	Local knowledge holders
	Eepe	Kupe	Consumed raw or delicious curry is made with vegetables, meat or fish	Regular	Roy et al. (2017)
	Eeku	Fresh minutely chopped bamboo shoots	Consumed raw or delicious curry is made with vegetables, meat or fish	Regular	Roy et al. (2017)
	Eep	Eeku	Consumed raw or delicious curry is made with vegetables, meat or fish	Regular	Roy et al. (2017)
	Agya	Fermented soybean	Used for making curry	Regular	Local knowledge holders

(continued)

Table 2.2 (continued)

Tribe	Name of fermented food	Main ingredients used	Use pattern/ storage	Use/cultural importance	Source
Monpa	*Chhurpi or durkha or churapi*	Yak milk is preferred for making this cheese; fresh milk may be used	Traditional mouth freshener	Regular	Rawat et al. (2018), Nehal (2013), Rai et al. (2016), Singh et al. (2007)
	Chhur chirpen	Yak milk and cut fruits of crab apple (thung)	Traditional mouth freshener	Regular	Rawat et al. (2018), Singh et al. (2007), Rai et al. (2016)
	Chhurpupu	The *chhurpi*, which becomes very old (4–5 years)	As medicine for curing stomach pain	A traditional medicine	Singh et al. (2007)
	Marchang	Chhurpupu, yak ghee and barley flour kongpu	Mix with *rakshi*	Culturally valuable	Singh et al. (2007)
	Libi chhurpi	Soybean seeds and zola or banana	Mixed with vegetables	Prepared during the winter, i.e. from October to February	Singh et al. (2007)
	Lip chhuro	Fermented soybean	Used for making curry	Regular	Local knowledge holders
Sherdukpen	*Chhuk chhoro*	Fermented soybean	Used for making curry	Regular	Local knowledge holders
	Moh	Milk	Used as flavouring agent with food	Regular	Local knowledge holders
Khambas	*Churapi*	Yak milk is preferred for making this cheese	Used as a traditional chocolate, as mouth freshener and also in preparation of vegetable and nonvegetable dishes in the form of paneer	Regular	Tiwari and Mahanta (2007), Rawat et al. (2018), Rai et al. (2016)
	Churkham	Yak milk and cut fruits of crab apple	Used as a traditional chocolate, as mouth freshener and also in preparation of vegetable and nonvegetable dishes in the form of paneer	Regular	Tiwari and Mahanta (2007), Rawat et al. (2018), Rai et al. (2016)

(continued)

Table 2.2 (continued)

Tribe	Name of fermented food	Main ingredients used	Use pattern/ storage	Use/cultural importance	Source
Nyishi	*Bamboo Tenga/ hitak*	Fresh shoots of *Dendrocalamus hamiltonii* hook (*choya bans*)	Used as pickles. It is one of the most favoured traditional foods	Regular	Tiwari and Mahanta (2007), Tamang et al. (2012), Rawat et al. (2018)
	Hikung/ ekung	Fresh small pieces of bamboo shoots and incubated under soil for 1 month	As an ingredient of different dishes such as boiled vegetables, cooked meat, chutney, etc. or as pickle	Regular	Roy et al. (2017)
	Hiyup/eup	Dry bamboo shoot incubated under soil for 1 month and then dried in sunlight	As an ingredient of different dishes such as boiled vegetables, cooked meat, chutney, etc. or as pickle	Regular	Tamang et al. (2012)
	Fermented soybean	Boiled soybean and dry in air/ sunlight	Used for making curry	Regular	Local knowledge holders
Different tribes	*Lukter*	Cooked dry meat and king chilli	Used as chutney	Regular	Local knowledge holders
	Pehak	Fermented soybeans and king chilli	Used as spicy chutney	Regular	Local knowledge holders
	Churasabji	Fermented cheese made by yak milk or cow milk, vegetable and king chilli	As soup	Regular	Local knowledge holders
Idu-Mishmi	*Anpo*	Fresh bamboo shoots	As vegetable	Regular	Local knowledge holders
	Anpo-shi-anjita	Fresh bamboo shoots and ginger	As garnishing ingredient of vegetables	Regular	Local knowledge holders

nutritional security. WEPs are extensively consumed by the local people in Arunachal Pradesh. WEPs are species that are neither cultivated nor domesticated but available from their natural habitat and used as sources of food and nutrition in the population (Das et al. 2016).

The fermentation products are utilised by all groups of people here, from poor to rich, rural to city dwellers, children to elders, and men and women. Fermentation activities not only are advantageous for producing food products but also produce useful by-products such as nutritious fodder for poultry and other livestock or serve as substrates for extended fermentation products. The lactic acid fermentation is the key process involved in the fermentation of plants or their parts as a preservation method for the production of finished and semifinished products.

Table 2.3 Fermented alcoholic beverage of different tribes of Arunachal Pradesh, India

Name of tribe	Name of fermented beverage	Main ingredients used	Use pattern/ storage	Cultural importance	Source
Almost all tribes	*Apong*	Rice, starter culture	Consumed as alcoholic beverage. Concentrated beverage can be stored for 4–5 days	For every purpose and regular use	Shrivastava et al. (2012); Tamang et al. (2012); Ray et al. (2016); Tiwari and Mahanta (2007)
Adi, Nyishi, Galo and *Mishmi*	*Opo (also called kala apong)*	Rice husk or rice, starter culture	Consumed as alcoholic beverage	For every purpose and regular use	Shrivastava et al. (2012)
Almost all tribes	*Madua apong*	Millet, starter culture, *Ektam* leaves	Consumed as alcoholic beverage	For every purpose and regular use	Shrivastava et al. (2012)
Monpa	*Themsing*	Millet or barley, starter culture	Consumed as alcoholic beverage	For every purpose and regular use	Shrivastava et al. (2012); Singh et al. (2007)
Monpa, Sherdukpen and *Idu-Mishmi tribes*	*Rakshi*	Millet or barley or rice	Consumed as alcoholic beverage	For every purpose and regular use	Shrivastava et al. (2012); Rawat et al. (2018); Sekar and Mariappan (2007); Singh et al. (2007)
Monpa tribe residing mainly in Tawang district	*Mingri, lohpani and bhangchang*	Finger millet/ barley/rice/ maize	Consumed as alcoholic beverage	For every purpose and regular use	Shrivastava et al. (2012); Singh et al. (2007)
Sherdukpen	*Phaak*	Finger millet/ barley/rice/ maize	Consumed as alcoholic beverage	For every purpose and regular use	Local knowledge holders
Idu-Mishmi	*Yu*	Rice, old starter cake	Consumed as alcoholic beverage	For every purpose and regular use	Local knowledge holders

(continued)

Table 2.3 (continued)

Name of tribe	Name of fermented beverage	Main ingredients used	Use pattern/ storage	Cultural importance	Source
Nyishi	*Poling*	Marua + locally made starter yeast	Consumed as alcoholic beverage	Any occasion	Local knowledge holders
	Pona	Dry boiled rice + locally made starter yeast, store in 1 week	Consumed as alcoholic beverage	Any occasion	Tamang et al. (2012)
Galo	*Nyongin*	Rice, opop (dried yeast tablets)	Consumed as alcoholic beverage	Mainly during Mopin, the main traditional festival of Galo	Local knowledge holders
	Pone	Rice husk, rice, opop (dried yeast)	Consumed as alcoholic beverage	Mainly during Mopin, the main traditional festival of Galo	Local knowledge holders
Apatani	*O/oh*	Rice mixed thoroughly with millet, rice beer starter	Consumed as alcoholic beverage	Any occasion	Tamang et al. (2012)
	Sira-o	Rice mixed thoroughly with millet, rice beer starter	Consumed as alcoholic beverage	Any occasion	Local knowledge holders

2.2.1 Fermented Food Products of Bamboo

Bamboo and bamboo-based products are closely associated with the culture and livelihood of the people of Arunachal Pradesh. The fermented bamboo is one of the main food items which are also considered to provide many health benefits. The varying agroclimatic conditions of the state have endowed with diverse and abundance of locally available bamboo species which play an integral part in their culture, tradition and custom. Of about 1200–1500 bamboo species belonging to 60–70 genera known worldwide (Wang and Shen 1987), around 125 species under 23 genera are available in India, and North Eastern states consist of about 84 bamboo species (Trivedi and Tripathy 1984). The common species of bamboo found in this

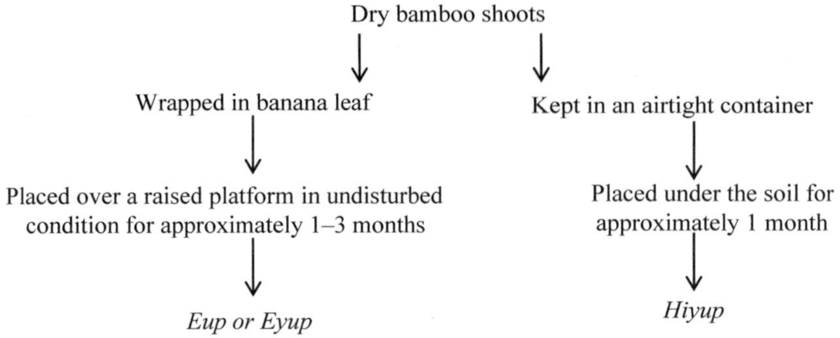

Fig. 2.1 Different stages of *eup or eyup and hiyup* preparation

region are *Bambusa tulda*, *B. pallida*, *Dendrocalamus hamiltonii*, *D. hookeri*, *Pseudostachyum polymorphum*, *Arundinaria* sp., etc. Of these, *B. tulda* and *D. hamiltonii* are most popular edible species (Angami et al. 2006). Fresh, tender and soft bamboo shoots are widely consumed as pickles and other fermented products by almost all ethnic communities of the state. Seasonally available bamboo shoots are fermented for different periods following various traditional processes to increase their shelf life and delicacy (Roy et al. 2017). The brief process technology of some important bamboo-based fermented products is elaborated below.

2.2.1.1 Eup, Eyup and Hiyup (Fig. 2.1)

Eup or *eyup* is a popular bamboo-based food product consumed by the people of *Adi* and *Nyishi* tribes of Arunachal Pradesh, India. The *hiyup* is also a fermented product prepared by *Nyishi* tribe only. These food products are used as one of the ingredients of different dishes such as boiled vegetables, as cooked meat and as pickle. They are also used to prepare pickle. All these preparations are made by using dry bamboo shoots; however *eup* or *eyup* is fermented directly by wrapping the bamboo shoots in banana leaves and keeping it in an undisturbed place preferably on a raised platform not reachable to members of family specially children, while *hiyup* is prepared by keeping the dried bamboo shoots in an airtight jar under the soil. Their fermentation period is about 1–3 months depending upon the season/weather. *Eup* word is derived from the *Nyishi* dialect (Tamang 2010). It is an acidic, sour sun-dried solid product which may be used to prepare curry. It is prepared by *Nyishi* and *Adi* tribes and its fermentation period is 1–3 months.

2.2.1.2 Iting, Bamboo Tenga or Hitak, Kupe, Hirring and Heccha (Fig. 2.2)

Fresh and long shoots of commonly available bamboos are used to prepare *iting* and kupe, whereas fresh shoots of *Dendrocalamus hamiltonii* are used for the preparation of *bamboo tenga* or *hitak*. The people of *Adi* tribe also prepare *iting* very often by using fresh bamboo shoots in place of dried shoots as used in the *eup* or *eyup*. *Kupe* and *bamboo tenga* (Plate 2.1) or *hitak* are the products of *Nyishi* and *Galo*

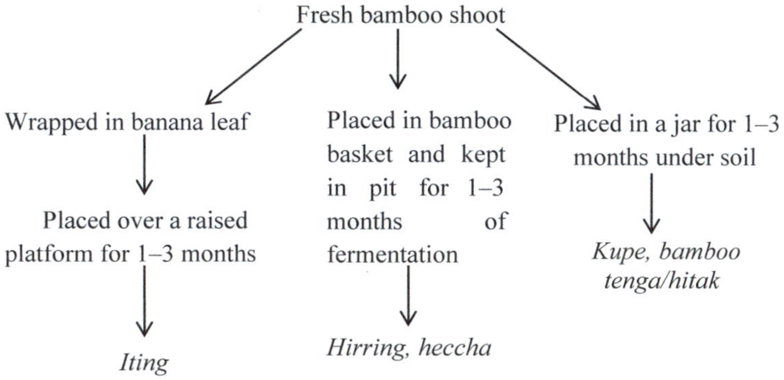

Fig. 2.2 Different stages *Iting, Hirring, Heccha, Kupe and Bamboo Tenga or Hitak* preparation

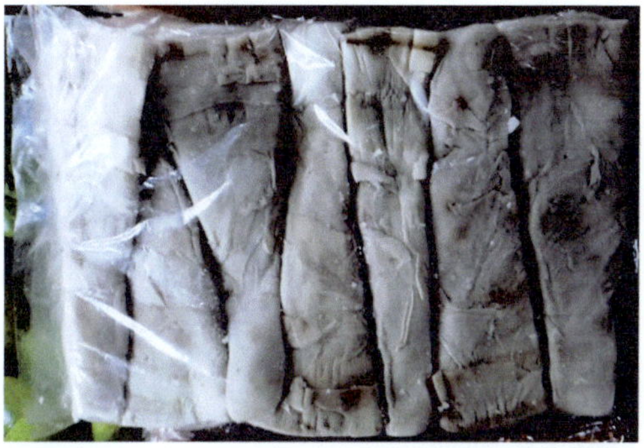

Plate 2.1 *Bamboo tenga*

tribes, respectively, where they use fresh bamboo shoots. Their preparation process is otherwise similar to that of *hiyup* of *Nyishi people*. These food items are also used as an ingredient of boiled vegetables and cooked meat or consumed as chutney or pickle. *Hirring* and *heccha* are other fermented foods made by using fresh and moist solid mass of bamboo shoots fermented for 1–3 months in a bamboo basket kept in pit under soil. *Hirring* is a food of *Nyishi* as well as *Apatani* tribes, whereas *heccha* is used only by *Apatanis.*

2.2.1.3 Ikhung or Ekung, Hikhung, Eeku, Hikhu and Hihi (Fig. 2.3)

In the preparation of *ikhung* or *ekung*, *hikhung* and *eeku*, fresh but chopped bamboo leaves are utilised; however, they are coarsely cut in case of *ikhung* and *hikhung* by the Adi people, whereas they are minutely chopped to prepare *eeku* by the people of *Galo* tribe. *Ikhung* or *ekung* is prepared by tightly wrapping the cut bamboo leaves in banana leaf and fermenting them, while bamboo leaves are fermented in an air-tight jar to prepare *hikhung* and *eeku* by *Nyishi* and *Galo* people, respectively. These bamboo products are also utilised in the similar way as that of *eup*, *hiyup*, *iting*, *kupe*, etc. All these food items are basically used as flavouring agents and to improve the taste of main food.

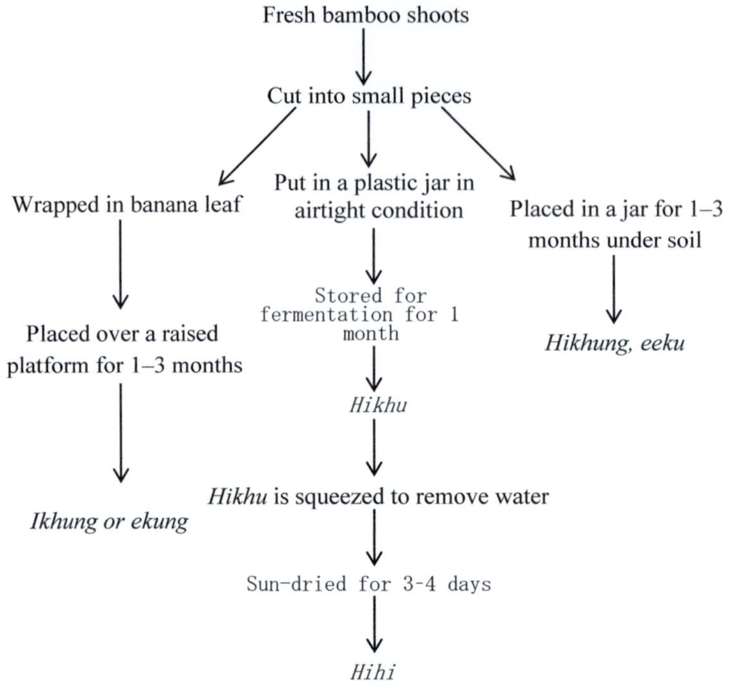

Fig. 2.3 Different stages of *Ikhung or Ekhung, Hikhu, Hihi, Hikhung and Eeku* preparation

Plate 2.2 *Hikhu*

Hikhu (Plate 2.2) is a bamboo shoot product of *Apatani* tribe in which fresh and cut bamboo shoots are kept in an airtight plastic jar for fermentation in a safe place for about 1 month or more. *Hikhu* serves as substrate for the preparation of *hihi*. The excess liquid is squeezed from *hikhu* and sun dried for 3–4 days to become hard. This is now called as *hihi.*

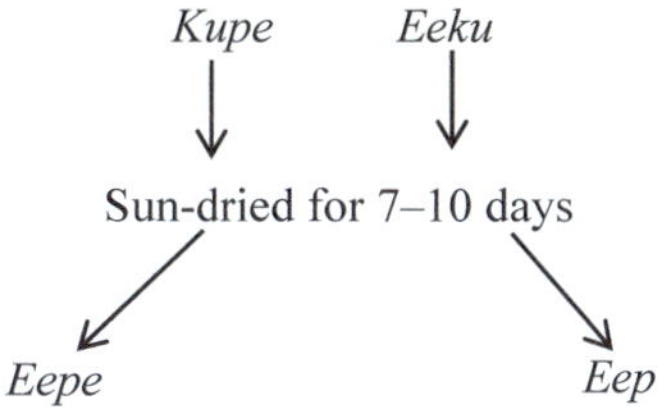

Fig. 2.4 Different stages of *Eepe and Eep* preparation

Plate 2.3 (**a**) *Eepe* and (**b**) *Eep*

2.2.1.4 *Eepe and Eep* (Fig. 2.4)

Eepe (Plate 2.3a) and *eep* (Plate 2.3b) are sun-dried forms of *kupe* and *eeku*, respectively, used by people of *Galo* community. Both are brown in colour but *eepe* is comparatively darker than *eep*. Both are pungent foods with enhanced shelf life owing to their dry nature.

2.2.1.5 Anpo and Anpo-Shi-Anjita (Fig. 2.5)

Anpo and *anpo-shi-anjita* are the two another bamboo-based fermented food products prepared by the *Idu-Mishmi* tribe of Arunachal Pradesh. In the preparation of *anpo*, fresh bamboo shoots are finely sliced in about 0.5–1 cm size. The sliced shoots are wrapped with wild *Colocasia* leaf and then pressed with heavy stones, keeping it airtight. It is then allowed to ferment for a period of about 1 month or so. The fermented product is ready for consumption and called as *anpo* (Plate 2.4a). It is consumed immediately after preparation normally within a period of 15 days.

The method of preparation of *anpo* and *anpo-shi-anjita* is as follows:

Anpo serves as the raw material for preparation of *anpo-shi-anjita*. *Anpo* after mixing with crushed ginger is sun dried till it is hard dry. The *anpo-shi-anjita* (Plate 2.4b) being a dried product has elongated shelf life than *anpo*. *Anpo* is used as vegetable after cooking, while *anpo-shi-anjita* is used for garnishing vegetables.

Fresh bamboo shoot

↓

Finely sliced in 0.5–1-cm-size pieces

↓

Wrapped with wild *Colocasia* leaf

↓

Pressed with heavy stones and left for 1 month for fermentation

↓

Anpo

↓

Crushed ginger is mixed and sun-dried till it is hard dry

↓

Anpo-shi-anjita

Fig. 2.5 Different stages of *anpo and anpo-shi-anjita* preparation

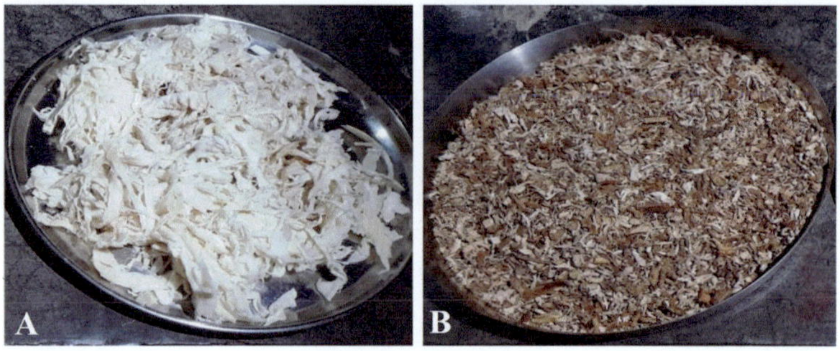

Plate 2.4 (**a**) *Anpo* and (**b**) *anpo-shi-anjita*

2.2.2 Fermented Food Products of Vegetables

The ethnic people of Arunachal Pradesh are utilising a number of locally available mostly wild species of plants to prepare fermented foods. The commonly used leafy vegetables include some *Brassica* spp., *Raphanus* sp., *Clerodendrum* spp., *Diplazium esculentum*, *Phragmites karka*, *Carica* sp., *Musa* spp., grasses, etc. Some commonly used vegetable-based fermented products prepared by the people of *Adi* and *Apatani* tribes only are described below.

2.2.2.1 Pila (Fig. 2.6)

Pila is a traditional food item of the *Apatanis* of Ziro valley, the ethnic residents of the Lower Subansiri district. Earlier, stem of a thorny wild cane, locally known as *lobyo' tare*, was used for *pila* preparation, but nowadays papaya trunk, banana peels

Fresh leaves of banana, *Clerodendrum* spp., fern (*Diplazium esculentum*, pepu (*Phragmites karka*), papaya trunk, banana peels, grasses, etc.

↓

Dried in sunlight for 2–3 days

↓

Semi-dried leaves are burnt

↓

Leaves' ashes are filtered using bamboo-made fennel

↓

Pila

Fig. 2.6 Different stages of *Pila* preparation

Plate 2.5 Types of *pila*

and a variety of other plants are used. First, leaves, stem or shoot portion of many plants including local grasses are semi-dried in the open, which imparts flavour and bite. Then, the semi-dried parts are burnt in a tin or on clean dry ground. The ash obtained after burn is collected and filtered using bamboo-made funnel called *piyu-khugyu*. The cold water is poured slowly over the ash and the filtrate is called as *pila* (Plate 2.5).

Pila is a clear ruby-coloured or a dark liquid, which can be stored for very long periods. Its use is also very versatile as a ready-made stock adding flavour to traditional foods like *hikhu pickey*, *salyo pikey*, *pike pila*, etc.

2.2.2.2 Tapyo (Fig. 2.7)

Tapyo is another popular fermented food prepared by *Apatani* people of Ziro valley, Lower Subansiri district of Arunachal Pradesh, India. The ingredients of *tapyo* are *pila* and rice starch. For preparing *tapyo*, first, a layer of rice starch is allowed to set in

A layer of rice starch is allowed to set in a flat, dry pan along with leaves of the selected plant over a low flame by burning bamboo

↓

Pila is then poured over this base and allowed to cook until the water is completely evaporated down to residual ash that is grey-black in colour unlike rock salt (tapyo)

Fig. 2.7 Different stages of *Tapyo* preparation

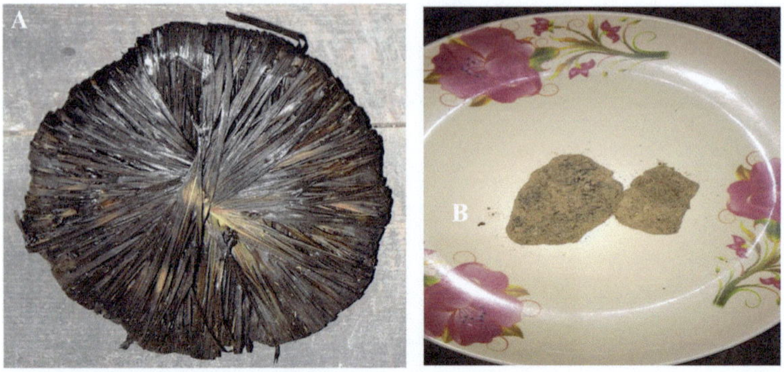

Plate 2.6 (**a**) *Tapyo* wrapped in *Tabo* leaves and (**b**) ready for use

a flat, dry pan over a low flame by burning bamboo. *Pila* is then poured over this base and allowed to cook until the water is completely evaporated down to residual ash that is grey-black in colour to obtain a thick cake. This cake is then wrapped in *Tabo* (*Tabo* is *Apatani* name of a wild plant whose scientific name is not known) leaves and kept over a raised platform in a smoky area till further use (Plate 2.6a). The *tapyo* (Plate 2.6b) may be used as a salt either fresh or stored for long period till consumed fully.

2.2.3 Fermented Food Products of Soybean

Soybean is one of the most common raw materials used for traditional preparation of both fermented and non-fermented foods in the Eastern Himalayan regions (Tamang 2015). In Arunachal Pradesh, the fermentation products of soybean are mainly utilised by *Galo*, *Apatani*, *Monpa* and *Nyishi* tribes only. Some of these food products are described below.

2.2.3.1 Peru-Yaan, Agya, Chhuk Chhoro, Lip Chhuro and Fermented Soybean (Fig. 2.8)

Fermented soybean is another food product among the tribes of Arunachal Pradesh. They are named differently such as *peru-yaan* by *Apatani* tribe, *agya* by *Galo* tribe, *chhuk chhoro* by *Sherdukpen* tribe, and *lip chhuro* by *Monpa* tribe, while *Nyishi*

Fresh soybean seeds

↓

Seeds are boiled in water for 1–2 h

↓

Seeds are cooled by draining the excess water

↓

Seeds are than packed in bamboo basket lined with ginger leaves/Niiji leaves

↓

Kept on wooden rack in the fireplace

↓

Fermented for 3–5 days

↓

Peru-yaan, agya, chhuk chhoro, lip chhuro, fermented soybean

Fig. 2.8 Different stages of *peru-yaan, agya, chhuk chhoro, lip chhuro* and *fermented soybean* preparation

Plate 2.7 (**a**) *Peru-yaan* and (**b**) *peru-yaan* in ball shape

people call it simply as fermented soybean, but their method of preparation is similar as depicted below:

Initially, the fresh soybean seeds are boiled in water for 1–2 h followed by its cooling by draining off the excess water. The boiled seeds are then packed in bamboo basket lined with ginger leaves or *Niiji* leaves (*Niiji* is Apatani name of a wild plant whose scientific name is not known) in an airtight condition. The basket is kept on wooden rack over the fireplace in specially designed fireroom called as *Chang-ghar* for fermentation in undisturbed condition for 3–5 days depending on the season of preparation. The fermented soybeans are packed tightly in banana leaves (Plate 2.7a) in Ziro area, or *Monpas* made it into a ball and wrap in the poly bag for sale in the market (Plate 2.7b). The fermented product is consumed as chutney after crushing with chilly by the *Apatani* people, while the *Galo* and *Nyishi* people used it in the preparation of curries.

Fig. 2.9 Different stages of *pehak* preparation

2.2.3.2 Pehak (Fig. 2.9)

Pehak is a food product used by *Adi* tribe of Arunachal Pradesh. It is prepared from fermented soybean and king chilli. Both the ingredients are crushed and mixed thoroughly to store. It is consumed as spicy chutney.

2.2.3.3 Libi Chhurpi (Fig. 2.10)

Libi chhurpi is prepared during the winter season, i.e. from October to February by the *Monpa* community of West Kameng district of Arunachal Pradesh. To prepare it, local varieties of soybean seeds are first boiled; excess water is drained out and spread on a locally made bamboo mat on an open raised platform called *charang*. Salt is added to the seeds and mixed properly when they are still lukewarm. After cooling, the seeds are put in a container called as *Shong* and covered with leaves of *zola* or banana. The mouth of *Shong* is tightly covered with its lid, is kept near the fire and is rotated regularly. When seeds become soft and threadlike and start giving off a characteristic smell, they are taken out and crushed carefully. These crushed seeds are then spread in small heaps on *Chang* for sun drying. They are again crushed and dried. The crushing process is repeated three times till seeds turn into a course powder. The *libi chhurpi* is now ready for consumption and can be stored in dried bottle gourd or bamboo containers for 1–4 years.

2.3 Fermented Products from Meat and Fish

The people of all tribes prepare and utilise fermented meat and fish products. Among all types of fermented foods, the meat and fish products play a very important role in the food habits of *Arunachalee* people. A number of fish and meat products have been fermented for their increased nutritional value and long-term edibility (Das et al. 2016). Salting, drying and cooking of meat are the most common and the most traditional methods employed for food preservation (Kabak and Dobson 2011). The most liked meats are pork meat and mithun meat; however, many other meat types are also consumed by local people. Some of them are briefed here.

Local varieties of soybean seeds

↓

Seeds are boiled and spread on bamboo mat

↓

Salt is added when the seeds are lukewarm and mixed thoroughly

↓

After cooling, seeds are put in Shong (container) and covered with *zola* or banana leaf

↓

Container is kept near the fireplace and rotated regularly

↓

When smell comes and seed becomes thread-like, they are taken out and crushed carefully

↓

Spread in small heaps on *charang* and sun-dried

↓

It is again crushed in to a coarse powder and dried

↓

Libi Chhurpi

Fig. 2.10 Different stages of *libi chhurpi* preparation

2.3.1 Pikey Pila (Fig. 2.11)

Pikey pila is a fermented food item of the *Apatani* tribe of Arunachal Pradesh prepared from pork fat. In its preparation, fresh pork (fat) is stored in smoked area for 3–6 months. The fermented meat is cooked with vegetables or bamboo shoot by adding chillies in excess amount. After that pila is added and thereby *pikey pila* is ready for consumption (Plate 2.8).

2.3.2 Lukter (Fig. 2.12)

Lukter is a meat-based preparation of the *Nyishi* tribe. The meat of any kind can be used for *lukter* preparation. Meat is first cooked followed by drying in sun for 2–3 days. King chilli is mixed with the dried meat. *Lukter* is now ready to use. It can be stored for long time or consumed when required.

Fig. 2.11 Different stages of *pikey pila* preparation

Fresh pork (fat) is stored in smoked area for 3–6 months
↓
Fermented pork meat + vegetables/bamboo shoot
↓
Mixed with chilli
↓
Pila is added
↓
Pikey pila

Plate 2.8 *Pikey pila*

Fig. 2.12 Different stages of *lukter* preparation

Cooked meat
↓
Dried at sunlight for 2–3 days
↓
Mixed king chilli
↓
Lukter

2.3.3 Dingkyo (Fig. 2.13)

Another preparation of fermented meat and fish used by *Nyishi* community is known as *dingkyo*. *Dingkyo* is mainly prepared from the mithun (*Bos frontalis*) meat while any other meat and fish may also be used. Cutting of fresh flesh to convenient size

Fig. 2.13 Different stages of *dingkyo* preparation

Fresh meat of mithun and any other meat and fish

↓

Properly washed and cut into convenient size

↓

They are kept hanging in smoky place

↓

Dingkyo

Fresh fish + mustard oil + salt

↓

Put into an airtight container and kept for 7 days for fermentation

↓

Ngiiyi-yaan

Fig. 2.14 Different stages of ngiiyi-yaan preparation

for hanging is the initial step for its processing. They are kept hanging in the smoky place for fermentation and consumed by cooking with *bamboo tenga* or any other leafy vegetable or used as chutney along with the chilli.

2.3.4 Ngiiyi-Yaan (Fig. 2.14)

Fermented fish product called as *ngiiyi-yaan* is prepared by the *Apatani* tribe of Arunachal Pradesh, India. During its preparation, the fresh fish is treated with mustard oil and salt. The oily and salty fish is then put into an airtight container and kept for about 7 days. This fermented fish product is called *ngiiyi-yaan*.

2.4 Fermented Food Products of Milk

The fermented milk products are important food used by *Monpa*, *Sherdukpen* and *Khamba* tribes. Rather, fermented milk products are considered very precious food and served during special occasions and to special guests (Singh et al. 2007) only. A good range of milk products are consumed; among them ghee and chhurpi (paneer) of yak milk are more popular. Fermentation plays a very important role in increasing shelf life of milk with the help of LAB since milk and its products are easily spoiled by pathogenic microorganisms due to their high nutritive value.

Yak milk

↓

Collected in 15–20-L container

↓

Left for about 10 days

↓

Extensive boiling

↓

Sieved to totally remove water

↓

Wrapped with animal's skin and sun-dried or placed over the fireplace

↓

Churapi, chhurpi or *durkha*

Fig. 2.15 Different stages of *Churapi* or *Durkha* or *Chhurpi* preparation

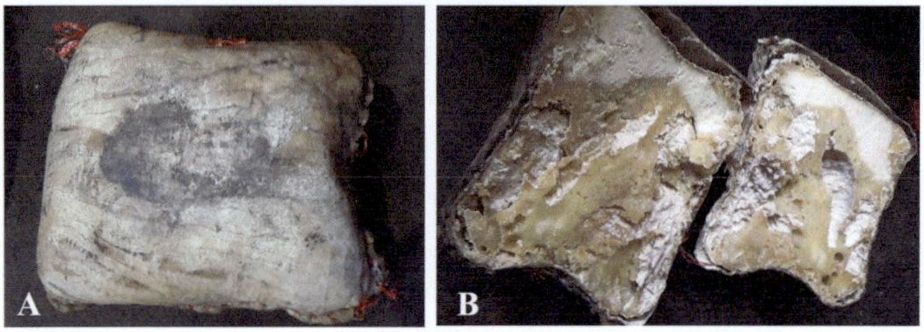

Plate 2.9 *Churapi*, (**a**) wrapped in animal's skin; (**b**) ready to use, *churapi*

Lactobacillus bacteria convert lactose sugar into lactic acid. In Arunachal Pradesh, yak milk is used for manufacturing a number of dairy products like *chhurpi*, *churapi* or *durkha*, *chhur singba*, *chhur chirpen*, *churkham*, *chhurpupu*, *marchang*, etc. The process of preparation of these milk products is detailed below:

2.4.1 Churapi/Chhurpi/Durkha (Fig. 2.15)

The people of *Monpa* tribe of Arunachal Pradesh prepare popular fermented food products from the yak milk named differently as *churapi* or *durkha* or *chhurpi*. Initially yak milk is collected in a container of 15–20 L in volume for a few days.

Chhurpi

↓

Stored for 4–5 years

↓

Chhurpupu

Fig. 2.16 Different stages of *chhurpupu* preparation

Yak milk is boiled

↓

Added cut fruits of *thung* to the lukewarm milk to facilitate the fermentation

↓

Water is removed after some time

↓

Remaining material is placed in cotton cloth and kept for 3–4 days

↓

Pressed with stone to give proper shape and kept on a bamboo mat placed over the fireplace in the kitchen

↓

Chhur chirpen

Fig. 2.17 Different stages of *chhur chirpen* preparation

This container is kept undisturbed for about 10 days. The material is extensively boiled followed by sieving to remove water totally. Then the remaining material is wrapped with cloth and allowed to dry under sunlight or by keeping over the fireplace. This is called *churapi* or *durkha* or *chhurpi* (Plate 2.9).

2.4.2 Chhurpupu (Fig. 2.16)

Chhurpi, which becomes very old (4–5 years), is called *chhurpupu*. The *chhurpi* is kept in yak skin (*mongnang*) and after sealing it can be kept for 3–20 years.

2.4.3 Chhur Chirpen (Fig. 2.17)

The people of *Monpa* tribe prepare another fermented food called *chhur chirpen* from the yak milk. Initially, the yak milk is boiled, and when it cools down to lukewarm, the cut fruits of *thung* (local name, crab apple; scientific name, *Malus sylvestris*) are

White cheese produced during preparation of *churapi* is mixed with pieces of old stored *churpi*

↓

Added little fresh milk to make a paste

↓

Shaped with hands

↓

Large cakes are packed in yak skin and sun-dried

↓

Churkham

Fig. 2.18 Different stages of *churkham* preparation

Plate 2.10 *Churkham*

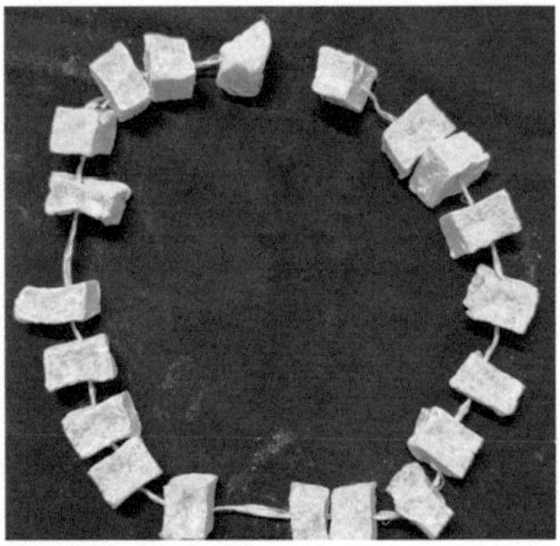

added to it to facilitate fermentation. Watery liquid is removed after some time, and the remaining solid matter is placed in cotton cloth for 3–4 days till its colour turns to creamy yellow followed by pressing with heavy stone to give proper shape. It is then kept on a bamboo mat and placed over the fireplace in the kitchen. There is gradual change of creamy yellow colour of churpi into light yellowish brown. This is known as *chhur chirpen.*

2.4.4 Churkham (Fig. 2.18)

For the preparation of *churkham*, the pieces of stored *churpi* are utilised. White cheese produced during the preparation of *churapi* is mixed with a little fresh milk to prepare a smooth paste, shaped with hands and cut into the pieces of desired size. Usually large cakes of this soft cheese are packed in yak skin and sun dried. The

Fig. 2.19 Different stages of *marchang* preparation

Chhurpup
↓
Fried in yak *ghee*
↓
Kongpu flour is added to it and mixed properly
↓
Marchang

Fig. 2.20 Different stages of *churasabji* preparation

Fermented cheese made by yak or cow milk
↓
Fresh vegetables are added
↓
Mixed with king chilli
↓
Churasabji

product thus prepared is called *churkham*. Generally, *churkham* cakes of 5–10 kg weight are packed and stored for 2–12 months or more, as taste improves with storage. The raw butter (Mar) formed during *churapi* making is used in the preparation of traditional *Monpa* salty butter tea. The clarified butter called *Jimar* is used to flavour rice and is also added to vegetable and meat dishes in small amounts. Further, *churapi* and *churkham* are used as a traditional chocolate (Fig. 2.2) and mouth freshener and also in the preparation of vegetable and nonvegetable dishes in the form of paneer (Plate 2.10).

2.4.5 Marchang (Fig. 2.19)

Chhurpupu is fried in yak *ghee* to remove the unpleasant odour, and then *kongpu* flour is added to it and mixed properly. This dish is known as *marchang*. *Ghee* and *chhurpupu* are mixed in old seasoned *rakshi* and are given to cure body ache.

2.4.6 Churasabji (Fig. 2.20)

This is a fermented food prepared by different tribes of Arunachal Pradesh. It is prepared from the fermented cheese made by yak milk or milk. Fresh vegetables are added to the cheese and mixed with king chilli. Then this mixture is stored in a jar until used.

Yak/cow milk

↓

Kept for 2–3 days is added

↓

The milk is masticated in a pot locally called as *jopuh* for 20–30 min

↓

Approximately 1 L of warm water is added

↓

Mastication continued

↓

Fat accumulates over the liquid that is collected

↓

Fat is poured in a pot and dipped in cold water to become hard

↓

Hard fat is *moh*

Fig. 2.21 Different stages of *moh* preparation

2.4.7 Moh (Fig. 2.21)

Moh is a traditionally prepared milk product of *Sherdukpen* tribe residing in the West Kameng district of Arunachal Pradesh. It is prepared from yak or cow milk. The milk is stored in the traditionally designed container locally called as *jopuh* (Plate 2.11) for 2–3 days. The milk may be collected in 2–3 days' time to obtain about 15–20 L at a time. The milk is then masticated for 20–30 min after which approximately 1 L of warm water is added to it and mastication continued. The fat will accumulate over the warm water which is collected and poured in pot dipped in cold water. It results in its hardening. The product so prepared is called as *moh.* It is consumed on regular basis.

2.5 Fermented Alcoholic Beverages

2.5.1 Apong (Fig. 2.22)

Apong is the most popular cereal-based fermented alcoholic beverage prepared from rice and consumed by almost all tribes of Arunachal Pradesh. The method of preparation of *apong* is quite similar in all tribes with minor variation.

The final product of *apong* is a clean and milky liquor. It is prepared from *Khamtip* (mixture of cooked rice and starter) by keeping it in either earthen pot,

Plate 2.11 Preparation of *Moh*

Fig. 2.22 Different stages of *Apong* preparation

Rice
↓
Cooking of rice to become soft
↓
Cooling of rice on bamboo mat
↓
Mixing of amylolytic starter culture with cooked rice
↓
Transferring whole material in pots
↓
Fermentation (4–6 days in summer or 7–9 days in winter)
↓
Mixing of water
↓
Filtration with muslin cloth
↓
Apong

wooden containers, teak or banana leaves for fermentation in traditional way which is now being replaced by metallic containers with lid. People of *Sulung* tribe use wild banana (*Phrynium capitatum*) leaves locally known as *etkam* to cover the earthen pots during fermentation. After fermentation, the solid material is transferred into another vessel and liquid is extracted. This concentrated liquid is pale yellow in colour and bears good alcoholic flavour and pleasant smell and taste. It

Plate 2.12 (**a–c**) Different stages of *apong* preparation, (**d**) *apong* ready to consume

can be stored for 4–5 days without spoilage as it does not contain any water contents; however, by mixing water before extraction, the drink becomes diluted; thus shelf life is also reduced to 1–2 days only (Plate 2.12).

Its traditional process technology is elaborated below:

The people of *Karbi* tribe use a special apparatus called *Kodtak* for the extraction of the *apong*. After fermentation, enough quantity of fermented stock is transferred in a vessel, and cold water (lukewarm water during winter) is added at a rate of 4–5 L per kg of rice. This mixture is then kept for 8–9 h for further fermentation. The beverage is filtered using a bamboo pole locally called *Hiru*, which is open from one side and has a small hole on the other near the closed end. This end is inserted vertically into the vessel having fermented material. Due to capillary action, the beer enters and collected inside the bamboo cavity. To collect the beverage from bamboo, a dried *Lngjak* fruit (*Luffa acutangula* Roxb.) is dipped into it. The absorbed liquid is then squeezed out in a container which is used as beer.

The people of *Tagin* tribe residing in Upper Subansiri district of the state prepare *apong* by filtering the fermented stock through a bamboo-made sieve.

2.5.2 Opo (Fig. 2.23)

Opo is also a rice-based beverage prepared by the *Adi*, *Nyishi*, *Galo* and *Mishmi* tribes of Arunachal Pradesh especially those residing in East Siang, Papumpare and Lohit districts. The fermented mixture of this preparation is locally called as pone which consists of cooked rice and burnt ashes of rice husk and amylolytic starter culture (Plate 2.13). Pone is allowed to ferment for 4–7 days. A bamboo-made funnel is used for extracting the drink from this mixture. The funnel is filled with pone, and boiled water is poured slowly through the funnel till the beer is extracted completely. A clean blackish-coloured beverage is collected in a container. People of Galo tribe, however, use boiled water only for the first preparation, and for succeeding extractions, normal water is used. This beverage is also called kala apong.

The detailed method of preparation is given below:

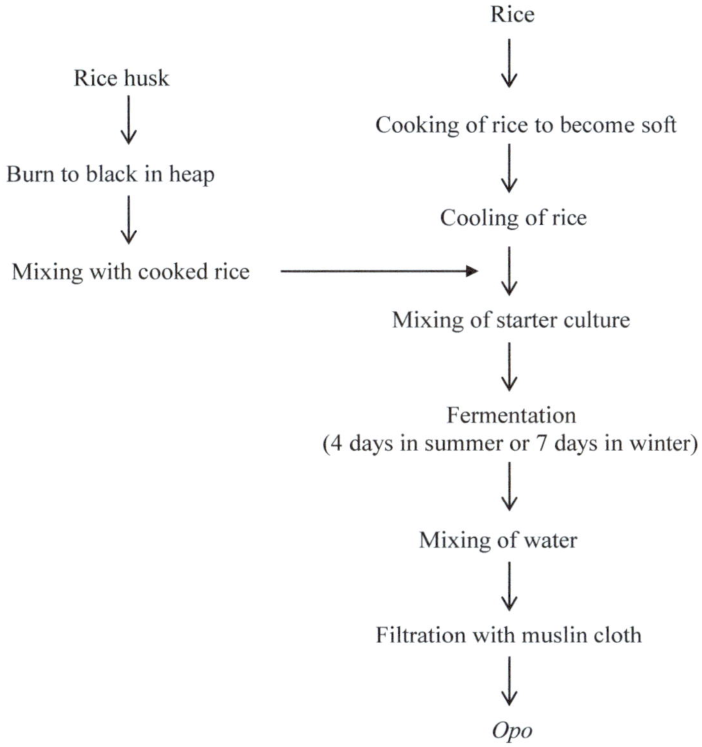

Fig. 2.23 Different stages of *Opo* preparation

Plate 2.13 (**a**) Pot used for preparation of *opo* and (**b**) *burning of rice husk*

2.5.3 Madua Apong (Fig. 2.24).

Madua apong is also one of the most popular and commonly used alcoholic beverages of different tribes of Arunachal Pradesh, India. It is a traditional alcoholic beverage prepared from finger millets. The best-quality *madua apong* is golden yellow in colour and sweet in taste and emits sweet alcoholic aroma during saccharification. This is an indicator of accomplishment of fermentation process. The fermented mass of finger millet is then transferred into a perforated bamboo basket, and lukewarm water is poured slowly at a rate of 1 l/h onto it. The filtrate so collected is known as *madua apong* (Plate 2.14).

The stepwise method of preparation of *madua apong* is depicted in flow diagram given below:

Fig. 2.24 Different stages of *madua apong* preparation

Millet

↓

Roasting of millet for 30 min

↓

Cooling of millet

↓

Cooking of millet to become soft

↓

Mixing of starter culture with millet

↓

Kept in perforated basket covered with *Ektam* leaves

↓

Fermentation
(4 days in summer/7 days in winter)

↓

Pouring of hot water from top

↓

Collection in a container

↓

Madua apong

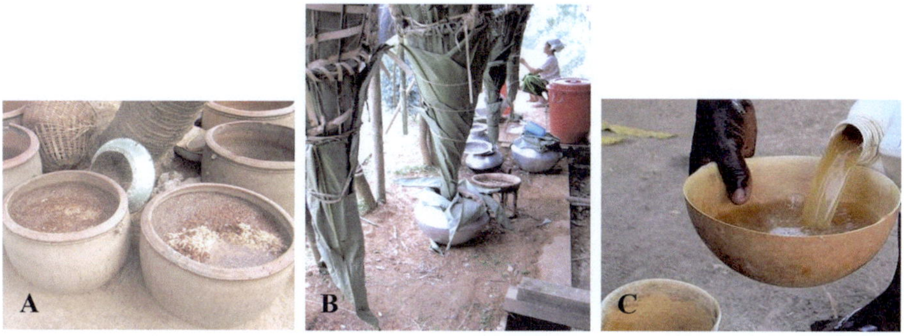

Plate 2.14 (**a–c**) Different stages of *madua apong* preparation

2.5.4 Themsing (Fig. 2.25)

Themsing is an alcoholic beverage consumed by *Monpa* tribe of Tawang district of Arunachal Pradesh only. It is prepared either from finger millet (locally called *kongpu*) or from barley (locally known as *bong*) or from the mixture of both. The detailed preparation of *themsing* is provided below:

For *themsing* preparation, cooked and cooled millet or barley grains taken, small amount of ash or charcoal and amylolytic starter are thoroughly mixed and kept for fermentation for 7–10 days. The fermented stock is transferred to special container fitted with a PVC pipe having a diameter of approximately 1 cm at the bottom. The drink sips out from the pipe which is collected in another container. The extraction of *themsing* may also be done by transferring the fermented material into a perforated bamboo basket, slowly adding the lukewarm water and collecting the filtrate/liquor. The filtrate has a golden colour and good aroma.

2.5.5 Rakshi, Aara, Aaro (Fig. 2.26)

Rakshi, *aara* and *aaro* are types of traditional distilled beverages prepared by the people of *Monpa*, *Sherdukpen* and *Idu-Mishmi* communities of Tawang and West Kameng districts of Arunachal Pradesh. Either finger millets, rice or barley grains are used to prepare it. The traditional processing technology of *rakshi* (Plate 2.15) is elaborated as follows:

The grains of finger millets, rice or barley are first cooked to make them soft and ready for fermentation. After boiling, they are spread on a bamboo mat *charang*; starter culture is mixed and left for fermentation for 1–2 weeks depending on season. On completion of fermentation, *rakshi* is distilled by adding water to the fermented stock, and a large vessel containing fermented material is kept on fire. A small metallic container is kept inside the large vessel above a triplet stand to collect the distillate. A wide vessel containing cold water is kept above the large vessel as condenser whose water is changed at frequent intervals to keep the water continuously cool. The drink prepared in this method has good alcoholic aroma and a very strong ethanolic taste.

Finger millets/barley

↓

Cooking of grains to become soft

↓

Cooling of grains on bamboo mat

↓

Mixing of starter culture with cooked grains

↓

Adding a pinch of ash/charcoal

↓

Fermentation in wooden containers by tightly covering with *zol*a/banana leaves (at this stage, mixture can be stored for 1–3 years)

↓

Adding of lukewarm water

↓

Themsing

Fig. 2.25 Different stages of *Themsing* preparation

Plate 2.15 (**a**) Distillation of *rakshi* and (**b**) *rakshi* ready to consume

2.5.6 Mingri or Niingri, Lohpani and Bhangchang or Phaak (Fig. 2.27)

These are traditional alcoholic beverages very unique to *Monpa* and *Sherdukpen* tribes residing mainly in Tawang and West Kameng districts of Arunachal Pradesh. These drinks are prepared from finger millet, rice, maize or barley. These three

Fig. 2.26 Different stages of *rakshi*, *aara* or *aaro* preparation

Finger millets/barley/rice

↓

Cooking of grains to become soft

↓

Cooling of grains by spreading on *charang* (bamboo mat)

↓

Mixing of starter culture with cooked grains

↓

Fermentation
(1 week in summer or 2 weeks in winter)

↓

Distillation of fermented stock

↓

Rakshi, aara or *aaro*

Fig. 2.27 Different stages of *Mingri/Lohpani/Bhangchang* preparation

Finger millet/barley/rice/maize

↓

Cooking of grains to become soft

↓

Cooling of grains

↓

Mixing of amylolytic starter culture with cooled grain

↓

Fermentation for 4–5 days

↓

Filtration by bamboo sieve

↓

Mingri/niingri

↓

Further fermentation of filtrate for 2–3 days

↓

Lohpani

↓

Further fermentation of filtrate for 4–5 days

↓

Bhangchang/phaak

Plate 2.16 *Lohpani*

types of alcoholic beverages are obtained at three different stages of fermentation of the same raw material. The cereal grains of finger millet, barley, rice or maize are taken, cooked till they become soft and cooled; amylolytic starter is mixed and kept for fermentation. After 4–5 days, the first lot of beer is collected by sieving through bamboo mat called *Khajir* without adding any additional water. This is called as *mingri* by *Monpas* and niingri by *Sherdukpen*. It is richer in alcoholic contents and light yellowish in colour and emits pleasant strong aroma being rich in alcoholic contents. The leftover filtrate is further fermented for 2–3 days more. As a result, comparatively darker yellowish beverage with less alcoholic contents and aroma is collected and called as *lohpani* (Plate 2.16). The remaining material is again allowed to ferment for 4–5 days more to obtain *bhangchang* (in *Monpa* tribe) or *phaak* (in *Sherdukpen* tribe). The product *phaak* is being reported for the first time. These beverages are collected by adding a little water into the mixture which is then squeezed out and collected which has least alcoholic contents, is whitish in colour and has a very weak aroma.

The brief process of preparation of these three types of alcoholic beverages is elaborated in following flow chart:

2.5.7 Yu (Fig. 2.28)

This is a traditional beverage prepared by the *Idu-Mishmi* tribe occupying the north-eastern tip of the central Arunachal Pradesh in Upper and Lower Dibang Valley, Lohit and Anjaw districts. *Idu-Mishmi* is a subdivision of Mishmi or Deng people of Tibet and Arunachal Pradesh, emerged due to the geographical distribution, but racially all the three groups [*Idu-Mishmi* (*Idu Lhoba*), *Digaro* tribe (*Taraon*, *Darang Deng*) and *Miju Mishmi* (*Kaman Deng*)] are of the same stock.

Fig. 2.28 Different stages of *yu* preparation

Rice is cooked

↓

Cooked rice is cooled

↓

Old starter cake is mixed with the cooled rice

↓

Kept in airtight pan for fermentation

↓

Filtration

↓

Yu

This beverage is prepared from rice. The cooked rice after cooling is mixed with old amylolytic starter cake and left for fermentation in an airtight pan. The drink is extracted by filtration when a fruity smell of alcohol is experienced. This drink is used as alcoholic consumption at different occasions and regular use. The process technology of 'yu' is detailed below:

2.5.8 Poling (Fig. 2.29)

It is an alcoholic beverage traditionally prepared by the people of *Nyishi* tribe. The largest ethnic group in Arunachal Pradesh spread across seven districts: Kra Daadi, Kurung Kumey, East Kameng, West Kameng, Papum Pare, parts of Lower Subansiri and Kamle districts.

This is a product obtained from *madua* (millets). Locally made starter yeast is mixed with boiled and cooled marua grains and left for fermentation in airtight container. When the fruity alcoholic aroma is observed to occur, it is then filtered. The drink so obtained is called *poling*. The poling is an alcoholic beverage consumed on regular as well as on some special occasions.

2.5.9 Pona (Fig. 2.30)

Another traditional liquor prepared by the *Nyishi* tribe of the state is named as *pona*. This is used as alcoholic beverage in many occasions, which is a fermentation product prepared from rice. The preparation method is started with cooking of rice followed by its drying under sunlight. When it is completely dried, locally made yeast starter culture is mixed with it and left for fermentation in airtight containers. It is stored as such for a week, followed by filtration by mixing sufficient amount of water, and the filtrate so obtained is the drink termed as *pona* (Plate 2.17). The *pona* is prepared as follows:

Fig. 2.29 Different stages of *poling* preparation

Madua
↓
Boiled and cooled by spreading on a mat
↓
Locally made starter yeast is mixed
↓
Left in airtight container for fermentation
↓
Filtration with water
↓
Poling

Fig. 2.30 Different stages of *Pona* preparation

Rice
↓
Cooking of rice
↓
Drying of boiled rice in the sun
↓
Mixed with locally made amylolytic starters
↓
Left in airtight container for fermentation
↓
Stored for 1 week
↓
Filtration with water
↓
Pona

2.5.10 Nyongin (Fig. 2.31)

Galo is also one of the major tribes of Arunachal Pradesh, with members who are descendants of *Abo Tani* and speak the *Tani* language *Gallong*. The *Galo* people primarily inhabit West Siang, Lepa Rada and Lower Siang districts of modern-day Arunachal Pradesh state in North Eastern India, but are also found in the southwestern side of East Siang district, the southeastern side of Upper Subansiri district, as well as in some small pockets in Itanagar, Lower Dibang Valley and Changlang districts.

Plate 2.17 Fermented rice in the preparation of *pona*

Fig. 2.31 Different stages of *nyongin* preparation

Rice is allowed to cook

↓

The cooked rice grains properly spread on a clean mat

↓

Mixing it with *opop*

↓

Kept in an airtight container for fermentation

↓

Filtration with water

↓

Nyongin

The people of *Galo* tribe used to prepare alcoholic beverage named as *nyongin*, which is used specially during *Mopin*, one of their most important festivals celebrated on a big way. It is again a rice fermentation product. During its preparation, rice is first cooked completely and, once it is cooked, is spread properly on a clean mat. *Opop*, an amylolytic yeast starter, is then mixed with the spread rice and the mixture put in airtight container for fermentation. After fermentation, it is filtered and the filtrate is the alcoholic drink called as *nyongin*.

2.5.11 Pone (Fig. 2.32)

Pone is traditional liquor used by the people of *Galo* tribe mainly during their festival called *Monpi*. Its preparation is initiated by burning the rice husk till it is completely black followed by the steaming of rice till they are properly cooked.

Fig. 2.32 Different stages of *pone* preparation

Burn the rice husk till it is completely black

↓

Steam rice till it is properly cooked

↓

Cool and then mix the burnt rice husk

↓

Add the boiled rice along with *opop* properly

↓

Pack them in leaves

↓

Leave in airtight container for fermentation

↓

Filtration

↓

Pone

The steamed and cooked rice is then cooled and mixed with the burnt husk. Mixing of *opop* to the mixture of cooked rice and ash of rice husk follows the process. It is then packed in *ekkam* leaves (*Phrynium pubinerve*), and the packed materials are put in airtight container for fermentation. After fermentation it is filtered and filtrate so obtained is the alcoholic drink *pone*.

2.5.12 O and Sira-O (Fig. 2.33)

O and *sira-o* are two cereal-based traditional beverages prepared by the *Apatani* of Arunachal Pradesh. The chief ingredients of these products are rice and millet. The preparation is initiated by thoroughly mixing rice and millet grains followed by cooking the mixture properly. The excess water is drained off, and cooked grains are cooled by spreading over a mat. The old amylolytic starter is added to the cooled mixture and transferred in a bamboo basket and kept over a raised platform normally on a false ceiling over the fireplace for one night.

Next day, the mixture is transferred to an airtight metal pot for fermentation for 4–5 days. This fermented product is called *pona*. The filtered product from *pona* is called *O* (Plate 2.18). Another beverage which is prepared from pona by distillation is called *sira-o*. The distillation of *sira-o* is done following the traditional distillation method as described in the preparation of *rakshi* produced by *Monpa*, *Sherdukpen* and *Idu-Mishmi* tribes. *Sira-o* is considered as special drink used during the social ceremonies or occasions.

Rice mixed thoroughly with millet

↓

Cooked properly

↓

Spread over a mat and cooled

↓

Amylolytic starter is added

↓

Mixture is transferred in a bamboo basket and kept over the fireplace for a night

↓

Transferred to a metal pot and kept for fermentation for 4–5 days

↓

Fermented product is termed as *pona*

↓ Filtered with water ↓ Filtered drink is called *o*

↓ Distillation of fermented stock in traditional way ↓ Distilled drink is *sira-o*

Fig. 2.33 Different stages of *o* and *sira-o* preparation

Plate 2.18 '*O*' ready to consume

2.6 Conclusion

The present survey and literature review indicate that, even with the shift of rural and urban food culture in recent years towards modern west-oriented fast foods, the traditional fermented food products are still very much preferred food items among the people of Arunachal Pradesh. Many of these foods are prepared and cooked in households regularly. The *Arunachalee* people who are staying away from home greatly experience shortage or absence of such traditional foods. The high diversity of fermented food products in the state is correlated to the diverse tribal communities, their varied food culture, customs, religious practices, etc. that might have evolved parallel with time. The remote geography and isolated areas from mainland in the state might have also contributed towards increased variety of fermented foods. The *Apatani*, *Galo* and *Monpa* communities of the state are consuming more fermented products as compared to *Nyishi*, *Adi*, *Idu-Mishmi* and *Khamba* tribes. It may be concluded that bamboo-based foods are the most preferred fermented foods consumed throughout the state and by the people of all tribes. The preparation and process technology of some fermented foods are similar, however known by different local names in different tribes. This may be attributed to diversity in local dialects and languages among them and sometimes mode of pronunciation or accent. On the other hand, there is little variation in food preparation methodology or the way bamboo material is used in different tribes. It is observed that people of *Nyishi* tribe ferment the bamboo-based food items like *bamboo tenga* or *hitak*, *hikung*, *hiyup*, etc. by keeping in airtight vessel under the soil, while people of other tribes ferment it either by wrapping in leaf or keeping on a raised platform as the case of *eup*, *iting*, and *ekung* of *Adi* tribe and *hikhu* and *hirring* of *Apatani* tribe. Not only the use of dried (*eup*, *eyup* and *hiyup*) and fresh leaves but the use of large (*iting*, *hirring, heccha*, *kupe*, *bamboo tenga* and *hitak*), coarsely cut (*ikhung*, *ekung*, *hikhung*, *eeku*, *hikhu* and *hihi*) and finely cut (*anpo* and *anpo-shi-anjita*) bamboo leaves also makes differences.

Acknowledgement The authors are grateful to the local traditional knowledge holders of different tribal groups from various districts of Arunachal Pradesh, India, specially Ms. Nirigi Linggi, Ms. Tage Yama, Ms. Binu Nguso, Ms. Hage Sumpi, Mr. Khyoda Rajan, Ms. Biki Dene, Ms. Nabam Sangte and Ms. Lucky Tamuk, and the authors are also thankful to Mr. Lobsang Tashi Thungon, a member of the BMC, Shergaon Village, and Tashi Tsering, Chairman of the BMC, Sanglem Village, for sharing the valuable information with us on various indigenous foods and beverages, their preparation methods and ingredients. We are highly indebted to the people of *Apatani*, *Galo*, *Idu-Mishmi*, *Sherdukpen* and *Nyishi* communities for their help and support to gather the valuable information.

References

Angami A, Gajurel PR, Rethy P, Singh B, Kalita SK (2006) Status and potential of wild edible plants of Arunachal Pradesh. Indian J Tradit Knowl 5(4):541–550

Anupama A, Pradhan P, Sha SP, Tamang JP (2018) Traditional skill of ethnic people of the Eastern Himalayas and North East India in preserving microbiota as dry amylolytic starters. Indian J Tradit Knowl 17(1):184–190

Asati BS, Yadav DS (2003) Diversity of horticultural crops in North Eastern Region. ENVIS Bull Himalayan Ecol 12:1–10

Chatterjee S, Saikia A, Dutta P, Ghosh D, Pangging G, Goswami AK (2006) Background paper on biodiversity significance of North East India for the study on natural resources, water and environment nexus for development and growth in North Eastern India. Background paper no. 13. WWF-India, New Delhi. p 1–71

Das AJ, Deka SC (2012) Mini review fermented foods and beverages of the North-East India. Int Food Res J 19(2):377–392

Das G, Patra JK, Singdevsachan SK, Gouda S, Shin H (2016) Diversity of traditional and fermented foods of the seven sister states of India and their nutritional and nutraceutical potential: a review. Front Life Sci 9(4):292–312

Dhar ON, Nandargi S (2004) Rainfall distribution over the Arunachal Pradesh Himalayas. Weather 59(6):155–157

Kabak B, Dobson ADW (2011) An introduction to the traditional fermented foods and beverages of Turkey. Crit Rev Food Sci Nutr 51:248–260

Konsam S, Thongam B, Handique AK (2016) Assessment of wild leafy vegetables traditionally consumed by the ethnic communities of Manipur, Northeast India. J Ethnobiol Ethnomed 12:15

Mao AA, Hynniewta TM (2000) Floristic diversity of North East India. J Assam Sci Soc 41:255–266

Medhi P, Kar A, Borthakur SK (2013) Medicinal uses of wild edible plants among the Ao Nagas of Mokokchung and its vicinity of Nagaland, India. Asian Res 2:64–67

Nehal N (2013) Knowledge of traditional fermented food products harbored by the tribal folks of the Indian Himalayan Belt. Int J Agric Food Sci Technol 4(5):401–414

Nongdam P (2015) Traditional fermented bamboo shoot foods of North-East India and their characteristic natural microbial flora, 10th World Bamboo Congress, Korea 2015, p 1–13

Rai R, Shangpliang HNJ, Tamang JP (2016) Naturally fermented milk products of the Eastern Himalayas. J Ethnic Foods 3:270–275

Rawat K, Kumari A, Kumar S, Kumar R, Gehlot R (2018) Traditional fermented products of India. Int J Curr Microbiol App Sci 7(4):1873–1883

Rinya P (2016) Traditional food and health implication with reference to Apatani tribe of Arunachal Pradesh. Int Res J Soc Sci 2:236–241

Ray M, Ghosh K, Singh S, Mondal KC (2016) Folk to functional: an explorative overview of rice-based fermented foods and beverags in India. J Ethn Foods 3:5–18

Roy A, Roy S, Rai C (2017) Insight into bamboo-based fermented foods by Galo (Sub-tribe) of Arunachal Pradesh. Int J Life Sci Sci Res 3(4):1200–1207

Sathe GB, Mandal S (2016) Fermented products of India and its implication: a review. Asian J Dairy Food Res 35:1–9

Sekar S, Mariappan S (2007) Usage of traditional fermented products by Indian rural folks and IPR. Indian J Tradit Knowl 6:111–120

Sha SP, Suryavanshi MV, Jani K, Sharma A, Shouche Y, Tamang JP (2018) Diversity of yeasts and molds by culture-dependent and culture-independent methods for mycobiome surveillance of traditionally prepared dried starters for the production of Indian alcoholic beverages. Front Microbiol 9(2237):1–15

Shrivastava K, Greeshma AG, Srivastava B (2012) Biotechnology in tradition—a process technology of alcoholic beverages practiced by different tribes of Arunachal Pradesh, North East India. Indian J Tradit Knowl 11(1):81–89

Singh RK, Singh A, Sureja AK (2007) Traditional foods of Monpa tribe of west Kameng, Arunachal Pradesh. Indian J Tradit Knowl 6(1):25–36
Singh RK, Srivastava RC, Adi Community and Monpa Community (2010) Bioculturally important plant diversity of Arunachal Pradesh: learning from Adi and Monpa communities about future crops of India. Indian J Tradit Knowl 9(4):754–759
Sonar NR, Halami PM (2014) Phenotypic identification and technological attributes of native lactic acid bacteria present in fermented bamboo shoot products from North-East India. J Food Sci Technol 51(12):4143–4148. https://doi.org/10.1007/s13197-014-1456-x
Tamang JP (2010) Himalayan fermented foods: microbiology, nutrition, and ethnic values. CRC Press, Taylor & Francis Group, New York, p 295. ISBN: 9781420093247
Tamang JP (2015) Naturally fermented ethnic soybean foods of India. J Ethnic Foods 2:8–17
Tamang B, Tamang JP (2009) Traditional knowledge of biopreservation of perishable vegetables and bamboo shoots in Northeast India as food resources. Indian J Tradit Knowl 8(1):89–95
Tamang B, Tamang JP, Schillinger U, Franz CMAP, Gores M, Holzapfel WH (2008) Phenotypic and genotypic identification of lactic acid bacteria isolated from ethnic fermented bamboo tender shoots of North East India. Int J Food Microbiol 121:35–40
Tamang JP, Chettri R, Sharma RM (2009) Indigenous knowledge on north-east women on production of ethnic fermented soybean foods. Indian J Tradit Knowl 8(1):122–126
Tamang JP, Tamang N, Thapa S, Dewan S, Tamang B, Yonzan H, Rai AP, Chettri R, Chakrabarty J, Kharel N (2012) Microorganism and nutritive value of ethnic fermented foods and alcoholic beverages of North-East India. Indian J Tradit Knowl 11:7–25
Tiwari SC, Mahanta D (2007) Ethnological observations on fermented food products of certain tribes of Arunachal Pradesh. Indian J Tradit Knowl 6(1):106–110
Trivedi S, Tripathy RS (1984) Bamboo as an important resource of Northeast India. In: Tripathy RS (ed) Resource potentials of North East India, Vol II, living resources. Meghalaya Science Society, Shillong, pp 9–15
Wang D, Shen SJ (1987) Bamboos of China. Christopher Helm, Bromley, Kent, p 167

Ethnic Fermented Foods and Beverages of Assam

3

Madhumita Barooah, Sudipta Sankar Bora,
and Gunajit Goswami

Abstract

The human migration waves from varied places over long intervals constitute the ancestors of the present-day ethnic communities of Assam in Northeast India. During the course of assimilation, people of this region developed the art of fermenting various perishable, inedible raw agricultural commodities into palatable, preserved and organoleptically superior foods and beverages. Locally available biological resources such as cereals, legumes, bamboo shoots, milk and vegetables are commonly fermented by most of these communities. Ethnic fermented foods such as *kharoli*, *kahudi*, *panitenga*, *khorisa*, *sukoti*, *namsing*, etc. and rice-based alcoholic beverages like *xaj* and *sai-mod* are unique to Assam and carry the signatures of rich culture and traditions. Abundance of functionally important indigenous microflora such as *Amylomyces rouxii*, *Rhizopus* sp., *Lactobacillus plantarum*, *Lactobacillus brevis*, *Saccharomyces cerevisiae*, *Meyerozyma guilliermondii*, etc. bears the testimony of people's knowledge in customary microbiology. Preparation of most of the fermented products is a women-associated activity that has remained mostly home-based linking with and local market.

Keywords

Assam · Fermentation · Food · Microorganisms · Nutrition

M. Barooah (✉)
Department of Agricultural Biotechnology, Assam Agricultural University,
Jorhat, Assam, India

S. S. Bora · G. Goswami
DBT-North East Centre for Agricultural Biotechnology, Jorhat, Assam, India

J. P. Tamang (ed.), *Ethnic Fermented Foods and Beverages of India: Science History and Culture*, https://doi.org/10.1007/978-981-15-1486-9_3

3.1 Introduction

The state of Assam located in Northeast India extends between 89° 42′ E to 96° E longitude and 24° 8′ N to 28° 2′ N latitude covering an area of 78,438 km^2. The Brahmaputra, an antecedent river, older than the Himalayas, is regarded as the lifeline of the state. The river with steep gorges and rapids in Arunachal Pradesh on entering Assam becomes a braided river; and with tributaries, it creates the Brahmaputra Valley. The valley is approximately 80–100 km wide and almost 1000 km long (Singh 1993). It covers areas of clay soil rich in the nutrients of the floodplain, which is very suitable for tea plantation. In the south, another major river, the Barak, originating in the Barail Range (Assam-Nagaland border) creates another valley, the Barak Valley (25–30 miles wide). The river flows through the Cachar District and bifurcating into Surma and Kushiyara Rivers inflowing Bangladesh (Das 2012). Climate of Assam is temperate (summer max. at 35–38 °C and winter min. at 6–8 °C). With heavy rainfall and high humidity, the state enjoys the "Tropical Monsoon Rainforest Climate" (Singh 1993). The vegetation of the area is dominated by tropical rainforests, deciduous forests, riverine grasslands, bamboo orchards and numerous wetlands (National Mission on Bamboo Applications 2004; Sharma 2003). People of different ethnic races have been migrating into the land of Assam and assimilating in a common harmonious whole since time immemorial. The mongoloid migration to Assam took place at long intervals and from widely divergent locations. They, in general, belonged to the Tibeto-Burman family of the Indo-Chinese group. The early waves of this group constituted the ancestors of the present-day *Kachari*s, *Dimasa*s, *Bodo*s, *Rabha*s and *Lalung*s, as also most of the tribes living in the hills neighbouring modern Assam (Kakati 1941). The later waves included the *Ahom*s from the northern and eastern hill tracts of Upper Burma and Western Yunnan and the *Mishing*s and the *Karbi*s of the Tibeto-Burman stock (Gait 1906).

The Ahoms of the Tai-Shan family came from Burma across the Patkai Range and entered Assam under an adventurous leader named, Chaolung Sukaphaa. The special section to which they belonged, or the Shans proper, occupied the northern and eastern hill tracts of Upper Burma and Western Yunnan, where they formed a group of states. The Ahoms subdued the various local chiefs through a series of determined and skilful moves and very soon firmly entrenched themselves as the masters over a long tract (Kakati 1941). The Ahoms are known to prepare fermented rice alcoholic beverage called as *xaj pani* or *koloh pani* in their own traditional way (Das et al. 2012).

Traditional fermented foods have occupied an important place in the Assamese culture. Varieties of fermented foods and beverages are prepared by the Assamese people. Some of the popular fermented foods and beverages that are unique to Assam are *khorisa* (fermented bamboo shoot), *kahudi* (fermented rapeseed produced by acidic fermentation), *kharoli* (fermented rapeseed produced by alkaline fermentation), *sukati/namsing* (fermented fish) and *poita bhat* (fermented rice). Besides catering to the growing demand of nutrition, these fermented foods also have gained an irrevocable place in the social fabric (in terms of local festivals, marriages and special occasions) of Assam (Das and Deka 2012).

A list of ethnic fermented foods and beverages of Assam is given in Table 3.1.

3.2 Fermented Bamboo Shoot Products

3.2.1 Khorisa

Khorisa is an ethnic fermented bamboo shoot food of Assam produced from succulent bamboo shoots (*Dendrocalamus hamiltonii*, *Bambusa balcooa*, *B. vulgaris* and *B. tulda*) during sprouting seasons of June–September. For traditional preparation of *khorisa*, the tips of mature bamboo shoots are collected and chopped into small pieces after removing the outer castings and the lower portions. The pieces are sun-dried for 2–3 days after which these are pounded finely in a *dheki* or *ural* (traditional grinder). A little water is added along with two to three chillies and small pieces of *Thekera* (dried mangosteen, scientific name *Garcinia pedunculata*; family Clusiaceae), which gives a characteristic flavour and odour. The mixture is kept in an air-locked container or an earthen pot for 5–6 days for maturation (submerged fermentation) at room temperature (Fig. 3.1).

Flow sheet depicting the traditional method of *khorisa* preparation in Assam is presented in Fig. 3.2.

Table 3.1 Ethnic fermented foods and beverages of Assam

Food	Substrate	Starter	Sensory property of product	Major ethnic consumers	Region/district
Khorisa	Young bamboo shoot	Autochthonous fermentation	Acidic, sour in taste, curry/pickle	Most of the ethnocultural groups	Upper, middle and lower Brahmaputra valley districts
Kharoli	Black or white mustard seeds; Rapeseeds		Alkaline, *chutney*		
Kahudi			Acidic, sour in taste, tangy *chutney*		
Panitenga			Acidic, sour in taste, tangy *chutney*		
Sukati/ namsing	Freshwater small- and medium-sized fishes		Consumed as an appetizer along with rice	Deori/ Mishing	Upper Brahmaputra valley districts

(continued)

Table 3.1 (continued)

Food	Substrate	Starter	Sensory property of product	Major ethnic consumers	Region/district
Xaj	Gelatinized rice preferably glutinous varieties	*Xaj-pitha*	Sweet and alcoholic beverage	Ahom	Upper Brahmaputra valley districts
Zu/ Judima		*Umhu* or *Humao*		Dimasa	Dima Hasao, Jatinga valley and lower portion of Karbi Anglong district (East), Nagaon, Cachar
Jou Bishi		*Angkur*		Bodo	Kokrajhar, Chirang, Udalguri and Baksa
Sujen		*Perok kushi*		Deori	Upper Brahmaputra valley districts
Aapong		*Apop pitha*		Mishing	Upper Brahmaputra valley districts
Haria		*Mod pitha*		Adivasi	Upper and lower Brahmaputra valley districts
Arak/ Hor-Alank		*Thap*		Karbi	Karbi Anglong, Dima Hasao, Nagaon, Golaghat, Karimganj, Sonitpur and Biswanath Chariali
Laopani		*Bhekur pitha*		Lalong	Middle Brahmaputra valley
Poita bhat	Cooked rice	Autochthonous fermentation	Acidic, sour in taste	Ahom	Upper, middle and lower Brahmaputra valley districts

Another type of *khorisa* preparation, called as *xukan khorisa*, slightly differs from that of *poka khorisa*. In this process, the chopped pieces of bamboo shoot are left for fermentation in an earthen pot for ~10–15 days. Sour water from the fermenting product is squeezed out, and the residue is left for sun-drying. The drying residual mass is mixed intermittently with the previously collected sour water in small increments and again exposed for sun-drying. This step is repeated for several times until the water gets exhausted. The product, after mixing with mustard oil and salt, undergoes a final sun-drying and is kept in a glass container. *Xukan khorisa* can be served with other dishes preferably with fish items.

Culinary and mode of consumption: *Khorisa* is commonly used as *chutney* (a condiment used in South Asian cuisine that contains spices and vegetables). *Khorisa* is a part of the diet of both rural and urban people and is extensively used as an ingredient in *pitika* (mashed potato or other vegetables mixed with chopped onion, mustard oil and fresh chillies). It can also be used to enhance flavours during the preparation of different food items like meat and fish curries, pickles, etc. (Sharma and Barooah 2017).

Socio-economy and ethnical or religious values: Although *khorisa* bears the Tibeto-Burman heritage, the term has an Austric language origin indicating that the condiment was thoroughly enjoyed by the people of ancient Kamarupa (Sen 2004). In the tribal societies of Assam, sour foods including *khorisa* are generally avoided during pregnancy for the fear of flatulence, miscarriage and bleeding. *Khorisa* is also an important ingredient of the therapeutic remedy for patients suffering from measles, a disease which is entirely treated at home. A special diet called *sukoti* where *magur* fish (*Clarias batrachus*) flavoured with *khorisa* is served on fresh plantain leaf along with rice to the patients on the 21st day of post-symptomatic period (Joshi 2015). Presence of phytosterols, flavonoid and antioxidants (Choudhury et al. 2012; Nongdam and Tikendra 2014; Nirmala et al. 2014) in *khorisa* warrants scientific investigation behind this practice.

Microorganisms: Based on sugar fermentation pattern and RAPD-PCR profile, the most dominating microflora were found to be members of the group lactic acid bacteria (LAB) with isolates, viz. *Lactobacillus brevis*, *Lactobacillus plantarum*, *L. paracasei* subsp. *paracasei*, *L. pentosus* and *L. collinoides* (Sharma and Barooah 2017). Culture-dependent microbiological analysis of *Mesu*, a similar non-salted fermented bamboo shoot product, consumed by the people of the Darjeeling hills and Sikkim in India revealed the prevalence of lactic acid bacteria, viz. *Lactobacillus plantarum*, *L. brevis* and *Pediococcus pentosaceus* (Tamang and Sarkar 1995). It was also observed that lactic acid contents increased significantly in the samples of *khorisa* from 0.32% to 2.74% which effectively inhibited the growth of spoilage and pathogenic bacteria. All the LAB isolates also found to produce bacteriocin, thereby

Fig. 3.1 (**a**) Bamboo shoot (*Bambusa balcooa*). (**b**) *Khorisa* in its final form

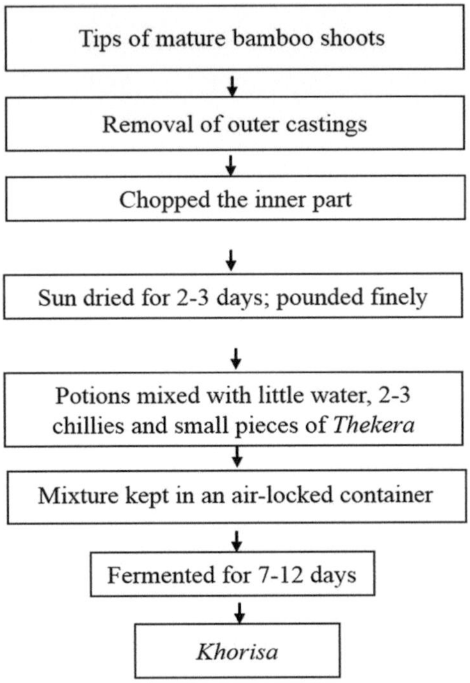

Fig. 3.2 Traditional method of *khorisa* preparation in Assam

showing antimicrobial activity against pathogenic bacteria such as *Listeria monocytogenes*, *Listeria innocua* and *Staphylococcus aureus* (Sharma and Barooah 2017).

Nutritional value: Biochemical evaluation of important parameters revealed that 100 g of mature *khorisa* contains 1.85 mg of total sugar, 3.835 mg of ascorbic acid, 1.23 g of protein, 9.90 g of crude fibre, 2007.25 mg of calcium, 1.715 mg of iron and 10.748 mg of phosphorus (Sharma and Barooah 2017). The value of pH in *khorisa* made from *bhuluka* was observed to decrease from an initial 5.57 on the 1st day of fermentation to 3.12 on the 10th day of fermentation. Total acidity of *khorisa* was found to be 0.63 on the 10th day of fermentation which was initially recorded at 0.41 on the 1st day of fermentation (Sharma and Barooah 2017).

3.3 Fermented Oil Seed Products (*Kharoli, Kahudi* and *Panitenga*)

Indigenous oil seed products, *viz. kharoli*, *kahudi* and *panitenga*, are traditionally consumed by most of the communities of Assam. *Khar* and *Thekera* are the two basic ingredients of these products which determine the type of fermentation (acidic or alkaline) of the oilseeds. *Khar* is an alkaline solution that is prepared preferably from matured dry banana (*Musa paradisiaca*, family Musaceae; vernacular name *bheem kol*) peels, although other substrates like banana stems, dried leaves of vegetables such as papaya (*Carica papaya*; family Caricaceae; vernacular name *Omita*), bottle gourd (*Lagenaria siceraria*; family Cucurbitaceae; vernacular name *Jatilao*), leafy vegetables such as Indian mustard (*Brassica juncea*; family Brassicaceae; vernacular name *Lai xaak*), thumbai (*Leucas aspera*; family Lamiaceae; vernacular name *Durun xak*), helencha (*Enhydra fluctuans*; family Asteraceae; vernacular name *Helosi-xak*), brahmi (*Bacopa monnieri*; family Plantaginaceae; vernacular name *Brahmi xak*), *sukuta* (*Corchorus olitorius*; family Malvaceae; vernacular name *sukuta*), etc. are also regularly used for *khar* preparation. Commonly, the ashes of sun-dried and burnt mature banana peels are collected and dipped in a cup of water. *Khar* has a black tea-like colloidal appearance (pH 8.5–10.0).

Kharoli involves basic (pH 8.0–8.5) solid-state fermentation owing to the alkaline nature of *khar.* For *kharoli* preparation, the seeds of black or white mustard (*Brassica rapa* subsp. *oleifera*, family Brassicaceae) are soaked. After draining the water and sun-drying, these are thoroughly pounded, with occasional addition of mustard oil during the process. Then the oily batter is mixed with final potions of mustard oil and *khar* with addition of a little salt. The concoction is then shaped into small balls and packed in banana leaves individually fermenting for 5–11 days at an ambient temperature (about 30 °C).

Flow sheet of traditional method of *kharoli* preparation is presented in Fig. 3.3.

Culinary and mode of consumption: *Kharoli* is served with rice meals or fermented rice (local name: *poita bhat/korkora bhat*).

Socio-economy and ethnical or religious values: *Khar*, the basic ingredient of *kharoli*, is beneficial for patients with hypertension condition. However, excessive *khar* may increase the blood pressure level. So, tribal pregnant women use to avoid *khar* and its derivative *kharoli* during period of pregnancy (Sen 2004).

Microorganisms: Unknown

Nutritional value: *Khar* is an effective salt substitute with nearly 46% of potassium, 2.6% of sodium and beneficial trace elements (Neog and Deka 2013). With limited supply of rock salt or table salt during the pre-British rule period, the commoners had to be satisfied with home-made *khar*. Exact nutritional value and pharmaceutical properties of *kharoli* are unknown and need further research.

3.3.1 Kahudi

The preparation of another mustard seed fermented product named *kahudi* is very similar to *kharoli* except that *Thekera* (*Garcinia pedunculata*) replaces *khar* in its preparation (Fig. 3.4). *Panitenga* has the exact ingredients as in *kahudi*, but instead of making into balls, the entire batter is packed tightly in a banana leaf (preferably leaves of *Musa paradisiaca*; family Musaceae) so as to facilitate anaerobic fermentation for 3–4 days under ambient temperature. The product with a small addition of water is finally kept in a tightly sealed glass container.

Culinary and mode of consumption: *Kahudi* is served with rice meals or fermented rice (local name: *poita bhat/korkora bhat*).

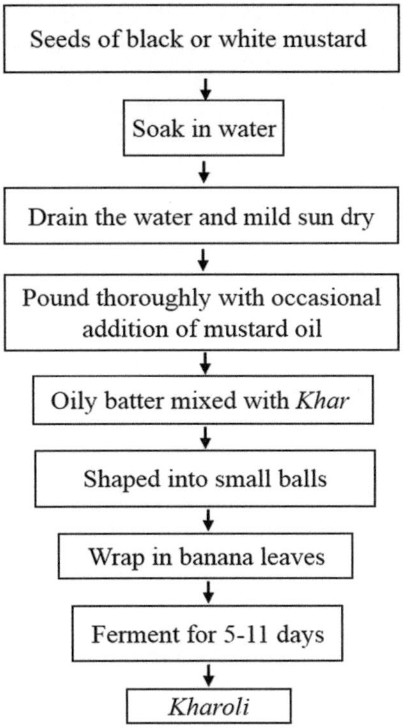

Fig. 3.3 Flow diagram depecting the traditional method of *kharoli* preparation

Socio-economy and ethnical or religious values: Sour food items like *kahudi* and *khorisa* are generally avoided during pregnancy in the tribal societies (Sen 2004).

Microorganisms: The 11-day-old ready-to-eat *kahudi* was recorded to have a pH of ~3.95 that had receded from an initial value of pH 6.8 through the process of fermentation. The initial microbial population recorded after 24 h ranged between 1 and 3 log CFU/g. It was observed that the population of LAB substantially increased (8.74 ± 0.163 log CFU/g) after 96 h (4th day) of fermentation as compared to yeasts and other bacteria. The drop in pH due to the lactic acid produced by LAB could have impeded the growth of other microorganisms such as yeasts and other bacteria. In fact, the LABs were found to be the most dominating (8 log CFU/g) microflora at the end of 11th day of fermentation. Coliform or other enteric bacteria and mycelial moulds were not detected during the fermentation period. Twelve isolates of LAB were identified as *Enterococcus durans*, *Lactobacillus casei*, *Lactobacillus fermentum* and *L. plantarum*. Fermentation of mustard seed batter appears to have a significant effect on the increase of all essential amino acids and in the reduction of antinutrients (such as phytic acid), enzyme inhibitors and flatus sugars (Goswami et al. 2017).

Nutritional value: Unknown

Fig. 3.4 Traditional *kahudi* preparation. (**a**) Powdered mustard seeds dried under low flame. (**b**) *Thekera* (*Garcinia pedunculata*) pieces soaked in water. (**c**) Concoction of mustard seeds with *Thekera*. (**d**) Shaped balls of *kahudi*. (**e**) *Kahudi* wrapped in banana leaves, left for fermentation

3.4 Ethnic Fermented Fish Products (*Namsing* and *Sukoti*)

Ethnic people of Assam depend upon the Brahmaputra and its tributaries and other small water bodies like *beels*, ponds, etc. to catch fish. In absence of adequate storage facility and commercial exportation, huge amount of catch just perish in a short duration. Usually, large fishes are consumed fresh; other small fishes undergo traditional processing such as sun-drying, smoking or fermentation, which are the principal methods of fish preservation in this region. *Namsing* and *sukoti* are traditional fermented fish pastes which are consumed by the people of *Mishing* and *Deori* communities of Assam, respectively.

The process of *sukoti/namsing* preparation starts with degutting and thorough washing of small- and medium-sized fishes like *karoti* (*Gudusia chapra*), *selkona* (*Oxygaster gora*), *darikana* (*Esomus danricus*), *muwa* (*Amblypharyngodon mola*), *puthi* (*Puntius chola*), *botiya* (*Nemacheilus corica*), *singara* (*Mystus bleekeri*), *magur* (*Clarias batrachus*), *chengeli* (*Channa orientalis*), *goroi* (*Channa punctatus*), *bariyala* (*Aspidoparia morar*), etc. These are then sun-dried (alternatively, smoking them over fireplace) and crushed to powder. The sun-dried fish powder is further ground thoroughly with some garlic and dry red chillies in *dheki* or *ural* (traditional grinder). The petioles of *Bor-kosu* (*Colocasia macrorhiza*) or *Kosu* (*Colocasia esculenta*) are cut into pieces, washed and exposed briefly to hot ash for few minutes. An equal proportion of cut pieces are then crushed along with fish powder and lemon juice to make a paste. The paste is now tightly stuffed into a bamboo (*Bambusa balcooa*) tube of 40–50 cm length tightly sealed with banana leaves (it is believed that production of good-quality *sukoti/namsing* can be achieved by tightly stuffing the contents and creating an airtight environment inside the tube). The bamboo tubes are kept on *dhuwa chang* (traditional rack above fireplace in kitchen) for incubation of up to 1–2 months. The preparation process may spatially differ depending on the availability of fish species, plant material and spices and condiments like ginger, garlic, chilli and turmeric (Chowdhury et al. 2019). Flow sheet of traditional method of *sukoti/namsing* preparation is presented in Fig. 3.5.

Fermented fish products such as *napham*, *nakham* and *nichaow* are also prepared by the ethnic communities of Assam. The underlying principle of fermentation involved in these products is very much similar to that of *namsing* or *sukoti*; however, spatial factors such as availability of biological resources (local fishes and herbs) and seasonality add minor varietal differences in the texture of these products (Narzary et al. 2016).

Dried fish products such as *bordia*, *karati* and *lashim* are very much common in the local markets of Assam. The preprocessing before sun-drying is very much similar in these products. However, the main differences are attributed through the preferential usage of some available local fishes [*bordia*, *Pseudeutropius atherinoides* Bloch (vernacular name *Bohduah* and *Patasi*); *karati*, *Gudusia chapra* Hamilton (vernacular name *Koroti*); and *lashim*, *Cirrhinus reba* Hamilton (vernacular name *Lashim*)] (Thapa et al. 2007).

Culinary and mode of consumption: It is consumed as an appetizer along with rice or added to boiled vegetables to enhance the flavour. It is also consumed as supplementary ingredient by frying with fishes and other dishes or as chutney by mixing with salt and green chillies (Chowdhury et al. 2019).

Nutritional value: The availability of nutrients like phosphorus, potassium, calcium, magnesium, zinc, iron, manganese, copper and sulphur makes *namsing* a good source of minerals for fulfilling the daily recommended need (Chowdhury et al. 2019).

Socio-economy and ethnical or religious values: *Sukoti* is avoided during the period of menstruation as it may produce foul smell from the body (Sen 2004).

Microbiology: The pH of *namsing* increased from an initial 6.1 on the 1st day to pH 8.4 on the 7th day and then slightly decreased to pH 7.0 towards the end of fermentation on the 28th day. The initial microbial population of *namsing* fermentation recorded ranged between 2 and 3.5 log CFU g^{-1} on the 1st day which increased to 6–8.2 log CFU g^{-1} on the 7th day and gradually decreased to 2.5–3 log CFU on the 28th day. The microflora of *namsing* was found to be dominated by the genus *Bacillus* (40.7%) followed by *Lactobacillus* (7.4%), *Kurthia* (14.8%), *Providencia* (11%), *Wohlfahrtiimonas* (7.4%), etc. Isolates from *namsing* were identified to be *Bacillus subtilis*, *Bacillus tequilensis*, *B. siamensis*, *B. cereus*, *B. thuringiensis*, *B. megaterium*, *Lactobacillus plantarum*, *Kurthia gibsonii*, *Providencia alcalifaciens*, *Wohlfahrtiimonas chitiniclastica*, *Staphylococcus xylosus*, *Klebsiella pneumonia*, *Vagococcus fluvialis* and *Paenibacillus polymyxa* (Chowdhury et al. 2019). The dominance of *Bacillus* and LAB in several other fermented fish products of Northeast India has been reported previously (Thapa et al. 2007; Tamang et al. 2012).

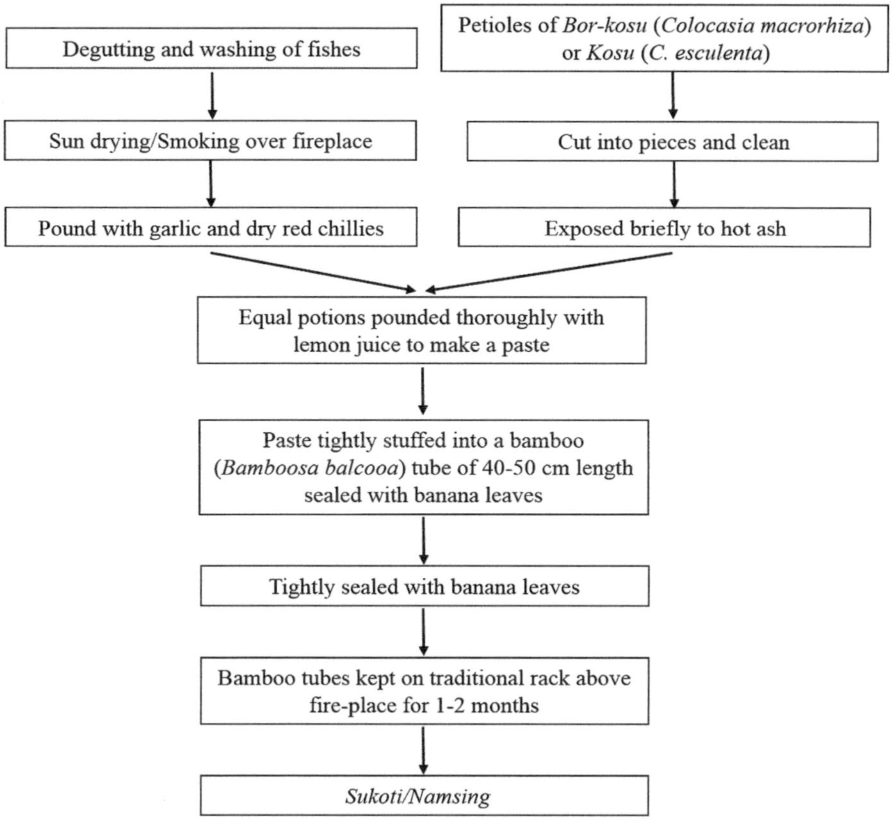

Fig. 3.5 Traditional method of *sukoti/namsing* preparation

Microbiological analysis of *bordia*, *karati* and *lashim* revealed that lactic acid bacteria were most prevalent (average Log cfu/g: *bordia* 5.3, *karati* 4.2, *lashim* 5.8) followed by *Bacillus cereus* (average log cfu/g: *bordia* 2.4, *karati* 2.6, *lashim* 2.2), *Staphylococcus aureus* (average log cfu/g: *Bordia* 2.0, *karati* 1.8, *lashim* 2.1), *Enterobacteriaceae* (average log cfu/g: *bordia* 3.2, *karati* 2.8, *lashim* 3.1) and yeasts (average log cfu/g: *bordia* 2.2, *karati* 3.1, *lashim* 3.0) (Thapa et al. 2007)

3.5 Ethnic Rice-Based Fermented Non-alcoholic Food (*Poita bhat*)

Poita bhat is fermented rice that is partaken after fermenting the cooked rice overnight. Usually leftover cooked rice is soaked in water and allowed to ferment overnight and consumed the next morning as a breakfast along with salt, chilly and lemon. *Poita bhat* is also prepared to prevent the spoilage of leftover rice.

Culinary and mode of consumption: The dish, slightly sour in taste with soft texture, is considered to be a coolant during the summer and well regarded for its nutritional and therapeutic value.

Nutritional value: In the course of *poita bhat* fermentation, it was observed that amount of phytic acids of initial cooked rice gradually decreased from a value of 1.255 ± 0.002 mg/g to 0.313 ± 0.004 mg/g after 12 h period of fermentation. With the decrease in phytic acids, a concomitant increase in minerals contents was observed. Iron (Fe) content of the cooked rice was 0.453 ± 0.008 mg/g, and after 12 h of fermentation, it increased to 1.351 ± 0.012 mg/g. Fermentation also increased the contents of other minerals, viz. phosphorus, magnesium, potassium, calcium and zinc (phosphorus initial 0.410 ± 0.008 mg/g, final 0.804 ± 0.004 mg/g; magnesium initial 2.6 mg/g, final 5.2 mg/g; potassium initial 0.51 mg/g, final 1.85 mg/g; calcium initial 2.0 mg/g, final 4.5 mg/g; and zinc initial 0.02 mg/g, final 0.03 mg/g) after 12 h of fermentation.

Microorganisms involved: Four isolates of *Lactobacillus plantarum* and three isolates each of *L. pentosus* and *L. fermentum* were identified from *poita bhat.* An isolate, *Lactobacillus plantarum* AAU-P-3, showed the highest phytase activity (4.0 U/mL). These findings suggested that *poita bhat* would serve as good source of minerals including iron, phosphorus, magnesium, potassium, calcium and zinc. Four isolates of *Lactobacillus plantarum* and three isolates each of *L. pentosus* and *L. fermentum* were identified from *poita bhat* (Barooah and Goswami 2011; Goswami et al. 2016).

Socio-economy and ethnical or religious values: *Poita bhat* is very similar to dish items like *Panta bhath* in Bengal, *Pokhalo* in Orissa, *Pazhankanji* or *Vellachoru* in Malayalam, etc. These food items are regularly prepared and consumed by the rural populace of Indian subcontinent and are believed to cool down stomach during hot summer (Goswami et al. 2016).

3.6 Fermented Milk Products

Doi or *dahi* has a white to mild yellow gel-like appearance which is similar to yogurt or *dahi.* It is prepared from the milk of cows or buffaloes without the addition of any starter culture. During *Doi* preparation cow or buffalo milk is not boiled; instead, the raw milk is set in an earthenware pot called *tekeli* and allowed to ferment naturally for 2–3 days without adding any inoculums or starter culture. *Doi* prepared in such way is known as *tekeli doi.* Gradual absorption and evaporation of water through the porous walls of *tekeli* not only thicken the *doi* but also produce the right conditions for the growth of the fermenting microflora. Some people also prepare *Doi* by incubating in bamboo tube, popularly known as *Bahor Chungar Doi.*

Culinary and mode of consumption: Local *doi* is an integral serving of *jolpan* or breakfast or mini meal. It is commonly mixed with *Chira* (dehusked and flattened rice) and *Goor* (jiggery) and served in a *Kanhor Baan Bati* (brass goblet).

Nutritional value: Unknown

Socio-economy and ethnical or religious values: In Assam, traditionally prepared *Doi* is a delicacy among the locals and has been preferred over milk since time immemorial. It is also regularly consumed during traditional *bihu* festivals or on special occasions like marriage or *shraddha* ceremonies.

Microbiology: Based on the phenotypic and biochemical characteristics as well as sugar fermentation pattern, the predominant microflora were identified to be *Lactobacillus plantarum*, *L. brevis*, *L. paracasei*, *L. collinoides*, *L. acidophilus* and *L. pentosus*. Among the isolates, *Lactobacillus plantarum* was the predominant species of LAB occurring in *Doi* and also had the best probiotic attributes such as highest survival to low pH (up to 2.5) and bile salts, bile tolerance, exopolysaccharide production, etc. All the LAB isolates inhibited the growth of microorganisms commonly found in fermented foods which indicated that the presence of LAB microorganisms in *Doi* might aid in the reduced survival of bacterial pathogens despite the simple preparation process (Fatima et al. 2018).

3.7 Rice-Based Traditional Alcoholic Beverages

Alcoholic beverages have played an important role in the spiritual and cultural life of some ethnic groups of the state. In the northeastern states of India, rice-based alcoholic beverages are called by various names such as *Zu/Judima* (Dimasa tribe), *Xaj* (Ahom and Tiwa community), *Haria* (Adivasi community), *Jou Bishi* (Bodo), *Apong* (Mishing), *Photika* (Kachari), *Laopani* (Lalong), *Hor-Alank* (Karbi) and *Tsa-pe* (Singpho) (Sarma 2017).

Mod pitha/xaj-pitha is a natural starter similar to several starter cultures available in northeast states of India and other Asian countries. These include *Bakhar Marcha*, *thiat*, *dawdim*, *hamei*, *humao*, *khekhrii*, *chowan* and *phut* (India), *bubod* (Philippines), *Chu* (China), *Look-pang* (Thailand), *Men ruou* (Vietnam), *Marcha* (Nepal), *Nuruk* (Korea), *Ragi* (Indonesia) and *Tapai* (Malaysia) (Steinkraus 1989, 1997; Nout 1992; Tamang and Sarkar 1995; Basuki et al. 1996; Lee and Fujio 1999; Anupma et al. 2018; Sha et al. 2018). These are in the form of starchy tablets containing mixed cultures of starch degrading microbes and fermenting yeasts. These mixed culture starters are used to convert starchy materials to sugars and subsequently to alcohol and organic acids. Like other traditional fermentation process, the *Xaj* brewing process also involves spontaneous fermentation and very little is known about the microorganisms involved in the process or about the eco-physical importance of these microorganisms. The initial step of rice wine fermentation begins with a solid state fermentation where the steamed rice is mashed through liquefaction and saccharification process brought about by moulds. During this stage, rice starch is converted into limit dextrins by the action of α-amylase, and later by the action of glucoamylase, these fragmented oligosaccharides are hydrolysed into glucose. Moulds like *Aspergillus*, *Amylomyces rouxii*, *Rhizopus oligosporus* and *Rhizopus oryzae* present in the starter culture are considered to be the major enzyme source. The importance of moulds during this initial phase can be gauged by the fact that moulds are the dominant microflora during this period. Decomposition of rice starch creates an internal anaerobic environment which is conducive for the growth of lactic acid bacteria (LAB) and *Bifidobacterium*. Production of lactic and acetic acid marks the onset of the transient second step of fermentation, a key determinant of the dynamics and outcome of the overall fermentation process. In the third and final step of fermentation, alcohol-producing yeasts such as *Saccharomyces cerevisiae*, *Saccharomycopsis burtonii*, *Saccharomycopsis fibuligera*, etc. succeed over other microbes; enrich the ferment with various vitamins, amino acids, etc.; and aid specific flavour and aroma (Ghosh et al. 2015).

3.8 Traditional Rice Wine Preparation Process

The traditional fermentation process of preparing *Xaj* involves physical, biochemical and microbiological operations including steaming of rice (preferably glutinous rice) into which a starter called *xaj-pitha* is sprinkled and mixed. The mixture is then put in narrow necked earthen pots for incubation under ambient conditions and allowed to ferment. The initial fermented product is called *Rohi* which is subsequently diluted with water. The fresh rice wine is decanted, filtered and served. The duration of fermentation varies on the requirement of the alcoholic percentage (Fig. 3.6).

Culinary and mode of consumption: *Xaj* is very much popular among the Ahom people during festive occasions and also as refreshing drink after daily hardship (Saikia et al. 2007).

Socio-economy and ethnical or religious values: Rice-based alcoholic beverages play an important role in the spiritual and cultural lives of the indigenous communities. These are regularly prepared and consumed by different ethnic communities of Assam during harvest celebrations, religious rituals and also in *shraddha* and marital ceremonies.

Microorganisms: The microbial diversity of *xaj-pitha*, the starter culture involved in traditional rice wine fermentation was assessed through a metagenomics approach involving Illumina-based whole genome shotgun (WGS) sequencing method. Metagenomic DNA was extracted from rice wine starter culture and analysed through a MiSeq® system (Illumina, USA). Dominant microflora associated with the starter culture were identified to be amylolytic fungal species, viz. *Rhizopus delemar*, *Mucor circinelloides* and *Aspergillus* sp., and saccharifying yeast species such as *Meyerozyma guilliermondii*, *Wickerhamomyces ciferrii*, *Saccharomyces cerevisiae*, etc. The genus *Lactobacillus* was found to be the most abundant microbial species (7.2%) (Fig. 3.7). The bacterial microflora was found to be dominated by LAB with members like *Lactobacillus plantarum*, *Lactobacillus brevis*, *Leuconostoc lactis*, etc. (Bora et al. 2016). Similar type of microorganisms is also reported to be present in other traditionally prepared starter cultures available in northeast states of India (Sha et al. 2018).

Nutritional value: Three-month-old *Xaj* samples were found to contain 12–13% (v/v) alcohol, 1.85–2.51% (w/v) crude protein, 1.163–1.675% (w/v) soluble protein and significant amounts of minerals like calcium (10.93–34.37 mg/mL), sodium

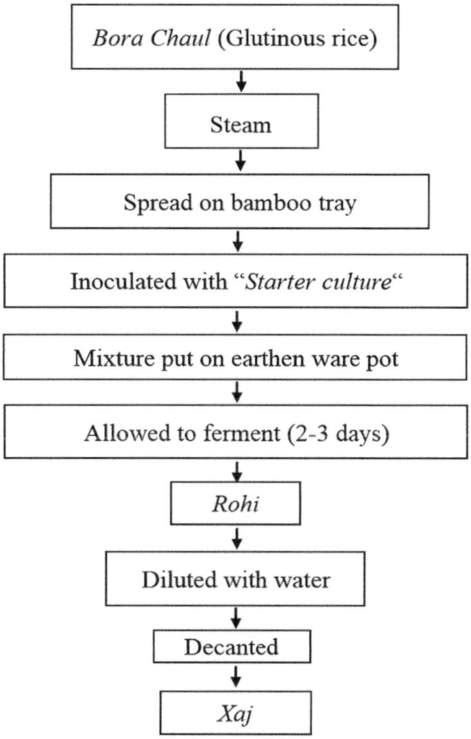

Fig. 3.6 Traditional method of *Xaj* preparation

(1.27–3.49 mg/mL), potassium (4.76–9.63 mg/mL), iron (0.296–3.069 mg/mL) and phosphorus (19.123–78.108 mg/mL) (Bhuyan et al. 2014).

3.9 Traditional Process of Making the Starter Culture

The starters are always in the form of a dry product, either as small flattened cakes, balls of a few centimetres diameter depending upon various communities. Rice is thoroughly mixed with spices and herbs that are believed to prevent the growth of undesirable microorganisms and have other organoleptic properties. Water is added to make a dough-like material which is shaped into small balls or cakes of about 4 cm in diameter and 1 cm thick. Dry powdered starter from previous batches is sprinkled over the cakes, placed over a bamboo tray overlaid with hay and incubated at ambient temperature for 2–5 days. These are then air- or sun-dried and stored. These have a shelf life of several months. Owing to the variation of the concoction practices of making the starter cultures and the process, the resultant product differs in quality and often has a short shelf life (Fig. 3.8).

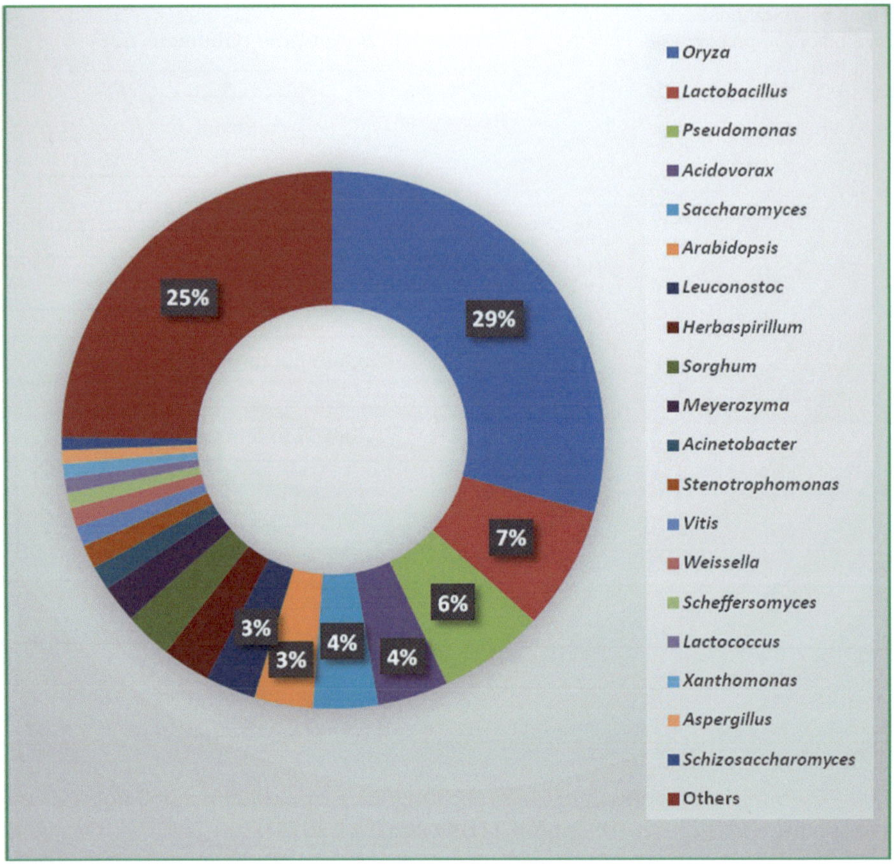

Fig. 3.7 Genus-level abundance of prevalent taxa in Assamese rice wine starter culture *xaj-pitha*

3.10 Preparation of Sai-mod by Mishing Community

Sai-mod is a traditional alcoholic beverage with a pleasant taste and appealing appearance produced by the people of *Mishing* community residing in the upper Brahmaputra valley region of Assam. The ash after burning rice bran or hay is collected and concocted with half-cooked rice which is finally used as substrate for fermentation. About 10 starter cultures (*Apong kusure*) are mixed with 1 kg of steamed rice along with other herbs commonly believed to have medicinal and other organoleptic properties. The mixture is placed in a rice straw sealed cone-shaped bamboo container and kept for incubation at an ambient temperature. On the 2nd day of incubation, small amount of water is added. From the 5th day (which can continue up to 10–12 days) of incubation, the poured-out *sai-mod* is collected at the other end of the container and is served (Figs. 3.9 and 3.10).

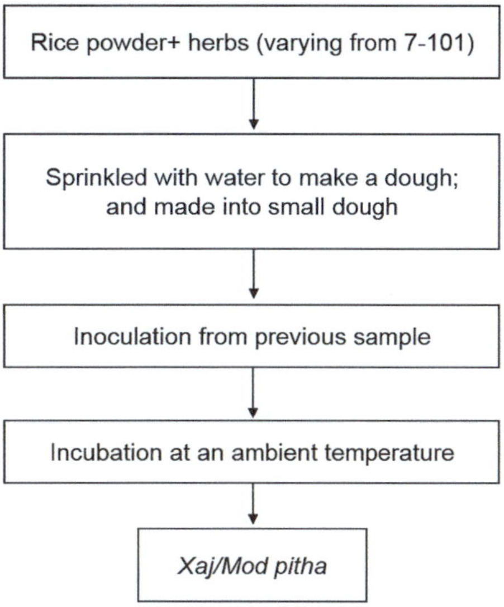

Fig. 3.8 Traditional method of *Xaj/Mod pitha* preparation

Fig. 3.9 Traditional *sai-mod* preparation (**a**) burning of rice bran, (**b**) spreading of rice and ash on bamboo tray, (**c**) concoction of rice and ash, (**d**) starter culture *Apong kusure* on substrate, (**e**) thorough mixing of starter culture and some herbs with substrate, (**f**) fermentation takes place in a bamboo made container and (**g**) secretion and collection of *sai-mod*

3.11 Conclusion

In terms of functionality, fermented foods and beverages are quite different from the raw, unpalatable starting materials. Most of these products are yet to be screened for their beneficial as well as other physical, chemical or microbiological hazards. These products are bio-enriched with protein, essential amino acids and vitamins

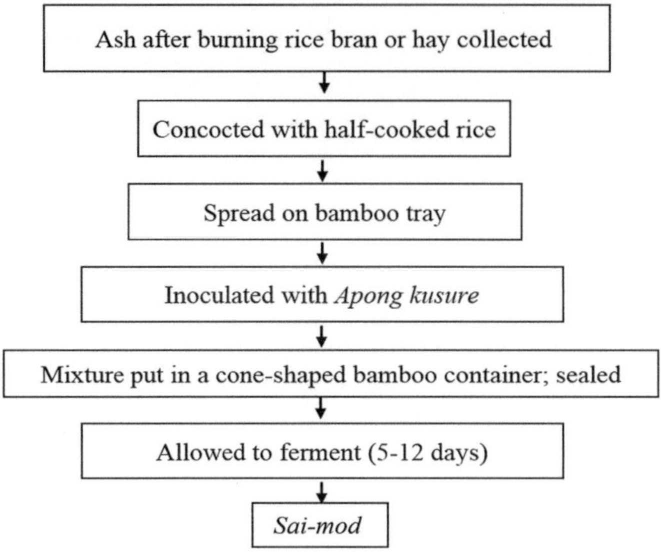

Fig. 3.10 Traditional method of *Sai-mod* preparation

with enhanced nutritive values. This has high significance for economically marginalized population of the developing countries where the majority cannot afford to have commercially available expensive fortified nutritive foods. However, the main obstacle in commercializing fermented foods and beverages has been the inability to produce uniform product and quality. The production process along with the wide varied fermenting microflora introduces spatial and temporal variations in the microbial and biochemical compositions that significantly influence organoleptic characteristics. In addition to that, lack of control on fermentation and aseptic conditions and occurrence of possible harmful microbes and by-products are some of the major constraints associated with traditional fermentation process. Therefore, upgrading the quality and safety of fermented products, while reducing their production cost and maintaining their authenticity and uniqueness, is the need of time. In this regard, intensive, multi-institutional collaborative research efforts to develop low-cost production technologies will benefit the low-income groups of Assam.

Acknowledgement The authors kindly acknowledge the Head of the Department of Agricultural Biotechnology and Director, DBT-NECAB Centre, Jorhat, Assam Agricultural University for their support and encouragement of the research.

References

Anupma A, Pradhan P, Sha SP, Tamang JP (2018) Traditional skill of ethnic people of the Eastern Himalayas and North East India in preserving microbiota as dry amylolytic starters. Indian J Tradit Knowl 17:184–190

Barooah M, Goswami G (2011) Enzyme activity of lactic acid bacteria isolated from poita-bhat—a potential for removing antinutritional factors from fermented foods. In: National seminar on 'Recent advances in the development of fermented foods'. Banaras Hindu University, Varanasi, p 231

Basuki T, Dahiya DS, Gautan Q, Jackson H, Ko SD, Park KI, Steinkraus KH, Uyerco FR, Wong PW, Yoshizawa K (1996) Indigenous fermented foods in which ethanol is a major product. In: Steinkraus KH (ed) Handbook of indigenous fermented foods. Marcel Dekker, New York, pp 363–508

Bhuyan DJ, Sakia Barooah M, Bora SS, Singaravadivel K (2014) Biochemical and nutritional analysis of rice beer of North East India. Indian J Tradit Knowl 13:142–148

Bora SS, Keot J, Das S, Sarma K, Barooah M (2016) Metagenomics analysis of microbial communities associated with a traditional rice wine starter culture (Xaj-pitha) of Assam, India. 3 Biotech 6:153. https://doi.org/10.1007/s13205-016-0471-1

Choudhury D, Sahu JK, Sharma GD (2012) Bamboo shoot: microbiology, biochemistry and technology of fermentation-a review. Indian J Tradit Knowl 11:242–249

Chowdhury N, Goswami G, Hazarika S, Sharma-Pathak S, Barooah M (2019) Microbial dynamics and nutritional status of *namsing*: a traditional fermented fish product of Mishing community of Assam. Proc Natl Acad Sci India Sect B Biol Sci 89:1027–1038. https://doi.org/10.1007/s40011-018-1022-9

Das P (2012) Meandering nature of Barak river in subtropical climate of Southern Assam, Northeast India—a geospatial analysis. Int J Environ Sci 2:2110–2119. https://doi.org/10.6088/ijes.00202030094

Das A, Deka S (2012) MiniReview fermented foods and beverages of the North-East India. Int Food Res J 19:377–392

Das A, Deka S, Miyaji T (2012) Methodology of rice beer preparation and various plant materials used in starter culture preparation by some tribal communities of North-East India: a survey. Int Food Res J 19:101–107

Fatima J, Goswami G, Bora SS, Sarmah-Pathak S, Boro RC, Barooah M (2018) "Doi", a naturally fermented milk product of Assam contains indigenous Lactobacillus spp. with probiotic attributes. Res J Biotechnol 13:1–15

Gait EA (1906) A history of Assam. Thacker, Spink & Co., Calcutta, p 383

Ghosh K, Ray M, Adak A, Dey P, Halder SK, Das A, Jana A, Parua (Mondal) S, Das Mohapatra PK, Pati BR, Mondal KC (2015) Microbial, saccharifying and antioxidant properties of an Indian rice based fermented beverage. Food Chem 168:196–202

Goswami G, Baruah H, Boro RC, Barooah M (2016) Fermentation reduces anti-nutritional content and increases mineral availability in *Poita bhat*. Asian J Chem 28:1929–1932

Goswami G, Bora SS, Parveen A, Boro RC, Barooah M (2017) Identification and functional properties of dominant lactic acid bacteria isolated from Kahudi, a traditional rapeseed fermented food product of Assam, India. J Ethnic Foods 4:187–197

Joshi VK (ed) (2015) Indigenous fermented foods of South Asia, 1st edn. CRC, London, pages 886. ISBN: 9780429067402

Kakati B (1941) Assamese: its formation and development, 1st edn. Government of Assam in the Department of Historical and Antiquarian Studies, Narayani Handiqui Historical Institute, Guwahati, pages 440

Lee AC, Fujio Y (1999) Microflora of banh men, a fermentation starter from Vietnam. World J Microbiol Biotechnol 15:51–55. https://doi.org/10.1023/A:1008897909680

Narzary Y, Brahma J, Brahma C, Das S (2016) A study on indigenous fermented foods and beverages of Kokrajhar, Assam, India. J Ethnic Foods 3:284–291. https://doi.org/10.1016/j.jef.2016.11.010

National Mission on Bamboo Applications (2004) Department of Science and Technology, Government of India. http://dst.gov.in/whats_new/press-release10/pib_5-5-2010_3.htm. Accessed 27 Jan 2019

Neog SR, Deka DC (2013) Salt substitute from banana plant (*Musa-balbisiana* Colla). J Chem Pharm Res 5:155–159

Nirmala C, Bisht MS, Laishram M (2014) Bioactive compounds in bamboo shoots: health benefits and prospects for developing functional foods. Int J Food Sci Technol 49:1425–1431. https://doi.org/10.1111/ijfs.12470

Nongdam P, Tikendra L (2014) The nutritional facts of bamboo shoots and their usage as important traditional foods of Northeast India. Int Sch Res Notices 2014:679073. https://doi.org/10.1155/2014/679073

Nout MJR (1992) Ecological aspects of mixed culture food fermentations. In: Wicklow GCC (ed) The Mycota, the fungal community: its organization and role in the ecosystem, vol VI, 2nd edn. Marcel Dekker, New York, pp 817–851

Saikia B, Tag H, Das AK (2007) Ethnobotany of foods and beverages among the rural farmers of Tai Ahom of North Lakhimpur district, Asom. Indian J Tradit Knowl 6:126–132

Sarma R (2017) Rice beer and tribes of Assam. IOSR Int J Human Soc 22:13–16. https://doi.org/10.9790/0837-2207021316

Sen CT (2004) Food culture in India (food culture around the world). Greenwood Press, Westport, pages 232. ISBN: 0313324875

Sha SP, Suryavanshi MV, Jani K, Sharma A, Shouche Y, Tamang JP (2018) Diversity of yeasts and molds by culture-dependent and culture-independent methods for mycobiome surveillance of traditionally prepared dried starters for the production of Indian alcoholic beverages. Front Microbiol 9:2237. https://doi.org/10.3389/fmicb.2018.02237

Sharma P (2003) An overview on wetlands in Assam, ENVIS Assam. Assam Science Technology and Environment Council, Assam, pp 1–8

Sharma N, Barooah M (2017) Microbiology of *khorisa,* its proximate composition and probiotic potential of lactic acid bacteria present in *Khorisa*, a traditional fermented Bamboo shoot product of Assam. Indian J Nat Prod Resour 8:78–88

Singh R (1993) India, a regional geography, 1st edn. National Geographical Society of India, Varanasi, p 992

Steinkraus KH (1989) Industrialization of indigenous fermented foods. Marcel Dekker, New York, pages 600. ISBN: 9780824747848

Steinkraus KH (1997) Classification of fermented foods: worldwide review of household fermentation techniques. Food Control 8:311–317

Tamang JP, Sarkar PK (1995) Microflora of murcha: an amylolytic fermentation starter. Microbios 81:115–122

Tamang JP, Tamang N, Thapa S, Dewan S, Tamang B, Yonzan H, Rai AK, Chettri R, Chakrabarty J, Kharel N (2012) Microorganisms and nutritional value of ethnic fermented foods and alcoholic beverages of North East India. Indian J Tradit Knowl 11:7–25

Thapa N, Pal J, Tamang JP (2007) Microbiological profile of dried fish products of Assam. Indian J Fish 54:121–125

Ethnic Fermented Foods and Beverages of Bihar and Jharkhand

4

Usha Singh, Seema Singh, and Sunita Kumari Kamal

Abstract

Bihar and Jharkhand are the two Indian states which were united until the year 2000. The food, culture and festivals among the people of these states are almost the same except those celebrated by tribal communities. Since both states come under rice-wheat cropping system, the staple food of the people is rice and wheat which cover the major proportion of the food products/dishes prepared at different occasions. The use of fermented foods has been preferred due to improved shelf life, better digestibility and palatability. Fermented foods like *pua*, *jalebi*, *dhuska*, *kadhi badi*, *dahi vada*, *baskarel*, etc. are prepared at different occasions. Products like *jalebi* and *dhuska* are commercially prepared by the rural people for sale, and these can be seen at different food stalls in the market. The popular beverage of tribal community of Jharkhand is *Hadia* prepared from rice or *mahua*. In Bihar, the consumption of *toddy* was popular among people of lower socio-economic status. But, after ban from Government of Bihar on use of alcoholic beverages, its consumption has drastically reduced. Since the people have the technique of preparation of these nutritious foods, the commercial production of these foods must be promoted under organized system giving them the employment and providing the common people the nutritious food.

Keywords

Ethnic fermented foods · Alcoholic beverages · Bihar and Jharkhand · Nutritional value

U. Singh (✉)
Department of Food and Nutrition, Dr. Rajendra Prasad Central Agricultural University, Pusa, Bihar, India

S. Singh
KVK, Dumka, Birsa Agricultural University, Ranchi, Jharkhand, India

S. K. Kamal
KVK, Palamu, Birsa Agricultural University, Ranchi, Jharkhand, India

J. P. Tamang (ed.), *Ethnic Fermented Foods and Beverages of India: Science History and Culture*, https://doi.org/10.1007/978-981-15-1486-9_4

4.1 Introduction

Bihar is surrounded on three sides by Indian states (to the east by West Bengal, to the south by Jharkhand and to the west by Uttar Pradesh) and one side by an independent country Nepal. Being the third largest populous state in India after Uttar Pradesh and Maharashtra, 88.70% of people of Bihar live in rural areas with agriculture as the mainstay of their livelihood. The diverse climate, highly fertile soil and available ground water in the state give rise to a range of agricultural produce such as cereals, pulses, oilseeds, fruits and vegetables (Kumar and Maulick 2016). Rice, wheat and maize are the major cereal crops, whereas red gram, black gram, green gram, Bengal gram, pea and lentils are the main pulse crops in the state. Bihar is the largest producer of vegetables such as potato, onion, brinjal, cabbage, cauliflowers, etc. besides green leafy vegetables. Litchi, mango, pineapple, banana, guava, etc. are the main fruits being grown in the state. Sugarcane and jute are the two other major cash crops of Bihar.

Bihar's major ethnic group is Indo-Aryan including Bhojpuris, Maithils and Magahis which further follow a specific caste system (Barik 2006). According to the Census of India (2011), scheduled caste and scheduled tribes (Gond, Santhal and Tharu communities) constitute 15 and 1.3% of Bihar populations, respectively. The diverse group follows different cultures and celebrates different festivals. Some of the most important festivals are *Chhath puja*, *Ram Navami*, *Makar Sankranti*, *Durga Puja*, *Deepawali*, *Saraswati puja*, *Holi*, *Id*, *Bakrid*, *Christmas*, etc.

The staple food of the people of Bihar is *bhat* (cooked rice), *dal* (pulses), *roti* (wheat flour), *tarkari/sabji* (vegetables) and *achar* (pickles). Mustard oil is generally used for cooking purposes. In the winter season, this oil is applied on the body also. *Khichdi* prepared from rice and pulse seasoned with spices is one of the popular dishes being relished with *chokha* (smashed potato), curd and pickles. Sometimes vegetables are added in *khichdi* to make it more tasty and nutritious. Other favourite foods of the people of Bihar are *litti* (made of *sattu*) and *chokha* (made of smashed potato, tomato and brinjal) which can widely be seen as street foods and among the dishes being served in party/functions.

Surrounded on all sides by Indian states (West Bengal to the east, Odisha to the south, Chhattisgarh to the east, Uttar Pradesh to the northwest and Bihar to the north), Jharkhand is a land of forest which was a part of Bihar until 2000. In Jharkhand, the predominant population is tribal. No doubt, a large proportion of population from other groups has migrated to Jharkhand and became inhabitants. The Mundas were the earliest prominent tribal settlers, and the Santhals were the last. The other tribal communities of Jharkhand are Oraon (Kurukh), Kharia and Ho who constitute a significant proportion of tribal population in Jharkhand. People of these communities strictly adhere to their culture and tradition.

The food habit and festivals in Jharkhand resemble those of Bihar, except the tribal communities whose food habits and culture are different. The tribal community celebrates festivals in their own style and enthusiasm. Some of the important ones are *Karma*, *Sohrai* and *Sarhul*. Since Bihar and Jharkhand are based on rice-wheat cropping system, the use of rice and wheat is reflected during festivals; also, rice- and wheat-based food products are prepared mainly by different methods. Fermented products of milk, cereals, pulses, and alcoholic beverages from fruits, cereal grains, etc. had been quite popular among the people of Bihar and Jharkhand. Fermented foods like *pua*, *jalebi*, *dhuska*, *kadhi badi*, *dahi vada*, *baskarel*, etc. are prepared at different occasions. The popular beverage of tribal community, *Hadia*, is a rice-based fermented beverage relished on different occasions.

4.2 Fermented Foods

The fermented products prepared most often in Bihar and Jharkhand from the beginning are *pua*, *jalebi*, *dhuska*, *kadhi badi*, *dahi vada* and *baskarel.* Nowadays, some recipes of South India, viz. *dosa*, *idli* and *uttappam*, have become quite popular in Bihar and Jharkhand.

4.2.1 Pua

Pua is a popular fermented dish available in every plate on the occasion of Holi in Bihar and Jharkhand. Sweet smell of *pua* can be felt in every household on the eve of *Holi*. It is a famous delicacy in all the states across India but actually is a part of the tradition of Bihar and now Jharkhand.

Traditional method of preparation: The high-energy *pua* is prepared from refined flour and sugar (Fig. 4.1). Sometimes, suji or whole wheat flour is added for the preparation of *pua*. Banana and milk are sometimes added in *pua* giving the name *malpua.*

Culinary and mode of consumption: *Pua* is a popular food which is synonymous with Holi in Bihar and Jharkhand. It is served with *puri*, *sabji* and *dahi*. Sometimes *pua* after preparation is immediately dipped in milk and served as soft and tasty *pua*. The nonvegetarians enjoy *pua* with mutton curry on the occasion.

Socio-economy and ethical or religious values: *Pua* is a product commonly relished by Hindu population during the occasion of Holi and even in Saraswati puja. It is being used as *Prasad* also that is offered to God and then distributed among the people for consumption.

Microbiology: Not known

Nutritional value: Not available

Optimization and commercialization: Not available

Mix refined flour and sugar (2:1).

↓

Add water just to make a soft batter.

↓

Leave overnight.

↓

Mix thoroughly the batter; add dry fruits and water to make the consistency thin.

↓

Heat refined oil in a frying pan.

↓

Put the batter in hot refined oil for deep-frying.

↓

Fry till golden brown colour is obtained.

↓

Then remove from the oil and put on blotting paper to get extra oil absorbed.

↓

Serve *pua.*

Fig. 4.1 Flow diagram of the preparation of *pua*

4.2.2 Kadhi Badi

Kadhi badi is a popular dish in Bihar and Jharkhand. *Kadhi* is a curd- and besan-based stew made with mixing curd and gram flour and tempered with aromatic spices. Badis are the pakoras made from gram flour added into kadhi while boiling. Sometimes badis are consumed as such as snack.

Traditional method of preparation: The traditional method for the preparation of *kadhi badi* has been presented in Fig. 4.2.

Take 11/2 cup of besan (gram flour).

↓

Add salt, powdered masala and water (half cup).

↓

Make a smooth batter so that badi can be made.

↓

Heat oil in a pan; drop badis batter in hot oil.

↓

Fry by keeping the heat low until the 'badis' are golden.

↓

Drain and keep *badi* aside.

↓

For the kadhi, beat the *curd* until smooth. Keep aside.

↓

Put 2 cups of water to the remaining 1/2 cup of besan and make a watery solution out of it.

↓

Now add the beaten curd and mix thoroughly the solution of curd and besan.

↓

Heat two tablespoons of oil in the pan.

↓

Add spices, chilli, garlic, onion, etc. for *tadka*.

↓

Add besan solution to it and keep stirring so that besan does not stick to the base.

↓

Cook for 10–15 min and then keep warm.

↓

Add *badis* to the *kadhi*.

Fig. 4.2 Flow diagram for *kadhi badi*

Culinary and mode of consumption: *Kadhi badi* is consumed with rice and *pakoda/pakora* (fried batters) on the eve of Holika Dahan which is celebrated 1 day before Holi. *Badi* is consumed as snack items also.

Socio-economy and ethical or religious values: *Kadhi badi* is a dish being consumed by people of every socio-economic status. But the poor people prepare only *kadhi* because of its low cost and consume it with rice or roti. The poor take up the *badi* as small business purposes. They prepare and sell in village *hat* or small *bazaar* along with *murhi* called as *murmure* also (puffed rice).

Microbiology: Not available

Nutritional value: Not available

Optimization and commercialization: Scientific data are not available for optimization and commercialization at large scale.

4.2.3 Jalebi

Jalebi is a popular Indian dessert made with maida (refined flour) and sugar syrup served during festivals and special occasions. It can be seen in every hotel and motel where it is served during breakfast along with *puri* and *sabji*.

Traditional method of preparation: The method of preparation of *jalebi* has been presented through the flow chart in Fig. 4.3.

Culinary and mode of consumption: *Jalebi* is relished during the party or function as such or as a dessert served with cream. *Puri*, *sabji* and *jalebi* are very common breakfast in both rural and urban areas.

Socio-economy and ethical or religious values: *Jalebi* is a popular sweet product which is commonly preferred by people of all socio-economic strata. After getting any good news, people go home by carrying *garam jalebi* (hot *jalebi*). The circular bizarre shape, crisp texture and juicy mouthfeel are the characteristics making *jalebi* quite popular among the people of India (Chakkaravarthi et al. 2009; Balaswamy et al. 2012).

Make a smooth batter of refined flour with the help of water and leave overnight.

↓

Mix the batter thoroughly till smooth and semi-liquid consistency is obtained.

↓

Now make sugar syrup.

↓

Put the frying pan on gas stove.

↓

Heat refined oil in frying pan.

↓

Take a thick cloth and make small hole.

↓

Put the batter on cloth and hold it properly.

↓

Now squeeze the cloth of batter above refined oil to make *jalebi*.

↓

Take out the *jalebi* from the pan and put it in sugar syrup for 5 minutes.

↓

Remove the *jalebi* from sugar syrup and put it on a serving plate.

Fig. 4.3 Flow diagram for the preparation of *jalebi*

Microbiology: *Lactobacillus fermentum* (6×10^8/g), *L. buchneri* (3.2×10^8/g), *Streptococcus lactis* (6×10^8/g), *S. faecalis* (6×10^8/g) and *Saccharomyces cerevisiae* are found in the fermented batter. The pH decreases from 4.4 to 3.3, whereas the volume of the batter is increased by 9%. Both amino nitrogen and free sugar decrease during fermentation (Steinkraus 1996).

Nutritional value: The nutritional value of *channa jalebi* is moisture, 20.23%; fat, 12.53%; protein, 5.71%; sucrose, 40.21%; ash, 0.29%; and carbohydrate, 67.11% (Panneerselvam et al. 2015).

Optimization and commercialization: Though there is no organized sector for commercial production of *jalebi* in Bihar, the product is being sold everywhere whether it is rural area or urban area

Take black gram dal, clean it and soak them overnight.

↓

Drain the materials and grind them.

↓

Keep aside for 3–4 h.

↓

Mix and make a thick batter with water as desired.

↓

Add chopped ginger and green chilli in the batter.

↓

Heat the frying pan and put refined oil.

↓

Now deep fry vada till it becomes golden brown.

↓

Soak the vada in water for 15–20 min.

↓

Drain the vada and put all vada in curd solution prepared separately.

↓

Garnish it with black salt, cumin seed powder, red chilli powder and tamarind chutney and serve.

Fig. 4.4 Flow diagram of dahi vada

4.2.4 Dahi Vada

Dahi vada is a gram flour- and *dahi*-based recipe quite popular in Bihar. It is a recipe which can be seen in every household and family in Holi and marriage ceremony especially in reception party. Nowadays, it has become a favourite recipe for working lunch served in seminar and symposia.

Traditional method of preparation: The flow diagram for the preparation of *dahi vada* can be seen in Fig. 4.4.

Culinary and mode of consumption: *Dahi vada* is consumed at the start of the meal. Sometimes, it is consumed after having complex food on the occasion of festival.

Socio-economy and ethical or religious values: *Dahi vada* is considered to have good digestion property and cooling effect for the stomach.

Microbiology: Not known

Nutritional value: Not available

Optimization and commercialization: There is neither organized nor unorganized sector for the commercial production of *dahi vada*. Rather it is prepared at home or ordered for its preparation at large scale for the food served in party/functions.

4.2.5 Dhuska

Dhuska is fermented rice-lentil product and is one of the most popular items in Jharkhand and is a part of every household. It is usually consumed as a breakfast item.

Traditional method of preparation: For the preparation of *dhuska*, rice and lentil are washed, soaked and ground. After grinding, the batter is left for fermentation. From the batter, small balls are taken, flatten and fried later.

Culinary and mode of consumption: It is normally served with traditional *ghugni*, which is a simple curry made with black chickpeas. *Dhuska* and *ghugni* is a delicacy which is loved even by the tourists.

Socio-economy and ethical or religious values: The preparation and sale of *dhuska* and *ghugni* by the people provide opportunity for income generation.

Microbiology: Not known

Nutritional value: Not available

Optimization and commercialization: *Dhuska* is commercially available at very small scale.

4.2.6 Baskareel

Bamboo shoot locally known as *baskareel* in Jharkhand is a forest plant which is normally consumed by elephants. But rural people of Jharkhand ferment it and use in the form of *vegetables*, *sherbet*, *sukhauta* (dried form), pickles, etc.

Traditional method of preparation: The processing method of *baskareel* has been presented in Fig. 4.5.

Culinary and mode of consumption: *Baskareel* after processing is being used for the preparation of vegetables, pickles and other products.

Socio-economy and ethical or religious values: These are being sold in the market also for which rural women got a very good price.

Microbiology: Not known

Nutritional value: Not available

Optimization and commercialization: Not available

Take *baskareel* and cut it into small pieces of 1 in. × 2 in. size.

↓

Soak them in 1% salt water and cover them.

↓

Leave for 3 days for fermentation.

↓

After 3 days, these pieces are ready for making vegetables, pickles and other products.

Fig. 4.5 Flow diagram of dahi vada

Clean the bark and roots of plants.

↓

Sun-drying of barks and roots and grind into powder.

↓

Moisten sun-dried rice.
(Boiled and dried rice also can be taken.)

↓

Convert it into flour.

↓

Add rice flour in the mixture of bark and root powder.

↓

Evenly mix and the dough is made.

↓

Roll the dough into small balls.

↓

Scatter the balls between the layers of straw and leave for 3 days.

Fig. 4.6 Flow diagram for the preparation of ranu tablet. Plant ingredients such as Basok leaves (**a**); Bhetkuna (**b**); Jiyati (**c**); Senwar leaves (**d**) and ranu tablet (**e**)

Fig. 4.6 (continued)

Boil rice with water.

↓

Break ranu tablet into pieces (for 1/2 kg rice: 1 tablet).

↓

Mix them with boiled rice.

↓

Allow them for fermentation in earthen pot (Handi) for 2–3days with mouth slightly open.

↓

After fermentation squeeze the materials slowly through a sieve for filtration.

↓

The filtrate is cream-coloured drink known as Hadia.

a b c

Fig. 4.7 Flow diagram for the preparation of rice-based *Hadia*: distillation and condensation (**a**); *Hadia*-ready for use (**b**) and *Hadia* in market (**c**)

4.3 Fermented Beverages

4.3.1 Hadia

Hadia is a fermented drink being used by tribal people during festival. The word *Hadia* came from its original word 'Handi', which is made from earthen material and widely used as cooking vessel from ancient times. For the preparation of fermented beverages also, this *Handi* has been used giving the name of beverage as *Hadia*.

After cleaning, *mahua* is soaked in water in clay utensil (*Handi*).

↓

It is left for fermentation for 7 days under the soil.

↓

The soaked mahua is transferred from clay utensil to aluminium utensil.

↓

Aluminium utensil filled with *mahua* is put on stove.

↓

Wooden utensils (*kathot*) fitted with outlet pipe is placed over aluminium utensil.

↓

Another aluminium utensil filled with cold water is placed over wooden utensil for condensation process.

↓

The liquid material coming out from pipe drop by drop is *mahua*-based *hadia*.

Fig. 4.8 Flow diagram for the preparation of *mahua*-based *Hadia*

Traditional method of preparation: *Hadia* is generally prepared from rice. But in some areas, mahua also is being used for the preparation of *Hadia.* In traditional fermenting culture, ranu (tablets/powder) is used for rice-/mahua-based *Hadia* preparation.

Ranu tablets locally known as *Ranu goti* or *Bakhar* are prepared by tribal women for preparation of *Hadia.* Ranu helps in fermentation of beverages. It is prepared by mixing different plant ingredients (roots, barks, rhizomes, leaves of about 20–25 plant species) with boiled rice powder and water and shaping them into small round balls. The balls are then wrapped in leaves, allowed to ferment and dried continuously in the shade for 3 days (Fig. 4.6).

Handia is generally prepared by women. The preparation of rice-based *Hadia* usually takes 3 days. Rice is boiled with water in such a way that rice is soaked through with water. Then the *ranu* tablet is broken into pieces and mixed in boiled rice. The mixture is left untouched for 2–3 days. During this time, the mixture gets fermented, and a sour taste is developed. To extract the juice, the mixture is squeezed through a sieve (chaluni) for filtration (Fig. 4.7).

Mahua-based *Hadia* is prepared by cleaning, soaking in water, fermentation, distillation and condensation (Fig. 4.8). Ten skg of *mahua* produces 5 L of concentrated *hadia.*

Culinary and mode of consumption: *Hadia* is considered as a sacred drink among Munda and Santhal tribes. During festival, this drink is offered to God and dead ancestors before consumption. It is used on the occasion of marriage and birth anniversaries also.

Socio-economy and ethical or religious values: *Handia* had got a specific role in the tribal community socially, culturally and economically. From social and cultural point of view, *Handia* binds the tribals together like a string of thread. *Firstly,* during social meetings and social functions (i.e. marriage, birth and death rituals), the tribals greet each other with *Handia*. *Secondly*, while going to friends' or relatives' houses, they take *Handia* with them as a gift. It indicates the status, love and affection of the guests. Similarly, the host also welcomes them with *Handia*. *Thirdly*, at the time of common rituals and cultural functions, the tribal people drink *Handia*, dance and enjoy together.

The tribal women take part in the preparation of *Handia*. They take *Handia* even in business mode and earn significant income for the household. By promoting *Handia* preparation and sales, the tribal women have been able to make economic gains. *Handia* is used for two purposes—consumption and business. Previously, tribal people used *Handia* only for consumption, but during the last 30 years, it has also been used for business purposes. When the Munda tribes from Bihar migrated to Orissa and settled in different parts of Keonjhar and other districts, they initiated the *Handia* business, and gradually it spread to the tribes in Orissa, who were attracted by the *Handia* practices (Munda and Santhal tribes). It is a secondary source of livelihood for most of the tribals. Some tribals accept the business as a primary source of income. Most Munda and some Mahanta and Majhi tribal women prepare and sell *Handia* either in neighbours or in the market. There are four categories of households engaged in the *Handia* business: *ranu* preparation and sale at the market; *Handia* business who purchase *ranu* from others; *ranu* and *Handia* preparation and business; and collection of roots from the forest and selling them at the market.

The Munda women prepare a special *Handia* for religious functions. Before preparing the *Handia*, they cleanse themselves by bathing, put on clean clothes and wash the 'dekchi' (big aluminium pot). They don't take any food while preparing *Handia*. This *Handia* is first offered to God, and only then other members of the family can consume it. Munda women prepare Handia two or three times a week. But in the summer season, most of them prepare it more often. As with production, the Munda women also have the responsibility for marketing because of better credibility towards work and family as compared to male counterparts. They sell *Handia* at home, at daily markets near the roadside, at the *Saptahik Hat* (weekly market), by order and at *Jatra* (festivals).

Microbiology: Not known

Nutritional value: Not available

Optimization and commercialization: From 5 kg of rice, one can produce 12 *mana* (6 kg) *ranu* tablets. For the preparation of *ranu* tablets, the required roots and barks are available in the forest mainly in the rainy season. So the tribal women collect these roots and barks in the season and keep them in stock for the whole year. Some tribal women have expertise in preparation of *ranu* at home. They collect the raw material and prepare and market ranu tablet to others who wish to buy for the preparation of *hadia*. For one mana or 1/2 kg rice, one can use one tablet. The *Handia* can be hard, medium and soft, depending on how the *ranu* is used. From 6 kg rice, the tribal women produce 18 L *Handia*. They sell the *Handia* in one-*gina* amounts (a *gina* is a local measuring pot with a capacity of 250 mL).

4.3.2 Toddy

Toddy had been a popular fermented drink among the people of lower economic status of backwards and scheduled caste category. The earlier stage of toddy is *neera* which is a sap extracted from palmyra tree (*Borassus* spp.).

Traditional method of preparation: *Neera* extraction is generally performed before sunrise. It is sweet and translucent in colour. It is susceptible to natural fermentation at ambient temperature within a few hours of extraction. Once fermented, neera becomes *toddy* with 4% alcohol. *Neera* is obtained by scraping the tendermost part, just below the crown of palmyra (*Borassus flabellifer*).

Culinary and mode of consumption: Generally, toddy alone or with certain snacks is being used by male members. Women always oppose the use of toddy by their male counterparts.

Socio-economy and ethical or religious values: The consumption of toddy is quite common among the people of lower socio-economic groups.

Microbiology: Not known

Nutritional value: Neera is rich in carbohydrates, mostly sucrose, and has a nearly neutral pH. Besides higher nutritional value, *neera* had not been the choice of people. *Toddy* has taken better place.

Optimization and commercialization: Toddy had been sold in the rural area at different chowks. But after the ban on the use of alcoholic beverages, its sale has been reduced.

4.4 Conclusion

In Bihar and Jharkhand, the food, culture, festivals and even cropping system are almost the same except those celebrated by tribal communities. Rice and wheat being the staple food in the area cover a large proportion of food plate from rice- and wheat-based food products/dishes. Some of these food products/dishes are being prepared by fermentation technique in order to improve shelf life and for better digestibility and palatability. Fermented foods like *pua*, *jalebi*, *dhuska*, *kadhi badi*, *dahi vada*, *baskarel*, etc. are prepared and relished at different occasions. Products like *jalebi* and *dhuska* are prepared by the rural people for sale, and these can be seen at different food stalls in the market. The popular beverage of tribal community of Jharkhand is *Hadia* which is prepared mainly from rice and from mahua for change of flavour. In Bihar, the consumption of *toddy* was quite popular among people of lower socio-economic status. But, after ban from Government of Bihar on use of alcoholic beverages, its consumption has drastically reduced. Since the people especially women have the technique of preparation of these nutritious foods, the commercial production of these foods can be a good opportunity for their employment.

References

Balaswamy K, Prabhakara Rao PG, Prabhavathy MB, Satyanarayana A (2012) Application of annatto (Bixa orellena L.) dye formulations in Indian traditional sweetmeats: jilebi and jangri. Indian J Tradit Knowl 11:103–108

Barik R (2006) Land and caste politics in Bihar. Shipra Publications, New Delhi. ISBN 9788175413054

Chakkaravarthi A, Punil kumar HN, Bhattacharya S (2009) Jilebi—an Indian traditional sweet: attributes, manufacturing practices and scope for large scale production. Indian Food Indust 28:30–36

Kumar A, Maulick BG (2016) Agriculture in Bihar: the latent sector of development. Int J Humanit Soc Sci Invent 5:09–20

Panneerselvam G, Arivazhagan R, Periyar Selvam S, Ida IM (2015) Process standardization, characterization and shelf life studies of Chhana jalebi—a traditional Indian milk sweet. Int Food Res J 22:155–162

Steinkraus KH (1996) Handbook of indigenous fermented foods. Marcel Dekker, New York

Ethnic Fermented Beverages and Foods of Chhattisgarh

5

Shubhra Tiwari, S. K. Jadhav, Esmil Beliya, Jai Shankar Paul, and G. D. Sharma

Abstract

Chhattisgarh is known as "rice bowl of India" due to enormous production of rice. Rice is the major ingredient of the ethnic fermented food and beverages of Chhattisgarh. Apart from rice, traditional dishes are also made from wheat, barley, and different lentils. Fermentation process has an impact on food's aroma, flavor, texture, and nutritional content besides preservation. The ethnic food of Chhattisgarh serves a wide range of mouthwatering dishes that are enriched with flavor and exceptional taste. With nutritional values, different fermented foods are lays an important ethic place in tribal people's life. Beverages play an important role in the life of tribal people of Chhattisgarh state. Fermented beverages are consumed in festivals, marriages, funeral feast, and other ceremonies cerebrated by tribal peoples. Fermented beverages and foods have several health benefits. Main ethnic beverage of Chhattisgarh is *handia*, *salfi*, and *mahua*. *Handia* is made up of cooked rice, *mahua* is made up of mahua flowers, and *salfi* is obtained from trunk sap of *salfi*. Some important fermented foods of Chhattisgarh are *boree basi*, *bara*, *dehori*, *aeersa*, *pidiya*, *bafaur*, *bijori*, *rakhiya bari*, etc. Most of the dishes are made up mainly from rice or rice flour and consumed during the festivals, marriage ceremony, and special occasions. Fermented food and their recipes give a real treat to the taste buds of people. These ethnic fermented Chhattisgarh beverages and dishes are healthy, cheap, and easy to digest.

Keywords

Chhattisgarh · Ethnic food · Fermented beverages · Tribes

S. Tiwari · S. K. Jadhav · E. Beliya · J. S. Paul
Pt. Ravishankar Shukla University, Raipur, Chhattisgarh, India

G. D. Sharma (✉)
Atal Bihari Vajpayee University, Bilaspur, Chhattisgarh, India

J. P. Tamang (ed.), *Ethnic Fermented Foods and Beverages of India: Science History and Culture*, https://doi.org/10.1007/978-981-15-1486-9_5

5.1 Introduction

Chhattisgarh was established on 1 November 2001 as the 26th state of India. State exists on the tenth position of the country on area wise and covers 135,192 km^2 (4.11%) of total area and is geographically situated at 17°46′ to 24°5′ North latitude and 80°15′ to 84°24′ East longitude. Chhattisgarh has four types of geology; the state mostly has oldest Archean rocks; upper region of state has Gondwana rocks, which have major coal deposition of country; middle region has Cuddapah rocks; and lower plateau consists of Dharwar rocks, which is rich in iron ore. As per census of India 2011, total population of Chhattisgarh is more than 25.54 million, which is 2.11% of the country (https://cgstate.gov.in/web/guest/census). Majority of the population (19,607,961) of state lives in rural areas. More than 30% of the populations are indigenous people with 42 different ethnic group types, and majority of them live in northern and southern region of the state. Some of the major tribes are *Abhujmariya*, *Gond*, *Baiga*, *Muria*, *Mariya*, *Uraon*, *Kamar*, *Korwa*, *Halba*, *Birhor*, etc. State climate is basically tropical and depends on monsoon rainfall. Chhattisgarh covers 44% of forest region, and most of tribal peoples depend on forest and agriculture for their basic needs and livelihood. Nearly 50% of the land is used for agriculture purposes, and major crop of state is paddy (https://cgstate.gov.in). Hence, Chhattisgarh is known as "rice bowl of India." So, the staple food of tribal peoples is rice, and different dishes made up of rice are the main part of their meal (http://agriportal.cg.nic.in/PortHi/). Many traditional dishes are made of rice and rice flour. Apart from rice, traditional dishes are also made from wheat, barley, and different lentils. The Chhattisgarh cooking is far influenced by the cooking of the neighboring states. The folks of the various tribal population of Chhattisgarh have their own distinct cooking, referred to as Chhattisgarh tribal cuisine. Few dishes of Chhattisgarh that represent experience the diversity of food culture in India are narrated. Traditional fermented foods of Chhattisgarh are divided into two categories, i.e., fermented beverages and fermented food. With nutritional values, different fermented foods are lays an important ethic place in tribal people's life. Fermented foods are major part of tribal people's emotions; they like to serve fermented foods in different festivals, in marriage ceremony, in religious event, and also in the death ceremony (teravih, barsi), so fermented foods and beverages are also ethnically important. Some fermented food stuffs like *bara*, *bari*, *salfi*, *mahua*, etc. provide economic support to the tribal peoples. *Bari* and *bara* are sold in local market of Chhattisgarh. *Salfi* drink extracted from salfi (*Caryota urens*) tree is very popular in Chhattisgarh and helps in revenue generation. This chapter emphasizes different traditional ethnic fermented foods and beverages of Chhattisgarh (Table 5.1).

5.2 Fermented Food

Traditional foods basically depend upon their indigenous crop availability as well as their climatic conditions favorable to microbial flora which directly or indirectly related to the health of native people. Thus, they are encouraged to preserve

Table 5.1 Ethnic fermented foods and beverages of Chhattisgarh

Food	Substrate	Sensory property of product	Major ethnic consumers
Boree basi	Rice	Fermented rice, breakfast	Gond, Halba, Kurmi, Baiga, etc.
Bara	Black lentils	Soaked lentils paste, fried round cake like, snack	Kurmi, Muria, Gond, Halba, Teli, Uraon, etc.
Dehori	Rice-*dahi* mixture	Round, fried dough into jaggery syrup	All tribes
Rakhiya bari	Black lentils, flesh of ash gourd/winter melon/white pumpkin	Fermented, dried round nuggets	Uraon, Halba, Kurmi, Gond, Baiga, etc.
Aeersa	Rice and jaggery	Fried in oil or ghee, rounded sweet dish; dessert	Uraon, Kurmi, Baiga, Halba, etc.
Pidiya	Rice	Sweet dish, small cylindrical shaped; snack	All tribes
Bijori	Black lentils	Sun dried, fried in oil. Flat round in shape; snack	Kurmi, Baiga, Teli, Gond, etc.
Bobra	Rice, wheat, jaggery	Sweet dish; dessert	All tribes
Hatphodva (Chhattisgarh idli)	Rice batter	Fermented batter, spongy-like *idli*; breakfast	All tribes
Dhuska bara	Chickpea, black gram, rice-*dahi*	Deep fried; snacks	Uraon and Korwa
Toddy	Sap of tad (*Borassus flabellifer*)	Mild alcoholic beverage	Halba, Gond, Mariya, Muria, etc.
Chhind	Extracted sap of chhind (*Phoenix sylvestris*) tree	Mild alcoholic drink. Fresh sap for jaggery making	Halba, Abhujmariya, Gond, etc.
Salfi	Trunk sap of salfi (*Caryota urens*)	Milky alcoholic juice	Abhujmariya
Mahua	Flower of *Madhuca longifolia*	Alcoholic beverage	Gond, Halba, Muria, Mariya, Baiga, Abhujmariya, etc.
Handia/landa	Rice, starter-*ranu*	Alcoholic beverage	Uraon, Kanwar, Halba, Gond

their indigenous culture and inherit it from one generation to other generation (Das et al. 2017).

5.2.1 Boree and Basi

For the people of Chhattisgarh, especially for farmers, their main breakfast is stale food. The most popular, cheap, and tasty stale dish of tribes is *boree basi*. There is a slight difference between *boree* and *basi*, but both are made up of cooked rice. *Boree* and *basi* is consumed with salt, chutney, and onion as a breakfast. In this way the tribal people preserve and reuse the cooked rice. *Boree* and *basi* both are eaten by laborers, by farmers, and even by local common people of any class in all over the Chhattisgarh state (Figs. 5.1, 5.2, and 5.3)

Socioreligion and nutritional values: This food is rich in carbohydrate. Cooked rice is soaked in water, so it avoids the problem of dehydration and controls the body temperature. Mostly *boree* and *basi* is consumed during summer season.

Microorganisms: Unknown

5.2.2 Dehori

It is a sweet dish made up of rice, curd, and jaggery (Figs. 5.4 and 5.5).

Socioeconomic and religious values: It is occasionally formed by tribal people of Chhattisgarh especially in marriage ceremony and festivals. It is a main sweet dish made during the Holi festival in Chhattisgarh.

Microorganisms: Unknown

Nutritional values: Due to fermentation, this dish is easy to digest and good for stomach. It is a rich source of carbohydrate. Jaggery is used as sweetener, so it is healthy to consume as compared with processed sugar.

Fig. 5.1 *Boree*

Fig. 5.2 Preparation of *boree*

Fresh cooked rice

↓

Cooled and soaked in water for few hours

↓

Rice with water is called *boree*

Remaining/old cooked rice

↓

Soaked in water for overnight

↓

Basi is ready

↓

Consumed at next morning

Fig. 5.3 Preparation of *basi*

Fig. 5.4 *Dehori*

Water soaked rice is grinded

↓

mixed with curd

↓

fermented for an hour

↓

Dough is made from this

↓

Round shaped *dehori* is made from dough

↓

deep fried in oil

↓

Dipped in the jaggery syrup

Fig. 5.5 Preparation of *dehori*

5.2.3 Bara

Bara is the most famous and delicious of all Chhattisgarh dishes. *Bara* is prepared and taken up by different regional classes of people, like Kurmi, Gond, Halba, Teli, Uraon, etc., all over the Chhattisgarh region (Figs. 5.6 and 5.7).

Socioeconomic and religious values: According to the old rituals, it is mandatory to prepare bara in pitru moksha amavasya festival to please the soul of dead family members. In Bastar region, it is necessary to serve guest with bara in any kind of celebrations, marriages, and funeral ceremony.

Microorganisms: Unknown

Nutritional values: Due to the use of pulse (black lentils (urad)), it is a good source of protein. According to nutrition one piece of *urad bara* gives 240 calories, 18 g fats, 16 g carbs, and 6 g of protein (https://platform.fatsecret.com/api).

Microorganisms: Unknown

5.2.4 Rakhiya Bari

Rakhiya bari (*rakhiya nugget*) is a traditional dish of Chhattisgarh. It is popular in all regions of the state, made up of fermented *urad dal* (black lentils) and white flesh of rakhiya (ash gourd/winter melon/white pumpkin). This *rakhiya bari* is cooked with vegetables or dal to enhance the flavor and taste, and it is especially used in rainy and summer season (Figs. 5.8 and 5.9).

Fig. 5.6 *Bara*

Fig. 5.7 Preparation of *bara*

Black lentils is soaked and fermented for 6 hours

↓

After that it is grinded

↓

Mixed with salt, chilies, coriander, onion

↓

Then it is made round and flat shaped

↓

Then deep fried in oil

Fig. 5.8 *Rakhiya bari*

Fig. 5.9 Preparation of *rakhiya bari*

Soak urad daal in water for 4 to 5 hours
↓
It is grinded smooth
↓
Peel off rakhiya then grate it
↓
Mixed and whipped with grinded daal
↓
Make small round shaped bari
↓
Keep it for sundry for 10-15 days

Nutritional values: It protects the body from heat during summer season. *Rakhiya* is also a rich source of vitamins like vitamin B and vitamin C and minerals like iron, calcium, zinc, magnesium, and potassium. This food item also provides a good source of carbohydrates, proteins, and dietary fiber (https://ndb.nal.usda.gov/ndb/foods/show).

Microorganisms: Unknown

5.2.5 Aeersa

It is a famous sweet dish of Chhattisgarh prepared from rice and jaggery (Figs. 5.10 and 5.11). *Aeersa* is most popular in Chhattisgarh people like Uraon, Kurmi, Baiga, Halba, etc., and it is also very popular in central region of Chhattisgarh.

Socioreligious and nutritional values: It is a famous and delicious sweet dish of Chhattisgarh state. It is made at the time of festivals. Due to containing rice, it is a good source of calorie to get instant energy. It is good for the digestive system.

Microorganisms: Unknown

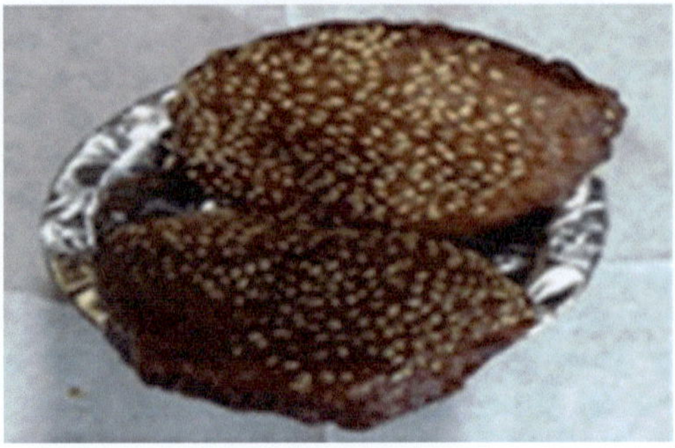

Fig. 5.10 *Aeersa*

Fig. 5.11 Preparation of *aeersa*

Soak rice for 6-7 hours in water
↓
Drain and spread on an absorbent cloth
↓
Grind the rice and sieve it
↓
Add jaggery and ghee to make dough
↓
Cover it and rest for 5 hours
↓
Shape them into small round balls
↓
Deep fry in oil

5.2.6 Pidiya

Pidiya is an expensive sweet dish famous in marriage ceremony and festivals of Chhattisgarh state. It is laborious to prepare but delicious in taste (Figs. 5.12 and 5.13).

Socioeconomic and religious values: It is a famous sweet dish of Chhattisgarh state. It is made at the time of festivals especially at the time of Diwali.

Microorganisms: Unknown

5.2.7 Bijori

Bijori is made from black lentils and used in lunch or dinner as complementary food, to enhance the taste of food. It is deep fried at the time of consumption (Figs. 5.14 and 5.15). *Bijori* is mostly prepared by major tribes and local peoples in central Chhattisgarh region. It is a traditional and side dish served with main course.

Nutritional values: *Bijori* is made from lentils, so it is highly proteinaceous.

Microorganisms: Unknown

Fig. 5.12 *Pidiya*

Fig. 5.13 Preparation of *pidiya*

Rice is soaked for 5-6 hours

↓

After that grind it and make flour from it

↓

Add curd and make dough, do not add water

↓

Fill a portion of dough in mold and make strings

↓

Deep fry these strings in oil

↓

Cool down the strings

↓

Then grind these strings and add powdered sugar in it

↓

Small cylindrical shaped of *pidiya* is made from this

↓

It is coated with sugar syrup

Fig. 5.14 *Bijori*

Fig. 5.15 Preparation of *bijori*

Black lentils are soaked in water for 8 hours
↓
Then finely grinded
↓
Whipped to make smooth paste
↓
Add salt and sesame
↓
Make round ball from batter
↓
Keep it on a cloth and gently pressed
↓
Dry under the sunlight for 10-15 days

5.2.8 Bobra

Bobra is made of rice or wheat flour and jaggery. It is sweet in taste; people love to eat *bobra* in Madai and Mela of tribal region of Chhattisgarh. *Bobra* basically is prepared in Bastar and southern region of Chhattisgarh by Gond, Muria, Mariya, Halba, etc. (Fig. 5.16)

Nutritional values: *Bobra* is rich in carbohydrates. It is easy to digest and good for the health.

Microorganisms: Unknown

5.2.9 Bafauri

It is a famous dish of the Chhattisgarh which is made by red lentil, black lentil, and split mung bean. Sometimes vegetables are also added to enhance taste (Fig. 5.17).

Nutritional values: It is a very healthy dish with high protein content.

Microorganisms: Unknown

Fig. 5.16 Preparation of *bobra*

Rice is soaked in water for 5-6 hours
↓
Grind the rice
↓
Add grated jaggery and cardamom
↓
Make batter, little thicker and coarser than the idli batter
↓
Fry the *bobra* by splashing ghee

Fig. 5.17 Preparation of *bafauri*

Different pulses are soaked in water for 8-10 hours

↓

Batter is made

↓

Batter is steamed in pot

↓

After steaming, then fried mildly

5.2.10 Hatphodva

Hatphodva is also called as Chhattisgarh *idli*. It is made up of rice batter poured in earthen pan and cooked in steam. It can be served with chutney or milk having both sweet and salty flavors separately (Fig. 5.18).

Microorganisms: Unknown

5.2.11 Duska Bara

Duska bara is the favorite dish of tribal people (Uraon, Korwa) mostly living in Jashpur and Sarguja district of Chhattisgarh. The ingredients of *duska bara* are rice, chana dal (split chickpeas), and urad dal (black lentils); all these ingredients are mixed and soaked in water for 2–3 h, and then the mixture is battered after grinding it and then added with curd to ferment and is fried in oil. It is delicious in taste and easily digestible. People made and eat it as snacks in special occasions (Fig. 5.19).

Microorganisms: Unknown

Fig. 5.18 Preparation of *hatphodva*

Soaked rice overnight

↓

Make batter

↓

Steam it and fry

Fig. 5.19 Preparation of *duska bara*

Rice, chana dal and urad dal mixed in a pot

↓

Soaked in water for 2-3 hours

↓

Grind it and make batter

↓

Add some curd to ferment it

↓

Make round shape ball and deep fry

5.3 Fermented Beverages

Beverages play an important role in the life of tribal people of Chhattisgarh state. *Handia, mahua*, and *salfi* are the important fermented beverages consumed by tribal people of Central India including Chhattisgarh state. It is consumed in festivals, marriages, funeral feast, and other ceremonies cerebrated by tribal people (Kumar and Rao 2007). There is a strong believe among the tribal people for serving the purpose of beverages sometimes acts as a remedy for various acute and chronic diseases (Das et al. 2017).

5.3.1 Handia/Landa

Handia or *landa* is a traditional drink of local tribal people of Chhattisgarh. *Handia* is basically used by Uraon and Kanwar in Chhota Nagpur region of Chhattisgarh, and in Bastar region of Chhattisgarh, *handia* is named as *landa*. A brown rice variety of Karahani and Gora are used for *handia* making (Fig. 5.20).

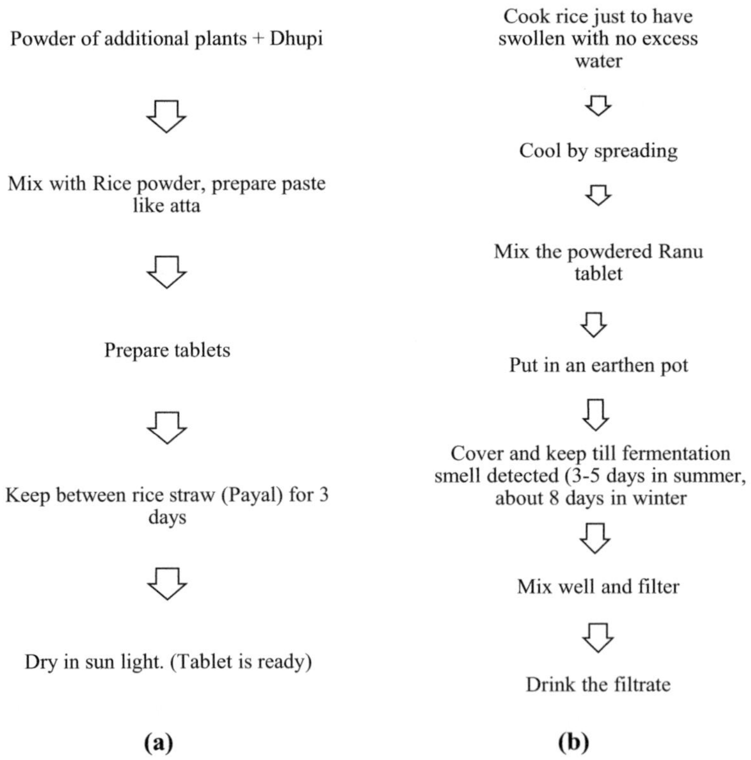

Fig. 5.20 Traditional preparation process of (**a**) *Ranu tablet* and (**b**) *Handia*

5.4 Culinary and Mode of Consumption

Ranu is prepared mainly from three plants *Holarrhena antidysenterica* (Koriya), *Casearia graveolens* (Ruchmuchia), and *Diospyros montana* (Patwan) and called as dhupi. Some tribal people add additional whole plants, roots, or tuber for the preparation of ranu besides dhupi. These plants are Jurbula (*Ruellia tuberosa*) whole plant, Kaisago (*Asparagus racemosus*), seet (climber) root bark, Tara Phool (*Vernonia cinerea*) whole plant, Meetha Jhad (*Scoparia dulcis*) whole plant, Runpawan (*Smilax zeylanica*) root (tuber), and Parhi (*Cissampelos pareira*) tuber (Kumar and Rao 2007). *Handia* is also known as rice beer or rice alcohol. *Handia* is offered in different social occasions. Tribal peoples mostly drink *handia* during summer season (Fig. 5.21).

Fig. 5.21 *Handia* and their ingredients. (**a**) Dhupi, (**b**) Ranu, (**c**) Ranu selling at local market, (**d**) Ranu and dhupi selling, and (**e**) Handia

Socio-economy and ethnical or religious values: The use of *Handia* is common in marriages and festivals. At the time of marriage celebration, use of *handia* drink is a must by Uraon.

Microbiology: Unknown

Nutritional values: Due to the presence of rice, it is rich in carbohydrate. It is believed that the consumption of *handia* controls the thermal regulation of body and makes the body cool which helps in avoiding sun stroke during summer season (https://www.bastariya.com/bastar-drinks; IGKV, Raipur).

5.4.1 Mahua Mand

Mahua mand is an important traditional alcoholic drink of tribal people of Bastar and other regions of Chhattisgarh. It can easily be available in their local market. People enjoy this drink in a group during festival and special occasions. It can be served in a cup made up of leaves known as chipdi (Figs. 5.22 and 5.23).

Socio-economy and ethnical or religious values: The mahua drink is having a socioreligious importance in tribal people and enjoyed by men and women in the evening after their hard day activities. This is unique in the traditional culture of Bastar (https://www.hindustantimes.com/photos/india-news/photos-making-mahua-wine-from-a-flower-sacred-to-chhattisgarh-tribes/ph).

Microorganisms: Generally mahua drinks are produced from yeast *Saccharomyces cerevisiae* (Gavankar and Chemburkar 2016). It is found that the presence of yeast species on the flower of *mahua* (Rao et al. 1961) and some indigenous species of microorganisms found in Mahua flowers, namely, *Kloeckera*, *Candida*, *Torulopsis*, *Pichia*, *Saccharomyces*, and unidentified genera (Ethiraj et al. 1980), is helpful in the fermentation process for the preparation of mahua drink.

Nutritional values: Local tribes of Bastar consume fresh flower of mahua. Fresh flower of mahua is juicy and rich in carbohydrate.

Dried mahu a flower kept in earthen pot with some water

↓

In pot some ranu tablets and leaf juice of *Buchanania lanzan* Spreng is added

↓

Pot is tightly closed and put for fermentation for 3-5 days

↓

After fermentation, the pot is kept for distillation over stove or chullah

↓

Bottom one is fermented pot, middle dome pot and tube is for alcohol vapour collection and supply, and third pot is shower with cold water as condenser and collection.

Fig. 5.22 Preparation process of *mahua mand*

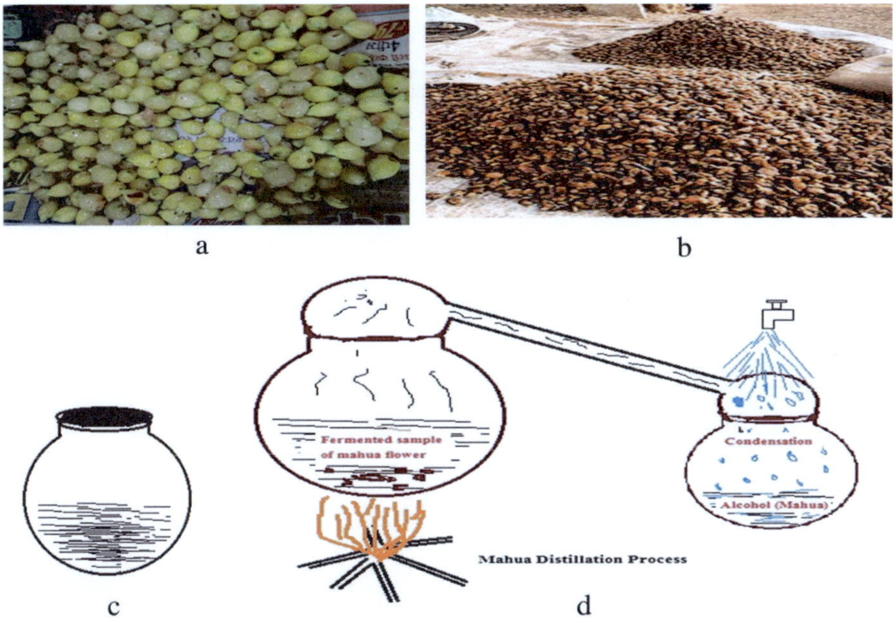

Fig. 5.23 Preparation process of *mahua mand*. (**a**) Fresh mahua flower, (**b**) dried mahua flower, (**c**) fermentation of mahua flower kept in earthen pots, and (**d**) boiling of fermented sap

5.4.2 Salfi

Salfi is also known as Bastar "beer" of Chhattisgarh. It is a local drink obtained from trunk sap of salfi tree (*Caryota urens*) after 9–10 years of age obtaining a height of approximately 40 ft. *Salfi* juice is sold at low cost of Rs. 40–50 rupees in the local market. Therefore, for the tribals of Bastar, salfi tree is a money tree. As salfi tree is an extinct category, the tribal people are using the sap of chhind (*Phoenix sylvestris*) family Arecaceae tree as the replacement of salfi (Figs. 5.24 and 5.25).

Hanging pot fixed in the trunk of salfi tree

↓

A cut is made on the trunk with a sharp knife

↓

Fresh milky liquid juice '*salfi*' is collected

↓

Collected *salfi* is kept overnight for fermentation to make alcoholic

Fig. 5.24 Extraction and preparation process of *salfi*

Fig. 5.25 Salfi tree and *salfi* beverage. (**a**) Salfi tree, (**b**) salfi drink, and (**c**) Salfi selling and drinking by tribals

(https://www.bbc.com/hindi/india/2015/05/150522_sulfi_chhattisgarh_tribal_an).

Socio-economy and ethnical or religious values: *Salfi* can be indigenously consumed by tribal people of Bijapur, Dantewada, Bastar, and other districts of Chhattisgarh as a means for natural mood elevator instead of wine. *Salfi* is one of the most important enjoyable drinks of Abhujmariya people. It is a natural alcoholic drink taken by Bastar tribes of Chhattisgarh on various occasions.

Microorganisms: Unknown

Nutritional values: *Salfi* is a rich source of carbohydrates, and fresh juice is taken as natural cold drink. This beverage is fresh collected sap of tree, so it contains several vitamins and proteins also.

5.4.3 Chhind

Chhind or *chhindras* is one more common famous beverage in tribes of Chhattisgarh. Generally, it is extracted sap of *chhind* tree, and the botanical name of this plant is *P. sylvestris* belonging to family Arecaceae. This beverage is used by various tribes of Chhattisgarh such as Halba, Gond, Mariya, Muria, Abhujmariya, etc. (Fig. 5.26)

a b c

Fig. 5.26 Extraction of chhind and tadiras from (**a**) chhind tree, (**b**) tadi tree, and (**c**) Salfi, chhind, and tadi tree in a villager's house

5.4.4 Toddy/Tadi

It is a sour sap beverage extracted from the tad tree *Borassus flabellifer* family Arecaceae. It becomes a moderate alcoholic beverage after placing in fermentation for 2–3 days. This beverage is also used by various tribes of Chhattisgarh such as Halba, Gond, Mariya, Muria, etc.

Microorganisms: Unknown

5.5 Conclusion

Ethnic fermented and beverages of Chhattisgarh are healthy, easy to prepare, cheap, and easy to digest. Indigenous people of Chhattisgarh have invented the traditional knowledge of food fermentation and also developed the traditional culinary skill with different recipes in local diets. However, no studies on microbiology, nutrition, and health benefits of these unique ethnic fermented foods and beverages of Chhattisgarh have been conducted yet.

References

Das M, Kundu D, Singh J, Rastogi A, Banerjee R (2017) Physiology and biochemistry of indigenous tribal liquor Haria: a state of art. Adv Biotechnol Microbiol 6(2):1–5

Ethiraj S, Onkarayya H, Suresh ER (1980) A note on the nature and sequence of yeasts during fermentation of apples grown in India. J Appl Bacteriol 48:97–100

Gavankar R, Chemburkar M (2016) Isolation and characterization of native yeast from Mahua flowers. Int J Curr Microbiol App Sci 5(11):305–314

Kumar V, Rao RR (2007) Some interesting indigenous beverages among the tribals of Central India. Indian J Tradit Knowl 6(1):141–143

Rao RTN, Dwaraknath CT, Johar DS (1961) Isolation of yeast strain from Mahua flower and its use for fermentation studies on the flower extract. Food Sci 10:88–89

Further Reading

http://agriportal.cg.nic.in/PortHi/

https://cgstate.gov.in

https://cgstate.gov.in/web/guest/census

https://ndb.nal.usda.gov/ndb/foods/show

https://platform.fatsecret.com/api

https://www.bastariya.com/bastar-drinks; IGKV, Raipur.

https://www.bbc.com/hindi/india/2015/05/150522_sulfi_chhattisgarh_tribal_an

https://www.hindustantimes.com/photos/india-news/photos-making-mahua-wine-from-a-flower-sacred-to-chhattisgarh-tribes/ph

Ethnic Fermented Foods and Beverages of Goa

6

Irene Furtado and Sheryanne Velho-Pereira

Abstract

The natives, through centuries, followed indigenous traditional processes of food and beverage preparations which can be viewed under three categories, based on the key ingredients used: (i) plant saps/juices; (ii) lentil, i.e., urid dal; and (iii) crude solar salt. These processes possibly involve natural spontaneous uncontrolled fermentations, initiated and sustained by mixed microbial cultures that grow either simultaneously or succession. These microbial cultures are inherent microflora of coconut palm sap *sur*, cashew apple juice *nero*, lentil *urid dal*, raw tender mango *tor*, and fresh mackerel *bangde* and additionally the microbes present in Goan crude salt g*aunti mith* which is added to the processes as preservative or flavoring. Unknowingly, the indigenous people particularly of Reinder, Kazkar, Ramponkar, and other communities have been carrying the food and beverage preparations, exploiting the potential of inherent microflora to produce *madda feni*, *caju feni*, salted fish, and other indigenous foods. In this chapter we have documented the ethnic fermented foods and beverages of Goa.

Keywords

Foods · Goans · Beverages · Fermentation

I. Furtado (✉)
Department of Microbiology, Goa University, Taleigao, Goa, India
e-mail: ijfurtado@unigoa.ac.in

S. Velho-Pereira
Department of Microbiology, St. Xavier's College, Mapusa, Goa, India

J. P. Tamang (ed.), *Ethnic Fermented Foods and Beverages of India: Science History and Culture*, https://doi.org/10.1007/978-981-15-1486-9_6

6.1 Introduction

Geographically Goa is located on midwest coast of India at 15.2993°N, 74.124°E. It has approximately 105 km coastline, is 6.2 km east to west, and covers approximately 3702 km^2 with a total population of 1,478,545 (Goa population census of 2011). Historically, Goa has been shaped by the Bhojas, Shilaharas, Rashtrakutas, Kadambas, Vijayanagar rulers, Adilshashi, and Sephardic Jewish settlers and finally occupied and ruled by the Portuguese for 451 years from 25 November 1510 to 19 December 1961 (Mitragotri 1999; Dhume 1986). Early, Indo-Aryan and Dravidian migrants amalgamated with native Gawda, Kunbi, Velips, and Dhangar communities who had agriculture and fishing as their occupations and followed unique cultural setup (De Souza 1990).

The preparation and preservation method of Goan Food and Beverages based on one key ingredient are viewed under following three categories:

1.
   (a) Coconut palm sap "*sur*" used as source of leaven in making *pao/undoh* (bread), *sannas* (red rice bread), *kazara-bollo* (wheat/wheat and finger millet), and *madda goddh* (coconut sugar/jaggery);
   (b) Beverage(s), directly as "sur," natural conversion of *sur* to sour acetic acid—*vinagre* (vinegar), distillate of alcoholic fermentation of *sur-madda feni* (coconut feni); and
   (c) Cashew apple juice, *neero* directly as beverage or as distillate of alcoholic fermentation of *neero-caju feni*
2. Use of lentil *urid dal* as leaven for *hitlli/idli* (red rice cakes), *amboli*, and *ambil*
3. *gaunti mith* (local Goan salt) in preservations, *chepni tor* (tender mango pickle) and *suke* and *chala bangde* (dry and wet salted mackerels).

Feni, particularly *caju feni* being exclusively of Goan origin, was awarded geographical indicators registration as a "heritage brew/spirit" in 2009 and hence cannot be made and sold outside Goa (Chamaria 2018). All of these above indigenous, cost-effective preparations of Goa have contributed toward economy of the state, right through colonial times. Yet, there are no basic nor dedicated studies on these indigenous processes. This chapter records the steadfast, meticulous indigenous processes practiced by the Goan community with a hope that scientific community will take up exploratory and evaluator studies of these natural, spontaneous uncontrolled fermentations brought about by the inherent flora of plant or marine ingredient involved, therein. The preparation methods used are described using the colloquial terminologies and names for these indigenous practices.

6.2 Various Indigenous Food and Beverage Preparation of Goa

The food and beverage preparations and various preservation methods based on one key ingredient used are classified for convenience as shown in Fig. 6.1 and Plate 6.1.

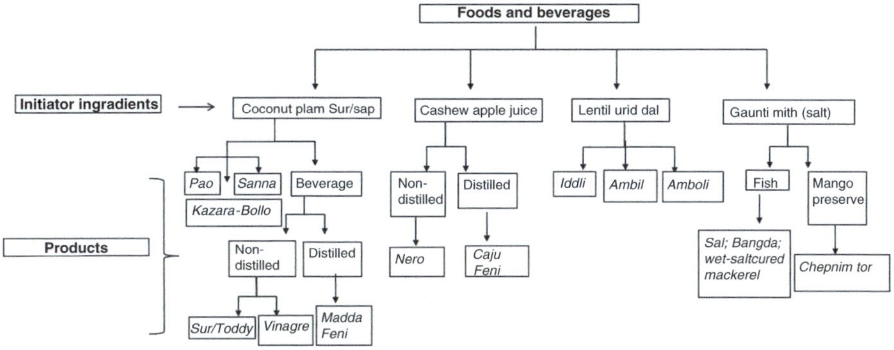

Fig. 6.1 Classification of indigenous foods and beverages of Goa, India, based on key ingredients used for natural fermentation

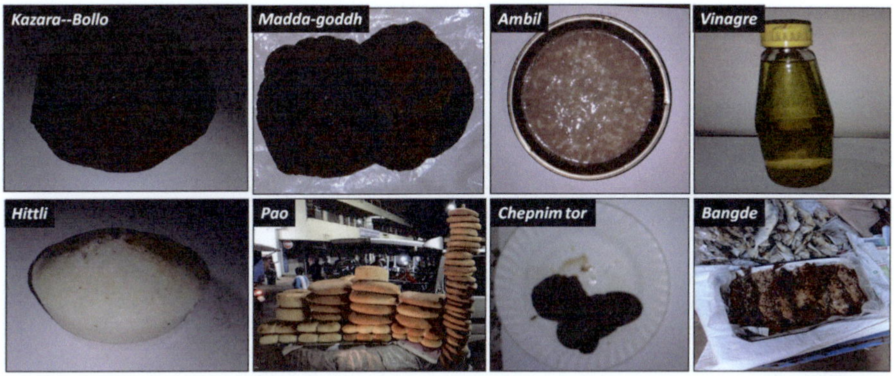

Plate 6.1 Photos of indigenous Goan—foods

6.3 Foods and Beverages Prepared Using Coconut Palm (*Cocos nucifera* L.) Sap (*Sur*)

Sur is tapped by incising the tender part of the coconut spadix or inflorescence (Gomes 2009) and is detailed in flow sheet (Fig. 6.2; Plate 6.2). The understanding is that s*ur* starts undergoing changes immediately, due to the presence of microbes, and increasingly attains an acidic taste past 2 h of collection and standing at ambient temperature. This weekly fermented s*ur* as well as the s*ur* oozing out into the damnnem, the collector pot, is (i) consumed directly as a healthy nutrition beverage and (ii) used as leaven for making *pao* as in Fig. 6.3; *sannas*, as described in Fig. 6.4; and *kazara-bollo* of wheat as detailed in Fig. 6.4. "Sur" is set aside after addition of edible lime for preparation of *madda goddh* and jaggery (Fig. 6.5) and while in its natural form is allowed to stand for 22 days to obtain *vinagre* (vinegar) (Fig. 6.6). For preparation of *madda feni*, however, the natural fermentation process of s*ur* is arrested by distilling the fermented product on fourth day and collecting the

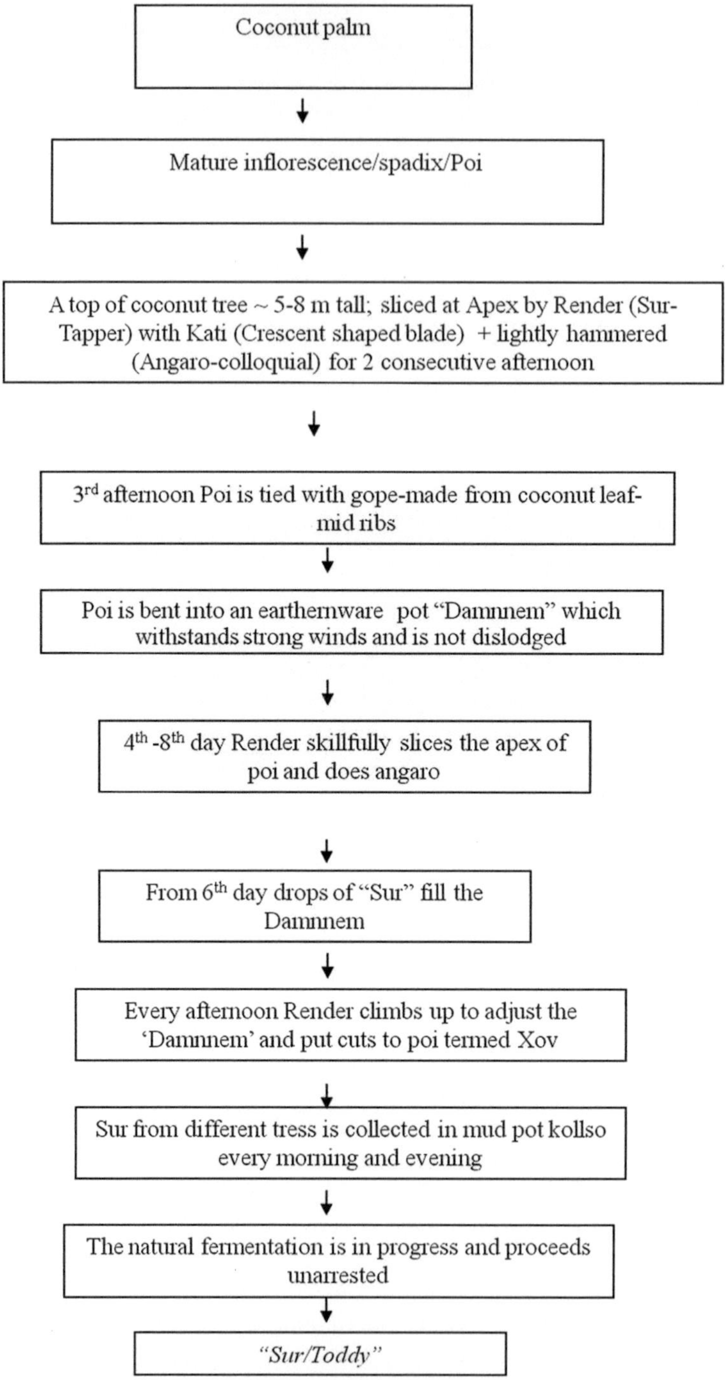

Fig. 6.2 Tapping of "sur/toddy" in Goa from coconut palm

Plate 6.2 Photos of Reinder—the local coconut sap tapper and "sur"

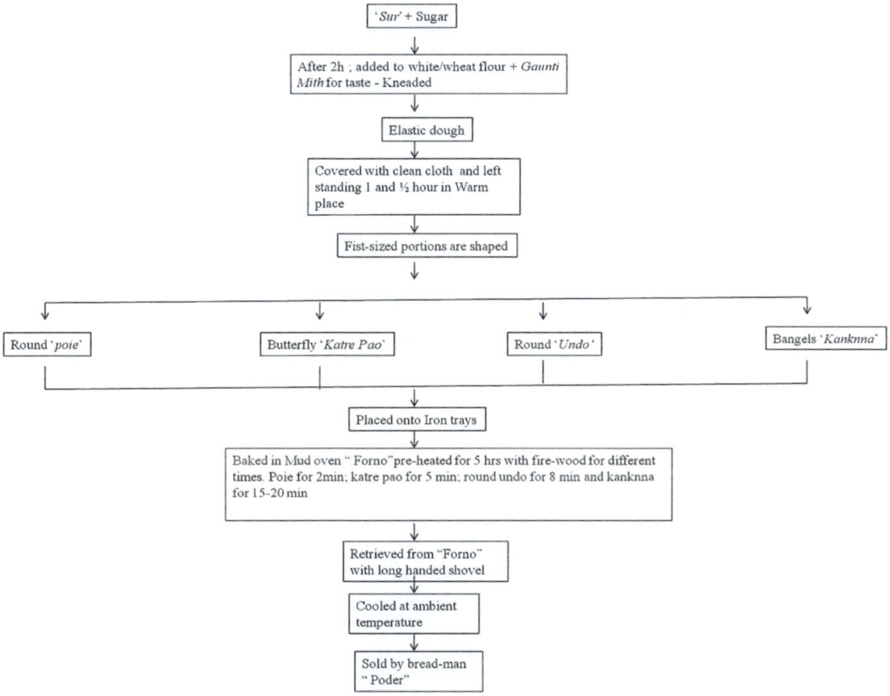

Fig. 6.3 Preparation of "pao"

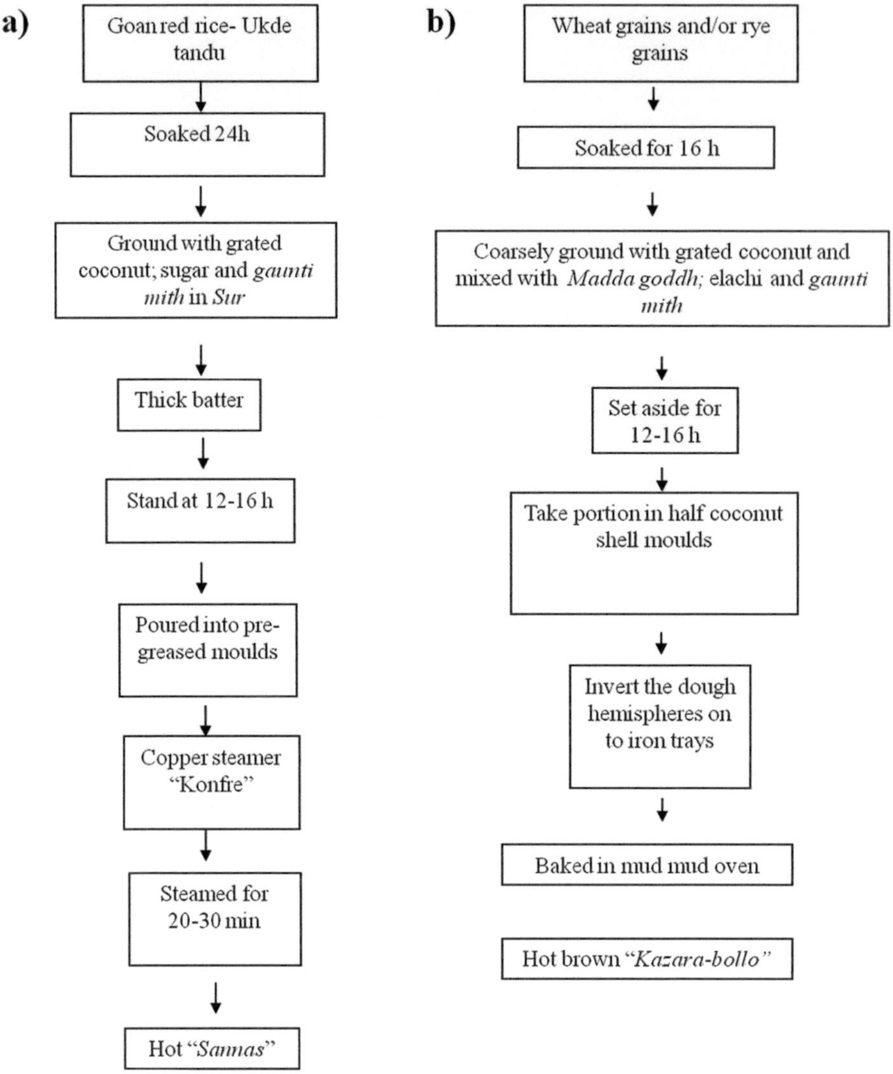

Fig. 6.4 Preparation of (**a**) "*sannas*" and (**b**) "kazara-bollo"

distillate called *molop* which is 30 proof [a measure of ethanol content by volume (abv)] having 15% abv and which is redistilled after mixing with s*ur* to get *madda feni* with 42–45% abv (Fig. 6.7); the distillation art is believed to be introduced by the Portuguese.

A very basic preliminary study, carried out as a student project by Caldeira et al. (2018), under mentorship of Arina Frank, merely enumerated total microbial counts on specific media such as MRSA and Sabouraud and reported total reducing sugars, proteins, and vitamin C of 22.25 μg/mL, 0.33 μg/mL, 0.25 μg/mL, and 0.312 μg/mL in "sur" collected from Siolim, a village in Goa.

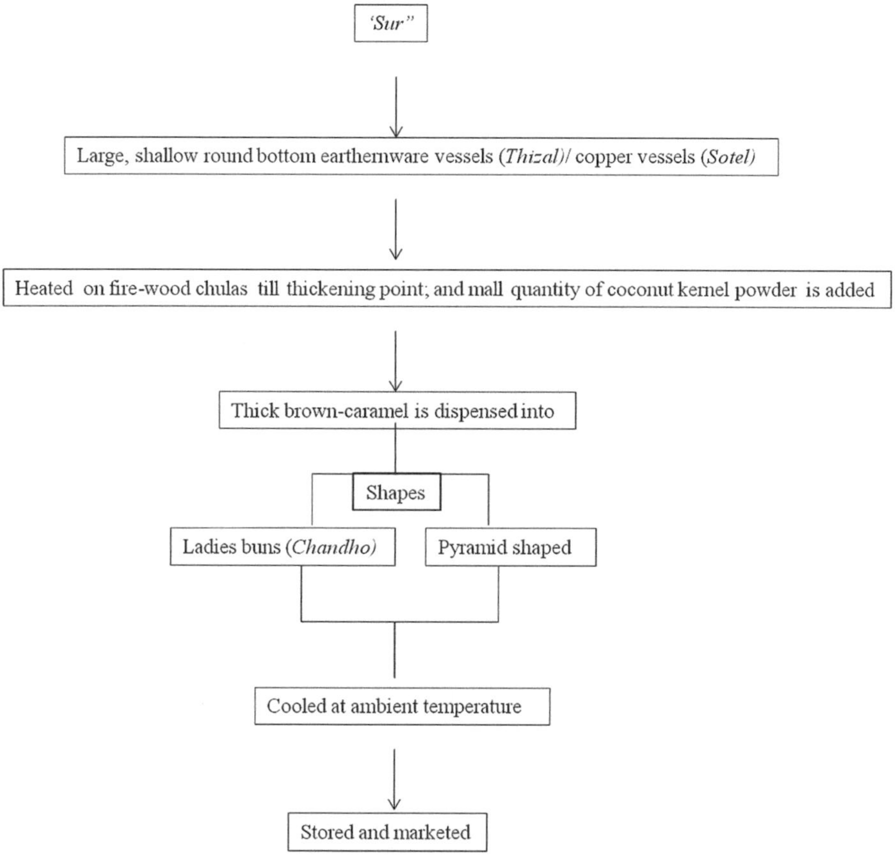

Fig. 6.5 Preparation of Goan "madda goddh" from coconut "sur"

Knowledge of toddy an equivalent of *sur* fermentation from other states of India has been accrued pertaining to the microflora, biochemistry, and chemical content (Atputharajah et al. 1986; Sekar and Mariappan 2007; Kadere and Kutima 2012; Shetty et al. 2017; Chandrasekhar et al. 2012).

Locally prepared *pao* are baked in mud ovens called *forno* heated with wood (Plate 6.3). Pao are named based on their shape *katre pao*, *pao*, *poie*, and *kanknna*. There is no data available on microbes present dominating the rising of dough or on microbes that are acquired during cooling and storage, in open environment. Additionally no information is available on the biochemical process involved or changes happening from the time dough is prepared with *sur* and *gaunti mith* which are added for taste, allowed to rise, and baked nor on the microflora that establishes during cooling and marketing. Information, however, is available on bread leavened in other countries, using baker's yeast. Since indigenous Goan *pao* (bread) as well as *sannas* (rice cake) and *kazara-bollo* (coarse wheat/wheat and rye cake) are natural fermentation leavened with s*ur*, a dedicated study of these is recommended,

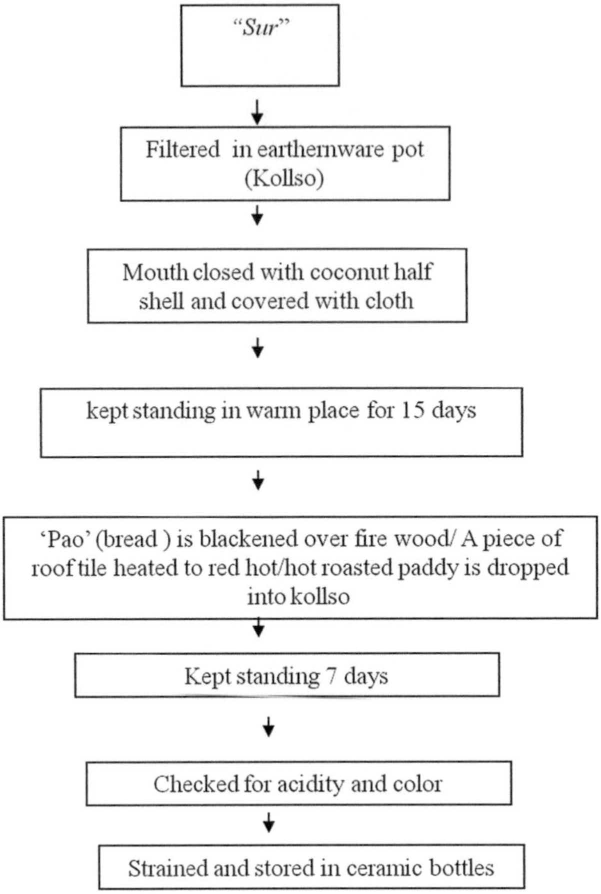

Fig. 6.6 Preparation of "vinagre"

before availability of *sur* becomes extinct due to the stress and risk involved in the *sur* tapping profession of Reinder, which is on a fast decline. Similarly, there are no microbial studies on other natural indigenous fermentation using urid dal lentil described in Fig. 6.8a–c and those using *gaunti mith* (locally farmed Goan salt) used for preservation of raw tender mangoes and fish that are schematically detailed in Figs. 6.9 and 6.10. It may be noted that these products are exported. The conventional fermented fish preparations of Goa are legacy of Goa's Portuguese past and its trade association with Arabia, Africa, and China.

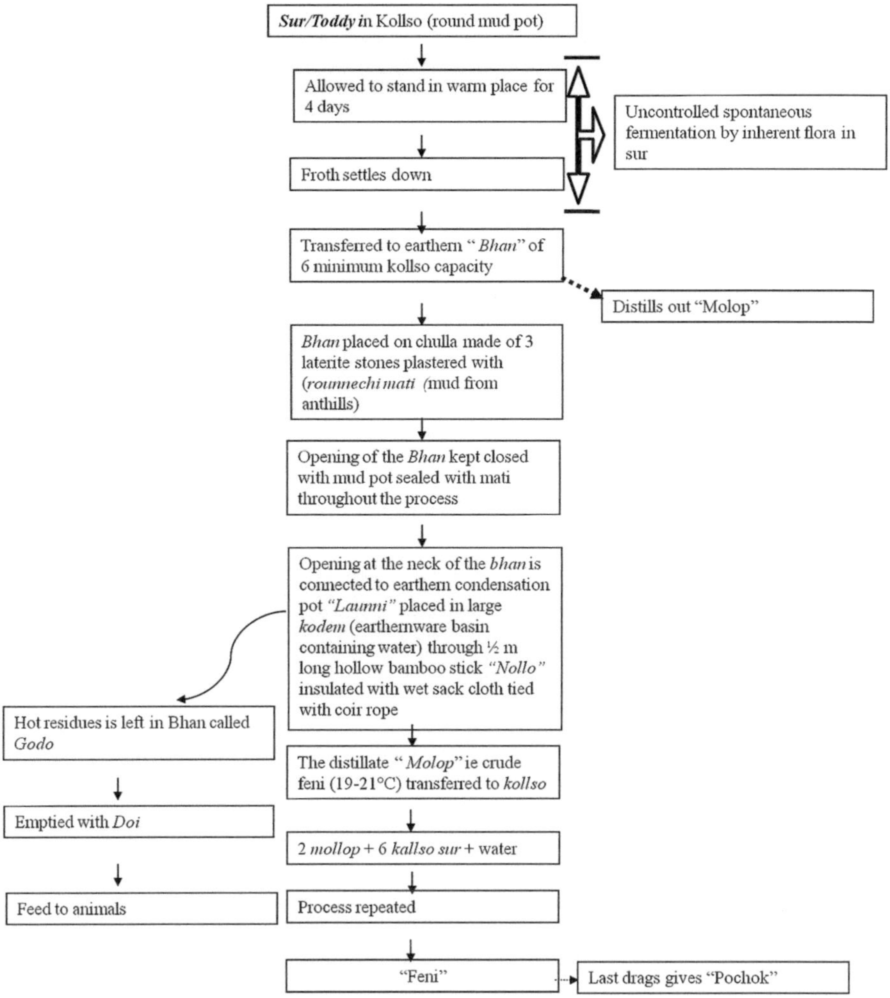

Fig. 6.7 Production of Goan "madda feni" in "bhatti"—an ancient distillation unit

6.4 Production of *Caju Feni* from Cashew (*Anacardium occidentale*) Apple Juice

Caju tree, *Anacardium occidentale*, was brought into by the Portuguese from Brazil. Portuguese adapted its Brazilian name acaju to *caju*. The cashew apple botanically is the succulent stem. Mostly the Velips possess SAUADS (community cashew plantations). Cashew plantations are present in every taluka of Goa. The community involved in cashew *feni* industry is called *kazkar.* The Goa government organizes a *pauni* (auction) for setting of distillation unit; the highest bidder gets the license and all *kazkar* from that taluka, compulsorily distil at this unit.

Plate 6.3 (**a**) Goan mud oven fired with wood used for baking "pao," "kazara-bollo"

The step-by-step procedure for preparation of *caju feni* is laid out in Fig. 6.11. Various earthenwares used in the indigenous native process such as *bhann*, *launni*, *kodem*, and *kollso* are replaced by copper/steel vessels, while cooling vessel basins are replaced by cement tanks. Instead of stomping, a rotating cage "pingre" is used for extraction and recovery of juice. A picture of earthenware bhann used for distillation and collection of distillate is shown in Plate 6.4.

The word *feni* is from the Sanskrit word "phena" meaning froth. Approximately 35 L of "urrack" containing 15% abv strength is recovered as distillate from 50 L of fermented juice. Redistillation of "neero" mixed with "urrack" in 1:2 ratio gives "cazulo" or "cajulo," a distillate having 40% abv strength. Redistillation of Cazulo gives feni of 45–48.2% abv strength. Barbuddhe et al. (2011) reported *Saccharomyces cerevisiae*, *Pichia kudriavzevii*, *Candida ethanolica*, *Lachancea fermentati*, and *Pyrenochaeta nobilis* as dominant yeast and as the dominant natural flora of cashew apple. Following this, Khorjuvenkar (2016) examined the natural fermentation of cashew apple juice and reported the different species belong to six different genera and used *S. cerevisiae*, *P. galeiformis*, *P. kudriavzevii*, and *L. orientalis* to obtain quality products. Use of non-*Saccharomyces* yeast strains of *Pichia*, *Rhodotorula*,

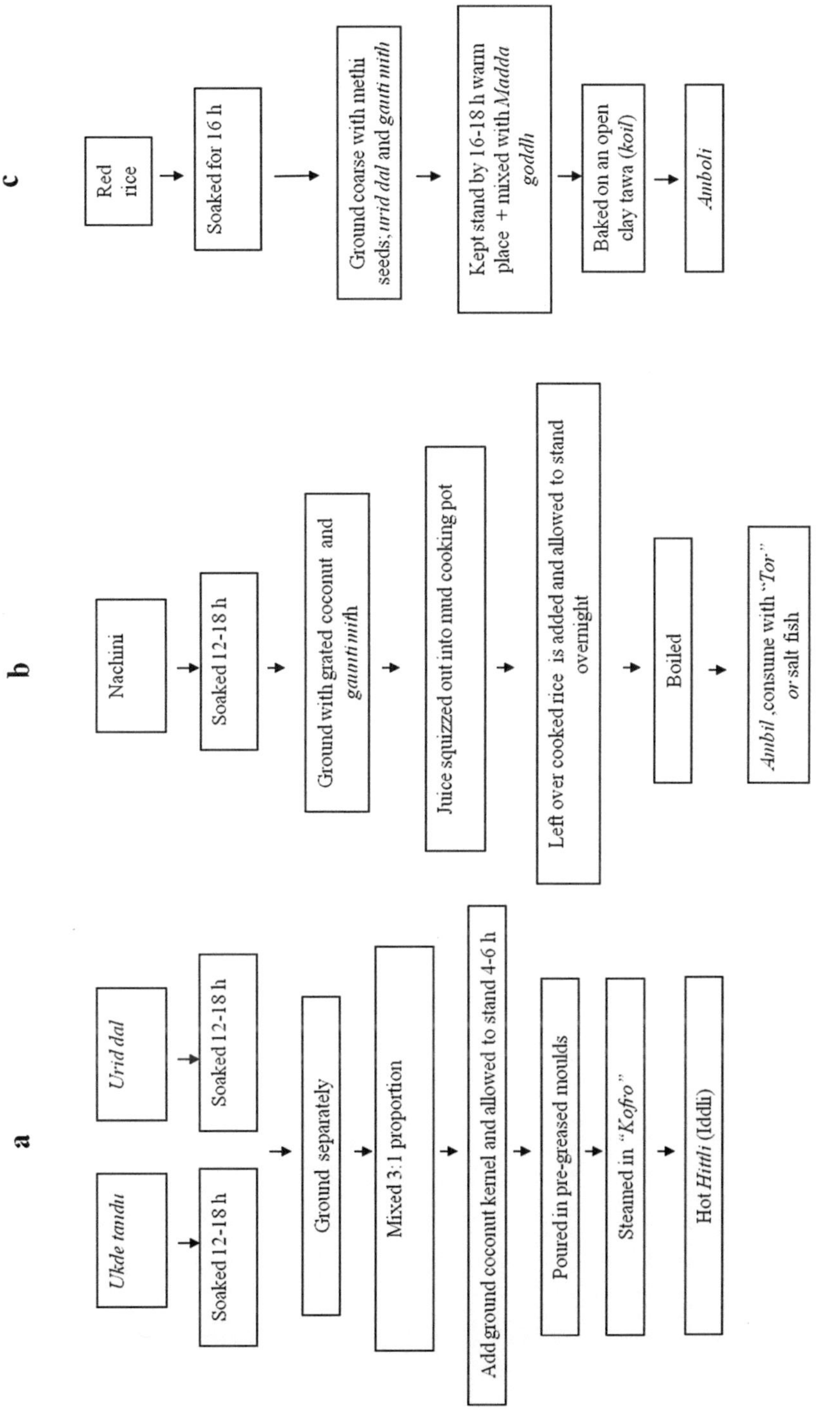

Fig. 6.8 Layout of *urid dal* fermented preparations of **(a)** *hittli* (idli), **(b)** *ambil*, and **(c)** *amboli*

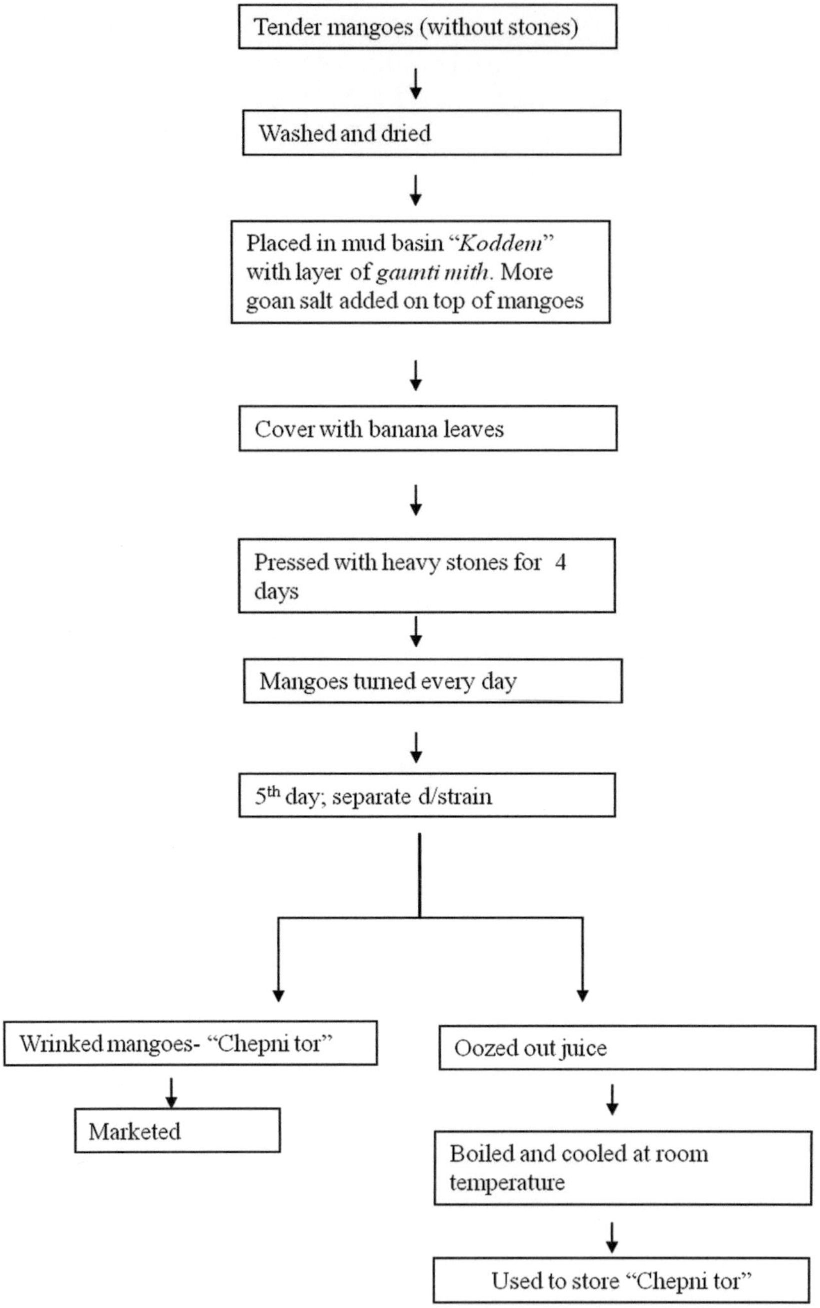

Fig. 6.9 Fermentation of tender mangoes to c*hepnim tor* using *gaunti mith*

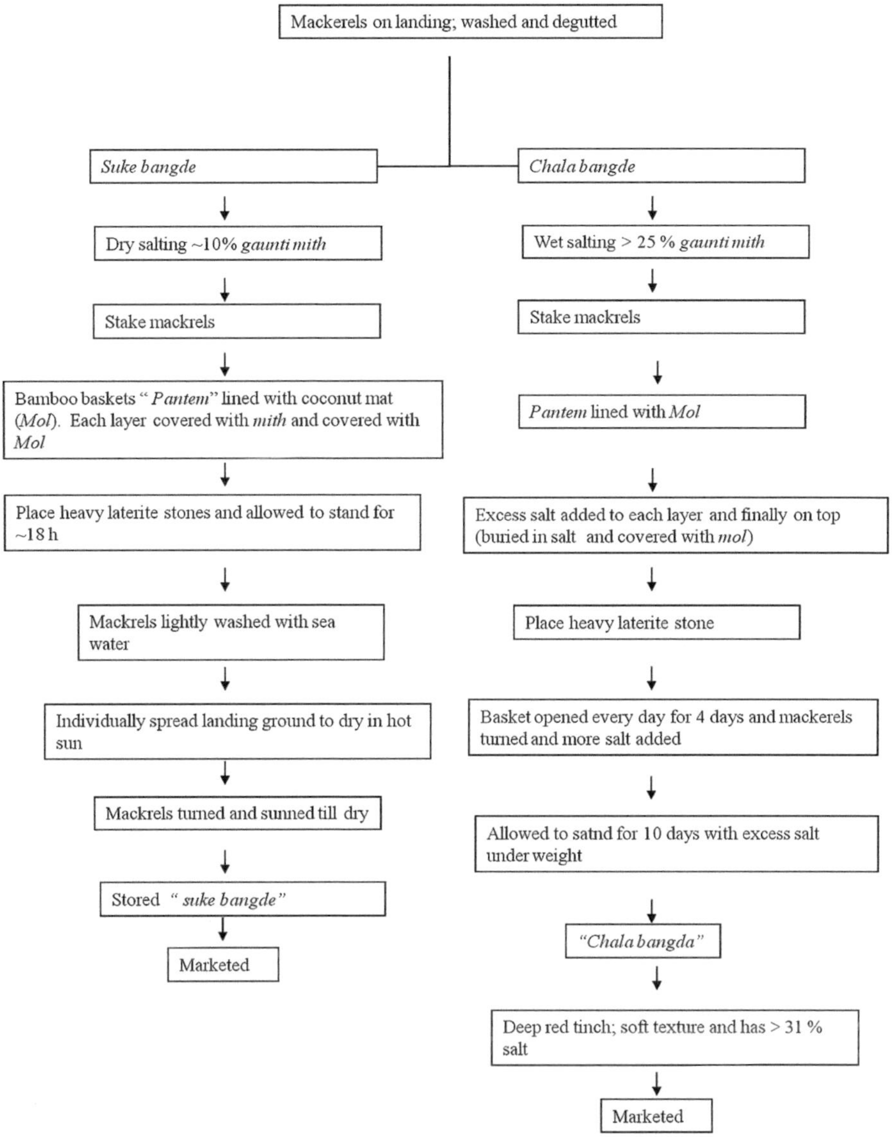

Fig. 6.10 Preparation of indigenous salted mackerel by Goan Ramponkar

Candida, and *Lanchancea* were evaluated as mixed cultures for high alcohol tolerance and improvement of product through manipulation of these microbial cultures.

The natural fermentation brought about by native microflora contained total acid, volatile acids, higher alcohol, and their esters which contribute to the aroma. Importantly, methanol formation was not observed and totally absent. *Caju feni* obtained through these cultures was within the requirements laid down for *caju feni* by Bureau of Indian Standards Specifications.

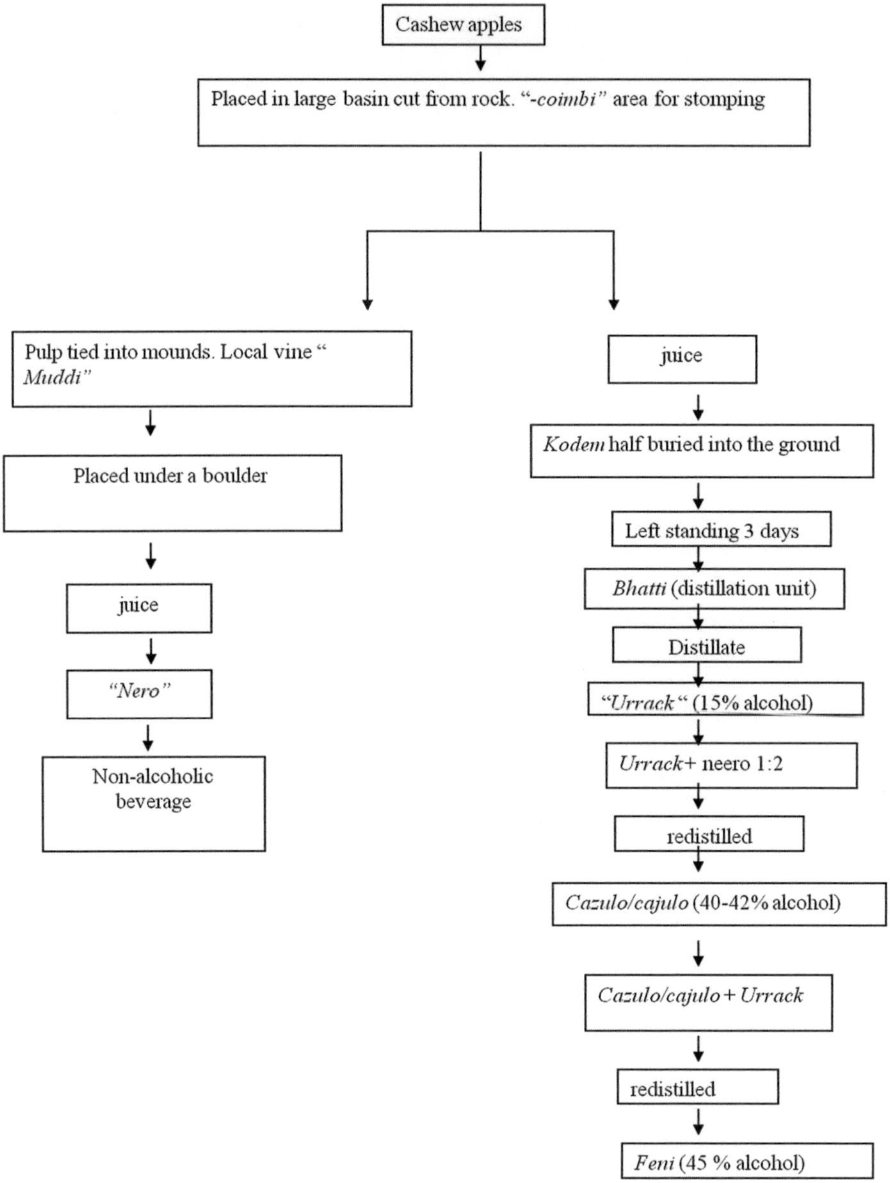

Fig. 6.11 Production of "cashew feni" by kazkar of Goa

Plate 6.4 Bhann employed for distillation and collection of distillate

6.5 Indigenous Method of Alcohol Evaluation in *Madda Feni* and *Caju Feni*

In the bygone era, in the absence of alcohol meter, the Reinder and Kazakar of Goa evaluated the alcohol content of *feni* using an indigenous method wherein the extent of effervescence produced by the l liquor was taken as indicator of alcoholic strength. Liquor was vigorously stirred with a *chuti* (coconut palm leaflet) or rapidly transferred from one glass container/coconut shell to another and examined for frothing or fizzle. Strong froth was attributed to strong liquor and higher content of alcohol. A low fizzle was attributed to a very low alcoholic content and the liquor as mildly alcoholic.

6.6 Socioeconomy and Ethnical or Religious Values

All products using *sur* as leaven are marketed as indigenous Goan: food, jaggery, vinegar, mango, and mackerel preserve by family-based industries in local markets. Vinegar, coconut feni, cashew feni, and salted fish are exported to other states of India and overseas.

Sur and distillate of fermented *sur*, *madda feni*, is used in traditional celebrations and ceremonies such as marriages and burials. Natives believe that sur acts as a mild laxative and helps to relieve constipation. Goans are known to use *madda feni* in treatment of colds and also use it to control soaring of temperature in fever. Similarly, Goans claim health benefits of *caju feni* in treatment of gout and rheumatism where it is rubbed in the joints. Some Goan natives dissolve sugar in feni and set it ablaze and then drink the residual mixture as medicine for coughs and colds. In addition to coconut vinegar, *feni* is also used as a preservative, presumably to enhance flavor (Chamaria 2018).

6.7 Conclusion

Goa and Goans have a rich heritage of indigenous, traditional foods and beverages obtained through natural fermentations using coconut palm sap "sur," caju juice, *urid dal*, and local Goan salt—*gaunthi mith. Caju feni* has earned the award as geographical indicator drink. Although all of these products actively help in the betterment of economy of the state of Goa and are well appreciated in overseas markets, there is no knowledge on microbiology, biochemistry, enzymology, and chemistry that is involved in these natural fermentation processes. Such knowledge along with the use of microbial cultures retrieved is expected to shorten the fermentation processes and increase flavor and quality of products. Furthermore, retrieved microbial cultures could replace "sur" and yet maintain the quality of fermentation as well as the product. A dedicated, detailed microbiological study is much desired and recommended.

Acknowledgments Furtado I is very grateful to her PhD student, Mr. Sanket Gaonkar, Department of Microbiology, Goa University, for typing all the layouts in figures, into computer sketches and organizing the photos in plates.

References

Atputharajah JD, Widanapathirana S, Samarajeewa U (1986) Microbiology and biochemistry of natural fermentation of coconut palm sap. Food Microbiol 3(4):273–280

Barbuddhe SB, Desai R, Doijad SP, Singh NB (2011) Diversity of yeast isolated from fermented cashew apple juice. Paper presented at National Symposium on Microbial Diversity and its applications in health, Agriculture and Industry. Organized by ICAR Research Complex for Goa, Old Goa, 4–5 March 2011

Caldeira J, Dias B, Lobo J, Shetye N, Gaonkar N (2018) Fermentation of Sanna—a traditional food from Goa. BSc project. St. Xaviers College; Mapusa-Goa; affiliated to Goa University

Chamaria P (2018) The spirit of Goa—"TheStatesman" Indian English-language broadsheet newspaper of April 2018. Delhi

Chandrasekhar K, Sreevani S, Seshapani P, Pramodhakumari J (2012) A review on palm wine. Int J Res Biol Sci 2(1):33–38

De Souza TR (1990) Goa through ages: an economic history. Concept Publishing Company, New Delhi. ISBN 978-81-7022-259

Dhume RAS (1986) The cultural history of Goa from 10000 BC-1352 AD. University of Michigan, pp 100–150. https://books.google.co.in/books/about/The_Cultural_History_of_Goa_from_10000_B.html?id=YtALAAAAIAAJ

Gomes V (2009) Three cheers for toddy Goa Chitra. Preserving the past to enrich the future. www.goachitra.com/research.html
Kadere TT, Kutima PM (2012) Isolation and identification of lactic acid bacteria in coconut toddy (MNAZI). J Asian Sci Res 2(12):807–819
Khorjuvenkar (2016) Diversity among yeast isolated from naturally fermented cashew apple juice. PhD thesis, Goa University
Mitragotri VR (1999) A socio-cultural history of Goa from the bhoja to Vijayanagar rulers. Institute Menezes Braganza, Panaji
Sekar S, Mariappan S (2007) Usage of traditional fermented products by Indian rural folks and IPR. Indian J Tradit Knowl 6(1):111–120
Shetty P, D'Souza A, Poojari S, Narayana J, Rajeeva P (2017) Study of fermentation kinetics of palm sap from *Cocos nucifera*. Int J Appl Sci Biotechnol 5(3):375–381

Ethnic Fermented Foods and Beverages of Gujarat and Rajasthan

7

V. Sreeja and Jashbhai B. Prajapati

Abstract

Traditional fermented foods prepared using milk, cereals, and pulses occupy a pivotal position in the dietary practice of people of India's western states Gujarat and Rajasthan. Fermented foods such as *dahi*, *chhash*, *raita*, *shrikhand, dhokla*, *khaman*, *handvo* and *khandvi* are consumed routinely and are associated with the traditional customs and beliefs of people of these two states. *Dahi*, *chhash*, *raita*, and *shrikhand* are fermented milk products, whereas *dhokla*, *khaman*, *handvo*, *raabadi* and *khandvi* are prepared using cereals and pulses. These popular traditional foods are valued for their high nutrition and possible therapeutic value and relished for their characteristic flavour and palatability. Lactic acid bacteria and yeasts are the major groups playing a crucial role in the fermentation of these foods. Production of most of these products is still largely confined to home and by unorganized sector, mainly by small dairy farms. Recently several large cooperative dairies and organized sectors have started manufacturing *dahi*, buttermilk, and *shrikhand* on large scale and are having good market. The limited shelf life of these foods at room and refrigeration temperature is a major constraint in their effective marketing. Researchers have tried various techniques for preparation and extending the shelf life of some of these ethnic foods.

Keywords

Fermented foods · Traditional · Gujarat · Rajasthan · Probiotics · Lactic acid bacteria

---

V. Sreeja · J. B. Prajapati (✉)
SMC College of Dairy Science, Anand Agricultural University, Anand, Gujarat, India
e-mail: jbprajapati@aau.in

J. P. Tamang (ed.), *Ethnic Fermented Foods and Beverages of India: Science History and Culture*, https://doi.org/10.1007/978-981-15-1486-9_7

7.1 Introduction

Traditionally fermented foods were prepared to preserve the surplus local produce, and gradually these products became a part of the traditional diet of local population (Prajapati and Sreeja 2013). Many of such foods were taken as home remedies to cure ailments particularly those related to stomach and gastrointestinal tract (Prajapati and Sreeja 2013). These fermented foods were known for their improved nutritional status and ease of digestion and hence were often given to infants and individuals in specific conditions. Gujarat and Rajasthan are western states of India well-known for milk production. Hence majority of the fermented foods of these two states comprises fermented milk products such as *dahi*, *chhash*, *raita*, *shrikhand*, and *chakka*. The fermented cereal products of these states include *dhokla*, *khaman*, *handvo*, *raabadi*, and *khandvi*. The microorganisms associated with such fermentation usually belong to lactic acid bacteria and yeast. Production of most of these products is now commercialized. This chapter encompasses the technological, microbiological, and nutritional and health aspects of traditional fermented foods of Gujarat and Rajasthan.

7.2 Gujarat

Gujarat is one of the western states in India. The name of the state came from the Gujjars, who ruled the area during the 1700s and 1800s. Archaeological evidence shows the region had cities as early as 2000 BC. Muslims conquered Gujarat in the thirteenth century AD and ruled for the next 450 years. After that British East India Company ruled in 1818. After India's independence in 1947, Gujarat was incorporated into Bombay state. In 1960, the Gujarati-speaking areas of Bombay were split off to form the present-day Gujarat. The state has three broad geographic divisions, viz. mainland Gujarat, the Saurashtra Peninsula, and Kachchh. Mainland Gujarat consists of coastal plains. These merge with lowlands around Ahmadabad and northern Gujarat. Fringing this area on the north and east are the uplands of the southern Aravallis, the western Vindhya and Satpura Ranges, and the Western Ghats. The southern areas are good for farming, even though most of the state is dry. The Saurashtra (also known as Kathiawar) region consists of a peninsula bounded by the Gulf of Cambay, the Arabian Sea, and the Gulf of Kachchh. Broad coastal plains surround low plateaus and hills. One of these, Gir Range, is home to a wildlife sanctuary for the last Asian lion population in the world. The Rann, a vast expanse of tidal mud flats and salt marshes, takes up much of Kachchh. People of Gujarat speak Gujarati. These include Kachchi, Kathiawadi, and Surati. Bhili, a language similar to Gujarati, is spoken by tribal groups in northern and eastern Gujarat. Gujaratis have a cultural heritage that can be traced back to a civilization that existed 3000 years ago. Many groups contribute to Gujarati culture. From the Vaishnavas come the legends and mythology of Krishna, to whom are ascribed the popular Raas and Garba folk dances. Jains influenced temple architecture and developed a distinctive style of painting. Muslim architecture in Gujarat combined Hindu elements with its own styles (https://www.everyculture.com/wc/Germany-to-Jamaica/Gujaratis.html).

Out of total population of Gujarat, 42.60% people live in urban regions. Average literacy rate in Gujarat for urban regions was 86.31% in which males were 90.98% literate while female literacy stood at 70.26%. Hinduism is majority religion in state of Gujarat with 88.57% followers. Islam is second most popular religion in state of Gujarat with approximately 9.67% following it. In Gujarat state, Christianity is followed by 0.52%, Jainism by 0.96%, Sikhism by 0.10%, and Buddhism by 0.10% (https://www.census2011.co.in/census/state/gujarat.html). Gujarati cuisine is mostly vegetarian, reflecting the strong influence of Jains and the Vaishnavas in the region. Wheat and the two kinds of millet (*jowar, bajra*) are the main staples. Flour is made into unleavened bread called *roti*. This is eaten with a variety of vegetable dishes. The villager takes a light breakfast of roti and milk or curds before setting out for the fields. Lunch is usually roti and buttermilk. The main meal is eaten in the evening and consists of rice, split peas (*dal bhat*), and vegetables. Meals are served on a *thali*, a metal tray on which roti, rice, and small bowls are placed. The bowls may hold vegetables such as eggplant, potatoes, beans, *dal* (lentils), and *dahi* (curds). *Kadhi*, a savoury curry of curd made from pulses (legumes), is a popular dish. *Ghee* (clarified butter) also forms part of the menu. Milk-based desserts are common. *Shrikhand* is a rich dessert made with curd and spiced with saffron, cardamom, nuts, and fruit.

7.3 Rajasthan

This state is located in the north-western part of the Indian subcontinent. Rajasthan, meaning "The Abode of the Rajas", was formerly called *Rajputana*, "The Country of the Rajputs" (sons of rajas). Before 1947, the state comprised some two dozen princely states and chiefships, the small British-administered province of Ajmer-Merwara, and a few pockets of territory outside the main boundaries. After 1947 the princely states and chiefships were integrated into India in stages, and the state took the name Rajasthan. It assumed its present form on November 1, 1956, when the States Reorganization Act came into force.

Most of Rajasthan's population consists of Indians of various social, occupational, and religious backgrounds. The Rajputs (various clans of landowning rulers and their descendants), though representing only a small percentage of Rajasthan's residents, are perhaps the most notable section of the population. In terms of caste structure, the Brahmans (highest caste) are subdivided into many *gotra*s (lineages), while the Mahajans (trading caste) are subdivided into a bewildering number of groups. In the north and west, the Jats (peasant caste) and Gujars (herding caste) are among the largest agricultural communities. Tribal peoples constitute more than one-tenth of the population of Rajasthan. In the eastern part of the state, those groups include the Mina, most of whom are farmers; the Banjara, who have been known as travelling traders and artisans; and the Gadia Lohar, another historically itinerant tribe, who traditionally have made and repaired agricultural and household implements. The Bhil, one of the oldest communities in India, generally inhabit southern Rajasthan and have a history of possessing great skill in archery. The Grasia and Kathodi also largely live in the south, mostly in the Mewar region.

Sahariya communities are found in the southeast, and the Rabari, who traditionally are cattle breeders, live to the west of the Aravallis in west-central Rajasthan (https://www.britannica.com/place/Rajasthan).

Out of total population of Rajasthan, 24.87% people live in urban regions. Average literacy rate in Rajasthan for urban regions was 79.68% in which males were 87.91% literate while female literacy stood at 63.81%. Hinduism is majority religion in state of Rajasthan with 88.49% followers. Islam is second most popular religion in state of Rajasthan with approximately 9.07% following it. In Rajasthan state, Christianity is followed by 0.14%, Jainism by 0.91%, Sikhism by 1.27%, and Buddhism by 1.27%. Around 0.01% stated "other religion", approximately 0.10% stated "no particular religion" (https://www.census2011.co.in/census/state/rajasthan.html).

Rajasthani cuisine is influenced by the state's warring life style as well as the availability of the ingredients. Hence, dishes that could be stored up for several days without heating were preferred. Geographically, the state experiences scarcity of water and fresh green vegetables due to its hot and arid climate, which in turn has influenced the cuisine of this land. The cooks in the desert belt of Bikaner, Barmer, and Jaisalmer prefer to use clarified butter (ghee), milk, and buttermilk, with the least possible amount of water. Rajasthani cuisine also has elements of the royal era of the Rajput rulers. The natives of this region prefer to have a wide variety of chutneys made from local spices like mint, coriander, turmeric, and garlic. An integral part of the cuisine of Rajasthan is the different forms of sweets. The passion of hunting in Royal Maharajas of Rajasthan also shaped the culinary of Rajasthan. Apart from all these, there is vegetarian cooking of the Maheshwaris of Jodhpur. The use of garlic and onion is prohibited in their cooking as they believe that these excite blood. The Marwaris of Rajasthan were vegetarian too, but their cuisine was richer in its method of preparation, similar to that of Rajputs. Then there were the Jains, who apart from being vegetarian, would not eat after sundown. Their food had to be devoid of the important ingredients of Rajasthani cuisine, garlic, and onion. The Bishnois, who were known to conserve animal and plant life, were vegetarians and so were the Vaishnavas, followers of Lord Krishna. Even there were few royal Rajput kitchens where only vegetarian meals were cooked.

(https://www.ohmyrajasthan.com/rajasthani-cuisine-history).

7.4 Classification of Fermented Foods and Beverages of Gujarat and Rajasthan

Fermented foods are generally classified on the basis of raw material used, type of fermentation, incubation conditions, type of microorganisms, form and state of the products, and many other conditions. The typical fermented foods of Gujarat and Rajasthan use mainly milk, cereals, and pulses as the raw materials. Fermented foods based on fruits, vegetables, fish, meat, and tubers are not generally found in these two western states of India. However, this excludes pickles which are traditionally made by the housewives of Gujarat and Rajasthan as a means of preservation of several fruits, vegetables, and tubers. The classical fermented foods of

Table 7.1 Ethnic fermented foods and beverages of Gujarat and Rajasthan

Food	Substrate	Sensory property of product	Major ethnic consumers	State
Dhokla	Cereal, spices	Steamed pancake-like consistency, fermented slightly salty-sour taste	People of Gujarat	Gujarat
Khaman	Cereal	Soft and spongy texture	People of Gujarat	Gujarat
Handvo	Cereal	Savoury and spicy with cake-like texture	People of Gujarat	Gujarat
Raabadi	Cereal	Beverage type with pleasant taste	People of Rajasthan	Rajasthan
Khandvi	Cereal	Soft, slippery texture, mild sour taste	People of Gujarat	Gujarat
Dahi	Cow/buffalo milk	Semi-solid, soft texture, sour taste	People of both Gujarat and Rajasthan	Gujarat/ Rajasthan
Chhash	Cow/buffalo milk/mixed milk	Beverage type with pleasant diacetyl aroma and mildly sour taste	People of both Gujarat and Rajasthan	Gujarat/ Rajasthan
Raita	Cow/buffalo milk	Viscous custard-like consistency and pleasant taste of added ingredients	People of both Gujarat and Rajasthan	Gujarat/ Rajasthan
Shrikhand	Cow/buffalo milk	Paste-like consistency, sweet-sour in taste	People of Gujarat	Gujarat

Gujarat and Rajasthan are basically of two categories: (1) milk-based products such as *dahi*, *chhash* (buttermilk), and *shrikhand* and (2) cereal- and pulse-based products such as *dhokla, khaman, handvo, raabadi, and khandvi*. The ethnic fermented foods and beverages of Gujarat and Rajasthan are shown in Table 7.1.

7.5 Milk-Based Fermented Foods

7.5.1 Dahi

Dahi is said to be one of the oldest traditional fermented dairy product. It is obtained by fermenting cow or buffalo milk using homofermenting lactic starter cultures.

Traditional method of preparation: For the traditional preparation of *dahi* in the household, milk is boiled, cooled to ambient temperature, and inoculated with spoonful of (roughly 0.5–2.0%) starter (previous day's *dahi* or buttermilk) and allowed to set overnight. It is then stored under refrigeration and consumed (Fig. 7.1). In cooler weather the *dahi* setting vessel is usually wrapped in a woollen cloth to maintain warmth. Homemade *dahi* is shown in Plate 7.1. In shops, the method is more or less the same except that milk is concentrated somewhat before inoculation, and the dahi is usually set in a shallow circular earthen pot, which helps in the absorption of any whey that may ooze out. In *halwais's* method, which is usually followed in restaurants and sweetmeat shops, *dahi* is prepared by the short set

Cow/buffalo milk

↓

Boiling for few minutes

↓

Cooling to body temperature

↓

Addition of previous day's *dahi* (0.5–5.0%)

↓

Incubation at ambient temperature (usually overnight, done at a cool place in summer and warm place in winter)

↓

Dahi

Fig. 7.1 Traditional method of preparation of *dahi*

Plate 7.1 *Dahi*

method (curd is ready within 4–6 h). They use symbiotic cultures added at the rate of 2–4% followed by incubation at 42–45 °C till setting of the curd (Singh 2006).

Traditionally, in Gujarat and Rajasthan, earthen pots or *kulhars* and dried *dhak* (*Butea frondosa*) leaf cups were used as packaging material for dahi (Goyal 1974; Yadav et al. 1989). Dahi packaged in earthen pots display a firm body and little of whey separation which is highly preferred by the people. These traditional packaging materials are still used for packaging dahi in rural area. Such packages are eco-friendly and easily degradable after disposal, but these are heavy and breakable and cause excessive shrinkage of the product during storage due to moisture seepage through pores. There are also problems of maintaining high standards of hygiene and proper sealing.

Culinary and mode of consumption: *Dahi* is usually consumed as a part of daily diet with meals and/or as a refreshing beverage. It is consumed with rice or wheat preparations in Gujarat and Rajasthan. *Dahi* is also used as a dessert. Surplus

Table 7.2 *Dahi*-based products popular in Gujarat and Rajasthan

S.no	Popular name	Short description
1	*Chach/chhash/ buttermilk/ mattha*	*Chhash* is a by-product obtained from *dahi* during the preparation of *makkhan* by the traditional churning process
2	*Chakka*	Concentrated product obtained by partial draining off whey from dahi. Usually not consumed as such and used for preparation of *shrikhand*
3	*Shrikhand*	*Shrikhand* is a concentrated sweet-sour fermented milk product prepared from *dahi*
4	*Raita*	It consists of various vegetables and fruits such as shredded cucumber and carrots; diced tomato or onions; boiled and diced potato, spinach, and beets; diced fruits such as banana and grapes mixed into plain dahi. This mix is seasoned with various spices and herbs
5	*Dahi vada/dahi bhalla*	*Dahi vada* is a popular snack in India. It is prepared by soaking *vadas* in thick *dahi*
6	*Kadhi*	This product is made from *chhash* by a recipe which varies from region to region. It is then served hot with rice. Sometimes, small balls made out of *besan* (Bengal gram flour) dough and fried in oil are added to *kadhi* and served as a curry (FAO 1990)

dahi is used as the intermediate product and churned into *makkhan* (butter), while the whey, popularly called *chhach* or *mattha*, is consumed as a refreshing beverage or converted into *kadhi*, a spicy curry served hot with rice. *Dahi* is used as a base for producing other products such as *chakka*, *shrikhand*, *raita*, *chaas*, and *lassi*. It is also used in the preparation of gravies of a number of dishes. Table 7.2 shows the popular *dahi*-based products consumed in Gujarat and Rajasthan.

Socio-economy and ethnical or religious values: *Dahi* has been associated with the traditional beliefs, customs, tales, and rituals of Hindu community in India. It is mentioned in ancient Indian scriptures like *Vedas* and *Upanishads*, and various hymns for which exact date is not known, but has been referred in about a 10,000-year-old era of Rama. During the period of Lord Krishna (ca. 3000 BC) *dahi*, buttermilk and *makkhan* were highly regarded. Aryans probably used *dahi* in their daily diet as it provided an acidic, refreshing taste and reduced putrefactive changes. *Dahi* is also used as an ingredient of *panchamrut* (five nectars), an article in every Hindu rituals. It is also used for giving bath to Idols of God's made up of metals before the routine worship in temples.

Microbiology: Microbiota in *dahi* can be broadly classified into two categories: (1) essential and (2) contaminants. Essential comprises the lactic acid bacteria used as starter cultures, while the contaminants are mainly coliforms, yeasts, and aerobic spore formers. Microbiological analysis of dahi samples collected from various sources in different seasons revealed a predominance of *Streptococcus thermophilus* and *Lactobacillus bulgaricus* (Yadav et al. 1993). Other than several species of lactic acid bacteria, lactose fermenting yeasts (Prajapati and Sreeja 2013; Tamang et al. 2016), spore formers (Jayaram and Gandhi 1987), spores (Ramadasappa et al. 1968), and staphylococci (Hosona et al. 1989) have been reported as microbiota of dahi. *Lc. lactis* subsp. *lactis*, *Lc. lactis* subsp. *diacetylactis*, *Lc. lactis* subsp. *cremoris*, *S.*

thermophilus, and *Lb. delbruecki* subsp. *bulgaricus* were reported to be isolated from shrikhand samples (Tamang et al. 2016). Yeasts and moulds are considered as the most common contaminants in *dahi*. Low-pH environment of dahi favours the growth of these organisms. Yeasts and moulds get entry into the product from the utensils, human hands, and environment under natural conditions (Lembhe and Ranganathan 1968; Jayaram and Gandhi 1987; Ghosh and Rajorhia 1987; Hosona et al. 1989; Sharma et al. 1993). Yeasts can multiply in *dahi* producing gassiness and flavour defects associated with lipolysis of milk fat, if the product is held too long at ambient temperatures. Different pathogenic organisms have been found to survive in *dahi* even though the conditions present in *dahi* are not favourable for their growth. The organisms, due to their variable degree of survival, viz. *S. aureus* (Saha and Ganguli 1957; Ghosh and Chattoraj 1963; Kahlon and Grover 1984), *E. coli* (Prasad et al. 1984; Ghosh and Rajorhia 1987), *Enterobacter aerogenes* (Prasad et al. 1980), and *Salmonella paratyphi* (Tiwari and Singh 1964, 1966), can be a potential threat.

Microbiological examination of *dahi* and *shrikhand* includes checking for the survival of the starter organisms, as well as for the presence of undesirable spoilage or pathogenic organisms. Selective agar media, viz. deMan Rogosa Sharpe Agar, M17 Agar, Violet Red Bile Agar/MacConkey's agar, and Potato Dextrose Agar, are employed for enumeration of lactobacilli, streptococci, coliforms and yeast, and moulds, respectively. Microscopic examination of product using Gram's staining can be done for rapid checking of contamination as well as for checking the ratio of starter cultures present in the products (Prajapati and Sreeja 2013).

Nutritional and therapeutic value: *Dahi* contains all the valuable constituents of milk. Its nutritional aspects further include the changes caused as a result of heat treatment of milk, starter growth, and fermentation processes. High heat treatment of milk affects the availability of lysine to some extent and also causes some damage to sulphur amino acids, vitamin C, and some of the B complex vitamins. High heat treatment also causes denaturation of whey proteins, release of reducing substances. Uniform fat distribution due to homogenization results in soft curd. Nutrients like lactose, protein, fat, and some vitamins decrease, while nutrients like lactic acid, peptides, amino acids, volatile flavour compounds, some vitamins, enzymes, and bacterial proteins increase. Digestibility of the proteins and availability of minerals increases. The unutilized amount of lactose after its partial use by lactic acid bacteria for production of lactic acid and other compounds contribute to the calorific value of *dahi* and may help in alleviating the harmful effects of lactose intolerance (Ayebo and Shahani 1980). The protein quality of *dahi* is the sum of quality of proteins derived from milk, microbial cell protein built up as a result of growth of starter bacteria and the free amino acids and peptides released by the proteolytic activities of the organisms. The native milk proteins are converted into a soft curd containing finely dispersed casein particles due to bacterial action during the preparation of *dahi*. As a result, the proteins in *dahi* are more easily digested and assimilated than milk proteins and are therefore, particularly useful to children, old people and persons suffering from stomach ulcers who cannot tolerate the milk proteins in its original form. Additionally, *dahi* is considered as a potential source of B complex vitamins, folic acid, and riboflavin (Sarkar et al. 2015).

The therapeutic value of *dahi* and its related products in the treatment of gastrointestinal disorders and other ailments has been recognized in the Ayurvedic system of medicine in India quite a long time ago. Various properties of cow and buffalo milk *dahi* and its therapeutic aspects are being discussed and emphasised in *Charaka Samhita* and *Sushruta Samhita* (Prajapati and Nair 2008). The beneficial effect of fermented milks in the treatment of intestinal diseases may be ascribed partly to the effect of lactic acid and antibacterial substances present in them and partly to the creation of favourable condition for the growth of *Lactobacillus acidophilus*, which is a normal inhabitant in the intestine and plays a key role in controlling the microbial population there. Also, the consumption of fermented milk has the potential to increase gut bacterial colonization (Fernandez et al. 2003). Several in vitro tests and in vivo animal studies, especially in mice and rat models, are being done on the therapeutic aspects of *dahi* and its related products. Some of these are mentioned in Table 7.3.

Optimization and commercialization: The recent decade saw many organized dairies entering into the arena of *dahi* production adopting a standardized method. In this method, fresh, good-quality milk is preheated and subjected to filtration and clarification, standardized to desired fat and SNF level, homogenized and heat treated followed by cooling to incubation temperature, and inoculated with specific *dahi* starter culture. It is then filled in suitable containers (cups) of the appropriate size and incubated. When a firm curd is formed and the acidity reaches to about 0.7% lactic acid, *dahi* cups are transferred to cold room maintained at about 4–5 °C and stored at that temperature till consumption. The procedure for commercial production of *dahi* is depicted in the Fig. 7.2. Many organized dairies are now preparing *dahi* adopting mechanized and standardized method. The starter cultures for *dahi* making are now commercially available in the form of liquid, concentrated, freeze-dried cultures. Alternatively a Direct Vat Inoculum (DVI)/Direct Vat Set (DVS) concentrated culture can be used directly in fermentation vat for the preparation of *dahi*.

Dahi manufactured by commercial dairy plants is packaged in low-density polyethylene (LDPE) cups and heat sealed with aluminium foil. Pouches are the most popular, convenient, and accepted packaging for *dahi*. Dairy plants fill the inoculated milk in pouches and transfer them to the incubation room for setting the *dahi*, and this is then marketed. Form-fill-seal (FPS) machines are now widely used for packaging of *dahi* in pouches. Cup filling and sealing machines are industrially used for filling and sealing *dahi* in preformed cups. These may be of rotary or straight through type. Cup thermo-fill and sealing machine does thermoforming of cups from the plastic film, filling of product, and sealing of lid in a continuous sequence of operation.

Value addition to dahi: A number of ingredients and methods are being tried for value addition to *dahi* and its related products. One of the most important functional ingredients tried is probiotics. Probiotic *dahi* is a regular curd with added probiotic cultures such as *Lactobacillus acidophilus*, or *Bifidobacterium* sp. *Dahi*, which is already considered good for digestion, is the most appropriate product to introduce probiotics. At Dairy Science College, Anand (Gujarat), India, several value-added

Table 7.3 Research studies highlighting the therapeutic aspects of *dahi* and related products

Proposed therapeutic effect	Cultures used for dahi/related products	Model used	References
Activates the non-specific immune system; protects against enteric infection by *Shigella dysenteriae*	Dahi made using *Leuconostoc citrovorum* and *Lactococcus lactis*	Mice	Singh and Kansal (2003)
Modulation of immune response against *Salmonella enteritidis*	Dahi containing probiotic *Lactobacillus casei*	Mice	Jain et al. (2009)
Anti-allergic effect	Dahi containing probiotic strains *L. acidophilus* NCDC14, *L. casei* NCDC19	Mice	Jain et al. (2010)
Protective effect against intestinal infection by *Shigella dysenteriae*	Buffalo milk probiotic dahi containing selected strains of *Lactobacillus acidophilus* and *Bifidobacterium bifidum* and using dahi culture (*Lactococcus lactis* ssp. *cremoris* and *Lactococcus lactis* ssp. *lactis* biovar. *diacetylactis*)	Mice	Rajpal and Kansal (2009a, b)
Maintenance of balance in cytokine production by increasing the production of IL-10	Dahi supplemented with leaf protein concentrate	Children	Dewan et al. (2009)
Immunostimulating effect	Lassi prepared using *S. thermophilus* MTCC 5460 and probiotic *L. helveticus* MTCC 5463	Geriatric volunteers	Senan et al. (2015)
Reduce the duration of diarrhoea	Dahi containing 10^8 of each *Lactococcus lactis*, *Lactococcus lactis cremoris*, and *Leuconostoc mesenteroides cremoris* per gram	Children (6 months to 5 years of age)	Agarwal and Bhasin (2002)
Blood pressure reduction	Dahi containing *Lactobacillus delbrueckii* subsp. *bulgaricus*, *Streptococcus salivarius* subsp. *thermophilus*, and *Lactococcus lactis* biovar. *diacetylactis*	Adult humans (35–65 years)	Ashar and Chand (2004)
Cholesterol lowering property	Probiotic dahi containing selected strains of *Lactobacillus acidophilus* and *Bifidobacterium bifidum*	Rats	Rajpal and Kansal (2009a, b)
Antihypercholesterolemic effect	Probiotic *dahi* containing *L. acidophilus* LaVK2 and dahi culture *Lactococcus lactis* ssp. cremoris NCDC-86 and *L. lactis* ssp. *lactis* biovar. *diacetylactis* NCDC-60	Rats	Mohania et al. (2014a, b)

(continued)

Table 7.3 (continued)

Proposed therapeutic effect	Cultures used for dahi/related products	Model used	References
Reduced serum levels of LDL, TC/HDL ratio, and LDL/HDL ratio	*Lassi* prepared using *S. thermophilus* MTCC 5460 and probiotic *L. helveticus* MTCC 5463	Geriatric volunteers	Senan et al. (2015)
Antidiabetic effect	*Dahi* containing *L. lactis*	Rats	Yadav et al. (2006)
	Probiotic *dahi* containing *Lb. acidophilus* and *Lb. casei* with dahi cultures	Rats	Yadav et al. (2008a, b)
Reduction of gastrointestinal cancer	*Dahi* containing selected strains of *Lactobacillus acidophilus* and *Bifidobacterium bifidum* and using dahi culture (*Lactococcus lactis* ssp. *cremoris* and *Lactococcus lactis* ssp. *lactis* biovar. *diacetylactis*)	Rats	Rajpal and Kansal (2008)
Antiatherogenic effect	Probiotic *dahi* prepared using culture with probiotic *Lactobacillus acidophilus* and *Lactobacillus casei*	Rats	Sinha and Yadav (2007)
Anti-oxidative effects	Probiotic *dahi* prepared by coculturing selected strains of *Lactobacillus acidophilus* and *Bifidobacterium bifidum* and *dahi* culture in buffalo milk	Rats	Rajpal and Kansal (2009a, b)
Antineoplastic and antiproliferative activities	Probiotic *dahi* containing *L. acidophilus* and *B. bifidum*	Rats	Mohania et al. (2014a, b)
Anticarcinogenic effect	Probiotic *dahi* containing *Lactobacillus acidophilus* LaVK2 and *Lactobacillus plantarum* Lp9	Rats	Mohania et al. (2013a, b)
Chemoprotective effect	Probiotic *dahi*	Rats	Shruti and Kansal (2011)
Ameliorative potential against colitis	*Dahi* containing selected strains of *Lactobacillus acidophilus* LaVK2 and *Bifidobacterium bifidum* BbVK3	Mice	Jadhav et al. (2013)
Immunoprotective effect against ulcerative colitis	Probiotic *dahi* containing *L. acidophilus* LaVK2 and *B. bifidum* BbVK3	Mice	Jadhav et al. (2012)

Raw milk

↓

Preheating (35–40°C)

↓

Filtration/clarification

↓

Standardization (fat 4.5% and SNF 10–14%)

↓

Preheating to 60°C

↓

Homogenization (100 kg/cm^2) (this step is optional)

↓

Heat treatment (90°C /10 min)

↓

Cooling to incubation temperature which usually ranges from 30 to 45°C

↓

Inoculation with active starter culture

↓

Package in containers (cups)

↓

Incubation till desired acidity of about 0.7% lactic acid

↓

Dahi

↓

Cooling and storage (5–7°C)

Fig. 7.2 Method for commercial production of *dahi*

probiotic fermented milk products have been developed. A synbiotic dahi was developed using probiotic culture *L. helveticus* MTCC 5463 and functional ingredient inulin. The synbiotic dahi was stable during storage of 28 days at 4 °C. Several other dahi-based synbiotic products like synbiotic raita, synbiotic lassi, synbiotic whey drink, herbal probiotic lassi, and carbonated probiotic milk have also been developed. All the products were organoleptically acceptable up to 3–4 weeks at refrigeration temperature and had optimum dose of probiotic lactobacilli (Prajapati and Sreeja 2013). Some of the research developments related to value addition to *dahi* are enlisted in Table 7.4.

Table 7.4 Research attempts made on value addition to *dahi*

Research attempts	Purpose	References
Probiotic cultures	Added health benefits	Yadav et al. (2005), Prajapati and Sreeja (2013), Panjagari et al. (2016)
Probiotic culture and inulin	Added health benefits	Gawai and Prajapati (2012)
Fruits	Improve nutritional as well as therapeutic values through phytonutrients and dietary fibres	Khurana and Kanawjia, (2007), Panjagari et al. (2016), Prajapati and Sreeja (2013)
Mineral fortification	Improve calcium content	Singh et al. (2005), Ranjan et al. (2006)
Beta-carotene-rich sweet potato	Vitamin A and dietary fibre enrichment	Sivakumar et al. (2008)
Aloe gel powder enrichment	Improve product quality through use as biothickener, improving nutritional and therapeutic potency	Pushkala and Srividya (2011)
Incorporation of cereal and fruits	Improve nutritional as well as therapeutic values through phytonutrients and dietary fibres	Kale et al. (2011)
Incorporation of artificial sweeteners such as sac-sweet or sucrol	Dahi meant for diabetic patients	Islam et al. (2010)
Sweet corn milk blended with buffalo milk	For preparation of low-cost nutritious dahi	Padghan et al. (2015)
Nisin-producing bacterial strain *Lactococcus lactis* W8	Biopreservation/extended shelf life	Mitra et al. (2010)
EPS-producing mesophilic strains	Reduce whey separation and improve sensory aspects	Behare et al. (2009)

7.5.2 Shrikhand

This is a fermented dairy product prepared by straining of fresh curd through a muslin cloth to cause partial removal of whey to obtain a concentrated curd mass called *chakka*. This *chakka* is then finely mixed with sugar and flavourings to get a semisoft consistency and sweetish-sour taste. *Shrikhand* is known for its high nutritive value, characteristic flavour, palatable nature, and possible therapeutic value (Nigam et al. 2009, Singh and Singh 2014).

Traditional method of preparation: The traditional method of *shrikhand* production (Fig. 7.3) involves preparation of *dahi* from milk of cow and buffalo or mixed milk, draining off whey from the curd by hanging in a muslin cloth for 10–12 h to obtain *chakka* or *maska*, mixing *chakka* with the required amount of sugar, flavour, colour, and other additives such as nuts, seasonal fruits, spices, raisins, etc. Starter cultures used are generally the same as those used for making *dahi*. The finished product is packaged in polystyrene cups.

Milk (cow/buffalo/mixed)
↓
Boiling
↓
Cooling to room temperature
↓
Inoculation with *dahi* culture
↓
Incubation at room temperature till firm curd is formed
↓
Hanging the curd in a muslin cloth to drain off whey (10–12 h.)
↓
Chakka/maska
↓
Addition of sugar (usually 50–100% of curd quantity), flavour, colour, spices,etc.
↓
Shrikhand
↓
Storage at refrigerated temperature

Fig. 7.3 Traditional method of preparation of shrikhand

Culinary and mode of consumption: *Shrikhand* is a very refreshing product and consumed as such or along with puri/roti or used as a dip/spread also.

Socio-economy and ethnical or religious values: *Shrikhand* derives its name from the Sanskrit word *Shikharani* meaning a curd prepared with added sugar, flavouring agents (saffron), fruits, and nuts (Karche et al. 2015). It is said to have originated in Persia, where the product was made using Frasi-shir (milk) and *khand* (sugar). *Shrikhand* was later brought to Gujarat by the Parsi-Zoroastrian settlers. Other than Gujarat, its popularity spans the states of Maharashtra and Karnataka.

Microbiology: *Shrikhand* quality is largely defined by the type and quality of starter cultures used for *dahi* preparation. A mixed culture containing *Lactococcus lactis* subsp. *lactis*, *Lactococcus lactis* subsp. *diacetylactis/Leuconostoc* spp., and *Lactococcus lactis* subsp. *cremoris* in the ratio of 1:1:1 may be used. Other recommended cultures are those containing *Streptococcus thermophilus* and *Lactobacillus delbrueckii* subsp. *bulgaricus* (1:1), a culture comprising *Lactococcus lactis* subsp. *lactis* and *Lactococcus lactis* subsp. *diacetylactis.* As *dahi* serve as the base material for *shrikhand* making, the type and quality of *dahi* has a huge influence on *shrikhand* quality.

Nutritional value: *Shrikhand* typically contains 39.0% moisture and 61.0% of total solids of which 10.0% is fat, 11.5% is proteins, 78.0% is carbohydrates, and 0.5% is ash, on a dry matter basis with a pH of about 4.2–4.4 (Boghra and Mathur 2000, Kulkarni et al. 2006). Some researchers have attempted to improve the sensory and nutritive characteristics of *shrikhand* by adding fruit pulp. In addition to

Plate 7.2 *Shrikhand*

fresh fruit pulps and fruit pieces, cardamom, nutmeg, dry fruits, charoli, saffron, and almond are added to improve the taste and nutritional quality of *shrikhand* (Plate 7.2).

Optimization and commercialization: Several commercial dairy plants are now manufacturing *shrikhand*. A small part of the total production is exported too.

Fully mechanized continuous process has been used for industrial production of *shrikhand*. The process consists of centrifugal separation of whey from the curd and the mechanical mixing of *chakka*, sugar, and flavourings, making the entire process hygienic as well as labour and time saving. Figure 7.4 shows the industrial method of *shrikhand* manufacture. Normally skim milk is preferred for preparing *dahi* for the manufacture of *shrikhand*. Use of skim milk, not only eliminates fat losses, but ensures faster moisture expulsion and less moisture retention in the curd. Skim milk is usually heated to 85 °C for 30 min. Such heat treatment results in denaturation of β-lactoglobulin, which in turn interacts with casein on acid coagulation, thereby increasing the yield. Also high temperatures are reported to result in the alteration of proteins in milk, which improves the growth and activity of starter culture. Such high heat treatment to milk causes destruction of certain heat labile inhibitors as well as many of the competing microorganisms that are present in raw milk. The heated milk is then cooled to incubation temperature and inoculated with *dahi* culture of mixed strains. Incubation is done at the optimum temperature of the starter culture till the required acidity (0.8–1.0% lactic acid) is achieved. This curd is then further processed to separate the whey from *dahi* to obtain *chakka/maska*. At industrial level, rapid whey drainage is recommended to control the level of acidity in the product. Hence mechanized means of whey removal have been favoured, which may include use of basket centrifuge or batch-type vacuum whey separator (IDF 2007). The process has been standardized using a quarg separator. Other technological developments include use of ultrafiltration to concentrate the curd to improve the yield of *chakka* over traditional methods (Sharma and Reuter 1998) and preconcentration of skim milk using ultrafiltration (Md-Ansari et al. 2006).

On industry level, polystyrene cups are widely used for the packaging of *shrikhand* (Goyal and Rajorhia 1991). Cup filling and sealing machines are

Skim milk

↓

Heat treatment (90°C/15 s)

↓

Cooling to 30°C

↓

Inoculation with *dahi* culture (0.5–2%)

↓

Incubation till desired acidity (0.8–1.0% lactic acid)

↓

Whey separation using a basket centrifuge (1100 rpm)/quarg separator

↓

Chakka

↓

Addition of sugar (@ 80% of amount of *chakka*) and plastic cream (to adjust fat to min. 8.5%) in a scraped surface heat exchanger

↓

Packaging in polystyrene cups/tubs

↓

Refrigerated storage (<5°C)

Fig. 7.4 Industrial method for *shrikhand* manufacture

industrially used for this purpose. *Shrikhand* is packaged in polystyrene cups of various sizes ranging from 100 g to 1.0 kg.

Normally the shelf life of *shrikhand* is limited to 2–3 days at ambient temperature and 2–3 weeks at refrigerated temperature (Prajapati and Sreeja 2013, Senapati et al. 2016). The limited shelf life of *shrikhand* at room and refrigeration temperature is a major constraint in its effective marketing. This limited shelf life is ascribed mainly to two factors: (1) excessive souring due to continuous growth of starter and (2) spoilage owing to presence of contaminants such as yeasts and moulds. These organisms get entry into the product mainly from the atmosphere. These organisms are capable of growing at the high osmotic pressure conditions existing in *shrikhand* due to presence of sugar (Dave et al. 1991; Sharma et al. 1993; Kumbhar et al. 2010). Kilara and Chandan (2013) reported that *shrikhand* remains stable for 40 days under refrigerated condition (4 °C). Narayanan and Lingam (2013) have reported that banana (20%) blended shrikhand, packed in polythene cups, could be preserved for 2 weeks (14 days) under refrigeration (4 ± 1 °C) with non-significant difference in sensory attributes from ordinary *shrikhand*. The limitations in shelf life of these products are being tackled through the methods of suppression of starter culture and prevention and control of contaminants, especially yeast and moulds after the products are ready. Researchers have tried various techniques for extending

Table 7.5 Methods tried to improve shelf life of dahi and shrikhand

Treatments	Details	References
Thermization	Heating cultured milk products below pasteurization temperature to extend their shelf life	Chander et al. (1989, 1992), Sarkar et al. (1992), Aziz (1985)
Carbonation	Carbonation process is economical and safe and does not have any negative effect on cultured dairy products. Carbonation of heat treated milk prior to inoculation with starters is tried to extend the shelf life of *dahi*	Singh and Anantharamaiyya (1968), Singh et al. (1970); Shah and Prajapati (2014)
Addition of preservatives	Use of nisin	Gupta and Prasad (1989)
	Potassium sorbate	Rajmohan and Prasad (1994)
	Sodium benzoate, potassium sorbate	Sanyal et al. (1990)
Drying	Preparation of freeze-dried *dahi* *Shrikhand* powder *Shrikhand vadi*	Rathi et al. (1990), De and Patel (1990), Prajapati and Sreeja (2013)
Biopreservation	Nisin producing *Lactococcus lactis* W8	Sarkar (2008), Mitra et al. (2010)

the shelf life, viz. heat treatments, addition of preservatives, use of gas, effective packaging, microwave treatment, irradiation, hurdle technology, biostabilization, etc. (Behare and Prajapati 2004). Some of the methods tried to improve shelf life of *dahi* and *shrikhand* are mentioned in Table 7.5.

Value addition to shrikhand: Some of the research efforts towards value addition to *shrikhand* are enlisted in Table 7.6.

7.5.3 Shrikhand Vadi

It is a desiccated form of *shrikhand*. Under ambient storage conditions, *shrikhand* spoils within 2–3 days. Refrigerated storage can increase its shelf life up to 40 days. To further extend the shelf life of *shrikhand*, *shrikhand vadi (Plate 7.3)* is made.

Traditional method of preparation: It is prepared by mixing an equal amount of dewatered *dahi* and sugar by weight and then dried in an open pan at low heat. When the mass begins to harden, it is tested for stickiness. The nonsticky product is flavoured and coloured. Powdered sugar is further added as desired. The product is mixed, rolled, and cut into shapes and packed like biscuits. The product yield is about 50% of the original weight of milk (Boghra and Mathur 2000; Prajapati and Sreeja 2013).

Table 7.6 Research attempts made on value addition to shrikhand

Research attempts	Purpose	References
Mango pulp	Improve sensory and nutritive characteristics	Bardale et al. (1986)
Soy fortification	Cost reduction and enhancement of nutritive and therapeutic value	Singh et al. (2014)
Blending with sapota pulp	Improve nutritional value in terms of proteins and minerals	Karche et al. (2015)
Papaya pulp addition	Improve sensory and nutritive characteristics	Nigam et al. (2009)
Addition of *Ashwagandha* (*Withania somnifera* L.) powder	Added health benefits	Landge et al. (2011)
Apple pulp with celosia powder	Improve product sensory quality and nutritional value	Kumar et al. (2011)
Guava powder	Improve product quality and nutritional value in terms of vitamin C, dietary fibres, and therapeutic potency	Chauhan et al. (2015)
Cocoa powder and papaya pulp	Improve product sensory quality and nutritional value	Vagdalkar et al. (2002)
Strawberry pulp	Improve product sensory quality and nutritional value	Sonawane et al. (2007)
Gulkand and rose petal powder	Improving product sensory quality and nutritional value	Nadaf et al. (2012)
Probiotic *shrikhand*	Possible health benefits	Devshete et al. (2012)
Walnut powder	Omega-3 enrichment and possible health benefits	Devshete et al. (2013)
Stevia powder	Partial replacement of sugar	Mehrotra et al. (2014)
Orange pulp and chiku pulp	Improve product sensory quality and nutritional value	Para et al. (2014)
Low-fat sugar-free mango *shrikhand*	Reduce fat content, improve flavour, and reduce calorie content	Shelke et al. (2014)
Sweet corn milk blended with buffalo milk	For preparation of low-cost nutritious shrikhand	Mane et al. (2015); Padghan et al. (2015)
Saffron	Improve colour, appearance, and flavour	Mehrotra et al. (2014)
Saffron and cardamom	Enhance the shelf life of shrikhand	Dandile et al. (2014)
Basil (*Ocimum basilicum*) extract	Herbal shrikhand with potential health benefits	David (2015)

7.5.4 Chhash

Chhash, also known as chaach or chaas, is a by-product obtained from *dahi* during the preparation of *makkhan* by the traditional churning process. It is similar to that of cultured buttermilk (Plate 7.4).

Plate 7.3 *Shrikhand vadi*

Plate 7.4 *Chhash*

Traditional method of preparation: Traditionally, *chaas* is made by churning *dahi* and cold water together in a pot, using a hand-held instrument called *madhani* with an aim to separate the milk fat to prepare *makhan.* An earthen pot is used to prepare and store *chaas*.

Culinary and mode of consumption: *Chhash* is used as a refreshing beverage, especially in the summer season and an accompaniment to main meal. It is preferred for its pleasant aroma and mildly sour taste. It is consumed with or without added salt and spices.

Socio-economy and ethnical or religious values: It is a widely consumed product throughout Gujarat and Rajasthan.

Microbiology: As *dahi* is the base material for preparation of this product, its microbiology is almost similar to that of *dahi*.

Nutritional value: It has less fat content and fewer calories compared with regular milk or *dahi*. It is easier to digest and, with the presence of live cultures, helps in intestinal health. It is advised in the Ayurveda for curing intestinal disorders (Pushpangadan et al. 2012; Prajapati and Sreeja 2013).

Skim milk/low fat milk

↓

Heat treatment 85° C/20 min or 90°C/10 min

↓

Cooling to inoculation temperatutre

↓

Inoculation with starter culture (DVS cultures are used nowadays)

↓

Incubation (22–25°C for mesophilic starters)

↓

Cooling and agitation to break the curd

↓

Packaging

Fig. 7.5 Method of preparation of cultured buttermilk

Optimization and commercialization: Manufacturing of cultured buttermilk on industrial scale involves selection of good-quality raw material, standard cultures, and optimized process of fermentation, packaging, and storage. Cultured buttermilk is prepared with the help of mesophilic lactic acid bacteria comprising of both acid producers and diacetyl producers such as *Lactococcus lactis* subsp. *lactis*, *Lactococcus lactis* subsp. *cremoris*, *Lactococcus lactis* subsp. *lactis* biovar. *Diacetylactis*, *Leuconostoc mesenteroides* subsp. *cremoris*, and *Leuconostoc mesenteroides* subsp. *dextranicum*. Flow diagram for preparation of cultured buttermilk is shown in Fig. 7.5. Polyethylene pouches/bottles/cartons are used for packaging buttermilk.

7.5.5 Kadhi

Kadhi is a product made from *chhash* by a recipe which varies from region to region. In Gujarat, it is made using either sour *dahi* or *sour chhash*.

Traditional method of preparation: For the preparation of *kadhi (Plate 7.5)*, a blend of spices comprising of ingredients such as black pepper, green chillies, turmeric, salt, and ground cumin are added with a small amount of Bengal gram flour, which is then added to an appropriate quantity of *chhash* and the mixture is brought to boiling point (Prajapati and Sreeja 2013). The consistency of *kadhi* varies widely from home to home, and particularly in Gujarat, *jaggery* or sugar is also added. It is sour-sweet and spicy in taste.

In Gujarat, *kadhi* is usually served hot with rice or *khichdi*, while in Rajasthan, it is also served with *bajre ki roti*.

Plate 7.5 *Kadhi*

7.5.6 Raita

Raita is a side dish made using *dahi* containing vegetables/fruits and spices.

Traditional method of preparation: Raita is prepared instantly using various seasonal vegetables and fruits such as shredded cucumber and carrots; diced tomato or onions; boiled and diced potato, spinach, and beets; and diced fruits such as banana and grapes mixed into plain dahi. Sometimes *boondi* (a water droplet-sized deep-fried crispy balls prepared from gram flour) is incorporated in dahi to prepare *boondi ka raita. Raita* is usually seasoned with various spices and herbs such as cilantro, cumin, mint, cayenne pepper, basil, black salt, and black pepper.

Culinary and mode of consumption: *Raita* is served as an accompaniment to meals in Gujarat and Rajasthan. These diverse raitas are virtually daily fare, especially among vegetarians, who rely on the milk and fermented milks for protein. *Raita* is served chilled or at room temperature. It is excellent with the regular meals as it balances the spiciness in the food (Prajapati and Sreeja 2013).

Nutritional value: Raita combines the goodness of both fermented milks and vegetables/fruits as well as spices. These are cooling, healthy, nutritious, and easy to rustle up.

7.6 Cereal and Pulses Based Fermented Foods

7.6.1 Rabdi

Rabdi or *raabadi* is a famous indigenous cereal-based fermented dairy product, prepared in the north-west part of India especially in Punjab, Haryana, Uttar Pradesh, and the western region of Rajasthan.

Traditional method of preparation: *Rabdi* is prepared by mixing and fermenting the flour of pearl millet, barley, wheat, sorghum, or maize with homemade buttermilk in earthen and metallic vessel in hot summer days in open sun or at room temperature (35–45 °C), kept for 4–6 h, followed by boiling, salting to taste, cooling, and consuming (Kumar et al. 2013; Sathe and Mandal 2016).

A couple of research studies have been carried out to standardize the procedure for preparation of different types of *raabadi* using barley (Gupta et al. 1992; Gupta and Khetarpaul 1993) and pearl millet (Dhankher and Chauhan 1987) and packaging conditions for enhancing the shelf life.

7.6.2 Dhokla

Dhokla is an indigenous fermented food found mainly in Gujarat. It is a cereal- and pulse-based fermented food.

Traditional method of preparation: It is prepared by fermentation of *Bengal gram* and rice.

Culinary and mode of consumption: *Dhokla* is consumed as a breakfast dish/main course/side dish or as a snack.

Microbiology: The organisms involved in fermentation come mainly from the substrate. Both lactic acid bacteria and yeast are involved in the fermentation and their numbers increase significantly during fermentation with an associated increase in the volume and decrease in pH of the batter. The lactic acid bacteria involved in the fermentation were identified as *Lactobacillus fermentum*, *Lb. lactis*, *Lb. delbrueckii*, *Leuconostoc mesenteroides*, and the yeast *Hansenula silvicola* (Joshi et al. 1989). During fermentation, lactic acid bacteria produce acid and contribute to sour taste of the product and improved flavour. Yeast produces gas and helps in making sponginess in food by increasing the batter volume.

Nutritional value: An increase in vitamins, viz. thiamine, riboflavin, and niacin content, in the range of 15–100% for the *dhoklas* prepared from wheat and Bengal gram has been reported. The amount of tocopherols has also been shown to increase during fermentation. Yeasts are reported to produce folic acid during fermentation. Joshi et al. (1989) reported that *dhoklas* prepared from fermented mixture of Bengal gram and curds which exhibited the best sensory qualities comprised several volatile acids including propionic acid, isobutyric acid, isovaleric acid, and acetoin. The antioxidant property of fermented *dhokla* batter is reported to be high, and this helps in curing age-related diseases and oxidative stress-induced degenerative diseases (Moktan et al. 2011). So, *dhokla* can be a good food item for the diabetic patients.

Although a number of cereals and pulses are fermented, very few reports are available on the antimicrobial effect of these fermenting substrates. Roy et al. (2009a, b) studied the survival and growth of food-borne bacterial pathogens in fermenting batter of *dhokla* and reported that though the fermentation is capable of holding the growth of common pathogenic bacteria in check, steaming is essential to produce microbiologically safe *dhokla* (Plate 7.6).

Plate 7.6 *Dhokla*

Plate 7.7 *Khaman*

7.6.3 Khaman

Khaman is the traditional fermented food of Gujarat. It is similar to *dhokla*, but made wholly of Bengal gram dal. It has yellow colour, soft, spongy texture, and high nutritive value.

Traditional method of preparation: *Khaman* is prepared by steaming the leavened and acidified batter (Plate 7.7). Two significant changes occurring in *khaman* fermentation are leavening and acidification of the batter. The organisms involved in the fermentation are important to give rise to high product quality.

Microbiology: During fermentation, lactic acid bacteria contribute to sour taste due to acid production in the product and improve the flavour. Yeasts present will produce gas and contribute to sponginess in product, further it will also produce folic acid (Kanekar and Joshi 1993). *Khaman* must be prepared and consumed within a day due to its poor keeping quality (Sekar and Mariappan 2007; Shrestha et al. 2017).

7.6.4 Handvo

Handvo/handwa is a fermented household product eaten as snack or light meal by Gujarati people. It is closely related to sourdough bread used in the western world.

Wash and soak the rice and all the dals together in enough water for 3–4 h, fine grind the soaked rice and dals in mortar, and prepare a fine paste. Add the *dahi/chaash* in this paste along with salt and blend it well.

↓

Transfer this handvo batter in a large bowl covered with a lid, and allow it to ferment for 8–10 h.

↓

Add grated bottle gourd, ginger paste, a little oil, green chilli paste, and a little turmeric powder. Mix all the ingredients well, and make sure that the fermented *handvo* batter has a thick consistency.

↓

Add baking soda along with lemon juice in it if more sponginess is required. Bake the mixture for 20–30 min until golden brown.

↓

Garnish with mustard seeds, cumin seeds, black sesame seeds, curry leaves, and asafoetida crackled in oil.

Fig. 7.6 Method of preparation of *handvo*

Traditional method of preparation: It belongs to the category of foods which are prepared by combining the cereals and pulses, involving coarse grinding of cereals and pulses, batter preparation from the ground mix, fermentation of the batter followed by its baking. It is prepared using fermented batter of rice and mixture of dals such as *urad dal*, *chana dal*, and *moog dal*. The batter is mixed with *dahi* or *chhash*; spices such as turmeric, chilli, and ginger paste; and seasonal vegetables such as bottle gourd, carrot, peas, etc. During the natural fermentation, the leavening is caused due to the action of bacteria and yeast. The fermented steamed product is usually garnished with black sesame seeds, mustard seeds, cumin seeds, asafoetida, and sometimes curry leaves fried in oil. Wide variations in the baking devices and source of heat ranging from the use of electric ovens to the use of coals have been found in the preparation of *handvo*. Household practice appears to be a short high temperature preheating for 5–10 min followed by uniform lower temperature heating for 50–80 min (Gajjar et al. 2014). The method of preparation of *handvo* is depicted in Fig. 7.6.

Two major steps in *handvo* preparation, viz. fermentation of the batter and baking of the fermented batter mix, greatly influence the ultimate product quality such as body, texture, flavour, colour, appearance, and nutrient content. The components of *handvo* batter include water, protein, fat, starch, and mineral contents of cereal-pulse flour mix and fat, sugar, salt, and spices added to the mix.

Nutritional value: The ultimate quality of *handvo* is dependent on the changes taking place in these ingredients during baking, although little is reported in the literature regarding such changes taking place in the *handvo* batter. Even though no specific literature can be traced with particular reference to *handvo*, similar changes like in *idli* batter are likely to occur during the fermentation of a typical *handvo* batter. However the net effect of their beneficial role in *handvo* has to be assessed in

Plate 7.8 *Handvo* cooker and *handvo*

Plate 7.9 *Khandvi*

terms of their surviving the baking treatment. Being a household item, *handvo* does not have a very uniform product profile, and there is a wide scope for accepting large variations of *handvo*, which may suit to the regional, socio-economic, seasonal, and personal requirements of different population groups (Plate 7.8).

7.6.5 Khandvi

It is a traditional Gujarati snack prepared with gram flour and buttermilk and seasoned with sautéed sesame seeds and other spices.

Traditional method of preparation: For the preparation of *khandvi*, gram flour, buttermilk, turmeric powder, and salt are mixed properly to make a smooth batter. The batter is then cooked in a heavy bottomed pan or a *kadai* on low flame to make a smooth paste-like consistency. It is then spread on a previously greased plate to form a thin layer. The layer is allowed to cool. It is then cut into strips, and these are rolled into a Swiss roll form. It is then seasoned with mustard seeds, cumin seeds, and curry leaves crackled in oil. It can be further garnished with shredded coconut and chopped coriander leaves (Plate 7.9).

7.7 Conclusion

Ethnic fermented foods and beverages of Gujarat and Rajasthan are a reflection of the life style, customs, beliefs, and sentiments of the people of these states. These fermented foods, viz. *dahi*, *shrikhand*, *chhash*, *raita*, *dhokla*, *handvo*, *khaman*, *khandvi*, etc., play a significant role in the dietary habits of people owing to good taste, high nutritive and therapeutic value, and, most importantly, increased keeping quality. At present most of these traditional fermented foods are being produced and marketed at commercial scale. In spite of the changing life styles and food habits with time, these foods have maintained its popularity in diet of people from Gujarat and Rajasthan. The demand for these products is continuously increasing offering ample opportunities to the modern dairy and food sector to scale up the production as well as to do value addition to these products to put them in a modern perspective. Well-known for their nutritional as well as therapeutic aspects, these products can be very well be explored for functional food categories and personalized nutrition. Another area which needs attention is improving the shelf life of these products. Modern completely automated production facilities combined with aseptic packaging and improved cold chain facilities throughout the marketing may be a solution to great extent. Further, the scientists and technologists need to seriously explore the possibilities of alternate preservation techniques and extending the shelf life of these valuable products which contributes not only to the economy of both rural and urban community but also to the physical and emotional well-being associated with traditions and beliefs of people in Gujarat and Rajasthan.

References

Agarwal KN, Bhasin SK (2002) Feasibility studies to control acute diarrhea in children by feeding fermented milk preparations actimel and indian dahi. Eur J Clin Nutr 56:S56–S59

Ashar MN, Chand R (2004) Fermented milk containing ACE-inhibitory peptides reduces blood pressure in middle aged hypertensive subjects. Milchwissenschaft 59:363–366

Ayebo AD, Shahani KM (1980) Role of cultured dairy products in diet. Cult Dairy Prod J 15:21–29

Aziz T (1985) Thermal processing of dahi to improve its keeping quality. Indian J Nutr Diet 22:80–87

Bardale PS, Waghmare PS, Zanjad PN, Khedkar DM (1986) The preparation of Shrikhand like product from skim milk chakka by fortifying with fruit pulps. Indian J Dairy Sci 39:480–483

Behare PV, Prajapati JB (2004) How to extend the shelf life of fermented milks? Proc Food Ind 2:16–19

Behare PV, Singh R, Singh RP (2009) Exopolysaccharides-producing mesophilic lactic cultures for preparation of fat-free dahi- an Indian fermented milk. J Dairy Res 76:90–97

Boghra VR, Mathur ON (2000) Physico-chemical status of major milk constituents and minerals at various stages of shrikhand preparation. J Food Sci Technol Mysore 37:111–115

Chander H, Batish VK, Mohan M, Bhatia KL (1989) Effect of heat treatment on the physico chemical properties of dahi—an Indian fermented dairy product. Cult Dairy Prod J 24:11

Chander H, Batish VK, Mohan M, Chand R, Singh RS (1992) Effect of heat processing on bacterial quality of dahi—an Indian fermented dairy product. Cult Dairy Prod J 27:8–9

Chauhan AK, Singh S, Singh RP, Singh SP (2015) Guava-enriched dairy products: a review. Indian J Dairy Sci 68:1–5

Dandile UM, Pawar BK, Choudhari DM (2014) Sensory quality of Shrikhand prepared by using cardamom and saffron. Res J Animal Husband Dairy Sci 5:1–5
Dave RI, Dave JM, Sannabhadti SS (1991) Microbiological quality of some market and household dahi samples. Asian J Dairy Res 10:111–114
David J (2015) Preparation of herbal Shrikhand prepared with basil (*Ocimumbasilicum*) extract. Pharma Innovation J 4:81–84
De A, Patel RS (1990) Technology of Shrikhand powder. Cult Dairy Prod J 25(2):23–24
Deodhar AD (1984) Nutritional role of dahi and yoghurt. Indian Dairyman 36:325
Devshete NG, Sontake AT, Patil HD, Watharkar RB (2012) Sensory quality of probiotic shrikhand using yoghurt culture. Int J Proc Post Harv Technol 3:212–214
Devshete NG, Hembade VR, Mandwade KG, Burbade R (2013) Preparation of omega-3 enriched probiotic Shrikhand using walnut powder. Int J Proc Post Harv Technol 4:74–78
Dewan P, Kaur IR, Faridi MMA, Agarwal KN (2009) Cytokine response to dietary rehabilitation with curd (Indian dahi) and leaf protein concentrate in malnourished children. Indian J Med Res 130:31–36
Dhankher N, Chauhan BM (1987) Effect of temperature and period of fermentation on protein and starch digestibility (in vitro) of Raabadi: a pearl millet fermented food. J Food Sci 52:489–490
FAO (1990) The technology of traditional milk products in developing countries, in Series on FAO Animal Production and Health Paper, 85. ISBN: 9251028990. http://www.fao.org/docrep/003/T0251E/T0251E04.htm
Fernandez MF, Boris S, Barbes C (2003) Probiotic properties of human lactobacilli strains to be used in the gastrointestinal tract. J Appl Microbiol 38:449–455
Gajjar DH, Joshi SD, Modi HA (2014) Nutritional upliftment studies of microbially fermented probiotic food (handvo) prepared from indigenous food ingredients. Horiz Holist Ed 1:124–133
Gawai K, Prajapati JB (2012) Status and scope of dahi industry in India. Indian Dairyman 64:46–50
Ghosh DN, Chattoraj SB (1963) Staphylococcal food poisoning from dahi. Indian J Public Health 7:1–4
Ghosh J, Rajorhia GS (1987) Chemical, microbiological and sensory properties of Mistidahi sold in Calcutta. Asian J Dairy Res 6:11–16
Goyal GK (1974) Packaging of dairy products in India. I. Milk and indigenous dairy products. Packaging India 6:23–24
Goyal GK, Rajorhia GS (1991) Role of modern packaging in marketing of indigenous dairy products. Indian Food Ind 10:32–34
Gupta M, Khetarpaul N (1993) Effect of Raabadi fermentation on phytic acid and in vitro digestibility of barley. Nahrung 37:141–146
Gupta RK, Prasad DN (1989) Incorporation of nisin in the stirred yoghurt. II. Effect of biochemical activities during storage. Cult Dairy Prod J 24:9–10
Gupta M, Khetarpaul N, Chauhan BM (1992) Preparation, nutritional value and acceptability of barley Raabadi-an indigenous fermented food of India. Plant Foods Human Nutr 42:351–358
Hosona A, Wardojo R, Otani H (1989) Microbial flora in 'dadih'a traditional fermented milk in Indonesia. LWT-Food Sci Technol 22:20. Cited from Dairy Sci. Abstr. 51: 4337
Hussain SK (2008) Studies on shelf life extension of cereal based fermented milk beverage. MSc thesis, NDRI Deemed University, Karnal, India
IDF (2007) Asian indigenous dairy products. Bulletin of the International Dairy Federation, Brussel, pp 1–20. ISSN0250-5118
Islam MN, Akhter F, Masum AKM, Khan MAS, Asaduzzaman M (2010) Preparation of dahi for diabetic patient. Bangladesh J Animal Sci 39:144–150
Jadhav SR, Shandilya UK, Kansal VK (2012) Immuno protective effect of probiotic dahi containing *Lactobacillus acidophilus* and *Bifidobacterium bifidum* on dextran sodium sulfate-induced ulcerative colitis in mice. Probiotic Antimicro Prot 4:21–26
Jadhav SR, Shandilya UK, Kansal VK (2013) Exploring the ameliorative potential of probiotic Dahi containing *Lactobacillus acidophilus* and *Bifidobacterium bifidum* on dextran sodium sulphate induced colitis in mice. J Dairy Res 80:21–27

Jain S, Yadav H, Sinha PR (2009) Probiotic dahi containing *Lactobacillus casei* protects against *Salmonella enteritidis* infection and modulates immune response in mice. J Med Food 12:576–583
Jain S, Yadav H, Sinha PR, Kapila S, Naito Y, Marotta F (2010) Anti-allergic effects of probiotic dahi through modulation of the gut immune system. Turk J Gastroenterol 21:244–250
Jayaram D, Gandhi DN (1987) Role of market dahi as inoculum for rapid preparation of dahi. Indian J Dairy Sci 40:374–376
Joshi N, Godbole SH, Kanekar P (1989) Microbial and biochemical changes during dhokla fermentation with special reference to flavour compounds. J Food Sci Technol 26:113–115
Kahlon SS, Grover NK (1984) Incidences of staphylococci in milk products sampled from Ludhiana. Indian J Dairy Sci 37:381–383
Kale AK, Dhanalakshmi B, Kumar U (2011) Development of value added dahi by incorporating cereal and fruits. J Food Sci Eng 1:379–385
Kanekar P, Joshi N (1993) *Lactobacillus fermentum*, *Leuconostoc mesenteroides* and *Hansenula silvicola* contributing to acetoin and folic acid during dhokla fermentation. Indian J Microbiol 33:111–117
Karche RV, Thakare VM, Bhagat AV, Shirsath SA (2015) Microbiological quality of cow milk shrikhand blended with sapota pulp. Int J Food Agric Vet Sci 5:18–22
Khurana HK, Kanawjia SK (2007) Recent trends in development of fermented milks. Curr Nutr Food Sci 3:91–108
Kilara A, Chandan RC (2013) Greek-style yoghurt and related products. In: Chandan RC, Kilara A (eds) Manufacturing yogurt and fermenting milks, 2nd edn. Wiley-Blackwell, Ames, IA, pp 305–315
Kulkarni C, Belsare N, Lele A (2006) Studies on shrikhand rheology. J Food Eng 74:169–177
Kumar S, Bhat ZF, Kumar P (2011) Effect of apple pulp and celosia argentea on the quality characteristics of shrikhand. Am J Food Technol 6:817–826
Kumar RS, Kanmani P, Yuvaraj N, Paari KA, Pattukumar V, Arul V (2013) Traditional Indian fermented foods: a rich source of lactic acid bacteria. Int J Food Sci Nutr 64(4):415–428. https://doi.org/10.3109/09637486.2012.746288
Kumbhar SB, Ghosh JS, Samudre SP (2010) Occurrence of disease causing organisms including bacteriophages in indigenous fermented milk products. Adv J Food Sci Technol 2:196–199
Landge UB, Pawar BK, Choudhari DM (2011) Preparation of shrikhand using ashwagandha powder as additive. J Dairy Foods Home Sci 30:79–84
Lembhe AF, Ranganathan B (1968) Occurrence of lactose fermenting yeasts in milk and milk products. Annual Report. National Dairy Research Institute, Karnal, p 32
Mane RR, Padghan PV, Patil S, Kamte R (2015) Studies on preparation of shrikhand from sweet corn milk blended with buffalo milk. Int J Food Nutr Sci 4:3
Md-Ansari IA, Rai P, Sahoo PK, Dutta AK (2006) Manufacture of shrikhand from ultra filtered skim milk retentates. J Food Sci Technol 43:49–52
Mehrotra R, Singh D, Tiwari A (2014) Physico-chemical analysis of low calorie high protein shrikhand prepared using stevia leaf powder. Innov J Food Sci 2:26–28
Mitra S, Chakrabartty PK, Biswas SR (2010) Potential production and preservation of dahi by *Lactococcus lactis* W8, a nisin-producing strain. Food Sci Technol 43:337–342
Mohania D, Kansal VK, Kumar M, Nagpal R, Yamashiro Y et al (2013a) Modulation of expression of Programmed Death-1 by administration of probiotic Dahi in DMH-induced colorectal carcinogenesis in rats. Acta Biomeica 84:102–109
Mohania D, Kansal VK, Sagwal R, Shah D (2013b) Anticarcinogenic effect of probiotic dahi and piroxicam on DMH-induced colorectal carcinogenesis in Wistar rats. Am J Pharmacol Toxicol 8:24
Mohania D, Kansal VK, Kruzliak P, Kumari A (2014a) Probiotic dahi containing *Lactobacillus acidophilus* and *Bifidobacterium bifidum* modulates the formation of aberrant crypt foci, mucin-depleted foci, and cell proliferation on 1,2-dimethylhydrazine-induced colorectal carcinogenesis in wistar rats. Rejuvenation Res 17:325–333

Mohania D, Kansal VK, Sagwal R, Kruzliak P (2014b) Probiotic LAVK2 dahi improves lipid profiles in hypercholesterolemic rats. Adv Dairy Res 2:113

Moktan B, Roy A, Sarkar PK (2011) Antooxidant activities of cereal–legume mixed batters as influenced by process parameters during preparation of dhoka and idli, traditional steamed pan cakes. Int J Food Sci Nutr 62(4):360–369

Nadaf NF, Patil RS, Zanzurne CH (2012) Effect of addition of gulkand and rose petal powder on chemical composition and organoleptic properties of Shrikhand. Recent Res Sci Technol 4:52–55

Narayanan R, Lingam J (2013) Sensory analysis of banana blended shrikhand. Afr J Agric Res 8:5518–5521

Nigam N, Singh R, Upadhayay PK (2009) Incorporation of chakka by papaya pulp during the manufacture of shrikhand. J Dairy Foods Home Sci 28:115–118

Padghan PV, Patil S, Jaybhaye RV, Katore VD, Deshmukh N (2015) Studies on cost of production of sweet corn milk and its blended milk products. J Ready Eat Food 2:51–55

Paltani IP, Goyal GK (2007) Packaging of dahi and yoghurt—a review. Indian J Dairy Sci 60:1–10

Panjagari NR, Singh RRB, Singh AK (2016) Chapter 7. Indian traditional fermented dairy products. In: Kristbergsson K, Oliveira J (eds) Traditional foods: general and consumer aspects. Springer, New York, p 105

Para PA, Razvi R, Nisar NA (2014) Effect of orange pulp and chiku pulp in combination (1:1) on the quality characteristics of shrikhand. World J Dairy Food Sci 9:135–137

Prajapati JB, Nair BM (2008) The history of fermented foods. In: Farnworth ER (ed) Hand book of fermented functional foods. CRC Press, Boca Raton, pp 2–6

Prajapati JB, Sreeja V (2013) Dahi and related products. New Delhi Publishers and SASNET-Fermented Foods, New Delhi. ISBN: 978-93-81274-31-6

Prasad G, Khan BL, Kulshreshtha DC (1980) Survival of *Escherichia coli* and *Enterobacter aerogenes* in dahi. I. Contamination of milk before preparation of dahi. Indian J Dairy Sci 30:497

Prasad G, Khan BL, Kulshreshtha DC (1984) Survival of *Escherichia coli* and *Enterobacter aerogenes* in dahi. II. Contamination after curdling of milk. Indian J Dairy Sci 37:261

Pushkala R, Srividya N (2011) Influence of aloe gel enrichment on the physicochemical, textural and sensory characteristics of dahi. J Food Sci Eng 1:141–153

Pushpangadan P, Dan VM, Ijinu T, George V (2012) Food, nutrition and beverage. Indian J Tradit Knowl 11:26–34

Rajmohan S, Prasad V (1994) Potassium sorbate as an antimycotic agent in enhancing the keeping quality of dahi. Cheiron 23:26–31

Rajpal S, Kansal VK (2008) Buffalo milk probiotic dahi containing *Lactobacillus acidophilus*, *Bifidobacterium bifidum* and *Lactococcus lactis* reduces gastrointestinal cancer induced by dimethyl hydrazine dihydrochloride in rats. Milchwissenschaft 63:122–125

Rajpal S, Kansal VK (2009a) Probiotic dahi containing *Lactobacillus acidophilus*, *Bifidobacterium bifidum* stimulates immune system in mice. Milchwissenschaft 64:147–150

Rajpal S, Kansal VK (2009b) Probiotic dahi containing *Lactobacillus acidophilus*, B*ifidobacterium bifidum* stimulates antioxidant enzyme pathways in rats. Milchwissenschaft 64:287–290

Ramadasappa EG, Jaganath EJ, Anantharamaih SN (1968) Non-lactic contamination in dahi. Annual Report. National Dairy Research Institute, Karnal, p 81

Ranjan P, Arora S, Sharma GS, Sindhu JS, Singh G (2006) Sensory and textural profile of curd (dahi) from calcium enriched buffalo milk. J Food Sci Technol 43:38–40

Rathi SD, Deshmukh DK, Ingle UM, Syed HM (1990) Studies on the physicochemical properties of freeze dried dahi. Indian J Dairy Sci 43:249–251

Roy A, Moktan B, Sarkar PK (2009a) Survival and growth of foodborne bacterial pathogens in fermenting batter of dhokla. J Food Sci Technol 46(2):132–135

Roy A, Moktan B, Sarkar PK (2009b) Survival and growth of foodborne bacterial pathogens in fermenting batter of Dhokla. J Food Sci Technol 46(2):132–135

Saha AL, Ganguli NC (1957) An outbreak of staphylococcal food poisoning from consumption of dahi. Indian J Public Health 1:22–26

Sanyal MK, Yadav PL, Dubey PC (1990) Use of preservatives for improving shelf life of curd (dahi). J Food Sci Technol 27:388–389
Sarkar S (2008) Innovations in Indian fermented milk products—a review. Food Biotechnol 22:78–97
Sarkar SP, Dave JM, Sannabhadti SS (1992) A note on the effect of thermization of mistidahi on the acid producers count. Indian J Dairy Sci 45:131–134
Sarkar P, Kumar LDH, Dhumal C, Panigrahi SS, Choudhary R (2015) Traditional and ayurvedic foods of Indian origin. J Ethnic Foods 2:97–109
Sathe GB, Mandal S (2016) Fermented products of India and its implication: a review. Asian J Dairy Food Res 35(1):1–9. https://doi.org/10.18805/ajdfr.v35i1.9244
Sekar S, Mariappan S (2007) Usage of traditional fermented products by Indian rural folks and IPR. Indian J Tradit Knowl 6(1):111–120
Senan S, Prajapati JB, Joshi C, Sreeja V, Gohel M, Trivedi S, Patel R, Pandya H, Singh U, Phatak A, Patel H (2015) Geriatric respondents and non-respondents to probiotic intervention can be differentiated by inherent gut microbiome composition. Front Microbiol 6:944
Senapati AK, Rosma A, Nadiah AHS et al (2016) Quality and safety of indigenous fermented foods. In: Joshi VK (ed) Indigenous fermented foods of South Asia. CRC Press, New York, p 289
Shah NP, Prajapati JB (2014) Effect of carbon dioxide on sensory attributes, physico-chemical parameters and viability of probiotic *L. helveticus* MTCC 5463 in fermented milk. J Food Sci Technol 51:3886–3893
Sharma DK, Reuter H (1998) Ultrafiltration technique for shrikhand manufacture. Indian J Dairy Sci 45:209–213
Sharma NK, Gill JPS, Joshi DV, Kwatra MS (1993) Microflora of Indian fermented milk products (dahi). Indian J Dairy Sci 46:85–87
Shelke PA, Shegokar SR, Shelke RR, Kahaate PA, Chavan SD (2014) Studies on preparation of low fat, sugar free mango shrikhand. Res J Animal Husband Dairy Sci 5:122–125
Shrestha R, Mehta M, Shah G (2017) Isolation and characterization of Khaman fermenting microorganisms. Trends Biosci 10(8):1574–1576
Shruti S, Kansal VK (2011) Effect of feeding probiotic dahi *containing Lactobacillus acidophilus* and *Bifidobacterium bifidum* on enzymes that catalyze carcinogen activation and detoxification in rats. Milchwissenschaft 66:244–247
Singh R (2006) Characteristic and technology of traditional Indian cultured dairy products. Indian Dairyman 58:49–56
Singh BH, Anantharamaiyya SN (1968) Manufacture and storage of dahi and fermented milks. Annual Report. NDRI, Karnal
Singh R, Kansal VK (2003) Augumentation of immune response in mice fed with dahi—a fermented milk containing *Leuconostoc citrovorum* and *Lactococcus lactis*. Milchwissenschaft 58:480–482
Singh D, Singh J (2014) Shrikhand: a delicious and healthful traditional Indian fermented dairy dessert. Trends Biosci 7:153–155
Singh BH, Jagannath EG, Anantharamaiyya SN, Ananthakrishnan CP (1970) Studies on the preparation and storage of dahi and fermented milks at room temperature. XVII Int Dairy Cong 16:410. Cited from *Dairy Sci. Abstr.* 32, 4649
Singh G, Arora S, Sharma GS, Sindhu JS, Rajan P, Kansal V (2005) Sensory and textural profile of dahi from calcium enriched cow milk. Indian J Dairy Sci 58:23–26
Singh D, Singh J, Kumar S, Verma T (2014) Microbiological evaluation of soy fortified shrikhand by using response surface methodology. Int J Appl Biol Pharm Technol 5:1–7
Sinha PRV, Yadav H (2007) Antiatherogenic effect of probiotic dahi in rats fed cholesterol enriched diet. J Food Sci Technol 44:127–129
Sivakumar PS, Panda SH, Ray RC, Pradhan DC, Sivaramane N (2008) Modeling consumer acceptability of b-carotene rich sweet potato curd. J Sens Stud 23:791–803
Sonawane VK, Chavan KD, Pawar BK (2007) Effect of levels of strawberry pulp and sugar on chemical composition during storage of Shrikhand. J Dairy Foods Home Sci 26:153–158

Tamang JP, Watanabe K, Holzapfel WH (2016) Review: diversity of microorganisms in global fermented foods and beverages. Front Microbiol 7:377
Tiwari MP, Singh IP (1964) Survival of pathogens in dahi. I. Contamination of milk before the preparation of dahi. Indian J Dairy Sci 17:97–101
Tiwari MP, Singh IP (1966) Survival of pathogens in dahi. II. Contamination after the curdling of milk. Indian J Dairy Sci 19:162–164
Vagdalkar AA, Havan BR, Morkile VM, Thalkari BT, Landage SN (2002) A study on preparation of Shrikhand by using cocoa powder and papaya pulp. Indian Dairyman 54:49–51
Yadav JS, Grover S, Batish VK (1989) Role of marketing and related factors in the microbiological quality of milk and milk products. Indian Dairyman 41:527–534
Yadav JS, Grover S, Batish VK (1993) A comprehensive dairy microbiology. Metropolitan Pub, New Delhi, p 764
Yadav H, Jain S, Sinha PR (2005) Preparation of low fat probiotic dahi. J Dairy Foods Home Sci 24:172–177
Yadav H, Jain S, Sinha PR (2006) Effect of dahi containing *Lactococcus lactis* on the progression of diabetes induced by a high fructose diets in rats. Biosci Biotechnol Biochem 70:1255–1258
Yadav H, Jain S, Sinha PR (2008a) Oral administration of dahi containing probiotic *Lactobacillus acidophilus* and *Lactobacillus casei* delayed the progression of streptozotocin-induced diabetes in rats. J Dairy Res 75:189–195
Yadav H, Jain S, Sinha PR (2008b) The effect of probiotic dahi containing *Lactobacillus acidophilus* and *Lactobacillus casei* on gastropathic consequences in diabetic rats. J Med Food 11:62–68

Further Reading

Links

https://www.everyculture.com/wc/Germany-to-Jamaica/Gujaratis.html
https://www.census2011.co.in/census/state/gujarat.html
https://www.britannica.com/place/Rajasthan
https://www.census2011.co.in/census/state/rajasthan.html
https://www.ohmyrajasthan.com/rajasthani-cuisine-history

Ethnic Fermented Foods and Beverages of Himachal Pradesh

8

S. S. Kanwar and Keshani Bhushan

Abstract

Himachal Pradesh is situated in the lap of the Western Himalayas with a diverse culture and cuisine. Mostly preparation and usage of fermented foods and alcoholic beverages are associated with tribal areas of the region. Some ethnic fermented foods, commonly prepared in Himachal Pradesh, are *bhaturu*, *manna*, *marchu*, *siddu*, *dosha*, *chilra (lwar)*, *bedvin roti*, *gulgule*, *seera*, *pinni/bagpinni*, *pakk*, *thuktal*, *borhe*, *sepubari*, *bari*, *churpa*, and *aska*. Mostly these fermented foods are prepared from cereals, viz., wheat, barley, and buckwheat, but some are prepared from legumes and milk too. Traditional fermented alcoholic beverages of Himachal Pradesh include *sura*, *chhang/lugri*, *angoori/kinnauri*, *daru/chakti*, *chulli*, *behmi*, and *arak/ara* which are prepared from cereals (rice/barley), fruits (apple/wild apricot/grapes), jaggery, and millets. These ethnic fermented foods and alcoholic beverages are especially used during the local festivals, religious ceremonies, and marriage functions in the rural and urban areas of the state and are an important part of the day-to-day life of people of Himachal Pradesh.

Keywords

Himachal Pradesh · *Saccharomyces cerevisiae* · Ethnic fermented foods and alcoholic beverages

S. S. Kanwar (✉)
Department of Microbiology, CSK Himachal Pradesh Agriculture University, Palampur, Himachal Pradesh, India

K. Bhushan
Department of Microbiology, Punjab Agricultural University, Ludhiana, Punjab, India

J. P. Tamang (ed.), *Ethnic Fermented Foods and Beverages of India: Science History and Culture*, https://doi.org/10.1007/978-981-15-1486-9_8

8.1 Introduction

The Western Himalayan region covers Jammu and Kashmir, Himachal Pradesh, and Uttaranchal hills. Himachal Pradesh, situated in the lap of Western Himalayas, has majestic mountains, fertile valleys, perennial rivers, precious forests, invaluable flora and fauna, tremendous wealth of resources, minerals, very rich culture, and diverse customs and traditions. The state of Himachal Pradesh is situated between 30° 22′ 44″ and 33° 12′ 40″ N latitude and 75° 45′ 55″ to 79° 04′ 20″ E longitude. The geographical area of the state is 55,673 km^2, while the population is 68.64 lakhs (Census 2011). Himachal Pradesh is a hilly state with a general increase in elevation from west to east and south to north ranging from 350 m to 6975 m (Kayastha 1971). Its one-third area remains snow covered for about 7 months in a year. Due to its difficult terrains, it is mostly remained untouched by external customs.

Apart from national fairs and festivals, numbers of other festivals are celebrated nearly in every region of Himachal Pradesh which are of great significance to the local people. During these festivals, indigenous fermented foods and beverages are prepared by locals to mark the celebrations. These products are prepared by native people using their inherited traditional knowledge from locally available plant or animal raw materials. Fermented product is prepared either naturally or by adding starter culture/local inoculum to transform the substrate into an edible product that is ethnically and socially conventional to the local people (Rawat et al. 2018).

8.2 Food Culture and Ethnic Fermented Foods of Himachal Pradesh

The day-to-day food of *Himachalis* is very similar to the rest of the North India; however, there are certain special fermented foods that are prepared by different ethnic tribes of the state which are specific to a particular region. Some of the specialties of Himachal Pradesh are *sepubari*, *chilra*, *seera*, *bhaturu*, *babroo*, *mukund bari*, etc. Inhabitants of tribal areas of Himachal Pradesh routinely consume these indigenous products which are mainly of cereal-based (Kanwar et al. 2007). Savitri and Bhalla (2007) studied a wide range of traditional fermented foods and beverages of Himachal Pradesh (Kinnaur) which constitute a part of staple food consumed during marriages, local festivals, and special occasions. *Bhaturu*, *siddu*, *marchu*, *seera*, *chilra*, *manna*, *aenkadu*, *sepubari*, *patande*, *doo*, *baari*, *dosha*, *malpude*, *babroo*, *bedvin roti*, *madrah*, *tchati*, *churpa*, *sura*, *chhang*, *kinnauri*, *angoori*, *chulli*, *lugri*, *arak/ara*, *rak*, *chukh*, and pickles made from different fruits, vegetables, and cereals are some of the popular traditional products in tribal areas of Himachal Pradesh.

These fermented foods and beverages are important constituents of diet of rural people especially in tribal belts of Lahaul and Spiti, Kinnaur, Chamba, Kullu, Shimla, Mandi, and Kangra districts of Himachal Pradesh. Mostly fermented food products are acidic in nature. Microbiological examination of these fermented food products and their sources of inocula reveals the dominance of yeasts mainly from the genera *Saccharomyces*, *Debaryomyces*, and *Schizosaccharomyces* and bacteria from the genera *Lactobacillus*, *Lactococcus*, and *Leuconostoc* (Kanwar et al. 2007). These organisms are generally contributed by the food ingredients, workers, and/or surroundings (Venkatasubbaiah et al. 1984). Organisms associated with fermentation of these products may produce desirable number of enzymes to degrade antinutritive compounds to enhance their acceptability with improved flavor and aroma (Kumari et al. 2015). As the source of inoculum is not apparent, considerable attention has been attracted toward exploring the microbes actually associated with these fermentations over the years. In one of the studies, Pathania et al. (2010) did the microbiological exploration of various fermented food products (*babru*, *bhaturu*, and *chilra*), alcoholic beverages (*chhang*, *lugari*, *aara*, *chiang*, *apple wine*, and *chulli*), and traditional inocula (*khameer*, phab, and *dhaeli*) of tribal areas of Himachal Pradesh representing regions of Lahaul and Spiti, Sangla, Bharmour, Pangi, and Kinnaur. They showed the dominance of yeasts in the microflora of these products. As these isolates were of different geographical origin, and from different sources, then Keshani et al. (2015) further explored these yeasts isolates (*S. cerevisiae*) with respect to existence of strain level differences as well as functional diversity. These yeast strains also showed variability during the production of apple cider (Kanwar and Keshani 2016). Tamang et al. (1996) highlighted the nutritional importance of these foods in providing calories, proteins, vitamins, and minerals to the local people.

Most of the traditionally fermented alcoholic beverages are prepared by cooking grains and then inoculating them with traditional inocula *phab/dhaeli*. There are number of different traditional alcoholic beverages prepared in tribal belt of Himachal Pradesh, viz., *chhang*, *lugari*, *aara*, *chakti*, *daru*, chiang, etc. In a study conducted by Kanwar et al. (2011), the chemical analysis of some of the traditional alcoholic beverages of Lahaul and Spiti area of Himachal Pradesh revealed the acidic nature of these beverages. The ethanol content ranged from 5% to 12% (v/v) in case of undistilled and 13% to 19% in case of distilled beverages. Some other chemical constituents, viz., acetaldehyde, methanol, ester, *n*-propanol, etc., were also observed in distilled alcoholic beverages. Microbiological examination of these beverages and their source of inocula revealed the dominance of yeasts mainly from genera *Saccharomyces* and *Endomyces* and bacteria from genera *Lactobacillus*, *Acetobacter*, and *Bacillus*. In cold desert region of Himachal Pradesh, the alcoholic beverages are believed to be good tonic. Apart from having ethnic value, these beverages do have some religious touch also.

In spite of scientific and technological revolution, the art of fermentation practiced by common man has continued but mostly in rural areas due to several reasons like (1) inaccessibility of the industry-made products in these secluded areas, (2i) accepted taste of traditional fermented products by the people, and (3) their sociocultural linkages with such products (Thakur et al. 2004). Specific microflora is associated with the production of indigenous fermented foods which involves a combined action of bacteria, yeast, and fungi; however, this may vary from region to region, village to village, and family to family (Kanwar and Keshani 2014). Indigenous fermented foods are an intrinsic part of diet of the ethnic tribes in the

Himalayan belt of India, being the oldest and most economic methods for biological enrichment of food products by the manipulation of different microbial populations (Nehal 2013). Most of the traditional fermented foods are prepared by processes of solid substrate fermentation in which the substrate is allowed to ferment either naturally or by added starter cultures. The consumption of fermented foods and beverages has been the integral part of life of people living in the rural and tribal areas of Himachal Pradesh (Thakur et al. 2003). The know-how of processes involved in the preparation of fermented products is being transferred from generations among tribal and rural people. The locally available raw materials and traditional equipment are being utilized by folk people as these are affordable and easy to maintain (Thakur et al. 2004).

People believe that these traditional fermented food products possess the potential to improve their health apart from providing nutrients. The positive role of fermented foods in modulating complex gastrointestinal functioning and imparting physiological benefits is chiefly ascribed to their microbial components (Ghosh 2012). A variety of indigenous fermented food products and beverages of Northwestern Himalayas have been documented with respect to their substrates and probiotic diversity (Sourabh et al. 2010, 2011). The property of probiotic microorganisms to act as DNA-bioprotective agents is of great significance in terms of food quality and its relationship with health. In one of the studies conducted by Walia et al. (2014), 11 indigenous potential probiotic bacteria and 6 yeast isolates obtained from traditional fermented foods of the Northwestern Himalayas were screened for antigenotoxicity and antimutagenicity. These activities were very much comparable with the reference strain *Lb. rhamnosus* GG (ATCC 53103) used in their study. These strains were further studied for the effect of prebiotics on their growth behavior and for their adherence potential (Sharma and Kanwar 2017, 2018). The probiotic microorganisms also affect the gut microbiota and, thus, affect the host behavior in a positive way (Kanwar et al. 2015). Hence, the enrichment of traditional fermented foods with such type of indigenous probiotics can play a major role in imparting the health benefits to the locals. Some of the information pertaining to these foods is summarized in Table 8.1.

8.3 Ethnic Fermented Foods

Mostly the fermented foods are prepared from cereals, viz., wheat, barley, and buckwheat, and their cuisine is unique to Himachal Pradesh (Kanwar et al. 2007). In Himachal Pradesh, large variety of fermented foods are prepared in the tribal districts either daily or during particular occasions or for the consumption during journey (Sharma et al. 2013). Traditional starter cultures, viz., *malera* and *treh*, are used as inocula in these fermented foods. Though natural fermentation method is also used in the preparation of several products like seera, sepubari, borhe, etc. (Thakur et al. 2004) but mostly appropriate starter cultures are being used for the preparation of most of the fermented food products. Some of the cereal-based traditional fermented products are discussed as follows:

Table 8.1 Fermented beverages and foods of Himachal Pradesh

Product	Substrate	Nature with use	References
Aska	Rice flour	Dish served during festivals	Thakur et al. (2004)
Babru	Wheat or rice flour	A type of pancake, used in breakfast or as a snack	Kanwar et al. (2007)
Bagpinni	Roasted barley flour with *Chhang*	It is a ball-like snack	Savitri and Bhalla (2007)
Bedvin roti	Wheat flour, black gram, and walnut	Baked and staple food	Thakur et al. (2004)
Bhatabaru	Wheat flour, milk, sugar	Sweet, deep fried, oval shaped snack	Sharma et al. (2014)
Bhatooru/bhaturu	Wheat or barley flour	Baked and staple food	Savitri and Bhalla (2012)
Borhe	Black gram	Snack as well as staple food	Thakur et al. (2004)
Chilra	Wheat and Buckwheat flour	Like *dosa*; staple food	Thakur et al. (2004)
Dangal bari	Black gram, dangal, spices	Staple food	Sharma et al. (2013)
Gulgule	Wheat flour	Cooked as flat pancake, snack	Thakur et al. (2004)
Jhan Chang	Barley grains, *Phab* as inoculum	Halwa-like preparation, snack food	Kanwar et al. (2007)
Marchu	Wheat flour	Baked breads; staple food	Thakur et al. (2004)
Mash bari	Black gram, spices	Staple food	Sharma et al. (2013)
Pakk	Barley, butter, and lassi	Staple food	Thakur et al. (2004)
Seera	Wheat grains	Dried sweet dish	Thakur et al. (2004), Savitri et al. (2012)
Siddu	Wheat flour, walnut/opium seeds/black gram	Steamed bread, oval shaped; staple food	Thakur et al. (2004)
Tchog	Roasted barley flour, *chhang*	Solid ball-like, staple food	Savitri and Bhalla (2007)
Thuktal	Barley	Served as snack	Thakur et al. (2004)
Sepubari	Black gram	Fermented paste, staple food as well as side dish	Thakur et al. (2004)
Chhurpi (hard variety)	Cow or yak milk	Hard mass, masticator	Katiyar et al. (1991), Pal et al. (1996)
Kachauri	Wheat flour, urad dal	Staple as well as snack food	Sharma et al. (2014)
Khoru	Buttermilk and spices	Accompaniment to the main diet	Sharma et al. (2013)
Angoori/kinnauri	Grapes	Distilled drink	Thakur et al. (2004)
Ark/ara	Barley, apple, wild apricot	Light brown drink	Kanwar et al. (2007)

(continued)

Table 8.1 (continued)

Product	Substrate	Nature with use	References
Behmi	Behmi, apple, chulli	Alcoholic drink	Thakur et al. (2004)
Chakti	Jaggery	Filtrate drink	Thakur et al. (2004)
Chhang	Finger millet/barley and *phab*	Mild alcoholic, slightly sweet acidic beverage	Savitri and Bhalla (2007)
Chulli	Apricot	Filtrate drink	Thakur et al. (2004)
Daru	Cereal	Filtrate jaggery	Sekar and Mariappan (2007)
Ghanti	Apple and apricot	Distillate drink	Roy et al. (2004)
Lugri	Barley	Alcoholic beverage	Thakur (2013)
Rak	Apple, wild almond, apricot, and jaggery and *phab*	Filtrate drink	Thakur et al. (2004)
Soor	Fruits and *keem*	Distillate drink	Rana et al. (2004)
Sura	Finger millet	Alcoholic beverage	Thakur et al. (2004)
Malera/treh	Wheat flour	Starter to ferment *bhaturu/* chilra, etc.	Savitri and Bhalla (2007)
Phab	Rice and leaves of *Solanum khasianum*	Starter to ferment alcoholic beverages—*chhang*, *arak*, *kinnauri*	Angmo and Bhalla (2014)
Keem	Wheat and plants	Starter to ferment alcoholic beverages—*soor*	Rana et al. (2004)
Dhehli	Mixture of 36 herbs and roasted barley flour	Starter to ferment alcoholic beverage—*sura*	Thakur et al. (2004), Savitri and Bhalla (2007)

8.3.1 Siddu

It is also known as *khobli* in the capital of Himachal Pradesh (Shimla). *Siddu* is a traditionally fermented steam cooked dish, prepared as a delicacy in Kullu and Shimla districts.

Figure 8.1 flow sheet as follows:

Culinary and Mode of Consumption *Siddu* is dish up hot with *chutney* or *desi ghee* (Thakur et al. 2004).

Socioeconomy and Ethnical or Religious Values It is prepared as a special dish in Kullu, Manali, and Shimla districts.

Microorganisms Unknown

Wheat flour

↓

←Inoculum (*malera*)

↓

Dough preparation

↓

5–6 h dough fermentation

↓

Prepare dough balls stuffed with spices, mixed paste of walnut, opium seeds, and black gram

↓

Steam the *siddu* balls for 15 min

Fig. 8.1 Traditional method of preparation of *siddu* in Himachal Pradesh

Wheat/barley and buckwheat flour

↓ (Inoculum (*treh*))

Dough preparation

↓

5–6 h dough fermentation in traditional bucket shaped wooden vessel "Lwarenza"

↓

Spread on hot plate and baked on both sides

Fig. 8.2 Traditional method of preparation of *chilra* in Himachal Pradesh

Nutritional Value Opium seeds and ghee provide fat and maintain the body temperature during winters.

Optimization and Commercialization Familiarity of *Siddus* is limited as compared to other Indian cuisines. They are served in the local markets of Shimla and Kullu districts of Himachal Pradesh.

8.3.2 Chilra

Chilra is prepared through fermentation of wheat/barley and buckwheat flour. It is preferred due to its easy digestibility.

Figure 8.2 flow sheet as follows:

Culinary and Mode of Consumption *Chilra* is consumed in Lahaul and Spiti area of Himachal Pradesh. It is mainly consumed during festivals.

Wheat flour, black *jeera,* and salt

↓ (Inoculum (*malera*))

Dough preparation

↓

Fermentation

↓

Rotis are prepared with designed imprints on a wooden base with carving

↓

Deep fried in mustard oil

Fig. 8.3 Traditional method of preparation of *marchu* in Himachal Pradesh

Socioeconomy and Ethnical or Religious Values It is served during local festivals.

Microbiology Yeasts isolated from *Chilra* are *S. cerevisiae*, *Debaromyces hansenii*, and *Schizosaccharomyces* sp. (Kanwar et al. 2007).

8.3.3 Marchu

Marchu is baked staple bread prepared from wheat flour. It is mainly used in Lahaul and Spiti. It is served during the local festivals (*Phagli*, *Halda*).

Figure 8.3 flow sheet as follows:

Culinary and Mode of Consumption *Marchu* is salted *roties*, served with tea as a snack or breakfast in Lahaul and Spiti and Kinnaur districts of the state (Thakur et al. 2004).

Socioeconomy and Ethnical or Religious Values It is prepared during the local festivals, religious, and marriage ceremonies (*Phagli*, *Halda*) in Lahaul valley.

Microorganisms Unknown

8.3.4 Bhaturu

Bhaturu is an "indigenous bread" also known as *sumkeshi roti* in Lahaul region, which serves as a staple diet of the people of Himachal Pradesh especially in rural areas of Mandi, Kullu, Chamba, Shimla, and Kangra districts. The normal fermentation time is 2–3 h and 4–5 h in summers and winters, respectively. To reduce the fermentation time, people knead flour with warm water and more inoculum.

Figure 8.4 flow sheet as follows:

Fig. 8.4 Traditional method of preparation of *bhaturu* in Himachal Pradesh

Wheat flour

↓ (Inoculum (*malera*))

Dough preparation

↓

3–5 h dough fermentation

↓

Fermented dough makes into *rotis*

↓

Bake/deep fry after half an hour

8.3.5 Bhaturu

Culinary and Mode of Consumption *Bhaturu* serves as a staple diet for rural people, and they also take it during meals with vegetables or curries (Sharma et al. 2013).

Socioeconomy and Ethnical or Religious Values *Bhaturu* is prepared during marriage ceremonies especially at night and during religious days and local festivals. It is easily digestible.

Microbiology *S. cerevisiae* and *Lactobacillus* sp. have been isolated from *malera*, the traditional inoculum used for *bhaturu* (Thakur et al. 2004).

8.3.6 Seera

Figure 8.5 flow sheet as follows:

Soak wheat grains in water for 2–3 days

↓ Fermentation (natural flora)

Grains are ground, steeping is done, and bran is removed

↓

Grounded material is sun-dried along with the removal of starch and proteins

↓

This dried matter is "seera"

Fig. 8.5 Traditional method of preparation of *seera* in Himachal Pradesh

Culinary and Mode of Consumption Dried *seera* is first soaked in water to make slurry and is then poured into hot *ghee*. For sweetness, sugar is added, cooked, and served as snack. *Seera* is nutritious and easily digestible snack food.

Socioeconomy and Ethnical or Religious Values It is usually offered to the guests as a sweet dish in Mandi, Kullu, Kangra, and Chamba districts of Himachal Pradesh (Kanwar and Sharma 2011). People suffering from jaundice/hepatitis usually consume *seera*. It is also used during fast.

Microbiology The microflora isolated from *seera* comprised mainly of yeasts, viz., *Saccharomyces cerevisiae*, *Cryptococcus laurentii*, and *Torulaspora delbrueckii*; bacterial population includes *Lb. amylovorus*, *Cellulomonas* sp., *Staphylococcus sciuri*, *Weissella cibaria*, *Bacillus* sp., *Leuconostoc* sp., and *Enterobacter sakazakii* (Savitri et al. 2012).

Soak urd bean or seeds of black gram in water for 6–12 h

↓

Dewatered, dehulled, and ground into a smooth, mucilaginous paste

↓ (Add inoculum from previous batch)

Mix the dough with salt, and spices including asafetida, caraway, cardamom, clove, fenugreek, ginger, and red pepper

↓ (Fermentation for 1–3 days)

Molded into balls, and dried for 2–8 days on bamboo or palm mats

Fig. 8.6 Traditional method of preparation of *wari* in Himachal Pradesh

8.3.7 Wari

Wari is an ethnic Indian fermented black gram or urd bean product. It is a dry, hollow, brittle, spicy, and friable ball with 3–8 cm in diameter and 15–40 g in weight (Batra 1986).

Figure 8.6 flow sheet as follows:

Culinary and Mode of Consumption *Wari* is used as condiment and is mixed with vegetables as side dish; nowadays it is also taken as main dish with rice or *roti*.

Microbiology Yeast species of *Candida krusei*, *C. vartiovaara*, *Kluyveromyces marxianus*, *Trichosporon beigelii*, *Hansenula anomala*, *S. cerevisiae*, *Leuc. mesenteroides*, and *Lb. fermentum* have been isolated from *wari* (Batra and Millner 1976; Batra 1986).

Nutritional Value Increase in total acids, soluble nitrogen, free amino acids, thiamine, riboflavin, and cyanocobalamin has been observed during *wari* fermentation (Soni and Sandhu 1990).

8.3.8 Bhatabaru

It is a sweet deep fried oval-shaped snack, mostly prepared during religious ceremonies.

Figure 8.7 Flow sheet as follows:

Culinary and Mode of Consumption It is consumed as a sweet snack in different parts of Himachal Pradesh.

Socioeconomy and Ethnical or Religious Values *Bhatabaru* are prepared and served during religious and social ceremonies around the state.

Microorganisms Unknown

Certain other traditional fermented foods of Himachal Pradesh include as follows: the leaves of buckwheat are mixed with wheat flour and made into cakes, *akhtori*; *gulgule* are sweet snacks prepared from batter of wheat flour with sugar that are served during religious and marriage ceremonies both in rural and urban areas; *patrodu* is prepared from the leaves of *Colocasia esculenta*, the leaves are rolled with slurry of gram flour, steamed cooked, and then fried; and *kadu* or *kheeru* is a buttermilk-based dish that is prepared in most parts of the state. It is prepared by cooking the buttermilk or *chha* with spices and *besan*; *nudu* is a ceremonial food prepared by cooking wheat flour in milk with small amount of salt to be eaten with *ghee* (Savitri and Bhalla 2007). Kachauri is a modified form of bhaturu and is prepared by stuffing the dough with spicy paste of urad dal (Vigna mungo). *Kachauris* are simply baked or deep fried stuffed *roties*. These are usually taken as snacks with tea. Some modifications have been made in *kachauri* such as along with dal, opium seeds (*Papaver somniferum*); dry fruits and ground soybean (*Glycine max*)/kulath (*Macrotyloma uniflorum*) pulses are also added.

Wheat flour is kneaded into dough by adding milk, water, and sugar

↓ (Fermentation by using *malera* as inoculum)

Small oval shape balls are prepared from sweetened dough

↓

Deep fried until brown

Fig. 8.7 Traditional method of preparation of *bhatabaru* in Himachal Pradesh

Jhol is a curry prepared by adding spices and then boiling it. For its preparation, freshly chopped coriander leaves, cumin seeds, and salt are mixed with the curd/*lassi*/buttermilk. The mixture is constantly stirred and heated till the mixture changes color from white to pale yellow to bright yellow. It is a very healthy hot curd soup and is also used as medicine during cold and in some stomach ailments (Sharma et al. 2014). Microbiological exploration of inocula of these fermented foods as carried out by Sharma et al. (2014) and Thakur et al. (2004) reveals the presence of *Saccharomyces cerevisiae*, *Candida* sp., *Zygosaccharomyces bisporus*, and *Kluveromyces thermotolerance* along with bacteria dominated by *Leuconostoc* and *Lactobacillus*. Traditional fermentation of food substrates enriches them with protein, essential amino acids, and vitamins, thereby enhancing their nutritive value. Dough is enriched in vitamin B complex especially thiamine and riboflavin and antioxidants due to the growth of microflora especially yeast during fermentation (Sharma et al. 2013).

8.4 Traditional Alcoholic Beverages of Himachal Pradesh

There are a number of traditional alcoholic beverages popular among *Himachalis*. These are mostly prepared from cereals (rice and barley), fruits (grapes, wild apricot, and apple), jaggery, and millets. These alcoholic beverages are consumed on the regular basis and also during functions. Some of the common traditional alcoholic beverages are *sura*, *daru/chakti*, *chhang/lugri*, *chulli*, *angoori/kinnauri*, *behmi*, and *arak/ara* (Kanwar et al. 2011). *Phab* is used as inoculum for the fermentation and preparation of these fermented beverages except *sura*. *Sura* is a prepared by natural fermentation (Thakur et al. 2004). Some of the traditional alcoholic beverages are discussed as follows:

8.4.1 Chhang

Chhang is prepared from rice and is an indigenous beer. It is produced in the tribal belt of Lahaul and Spiti districts of Himachal Pradesh. Solid-state fermentation is carried out for the production of *chhang* since no additional water is added. It is also known as *lugri* in Lahaul region of Himachal Pradesh.

Figure 8.8 flow sheet as follows:

Fig. 8.8 Traditional method of preparation of *chhang* in Himachal Pradesh

Cooked rice/barley

↓ Fermentation (*phab*)

Fermentation for 1 week maximum

↓

Chhang is prepared

Culinary and Mode of Consumption The alcohol content in *chhang* is around 8.5% (Thakur et al. 2004). The conventional vessel made of stone or metal is used to store *chhang* and is known as "uthi" in Lahaul. It is served in traditional jugs called "chapkiayan." *Chhang* is mainly served during marriage ceremonies and traditional New Year of *Lahulis* called *Phagli.*

Socioeconomy and Ethnical or Religious Values It is sprinkled as *shagun* on guests (tribal custom) and thus has religious significance. In Kullu this fermented beverage is known as *jhol.* Its distilled form is known as *sra* in Lahaul valley. *Chhang* is rich in vitamins, amino acids, and sugars besides alcohol and, hence, serves as a tonic in winters. It is considered to provide protection against cold (Basappa 2002).

Microbiology Microorganisms identified from *chhang* are *Lb. plantarum*, *Lb. casei*, *Enterococcus faecium*, *Pediococcus pentosaceus*, *S. cerevisiae*, *S. fibuligera*, *Pichia kudriavzevii*, and *Candida tropicalis* (Kanwar et al. 2011; Thakur 2013).

8.4.2 Sura

Sura is a millet-based fermented beverage prepared from *kodra/kached* (*Eleusine coracana*) and is mostly produced in *Lug* valley of Kullu district. *Sura* is high in ethanol content having about 15% of ethanol (Thakur et al. 2004). This millet has long shelf life and is popular as the "famine grain." No specific inoculum is used for the preparation of *sura*. Natural microflora carries out the fermentation and starch-hydrolyzing activities during its production.

Figure 8.9 flow sheet as follows:

Method to Prepare *Dhehli* During fermentation *dhehli* is added, which is an herbal mix in *sattu* base. The preparation of *dhehli* is a community effort, in which elderly people go to the forests on 20th day of *Bhadrapada* month (usually 5 or 6th

Fig. 8.9 Traditional method of preparation of *Sura* in Himachal Pradesh

Kodra/kached (mixed with water)

↓

Natural fermentation

←Addition of *dhehli*

↓

Fermentation

↓

Sura

of September) and collect approximately 36 fresh herbs. After that, the herbs are crushed in a stone with a large conical cavity called "ukhal" using a wooden bar called "mussal." The extracts as well as plant biomass are added into the sattu and are roughly kneaded. This whole material is added into a wooden mold to give the shape of a brick, after that it is dried and is known as *dhehli*. It is evenly divided among the villagers and is used, whenever *sura* is to be prepared. The *dhehli* provides bioactive compounds as well as microflora for fermentation.

Culinary and Mode of Consumption *Sura* is high in ethanol content and is consumed in Chamba and Kullu districts specifically during the winter season.

Socioeconomy and Ethnical or Religious Values *Sura* is usually consumed during local festivals like marriage ceremonies and *shoeri saja* (Thakur et al. 2004).

Microbiology *S. cerevisiae* and *Zygosaccharomyces bisporus* yeasts have been isolated from *sura* (Thakur et al. 2004).

8.4.3 Daru/Chakti

Daru is a traditionally fermented beverage prepared from jaggery (ethanol content 5%) prepared and is consumed in rural belt of Kullu, Shimla, and other areas of Himachal Pradesh (Thakur et al. 2004). In Kullu district of Himachal Pradesh, it is also famous as *chakti*.

Figure 8.10 flow sheet as follows:

Culinary and Mode of Consumption *Daru* is served during marriage ceremonies and local festivals in the rural areas for entertaining guests. To give taste and aroma to *chakti*, a local babool wood called *kikar* (*Acacia nilotica*) is added during its preparation (Nehal 2013).

Fig. 8.10 Traditional method of preparation of *daru* in Himachal Pradesh

Gur (jaggery) mixed with water

↓ Fermentation

Fermented liquid

↓

Distillation

↓ Addition of burnt sugar for color

Daru

Socioeconomy and Ethnical or Religious Values *Daru* is one of the popular fermented alcoholic beverages of Himachal Pradesh and is especially used for entertaining guests during marriage ceremonies and local festivals.

Microbiology The yeasts isolated from *daru/chakti* are *S. cerevisiae*, *Candida famata*, *C. valida*, and *Kluveromyces thermotolerance* (Thakur et al. 2004).

8.4.4 Chulli

Chulli is prepared from the dried wild apricots (ethanol content 3%) and is indigenous to the tribal district of Kinnaur.

Figure 8.11 flow sheet as follows:

Culinary and Mode of Consumption *Chulli* is usually consumed during winter season especially in Kinnaur region of the state.

Socioeconomy and Ethnical or Religious Values *Chulli* forms an essential part of the social life of *Kinnaur*, as it is served to the guests during marriage ceremonies, fairs, and local festivals (Thakur et al. 2004).

Microbiology The main yeast associated with *chulli* fermentation is *S. cerevisiae* (Thakur et al. 2004).

Fig. 8.11 Traditional method of preparation of *chulli* in Himachal Pradesh

Crushed wild apricots mixed in water

↓(*Phab* inoculation)

Heat the mixture

↓ (Fermentation for 2–3 days)

Fermented liquid (*chulli*)

↓

Filtration

↓ (Fermentation for 10–15 days)

Distillation

↓

Distilled *chulli*

8.4.5 Angoori

Angoori is prepared from red and green grapes; in case of red grapes, the ethanol content of the beverage ranged about 5%, and in case of green grapes, it is about 3.5%. Different numbers of grape varieties are cultivated in Kinnaur and are used in the preparation of this local alcoholic beverage. *Angoori* is also called *kinnauri* in Kinnaur district.

Figure 8.12 Flow sheet as follows:

Socioeconomy and Ethnical or Religious Values *Angoori/kinnauri* is one of the very popular traditional fermented alcoholic beverages consumed during marriage and local festivals ceremonies by the people of that area (Thakur et al. 2004).

Microbiology *S. cerevisiae* is the main organism associated with fermentation of *angoori.*

8.5 Traditional Starter Cultures of Himachal Pradesh

Malera/khameer and *treh* are the traditional inocula used for preparation of some fermented foods. It is a portion of a previous leftover fermented dough or fermented slurry. Every time fermentation is carried out, some portion of dough/slurry is kept so that it can be used next time. The microbiological analysis of *malera* revealed that it is a consortium of microorganisms which mainly consisted of lactic acid bacteria and yeast. *Lb. plantarum* (MTCC8296), *Leuconostoc* sp. and *S. cerevisiae* (MTCC 7840) have been isolated from different samples of *malera* (Savitri and Bhalla 2012).

Phab/dhaeli is a starter culture or a traditional inoculum used for fermentation of alcoholic products. It is an indigenous inoculum of the Trans-Himalayan region. It

Fig. 8.12 Traditional method of preparation of *Angoori* in Himachal Pradesh

Grape juice
↓
Pasteurization
↓ (*Phab* inoculation)
Fermentation
↓
Fermented liquid
↓ (Distillation)
Brandy
↓ (Addition of burnt sugar)
Angoori

is mainly used for the fermentation of two barley-based alcoholic beverages, viz., *chhang* and *aarak*. It is prepared from barley/rice and leaves of *Artemesia* sp. *Phab* preparation is temperature dependent and involves microbial processes. *S. cerevisiae* and *Bacillus* sp. have been reported from *phab* (Thakur et al. 2004). An amylolytic yeast *Endomyces fibuliger* has also been isolated from *dhaeli* (*phab*) from Lahaul and Spiti area (Bhushan et al. 2012). It is generally prepared from mid-June to October when shrub *Artemesia* locally called *burnak* is fully grown. This shrub plays an important role in *phab* preparation; it provides the peculiar aroma to *phab* and has inhibitory effect on the growth of foodborne bacteria and allows the growth of only inherent microflora present in *phab*. For *phab* preparation, husked barley grains are roasted on fire and ground into coarse flour. The flour is kneaded into dough and rolled into small tablets, which are inoculated with powder from previous year *phab* and mixed properly, small amount of wheat flour is also mixed, and the tablets are kept between the leaves of *Artemesia*. These are kept as such for 2–3 days for fermentation and after that dried in shade for 5–7 days (Angmo and Bhalla 2014). *Phab* is commercially available as dried white cake or granule. Approximate weight of each granule (*phab*) is 3–5 g and of cake (*dhaeli*) is about 13–15 g (Kanwar et al. 2011).

8.6 Conclusion

Himachal Pradesh is a rich repository of microbial genetic diversity. Fermentation of traditional foods is beneficial in terms of food quality, preservation, and decontamination of toxins, often found in foods. Ethnic fermented foods, prepared by the rural and tribal people of Himachal Pradesh, include *bhaturu*, *manna*, *marchu*, *siddu*, *chilra (lwar)*, *dosha*, *seera*, *bedvin roti*, *gulgule*, *borhe*, *pinni/bagpinni*, *pakk*, *thuktal*, *sepubari*, *bari*, *churpa*, and *aska*. Most of these fermented foods are prepared from cereals, viz., wheat, barley, and buckwheat, but some are prepared from legumes and milk too. A number of indigenously made traditional fermented alcoholic beverages are also popular among the people of Himachal Pradesh, which are mostly prepared from cereals (rice/barley), fruits (apple/wild apricot/grapes), jaggery, and millets. Some of the common traditionally fermented alcoholic beverages of Himachal Pradesh are *sura*, *chhang/lugri*, *angoori/kinnauri*, *daru/chakti*, *chulli*, *behmi*, and *arak/ara*. The fermentation of these products enhances the nutritive and therapeutic values which play an important role in benefiting the human health. The know-how of the traditional processes and technologies involved in the production is being transferred from generation to generation as trade secrets. The raw materials needed are locally available cereals, pulses, vegetables, fruits, etc. These food products are prepared by using simple equipment, under primitive conditions, which result in low yield and poor quality. However, with the emerging technologies, the development of appropriate starter cultures and processing methods may ensure better quality of these products.

References

Angmo K, Bhalla TC (2014) Preparation of *Phabs*—an indigenous starter culture for production of traditional alcoholic beverage, Chhangin Ladakh. Indian J Tradit Knowl 13:347–351

Basappa SC (2002) Investigations on *Chhang* from finger millet (*Eleusine coracana* Gaertn.) and its commercial prospects. Indian Food Ind 21:46–51

Batra LR (1986) Microbiology of some fermented cereals and grains legumes of India and vicinity. In: Hesseltine CW, Wang HL (eds) Indigenous fermented food of non-western origin. J. Cramer, Berlin, pp 85–104

Batra LR, Millner PD (1976) Asian fermented foods and beverages. Dev Ind Microbiol 17:117–128

Bhushan K, Jain A, Sharma OP, Singh B, Kanwar SS (2012) α-Amylase production from *Endomyces fibuliger*—an indigenous yeast isolate of Western Himalayas. Int J Food Ferment Technol 2:63–69

Ghosh AR (2012) Appraisal of probiotics and prebiotics in gastrointestinal infections. Gastroenterology 3:WMC003796

Kanwar SS, Keshani (2014) Bio-prospecting of indigenous yeast isolates of fermented foods of North-Western Himalayas. In: Singh RS, Pandey A, Larroche C (eds) Advances in industrial biotechnology. IK International Publishing House Pvt. Ltd., New Delhi, pp 180–195. ISBN: 9789382332763

Kanwar SS, Keshani (2016) Fermentation of apple juice with a selected yeast strain isolated from fermented foods of Himalayan regions and its organoleptic properties. Front Microbiol 7:1012. https://doi.org/10.3389/fmicb.2016.01012

Kanwar P, Sharma N (2011) Traditional pre- and post natal dietary practices prevalent in Kangra district of Himachal Pradesh. Indian J Tradit Knowl 10:339–343

Kanwar SS, Gupta MK, Katoch C, Kumar R, Kanwar P (2007) Traditional fermented foods of Lahaul and Spiti area of Himachal Pradesh. Indian J Tradit Knowl 6:42–45

Kanwar SS, Gupta MK, Katoch C, Kanwar P (2011) Cereal based traditional alcoholic beverages of Lahaul and Spiti area of Himachal Pradesh. Indian J Tradit Knowl 10:251–257

Kanwar SS, Walia S, Sharma S (2015) Impact of probiotics and gut microbiota on host behavior. In: Garg N, Abdel-Aziz SM, Aeron A (eds) Microbes in food and health. Springer Publishers, New York, pp 29–42

Katiyar SK, Bhasin AK, Bhatia AK (1991) Traditionally processed and preserved milk products of Sikkimese tribes. Sci Cult 57:256–258

Kayastha SL (1971) Himachal region. In: Singh RL (ed) India—a regional geography. National Geographical Society of India, Varanasi, pp 390–442

Keshani, Sharma PN, Sharma KD, Kanwar SS (2015) Molecular and functional diversity of *Saccharomyces cerevisiae* strains of traditional fermented foods of the North-Western Himalayas. Ann Microbiol 65:2265–2275

Kumari A, Bhushan K, Kanwar SS et al (2015) Microbiology and biochemistry of indigenous fermented foods. In: Joshi VK (ed) Indigenous fermented foods of South Asia. CRC Press Taylor & Francis Group, London, New York, pp 107–226. ISBN: 9781439887905

Nehal N (2013) Knowledge of traditional fermented food products harbored by the tribal folks of the Indian Himalayan belt. Int J Agric Food Sci Technol 5:401–414

Pal PK, Hossain SA, Sarkar PK (1996) Optimization of process parameters in the manufacture of chhurpi. J Food Sci Technol 33:219–223

Pathania N, Kanwar SS, Jhang T, Koundal KR, Sharma TR (2010) Application of different molecular techniques for deciphering genetic diversity among yeast isolates of traditional fermented food products of Western Himalayas. World J Microbiol Biotechnol 26:1539–1547

Rana TS, Datt B, Rao RR (2004) Soor: a traditional alcoholic beverage in *Tons* valley, Garhwal Himalaya. Indian J Tradit Knowl 3:59–65

Rawat K, Kumari A, Kumar S, Kumar R, Gehlot R (2018) Traditional fermented products of India. Int J Curr Microbiol App Sci 7:1873–1883

Roy B, Kala CP, Farooquee NA, Majila BJ (2004) Indigenous fermented food and beverages: a potential for economic development of the high altitude societies in Uttaranchal. J Hum Ecol 15:45–49

Savitri, Bhalla TC (2007) Traditional foods and beverages of Himachal Pradesh. Indian J Tradit Knowl 6:17–24

Savitri, Bhalla TC (2012) Characterization of *bhatooru*, a traditional fermented food of Himachal Pradesh: microbiological and biochemical aspects. Biotechnology 3:247–254

Savitri, Thakur N, Kumar D, Bhalla TC (2012) Microbiological and biochemical characterization of seera: a traditional fermented food of Himachal Pradesh. Int J Food Ferment Technol 2:49–56

Sekar S, Mariappan S (2007) Usage of traditional fermented products by Indian rural folks and IPR. Indian J Tradit Knowl 6:111–120

Sharma S, Kanwar SS (2017) Adherence potential of indigenous lactic acid bacterial isolates obtained from fermented foods of Western Himalayas to intestinal epithelial Caco-2 and HT-29 cell lines. J Food Sci Technol 54:3504–3511

Sharma S, Kanwar SS (2018) Effect of prebiotics on growth behavior of *Lactobacillus plantarum* and their impact on adherence of strict anaerobic pathogens to intestinal cell lines. J Food Saf 38:e12384

Sharma N, Handa S, Gupta A (2013) A comprehensive study of different traditional fermented foods/beverages of Himachal Pradesh to evaluate their nutrition impact on health and rich biodiversity of fermenting microorganisms. Int J Res Appl Nat Soc Sci 1:19–28

Sharma N, Gupta A, Handa S (2014) An exploration of rich microbial diversity of rare traditional functional foods of Trans Himalayan state of India with proven additional probiotic effect. Int J Curr Microbiol App Sci 3:999–1014

Soni SK, Sandhu DK (1990) Biochemical and nutritional changes associated with Indian Punjab *Wari* fermentation. J Food Sci Technol 27:82–85

Sourabh A, Kanwar SS, Sharma PN (2010) Diversity of bacterial probiotics in traditional fermented foods of Western Himalayas. Int J Probio Prebio 5:193–201

Sourabh A, Kanwar SS, Sharma OP (2011) Screening of indigenous yeast isolates obtained from traditional fermented foods of Western Himalayas for probiotic attributes. J Yeast Fungal Res 2:117–126

Tamang JP, Thapa S, Tamang N, Rai B (1996) Indigenous fermented food beverages of Darjeeling hills and Sikkim: process and product characterization. J Hill Res 9:401–411

Thakur N (2013) Characterization of traditional fermentation processes used for the production of some alcoholic beverages (Chhang, Sura and Jau Chhang) in Himachal Pradesh. PhD thesis, Himachal Pradesh University, Summer Hill, Shimla

Thakur N, Kumar D, Savitri, Bhalla TC (2003) Traditional fermented foods and beverages of Himachal Pradesh. Invent Intell 2003:173–178

Thakur N, Savitri, Bhalla TC (2004) Characterization of some traditional fermented foods and beverages of Himachal Pradesh. Indian J Tradit Knowl 3:325–335

Venkatasubbaiah P, Dwarakanath CT, Sreenivasa MV (1984) Microbiological and physicochemical changes in idli batter during fermentation. J Food Sci Technol 21:59–62

Walia S, Keshani, Sood S, Kanwar SS (2014) Exhibition of DNA-bioprotective activity by microflora of traditional fermented foods of North-Western Himalayas. Food Res Int 55:176–180

Ethnic Fermented Foods and Beverages of Karnataka

9

Rwivoo Baruah, K. A. Anu Appaiah, and Prakash M. Halami

Abstract

In Karnataka a state of Southern India, there are numerous varieties of traditional food products, which are present due its vast variation in culture, geographical indication and variability in the raw material present in all three different zones, viz. north, south and coastal. Preservation of food by fermentation that extends shelf life of the food product is a common practice followed at household level. Fermentation by microorganisms in the raw material leads to the production of finished product having higher nutritive value. Most of the microorganisms involved in fermentation of these food products are known to have the Generally Regarded as Safe (GRAS) status and are known to promote human health. Fermented products are known to enhance the nutritional property as well as the therapeutic property of the food. This modification is seen especially in the case of dairy products. Karnataka is a home for few unique fermented products of cereal-legume-based product as well as dairy-based product and toddy-based beverages. Indigenous knowledge is being utilized to prepare fermented foods like *idli* and *dosa* and beverages like *toddy* and *neera*. This chapter provides methods of preparation, role of microorganism and health benefits of those fermented food products of Karnataka.

Keywords

Karnataka · Fermented food · Lactic acid bacteria · *Leuconostoc mesenteroides* · *Lactococcus lactis*

R. Baruah · K. A. A. Appaiah · P. M. Halami (✉)
Microbiology and Fermentation Technology Department, CSIR-Central Food Technological Research Institute, Mysuru, India

J. P. Tamang (ed.), *Ethnic Fermented Foods and Beverages of India: Science History and Culture*, https://doi.org/10.1007/978-981-15-1486-9_9

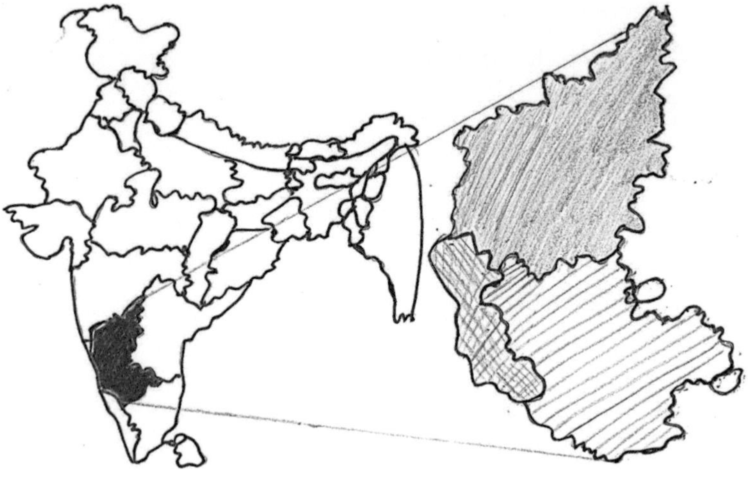

Fig. 9.1 Map of India and the state of Karnataka (depicted in three regions)

9.1 Introduction

The state of Karnataka in India is located at 11°30′ North and 18°30′ North latitudes and 74° East and 78°30′ East longitude. It is situated in the western part of the Deccan Peninsular region on a tableland where the Western and Eastern Ghat ranges converge into the complex. Historically, the state of Karnataka was created on 1 November 1956 by carving the old Mysore state and Bombay state (Ramachandra et al. 2017). In the west, Karnataka is bordered by the Arabian Sea also called as Karavalli of 320 km in length, the union territory of Maharashtra to the north, Goa to the northwest, Tamil Nadu to the southeast, Telangana and Andhra Pradesh to the east and Kerala to the southwest (Fig. 9.1). The state of Karnataka covers an area of 191,791 km^2 accounting for 5.83% of the total geographical area of India. This state is the seventh largest state in Indian zone, and according to the 2018 census, Karnataka has 69,371,567 inhabitants being the eighth largest state by population, comprising 30 districts. Karnataka has three main regions comprising of costal Karnataka, northern interior and southern interior regions (Geography of Karnataka, Wikipedia.org. 2019a).

Climate and Culture Karnataka has typical composition of most of the agro-climatic conditions in the country. It has been demarcated into ten agro-climatic zones. This classification was done based on humidity, rainfall, vegetation, soil structure, topography, elevation and other agro-climatic factors (Ramachandra et al. 2017). In the Indian subcontinent, the state of Karnataka falls under Zone X (Southern Plateau and Hilly region) and Zone XII (West Coast Plains and Ghat region). Since most food habit arises from the cultural group of the community, the main ethnic groups seen in Karnataka are Kannadiga, Tuluva people, Konkani people and Kodava people. In Karnataka, the dominant ethnic group are the Kannadigas,

Fig. 9.2 Typical Kannadiga meal, (**a**) thalli, (**b**) *ragi mudde*, (**c**) different types of chutneys

making up to 72% of the total population. In the district of Dakshina Kannada of Karnataka also referred as Tulu Nadu, the Tuluvas are a dominant ethnic community and are the native speakers of Tulu language. The regions of Uttara Kannada, Udupi and Dakshina Kannada have a large number of Konkani language speakers. Kodava people are native to the district of Kodagu (also known as Coorg). They are a martial race and are the native speakers of Kodava language.

The common agriculture produce of the state includes rice (*Oryza (O) sativum*), and millet or *ragi* (*Eleusine coracana*) forms the staple food in South Karnataka, whereas in North Karnataka, *jolada rotti*, a Sorghum-based *roti*, is staple to the people. Few of the popular food items in Karnataka are *masala dosa*, *benne dosa*, *ragi mudde*, *bisi bele bath*, *uppittu*, *jolada rotti* and *maddur vade*. Among popular sweets of the region are *mysore pak*, *amingad*, *belgaavi*, *dharwad pedha* and *kunda karadantu* of Gokak. *Dharwad pedha* has recently been acclaimed with the UNESCO tag of popular sweet dish. Other parts of Karnataka have their own distinctive cuisines such as the regions of coastal Karnataka and Kodagu. The cuisine of the coastal Karnataka region of Udupi is popular all over India.

The cuisine of Karnataka traces its origin to the Iron Age and is considered as one of the oldest surviving cuisines. The cuisine of the Karnataka with its varieties is responsible for influencing and being influenced from the cuisines of neighbouring states like Kerala, Andhra Pradesh and Tamil Nadu. Ragi has been mentioned by the great poet Adikavi Pampa seen in his historical works and by Sushruta Samhita, the ancient Sanskrit medical text known worldwide (Karnataka, Wikipedia.org. 2019b). After ignoring minor differences, a typical Kannadiga meal would include rice *palya*, *raita*, *gojju*, *chitranna*, *kosambari*, *tuppa*, *tovve*, kheer, pickle, salt, etc. served preferably on banana leaves. Cereals and millets are the staple food items for Karnataka especially in the south region of the state, whereas folks from coastal regions mainly consume fish curries (Fig. 9.2) (Karnataka, Wikipedia.org. 2019b).

9.2 Cereal-Based Fermented Foods

In the state of Karnataka, the best examples of staple foods are the cereal-based fermented food products. Commonly used cereals include rice, wheat flour (*Triticum (T)* spp.), *ragi* flour and barley flour (*Hordeum vulgare*), and pulses such as green gram, red gram and black gram dhal are among the principal components commonly used for the preparation of cereal-based fermented foods. Among these, popular foods that are consumed on the daily basis in the state are *idli*, *dosa*, *dhokla*, *ambali*, *nan*, *parotta*, *koozhu* and *pazhaiya soru*.

9.2.1 Idli

Idli or *idly* is a type of rice cake with savoury taste; it originated in the Indian subcontinent and is considered a popular breakfast food item consumed mainly in Southern India and northern Sri Lanka. The cakes are prepared using a fermented batter followed by steaming of the cakes. The key components of the batter include rice and black lentils (de-husked). Because of the fermentation, the native starches present get broken down resulting in the product to be easily metabolized by the body. Several variations of *idli* are found in Karnataka such as *rava idli* that is made from semolina and also regional variants such as sonna of Konkan and Mangalore region (Sarkar et al. 2015).

For the preparation of *idli*, uncooked rice (*idli* rice or parboiled rice) and de-husked whole black lentil (urad dal, *Vigna mungo*) in the ratio 4:1 are soaked separately for either 4–6 h or overnight in some cases. For additional flavour, spices such as fenugreek seeds can be added at the time of soaking. After soaking, the rice is coarsely grounded, and the lentils are grounded to a fine paste separately and then are combined. Subsequently, fermentation of the mixture is allowed to proceed overnight; this finally results in an increase of more than double of its volume. For the steaming process, the finished *idli* batter is put into greased moulds of an idli tray also referred to as "tree". The moulds are perforated so that the *idli* is cooked evenly. The tree kept for steaming over some boiling water in a covered pot. The time of steaming depends upon the size and number of *idli* kept. Instead of using moulds, the use leaves are common in the traditional method of *idli* preparation (Sekar and Mariappan 2007). For the considerable reduction of the soaking time, parboiled rice can be replaced with regular rice. Sour flavour may be incorporated in the unfermented batters with the addition of *dahi* (yoghourt). Besides the addition of fenugreek, other spices can also be used such as chilli peppers, cumin, ginger, coriander, mustard seeds, etc. The savoury taste of *idlis* can be replaced by the addition of sugar which makes the idli sweet. *Idli* can also be stuffed with a range of filling such as potato, onions, carrot, beans and masala. General method of preparation of cereal-legume-based fermented food is described in Fig. 9.3. Though traditionally *idli* batter preparation is done by spontaneous fermentation, studies have been made wherein back-sloping has been introduced to reduce the fermentation time from conventional 12 to 3 h. The batter obtained found to have increased levels of thiamine content and a reduction in antinutritive factors (Shrivastava and Ananthanarayan 2015).

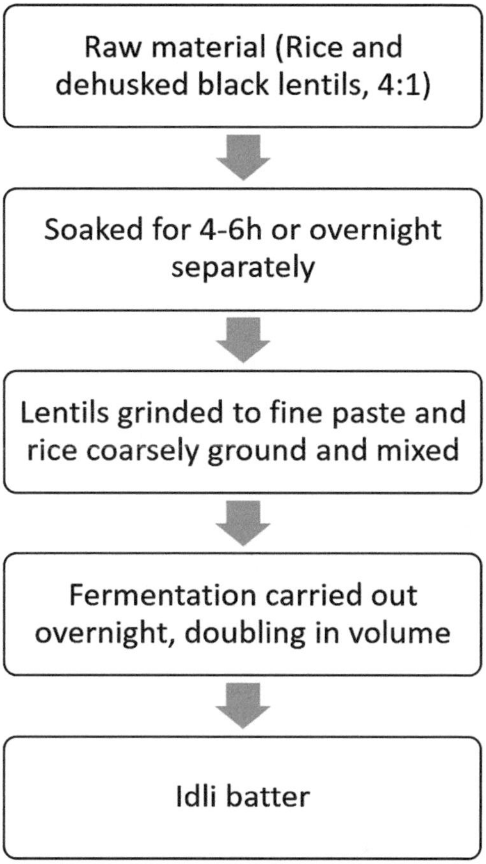

Fig. 9.3 Schematic representation of the preparation of cereal legume-based idli batter

Idli has been mentioned in several ancient Indian historical works; these were the precursors to the modern *idli*. *Iddalige*, prepared only from a black gram batter, has been mentioned by Shivakotiacharya in his 920 CE Kannada language work *Vaddaradhane*. This food is prepared by soaking black gram in buttermilk, then grinding it to a fine paste, and finally mixing it with the clear water of curd along with spices as described by Lokopakara (c. 1025 CE). In the encyclopaedia, *Manasollasa* (1130 CE), written by Someshwara III, the Western Chalukya king and scholar, who reigned in the area now called as Karnataka, the recipe of *idli* was also included which was described as *iḍḍarikā*. In present-day Karnataka, the *idli* prepared following this recipe is called *uddina idli* (Nair and Prajapati 2003). However, the recipe mentioned above lacked three key aspects found in the preparation of modern *idli* recipe: (1) the combined use of rice along with black lentils, (2) the overnight fermentation of the mixture and (3) the steaming of batter for desired fluffiness. It is only after 1250 CE that mentions of the modern recipe had appeared in the Indian works. The food historian K.T. Achaya had speculated that the origins of the modern *idli* recipe might have taken place in present-day Indonesia. His hypothesis includes that the steamed *idli* was invented by the cooks who were employed by the Hindu kings of the Indonesian kingdoms and had subsequently brought the recipe back to India during 800–1200 CE. An Indonesian dish similar to *idli* called *kedli* also exists (Achaya 1994, 2009) .

Idli has spread across the world as a result of the emigration of South Indians and Sri Lankans throughout the world, which led to the creation of many variations on *idlis*. Commercially different fusion recipes of *idli* are available such as *idly* manchurian, *idly* fry and chilly idly. *Tatte idli* is a plate-sized variant of *idli* from Karnataka. *Rava idli* (Sooji *idli*) is a specialty of Karnataka made with the addition of semolina. *Ragi idli* is ragi flour mixed with small portion of rice flour and urad dal. *Sannas* is a Goan variant of *idli* aka Hittli in Konkani. *Muday idli* is a Mangalorean variant of idli. Figure 9.4 represents variants of idli and its process of steaming in Mysore region. In present day, instant *idli* mixes are available in the market; these pre-packaged batter mixes are prepared and sold for the instant preparation of *idlis*. *Idli upma* can be prepared using leftover *idlis* that can be cut-up or crushed and then sautéed (Sankaran 1998). Curd and *idli* can be combined and then tempered with to prepare *dahi idli*. *Paddu* is prepared by using idli batter and steamed in a specially designed pan with small fissures; it can be made either sweet using jaggery or spicy using chillies. Modern cooking equipment for making *idlis* such as microwave and automatic electric *idli* steamers having convenient non-stick coating are also available in the market. Similarly, for batter preparation, manual rocking rock grinder can be replaced with electric grinders or blenders (mixies). *Idlis* are generally accompanied with red, green or white coconut chutney.

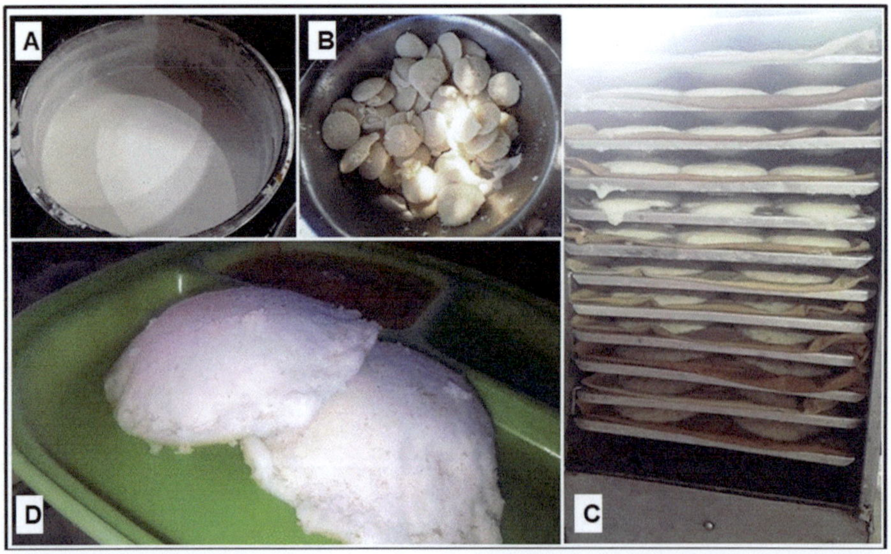

Fig. 9.4 (**a**) *Idli* batter, (**b**) finished *idlis*, (**c**) method of steaming in the preparation of *idli*, (**d**) *idli* served with sambar

During the fermentation of *idli* batter, the generation of carbon dioxide results in leavening, and production of acids (lactic acid) results in increase in acidity. LAB such as the heterofermentative strain *Leuc. mesenteroides* and the homofermentative strain *Enterococcus (E) faecalis* are the predominant microorganisms found during fermentation of batter. Heterofermentative LAB such as *Leuc. mesenteroides* possesses the ability to generate both lactic acid and carbon dioxide, whereas homofermentative LAB such as *E. faecalis* can only generate lactic acid (Satish Kumar et al. 2013; Agaliya and Jeevaratnam 2013). Urad dal is the source of both *Leuc. mesenteroides* and *E. faecalis* which gets included into the batter. The initial multiplication of the bacteria starts, while the grains are soaking and then continue to do so after grinding is completed. *Leuc mesenteroides* has the ability to tolerate high salt concentrations which is rare in most other bacteria. The high salt concentration in the batter and the ongoing generation of lactic acid both lead to the suppression of growth of other undesirable microorganisms (Vijayendra et al. 2010). Sridevi et al. (2010) have evaluated the suitability of three lactic acid bacterial starter cultures along with yeast culture in *idli* batter fermentation. The idly prepared using those starter cultures showed higher score compared to naturally fermented batter. Recently use of mustard essential oil as a bio-preservative has been investigated for shelf life improvement of idly batter. The decrease in growth of yeast and LAB in batter upon storage was evident in delay in deterioration, when idly batter was treated with 0.1% concentration of mustard essential oils (Regubalan and Ananthanarayan 2018). Similarly, curry leaves as a vehicle for bio-fortification and shelf life extension of *idli* batter have been studied. Incorporation of curry leaves powder at 5% in *idli* batter found to increase the shelf life up to 5 days when stored at 30 °C, suggesting the antimicrobial activity (Chelliah et al. 2016). Attempts have been made for the nutritional improvement of *idli* batter during fermentation, wherein the blend ratio of different combinations of ingredients was evaluated. The *idli* batter prepared using blend ratio of 3:1 (rice and black gram) was found to be better with reference to maximum nutrient content (Ghosh and Chattopadhyay 2011). With reference to in situ bio-fortification of *idli* batter, wherein defined starter cultures were used during fermentation, enrichment in the vitamin B produced by the starter culture comprising of *Lactococcus (Lc) lactis* and *Saccharomyces (S) boulardii* has been reported (Chandrasekar Rajendran et al. 2017). Studies have also been made to investigate the presence of biogenic amines in idli batter during storage. The predominant biogenic amines found were putrescine and cadaverine in both laboratory-prepared and market sample. However, the presence of such biogenic amines was under below the harmful limit (Regubalan and Ananthanarayan 2019).

A few popular variants and *idli* in the state of Karnataka are as follows.

9.2.2 Thatte Idli

Thatte idli is a popular breakfast dish, widely consumed in the southern part of Karnataka. The method of preparation has a slight difference than the regular *idli*. To make *thatte idli*, all the ingredients, i.e. rice and sabudana (tapioca), are washed and soaked separately for about 4–5 h and urad dal and beaten rice for 1–1.5 h. All ingredients are grinded into smooth batter and transferred to a big vessel. After the addition of salt, everything is mix well and kept aside for fermentation for 10–12 h. After fermentation, the mix is gently poured into greased plates and subsequently steamed for 10–12 min before serving. *Thatte idli* steaming is shown in Fig. 9.4.

9.2.3 Sannas/Mangalorean Steamed Rice Cakes

Sannas also known as Goan or Mangalorean steamed rice cakes is popular along the Konkan coast of South India. *Sannas* when compared to *idlis* are much more fluffy and porous making it ideal for absorbing gravies and curries. It is prepared by soaking rice and urad dhal separately for 6 h and then grinding it into a paste. Then everything is mixed with the addition of yeast, sugar and salt. The mixture is kept to ferment for 5 h and then steamed into cakes similar to idlis (Mascarenhas-Keyes 1977).

9.2.4 Dosa

Dosa or *dose* is a type of crèpe or pancake which is made from a fermented cereal legume batter that originates from the Indian subcontinent. Its main ingredients are rice and black lentils which are ground together in a fine, smooth batter with salt. *Dosas* is an iconic dish of Southern India and Sri Lanka and is currently popular all over the world. D*osas* are generally served hot along with a stuffing of potatoes, sambar and coconut chutney. According to P. Thankappan Nair, a historian, the dosa had originated in the town of Udupi in Karnataka. Different variants of *dosa* include *set dosa*, *davangere benne dosa*, *onion dosa*, *mixed dal dosa*, *whole wheat dosa*, *neer dosa*, *ridge gourd dosa*, *curd dosa* and *ragi dosa* recipes (Nair and Prajapati 2003; Sarkar et al. 2015).

Dosa is prepared using a mixture of rice and black lentils (in ratios 4:1 or 5:1) which are soaked in water and ground finely to form a batter. A handful of soaked fenugreek seeds are also added along with the rice. Overnight fermentation in the batter is allowed to take place. Subsequently, the batter is diluted with water to obtain the desired consistency. The batter is baked using a hot *tava* (griddle) that can be greased using vegetable oil or *ghee* (clarified butter). The shape of *dosa* is formed using the base of a ladle or bowl and then spread evenly to form a pancake, which can be thick, or it could be thin and crispy (Sathe and Mandal 2016).

Dosa batter fermentation also comprises a diversity of microflora predominantly lactic acid bacteria and *Bacillus* (Satish Kumar et al. 2013). Similarly, a potential probiotic strain of *Lb. plantarum* has been obtained from dosa batter (Gupta and Tiwari 2014). Studies pertaining to enhancing the nutritional value of dosa batter were performed by incorporation of finger millet and horse gram flour. The end product prepared found to have better sensory attributes and can be used to overcome protein energy malnutrition (Palanisamy et al. 2012). Food-grade bacilli-producing bioactive peptides such as fibrinolytic enzymes have been reported by Devaraj et al. (2018).

A few popular variants of *dosa* found in Karnataka are as follows.

9.2.5 Mysore Masala Dosa

Mysore masala dosa is a very popular breakfast recipe from Mysore and the southern part of Karnataka. In Karnataka, Mysore masala dosa is known as *masala dose*. An authentic Mysore masala dosa should be thick and crispy from outside and should be a little soft and spongy inside. For the preparation of Mysore masala dosa, red garlic chutney is applied as a coat in the inner layer accompanied by two tablespoons of potato filling (a masala prepared using potato and other spices) which is placed in the middle and is generally served along with sambar and coconut chutney (Masala dosa Wikipedia.org 2019c). The pictorial representation of Mysore masala dosa is shown in Fig. 9.5.

9.2.6 Davangere Benne Dosa

Benne dosa or *benne dose* is a type of *dosa* which origins can be traced from the city of Davangere in Karnataka, India. The term *benne dosa* in the English language corresponds to butter *dosa*. It differs from preparing the normal *dosa* by addition of generous amount of butter and is generally accompanied by coconut chutney. It is made out of rice batter and much butter and smaller in size compared to masala *dosa* or set *dosa*. While serving, liberal helpings of butter are sprinkled (Benne dose Wikipedia.org 2019d). The traditional Davangere *benne dosa* is presented in Fig. 9.6.

Fig. 9.5 (**a**) *Dosa* batter, (**b**) baking of *dosa* in *tava*, (**c**) pictorial representation of Mysore masala dosa

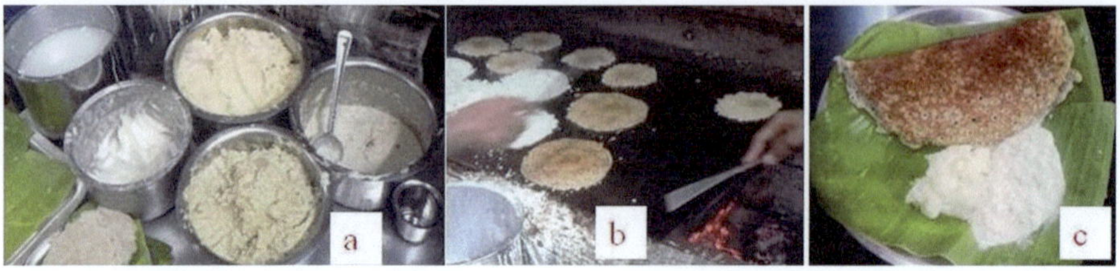

Fig. 9.6 *Benne dosa* of Davangere. (**a**) Batter and chutney, (**b**) mass baking of dosa and (**c**) serving dosa on banana leaves

9.2.7 Ambali

Ragi ambli or *ambali* is very popular across many regions of Karnataka. Thick batter is prepared using ragi-millet flour and water, which is then left to ferment for about 14–16 h. The fermented batter is combined with partially cooked rice by continuous stirring to complete the cooking. A*mbali* is ready to eat when yoghourt or curd is added to cooled cereal mixture (Ramakrishnan 1979). Microorganisms found in *ambali* fermentation were *Leuc. mesenteroides*, *Lb. fermentum* and *Strep. faecalis*. The acid production during fermentation decreased the pH from 6.4 to 4.0, and because of CO_2 production, the volume increases by about 20% (Ramakrishnan 1979). Ragi has numerous health benefits such as high amount of dietary fibre, and it helps in weight loss. It helps in reducing the body heat (Sarkar et al. 2015). Schematic representation of the preparation of ambali is shown in Fig. 9.7.

9.3 Milk-Based Fermented Foods

Karnataka Milk Federation (KMF) mainly controls the production and distribution of milk and milk products in Karnataka. It is the largest Cooperative Dairy Federation in South India, which is owned and managed by milk producers of the state of Karnataka. KMF has over 2.25 million milk producers in over 12,334 Dairy Cooperative Societies at village level, functioning under 13 District Cooperative Milk Unions in the state of Karnataka. According to the Department of Animal Husbandry, Dairying and Fisheries, the Ministry of Agriculture, Government of India, the state of Karnataka, produces 5997 tonnes of milk in the year 2013–2014 (Dairyknowledge.in 2019).

Milk and milk-based products are widely consumed due to their nutritive value. Fermentation of milk by LAB is preferred for preservation of nutrient quality of milk as milk gets easily curdled by pathogenic microorganisms. LAB can convert lactose, the sugar found in milk, into lactic acid, along with by-products such as vitamins and exopolysaccharides. Some strains of LAB can produce bacteriocin, an antibacterial substance that suppresses the milk curdling bacteria (Sarkar 2008). Fermented milk-based products are simply prepared by the addition of LAB into milk, which could

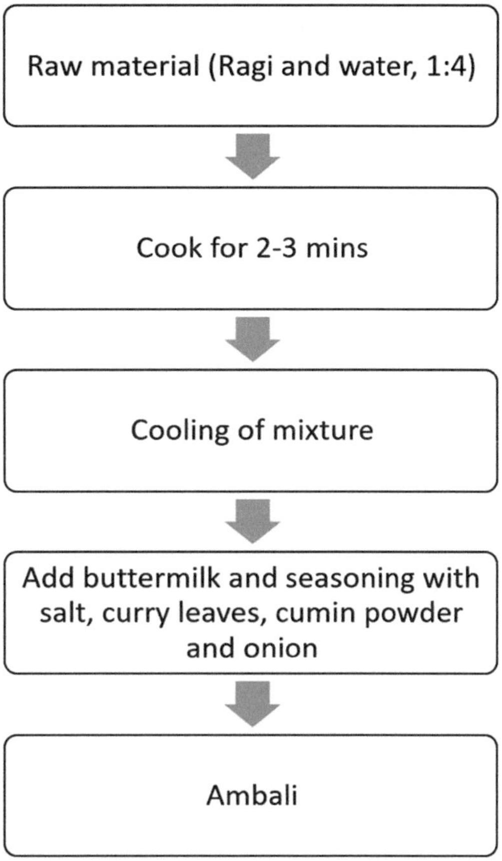

Fig. 9.7 Schematic representation of the preparation of cereal-based ambali

be from either cow, buffalo or yak, and allowed to ferment. *Dahi* or curd is the most popular and commonly used traditional Indian fermented product, and its description was found in texts as old as 700 BC (Satish Kumar et al. 2013).

9.3.1 Dahi

Dahi is prepared from boiled milk of either cow or buffalo, which is then soured using lactic cultures. The difference of *dahi* from yoghourt is the use of mesophilic lactococci as a mixed starter culture. The micro-organisms that are commonly used to make dahi are *Lc. lactis* spp. *lactis*, *Lc. lactis* spp. *cremoris* and *Lb. diacetylactis* which can be used separately or in combination. LAB found in *dahi* can demonstrate probiotic effect that can help in maintaining intestinal health and can also help in controlling diarrhoea in children (Agrawal et al. 2000; Shangpliang et al. 2018). Dahi is a potential source of B-complex vitamins folic acid and riboflavin (Sarkar et al. 2015). Flavours present in dahi are induced by metabolites such as diacetyl, which is supposed to be appreciated more by people of South Asian origin compared to the acetaldehyde flavour found in yoghourt (Yadav et al. 2007; Devi et al. 2015). Black slopping is a common practice of *dahi* preparation at household level. Bacteria present in fermented milk act as a starter culture (Tamang et al. 2016a, b).

9.3.2 Buttermilk

Buttermilk is a milk-based fermented dairy product that is prepared by the leftover liquid obtained when butter is churned out of cream or *dahi*; it can be consumed with or without added salt and spices. In some states of India, it is also known as *chaach*. Buttermilk consumption is very common in Karnataka. Live cultures of *Lb. acidophilus* present in buttermilk help in maintaining intestinal health. Compared to regular milk or dahi, buttermilk has less fat content and fewer calories. Buttermilk is a rich source of protein and calcium as like milk and is advised in the Ayurveda to be consumed with meals (Sathe and Mandal 2016). Method of preparation of buttermilk is presented in Fig. 9.8.

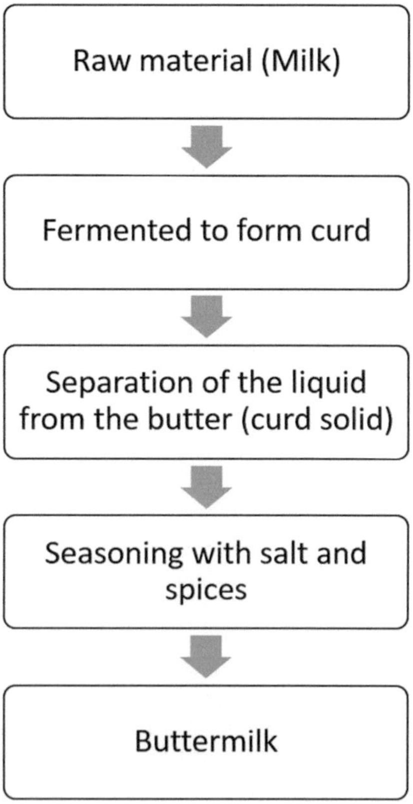

Fig. 9.8 Schematic representation of the preparation of milk-based fermented beverage buttermilk

9.3.3 Lassi

Lassi is a traditional fermented milk beverage prepared from dahi which is consumed mostly as a refreshment in summer seasons. Based on the use of ingredients in lassi, it is classified as salty lassi or sweet lassi. Dahi is blended with water, sugar, salt and spices to prepare *lassi*. The presence of spices decreases thirst. Due to the presence of *Lb. acidophilus* and *Strep. thermophilus* as active cultures, *lassi* is a probiotic product (Patidar and Prajapati 1998; Sarkar et al. 2015).

9.3.4 Ginna

It is also known as *gennu* in the state of Karnataka. *Ginna* is a colostrum-based Indian sweet which in different regions of India is also known as *junnu*, *posu* or *kharvas*. *Ginna* is prepared from the milk of a cow from the 1st day after calving; this thick milk obtained is called colostrum. The colostrum diluted using normal milk along with sugar/jaggery powder and cardamom can also be added for the taste. The final product is obtained after the cooking is complete, and subsequently the product is cut into pieces and can be served with *idli*, *dosa* or *roti*. Colostrum is a rich source of vitamin A and minerals along with immune components such as immunoglobulins and iron-binding lactoferrin protein (Sarkar et al. 2015).

9.4 Vegetable, BS and Unripe Fruit-Based Fermented Foods

As on 2014–2015, it is estimated that the state of Karnataka had produced 6941 tonnes of fruits and 8564 tonnes of vegetables (Neeraj et al. 2017). Though bamboo cultivation is done in some patches, it has not been used for preparing fermented foods in this state.

The lactic acid fermentation of vegetables can be applied as a preservation method. This method of preservation serves as an important technology which is responsible for the improvement of the nutritive value, microbial quality, acceptability, palatability and shelf life of the fermented products (Kingston et al. 2010). Considering the fact that the majority of rural people cannot afford canned or frozen foods or storage equipment, this is a very effective storage method in the absence of facilities such as cold storage or refrigeration especially for perishable vegetable. Fermented bamboo shoot (BS) products are generally consumed by the ethnic people of the north-eastern states of India as a traditional food (Tamang et al. 2009; Sonar and Halami 2014).

9.4.1 Pickles

Pickle is a fermented food product where brine or various spice mixes are used to preserve fruit and vegetables for a long period of time (Tamang et al. 2016a, b). In Karnataka, a typical meal consists of pickle which is a very common condiment in a daily basis. Pickles from various vegetables and fruits such as mango (*Mangifera indica* L.), lemon (*Citrus limon* L.) and *amla* (*Emblica officinalis* L.) are dietary supplements and used for culinary purposes in several parts of the world. Pickle contains a massive microbiome, which could differ from batch to batch (Tamang et al. 2009).

9.5 Meat-Based Fermented Foods

Drying, smoking and fermentation of meat are critical steps in the traditional processing of meat preservation, as meat is highly susceptible to microbial spoilage (Oki et al. 2011). However, no specific fermented meat product is prepared in Karnataka. The consumption of meat is mainly done in fresh form. As on 2015–2016, it was estimated that the total production of meat in state of Karnataka was 196,597 tonnes.

9.6 Fermented Fish Products

As on 2012–2013, it is estimated that the production of fish in state of Karnataka was 575.38 tonnes (iomenvis.nic.in 2019). Fish production in state and coastal area is important as a cheap protein-rich dietary component. The methods of preservation of fish by salt are an age-old technology and are popular in many developing countries. However, fermented food products are less known in Karnataka.

9.7 Fermented Beverages

In a basic level, a beverage (a drink) is a type of liquid that can be consumed for sustenance, energy or hydration. Historically, the idea of a beverage was most likely restricted to water or milk and perhaps juice squeezed from fruits. Popular beverages that are consumed by the population are various teas, coffees, sodas, wine, cocktails, cocoas and ciders (Rawat et al. 2018). Fermented beverages are classified as alcoholic and non-alcoholic (Marsh et al. 2014).

9.8 Non-alcoholic Beverages

A non-alcoholic beverage is a drink that contains no alcohol. However, in the western world, any beverage that contains alcohol less than 0.5% by volume can be termed as a non-alcoholic drink; these include apple cider and low-alcohol beer. Commercially available non-alcoholic beverages can include food products ranging from freshly squeezed orange juice to chemical-packed energy drinks to teas and coffees (Tamang and Kailasapathy 2010). Some of the commonly found ethnic fermented beverages are described below (Table 9.1).

Table 9.1 Ethnic fermented foods and beverages of Karnataka

Food	Substrate	Sensory property of product	Major ethnic consumers	State
Idli	Idli batter (rice, black gram dal)	Round steamed rice cakes	Kannadiga	Karnataka
Dosa	Idli batter (rice, black gram dal)	Crepe made using idli batter with/out potato fillings	Kannadiga	Karnataka
Adai Dosa	Idli batter (rice, black gram dal)	–	Kannadiga	Karnataka
Sannas	Idli batter (rice, black gram dal)	Very soft round steamed rice cakes	Kannadiga	Karnataka
Ambali	Fox millet Rice Kudo millet	Cereal porridge	Kannadiga, Konkani	Karnataka
Pickle	Fruit, salt, brine	Spicy semisolid condiment	Kannadiga	Karnataka
Sandigie	Dal, rice	–	Kannadiga	Karnataka
Dahi	Milk	Semisolid slurry	Kannadiga	Karnataka
Butter milk	Aceto milk	Non-alcoholic beverage consumed for refreshment	Kannadiga	Karnataka
Ginna	Colostrum milk	Solid sweets	Kannadiga	Karnataka
Lassi	Milk	Non-alcoholic beverage consumed for refreshment	Kannadiga	Karnataka
Beverages				
Neera	Palm sap Coconut palm Date palm Palmyra *Borassus*	Non-alcoholic beverage	Kannadiga	Karnataka
Palm vinegar	Toddy	Non-alcoholic beverage	Kannadiga	Karnataka
Toddy	Palm sap	Alcoholic beverage	Kannadiga	Karnataka
Cashew apple wine	Cashew apple	Alcoholic beverage	Kannadiga	Karnataka
Jamun wine	Jamun	Alcoholic beverage	Kannadiga	Karnataka
Kanji	Beetroot, carrots	Non-alcoholic beverage	Kannadiga	Karnataka
Wine	Grapes	Alcoholic beverage	Kannadiga	Karnataka

9.8.1 Neera

Neera is a popular non-alcoholic beverage which is prepared from the sap of several species of palm tree such as the date palms, coconut palms and palmyra. *Neera* or *pathaneer* is also called either palm wine or *toddy*, which is a sweet, non-alcoholic beverage that is derived from fresh sap, or *kallu*, which is a sour beverage made from fermented sap. The unfermented sap of palm trees is refrigerated, stored and distributed as neera (palm wine, Wikipedia.org 2019). For the prevention of any further fermentation, lime (calcium hydroxide) can be added to the sap. Fermentation of glucose present in the sap to alcohol is carried out by yeasts that are naturally present in palm sap (Chandrasekhar et al. 2012). The ideal time for the consumption of palm wine is said to be 1 day after extraction of sap, as the alcohol and vinegar content is minimal. *Neera* is relatively rich in potassium similar to fruit-juice products. Refrigeration is carried out to extend the shelf life of the beverage, and variety of spices are added that contribute to flavour (Law et al. 2011).

9.8.2 Palm Vinegar

Palm vinegar or sugar palm vinegar refers to vinegar made from palm or sugar palm sap. The yeast and acetobacter present in palm sap covert the free sugars into alcohol and then covert the alcohol to vinegar, respectively (Chandrasekhar et al. 2012). The method of preparation of palm vinegar is presented in Fig. 9.9.

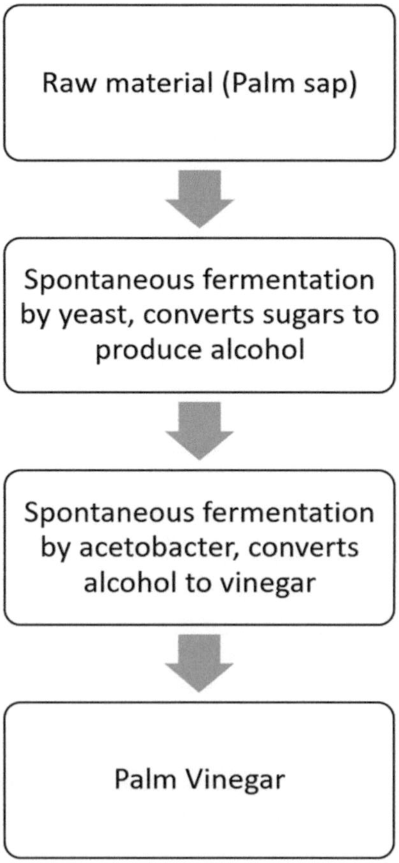

Fig. 9.9 Schematic representation of the preparation of fermented non-alcoholic beverage palm vinegar

9.8.3 Kanji

Kanji is a popular fermented drink, which is prepared using a deep purple-coloured carrot and then fermented along with crushed mustard seed, salt and hot chilli powder for 7–10 days. It is considered to have high nutritional value and cooling and soothing properties (Kingston et al. 2010). Beetroot can be used to make a similar drink, which is considered to have the potential to prevent infectious and malignant disease. LABs associated with this product are *Lb. acidophilus*, *Lb. casei* and *Lb. plantarum* (Vanajakshi et al. 2015).

9.9 Alcoholic Beverages

In western countries, numerous varieties of beverages are prepared by means of barley malt, as a source of starch and saccharifying agent (Marsh et al. 2014). In contrast, in the preparation of Indian traditional beverages such as ragi, rice and barley are used as a rich source of starch and saccharifying enzyme (Rawat et al. 2018). Many varieties of alcoholic beverages are manufactured and consumed by ethnic people of India. Traditionally different fermented beverages have been produced at different parts of India, depending upon the geographical condition, the climatic condition and the raw food material chiefly available in that locality of area commercially (Tamang and Kailasapathy 2010).

9.9.1 Toddy

Toddy is obtained by fermenting the sap from an excised tree species of palms, especially palmyra and coconut. The production of toddy is a restricted activity and requires permission. *Toddy* generally contains less than 5% alcohol. In Karnataka, the liquid collected from the coconut trees is named as *neera*. *Toddy* is extracted in a limited quantity in Mangalore, Udupi and Chitradurga districts in Karnataka (Sekar and Mariappan 2007). *Acetobacter* and *Gluconobacter* were among the dominant species found in coconut toddy (Kadere et al. 2008). *Toddy* consumption provides health benefits such as improved eyesight and mild laxative relieving constipation and helps in recovering from chicken pox, and toddy can act as a sedative (Sekar and Mariappan 2007).

9.9.2 Wine

Karnataka is India's second largest producer of grapes (*Vitis vinifera*). Regions such as Krishna Valley, Cauvery Valley and Nandi Valley are prominent grape-growing regions along with districts such as Mysore, Koppal and Chikmagalur. The climate of the state is also favourable for grape cultivation. The initiation of grape cultivation in the state can be traced back to the eighteenth century (Karnataka.com

Vineyards in Karnataka 2019). It was later during the nineteenth century that grape cultivation got popularized in the southern districts of the state by Christian missionaries. The annual production of the fruit is about 1.67 lakh tonnes. Karnataka has around 9700 ha of land under grape cultivation (Karnataka.com Vineyards in Karnataka 2019). Some of the popular wineries in the state include Elite Vintage Winery, Bagalkot, Myra Vineyards, Hampi Heritage Vineyards, Bijapur, Alpine Wineries, Mysore, Victoria Road, Bangalore, Grover Vineyards, Nandi Hills, Bangalore, Nandi Valley Winery, Yelahanka, Hobli, Yaana Wines and Bidar—premium wines from organic grapes (Karnataka.com Vineyards in Karnataka 2019). Fermentation in wine is generally carried out by *S. cerevisiae*. Multi-starter cultures that contain non-*Saccharomyces* (*Torulaspora delbrueckii*, *Hanseniaspora uvarum* and *Kluyveromyces thermotolerans*) and *S. cerevisiae* yeast strains have also been used for making wine (Ciani and Comitini 2015). A moderate consumption of wine can help in preventing coronary vascular disease (CVD) by increasing the levels of high-density lipoprotein cholesterol (HDL) in plasma, by restoring the endothelial function and by decreasing platelet aggregation, and it also has antioxidative effects (Saleem and Basha 2010).

9.9.3 Jamun Wine

Jamun (*Syzygium cuminii* (Linn.) Skeels) berry or black plum is an oblong, ovoid and shinning crimson black (rich in anthocyanin pigment, an anti-oxidant) when fully ripe. Clarified juice of jamun is inoculated with yeast, and fermentation is carried out in room temperature for 6 days. The jamun berry is used to prepare wine by using wine yeast, *S. cerevisiae* var. *bayanus*. Jamun fruits are universally accepted to be very good for medicinal purposes especially for curing diabetes because of its effect on the pancreas (Joshi et al. 2012).

A study by VenuGopal and Anu-Appaiah (2017) for the comparison of jamun juice fermentation two indigenous yeast strains were used. This study also evaluated the effect of seeds on fermentation and properties such as phenolic composition, chromatic properties and sensory attributes. Authors found that wines fermented with seeds had higher phenolic content, which later reduced during ageing. Due to the presence of seed phenolic and polysaccharides, the wines produced with seeds were prone to browning. Principal component analysis (PCA) revealed that the redness to jamun wines is imparted by the association of anthocyanin with younger wines. The overall composition of the wine affects the sensory perception of jamun wine, which was indicated by PLS analysis exhibiting both positive and negative correlation between various attributes (VenuGopal and Anu-Appaiah 2017).

9.9.4 Cashew Wine/Cashew Apple Wine

Cashew production in Karnataka is 53,000 tonnes with an average productivity of 461 kg/ha (Maruthi Prasad et al. 2011). Cashew wine is a light yellow alcoholic drink prepared from the fruit of the cashew tree. It contains an alcohol content of between 6% and 12%. The cashew apples undergo reverse osmotic process for juice extraction and then undergo the process of fermentation to produce the wine. Wine fermentation takes place with the help of wine yeast *S. cerevisiae*. Cashew juice has high nutritive values with high vitamin C content and high amount of minerals such as iron, calcium and phosphorus (Mohanty et al. 2006).

9.9.5 Garcinia Wine

Garcinia xanthochymus is a tree, which is endemic to the Western Ghats and Eastern Himalayas in India. The trees bear yellow fruit of 6–7 cm in diameter with juicy acidic yellow pulp containing up to two to three seeds. Wine from *Garcinia* can be prepared by fermenting 10% (w/v) garcinia pulp with wine yeast *Saccharomyces cerevisiae* for 21 days. *Garcinia species* possesses therapeutic properties, as they are a rich source of secondary metabolites such as xanthones, flavonoids, benzophenones, lactones, phenolic compounds and organic acids (Rai et al. 2010).

9.10 Health Benefits of Traditional Fermented Foods

The consumption of fermented foods by the local population is not only associated with its nutritive aspects but also by the array of health benefits these foods offer (Chaudhary et al. 2018). Traditional fermented foods are been known to have beneficial effect on various ailment and are used as a special diet or medicine for ages by the local people. Some of the fermented foods such as *idli* are often used as food for infant and persons with disabilities as they are easily digested. They are also prescribed for patients undergoing treatment in the hospitals as a part of their diet (Steinkraus et al. 1996). Some of the traditional fermented foods such as koozhu having high thiamine, riboflavin and niacin contents are included in the daily diet of rural and agricultural workers (Antony and Chandra 1997). This is also seen in the case of fermented bamboo plants where increased mineral content and has no cholesterol (Sonar and Halami 2014).

9.11 Conclusion

Ethnic fermented foods represent important components of diet in the people of Karnataka wherein some of the cereal-legume-based products and toddy-based beverages are originated. These food items safeguard traditional knowledge of indigenous people for generations to come. Several of the underutilized agricultural commodities like *ragi* have been utilized for the preparation of fermented food that eventually provides a microbial accessible (MAC) diet. Most of the fermented food also provide a diversity of microorganisms in the gut and help in curing many GI tract-related diseases. Traditional fermented foods also provide high level of nutrients and hence assure nutritional security and act as a functional food. Traditional fermented foods require large-scale production and commercialization to make them reach across the country. During the production of those traditional fermented foods, safety issues with reference to microbial contamination need to be addressed properly.

Acknowledgments The authors acknowledge the contributions of Ms. Ashwini and Dr. Abhishek for the photographs. Authors acknowledge kind approval of the Director CSIR-CFTRI Mysuru for this work.

References

Achaya KT (1994) Indian food: a historical companion. Oxford University Press, Delhi

Achaya KT (2009) The illustrated foods of India, AZ. Oxford University Press, New Delhi

Agaliya PJ, Jeevaratnam K (2013) Molecular characterization of lactobacilli isolated from fermented idli batter. Braz J Microbiol 44(4):1199–1206. https://doi.org/10.1590/S1517-83822013000400025

Agrawal R, Rati ER, Vijayendra SVN, Varadaraj MC, Prasad MS, Nand K (2000) Flavour profile of idli batter prepared from defined microbial starter cultures. World J Microbiol Biotechnol 16(7):687–690. https://doi.org/10.1023/A:1008939807778

Antony U, Chandra TS (1997) Microbial population and biochemical changes in fermenting finger millet (Eleusine coracana). World J Microbiol Biotechnol 13(5):533–537

Chandrasekar Rajendran SC, Chamlagain B, Kariluoto S, Piironen V, Saris PEJ (2017) Biofortification of riboflavin and folate in idli batter, based on fermented cereal and pulse, by *Lactococcus lactis* N8 and *Saccharomyces boulardii* SAA 655. J Appl Microbiol 122(6):1663–1671. https://doi.org/10.1111/jam.13453

Chandrasekhar K, Sreevani S, Seshapani P, Pramodhakumari J (2012) A review on palm wine. Int J Res Biol Sci 2(1):33–38

Chaudhary A, Sharma DK, Arora A (2018) Prospects of Indian traditional fermented food as functional foods. Indian J Agric Sci 88(10):1496–1501

Chelliah R, Ramakrishnan SR, Premkumar D, Antony U (2016) Bio-fortification and shelf-life extension of idli batter using curry leaves (*Murraya koenigii*). J Food Sci Technol 53(6):2851–2862. https://doi.org/10.1007/s13197-016-2264-2

Ciani M, Comitini F (2015) Yeast interactions in multi-starter wine fermentation. Curr Opin Food Sci 1:1–6. https://doi.org/10.1016/j.cofs.2014.07.001

Dairyknowledge.in (2019) Karnataka Milk Federation (KMF). http://www.dairyknowledge.in/link/karnataka-milk-federation-kmf

Devaraj Y, Rajender SK, Halami PM (2018) Purification and characterization of fibrinolytic protease from *Bacillus amyloliquefaciens* MCC 2606 and analysis of fibrin degradation product by MS/MS. Prep Biochem Biotechnol 48(2):172–180. https://doi.org/10.1080/10826068.2017.1421964

Devi SM, Archer AC, Halami PM (2015) Screening, characterization and in vitro evaluation of probiotic properties among lactic acid bacteria through comparative analysis. Probio Antimicrobial Prot 7(3):181–192. https://doi.org/10.1007/s12602-015-9195-5

Ghosh D, Chattopadhyay P (2011) Preparation of idli batter, its properties and nutritional improvement during fermentation. J Food Sci Technol 48(5):610–615. https://doi.org/10.1007/s13197-010-0148-4

Gupta A, Tiwari SK (2014) Probiotic potential of *Lactobacillus plantarum* LD1 isolated from batter of Dosa, a south Indian fermented food. Probio Antimicrobial Prot 6(2):73–81. https://doi.org/10.1007/s12602-014-9158-2

iomenvis.nic.in (2019) Statewise Fish Production in India. http://www.iomenvis.nic.in/index3.aspx?sslid=1172&subsublinkid=131&langid=1&mid=1

Joshi VK, Sharma R, Girdher A, Abrol GS (2012) Effect of dilution and maturation on physicochemical and sensory quality of jamun (Black plum) wine. Indian J Nat Prod Resour 3(2):222–227. http://nopr.niscair.res.in/handle/123456789/14423

Kadere TT, Miyamotoo T, Oniango RK, Kutima PM, Njoroge SM (2008) Isolation and identification of the genera acetobacter and gluconobacter in coconut toddy (mnazi). Afr J Biotechnol 7(16):2963–2971

Karnataka.com (2019) Vineyards in Karnataka. https://www.karnataka.com/vineyards

Kingston JJ, Radhika M, Roshini PT, Raksha MA, Murali HS, Batra HV (2010) Molecular characterization of lactic acid bacteria recovered from natural fermentation of beet root and carrot kanji. Indian J Microbiol 50(3):292–298

Law SV, Abu Bakar F, Mat Hashim D, Abdul Hamid A (2011) Popular fermented foods and beverages in Southeast Asia. Int Food Res J 18(2):475–484

Marsh AJ, Hill C, Ross RP, Cotter PD (2014) Fermented beverages with health-promoting potential: past and future perspectives. Trends Food Sci Technol 38(2):113–124. https://doi.org/10.1016/j.tifs.2014.05.002

Maruthi Prasad BN, Nataraja A, Yathindra HA, Aravinda Kumar JS, Mallikarjun-Gowda AP (2011) Prospects of cashew cultivation in Karnataka. Int Symp Cashew Nut 1080:83–87. https://doi.org/10.17660/ActaHortic.2015

Mascarenhas-Keyes S (1977) Catholic Goan food. In: Kuper J (ed) The anthropologists' cookbook. Routledge, London, pp 207–212

Mohanty S, Ray P, Swain MR, Ray RC (2006) Fermentation of cashew (*Anacardium occidentale* L.) "apple" into wine. J Food Process Pres 30(3):314–322. https://doi.org/10.1111/j.1745-4549.2006.00067.x

Nair BM, Prajapati JB (2003) The history of fermented foods. In: Handbook of fermented functional foods. CRC Press, Boca Raton, FL, pp 17–42

Neeraj, Chittora A, Bisht V, Johar V (2017) Marketing and production of fruits and vegetables in India. Int J Curr Microbiol App Sci 6(9):2896–2907. https://doi.org/10.20546/ijcmas.2017.609.356

Oki K, Rai AK, Sato S, Watanabe K, Tamang JP (2011) Lactic acid bacteria isolated from ethnic preserved meat products of the Western Himalayas. Food Microbiol 28(7):1308–1315

Palanisamy BD, Rajendran V, Sathyaseelan S, Bhat R, Venkatesan BP (2012) Enhancement of nutritional value of finger millet-based food (Indian dosa) by co-fermentation with horse gram flour. Int J Food Sci Nutr 63(1):5–15. https://doi.org/10.3109/09637486.2011.591367

Patidar SK, Prajapati JB (1998) Standardisation and evaluation of lassi prepared using *Lactobacillus acidophilus* and *Streptococcus thermophilus*. J Food Sci Technol 35(5):428–431

Rai AK, Prakash M, Anu Appaiah KA (2010) Production of Garcinia wine: changes in biochemical parameters, organic acids and free sugars during fermentation of Garcinia must. Int J Food Sci Technol 45(7):1330–1336. https://doi.org/10.1111/j.1365-2621.2010.02181.x

Ramachandra TV, Vinay S, Bharath S, Aithal BH (2017) Karnataka's Waterscape. Sahyadri E-News 60:1–83

Ramakrishnan CV (1979) Studies on Indian fermented foods. Baroda J Nutr 6:1–57

Rawat K, Kumari A, Kumar S, Kumar R, Gehlot R (2018) Traditional fermented products of India. Int J Curr Microbiol App Sci 7(4):1873–1883. https://doi.org/10.20546/ijcmas.2018.704.214

Regubalan B, Ananthanarayan L (2018) Shelf life improvement of idli batter by addition of mustard essential oil as bio-preservative. J Food Sci Technol 55(9):3417–3426. https://doi.org/10.1007/s13197-018-3247-2

Regubalan B, Ananthanarayan L (2019) Investigation of biogenic amines content in fermented idli batter during storage. J Food Sci Technol 56(4):1775–1784. https://doi.org/10.1007/s13197-019-03609-9

Roy A, Moktan B, Sarkar PK (2007) Characteristics of *Bacillus cereus* isolates from legume-based Indian fermented foods. Food Control 18(12):1555–1564. https://doi.org/10.1016/j.foodcont.2006.12.006

Saleem TM, Basha SD (2010) Red wine: a drink to your heart. J Cardiovasc Dis Res 1(4):171–176. https://doi.org/10.4103/0975-3583.74259

Sankaran R (1998) Fermented foods of the Indian subcontinent. In: Microbiology of fermented foods. Springer, Boston, MA, pp 753–789. https://doi.org/10.1007/978-1-4613-0309-1_24

Sarkar S (2008) Innovations in Indian fermented milk products—a review. Food Biotechnol 22(1):78–97. https://doi.org/10.1080/08905430701864025

Sarkar P, DH LK, Dhumal C, Panigrahi SS, Choudhary R (2015) Traditional and ayurvedic foods of Indian origin. J Ethnic Foods 2(3):97–109. https://doi.org/10.1016/j.jef.2015.08.003

Sathe GB, Mandal S (2016) Fermented products of India and its implication: a review. Asian J Dairy Food Res 35(1). https://doi.org/10.18805/ajdfr.v35i1.9244

Satish Kumar R, Kanmani P, Yuvaraj N, Paari KA, Pattukumar V, Arul V (2013) Traditional Indian fermented foods: a rich source of lactic acid bacteria. Int J Food Sci Nutr 64(4):415–428. https://doi.org/10.3109/09637486.2012.746288

Sekar S, Mariappan S (2007) Usage of traditional fermented products by Indian rural folks and IPR. Indian J Tradit Knowl 6(1):111–120

Shangpliang HNK, Rai R, Keisam S, Jeyaram K, Tamang JP (2018) Bacterial community in naturally fermented milk products of Arunachal Pradesh and Sikkim of India analysed by high-throughput amplicon sequencing. Sci Rep 8:1532. https://doi.org/10.1038/s41598-018-19524-6

Shrivastava N, Ananthanarayan L (2015) Use of the backslopping method for accelerated and nutritionally enriched *idli* fermentation. J Sci Food Agric 95(10):2081–2087. https://doi.org/10.1002/jsfa.6923

Sonar NR, Halami PM (2014) Phenotypic identification and technological attributes of native lactic acid bacteria present in fermented bamboo shoot products from North-East India. J Food Sci Technol 51(12):4143–4148. https://doi.org/10.1007/s13197-014-1456-x

Sridevi J, Halami PM, Vijayendra SVN (2010) Selection of starter cultures for idli batter fermentation and their effect on quality of idlis. J Food Sci Technol 47(5):557–563. https://doi.org/10.1007/s13197-010-0101-6

Steinkraus KH, Shapiro KB, Hotchkiss JH, Mortlock RP (1996) Investigations into the antibiotic activity of tea fungus/kombucha beverage. Acta Biotechnol 16(2–3):199–205

Tamang JP, Kailasapathy K (2010) Fermented foods and beverages of the world. CRC Press, Boca Raton, FL

Tamang JP, Tamang B, Schillinger U, Guigas C, Holzapfel WH (2009) Functional properties of lactic acid bacteria isolated from ethnic fermented vegetables of the Himalayas. Int J Food Microbiol 135(1):28–33. https://doi.org/10.1016/j.ijfoodmicro.2009.07.016

Tamang JP, Shin DH, Jung SJ, Chae SW (2016a) Functional properties of microorganisms in fermented foods. Front Microbiol 7(578). https://doi.org/10.3389/fmicb.2016.00578

Tamang JP, Watanabe K, Holzapfel WH (2016b) Diversity of microorganisms in global fermented foods and beverages. Front Microbiol 7:377. https://doi.org/10.3389/fmicb.2016.00377

Vanajakshi V, Vijayendra SVN, Varadaraj MC, Venkateswaran G, Agrawal R (2015) Optimization of a probiotic beverage based on Moringa leaves and beetroot. LWT-Food Sci Technol 63(2):1268–1273. https://doi.org/10.1016/j.lwt.2015.04.023

VenuGopal KS, Anu-Appaiah KA (2017) Seed incorporation during vinification and its impact on chemical and organoleptic properties in *Syzygium cumini* wine. Food Chem 237:693–700. https://doi.org/10.1016/j.foodchem.2017.05.160

Vijayendra SVN, Rajashree K, Halami PM (2010) Characterization of a heat stable anti-listerial bacteriocin produced by vancomycin sensitive *Enterococcus faecium* isolated from idli batter. Indian J Microbiol 50(2):243–246. https://doi.org/10.1007/s12088-010-0030-0

Wikipedia.org (2019) Palm wine. https://en.wikipedia.org/wiki/Palm_wine#cite_note-jak-1

Wikipedia.org (2019a) Geography of Karnataka. https://en.wikipedia.org/wiki/Geography_of_Karnataka

Wikipedia.org (2019b) Karnataka. https://en.wikipedia.org/wiki/Karnataka

Wikipedia.org (2019c) Masala dosa. https://en.wikipedia.org/wiki/Masala_dosa

Wikipedia.org (2019d) Benne dose. https://en.wikipedia.org/wiki/Benne_dose#cite_note-1

Yadav H, Jain S, Sinha PR (2007) Evaluation of changes during storage of probiotic dahi at 7°C. Int J Dairy Technol 60(3):205–210. https://doi.org/10.1111/j.1471-0307.2007.00325.x

Ethnic Fermented Foods and Beverages of Jammu and Kashmir

10

Rehana Akhter, F. A. Masoodi, Touseef Ahmed Wani,
Jeelani Raja, and Sajad Ahmad Rather

Abstract

The umpteen varieties of traditionally fermented foods of Jammu and Kashmir (J&K) deliver a wallop of savour, aroma, nutrients, probiotics and functional bioactive compounds. The volume of credible scientific evidence has made probiotics a crucial part of human nutrition and led to a demand for probiotics in functional foods and clinical application by demonstrating their beneficial effects on health and disease prevention. In Jammu and Kashmir state, there are nutrient-rich crops (cereals, vegetables) and animal products (milk) which are fermented and used as foods and beverages in households. Although traditionally fermented foods represent an important role in contributing to the livelihood of people of the J&K state through enhanced food security and income generation, there is still scarcity of information regarding their traditional production methods, microbiological and biochemical characteristics, nutritional value and safety. Research into the processing technologies of these fermented foods is still in its infancy stage. Thus, there is a need to document their traditional production methods, microbiology and biochemistry in order to evaluate their nutritional value and safety; standardize and industrialize them, where possible; and preserve them for future generations.

Keywords

Ethnic foods · Fermented foods · Beverages · Pickles · Cereal products

R. Akhter · F. A. Masoodi (✉) · T. A. Wani · J. Raja · S. A. Rather
Department of Food Science and Technology, University of Kashmir,
Srinagar, Jammu and Kashmir, India

J. P. Tamang (ed.), *Ethnic Fermented Foods and Beverages of India: Science History and Culture*, https://doi.org/10.1007/978-981-15-1486-9_10

10.1 Introduction

The state of Jammu and Kashmir is located in the northern Himalayas in India. Agriculture is the main occupation of the people of the state. The state is basically a mono-cropped and rainfed economy with 40% area in Jammu division and 60% in Kashmir division. Over 80% population of the state is engaged in agribusiness sector and about 60% of the state's revenue is earned by this sector only. Agribusiness therefore is a serious business when it comes to the state of Jammu and Kashmir (Chandra 2014). Those who are working in other industries also depend on agriculture for the requirement of raw material for their production. The major crops of the state include rice, maize and wheat. Recently the farmers have also started cropping cash crops and oil seeds. Production and export of fruits like apples, pears, cherries, plums, grapes, pomegranate, mulberry, peaches, apricot, walnut and almonds is common in the state. The horticulture business employs over 250,000 people in the state (Mansoor et al. 2018). A massive export potential is seen for fruits, vegetables, flowers, juices and pulps especially in the markets of Middle East, Europe, China and Australia.

Jean Anthelme Brillat-Savarin, a renowned gastronomist, quoted in his celebrated book, *The Physiology of Taste*, "Tell me what you eat and I will tell you who you are," to explore how food shapes us and our culture (Savarin 1825). Food is basic to all human life, so ubiquitous that it builds the identity of individuals associated with their culture and nations; likewise many people relate Italy with pizza and pasta, Korea with *kimchi*, England with tea, Ireland with potatoes and Kashmir with wazwan. Thus, food structures people's everyday world and defines relationships and hierarchies. The state of Jammu and Kashmir is famous all over the world for its ethnic/heritage and region-specific foodstuffs such as apple, saffron, baked products, dairy products, wazwan, etc. (Rather et al. 2016). These ethnic- and region-specific foodstuffs of Jammu and Kashmir have earned a distinction in international as well as domestic markets. Jammu and Kashmir has been described as an ancient region possessing a distinct ethnicity, character, language, dress, customs, rituals, cultural heritage and traditional food products (Lawrence 1886). In this subcontinent, fermentation is one of the oldest forms of food preservation. Fermentation has been used since ages as a traditional method of food preservation making them available in the harsh winters. In the state of Jammu and Kashmir, traditionally fermented foods like fruit and vegetable products including garlic, collard, yam, glue berry and lotus stem pickles and fermented cereal products including various flatbreads and pancakes are prepared. However traditional knowledge of fermented foods of the region is not documented properly.

Indigenous fermented foods are an intrinsic part of the diet of the ethnic tribes in the Himalayan belt of Jammu and Kashmir, being the oldest and most economic method for development of a diversity of aroma, flavour and texture, food preservation and biological enrichment of food products by the manipulation of different microbial population. The aim of this chapter is to present some important indigenous fermented food products of Jammu and Kashmir and update the existing knowledge about the methods of their production. The constraints and approaches

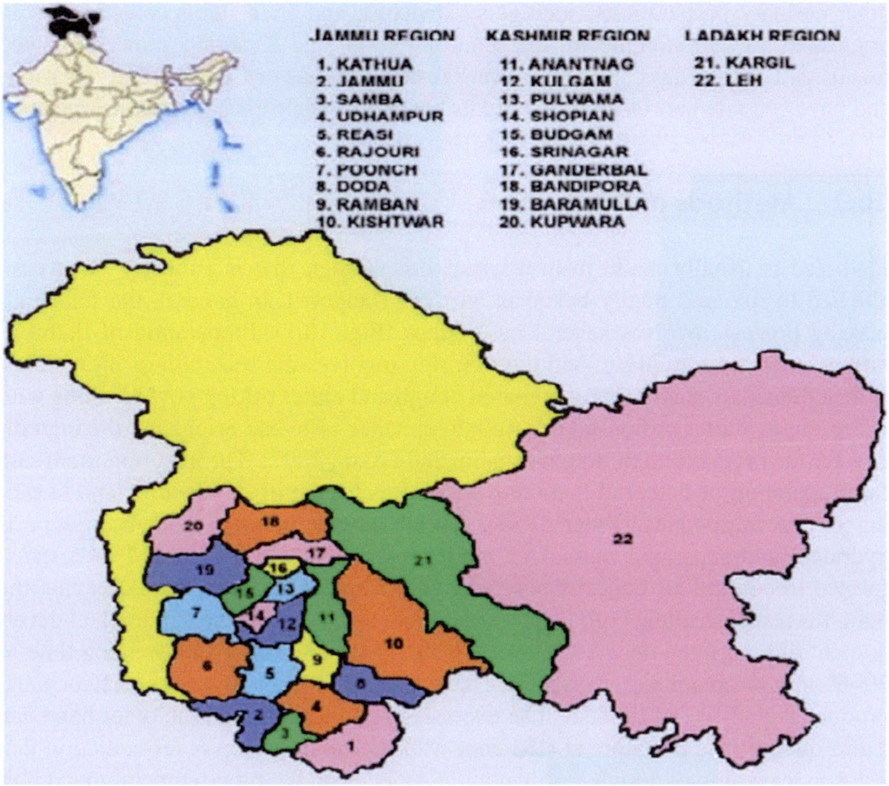

Fig. 10.1 Map of Jammu and Kashmir

in the manufacturing process of these fermented products to enhance their value in the functional food market were also reviewed (Fig. 10.1).

10.2 Cereal-Based Ethnic Fermented Foods

Grains are the staple nourishing food of India. The consumption of grains in different civilizations of India can be observed from the literature as they have good calorific values, perceived largely from starch and proteins. Wheat is commonly used to prepare loaves which are flat as staple diet in Jammu and Kashmir (Rehman et al. 2007). Generally, most of the cereal-based ethnic bread products of the state are prepared mainly from the fermentation process. Different forms of flatbreads are prepared depending on the composition of ingredients and process technology. Preparation involves the dough formation that is left at room temperature for fermentation. The rate of fermentation usually enhances by adding yeast to the dough. Traditionally flatbread is consumed in a variety of ways, principally served in various forms with any meal of the day. It can be eaten as a snack and used in other culinary preparations, such as sandwiches, and fried items coated in bread crumbs to prevent sticking. In the state of Jammu and Kashmir, some fermented breads and snacks are consumed during festive seasons or other special occasions and have become part of religious rituals beyond its importance as nourishment.

10.3 Methods of Production

Flatbread is usually made from a wheat flour dough that is cultured with yeast, allowed to rise and finally baked in an oven (tandoor). In general, the flatbread-making process involves several basic steps (Fig. 10.1). Preparation of flatbread differs from place to place, and the first step involves the selection of ingredients. Wheat flour, salt, traditionally prepared dough and either baking powder along with water, sugar, butter or non-fat dry milk to enhance taste and aroma are the ingredients for the preparation of dough (Shalini and Laxmi 2007). The next important step for preparation of flatbread is mixing which involves uniform blending and kneading of the ingredients (Pyler 1988). During dough mixing, the wheat flour is hydrated, and separated masses of gluten proteins are disrupted and then transformed to continuous cohesive viscoelastic gluten protein network. After that the dough is rested to allow complete flour particle hydration and relaxation and development of the gluten network. Normally the dough is given a short resting time of 30–60 min to obtain a more coherent form and achieve maximum consistency and produce a quality end product. The excessive resting period on the other hand can cause quality loss in flatbread (Hoseney 1988). The next step is fermentation that helps in leavening of dough. Fermentation depends on the time-temperature combination that usually varies from 0 to 3 h at room temperature (20 °C–35 °C) and is brought by baker's yeast which generates carbon dioxide and ethanol. Gas nuclei is retained, and dough is leavened that makes it lighter and extensible (Hoseney 1988). Sourdough fermentation has a detrimental effect on flatbread quality because it improves its flavour and texture, and thus it prolongs shelf life due to the formation of antifungal compounds and hence prevents staling (Corsetti et al. 1998). Moulding is an important step in flatbread production; whereby the dough is divided individually into loaf-sized pieces and then traditionally rounded between the palms of the hand. On a commercial level, the rounded dough is sheeted (rolled or flattened) into a circular shape in order to get a gas retention capability and good flattening (Boyacyoglu 1999). The dough is then placed for a short time followed by baking. During baking operation, physical, chemical and rheological changes such as loss of moisture content, volume increase, starch gelatinization, protein denaturation, crust formation browning, rupturing of gas cells, etc. occur. Time and temperature of the baking process determine the quality and shelf life of the bread. To achieve optimal baking conditions, the oven temperature should be kept in between 350 °C and 550 °C for good quality flatbreads. Sometimes, flatbreads are often baked in shorter time to retain more moisture and softness (Fig. 10.2).

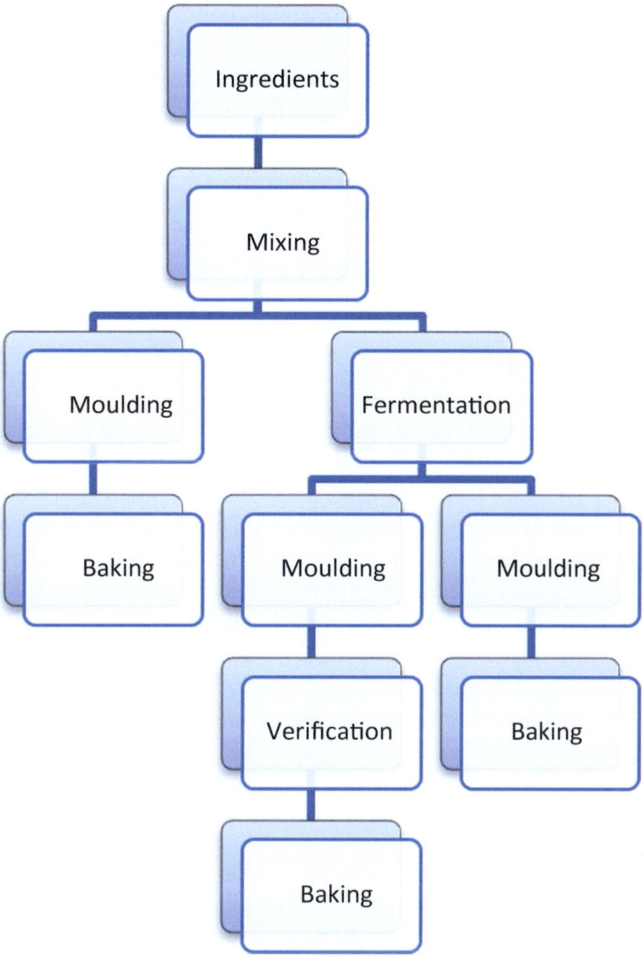

Fig. 10.2 Preparation of flatbread

10.4 Some Important Flatbreads Produced in Jammu and Kashmir

The state of Jammu and Kashmir has a long tradition of baking typical flatbreads in tandoor which are consumed generally at breakfast. Tandoor is an earthen oval in-ground oven, the walls of which are plastered with clay and are heated either by burning wood or natural gas. The sheeted dough is placed on a cloth pad and made to adhere to the internal wall of the tandoor for baking up to 60–90 s depending on the intensity of heat (Singh et al. 2000). These flatbreads are traditionally prepared by local bakers known as “Kandur”; while the sky is still tinged, they fire up wood-burning oven (tandoor) and start baking exquisite bread varieties. The first few hours of the morning are very hectic for the baker to cater to the demands of his customers.

These local breads are served as breakfast with butter along with salted green tea called "noon chai" and occasionally with omelettes. Some of the traditional breads consumed in J&K state include *chzot*, *katlam*, *kulcha*, *czhcwor*, *lavaas*, etc. which are made by these local bakers. These traditional flatbreads are commonly prepared from mixture of wheat flour, salt and water in different proportions; moreover, the difference lies in the use of other ingredients such as baker's yeasts, fat, skim milk powder and few food additives for improving quality and extension of shelf life (Qarooni 1996). The nature of flour and water of flatbreads is the most important ingredients for the development of texture, aroma and chewability (Zanoni and Peri 1993).

10.4.1 Girda/Chzot

Girda is a special Kashmiri traditionally baked disc-shaped flatbread with a crispy, furrowed exterior and a soft white pillowy interior often served as breakfast along with green tea. This is a single-layered flatbread consumed by local people on a daily basis with jam or butter and sometimes with omelettes. *Girda* is also known as *chzot* in Kashmiri language which is prepared from various ingredients including wheat flour, starter, baking soda, salt and water and is kneaded into a soft dough and allowed to ferment for 12 h. The fermented dough is rolled into small balls and sheeted and finally baked in tandoor. It has a golden-brown crust from the upper side due to polishing of its surface with skimmed milk paste and white colour on the lower side. The handmade indentations are created on the upper surface of *girda* with fingertips ranging from 7 to 9 in number (Mir 2011).

10.4.2 Ghyev Czhot

Ghyev czhot is another popular fermented baked flatbread of Kashmir. It is prepared in the same way as *girda* and is made from wheat flour dough that is fermented overnight and then baked in tandoor. A thin layer of ghee and sprinkling of poppy seeds is brushed on the upper side of the *ghyev czhot* that makes it different from *girda.* Due to the presence of ghee on the upper portions, it possesses light-brown crust, a soft pliable hand feel and squishy and slightly chewy texture and gets easily broken down in the mouth with typical pleasant taste and aroma. *Ghyev czhot* is popularly consumed during the month of Ramadhan and several special occasions and served with *wazwan* in social functions. These flatbreads can also be relished with an array of vegetarian and non-vegetarian gravy recipes.

10.4.3 Lavash/Lawaas

Lavash also known as *lawaas* is a kind of fermented, thin, large unleavened flatbread consumed by people on a daily basis with butter or jam. It is prepared from maida (finely milled wheat flour), water, yeast and sour starter. Traditional production of *lawaas* includes a well-developed sourdough starter, wheat flour, yeast, sugar and salt mixed to develop a cohesive dough and then allowed to ferment for 1–3 h. A small portion of the dough (100 g) is sheeted on a cotton pad, docked and baked at 480 °C for about 1.5 min. Typically, *lavash* is circular in shape and 60–70 cm long, 30–40 cm wide and 2–3 mm thick (Gargari et al. 2007). These unleavened flatbreads are well characterized by a papery thin blistered upper surface possessing a creamish white colour and soft or crispy texture like naan. *Lavash* is often consumed as a wrap for chickpeas and meat.

10.4.4 Czochwor

Czochwor is a kind of donut-shaped, fermented, slightly hard bread sprinkled with sesame seeds. It is a small, soft and round bread resembling that of bland bagel having a diameter of about 3 in. and a circumference of 6 in. *Czochwor* is also known as "teliwor" which is prepared from leavened dough of slight viscosity and is traditionally consumed in a variety of ways, principally as an evening and afternoon bread. For its preparation, flour is mixed with salt, water, shortening, sugar, bakers' yeast and sourdough starter and kneaded for 15–20 min until an elastic dough is obtained. The dough is first fermented for 50–60 min at 25 °C and divided in an appropriate weight and rounded. After that, it is proofed and again fermented for 30–40 min. Other ingredients such as sesame seeds are sprinkled on the upper portion of the bread that makes a crispy texture. It is then baked at a temperature of 320 °C for 15 min in the tandoor.

10.4.5 Kulcha

Kulcha can be defined as a circular, hard dry, crumbly fermented bread with a creamy yellow colour. For its preparation, flour is mixed with yeast solution, sourdough, oil, salt and milk to make a soft stiff dough and is left to ferment for 4 h at 30 °C. The dough is then scaled, divided, rounded and given a final proof of 15–20 min at 30 °C, and then the dough pieces are rolled to a thick round shape of about 6 cm wide. The disc-shaped dough is then sprinkled with poppy seeds and baked at a temperature of 320 °C for 15 min in the traditional tandoor. *Kulchas* are traditionally prepared by the local bakers during many festive occasions including religious and marriage ceremonies. Another type of *kulcha* is sweet *kulcha* or *khatai* that possesses a sweet taste and is bigger in size than salted *kulcha*. Similar ingredients are used to make the *khatai*, but sugar is added instead of salt.

10.4.6 Sheermal

Sheermal, also known as *krippe*, is a saffron-flavoured, dry, crumbly traditionally fermented flatbread with a long shelf life, and the word is a combination of sheer, which means "milk", and mal, which means loosely connoting "rich food". The origin of *sheermal* dates back to Mughal period where it was an upper-class food tracing its history to the Nawabi cuisine and also was a part of Awadhi cuisine. It is a mildly sweet flatbread common in Lucknow, Hyderabad and Kashmir in India and in other countries like Pakistan, Iran and Bangladesh. There are different varieties of *sheermal* made in Kashmir, and certain places like Pampore are famous for producing fresh and crispy *sheermals*. In Kashmir, both sweet and salt varieties of *sheermal* bread are made where the sweet type is generally served with "kehwa", a traditional tea made with water and mixed with saffron, sugar and almonds. The salty type on the other hand is consumed with green tea known as *noon chai*. Traditionally, *sheermal* is prepared by using different ingredients that include flour, sourdough starter, baker's yeast, milk and some other ingredients. For its production, all the ingredients are mixed together and kneaded into fully developed dough and fermented for 2 h. After that the dough is again kneaded for a minute in order to release any air pockets if present; then rolled, relaxed and spread into circular sheets; and finally brushed with saffron-flavoured milk, sprinkled with sesame seeds on the top surface, and baked for 10–12 min at a temperature of 199 °C.

10.4.7 Bakarkhani

Bakarkhani is another traditionally thick crispy-layered Kashmiri flatbread similar to puff pastry in appearance. Historical accounts have suggested that *bakarkhani* belongs to the Mughlai cuisine of the Indian subcontinent and is believed that this Mughlai bread has been originated from the Central Asia and arrived to the Indian subcontinent as part of Mughal dynasty. *Bakarkhani* is made from flour, semolina, sugar, sourdough, clarified butter, poppy seeds, etc. The dough is allowed to ferment for 2–3 h and finally stretched. The sheeted dough is folded, coiled and interleaved with ghee, sprinkled with poppy seeds and then baked in tandoor. Typically, *bakarkhani* is often prepared on certain religious occasions (Urs) at Sufi shrines where halva, a sweet dish, is wrapped in this bread which is called paratha. In addition, *bakarkhani* is served as accompaniments to chicken or rogan josh trays in social ceremonies like childbirth, wedding and engagements.

10.4.8 Roath

Roath is a traditional Kashmiri dry fruit pancake which is a usually one-meter-long, two-and-a-half-meter-wide large flatbread. *Roath* has a sociocultural importance and is usually prepared during the Kashmiri Hindu festival, "Pun", in the month of September, and puja is performed for invoking gods to seek blessings of prosperity and good luck.

10.4.9 Loochi

Loochi is a traditionally fermented savoury snack prepared from fine flour containing mild spices. *Loochi* has always been closely related with the cultural and religious practices of the Kashmiri Pandits, prepared domestically by the local population at the time of celebrating festive occasion including Kheer Bhawani Mela. This savoury snack is distributed at the end of congregational service often termed as "Bhog". It is prepared by mixing flour, carom seeds and whipped curd together to form a soft dough. The resulting dough is allowed to ferment for 2–3 h. The leavened dough is then rolled into discs, fried in oil and taken out from the pan quickly in order to retain colour and prevent crispness of the final product.

10.4.10 Yaji

Yaji is another popular, convenient ready-to-eat steamed dumpling-filled snack that is made from rice flour. It holds a great significance among Hindu folk where it is chiefly prepared during the rainy season known as vahrath. It is a traditionally fermented healthy snack that is eaten as Bhog or Prasad and is believed to improve digestion and eradicate some stomach ailments. Ethnically, it is prepared using a mixture of walnut, salt and water which is added to flour containing carom and cumin seeds. The resulting mixture is kneaded to form a soft dough and is left to ferment for a short period. The dough is then divided and rolled between the palms, moulded into the shape of a cup and either cooked with steam for about half an hour or deep-fried in vegetable oil until a golden-brown colour is obtained (Table 10.1).

10.5 Vegetable-Based Ethnic Fermented Products of Jammu and Kashmir

Vegetables are an important component of human diet due to their diverse nutrient profile and rich phytochemistry. According to the Asian Vegetable Research and Development Center (AVRDC 1992), vegetables are an edible plant or a portion of it is eaten with staples as a main course or supplementary food in cooked or raw form. They provide nutrients like vitamins, minerals, fibre and non-nutrient phytochemicals that may contribute to the normal functioning of the human body (Wettasinghe et al. 2002). Various processing methods have been applied to preserve and enhance the storage life of these vegetables. The capability to store vegetables and protect them from spoilage made fermentation an important method of preservation. Fermentation is defined as the biochemical process in which changes are brought about in organic substrates by the action of microorganisms resulting in the conversion of degradable food components into more stable forms (Caplice and Fitzgerald 1999). Fermentation not only inhibits the growth of pathogens but also improves its safety (Adams and Mitchell 2002). Fermentation of vegetables can occur in three ways, spontaneous fermentation by natural microflora, fermentation

Table 10.1 Ethnic fermented types of bread commonly consumed in Jammu and Kashmir

Product	Common name	Main ingredients	Type of fermentation	Traditional food uses
Chzot or girda	Flatbread	Wheat flour, sour starter, baking soda, salt with water	Fermentation followed by baking	Consumed throughout the year
Ghyev czhot	Crispy flatbread	Wheat flour, sour starter, baking soda, salt, *ghee* with water	Fermentation followed by baking	Mainly prepared during the fasting month of Ramadhan and other festive occasions
Lawaas	Flatbread	Wheat flour, sour starter, baking soda, salt, sugar with water	Fermentation followed by baking	Consumed throughout the year along with tea
Czochwor	Bagel bread	Wheat flour, water, shortening, sugar, bakers' yeast, sourdough starter and poppy seeds or sesame seeds with water	Fermentation followed by baking	Consumed throughout the year, especially as evening snack religious ceremonies
Kulcha	Round, dry crumbly bread	Wheat flour, sugar, salt, bakers' yeast, sourdough starter, ghee, milk, poppy seeds and water	Fermentation followed by baking	Served during weddings
Sheermal	Crispy flatbread	Flour, sourdough starter, baker's yeast, milk, saffron, sesame seeds and water	Fermentation followed by baking	Consumed throughout the year during several festive occasions
Bakarkhani	Puff-layered bread	Flour, semolina, sugar, sourdough, clarified butter, poppy seeds and water	Fermentation followed by baking	Consumed mainly during weddings and other festive occasions
Roath	Dry fruit-enriched pancake	Flour, baker's yeast, sugar, ghee, dry fruits, eggs and water	Fermentation followed by baking	Consumed during weddings and local festivals
Loochi	Leavened fried bread	Four, carom seeds, whipped curd, oil and water	Fermentation followed by frying	Consumed during the religious festive occasions and when making vows
Yaji	Steamed dumplings	Flour, walnuts, salt, cumin seeds, carom seeds, oil and water	Fermentation followed by frying or boiling	Consumed during rainy season (vahrath) and religious ceremonies

by starter cultures that are added into raw materials and fermentation of heat-treated materials by starter cultures (Hammes 1990). Fermentation occurs at an optimum temperature ranging from 20 °C to 30 °C (Poutanen et al. 2009). Moreover, fermentation also improves the nutrient content, bioavailability and organoleptic properties of foods (Poutanen et al. 2009). The predominant bacterial floras involved in fermentation are *Leuconostoc mesenteroides*, *Lactobacillus brevis*, *L. plantarum*, *L. casei*, *Pediococcus pentosaceus*, etc. (Fig. 10.3)

Fig. 10.3 Some of the popular traditionally fermented breads consumed in Jammu and Kashmir: (**a**) *ghyev czhot*, (**b**) *girda/chzot*, (**c**) *lavash/lawaas*, (**d**) *czochwor*, (**e**) *kulcha*, (**f**) *sheermal*, (**g**) *bakarkhani*, (**h**) *roath*, (**i**) *loochi*, (**j**) *yaji*

Even though the world has advanced technologically, e.g. preparing fast food with a tick of a clock, Jammu and Kashmir particularly Srinagar and Jammu districts stick to the traditional way of preparing pickles. Besides the vegetarian varieties, it includes pickling of chicken, mutton, fish, etc. Among the vegetables garlic, knol-khol, kale, lotus stem and bitter gourd locally known as *rohun*, *mounji*, *haakh*, *nadru* and *kareel achar*, respectively, are famous throughout the valley because of their nutraceutical properties. Pickle making has the potential to be developed as an industry in Jammu and Kashmir which can help in earning foreign exchange and in preserving the pickle-making art which forms an important part of the culture of Jammu and Kashmir.

10.6 Methods of Production

Pickles are a very important part of the Jammu and Kashmir meal that enhances the taste and increases the satisfaction after every meal. The most common spices that are the ingredients in making pickles include asafoetida, red chilli powder, turmeric, fenugreek, mustard seeds and carom seeds. Salt is generally used for imparting preservative effect and for enhancing taste. Even using the same main ingredients, pickles come in a wide variety of flavours due to differences in their spice formulation and preparation techniques, e.g. kadam (knol-khol) pickle from Jammu may taste very different from the one made in Kashmir. In Jammu region, sesame oil is generally preferred, while mustard oil is generally preferred in Kashmir for manufacturing of pickles. The traditional practice of pickling involves cleaning, cutting and blanching of vegetables. The vegetables are gathered together, tossed to mix them well and then sun-dried in order to remove the surface moisture. Spices like mustard and coriander seeds are lightly dry-roasted and coarsely crushed. The other spices are mixed in mustard oil and then added to the vegetables. Then the mixture is filled in earthen pots and is well pressed down on the top, till some of the oil starts seeping up through the top. The pushing and pressing develop a shear pressure that creates anaerobic environment for fermentation. Then the mixture is left to ferment for 2–3 weeks. Pickles prepared by fermentation contain many useful microbes (probiotics), but pickles prepared from vinegar do not possess probiotic effect. The general method for the preparation of pickles is shown in Fig. 10.4.

10.7 Some Indigenous Pickles Manufactured in J&K

10.7.1 *Garghal* (Lemon) Pickle

It is a traditional lemon pickle prepared in Jammu region especially in Dogra households. It is usually consumed with *paratha*, *chapati* or *khichdi*. Depending upon the taste, different ways are employed to prepare this kind of pickle. It is usually prepared in winter season as lemons are easily available during this time of the year. It helps in improving digestion. For its preparation lemons are washed, pat dried and

Fig. 10.4 Preparation of pickles

Fruits/vegetables
↓
Washing
↓
Peeling/coring/shredding
↓
Blanching
↓
Mixing of different spices in the form of paste
↓
Fermentation
↓
Packaging
↓
Final product (pickle)

cut into small pieces followed by the addition of roasted fenugreek seeds. The other ingredients added to the prepared mixture include aniseeds, carom seeds, turmeric powder and red chilli powder, which are previously shallow fried in mustard oil. Finally, salt is added to impart taste and flavour. The resulting mixture is then stored in clean, dried bottles till fermentation takes place which lasts for 10 days.

10.7.2 *Kasrod* (Fiddlehead Fern) Pickle

Fiddlehead fern pickle commonly known as *kasrod* pickle is another popular fermented product of Jammu region which is usually made in Dogra households. It can be served with chapati, paratha or bread. Fiddleheads are furled fronds of a young fern, harvested for vegetable use. Fiddleheads possess antioxidant activity, contain omega-3 and omega-6 fatty acids and have high iron and fibre content. However, some varieties of fiddlehead are believed to be carcinogenic. *Kasrod* pickle is prepared by shaving off the skin, and then it is cut into small pieces followed by blanching in water with addition of little turmeric powder. Once blanched, the pieces are thoroughly washed to remove any adhered dirt or unwanted impurities. The pieces are then sundried for 1–2 h and are well mixed with other ingredients which include red chilli powder, turmeric powder, salt, mustard seed powder and mustard oil. The resulting mixture is then kept in sterilized containers and left to ferment for about 10 days at room temperature (25 °C) to achieve complete fermentation.

10.7.3 *Zimikand* (Elephant Foot Yam) Pickle

Elephant foot yam (*Amorphophallus paeoniifolius*) is cultivated extensively in rainfed and hilly tribal regions of India (Singh et al. 2016). Besides being used as a vegetable, the tubers can also be utilized for making pickles. Ethnically known as *zimikand*, the pickle gains its popularity in Jammu regions as it is now cultivated locally. The pickle is usually prepared from starch-rich corm by lactic acid fermentation. Its preparation includes peeling, washing and removal of surface moisture, followed by cutting into small pieces, and then blanching for 20 min. The yam is mixed with 50% of total salt, turmeric powder and tamarind and left for 36 h. The remaining salt and other roasted spices (fenugreek seeds, asafoetida, carom seeds and mustard seeds) are well mixed with yam and stored in glass jars at room temperature for about 15 days till fermentation gets completed.

10.7.4 *Lasooda/Gunda* (Glue Berry) Pickle

Lasooda (*Cordia dichotoma*), also known as glue berry, is a tree seen all over the plains of India. The fruits of glue berry are oval, are sweet in taste and contain a solitary fruit. It is widely cultivated in the regions of Jammu division (Krishna and Sharma 2014). The fruits of *lasooda* are used in pickle preparations. *Lasooda* pickle is made in two ways, one with spices and the other without any spices. It is usually prepared by breaking the stems of *lasooda* followed by washing and boiling for 3–5 min. The porridge of *lasooda* is prepared and is mixed with other ingredients which include turmeric powder, salt, red chilli powder and mustard oil. The final mixture is packed in previously sterilized containers and allowed to ferment for about 10 days at room temperature. *Lasooda* porridge can be consumed either as a drink or can be fermented into *lasooda* pickle.

10.7.5 *Rohun* (Garlic) Pickle

Garlic (*Allium sativum*), also called as stinking rose, is an herb that has been used by humans as a source of antimicrobial agent. It was formerly classified in the lily family; now it is included as a member of the Alliaceae family (USDA 2006). Garlic in local language is known as "rohun". The pickle gains its popularity as it enhances the immune system and is used as an agent for prevention and treatment of cardiovascular disease, atherosclerosis, hyperlipidaemia, thrombosis and diabetes (Banerjee and Maulik 2002). Fermented garlic not only lowers down the cholesterol level but also triglycerides. The preparation of pickle involves the following steps: First the garlic bulbs are separated into individual cloves, wherein the small and defective cloves are discarded and the fresh cloves are peeled and washed under tap water. The cloves are blanched in water at 90 °C for 15 min (Raja et al. 2016). The surface moisture is removed by spreading the cloves on stainless steel trays for a few hours. It is then mixed with vinegar, salt, mustard seeds, fennel seeds, carom seeds and crushed red pepper. The resultant mixture is then boiled over low heat with continuous stirring until the salt is completely dissolved. The mixture is then poured in already sterilized jars and kept aside for 10 days to ferment.

10.7.6 *Mounji* (Knol-Khol) Pickle

Knol-khol (*Brassica oleracea*) is a round and tuberous vegetable in the *Brassicaceae* family and is native to Europe. It is widely grown and cultivated in J&K particularly in Kashmir region. It is grown all over the temperate climates for its succulent round-shaped modified stem as well as for its turnip-flavoured top greens. Two varieties of knol-khol exist in J&K, white and purple. Knol-khol stem is a rich source of vitamin C and contains health-promoting phytochemicals such as isothiocyanates. It also contains B-complex groups of vitamins that act as cofactors to enzymes during various metabolic functions inside the body. The swollen stem is generally used in the manufacture of pickle called "mounji" achar. The stem is thoroughly washed under tap water. The material is washed, peeled and shredded into pieces (2–4 cm in length and 1 cm thick). The spices like garlic and ginger are peeled, chopped and then coarsely pounded. Red chilli powder, mustard seeds and turmeric are used in powder form. However, aniseeds are used in intact form. Shredded vegetables are dried under sunlight for 2–3 h in order to remove the surface moisture with periodic turnings. The shredded material is mixed with specific quantity of spices and other ingredients. The salt in the concentration of 2.5–3% is added to the final mixture. The material so prepared is then filled in previously sterilized glass containers, and fermentation is allowed for 7–10 days at 21 °C. Moreover, knol-khol leaves (commonly called *mounji haakh*) can also be used in the pickle preparation as it is a rich source of vitamins, minerals and carotenes.

10.7.7 Haakh (Collard Greens) Pickle

Collard greens are loose-leafed cultivars of *Brassica oleracea*. They are part of the *Acephala* group of species, which includes kale and spring greens. According to the George Mateljan Foundation for the World's Healthiest Foods, collard greens probably descended from wild cabbages found in Asia. They eventually spread through Europe, Greece, Rome and the USA. Collard greens contain substantial amounts of vitamin K. They also contain rich sources of vitamin A, vitamin C and manganese and moderate sources of calcium and vitamin B6. Collard green is a daily dish in Kashmir and is one of the most popular garden vegetables. They are also grown and eaten regularly in many countries across the world. In Brazil, a dish known as *couve a mineira* is prepared by sauteing collard greens in olive oil and butter. In Portugal most families use collards in a soup known as *caldo verde* or green broth. Similarly, in Kashmir region of J&K, *haakh* is used as a daily dish and is incorporated into a traditional and elaborate multi-course feast called wazwan. Collard greens are famous in Kashmir in the form of pickles. In the preparation of pickle, *haakh* is thoroughly washed under tap water. It is then shredded into small pieces and dried under sunlight for 3–4 h with periodic turnings to remove the surface moisture. The leaves of *haakh* are then mixed with the paste of specific quantity of spices. The spices include mustard seeds, carom seeds, red chilli powder, turmeric powder, cumin seeds and fenugreek. For best results 2.5% of salt is added to the mixture, is packed in previously sterilized containers and is left to ferment for 10–15 days at optimum temperature of 21 °C.

10.7.8 *Nadur* (Lotus Stem) Pickle

Lotus stem (*Nelumbo nucifera*), an herbaceous perennial aquatic plant belonging to the *Nelumbonaceae* family, is considered to be a sacred plant by Asian cultures as it is able to unfold perfectly clean leaves out of muddy and dirty water (Muller et al. 2007). The rhizome of lotus plant locally known as "Nadur" is edible, is extensively used in Kashmiri cuisines and is used for pickling (*Nadur achar*). The stem of lotus is long and tender that is planted in the soil of the pond or river bottom and can stretch up to 4 ft. It has a wide range of health benefits owing to its rich nutritional composition such as lowering blood pressure, reducing stress, reducing cholesterol, improving digestion, boosting the immune system, preventing various forms of cancer, improving blood circulation and maintaining proper enzymatic activity in the body (Wang-Tsau et al. 1972). In the preparation of pickle, lotus stems are washed thoroughly, and the surface moisture is removed by a muslin cloth. The stems are chopped, blanched for 10 min, followed by straining water, and then sun-dried for 1–2 h. The chopped stems are further mixed with a specific quantity of spices (nigella seeds, fennel seeds, coriander seeds, kalonji seed, carom seeds, turmeric powder, red chillies). The mixture is heated with mustard oil, bottled in sterilized jars and allowed to ferment for 10 days at 21 °C (Fig. 10.5) (Table 10.2).

10.8 Ethnic Fermented Milk Products of Jammu and Kashmir

The Gujjars and Bakerwals of Jammu and Kashmir are defined as seminomads concentrated in specific pockets all over the state. Generally their economy depends on cattle rearing and some agricultural activities. For this purpose, they migrate from hilly areas to other cattle-grazing regions where availability of green fodder is abundant for their livestock and also to avoid severe cold and snowfall that affect the health, growth, reproduction, milk production, lactation, etc. of their breeding stock, which is their major source of livelihood (Magray 2003). Hence, pastoralism is the mainstay of these nomadic tribes who live in the elevated regions of Jammu and Kashmir. Milching animals reared in these regions include buffalo, cow, sheep and goat. Milk and milk products are used most commonly by these communities. Owing to its perishable nature with a very short shelf life, fresh milk possessing high nutrient density makes milk a suitable medium for microbial contamination and colonization by autochthonous, spoilage and pathogenic microorganisms.

Fig. 10.5 Some of the popular traditionally fermented pickles consumed in Jammu and Kashmir: (**a**) *garghal achar*, (**b**) *kasrod achar*, (**c**) *zimikand achar*, (**d**) *lasooda achar*, (**e**) *rohan achar*, (**f**) *mounji achar*, (**g**) *haakh achar*, (**h**) *nader achar*

Table 10.2 Traditionally fermented pickles consumed in Jammu and Kashmir

Product	Common name	Scientific name	Main ingredients	Type of fermentation	Traditional food uses
Garghal achar	Lemon pickle	*Citrus limon*	Lemons, fenugreek seeds, aniseeds, carom seeds, oil, salt, turmeric and chilli powder	Spontaneous	Served with chapati, paratha or khichdi
Kasrod achar	Fiddlehead fern pickle	*Athyrium esculentum*	Fiddleheads, oil, salt, turmeric, chilli powder and mustard seed powder	Spontaneous	Consumed with chapati, paratha or bread
Zimikand achar	Elephant foot yam pickle	*Amorphophallus paeoniifolius*	Yam, salt, turmeric, tamarind, fenugreek seeds, aniseeds, carom seeds and mustard seeds	Spontaneous	Consumed especially in summer seasons
Lasooda achar	Glue berry pickle	*Cordia dichotoma*	Lassora, turmeric, salt, mustard seeds oil and chilli powder	Spontaneous	Consumed as beverage
Rohun achar	Garlic pickle	*Allium sativum*	Garlic, vinegar, salt, mustard seeds, fennel seeds, carom seeds and crushed red pepper	Quick	Mainly consumed due to its therapeutic value
Mounji achar	Knol-khol pickle	*Brassica oleracea*	Knol-khol, ginger, garlic, mustard seeds, aniseeds, turmeric and chilli powder	Spontaneous	Consumed with meals (rice)
Haakh achar	Collard green pickle	*Brassica oleracea*	Collard, carom seeds, red chilli, turmeric, cumin seeds and fenugreek seeds	Spontaneous	Consumed with meals (rice)
Nader achar	Lotus stem pickle	*Nelumbo nucifera*	Lotus stem, nigella, fennel, coriander, kalonji seeds, aniseeds, turmeric salt and red chillies	Spontaneous	Consumed with meals (rice)

Hence for the purpose of improving digestibility and extending shelf life of milk, it is naturally fermented at ambient temperature or with the addition of starter cultures in the raw form or after pasteurization (Jans et al. 2017). Thus fermentation with the help of lactic acid bacteria is encouraged for extending shelf life of milk as well as bio-improvement in lactose metabolism (Tamang 2010). Milk-based fermented products are mostly prepared by the addition of lactic acid bacteria in the form of starter culture to milk (cow, buffalo, sheep or goat) and are allowed to ferment in order to convert lactose (milk sugar) into lactic acid (Rawat et al. 2018). Naturally fermented milk products are widely produced in different agroclimatic regions providing nutrition and income generation. There are various reports on the stimulation of health-promoting functions of LAB present in fermented milk products, like yoghurt and acidophilus-fermented milk. These products are effectively used in the treatment of gastrointestinal tract (GIT) disorders, including colitis, diarrhoea and constipation (Sanders 1993). Consumption of fermented milk products is often regarded as popular for thirst quenching, along with condiments consumed with stiff porridge or sometimes with gruel (Bille 2014).

Naturally fermented milk products are numerous and predominantly available in different parts of the world. Apart from cheese and yoghurt, examples include kefir (Russia), koumiss (Russia), laban (Lebanon), kurut (China), chhurpi (India), shrikhand (India), etc. (Ghatani and Tamang 2017). A wide variety of traditionally fermented milk products are prepared and consumed in the state of Jammu and Kashmir that includes *dahi* or curd, *lassi*, *kalari or kradi*, *phuh*, *kaenz*, etc. The method of preparation, fermentation, utilization, procedural nuances and multiplicity of appellation is quite unique to each fermented product that manifests their ethnic diversity.

10.8.1 Kalari or Kradi

Kalari or kradi traditionally known as "maeshkraj" is a semisoft variety of cheese found in Himalayan regions of Jammu and Kashmir, India (Lawrence 1886). Traditionally, it is manufactured primarily by the tribal population locally called as Gujjars and Bakarwals living in the hilly regions of Shopian, Kupwara, Pahalgam, Poonch and Rajouri Jammu and Kashmir, India. It is prepared from the acidified buttermilk released during churning of curd to release desi butter (Punoo et al. 2017). The buttermilk is then added to fresh buffalo milk in the ratio of 1:5 and heated at 45–60 °C. The heating of the mixture results in the separation of whey and casein coagulum. Leftover whey is separated out, and the coagulum is heat stretched until dough-like mass is formed. Finally, small disc-shaped portions of varying diameter ranging from 5 to 20 cm with a thickness of about 0.2–2.5 cm are made (Mushtaq et al. 2015). The cheese has also been documented as "mozzarella of Kashmir" because of its stretchability and meltability during heating or baking (Lawrence 1886). The cheese possesses high moisture content and exhibits bright, smooth and white colour surface with slightly sour flavour. This cheese has a tremendous market potential and is relished as a dairy delicacy, often served to the guests in festive occasions and marriage parties.

The major *Kalari* microbiota includes *mesophilic aerobes*, *aerobic psychrotrophs*, *thermophilic lactobacilli*, *mesophilic lactobacilli*, *lactococci* and *Enterobacteriaceae*, and a few yeast and mould species have been reported (Mushtaq et al. 2015). Lactics present in kalari perform a number of functions, which include texture development, enhancing flavour, taste and inhibiting the growth of spoilage and pathogenic bacteria. Other health benefits include antioxidant, opioid, antihypertensive and antithrombotic effects, due to the presence of bioactive peptides that are released after their in vivo or in vitro hydrolysis by microbial enzymes or by heat (Hafeez et al. 2014). The product has a shelf life of 5–7 days under normal conditions and can be improved by adopting innovative processes and suitable storage conditions (Conte et al. 2004).

10.8.2 *Phuh* (Colostrum Cheese)

Colostrum is defined as milk secreted from the mammary glands of cow or buffalo immediately after parturition (first 24 h) or a few days after calving (Tsioulpas et al. 2007). Bovine colostrum acts as an effective natural immune booster due to the presence of proline-rich bioactive polypeptides that perform several immunomodulatory and immunostimulatory functions (Houser et al. 2008). Recent studies have reported that colostrum contains essential nutrients. In addition to this, antibacterial, antiviral and antifungal properties of colostrum have also been documented (Moore et al. 2005). It has been estimated that on average, 5–10 L of colostrum is produced by a bovine, in which 27% of colostrum produced is fed to calves, while 33% gets wasted (Kim and Heo 2000). Different types of functional foods are available in the market that have been fortified with colostrum; an example includes Azaco cheese which is prepared from cow colostrum (Srivastava et al. 1984). In the state of Jammu and Kashmir, colostrum is locally known as Kheder which is traditionally converted into a coagulated lump of proteins known in local parlance as “phuh” along with fat and other constituents entrapped into it by the action of heat. *Phuh* resembles closely to paneer that is typically a heat- and acid-coagulated variety of cheese. In its preparation, colostrum milk is pasteurized and inoculated with 2% starter culture (mesophilic *Lactobacillus* species) of local origin and is allowed to stand for 3–4 h for acidity development. Then the above colostrum is subjected to heat coagulation at 70 °C for 10 min, and the resulting coagulum is separated from whey. It is consumed in the form of a culinary dish. Traditionally, it is distributed among family members, relatives and friends as a special dish for celebrating arrival of a new calf, a propitious sign in the family.

10.8.3 Dahi

Curd is popularly known as *dahi*, a well-known fermented and commonly used traditional dairy product of Jammu and Kashmir, India. It is a lactic acid-fermented product of cow or buffalo milk or combination of both. Generally, curd is prepared by boiling milk and allowed to cool. It is then inoculated with the previous batch of buttermilk or starter culture and incubated overnight to allow fermentation at room temperature. The increased acidity results in the coagulation of milk proteins (casein) to form casein gel of ordered network structure known as curd. The microflora involved in the technology of curd production is most commonly made up of strains of *Streptococcus cremoris, S. lactic, S. thermophilus, Lactobacillus bulgaricus, Lactobacillus acidophilus, Lactobacillus helveticus and Lactobacillus cremoris* (Mudhu et al. 2013). The mechanism of curd formation involves lactose-fermenting microorganisms that convert milk sugar lactose into lactic acid with the help of lactase enzyme. Lactose molecule is further broken down into simple sugars releasing glucose and galactose through glycolytic pathway with the formation of lactic acid as an end product. Eventually a portion of lactic acid combines with calcium of casein to form calcium lactate, thus releasing casein coagulum.

$\mathbf{C_{12}H_{22}O_{11} + H_2O}$ *Lactic acid bacteria* $\mathbf{C_6H_{12}O_6 + C_6H_{12}O}$
(Lactase)
$\mathbf{C_6H_{12}O_6}$ *Glycolytic pathway* $\mathbf{C_3H_6O_3}$
(Glucose) (Lactic acid)
Ca caseinate + lactic acid → Ca lactate = casein (curd form)

Dahi is an important fermented food which can be eaten directly in sweetened or salted form. Different types of curd such as sweet curd, sour curd and flavoured curd are available in the market. Sweet *dahi* is generally prepared from a mixed culture of *L. lactis, S. thermophilus* and *R. citrophilus* with the addition of sugar in the range of 8–12% volume of milk. Flavoured *dahi* is prepared by the addition of synthetic flavour or natural fruit juice (Sarker et al. 2018). According to the Prevention of Food Adulteration Act (2006), *dahi* is defined as "a semisolid product", obtained from pasteurized or boiled milk by souring using harmless lactic acid or other bacterial cultures. It should have the same minimum percentage of fat and solid not fat (SNF) as the milk from which it has been prepared (De 1980; Kumar and Mishra 2004). Curd is generally consumed by a large number of people and has been designated as a functional food due to the presence of balanced nutrients which include fats, carbohydrates, proteins, minerals and vitamins (Fig. 10.6). In addition, curd forms the richest source of probiotics containing live microorganisms which upon ingestion exert health benefits beyond inherent general nutrition (Sekar and Mariappan 2007). Consumption of curd is effective for treating various illnesses like dyspepsia, dysentery and other intestinal disorders due to the presence of probiotics that lowers intestinal pH, thus preventing the growth of undesirable organisms, and improves digestibility (Dewan and Tamang 2007).

Fig. 10.6 Traditional method of *kalari/kradi* preparation

Buffalo milk
↓
Pasteurization
↓
Cooling
↓
Admixing of starter (buttermilk)
↓
Heat coagulation
↓
Draining of whey
↓
Moulding/patting
↓
Packaging
↓
Surface drying
↓
Storage

10.8.4 Lassi

Lassi (stirred curd) is a ready-to-serve traditionally fermented milk beverage consumed in summer season which is mainly popular in North India. It is a white to creamy white viscous liquid with a sweetish or salty, rich aroma and mild acidic taste (Rasane et al. 2017). *Lassi* is not only refreshing and delicious but also a nutritious and easily digestible beverage; hence, it is quite popular amongst all age groups. Traditionally it is prepared from curd, by blending with water. To enhance the taste and depending upon regional and consumer preferences, lassi can be either flavoured with salt or sugar and other condiments and spices. *Lassi* is a probiotic product due to the presence of active cultures that offer beneficial and healthy microflora to the alimentary canal through diet without any risk of adverse effects (Sarkar et al. 2015) (Table 10.3).

Table 10.3 Ethnic fermented dairy products consumed in Jammu and Kashmir

Product	Common name	Types of product	Main ingredients	Type of fermentation	Traditional food uses
Maeshkraj	*Kalari/ Kradi*	Soft cheese	Buffalo buttermilk, raw buffalo milk	Fermentation followed by acid-heat coagulation	Consumed mainly by tribal population and festive occasions
Phuh	Colostrum milk	Soft cheese	Colostrum milk	Fermentation followed by acid-heat coagulation	Consumed at puerperium and as a remedy for body weakness and festive occasions
Zamuthdod	*Dahi*	Curd	Milk, starter culture	Spontaneous fermentation	Consumed throughout the year and festive occasions
Laes	*Lassi*	Diluted curd	Milk, starter culture, water, salt, condiments	Spontaneous fermentation	Consumed throughout the year as a refreshing drink

10.9 Nutraceutical Potential of Fermented Foods

The metabolic activity of microbial cells during the food fermentation process gives rise to various bioactive compounds that perform functional health-promoting antimicrobial, antioxidant, antihypertensive, antidiabetic, antimutagenic, anticancer and other activities in the human body. Lactic acid being the principal acid produced during fermentation reduces the pH to below 4.2 that facilitates inhibition of spoilage and pathogenic microorganisms, which are the primary cause of food-borne diseases and human illnesses (Giraffa 2004; Holzapfel 1997). Apart from producing lactic acid during the process of fermentation, the LAB produce bacteriocins that inhibit bacteria and other microorganisms. Various researchers have reported antimicrobial effects of fermented foods against different microorganisms, which include enteropathogenic *Escherichia coli*, *Salmonella typhi*, *Salmonella paratyphi* (Odugbemi et al. 1991), *Shigella dysenteriae* (Mbugua and Njenga 1992), *Campylobacter jejuni*, *Shigella flexneri* (Muyanja et al. 2003), *Staphylococcus aureus*, *Salmonella enterica* (Omar et al. 2006), *Klebsiella pneumoniae*, *Pseudomonas aeruginosa* (Todorov 2010), *Listeria innocua*, *Listeria monocytogenes*, *Clostridium butyricum*, *Clostridium perfringens*, *Bacillus cereus* (Mitra et al. 2010), *Enterococcus faecalis*, *Staphylococcus epidermidis* and *Yersinia enterocolitica* (Iranmanesh et al. 2014). The LAB controls/prevents mould growth and mycotoxin production. Inhibition of the growth of *Aspergillus flavus* and its aflatoxin B1 (AFB1) metabolism by divergent LAB strains has been confirmed by Roger et al. (2015). *Lactobacillus paracasei* inhibit several species of *Aspergillus*, *Penicillium* and *Fusarium* (Hassan and Bullerman 2008). Strains of *Lactobacillus brevis* suppress the growth of carcinogenic *Aspergillus niger*, *Aspergillus awamori* and *Penicillium claviforme* and partially inhibit mycelial growth and conidia germination of *Aspergillus flavus* (Tropcheva et al. 2014).

The fermentation process in many protein-rich foods results in the generation of bioactive peptides like angiotensin-converting enzyme inhibitor (ACE-I). The investigations on ACE-I peptide production are mostly on fermented milk products and legumes, which are reported to inhibit angiotensin-converting enzyme and cause vasodilation that lowers blood pressure (Adewumi 2018). Foods fermented with *Lactobacillus casei*, *Lactobacillus acidophilus*, *Lactobacillus bulgaricus*, *Enterococcus faecium*, *Bifidobacterium longum*, *Lactobacillus plantarum* 69 and *Lactobacillus rhamnosus* MTCC 5945 (NS4) are reported to have ACE-I activity (Tsai et al. 2008; Martinez-Villaluenga et al. 2012; Li et al. 2017; Chen et al. 2018; Solanki and Hati 2018). Gamma-aminobutyric acid (GABA) is another peptide compound with hypotensive activity. Foods fermented with *Lactobacillus plantarum* are reported to have GABA-associated hypotensive activity (Liu et al. 2011).

The fermentation process enhances antioxidant activity through production of phenols, flavonoids and bioactive peptides (Farvin et al. 2010; Xiao et al. 2015) and poses antidiabetic effect by the action of bioactive peptides through enhanced glucose-simulated insulin secretion and insulin-simulated glucose uptake (Adewumi 2018). *Bacillus subtilis* natto T-2 and *Bacillus subtilis* natto TK-1 are reported to produce cyclic lipopeptide and lipopeptide bio-surfactant, respectively, that induce apoptosis in human leukaemia cells and inhibit the proliferation of human breast cancer cells (Cao et al. 2009; Wang et al. 2007). *Bacillus subtilis* CSY 191 produces a surfactin-like compound that suppresses the growth of human breast cancer (MCF-7) cells (Lee et al. 2012). *Lactococcus lactis* and *Lactobacillus acidophilus* strains produce water-soluble extract ($\leq$3 kDa) that inhibits the proliferation of Caco2, MCF-7 and HELA carcinoma cell lines (Ayyash et al. 2018) (Fig. 10.7).

10.10 Conclusion

Fermentation undoubtedly represents the easiest and the most suitable way of enhancing nutritional and therapeutic potential of foods and is an ancient form of biopreservation that is common to all regions of the world. Lactic acid fermentation prolongs shelf life, increases digestibility, improves health benefits and assists in the destruction and detoxification of certain undesirable compounds, which may be present in the raw material. The documented traditionally fermented foods and beverages in this study represent the most popular products of Jammu and Kashmir that are consumed either occasionally or on a daily basis. Most of these traditionally fermented foods are cereal-based, plant-based and dairy-based with recipes and preparation methods handed down from generation to generation. The common traditional methods of these food preparations include fermentation followed by baking (e.g. flatbread), spontaneous fermentation (e.g. pickles) and fermentation followed by acid-heat coagulation (e.g. cheese and beverages). Moreover, these fermented products are embedded in the social practices and relations of the everyday life of common people, where traditional foodways have been preserved by the

Fig. 10.7 Some of the popular traditionally fermented dairy products consumed in Jammu and Kashmir: (**a**) *kalari/kradi*, (**b**) *phuh*, (**c**) *dahi*, (**d**) *lassi*

elderly, who are less influenced by the change of food habits toward Western food patterns. Nevertheless, there is an urgent need to conserve and promote traditionally fermented food values among the newer generations of the world because of the changing food habits and their positive correlation with the incidences of chronic diseases, such as heart disease, diabetes and cancer.

References

Adams M, Mitchell R (2002) Fermentation and pathogen control: a risk assessment approach. J Food Microbiol 79:75–83

Adewumi GA (2018) Health-promoting fermented foods. In: Reference module in food science. Elsevier, Amsterdam, pp 1–21

AVRDC (Asian Vegetable Research and Development Center) (1992) Introduction to vegetable and vegetable production systems. In: Vegetable production training manual. Asian Vegetable Research and Development Center, Shanhua, pp 1–24

Ayyash M, Al-Dhaheri AS, Al-Mahadin S, Kizhakkayil J, Abushelaibi A (2018) In vitro investigation of anticancer, antihypertensive, antidiabetic, and antioxidant activities of camel milk fermented with camel milk probiotic: a comparative study with fermented bovine milk. J Dairy Sci 101:900–911

Banerjee SK, Maulik SK (2002) Effect of garlic on cardiovascular disorders. Nutr J 1:4

Bille PG (2014) Descriptive consumer preference of omashikwa, traditional fermented buttermilk from Nambia. African J Food Agri Nutr Develop 14:8725–8734

Boyacyoglu H (1999) The factors affecting bread quality: IV. The effects of salt amount. Technol Flour Products 8:34–41

Cao XH, Liao ZY, Wang CL, Cai P, Yang WY, Lu MF, Huang GW (2009) Purification and antitumour activity of a lipopeptide biosurfactant produced by *Bacillus natto* TK-1. Biotechnol Appl Biochem 52:97–106

Caplice E, Fitzgerald GF (1999) Food fermentation: role of microorganisms in food production and preservation. J Food Microbiol 50:131–149

Chandra R (2014) Sustainability through organic agro-biotechnology with special reference to Jammu & Kashmir scenario. Int J Genetic Eng Biotechnol 5(2):169–178

Chen L, Zhang Q, Ji Z, Shu G, Chen H (2018) Production and fermentation characteristics of angiotensin-I-converting enzyme inhibitory peptides of goat milk fermented by a novel wild *Lactobacillus plantarum* 69. LWT Food Sci Technol 91:532–540

Chen W-T, Brace RA, Scott JB, Anderson DK, Haddy FJ (1972) The mechanism of the vasodilator action of potassium. Exper Biol Med 140:820–824

Conte A, Gammariello D, Di Giullo S, Attanasio M, Del Nobile MA (2004) Active coating and modified- atmosphere packaging to extend the shelf life of Fior di Latte cheese. J Dairy Sci 92:887–894

Corsetti A, Gobbetti M, Balestrieri F, Paoletti F, Russi L, Rossi J (1998) Sourdough lactic acid bacteria effects on bread firmness and staling. J Food Sci 63:347–351

De S (1980) Outlines of dairy technology: *dahi*, vol 1. Oxford University Press, New Delhi, pp 404–410

Dewan S, Tamang JP (2007) Dominant lactic acid bacteria and their technological properties isolated from the Himalayan ethnic fermented milk products. Antonie Van Leeuwenhoek 92:343–352

Farvin KHS, Baron CP, Nielsen NS, Jacobsen C (2010) Antioxidant activity of yoghurt peptides: part 1-in vitro assays and evaluation in ω-3 enriched milk. Food Chem 123:1081–1089

Gargari BP, Mahboob S, Razavieh SV (2007) Content of phytic acid and its mole ratio to zinc in flour and breads consumed in Tabriz, Iran. Food Chem 100:1115–1119

Ghatani K, Tamang B (2017) Assessment of probiotic characteristics of lactic acid bacteria from fermented yak milk products of Sikkim, India: *Chhurpi, shyow, khachu*. J Food Biotechnol 31:210–232

Giraffa G (2004) Studying the dynamics of microbial populations during food fermentation. FEMS Microbiol Rev 28:251–260

Hafeez Z, Cakir-kiefer C, Roux E, Perrin C, Miclo L, Dary- Mourot A (2014) Strategies of producing bioactive peptides from milk proteins to functionalize fermented milk products. Food Res Int 63:71–80

Hammes WP (1990) Bacterial starter cultures in food production. Food Biotechnol 4:383–397

Hassan YI, Bullerman LB (2008) Antifungal activity of *Lactobacillus paracasei* ssp. tolerans isolated from a sourdough bread culture. Int J Food Microbiol 121:112–115

Holzapfel W (1997) Use of starter cultures in fermentation on a household scale. Food Control 8:241–258

Hoseney RC (1988) Principles of cereal science and technology. AACC, St. Paul, MN, p 378

Houser BA, Donaldson SC, Kehoe SI, Heinrichs AJ, Jayarao BM (2008) A survey of bacteriological quality and the occurrence of salmonella in raw bovine colostrum. Foodborne Pathog Dis 5(6):853–858

Iranmanesh M, Ezzatpanah H, Mojgani N (2014) Antibacterial activity and cholesterol assimilation of lactic acid bacteria isolated from traditional Iranian dairy products. LWT Food Sci Technol 58:355–359

Jans C, Merz A, Johler S, Younan M, Tanner SA, Kaindi DWM (2017) East and West African milk products are reservoirs for human and livestock- associated *Staphylococcus aureus*. Food Microbiol 65:64–73

Kim SK, Heo KC (2000) Effects of preservative treatment, heating and freezing on compositional and microbiological changes of colostrum. J Ani Sci Technol 42:659–668
Krishna P, Sharma S (2014) Ethno-botanical importance of some tree species in Jammu District, J&K. Int J Sci Res 3(11):2795–2798
Kumar P, Mishra HN (2004) Yoghurt powder—a review of process technology, storage and utilization. Food Bioprod Process 82(2):133–142
Lawrence WR (1886) Livestock. In: The Valley of Kashmir. Kashmir Kitab-Ghar, Jammu, pp 360–366
Lee JH, Nam SH, Seo WT, Yun HD, Hong SY, Kim MK, Cho KM (2012) The production of surfactin during the fermentation of cheonggukjang by potential probiotic *Bacillus subtilis* CSY191 and the resultant growth suppression of MCF-7 human breast cancer cells. Food Chem 131:1347–1354
Li C, Kwok LY, Mi Z, Bala J, Xue J, Yang J, Ma J, Zhang H, Chen Y (2017) Characterization of the angiotensin-converting enzyme inhibitory activity of fermented milks produced with *Lactobacillus casei*. J Dairy Sci 100:9495–9507
Liu CF, Tung YT, Wu CL, Lee BH, Hsu WH, Pan TM (2011) Antihypertensive effects of *Lactobacillus*-fermented milk orally administered to spontaneously hypertensive rats. J Agric Food Chem 59:4537–4543
Magray MB (2003) Tribal geography of India: Jammu and Kashmir, vol 66. Oberoi Publication, Jammu
Mansoor A, Mehdi M, Stanzin L, Tsering D, Lobzang S, Permendra S, Sharafat H (2018) Status of area and production of apricot (*Prunus armeniaca* L.) in cold arid Ladakh. J Pharmacogn Phytochem 7(2):593–595
Martinez-Villaluenga C, Torino MI, Martin V, Arroyo R, Garcia-Mora P, Estrella-Pedrola I, Vidal-Valverde C, Rodriguez JM, Frias J (2012) Multifunctional properties of soy milk fermented by *Enterococcus faecium* strains isolated from raw soy milk. J Agric Food Chem 60:10235–10244
Mbugua SK, Njenga J (1992) The antimicrobial activity of fermented Uji. Ecol Food Nutr 28:191–198
Mir SA (2011) Flat breads of Kashmir Valley. MSc thesis, Islamic University of Science and Technology, Awantipora, Jammu and Kashmir
Mitra S, Chakrabartty PK, Biswas SR (2010) Potential production and preservation of dahi by *Lactococcus lactis*W8, a nisin-producing strain. LWT Food Sci Technol 43:337–342
Moore M, Tyler JW, Chigerwe M, Dawes ME, Midelton JR (2005) Effect of delayed colostrum collection on colostral IgG concentration in dairy cows. J Am Veter Med Assoc 226(8):1375–1377
Mudhu, Prakash SM, Neetu (2013) Yoghurt is excellent vehicle for travelling probiotics to public health. Int J Food Nutr Sci 2(1):126–137
Muller F, Michel W, Schlicht V, Tietze A, Winter P (2007) Self-cleaning surfaces using the lotus effect. In: Handbook of cleaning/decontamination of surfaces, vol 2. Elsevier Science B.V., Amsterdam, pp 791–811
Mushtaq M, Gani A, Shetty PH, Masoodi FA, Ahmad M (2015) Himalayan cheese *(Kalari/Kradi)*: effect of different storage temperatures on its physicochemical, microbiological and antioxidant properties. LWT-Food Sci Technol 63:837–845
Muyanja CMBK, Narvhus JA, Treimo J, Langsrud T (2003) Isolation, characterisation and identification of lactic acid bacteria from bushera: a Ugandan traditional fermented beverage. Int J Food Microbiol 80:201–210
Odugbemi T, Odujinrin OM, Akitoye CO, Oyerinde JP, Esumeh FI (1991) Study on the pH of ogi, Nigerian fermented weaning food and its effect on enteropathogenic *Escherichia coli*, *Salmonella typhi* and *Salmonella paratyphi*. J Trop Med Hygiene 94:219–223
Omar NB, Abriouel H, Lucas R, Martinez-Canamero M, Guyot JP, Galvez A (2006) Isolation of bacteriocinogenic *Lactobacillus plantarum* strains from ben saalga, a traditional fermented gruel from Burkina Faso. Int J Food Microbiol 112:44–50
Poutanen K, Flander L, Katina K (2009) Sour dough and cereal fermentation in an nutritional perspective. J Food Microbiol 7:693–699

Punoo HA, Patil GR, Bijoy RR (2017) Textural and microstructural properties of Kradi cheese (an indigenous cheese of Jammu and Kashmir, India). Int J Dairy Technol 71:372–380

Pyler EJ (1988) Baking science and technology. Sosland Publishing Company, Kansas, p 1346

Qarooni J (1996) Flat bread technology. Springer, Heidelberg, p 206

Raja J, Mir SA, Masoodi FA (2016) Effect of dry salt and brine on the fermentation and colour of blanched garlic. J Nutr Food Sci 6:484

Rasane P, Kailey R, Singh SK (2017) Fermented indigenous Indian dairy products: standards, nutrition, technological significance and opportunities for its processing. J Pure Appl Microbiol 11(2):1199–1213

Rather SA, Masoodi FA, Akhter R (2016) Ethnic meat products of Kashmiri wazwan: a review. J Ethnic Foods 3:246–250

Rawat K, Kumari A, Kumar S, Kumar R, Gehlot R (2018) Traditional fermented products of India. Int J Curr Microbiol App Sci 7(4):1873–1883

Rehman S, Paterson A, Piggot JR (2007) Chapatti quality from British wheat cultivar flours. LWT-Food Sci Technol 40:775–784

Roger T, Leopold TN, Mbofung CMF (2015) Effect of selected lactic acid bacteria on growth of *Aspergillus flavus* and Aflatoxin B1 production in kutukutu. J Microbiol Res 5:84–94

Sanders ME (1993) Effect of consumption of lactic acid cultures on human health. Adv Food Nutr Res 37:67–130

Sarkar PD, Huma LC, Panigrahi SS, Choudhary R (2015) Traditional and ayurvedic foods of Indian origin. J Ethnic Foods 2(3):97–109

Sarker MT, Prabakusuma AS, Islam MS (2018) Dahi (curd) preparation from milk with different levels of carrot *(Dacus carota)* juice. MOJ Food Proces Technol 6(1):66–71

Savarin B (1825) The physiology of taste. (Translated from the last Paris edition by Robinson, F.). Penguin Group, London, pp 1–136

Sekar S, Mariappan S (2007) Usage of traditional fermented products by Indian rural folks and IPR. Indian J Tradit Knowl 6(1):111–120

Shalini GK, Laxmi A (2007) Influence of additives on rheological characteristics of whole wheat dough and quality of chapati (Indian unleavened flat bread) part 1-hydrocolloids. Food Hydrocoll 21:110–117

Singh AK, Chaurasiya AK, Mitra S (2016) Assessment of nutritional composition in elephant foot yam (*Amorphophallus paeoniifolius* Dennst-Nicolson) cultivars. Int J Food Stud 5(1):146–157

Singh N, Kaur K, Singh H, Singh H (2000) Effect of starch-lipids inclusion complex formation on functional properties of our in tandoori roti. Food Chem 69:129–133

Solanki D, Hati S (2018) Considering the potential of *Lactobacillus rhamnosus* for producing Angiotensin I-Converting Enzyme (ACE) inhibitory peptides in fermented camel milk (Indian breed). Food Biosci 23:16–22

Srivastava S, Srivastava RP, Kulsrestha DC (1984) Studies on a cheese like product (Azaco cheese) made from cow-colostrums. Indian Dairyman 36:378–380

Tamang JP (2010) Diversity of fermented foods. In: Tamang JP, Kailasapathy K (eds) Fermented foods and beverages of the world. CRC Press/Taylor & Francis, Boca Raton

Todorov SD (2010) Diversity of bacteriocinogenic lactic acid bacteria isolated from boza, a cereal-based fermented beverage from Bulgaria. Food Control 21:1011–1021

Tropcheva R, Nikolova D, Evstatieva Y, Danova S (2014) Antifungal activity and identification of lactobacilli, isolated from traditional dairy product "katak". Anaerobe 28:78–84

Tsai J, Chen T, Pan BS, Gong S, Chung M (2008) Antihypertensive effect of bioactive peptides produced by protease-facilitated lactic acid fermentation of milk. Food Chem 106:552–558

Tsioulpas A, Grandison AS, Lewis MJ (2007) Changes in physical properties of bovine milk from the colostrum period to early lactation. J Dairy Sci 90:5012–5017

USDA (2006) National Genetic Research Programme. Germplasm Resource Information Network, Beltsville, MD

Wang CL, Ng TB, Yuan F, Liu ZK, Liu F (2007) Induction of apoptosis in human leukemia K562 cells by cyclic lipopeptide from *Bacillus subtilis* natto T-2. Peptides 28:1344–1350

Wettasinghe M, Bolling B, Plhak L, Parkin K (2002) Screening for phase enzyme- inducing and antioxidant activities of common vegetables. J Food Sci 67:2583–2588

Xiao Y, Wang L, Rui X, Li W, Chen X, Jiang M, Dong M (2015) Enhancement of the antioxidant capacity of soy whey by fermentation with *Lactobacillus plantarum* B1–6. J Funct Foods 12:33–44

Zanoni B, Peri C (1993) A study of the breading baking process. I: a phenomenological model. J Food Eng 19:389–394

Further Reading

Links

http://dialkashmir.com/blog/traditional-kashmiri-bread/

https://easyayurveda.com/2016/12/21/lasoda-cordia-dichotoma-shleshmataka/

https://en.wikipedia.org/wiki/Collard_greens

https://weblogtheworld.com/formats/featured/history-of-collard-greens-extends-far-beyond-north-america

https://www.researchgate.net/publication/324586272_Evaluation_of_antioxidant_properties_in_elephant_foot_yam_Amorphophallus_paeoniifolius_Dennst-Nicolson_pickles

https://www.swadesi.com/news/kashmiri-breads/

Ethnic Fermented Foods and Beverages of Kerala

11

K. Madhavan Nampoothiri, Nimisha R. Nair, and M. P. Soumya

Abstract

In this chapter, attempt has been made in providing an account about the fermented foods and alcoholic beverages of Kerala. Kerala, the southernmost state in India, is a region with rich cultural diversity and ethnic groups. Each ethnic group has its own method of fermenting food materials for the purpose of preservation, taste, and nutritional enhancement and has been carrying this tradition from time immemorial. Because of its rich trading heritage, over time various cuisines have blended with indigenous Kerala dishes with foreign dishes adapted to local tastes. Some of the most popular traditional fermented preparations were described here.

Keywords

Beverages · Fermented foods · Heterogeneous · Kerala · Microflora · Traditional

11.1 Introduction

Fermented foods including alcoholic beverages are an indispensable part of diet of man all over the world (Sanlier et al. 2019). India is known for culture and tradition, where food culture with diverse cuisine plays an important role based on indigenous

K. M. Nampoothiri (✉) · M. P. Soumya
Microbial Processes and Technology Division, CSIR—National Institute for Interdisciplinary Science and Technology (NIIST), Trivandrum, Kerala, India
e-mail: madhavan@niist.res.in

N. R. Nair
Department of Integrated BSc-MSc Biotechnology (IBC), Kerala Agriculture University Vellayani, Thiruvananthapuram, Kerala, India

J. P. Tamang (ed.), *Ethnic Fermented Foods and Beverages of India: Science History and Culture*, https://doi.org/10.1007/978-981-15-1486-9_11

knowledge of the rural women using their village art craftsmanship. Most of these remain majorly secretive in their preparation being passed on from generation to generation and tend to be regionalized with many of them being made only on the home scale by using back slopping. In short, fermented foods have heterogeneousness of traditions and cultural preferences found in the different geographical areas, where they are produced (Rolle and Satin 2002).

Kerala is located in South India on the Malabar Coast. The name "Kerala" has been derived from "Kera," the local Malayali word for coconut, a wonderful natural resource prized for its flesh, water, milk, oil, and even sugar. The state is wedged between the Lakshadweep Sea and the Western Ghats. Lying between northern latitudes 8°18′ and 12°48′ and eastern longitudes 74°52′ and 77°22′, the Malayalis are a group of people of mixed ethnic heritage who speak Malayalam, a Dravidian language; they constitute the majority of the population of Kerala. Kerala's residents belong to Hindu, Islam, and Christian communities, and Kerala also has small population of Jain, Sikh, Buddhist, and Jewish communities. Religion to a great degree controlled the food habitats as well. Kerala cuisine has a multitude of both vegetarian and non-vegetarian dishes prepared using fish, poultry, and meat. Culinary spices cultivated in Kerala are well-known globally, and rice is a dominant staple that is eaten at all times of day, and equally important is fermented food products as well.

Kerala's history, geography, trade links, and migrations have deeply influenced its cuisine. Kerala cuisine is extremely diverse with unique taste and flavor. Scientific investigations of some of these foods and beverages highlight the roles of the natural microflora as starter cultures possessing the innate ability to produce macro- and micronutrients, phytochemicals, and other health-boosting components during the fermentation process. Traditionally boiled rice is a staple diet with few fermented and non-fermented products in Kerala.

11.2 Ethnic Fermented Foods of Kerala

11.2.1 *Idli*

Idli is one of the non-excludable low-calorie fermented breakfast foods in South India (Blandino et al. 2003). It is a fermented product of rice, dehulled black gram or other dehusked pulses. *Idli* is a white-colored, small, round, and spongy product (Fig. 11.1). In 1485 and 1600 CE, *idli* was compared to the moon, which suggests that rice was in use. Use of rice along with pulses is necessary as a source of mixed natural microflora needed for efficient fermentation (Achaya 1994).

11.2.1.1 Traditional Method of Preparation

Traditionally, *idli* preparation is as follows: (i) rice and black gram were soaked separately (ii) after draining the water, rice and black gram were grind separately with occasional addition of water during grinding process, (iii) the rice and black gram batters were then mixed together with addition of a little salt, and (iv) allowed the mixture

Fig. 11.1 Morphological appearance of *idli*

to get fermented overnight at room temperature, and then the batter was dispensed into special *idli* pans and allowed for steaming for 5–8 min (Balasubramanium et al. 2006).

11.2.1.2 Flow Sheet of Preparation of *Idli*

Soak Rice (4) Soak Black gram (1)

Overnight at – 25-30°C

↓

Drain the water, grind rice and black gram separately

↓

Add enough water during grinding process

↓

Mix the rice and black gram batters together with addition of a salt (1% w/v)

↓

Allow the mixture to ferment overnight at room temperature (25-30°C)

↓

Dispense the fermented batter in special *idli* pans and allow for steaming for 5–8 min

Table 11.1 Approximate composition of optimized *idli*

Parameters (g% ± sd)	Rice (variety IR 20)	Black gram (variety DT3)	*Idli* (31.18)
Starch	79.50 ± 2.90	52.00 ± 1.21	75.00 ± 2.84
Amylose	32.00 ± 1.60	17.00 ± 0.74	31.00 ± 1.42
Amylopectin	47.50 ± 2.37	35.00 ± 1.62	44.00 ± 2.08
Total carbohydrates	84.00 ± 3.52	65.80 ± 3.02	81.60 ± 3.42
Protein	06.46 ± 0.32	24.16 ± 1.20	10.21 ± 0.50
Fat	00.27 ± 0.13	00.87 ± 0.02	00.10 ± 0.01
Crude fiber	00.20 ± 0.01	00.70 ± 0.03	00.28 ± 0.01

11.2.1.3 Microbiology

During fermentation of *idli* batter overnight, the naturally occurring microorganisms *Leuc. mesenteroides*, *Lb. delbrueckii*, *Lb. fermenti*, *Lb. coryniformis*, *Ped. acidilactis*, *Ped. cerevisiae*, *Streptococcus* sp., *Ent. faecalis*, *Lact. lactis*, *B. amyloliquefaciens*, *Cand. cacaoi*, *Cand. fragicola*, *Cand. glabrata*, *Cand. kefyr*, *Cand. pseudotropicalis*, *Cand. sake*, *Cand. tropicalis*, *Deb. hansenii*, *Deb. tamarii*, *Issatchenkia terricola*, *Rhiz. graminis*, *Sacch. cerevisiae*, *Tor. candida*, and *Tor. holmii* (Steinkraus et al. 1967a, b, c; Sridevi et al. 2010) in grains/legumes/utensils grow rapidly, outnumbering the initial contaminants and dominating the fermentation. These microorganisms produce lactic acid ($\geq$1.0%) and carbon dioxide that make the batter anaerobic and leaven the product.

11.2.1.4 Nutritional and Health Benefits

Idli is starchy and nutritious food with proteins, carbohydrates, and water (Teniola and Odunfa 2001; Blandino et al. 2003). It comprises protein (3.4%), carbohydrate (20.3%), moisture (70%), verbascose (1%), stachyose (0.2%), and raffinose. Table 11.1 shows an optimized composition of *idli* (Steinkraus et al. 1967a, b, c).

Fermentation increases the amount of enzymes such as amylase and proteinase, along with total acids; batter volume; soluble solids; essential amino acids like lysine, cysteine, and methionine; nonprotein nitrogen, soluble vitamins such as folate, vitamin A, vitamin B1, vitamin B2, and vitamin B12 content, with a significant reduction in anti-nutrient phytic acid (Hung 2005; Moktan et al. 2011).

Idli is rich in nutritive value that is almost equal to that of the unfermented mixture of rice and black gram originally used for preparing it. It is mild acidic, soft, moist, spongy pudding and staple breakfast (Ramakrishnan 1979; Steinkraus 1996a, b, c, d; Soni and Sadhu 1990) and is generally recommended for weight loss diet and for anti-obesity. As there is no or less oil in *idli*, it is recommended for people with the risk of cardiovascular diseases, high blood pressure, and stroke. *Idli* is rich with micronutrients like iron, zinc, folate, and calcium and hence can prevent anemia and facilitate the oxygenation of blood and nourishment of muscle and bone, and it is used as a dietary supplement to treat children suffering from protein calorie malnutrition and kwashiorkor. The carbohydrate as well as dietary fiber content promotes healthy digestion and formation of bulky stools (Blandino et al. 2003; Purushothaman et al. 1993).

11.2.2 Dosa

Dosa is unarguably one of the most popular food dishes in South India especially Kerala. Although originated in Southern India, over the decades it has acquired a pan-India status and is today enjoyed across almost every part of India. But not many know that *dosa* falls under "fermented food" category. This Southern Indian dish is considered foremost among the "healthy fermented food." Regional names of *dosa* differ across India.

11.2.2.1 Mode of Consumption

Dosa is a highly seasoned pancake and contains rice and black gram as primary ingredients. It can sometimes be crunchy wraps filled with vegetable fillings and is consumed as breakfast, dinner, or sometimes as teatime snack.

11.2.2.2 Traditional Method Preparation of *Dosa*

During the preparation of *dosa*, rice and black gram are soaked overnight, dewatered, and ground finely to form a batter. To enhance the taste and flavor, it is recommended to add a handful of fenugreek seeds soaked along with the rice. The proportion of rice to lentils is generally 4:1 or 5:1. The batter is allowed to ferment overnight. After the overnight fermentation, the batter is mixed with water to get the desired thickness. The batter is then ladled onto a hot tava (griddle) greased with oil or ghee (clarified butter). It is spread out evenly with the base of a ladle or bowl to form a pancake (Fig. 11.2). A *dosa* is served hot, either folded in half or rolled like a wrap. It is also usually served with chutney and sambar. *Dosa* can be stuffed with fillings of vegetables and sauces to make a quick meal. The most well-liked version is the masala *dosa*, with a filling of the potato masala.

11.2.2.3 Flow Sheet of Preparation of *Dosa*

Overnight at – 25-30°C

↓

Drain the water, grind rice and fenugreek and black gram separately

↓

Add enough water during grinding process

↓

Mix the rice and black gram batters together with addition of a salt (1% w/v)

↓

Allow the mixture to ferment overnight at room temperature (25-30°C)

↓

Lay the batter onto a hot tava (griddle) greased with oil or ghee

↓

Fold in half or rolls like a wrap and serve hot

11.2.2.4 Microbiology

Microorganisms reported in *dosa* batter are *Leuconostoc mesenteroides*, *Streptococcus faecalis*, *Lactobacillus fermentum*, *Bacillus amyloliquefaciens*, *L. lactis*, *L. delbrueckii, and Lactobacillus plantarum, along with S. cerevisiae*, *D. hansenii*, *T. beigelii*, *Torulopsis* sp., and *Trichosporon pullulans* (Soni et al. 1986).

11.2.2.5 Nutritional and Health Benefits

Dosa is high in carbohydrates and contains no added sugars or saturated fats (Table 11.2). To improve the nutritional quality of *dosa*, finger millet and horse gram can be used as primary ingredients (Palanisamy et al. 2012). During fermentation, the volume of the batter doubles, and as fermentation time increases, the protein content of batter increases. Vitamin B1 and B2, niacin, thiamine, folic acid, total acid

Fig. 11.2 Stages in *dosa* preparation

Table 11.2 Approximate composition of *dosa*

Parameter	Composition
Energy	400–450 kcal
Water (on dry matter basis)	35–40%
Carbohydrates	40–50%
Fats	15–20%
Fibers	5–8%
Ash	4–6%

solids, nonprotein nitrogen, free amino acids, amylase, proteinases, amino nitrogen, formation of diols, and antimicrobial and antioxidant substances have been found to increase during fermentation of *dosa* batter (Soni et al. 1985). The amount of anti-nutrients is reduced during fermentation, and also, it enhances the bio-accessibility of zinc and iron (Labana and Kawatra 1986). For people with wheat allergies and gluten intolerance, *dosa* is considered as an appropriate vegan diet. The prediabetic and postdiabetic conditions in humans can be controlled taking *dosa* in their regular diet taking into consideration the low glycemic load and glycemic index of *dosa*. It offers adequate energy for prolonged physical endurance. Some people consume *dosa* taking into regard their medicinal properties to increase fertility, weight of fetus, and breast milk production. It can also cure rheumatism and neural disorders to a certain extent (Blandino et al. 2003; Gupta and Tiwari 2014).

11.2.3 Uttapam

Uttapam is a *dosa*-like dish made from the same batter with an increased thickness, which is popular mainly in South India (Fig. 11.3) for breakfast or dinner. But unlike *dosa*, which is crisp and unlike relatively soft crepes, it is thick pancake mostly topped with diced onions, tomatoes, cilantro, or cheese. *Uttapam* is traditionally made with toppings such as tomatoes, onion, chillies, capsicum, carrots, beets, coriander, and grated coconut. *Uttapam* is sometimes characterized as an Indian pizza.

Uttapam batter is made of *Vigna mungo* and rice in the ratio 1:3. Rice can be a combination of parboiled rice and a regular variety such as basmati. The lentils and rice are soaked overnight, ground and fermented at room temperature until it rises. The fermented batter is poured on a hot greased pan and spread into a round shape, toppings spread on it, patted gently, and cooked in a low flame. Oil is dripped around it, flipped when it is golden-brown, oil dripped again, and another flip to make sure the second surface is baked too.

Fig. 11.3 *Uttapam*

11.2.3.1 Flow Sheet of Preparation of *Uttapam*

Soak vigno mungo (1) Soak Rice (3)

Overnight at – 25-30°C

⇩

Grind the soaked vigna mungo and rice in the ratio 1:3 to form a batter

⇩

Allow the mixture to ferment overnight at room temperature

⇩

Add 1% (w/v) of salt and mix the batter

⇩

Spread the batter onto a hot tava (griddle) into a round shape with ≈ 0.5cm thickness,

⇩

Add finely chopped tomatoes, onion, chilies, capsicum, carrots, beets, coriander and grated coconut on to the spread batter and grease with oil or ghee

⇩

Flip the *Uttapam* when it is golden- brown in color and spread some oil and pat gently

⇩

Flip again to make sure the second surface is baked and serve hot

11.2.3.2 Microbiology

LAB and yeast play vital roles in the fermentation of *uttapam*. *Lactobacillus pentosus* and *L. plantarum* sp. were isolated from *uttapam* batter. These microbes can survive in GI tract withstanding the high bile salt conditions, gastric secretions, and intestinal conditions. As they are capable of producing b-galactosidase, phytase, pectinase, and bile salt hydrolase, they inhibit the growth of pathogens and are able to adhere on Caco-2 cell surface.

11.2.3.3 Nutritional and Health Benefits

Uttapam is a zero trans-fat fermented food having approximately 160 Cal per 50 g serving, 0.4 g of fat, 34 g carbohydrate, 3.0 g dietary fiber, and 5.0 g of protein, with source calcium, ferrous, vitamin A and vitamin. Being a cholesterol-free food item, *uttapam* is a prescribed food for high-sugar and cholesterol patients. It is easily digestible and can reduce body weight and prevent obesity (Saraniya and Jeevaratnam 2014).

11.2.4 Appam

Appam or hoppers are a common type of food in South Indian cuisine especially in Kerala and Tamil Nadu. The appam batter is made by fermenting rice batter with sugar, coconut milk, and toddy, which is also a fermented drink made of coconut flower. The nonalcoholic version is made by using yeast in the batter. The final mixture is poured into an earthen mold, covered, and fried in low heat to obtain traditional *appam* (Fig. 11.4). Due to lactic acid production by bacteria, *appam* fermentation is acidic in nature even though the details of microbial composition in the batter are under exploration. An *appam* can give calorie of about 138.8, 3.7 g total fat, 0.1 g unsaturated fatty acids, 31.7 mg sodium, 13.5 mg potassium, 23.2 g total carbohydrate, 1.1 g dietary fiber, 2.1 g protein, with vitamin A/B complex, calcium, folate, iron, niacin, riboflavin, and thiamine. This is a healthy and easy digestible, nutritionally enriched food (Roy et al. 2007; Ray and Swain 2013).

Another form of *appam* is "kallappam," which looks like a pancake. The name originated from Kallu which means toddy and is used for fermentation. The circular pancake, dense at the center and light and crisp around the edges, had a peculiar tang.

Fig. 11.4 *Appam*

11.2.4.1 Flow Sheet of Preparation of *Appam*

Soak rice in adequate amount of water for 6-7 hours
At – 25-30°C

⇩

Grind the soaked rice to form a batter

⇩

Add 2% (w/v) of sugar, 100-200mL of coconut water and few grains of yeast

⇩

Allow the mixture to ferment overnight at 25-30°C

⇩

Add 1% (w/v) salt and mix the batter to get the desired consistency

⇩

The final batter is poured into a concave shaped pan covered, and fried in low heat

⇩

Serve hot

11.2.4.2 Pesaha Appam

It is the unleavened Passover bread made by the Saint Thomas Christians of Kerala, to be served on Passover night. It is served on Passover night of Maundy Thursday. The whitish *Pesaha appam* is a firm rice cake (Fig. 11.5). It's made from rice batter like *palappam* but is different from palappam in that it is not fermented with yeast in its preparation. Also, the brown "palkurukku" is made mainly using jaggery and coconut milk which usually served along with *Pesaha appam*.

11.2.5 Pazham Kanji

Pazham kanji is the gruel or porridge which is very popular among Keralites. Similar cuisines in other states are known as *poitabh* in Assam, *pantabhat* in Bengal, and *pokhalo* in Orissa and are consumed as breakfast or lunch (Tamang 2012). It was the staple breakfast in Southern India for centuries. The principal ingredients of Pazham kanji are rice soaked in water. To increase the taste, d*ahi* (Indian yogurt), salt, spices, and leafy vegetables are added. The cooked rice after cooling down to room temperature, adequate cold water is added to it, and the watery soaked rice is allowed to

Fig. 11.5 *Peshaha appam*

ferment overnight at room temperature. The fermented rice with water is consumed along with cooked vegetables/other ingredients (Blandino et al. 2003). Pazham kanji provides a cooling effect for the whole day, removes constipation, as it is a fibrous food, a good source of vitamin B12 and B6, eliminating skin allergy, helps to control blood pressure, hyper tension and cholesterol are some of known health benefits (Fig. 11.6).

11.2.5.1 Flow Sheet of Preparation of *Pazham Kanji*

Cool the cooked rice to room temperature

⇓

Add enough water to soak the rice

⇓

Allow it to ferment overnight at room temperature

⇓

Add adequate amount of curd, salt, crushed green chilies and raw onions

⇓

Mix well and serve

11.2.5.2 Microbiology

The microbes involved in the fermentation process include LAB like *Lactobacillus bulgaricus*, *Lactobacillus casei*, *Pediococcus acidilactici*, *S. faecalis*, *Streptococcus thermophilus*, *Microbacterium flavum*, and *Saccharomyces* sp. The fermentation process increases the amounts of vitamin B complex and vitamin K, and a 100 g of

Fig. 11.6 *Sour rice/fermented rice gruel*

sour rice is reported to have 73.91 mg iron, 303 mg sodium, 839 mg potassium, and 850 mg calcium.

11.2.5.3 Nutritional and Health Benefits

Fermented sour rice is an energy-rich food with an ability to rehydrate the body with the watery nature it possesses. It prevents constipation by fiber release in the stool and helps easy bowl movements. It also ensures to restore healthy intestinal flora in the GI tract and prevents gastrointestinal ailments like duodenal ulcers, infectious ulcerative colitis, Crohn's disease, irritable bowel syndrome, celiac disease, candida infection, etc. (Ray and Swain 2013; Choi et al. 2014).

11.2.6 Cereal-Based Fermented Sweets and Snacks of Kerala

These foods are consumed mostly during festival times and other special occasions. Wheat and rice flours are predominantly used cereals. Sugar, jaggery, or salt is added compulsorily in these fermented foods. Unlike *idli* or *dosa,* usually the fermentation time will be very less.

11.2.6.1 Unniyappam

It is a traditional sweet snack from Gods' own place, Kerala. In Malayalam, "unni" means small and "appam" means rice cake or "pancake." It is a small round snack made from rice, jaggery, banana, roasted coconut pieces, ghee, roasted sesame seeds, and cardamom powder fried in oil (Fig. 11.7). It is Kerala's version of sweet *kuzhi paniyaram.* It is also called "Karollappam."

Fig. 11.7 *Unniyappam*

Fig. 11.8 *Neyyappam*

11.2.6.2 Neyyappam

It is a sweet rice-based fritter fried in ghee (Fig. 11.8). Neyyappam has its origins in the Southern Indian state of Kerala. The name is derived from the words neyy meaning "ghee" (clarified butter) and appam meaning "pancake." Neyyappam is typically made of rice flour (alternatively, with rava or semolina), jaggery, ghee-fried coconut (pieces or grated), ghee, cardamom, and milk. It is served as a teatime snack usually in the evenings. Neyyappam is also served as offering in many traditional churches and Hindu temples in Kerala.

In traditional Kerala cuisine, neyyappam is cooked in a bronze pan called appakara. Rinse the white rice in water and keep it soaked for 3–4 h. Drain the water and grind the rice (using a mixer-grinder) with ripe banana and jaggery to form a thick paste. Fry dry coconut slices (or grated coconut) in ghee until golden brown. Add the ghee-fried coconut, cardamom seeds, and cumin seeds. Also add a pinch of dry ginger powder and mix well. Keep the batter for 30 min. Heat ghee or cooking oil in a bottom-deep pan, and pour a ladle full of batter into the pan, and fry till it become golden brown in color. Keep the fried neyyappams aside to drain off the excess oil and let it cool down before serving.

Flow Sheet of Preparation of *Unniyappam*

Soak parboiled rice in water for 3–4 hours

⇓

Grind the rice (in a mixer) with 2-4 ripe banana (Palayanthodan variety) and jaggery to form a thick paste

⇓

Cut dry coconut into very small pieces

⇓

Fry the dry coconut in ghee until its golden brown in color and mix it with the batter

⇓

Add ghee-fried coconut, cardamom seeds, cumin seeds & dry ginger powder to the batter

⇓

Mix well; keep the batter undisturbed for 30-45 min at 25-30 °C

⇓

Heat ghee in appakarrol pan and pour the batter to fill the pan

⇓

Flip it upside down and fry until it is golden brown in color

⇓

Drain off the excess oil and Serve

11.2.6.3 *Vettu* Cake/Fried Tea Cake

Vettu cake is fried sweetened dough and is a popular tea stall snack, available only in rural Kerala (Fig. 11.9). It is made with sugar, all-purpose flour, and eggs. Duck eggs are used as they have more fat content than the chicken eggs. In a bowl beat the eggs along with sugar until it dissolves. Mix the flour, semolina, baking soda, and beaten egg—mix sugar, milk, salt, and cardamom powder; warm under very low flame to make the mixture to smooth dough. Keep the dough covered for around 45 min. Shape the dough into a log shape and then cut them into 2 × 2.5 in. square pieces. Make an "X" (cross) mark in the center using a sharp knife. Boil oil in a pan and turn the flame to minimum. Fry each piece in minimum heat for about 4–5 min, flipping them in between to ensure uniform cooking.

Flow Sheet of Preparation of *Vettu Cake*

Mix the flour, semolina, a pinch of baking soda and beaten eggs (3-4)

⇓

Add 8-10% (w/v) sugar, milk, 0.5% (w/v) salt and cardamom powder (1%)

⇓

Warm under very low flame to make the mixture to smooth dough

⇓

Keep the dough covered for around 45 minutes at room temperature

⇓

Shape the dough into a log shape and then cut them into 2×2.5 inch square pieces

⇓

Make an "X" (cross) mark in the centre using a sharp knife

⇓

Boil oil in a pan and turn the flame to minimum

⇓

Fry each piece in minimum heat for about 4-5 minutes flipping them in between to ensure uniform cooking

11.2.6.4 Vada

Vada is an inevitable traditional South Indian fermented dish consumed during breakfast or even as an evening snack. It is prepared using rice, lentils, and other seasoning vegetables. For vada, lentils are soaked in water for 2 h and then blended along with a little water until a smooth paste is obtained. The batter is then left to ferment overnight in a warm place until frothy. The batter is then mixed with spices like red chillies, fresh peppercorns, cumin, finely chopped ginger, green chilies, fresh coriander leaves, coconut pieces, etc., to give *vada* its fascinating and

Fig. 11.9 *Vettu cake*/fried tea cake

Fig. 11.10 *Vada*

mouth-watering taste and flavor. A spoonful of batter is dropped in hot oil in a flattened donut shape and fried until golden. It is the process of fermentation that creates those fluffy and light layers. *Vada* is cooked by deep frying in edible oil (Fig. 11.10).

Microbiology and Nutritional Benefits

Pediococcus sp., *Streptococcus* sp., and *Leuconostoc* sp. are the prominently seen microbial flora that enhances the fermentation of the *vada* batter. Even if it is made by deep frying, it gives only up to 197 Cal, 505 g of fat, 20 mg of sodium, 350 mg of potassium, 39.6 g of total carbohydrates, 6.5 g of dietary fibers, 1.7 g sugars, and 7.6 g proteins and contains 2% vitamin C, 3% calcium, and 14% iron. This crispy snack is recommended for children as well as women for the health benefits it carries as stated and is rich in proteins, iron, and dietary fibers.

Flow Sheet of Preparation of *Vada*

Soak the lentils in water for 2 hours at 25-30°C

⇩

Grind with the addition of adequate amount of water to obtain a smooth paste

⇩

Leave the batter undisturbed overnight at room temperature to ferment

⇩

Mix the batter well.

⇩

Add finely chopped red chilies, peppercorns, cumin, ginger, green chilies, fresh coriander leaves.

⇩

Drop a spoonful of batter into hot oil in a flattened donut shape

⇩

Flip it upside down and fry until its golden color

⇩

Drain off the excess oil and Serve

11.3 Fermented Dairy Products

11.3.1 Dahi

The Indian staple *dahi* consumed with every meal has numerous health benefits. Fermentation of milk results in the breakdown of casein or milk protein, one of the most difficult proteins to digest. The color of *dahi* should be pleasing, attractive, and uniform, without showing any signs of visible foreign matter. The color ranges from creamish yellow for cow milk to creamish white for buffalo milk, but should not be brown, with a smooth and glossy surface, without the appearance of any free whey on top. It should have pleasant sweetish aroma and a mild, clean, acidic taste. A good pleasant diacetyl flavor is desired in *dahi*. It should however not show any signs of bitterness, saltiness, or other off-flavors.

11.3.1.1 Nutritional and Health Benefits

The nutritional value of *dahi* is as follows: water 84.8%, pH 4.2, acidity 0.73%, ash 4.7% DM, fat 24.5% DM, protein 5.7% DM, carbohydrate 22.5% DM, and food value 503.6 kcal/100 g DM (Dewan 2002; Tamang 2012). The chemical composition of *dahi* has been reported as fat ranging 3–7%, protein 3.3–3.4%, ash 0.75–0.79%, lactic acid 0.5–1.1%, calcium 100 mg, minerals 0.4%, phosphorus 50 mg, vitamin A 100 (IU), biotin 15 μg, and ascorbic acid 1 mg/100 mL. The factors affecting composition of *dahi* are the type of milk used, the extent of dilution during churning, and the efficiency of fat removal (Laxminarayana et al. 1952).

11.3.1.2 Microbiology

Reported microorganisms in dahi are *Lb. bifermentans*, *Lb. alimentarius*, *Lb. paracasei*, *Lb. lactis*, *Strep. cremoris*, *Strep. lactis*, *Strep. thermophilus*, *Lb. bulgaricus*, *Lb. acidophilus*, *Lb. helveticus*, *Lb. cremoris*, *Ped. pentosaceus*, *P. acidilactici*, *W. cibaria*, *W. paramesenteroides*, *Lb. fermentum*, *Lb. delbrueckii* subsp. *indicus*, *Saccharomycopsis* sp., and *Candida* sp. (Harun-ur-Rashid et al. 2007; Patil et al. 2010). Culturing restores many of the enzymes destroyed during pasteurization including lactase which helps digest lactose, which in turn helps the body absorb calcium and other minerals. Lactase produced during culturing process allows many people who are sensitive to milk to tolerate fermented milk products. Research has shown that regular consumption of cultured dairy products lowers cholesterol and protects against bone loss.

11.3.1.3 Flow Sheet of Preparation of *Dahi*

Boil the milk at 95°C

⇩

Allow it to cool to room temperature

⇩

Add a tablespoon of curd and mix well

⇩

Keep the mixture closed and undisturbed at room temperature overnight

⇩

Serve it the next day

11.3.2 *Sambaram* or the Buttermilk

Buttermilk is the liquid that is left over when butter is churned out of cream or *dahi*, consumed with or without added salt and spices. It has less fat content and fewer calories compared with regular milk or *dahi*. Buttermilk is as rich a source of calcium and protein as like milk. In some states of India, it is also known as *chaach*. It is easier to digest and, with the presence of live cultures, helps in intestinal health. Buttermilk is consumed with meals and is advised in the Ayurveda (Pushpangadan et al. 2012).

11.3.3 Sadhya

Sadhya is a feast consisting of a variety of traditional vegetarian dishes usually served on a banana leaf in Kerala and is considered as a perfect balanced diet. *Sadhya* means banquet in Malayalam. It is a vegetarian feast prepared by both men and women, especially when needed in large quantities, for weddings and other special events like Onam and other Hindu festivals. *Sadhya* consists of different types of dishes most of which are fermented (Fig. 11.11).

Kalan is made of yogurt, coconut, and any one vegetable like "nendran" plantain or a tuber-like yam. It is very thick and more sour and typically can last for a longer period owing to the lower water content.

Pachadi is sour curry made of yoghurt and usually cucumber or sliced ash gourd cooked in ground coconut with mustard seeds and ginger and seasoned with sautéed mustard seeds and curry leaves. It is somewhat similar to a raita, a common dish in North India.

Fig. 11.11 *Sadhya*

Sweet pachadi is a sweet form of *pachadi*, made with pineapple, pumpkin, or grapes in yoghurt. The gravy masala comprises ground coconut with cumin seeds and green chillies. Due to its sweetness, it is also called madhura (sweet) curry in some places.

Pulisseri is a sour, yellow-colored thin curry made with slightly soured yoghurt and cucumber. A sweet variant called "*Mambazha Puliseri*" replaces cucumber with a combination of ripe mangoes and jaggery.

Injipuli is a sweet pickle made of ginger, tamarind, green chilies, and jaggery, also called puli inji.

Spicy pickles of raw mango (mango pickle) or lemon also belong to the fermented foods. It is made with the addition of tender mango or lemon with red chilli powder, salt, and oil and is stored mostly in earthen air-tight containers called "barani" and are opened only whenever needed to avoid contamination.

Sambaram, as we have described earlier, is also referred to as "moru": It is a drink made from salted buttermilk with green chilli, ginger, and curry leaves. It is drunk to improve digestion and is typically served near the end of the meal.

11.4 Ethnic Fermented Beverages of Kerala

11.4.1 *Toddy* or Coconut Vinegar

Coconut vinegar, made from the sap or *toddy* of the coconut palm (Fig. 11.13), is produced and used extensively in South Asian countries, including India (Murooka et al. 2009). A cloudy, white liquid, coconut vinegar has a particularly sharp, acidic taste (4% acetic acid) with a slightly yeasty flavor. Coconut water vinegar is made from coconut water and contains between 3% and 4% (v/v) acetic acid and is an indispensable commodity in many households (Sanchez 1990). Sugar vinegar, cane sugar vinegar (sugarcane juice), and palm vinegars are popular in South India including Kerala. They range in color from dark yellow to golden brown and have a mellow flavor.

For toddy preparation, coconut trees are carefully selected, and the toddy tapper uses a coir rope tied with coconut husks at a 2 ft. interval which makes for a ladder. Then the unopened buds of the palm flower are cut open and with 95% left remaining on the tree. Then the bud is made to swell by beating it and pounding it with a hammer. Then clay pots are attached to catch the sap (Fig. 11.12) which results in a milky liquid being collected in the clay pots which are then taken down and segregated. Some are left to ferment lightly, and the longer it ferments, the more alcohol accumulates. It is then sold in glass bottles to licensed toddy shops where they are sold. Toddy has an alcohol content of about 8.1% and is considered natural alcohol and a health drink in Kerala (Fig. 11.13).

Fig. 11.12 Clay pot attached to coconut tree to collect the sap

Fig. 11.13 *Toddy*

11.4.1.1 Flow Sheet of Preparation of *Toddy*

Take a tender coconut and get the water in it

⇩

Strain the water using a fine sieve into a clean glass

⇩

Add two tablespoons of sugar (≈ 3% w/v) and ¼ tablespoon of dry yeast and mix well

⇩

Transfer to a clean glass container and cover it with double layer of cheesecloth

⇩

Allow the mixture to ferment at room temperature for 24 hours

⇩

Chill and consume toddy within a couple of days

11.4.2 Palm Vinegar

The palms most frequently tapped for sap are raffia palms (*Raphia hookeri* and *Raphia vinifera*) and the oil palm (*Elaeis guineensis*). The acetic acid concentrations of palm vinegar are around 4% (v/v) (Battcock and Azam-AIi 1998; Gonzalez and De Vuyst 2009).

11.5 Conclusion

There are diverse forms of traditional fermented foods and beverages prepared by different ethnic groups within a small state like Kerala (Table 11.3). In most cases, they have been exploring indigenous microflora for spontaneous conversion of raw materials into useful edible products. Currently, these products are processed for local consumption only. However, there is very high demand for a systematic evaluation of these fermented products with its health benefits proved through adequate scientific interventions for any global business and also it is high time that country should a proper preservation of these indigenous micro flora isolated from traditional foods belonging to various regions. Better starter cultures suitable to our needs and food habits can be easily developed if we systematically explore the microflora of these ethnic food preparations.

Table 11.3 Ethnic fermented foods and beverages of Kerala

Food	Substrate	Sensory property of product	Major ethnic consumers	Region/ district
Idli	Rice, black gram	White color, soft, spongy with fermented aroma	Brahmins/ vegetarian	South India
Dosa	Rice, black gram	Golden color, crisp, thin	Brahmins/ vegetarian	South India
Uttapam	Rice, black gram	Creamish white color, thick, soft, non-crispy	Brahmins/ vegetarian	Kerala and Tamil Nadu
Appam	Rice, black gram	White color, concave shaped, bulged at the center with crisp edges with fermented aroma	Keralites	Kerala
Pesaha appam		White color, soft, with aroma, 2–3 cm thick	Mostly Christians	Kerala
Pazham kanji	Rice and cold water	Watery, salty and sour taste	Especially farmers	Kerala
Unniyappam	Rice flour and jaggery	Golden brown color, soft, spongy, but compact, a half-cut spherical shape	Mostly Hindus	Kerala
Neyyappam	Rice and jaggery	Golden brown color, firm, spherical but flat, oily	Keralites	Kerala
Vettu cake	Rice flour and eggs	Golden yellow color, thick, firm, moderately hard	Christians	South Kerala
Vada	Black gram	Golden brown, donut shaped, oily crispy at the periphery and soft core	Mostly Brahmins	South India
Dahi	Milk	White color, semisolid/viscous, fermented aroma	Keralites	South India
Sambaram	Curd and water	White color liquid, salt and sour taste	Keralites	Kerala
Toddy	Coconut water	White color, liquid, sour taste, frothy, fermented aroma	Mostly gents	Kerala
Palm vinegar	Palm fruit/ palm juice	Liquid, sour taste, frothy, fermented aroma	Mostly gents	Kerala

Acknowledgments Corresponding author is thankful to DBT, New Delhi, and CSIR, New Delhi, for financially supporting the probiotic and nutraceutical activities in the division.

References

Achaya KT (1994) Indian food: a historical companion. Oxford University Press, Delhi

Balasubramanium S, Singh N, Ilys SM, Wanjari OD (2006) Effect of selected decorticated legumes protein on rheology of maize extrudate pastes. J Food Sci Technol 43:590–594

Battcock M, Azam-Ali S (1998) Fermented fruits and vegetables: a global perspective. FAO beverage of the Himalayas. Food Microbiol 21:617–622

Blandino A, Al-Asceri ME, Pandiella SS, Cantero D, Webb C (2003) Cereal based fermented foods and beverages. Food Res Int 36:527–543

Choi JS, Kim JW, Cho HR, Kim KY, Lee JK, Ku SK, Sohn JH (2014) Laxative effects of fermented rice extract (FRe) in normal rats. Toxicol Environ Health Sci 6:155–163

Dewan S (2002) Microbiological evaluation of indigenous fermented milk products of the Sikkim Himalayas Ph. D. thesis. Food Microbiology Laboratory. Sikkim Government College (affiliated to North Bengal University), Gangtok, India

Gonzalez A, De Vuyst L (2009) Vinegars from tropical Africa. In: Solieri L, Giudici P (eds) Vinegars of the world. Springer-Verlag, Milan, pp 209–221

Gupta A, Tiwari SK (2014) Probiotic potential of Lactobacillus plantarum LD1 isolated from batter of *dosa*, a south Indian fermented food. Probiotics Antimicrob Proteins 6:73–81

Harun-ur-Rashid M, Togo K, Useda M, Miyamoto T (2007) Probiotic characteristics of lactic acid bacteria isolated from traditional fermented milk "Dahi" in Bangladesh. Pak J Nutr 6:647–652

Hung CL (2005) Health food processing process using germinated rice to make health food containing natural eatable fibers, GABA, IP6, and probiotic. US Patent 6977093

Labana AK, Kawatra BL (1986) Studies on protein quality and availability of zinc from *dosa*. J Food Sci Technol 23:224–227

Laxminarayana H, Nambudripad VKN, Laxmi NV, Anantaramiah SN, Srinvasmurty V (1952) Studies on dahi. II. General survey of market dahi. Indian J Vet Sci Ani Husband 22:13

Moktan B, Roy A, Sarkar PK (2011) Antioxidant activities of cereal-legume mixed batters as influenced by process parameters during preparation of dhokla and *idli*, traditional steamed pancakes. Int J Food Sci Nutr 62:360e9

Murooka Y, Nanda K, Yamashita M (2009) Rice vinegars. In: Solieri L, Giudici P (eds) Vinegars of the world. Springer-Verlag, Milan, pp 121–133

Palanisamy BD, Rajendran V, Sathyaseelan S, Bhat R, Venkatesan BP (2012) Enhancement of nutritional value of finger millet-based food (Indian *dosa*) by co-fermentation with horse gram flour. Int J Food Sci Nutr 63:5–15

Patil MM, Pal A, Anand T, Ramana KV (2010) Isolation and characterization of lactic acid bacteria from curd and cucumber. Indian J Biotechnol 9:166–172

Purushothaman D, Dhanapal N, Rangaswami G (1993) Indian *idli*, *dosa*, dhokla, khaman, and related fermentations. In: Steinkraus KH (ed) Handbook of indigenous fermented food. Marcel Dekker, New York, NY

Pushpangadan P, Dan VM, Ijinu T, George V (2012) Food, nutrition and beverage. Int J Tradit Knowl 11:26–34

Ramakrishnan CV (1979) Terminal report of American PL 480-project Nr GF-IN-491.Study of Indian fermented foods from legumes and production of similar fermented foods from U.S. soybean. Biochemistry Department. Baroda University, Baroda

Ray RC, Swain MR (2013) Indigenous fermented foods and beverages of Odisha, India: an overview. In: Joshi VK (ed) Indigenous fermented foods of South Asia. CRC Press, Boca Raton, FL

Rolle R, Satin M (2002) Basic requirements for the transfer of fermentation technologies to developing countries. Int J Food Microbiol 75:181–187

Roy A, Moktan B, Sarkar PK (2007) Traditional technology in preparing legume based fermented foods of Orissa. Indian J Tradit Knowl 6:12–16

Sanchez PC (1990) Vinegar. In: Coconut Research and Development Foundation (ed) Coconut as food. Philippine Inc Publication, Manila, pp 151–161

Sanlier N, Gocksen BB, Zesgin AC (2019) Health benefits of fermented foods. Crit Rev Food Sci Nutr 59:506–527

Saraniya A, Jeevaratnam K (2014) Purification and mode of action of antilisterial bacteriocins produced by *Lactobacillus pentosus*SJ65 isolated from *uttapam* batter. J Food Biochem 38:612–619

Soni S, Sandhu D, Vilkhu K (1985) Studies on dosa an indigenous Indian fermented food: some biochemical changes accompanying fermentation. Food Microbiol 2:175–181

Soni SK, Sadhu DK (1990) Indian fermented foods: microbiological and biochemical aspects. Indian J Microbiol 30:135–157

Soni SK, Sandhu DK, Vilkhu KS, Kamra N (1986) Microbiological studies on Dosa fermentation. Food Microbiol 3:45–53

Sridevi J, Halami PM, Vijayendra SVN (2010) Selection of starter cultures for *idli* batter fermentation and their effect on quality of *idli*. J Food Sci Technol 47:557–563

Steinkraus KH (1996a) Handbook of indigenous fermented foods. Marcel Dekker, New York
Steinkraus K (1996b) Handbook of indigenous fermented foods. New York, Marcel Decker
Steinkraus KH (1996c) Handbook of indigenous fermented foods, vol 331, 2nd edn. Marcel Dekker, Inc., New York, p 776
Steinkraus KH (1996d) Handbook of indigenous fermented food, 2nd edn. Marcel Dekker, Inc., New York, NY
Steinkraus KH, Van Veen AG, Thiebeau DB (1967a) Studies on *Idli*- and Indian fermented black gram rice food. Food Technol 21:110–113
Steinkraus KH, Van Veen AG, Thiebean DB (1967b) Studies on idli—an Indian fermented black gram-rice food. Food Technol 21:916–919
Steinkraus KH, van Veer AG, Thiebeau DB (1967c) Studies on *idli*—an Indian fermented black gram-rice food. Food Technol 21:110–113
Tamang JP (2012) Plant-based fermented foods and beverages of Asia. In: Handbook of plant-based fermented food and beverage technology. CRC Press, Boca Raton, FL
Teniola OD, Odunfa SA (2001) The effects of processing methods on the levels of lysine, methionine and the general acceptability of Ogi processed using starter cultures. Int J Food Microbiol 63(1–2):1–9

Ethnic Fermented Foods and Beverages of Madhya Pradesh

12

Isha Sharma, Rupesh Kapale, and Naveen Kango

Abstract

Tribes of central India such as *Baiga*, *Bharia*, *Bhatra*, *Bhil*, *Bhujia*, *Bhillala*, *Birhor*, *Gond*, *Halba*, *Korwa*, *Kol*, *Maria*, *Oraon*, *Sahariya*, *Santhal*, etc. contribute around 20–30% of the total population of scheduled tribes. Most of the ethnic fermented foods such as *pej*, *handia*, *rabdi*, *mahua*, *pendum*, *basi*, etc. are prepared at household level. The ethnic fermented foods and beverages of these tribes confer several health-promoting benefits to the consumers due to functional microorganisms associated with them. Microbial diversity ranges from filamentous molds to Gram-positive and few Gram-negative bacteria, and the fermentations are associated with alcohol-producing yeasts along with some other obscure microorganisms. In order to validate and enhance our scientific knowledge regarding these ethnic fermented foods and beverages, systematic studies to decipher underlying mechanisms are required.

Keywords

Ethnic · Fermented foods · Madhya Pradesh · *Handia* · *Pej* · *Basi*

12.1 Introduction

Madhya Pradesh is located in the geographical center of India surrounded by other states of Maharashtra, Rajasthan, Gujarat, and Chhattisgarh. The state has assimilated ethnic cultures and traditions of each of these surrounding states.

I. Sharma · N. Kango (✉)
Department of Microbiology, Dr. Harisingh Gour Vishwavidyalaya, Sagar, Madhya Pradesh, India

R. Kapale
Department of Botany, Government PG College, Chhindwara, Madhya Pradesh, India

J. P. Tamang (ed.), *Ethnic Fermented Foods and Beverages of India: Science History and Culture*, https://doi.org/10.1007/978-981-15-1486-9_12

Madhya Pradesh was formed on November 1, 1956, comprising the 17 districts of Mahakoshal, 2 districts of Bhopal, 8 districts of Vindhya Pradesh, and 16 districts of Madhya Bharat. The Chhattisgarh region, comprising of 16 districts, was separated to form a new state as per the provisions of Madhya Pradesh Reorganization Act, 2000, and the reorganized State of Madhya Pradesh came into existence on November 1, 2000 (http://www.mp.gov.in/web/guest/home-history). The total geographical area of the state is now 308 thousand sq. km. The largest state of the country with more than 75 million human population is a multicolored, aesthetic, and warm place because the people here are hospitable and cordial (https://www.census2011.co.in/census/state/madhyapradesh.html). Rich in mineral resources, MP has the largest reserves of diamond and copper in India. More than 30% of its area is under forest cover. Its tourism industry has seen considerable growth, with the state topping the National Tourism Awards in 2010–2011. In recent years, the state's GDP growth has been above the national average.

A number of tribes in Madhya Pradesh tribes make a large part of the population among which many tribal communities live in deep forests. These tribes depend upon natural resources for addressing issues related to food and health. Over the decades, these communities have acquired exclusive understanding about the use of wild flora (Nath and Khatri 2010). The ethnic pilgrimages Ujjain, Amarkantak, Maheshwar, Mandleshwar, and Omkareshwar Temple located at the bank of Narmada River are the most popular religious centers of the state. Madhya Pradesh is famous for quite essential ethnic culture, rich flora and fauna, and religious sites.

Fig. 12.1 Map of Madhya Pradesh showing 43 districts after formation of Chhattisgarh

The traditional foods processed and prepared by tribes of the region are intimately connected to their sociocultural, ecological, spiritual life, and health. The processing and preparation of ethnic foods not only demonstrate the creativity and treasure of food heritage of tribal people but also their incremental learning to sustain the life and ecosystem as a whole. Tribal women of Central region have a wide range of variability in the ethnic foods made of soybean, rice, wheat, tree, and bean. In each state, the processing method of these foods is somewhat different based on the culture, variability in the materials used in the food, climate, and overall knowledge of the processing and preparation. The foods used in the dietary system are nutritionally rich and culturally important in various festivals and ceremonies. Ethnic foods prepared and consumed by tribals cannot be seen in the isolated mode; instead it is a complex dynamics in which nutrition, health, food security, culture, ethics, economy, and ecological sustainability are integral components (Fig. 12.1).

History says that earlier the Bhils were the warrior tribes who were concentrated more on the western part of the state and the eastern part being occupied by the Oraons tribes. Kols, Bhilalas, Murians, and Korkens contribute to the other minority of tribes in the state (https://www.indianetzone.com/5/culturemadhyapradesh.html). The state was initially ruled by the Gonds tribe who are seen in the jungles, south of Jabalpur. On the west live the Bhils, while the Oraons inhabit the eastern part of the state. The name of the tribal groups Bhils is derived from a Dravidian word for bow, which is the hallmark of the tribe (https://www.indianmirror.com/tribes/bhiltribe.html). The unique cultural relationship is proved by the peaceful living and mutual cooperation which has kept the people of the state unified. Add to it, the various tribal populations speckled around its regions, and the cultural vitality of the place stands out among all this state of India. The ethnic value of the people of Madhya Pradesh is contributed by the dominating tribal population who live apart from mainstream India. Though there was lot of influence of urbanization during the twentieth century, these tribal people were not much affected by the modernization. Apart from these tribal people, the state has housed a number of people belonging to different ethnicity settled in different parts of Madhya Pradesh. Some of the ethnic fermented foods and beverages prepared and consumed in Madhya Pradesh are listed in Table 12.1.

12.2 Fermented Foods

12.2.1 *Pej* (*Peja*)

One of the prime ethnic fermented foods of Madhya Pradesh is *pej* or *peja* that can be made from the water left from boiled rice or *makka* powder. Local people give testimony that this food is much better and healthier than other alternatives. The cooked rice is fermented with rice soup for whole night. Salt and other ingredients such as *dahi* or lemon and chili pieces are added in the fermented food. Onion pieces are also added optionally into the dish (Fig. 12.2). It is called to be the poor people's diet, and it is tastier during summer season (Kapale et al. 2013).

Table 12.1 Ethnic fermented foods and beverages of Madhya Pradesh

Food	Substrate	Sensory property of product	Major ethnic consumers (tribe)	Region/district
Rabadi	Wheat, barley, pearl, millet, or maize	Mild acidic thick, slurry-like product	Bhil	Jhabua, Ratlam, Dhar, Badwani, Neemuch, Burhanpur, Neemuch, Khargone
Haria	Rice	Source of protein, Carbohydrates	Bharia	Chhindwara, Betul, Seoni, Balaghat, Hoshangabad
Mahua	Dried corollas of *Madhuca longifolia* and ranu	Distilled liquor	Bharia, Bhil, Korku	Chhindwara, Betul, Seoni, Balaghat, Hoshangabad, Jhabua, Ratlam, Dhar, Badwani, Neemuch, Burhanpur, Neemuch, Khargone
Papad	Black gram	Circular wafers; snack	–	Mostly in all parts of Central India
Dahi	Cow/buffalo/ yak milk curd	Savory	–	Mostly in all parts of Central India
Handia	Rice and ranu	Alcoholic beverage	Gond, Bharia	Near rivers of Satpura, Vindhya, and Narmada Ranges, Chhindwara, Betul, Seoni, Balaghat, Hoshangabad
Ghee	Cow or buffalo milk	Butter	–	Mostly in all parts of Central India
Lassi	Cow or buffalo milk	Buttermilk; refreshing beverage	–	Mostly in all parts of Central India
Paneer	Whey of cow milk	Soft, cheese-like product; curry	–	Mostly in all parts of Central India
Ranu goti	Rice, herbs	Starter to ferment alcoholic beverages	Bharia, Bhil, Gond, Kol	Chhindwara, Betul, Seoni, Balaghat, Hoshangabad, Jhabua, Ratlam, Dhar, Badwani, Neemuch, Burhanpur, Neemuch, Khargone, Rewa, Jabalpur
Pendum	Rice	Alcoholic beverage	Bharia	Chhindwara, Betul, Seoni, Balaghat, Hoshangabad, Chhindwara
Basi	Rice	Alcoholic beverage	Agaria	Dindori, Sidhi, Mandla, Balaghat, Shahdol
Pej/ Peja	Rice	Staple food	Baiga	Balaghat, Sidhi, Mandla, Dindori

Rice is washed and cooked properly
↓
The cooked rice is fermented with rice soup for whole night
↓
Food gets fermented
↓
Salt and other ingredients such as curd/lemon and chili pieces are added
↓
Onion pieces can also be added optionally into the dish

Fig. 12.2 Traditional method of preparation of *pej or peja* preparation in Madhya Pradesh

12.2.1.1 Culinary and Mode of Consumption

It is a drink of the tribals to beat the heat. It helps in beating the heat of the summers, but it also contains plenty of nutrients; thus it acts as a complete meal diet for them. Tribals drink *pej* to reduce the expenditure of rice, and it is consumed by all ages of the family (Kapale et al. 2013).

12.2.1.2 Medicinal and Nutritional Value

Tribal healers prescribe this mixture to the patients suffering from diarrhea, constipation, gastrointestinal distress, and irritable bowel syndrome. Dangis prescribe a sort of this mixture containing almost same contents for the women who suffer from chronic skin infections. It also decreases prevalence of allergies. *Peja* used as a probiotic supplement for people taking antibiotics to replenish the "health-promoting microbes" in the intestinal tract (Kapale et al. 2013).

12.2.2 *Rabdi* (Maize Porridge)

Rabdi or maize porridge is another tribal dish of central India where freshly harvested maize is crushed and boiled in water till it gets cooked. Buttermilk is poured

Fig. 12.3 Traditional method of preparation of *rabdi* (maize porridge) in Madhya Pradesh

Clean the earthen pot
↓
Crush and boil the maize slurry in water
↓
Maize Meal
↓
Mix thoroughly
↓
Cool until well gelatinized (thin porridge)
↓
Cool to temperature (25-30 °C)
↓
Ferment, 25-30 ° C> 48 h, 6d
↓
Rabdi

Fig. 12.4 Image representing the preparation of *rabdi* (maize porridge)

Rice is washed thoroughly and cooked properly
↓
After, being cooked, surplus water is separated and allowed to ferment
↓
Fermented water (mand) is collected in an earthen pot
↓
Pej and *Mand* are mixed up
↓
Leave the mixture overnight
↓
Salt and other ingredients such as curd or lemon are added, optionally

Fig. 12.5 Traditional method of preparation of *basi* preparation in Madhya Pradesh

and salt is added before eating. The addition of buttermilk is subject to availability. Tribals also prepared maize curry from maize flour as a substitute for vegetable (Gupta et al. 1992) (Figs. 12.3 and 12.4).

12.2.2.1 Culinary and Mode of Consumption

On ceremonial occasions, mainly at the time of marriage, rice and boiled grams are the full-fledged meal of the Bhil tribe. They also prepare *lapsi* from wheat flour; they are much obsessed to liquor which is mainly prepared from the crushed and boiled maize or from the bark of babul, or molasses. At the time of festivals, marriages, and after-harvest, this drink cum meal is consumed excessively by members of all the ages of the family.

12.2.2.2 Microbiology of *Rabdi*

Rabdi is a cereal-based fermented food product made mainly from cooked maize flour and buttermilk (optionally). Several microorganisms involved in the fermentation process have been recorded. For example, *Pediococcus acidilactici* (3.6×10^5/g), *Micrococcus* sp. (7.9×10^5/g), and *Bacillus* sp. (1.1×10^6/g) have been isolated from fermented *rabdi* (Steinkraus 1996).

Fig. 12.6 Earthen pots used for processing of *basi* (fermented liquor)

Fresh grapes are harvested

↓

Grapes are then de-steamed and partial or whole berries are pumped in wine tanks

↓

Grapes are initially cold soaked

↓

Fermented at 25°- 30° C

↓

Fermented wine is pumped over daily to wet the cap to extract enough colour and tannin.

↓

Wine is dried, the skins are pressed gently and malolactic fermentation is carried out

↓

Wine is then aged new oak barrels for 8-10 months before being racked off and bottled in another 2-3 months time

Fig. 12.7 Traditional method of preparation of *grape wine* in Madhya Pradesh

12.3 Fermented Beverages

12.3.1 *Basi*

The diet consists of rice, wheat, pulses (*masoor dal* and red gram), and vegetables such as potatoes, Amaranthus, spinach, and methi. Kadon, kutki, maize, and seasonal vegetables and roots form a major diet of the hill tribals. They consume more quantity of rice in winter, but they consume more quantity of wheat in summer. Rice is cooked in excess of water. After being cooked, surplus water is removed, and it is collected in an earthen pot (Figs. 12.5 and 12.6). It is known as the mand. When the mand is not removed during cooking, the cooked rice is known as pej. When the *pej* and mand are mixed up and fermentation is allowed, it is known as "basi" (Nout et al. 2007).

12.3.2 *Grape Wine*

Grapes are cultivated on the hills of the Dindori vineyards that have the gentle slopes with gravel red soils rich in iron and calcium. These selected grapes are planted on the top of gentle slopes and are hand harvested. Grapes are de-steamed and must with partial and whole berries are pumped in wine tanks. The grapes are initially cold soaked and then fermented between 25 and 30 °C. The fermented wine is pumped over daily to wet the cap and extract enough color and tannin. After the wine is dried, the skins are pressed gently, and malolactic fermentation is carried out. The wine is then aged in new oak barrels for 8–10 months before being racked off and bottled in another 2–3 months' time. The wine is then bottle aged for nearly 3–4 months before release (Fig. 12.7).

12.3.3 *Mahua*

Mahua, a distilled beverage is prepared from dried flower bud (corollas) of *Madhuca indica* or *Madhuca longifolia* var. *latifolia* (Koen.) Macbr plant (Kumar and Roy 2007). For its preparation, two popular methods are used:

1. Closed method
2. Tube (nala) method

The main difference between the close and tube method is mainly their preference of sample collection. In Closed method, the liquor is collected in a pot, while in Tube (nala) method, a tube is coupled with the pot in which liquor comes out by

Flower buds (*Madhuca indica*)
↓
Dried corollas are kept in airtight pot and water is added
↓
Incubated for 4-6 days till a characteristic smell is observed
↓
Distillation using wood stove
↓
Vapour cools down, and droplets are collected in the middle pot

Closed method ↙ ↘ Tube Method

A small pot called *dolki* is placed in which liquor is collected | Vapour is passed through the tube, connected to the middle pot

↓
Collected *Mahua* is prepared and used as beverages

Fig. 12.8 Traditional method of preparation of *mahua* in Madhya Pradesh

Fig. 12.9 Preparation of *mahua* beverage from *Madhuca indica*. (**a**) Mahua flowers. (**b**) Sun-dried *mahua* flower for liquor preparation. (**c**) Processing of *mahua* liquor in earthen pots (Kapale 2011)

the help of tube and stored in pot or cane. The dried corollas are kept in pot and some water is added. The pot is closed by the air tight cloth and kept for 4–6 days or till it starts giving a characteristic smell. Then the pot is kept on wood stove for distillation. On this pot, another pot is placed, and above this another pot is placed in which the cold water is filled; the junctions of these pots are made air-tight using cloth. The vapor passed through the middle pot strikes the bottom of the upper pot in which the cold water is kept. The vapor cools down and droplets collect in the middle pot. In closed system, in middle pot (*Paina*), a small pot called *dokli* is placed in which the liquor is collected, whereas in tube (*nala*) method, the vapor is passed through the tube, which is connected to middle pot and collected in cane or pot. The water of upper pot is regularly changed to keep it at low temperature. The collected mahua liquor is prepared and used as beverages (Figs. 12.8 and 12.9). The potency of preparation of *mahua* beverage depends on changes of water in the upper pot (Kapale 2011). Apart from being a customary drink in festivals such as religious rituals, ceremonies, birth, marriage, and death, it is mainly used as a medicine for curing various diseases such as jaundice. This kind of usage of a beverage in traditional tribal culture might be a subject of scientific and medical investigation.

12.3.3.1 Socio-economy and Ethnical or Religious Value

The largest number of Baigas are found in baiga-chuk in Satpuda-Meikal forest range of Amarkantak. Beverages obtained from *mahua*, a popular local beverage, have a very important role in the sociocultural life of Baiga tribe of Amarkantak region. *Mahua* beverages are drunk in all their festive occasion of celebrations (Kapale 2011).Tribal people, men and women, consume this drink regularly. It is an obligatory item during celebrations and evening activities. *Mahua* beverage is not only popular drink of the Baiga tribe but is also considered to be pure and used as holy water or elixir by Bhil and Bharia tribal people (Kapale 2011).

12.3.3.2 Medicinal and Nutritional Value

Tribals not only consume *mahua* for inebriation, but also this beverage fulfills the need of around 5–10% of the daily requirements of the nutrients and plays a complementary role in the nutrition of the people. On the other hand, such beverages are also used as medicine for the treatment of different diseases or ailments. It is also used as light tranquillizer by Maria tribe of Bastar. Additionally, it is used to treat fever, diarrhea, dysentery, and gynecological complaints by Baiga, Gond, and Kol tribes of Surguja district. *Ranu* tablets are also used in treating cholera by Gond tribe of Surguja district (Kumar and Rao 2007).

12.3.4 *Pendum*

Pendum is the most popular homemade alcoholic rice beverage of the tribal community of the *Bharia* tribe of Madhya Pradesh; this beverage is prepared by fermentation of rice mixed with plant leaves and barks containing different phytochemicals (Arjun 2015). The starter material for fermentation is homemade starch mixed with

Starter material for fermentation is homemade starch mixed with number of plant parts
↓
Gurji (*Setaria italican*) is washed properly
↓
Boiled water and *Gurji* are added and a paste is obtained
↓
Paste is mixed with cold water in a mud pot
↓
Left for fermentation for 5-7 days
↓
Digestive tablets are added to the pot
↓
Fermentation continued upto 2-3 days

Fig. 12.10 Traditional method of preparation of *pendum* in Madhya Pradesh

a number of plant parts which are reported to carry medicinal properties (Das et al. 2012). The yeast culture is maintained in semi-sterilized medium made of ground rice. The fermentation temperature during indigenous process of wine making is generally maintained at around 30 °C. In the first step, the Gurji (*Setaria italican*) is washed properly and boiled for making a consistent paste. The paste is then mixed with cold water in a mud pot and left for fermentation for 5–7 days. After fermentation, digestive tablets made up of locally available plant parts are added to the pot, and fermentation is continued for couple of days (Fig. 12.10).

12.3.4.1 Nutritional Value

Pendum is consumed by the local people as to refresh their mood and get rid of anxiety and drowsiness. The traditional alcoholic drink prepared from aboriginal rice and finger millet is rich in protein and consumed most frequently by Bharia tribe. Its preparation and more frequent use as beverage might be on account of availability of preparing materials and cultural requirements.

12.3.5 *Handia*

The word *handia* finds its origin from the word handi in local language meaning large earthen pot. Handia occupies a key position in the social, cultural, and economic life of Santhal, Bharia, and Gond tribes and is popularly accepted as a traditional drink (Kumar and Roy 2007). It is prepared from rice along with some of the locally available plant parts using indigenous method.

Preparation of *handia* includes two distinct phases:

1. Preparation of ranu or bakhar tablets.
2. Production of *handia.*

12.3.5.1 Preparation of *Bakhar* or *Ranu*

Ranu or *bakhar* tablets act as starter and help in fermentation of the beverage. Ranu tablets are the mixtures of various plant parts (50%) and powdered unboiled rice (50%). A number of plant species and parts thereof are used for this purpose. Some species, viz., *Asparagus racemosus* Wild., *Cissampelos pareira* var. *hirsuta* Forman, *Clerodendrum serratum* (L.) Moon, *Coccinia grandis* (L.) Voigt, *Holarrhena pubescens* (Buch.-Ham.) Wall. ex. G. Don, *Madhuca longifolia* (Koenig) MacBride var. *latifolia* Roxb., *Smilax macrophylla* (Roxb.), *Woodfordia fruticosa* (L.) Kurz, and *Rauvolfia serpentina* (L.) Benth, are commonly used by the people of all localities. A total of 20 plant species belonging to 18 families are recorded to be used for preparing *bakhar* or *ranu*. The exact ratio of different plants used for *ranu* preparation could not be ascertained as the informants are usually reluctant to unveil the same. Mostly the roots are preferred followed by bark, flower, fruit, young shoot, and leaves. Preferred parts and plants vary from place to place. Roots and barks of the collected plants are dried under sun, chopped into pieces, and then powdered by husking paddles. Dried root, stem, and other parts used for the purpose, both as such

Parts of six different plants are collected, cleaned and dried

↓

Samples are powdered using grinding stone

↓

2 gm of powder is mixed with 200 gm of rice powder mixed in water

↓

After mulching for 10-15 min, small pellets are made out of it

↓

Small tablet form of these pellets are made

↓

Used to carry out fermentation process

Fig. 12.11 Traditional method of preparation of *bakhar* in Madhya Pradesh

and powdered, are abundantly and openly sold in the local markets. Powdered plant ingredients are mixed with rice (*Oryza sativa* L.) powder. A suitable amount of water is added to make dough. *Ranu* is prepared in the form of rounded tablets and spread over straw beds in layer over layer with a final thin layer of straw cover. After 3 days, the *ranu* tablets are picked up from straw beds and dried under sun for about 2 days and stored for use in fermentation of rice beverages. These tablets are not only used for fermenting rice beverage but also used for treatment of various ailments. A paste of *ranu* tablets with saliva is applied on mumps by the tribes to get relief. Santhals also disinfect tasar silkworm eggs during indigenous rearing (Roy et al. 2012) (Figs. 12.11 and 12.12).

12.3.5.2 Production of *Handia*

Boiling of Rice

During boiling, the crystal structure of rice starch is destroyed and moisture content is increased.

This leads to the favorable conditions for the growth of microorganisms as the process of boiling results into gelatinization of the starch. After boiling the rice,

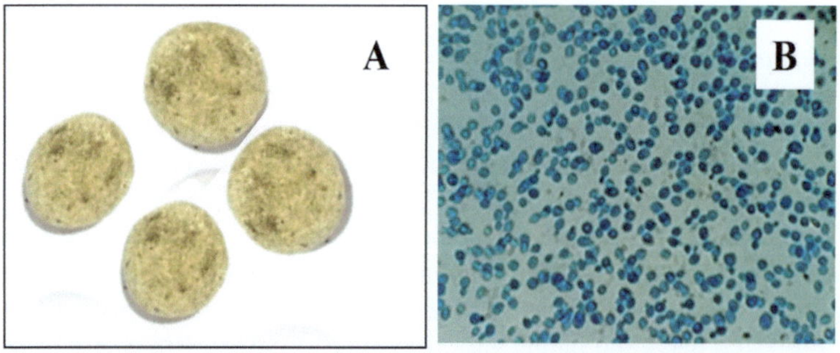

Fig. 12.12 Bakhar tablets used for liquor preparation. (**a**) Spherical tablets. (**b**) Yeast isolated from the bakhar tablet

excess water is drained off, and the steamed rice is immediately taken out from the pot and used for further processing.

Cooling of the Boiled Rice

After the excess water is drained off, steamed rice is evenly spread on a bamboo mat or on the mat prepared from date palm leaves for cooling. Then rice is turned over, breaking up large clumps using a wooden spade, by whacking it with the spade's flat blade.

Addition of *Bakhar* or *Ranu*

Generally 1 kg of rice is boiled and cooled, and then three small balls of *bakhar*, around 3 g of weight, are grinded and mixed well in accordance with the traditional process (Dhal et al. 2010). Traditional manufacturer generally uses earthen pot for fermentation. Fermentation is carried out for 2.5–3 days. Usually some *bakhar* powder is spread inside the earthen pot or the *handi*.

Saccharification and Fermentation

Bakhar mixed rice to be kept in earthen pot at room temperature for 2–3 days with lid slightly open to facilitate the fermentation process.

Separation of Spent Grains by Squeezing

Once fermentation is complete, the fermented mash becomes dense and mushy with cream white color and characteristic odor. The conventional process uses cloth

Earthen pot sterilization by washing, drying in sunlight and fumigation
↓
Substrate preparation
↓
Clean and remove impurities from rice grains
↓
Boil rice grains
↓
Cool and dry boiled rice
↓
Spread the mixture on mat
↓
Add bakhar tablets to rice grain mixture
↓
Put the mixture in a pot
↓
Incubate the pot in dark
↓
Fermentation continued upto 3-4 days
↓
Extraction and filtration

Fig. 12.13 Conventional method of preparation of *handia* preparation in Madhya Pradesh

called gamcha for filtering the liquid. A little amount of water (150 mL) is mixed during the separation of fermented liquid from the fermented mass by the traditional manufacturer (Fig. 12.13) (Roy et al. 2012).

12.3.5.3 Medicinal and Nutritional Value

Handia keeps the stomach cool, and it protects from extreme heat and is also intoxicating. Its preparation and selling is a source of livelihood for tribals, and some accept it as a primary occupation. The tribals believe it to possess medicinal value and often use it in curing jaundice, colic disorders, and dysentery.

12.3.6 *Haria*

Haria is classified as a fermented beverage made from rice which is liked by tribal and impoverished people of Central India and West Bengal (Ghosh et al. 2014). The liquor is mostly prepared using the indigenous knowledge of the tribal community. The eccentricity of the liquor lies in the chief component of this infusion, specifically low-quality boiled rice (*Oryza sativa* L.), which is prepared by mixing it with a traditional starter, called *bakhar*, an amylolytic starter culture followed by fermentation for 3–4 days in a heat-sterilized earthen pot. The fermentation is marked by the decrease in pH, which contributes to increased acidity, whereas the content of alcohol also increases up to 2–3% (v/v) in the consumable drink.

Although the preparation of *bakhar* varies from community to community, ~42 species of ethnobotanically important plants are mostly used. The microbial consortia are naturally outsourced from rice and herbs used for the preparation of *bakhar*. The amylolytic starter culture consistently generates malto-oligosaccharides, which is a low-calorie ingredient, less sweet, and viscous, capable of retaining sufficient quantity of water within it. Besides, serving the purpose of a beverage, *haria* can compensate the loss of water in the human system under extreme high temperature. Together with this, the drink also serves as a remedy for several acute and chronic diseases which may be due to the use of certain medicinal herbs as starter culture.

12.3.6.1 Haria Preparation

An uncontaminated, heat-sterilized earthen pot is used as a fermenter vessel. These are cleaned using a bundle of rice straw and then sterilized by direct heating and smoking by burning the readily available rice straw. Pots are then kept in standing and static condition throughout the process. Rice grains are charred by boiling in water followed by drying in open air. The *bakhar* (starter) is added to this, and then the mixture (rice grains + starter) is poured into the sterile earthen pots, and the lids are tightly closed. Now, the mixture is incubated at room temperature for 3–4 days in dark. Changes, such as change in texture of the rice grain, melting, and development of a thick gummy appearance, can be noticed during the course of fermentation. Lastly, the mixture is filtered with a clean cloth, and the creamy liquid so obtained is often supplemented with spicy vegetables and consumed. Any physical

Earthen pot is cleaned
↓
Rice grains are boiled in water (1:2)
↓
Mixture is spread over a mat for drying
↓
Mixture is supplemented with *bakhar* starter
↓
It is poured into the pot with the lid tightly closed
↓
It is incubated at room temperature in a dark room for 3-4 days
↓
Fermented rice grain develops a gelatinous appearance
↓
Extraction and filtration

Fig. 12.14 Traditional method of preparation of *haria* preparation

interruption during the course of fermentation is totally avoided. Various steps involved in traditional procedure of *haria* preparation are given in Fig. 12.14.

12.3.6.2 Substrate Preparation

After manual cleaning, the low-quality rice is used as the substrate for *haria* preparation. The grains are then subjected to vigorous boiling in suitable amount of water (1:2) in a large vessel until charring is achieved. The concept behind the vigorous boiling is that the charring kills the pathogenic and unwanted micro flora, which may extensively damage the infusion if not avoided. Also, this pre-boiling enhances the substrate availability by softening and increasing the surface area of the grains. The boiled product or *bhat* is then air-dried over a clean leaf mat.

12.3.6.3 Fermentation and Recovery

This inoculated boiled rice is tightly packed in the pot using an earthen lid so as to create anaerobic condition. This infusion is then incubated under static condition at room temperature for 3–4 days in dark. Once the fermentation is over, the boiled rice grains get gelatinized as a result of microbial actions. Then, the fermented material is transferred over a clean cloth. The thick gelatinous material is diluted using potable water and filtered through the clean cloth. The cream-colored, buttermilk like filtrate is then consumed by the local people. It is often supplemented with seasonal vegetables for spicing up the taste. The dilution and filtering is done in order to achieve homogeneity and eliminate solid particles and unfermented matter.

12.3.6.4 Microbiology of *Haria*

Scientific research unfold that *haria* is a combined product of liquefaction, saccharification, and fermentation processes. Mostly, aerobic bacteria, yeasts, molds, lactic acid bacteria (LAB), and *Bifidobacteria* work synergistically to complete the biochemical processes. At the beginning, an aerobic bacterium initiates the fermentation process; the hydrolytic activity of aerobic microbes sets the environment for the growth of the yeast and mold. In fact, LAB and *Bifidobacteria* are reported to be dominant throughout the fermentation process reflecting the fact that the microorganisms are outsourced from the medicinal herbs used for the fermentation. It is apparent to decode that amylase and glucoamylase are the enzymes responsible for the saccharification and liquefaction of the rice, where amylase converts complex polymers of amylose and amylopectin into limit dextrin and thereafter glucose is produced by the action of glucoamylase. This glucose is then utilized by yeasts for the production of alcohol. The synergistic enzymatic activity also leads to the formation of malto-oligomers including maltotriose, maltotetraose, and maltopentaose; these malto-oligomers are highly water soluble and thus responsible for the production of clear solutions, leading to the formation of the final product. It has been reported that, amylolytic yeast, *Saccharomycopsis fibuligera* in fermented rice is also responsible for the production of oligosaccharides. Microbiological studies reveal the presence of *Saccharomyces cerevisiae*, *Saccharomyces boulardii*, *Saccharomycopsis* sp., *Zygosaccharomyces cidri*, *Pichia*, *Candida tropicalis*, *Candida musa*, *Candida nitratophila*, and *Issatchenkia* in *haria* along with some of the species of lactic acid bacteria (Sha et al. 2012).

12.3.6.5 Medicinal and Nutritional Value

In tribal culture, *haria* is employed as a remedy for many contagious or transmissible diseases. It is believed to confer protection, particularly against several gastrointestinal disorders, including amoebic dysentery, appendicitis, amoebiasis, enteritis, and acid reflux. Moreover, the tribal practitioners also recommend it as a skin, eye, hair, and heart defensive agent.

12.4 Conclusion

Central India is rich in ethnics, and diversity of fermented foods is related to diversity of ethnicity. About 20–30% of the total population of scheduled tribe is found in central India. Conversely, rapid urbanization has affected and challenged the time-tested conventional technologies for preparation of ethnic fermented foods. If we continue to ignore the ethnic fermented foods and beverages, the time is not far when our indigenous traditional knowledge will vanish before it could be validated using scientific criteria and exploited further.

References

Arjun J (2015) Comparative biochemical analysis of certain indigenous rice beverages of tribes of Assam with some foreign liquor. J Biosci Biotechnol Res Commun 8:138–144

Das AJ, Deka SC, Miyaji T (2012) Methodology of rice beer preparation and various plant materials used in starter preparation by some tribal communities of North-East India: a survey. Int Food Res J 19:101–107

Dhal NK, Pattanaik C, Reddy S (2010) Bakhar starch fermentation—a common tribal practice in Orissa, India. Ind J Tradit Knowl 9:279–281

Ghosh K, Maity C, Adak A, Halder SK, Jana A, Das A, Saswati Parua SM, Das Mohapatra PK, Pati BR, Mondal KC (2014) Ethnic preparation of Haria, a rice-based fermented beverage, in the province of lateritic west Bengal, India. J Ethnobot Res Appl 12:39–49

Gupta M, Khetarpaul N, Chauhan BM (1992) Preparation, nutritional value and acceptability of barley rabadi—an indigenous fermented food of India. J Plant Foods Hum Nutr 42:351–358

Kapale R (2011) Beverages of Baiga tribe of Amarkantak Region. An ethnobotanical approach. J Adv Biores 2(2):14–17

Kapale R, Prajapati AK, Napit RS, Ahirwar RK (2013) Traditional food plants of Baiga Tribal's: a survey study in tribal villages of Amarkantak-Achanakmar biosphere, Central India. Indian J Sci Res Dent Tech 1:27–30

Kumar V, Rao RR (2007) Some interesting indigenous beverages among tribals of Central India. Indian J Tradit Knowl 6(1):141–143

Nath V, Khatri PK (2010) Traditional knowledge on ethno-medicinal uses prevailing in tribal pockets of Chhindwara and Betul Districts, Madhya Pradesh, India. Afr J Pharm Pharmacol 4:662–670

Nout MJR, Sarkar PK, Beuchat LR (2007) Indigenous fermented foods. In: Doyle MP, Beuchat LR (eds) Food microbiology: fundamentals and frontiers. ASM Press, Washington, DC, pp 817–835

Roy A, Khanra K, Bhattacharya C, Mishra A, Bhattacharyya N (2012) Bakhar-Handia fermentation: general analysis and a correlation between traditional claims and scientific evidences. J Adv Biores 3:28–32

Sha SP, Thakur N, Tamang B, Tamang JP (2012) Haria, a traditional rice fermented alcoholic beverage of West Bengal. Int J Agric Food Sci Technol 3:157–160

Steinkraus KH (1996) Handbook of indigenous fermented foods, 2nd edn. Marcel Dekker, New York, p 776

Web Resources

http://www.mp.gov.in/web/guest/home-history
https://www.census2011.co.in/census/state/madhyapradesh.html
https://www.indianetzone.com/5/culture_madhya_pradesh.html
https://www.indianmirror.com/tribes/bhiltribe.html

13 Ethnic Fermented Foods and Beverages of Maharashtra

Uday Shriramrao Annapure, Aarti Suryakant Ghanate, and Prabodh Shirish Halde

Abstract

Maharashtra state is geographically divided in five main regions such as *Konkan, Paschim* or *Madhya* Maharashtra, *Khandesh, Marathwada,* and *Vidarbha*. Food habits in all these regions are different with some overlaps. These differences are based on the availability of raw materials and the taste of the consumer in those specific regions. Maharashtra has a rich culture of different types of traditional fermented and non-fermented foods. Commonly consumed fermented foods and beverages in Maharashtra include milk-based products (*Shrikhand, Mattha*), cereal- and legume-based foods and beverages (*Ambil, Amboli, Olyafenya, Sandan, Anarase, Kurdai, Salpapdya, Bibdya, Kharodya, Dhapode,* and *Sandge*), dried fish (*Jawla, Karndi, Tengli, Sode, Sukke Bombil, Sukke Mandeli, Vaktya, Sukke Makul* and *Khara masa*), as well as fruits, vegetables, and meat-based pickles (*lonche*). Most of these foods are naturally fermented.

Keywords

Maharashtra · Fermented food · Cereals · Legumes · Dried fish · Pickles

13.1 Introduction

Among all the states and union territories in India, Maharashtra is located centrally in western region. Marathi is the native language of Maharashtra and the name Maharashtra is an imitative word from the Sanskrit words—'*Maha*' means '*Great*'

U. S. Annapure (✉) · A. S. Ghanate
Department of Food Engineering and Technology, Institute of Chemical Technology (ICT), Mumbai, Maharashtra, India

P. S. Halde
Marico Limited, Mumbai, Maharashtra, India

J. P. Tamang (ed.), *Ethnic Fermented Foods and Beverages of India: Science History and Culture*, https://doi.org/10.1007/978-981-15-1486-9_13

and *Rashtra* means '*Nation*'; hence it represents precisely great nation. Maharashtra is the third biggest state in India by area and located between latitudes 19.7515 N and longitudes 75.7139 N. Maharashtra state land occupies 307,713 km^2 area compared with allover India. Maharashtra has extended seashore prolonging up to 720 km beside the Arabian ocean. The western borders of Maharashtra (825 km) come into contact with Arabian Sea; eastern borderline touches the state of Madhya Pradesh and Chhattisgarh (750 km); the Southern frontier (1875 km) touches the states of Andhra Pradesh and Karnataka, and northern state line (1725 km) touches the states of Madhya Pradesh and Gujarat (https://www.maharashtratourism.gov.in). Maharashtra is one of the states in India which has five regions: Konkan, western or Madhya Maharashtra, Khandesh, Marathwada, and Vidarbha. They are, viz., Central Provinces, Old Berar region, and Vidarbha region that comprises Nagpur and Amravati divisions. Khandesh or northern Maharashtra comprises Nasik division, Marathwada region comprises Aurangabad division, western Maharashtra region involves Pune division, and Konkan region involves Konkan division which includes Mumbai City and Mumbai suburban areas (https://www.maharashtra.gov.in/1128/Districts).

Maharashtra is the second most populated state in India. As of the total population of India, the population of Maharashtra is more than 110 million (approximately 9.28%). Hinduism is the most popular religion in Maharashtra with 79.83% followers. Islam religion is second most popular with nearly 11.54% following it. In Maharashtra state, Jainism is followed by 1.25%, Christianity is followed by 0.96%, Buddhism by 0.20%, and Sikhism by 0.20%. Nearby 0.16% quantified 'other religion' and around 0.25% quantified 'no particular religion' (Census of India 2011).

The Maharashtrian people comprises castes including subcastes. There are about 250 castes, and within each caste there are several subcastes. **Marathwada and Vidarbha**: Kunbi (Tirole, Dhanoje, Khaire), Maratha, Brahman (Charak, Madhyandin), Gowari, Korku, Koshti, Halba or Halbi, Powar, Bawane Mahar, Kohali, Gond, Dheevar, and Kolam. **Khandesi**: Brahmin (Madhyandin, Rigvedi Deshastha), Vanjari, Leva Patidar, Leva Gujar, Bhill (Mavchi, Tadvi, etc.), Ahir Sonar, Bari, and Maratha. **Madhya Maharashtra**: Maratha, Brahmin, Phulmali, Mahar, Koli, Chambhar, Teli, Mang, Shimpi, Sali, Nhavi, Pareet, Gurav, Burud, Andha, Sonar, Bhavsar, and Dhangar (Hatkar and Khutekar). **Konkan**: Maratha, Vaishya Vani, Bhandari, Brahmin (Saraswat, Chitpawan, Karhade), Sonkoli, Pathare Prabu, Chandraseniya Kayastha Prabhhu, Daivadnya Brahmin-Sonar, Twashta Kasar, Pathare Kshatriya, Khatri, Somvanshi Vadval, Agri, Thakur, Varli, and Malhar Koli (Singh 1992).

The annual climate of Maharashtra state consists of three seasons, viz., four successive months of winter season followed up by 8 months of monsoon and summer seasons, respectively. The cold season is initiated at the beginning of November and continues till the end of February. The summer season starts at the beginning of March and continues till the end of May. The monsoon season reports approximately 85–90% of the yearly rainfall in the region, which normally commences in the first week of June and persists till the end of September. The post-monsoon season is the beginning of October and continues till the middle of November. October marks the transition from rainy season to winter.

Most of the fermented foods of Maharashtra are produced in the summer season, dried, and stored so that those can be used though out the year. The dried fermented food products are consumed mostly in the rainy and winter seasons. The Koli community of Maharashtra generally preserves fishes by applying salt and drying under natural sunlight during summer, and it consumed during the peak of monsoon when there is unavailability of sea foods as it is a breeding season for most of the fishes.

Agriculture is the mainstay of the state of Maharashtra. It is the major occupation of the people. Almost 82% of the rural population depends on the agriculture for livelihood. Due to the variations in the geographical, ecological, and weather conditions, variety of crops are grown in different parts of the state. Major crops cultivated in Maharashtra state include wheat, rice, sorghum, corn, millets, variety of pulses, groundnut, soybean, sugarcane, turmeric, etc. Major fruit cultivation includes mangoes, grapes, bananas, oranges, pomegranates, etc. Depending on the different environmental conditions, culture, ethnic groups, and occupations, a variety of cuisines are being developed as traditional foods for particular locality in the different parts of the state. Some of the traditional foods are also prejudiced by religion, cultural adoptions, and ethnicities.

In the coastal *Konkan* region, rice is the traditional staple food with fish being a major component, due to the abundant availability of variety of sea foods as it is in the proximity of seashore. Coconut is also a major part of most of the cuisines in Konkan in the form of wet coconut and coconut milk. Fish curry, fried fish, and many other cuisines prepared using fishes are popular in Konkan. *Kombdi vade* is also the most popular dish which is served with chicken curry. South Konkan has its own specific

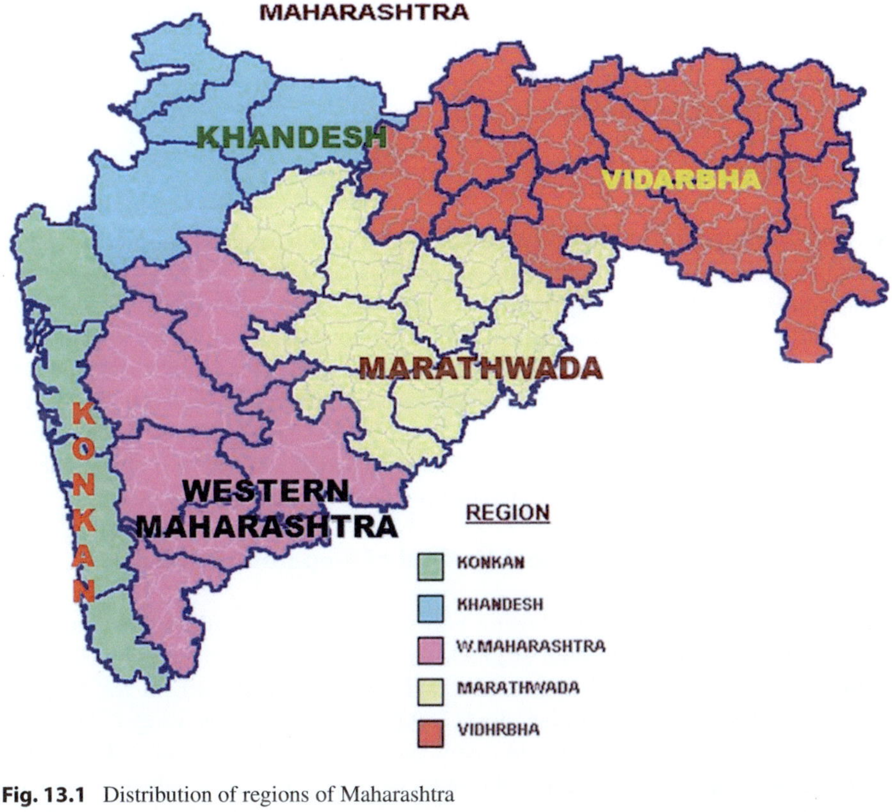

Fig. 13.1 Distribution of regions of Maharashtra

Table 13.1 Crop production statistics of Maharashtra 2016–2017

Crops	Production (in thousand tonnes)
Rice	3581
Wheat	2214
Sorghum	2538
Pearl millet	800
All cereals	12,646
All pulses	4584
All food grains	17,229
Sugarcane	54,237
Groundnut	420

Source: Economic Survey of Maharashtra 2016–2017

independent cuisine called *Malvani* cuisine, which is mostly non-vegetarian (fish curry with *Amboli).* The traditional staple food of Central (*Madhya*) Maharashtra and *Khandesh* and *Marathwada* region is mostly unleavened flat bread, *bhakri* (made by using a flour of *Jawri or bajari*) or *Chapati* (made by using a whole wheat flour), along with cooked spicy vegetables, vegetable curry, *dal*, and rice. In *Vidarbha, Marathwada*, and *Khandesh* regions, peanut is being widely used in daily preparations. *Savaji* is a speciality of *Vidarbha* in mutton and chicken dishes (Fig. 13.1; Tables 13.1 and 13.2).

13.2 Milk-Based Product

13.2.1 *Shrikhand*

Shrikhand is a traditionally homemade sweet dish in Maharashtra but now manufactured at a large scale by dairy industries across India. *Shrikhand* is a semisoft, sweetish-sour, whole milk product prepared from lactic fermented curd. The curd (*Dahi*) so obtained is strained through a muslin cloth to remove the whey to get a solid mass called *Chakka* which is the basic ingredient for *Shrikhand.* This *Chakka* is mixed with the required amount of sugar to yield *Shrikhand.* The *Shrikhand wadi* can also be prepared by desiccating *Shikhand* over an open pan. To enhance the flavour of *Shrikhand,* natural and synthetic flavours are added. *Shrikhand* added with mango pulp is called as an '*Aamrakhand*' which is mostly prepared in mango season (Fig. 13.2).

13.2.1.1 Modified Traditional Method of Preparation and Mode of Consumption

Shrikhand is generally prepared using buffalo milk (6% fat), pasteurized at 71 °C for 10 min, and allowed to cool at 28–30 °C. It is then inoculated with 1% lactic culture (*Str. Lactis*), mixed thoroughly, and incubated at 28–30 °C for 15–16 h (overnight) to set the firm curd (acidity 0.7–0.8% lactic acid). The whey is separated by placing the curd in a muslin cloth bag and hanging on a peg for 8–10 h. After draining the whey, the remaining solid mass is called as *Chakka*, which is the base

Table 13.2 Ethic fermented foods and beverages of Maharashtra

Food	Substrate	Sensory properties of product	Major ethnic consumers
Shrikhand	Milk	Acidic, sweet, viscous, consumed with puri or chapatti	All over Maharashtra
Mattha	Milk	Acidic, spicy, consistency like buttermilk	All over Maharashtra
Ambil	Cereals	Fermented beverage, thick consistency	All over Maharashtra
Amboli	Rice, black gram	Fermented pan cake like uttapam, consumed with black pea curry n chicken curry	Konkani specially Malvani
Olyafenya	Rice	Dosa like thin and round shape, breakfast	Konkani people
Sandan	Rice, Jackfruit	Idli like and round shape, sweet, breakfast	Konkani people
Anarase	Rice	Fried sweet snack, circular shape, crispy texture	All over Maharashtra
Salpapdya	Rice	Dried, fried crisp, salty, circular shape	All over Maharashtra
Bibdya	Sorghum	Dried, baked crisp, salty, circular shape	Khandesh
Kurdai	Sorghum, Rice	Dried, fried crisp, salty, circular shape	All over Maharashtra
Kharodya	Finger millets	Dried, fried crisp, salty	All over Maharashtra
Sandge	Finger millets, legumes	Dried	All over Maharashtra
Dhapode	Sorghum	Dried, circular shape	Khandesh

(continued)

Table 13.2 (continued)

Food	Substrate	Sensory properties of product	Major ethnic consumers
Jawla	Fish	Sun-dried shrimp product	Maratha, Vaishya Vani, Bhandari, Brahmin (Saraswat, Chitpawan, Karhade), Sonkoli, Pathare Prabu, Chandraseniya Kayastha Prabhhu, Daivadnya Brahmin-sonar, Twashta Kasar, Pathare Kshatriya, Khatri, Somvanshi Vadval, Agri, Thakur, Varli, Malhar Koli
Karndi	Fish	Sun-dried shrimp product	
Tengli	Fish	Sun-dried small fish product	
Sode	Fish	Sun-dried shrimp product	
Sukke bombil	Fish	Sun-dried fish product, salted, fried	
Sukke Mandeli	Fish	Sun-dried small fish product	
Khara masa	Fish	Sun-dried fish product, salted	
Vaktya	Fish	Sun-dried fish product	
Sukke Makul	Fish	Sun-dried fish product	
Pickles (Lonche)	Fruits and vegetables	Salty, sour	All over Maharashtra, prawns pickle is mainly prepared in Konkan and meat pickle prepared in Nagpur (Savaji community), *Mainmula* pickle specially prepared in Kolhapur region of Maharashtra

for *Shrikhand*. This *Chakka* is then mixed with ground sugar and well kneaded to insure uniform mixing. Colour and flavour may also be added as per the requirement (Fig. 13.3). The *Shrikhand* is usually consumed with *Puri* or *Chapatti*.

Microorganisms *Lactobacillus acidophilus*, *Lactobacillus sporogenes*, and *Lactobacillus rhamnosus* (Swapna and Chavannavar 2013).

Fig. 13.2 Photograph of *Shrikhand*

Fresh Buffalo milk (6% fat)
↓
Pasteurized (71°C / 10 min)
↓
Cooling to 28-30 °C
↓
Addition of 1 % curd cultures / lactic acid bacteria as starter culture.
↓
Incubation (28-30°C, 15-16 hours, till the acidity reaches 0.7-0.8 %)
↓
Curd formation
↓
Placed in muslin cloth & hung for 8-10 hours to remove the whey
↓
Chakka
↓
Add grounded sugar
↓
Mixing / whipping
↓
Shrikhand

Fig. 13.3 Flowsheet of *Shrikhand* preparation

Nutritional Composition The composition of *Shrikhand* is as follows: Moisture 34.48–35.66%, fat 1.93–5.56%, protein 5.33–6.13%, reducing sugar 1.56–2.18%, and non-reducing sugar (sucrose) 53.76–55.55% (Steinkraus 1983).

13.2.2 *Mattha*

Mattha is a popular traditional fermented beverage in Maharashtra. *Mattha* is prepared using curd with added spices. It is mainly consumed in the summer season. It is also a very popular beverage in the wedding and many other family functions of Maharashtra (Fig. 13.4).

Fig. 13.4 Photograph of *Mattha*

Fig. 13.5 Flowsheet of *Mattha* preparation

Cow/ buffalo Milk
↓
Curd
↓
Churn in earthen vessel and add cold water
↓
Light curd with consistency of buttermilk
↓
Add spices, salt and sugar
↓
Mattha

13.2.2.1 Modified Traditional Methods of Preparation and Mode of Consumption

Mattha is mildly acidic, little bit salty, and spicy in nature. Curd is the primary ingredient in *Mattha*. Traditionally, *Mattha* is prepared by churning the curd in earthen pot and then adding roasted cumin powder, graded ginger, and finely chopped green chilli and coriander leaves. They are mixed well and then water is added to adjust the desired consistency like buttermilk. Salt and sugar is added to the desired taste. It is then kept in the refrigerator for couple of hours and served cooled. *Mattha* is traditionally stored in earthen pots for cooling purpose (Lonkar et al. 2011).

Microorganisms *Lactobacillus acidophilus*, *Lactobacillus sporogenes*, and *Lactobacillus rhamnosus* (Fig. 13.5).

13.3 Cereal- and Legume-Based Fermented Food

13.3.1 *Ambil*

Ambil is a traditional fermented beverage that is very popular in Maharashtra. It is prepared by fermenting the rice/sorghum/ragi flour and then the slurry obtained was cooked to get the product *Ambil*. Generally, in Maharashtra it is made from rice and sorghum and from ragi in Karnataka. It is highly nutritious product in minimum cost.

Ambil as a gel like consistency is used as a coveted *prasad* (naivedya) for Mahalaxmi, the goddess of prosperity. In earlier times, it was used as a breakfast food. Today, *Ambil* is typically enjoyed as an appetizer. It is a common product prepared during *Yala-amavasya* (day of new moon) at large in Maharashtra (Fig. 13.6).

13.3.1.1 Modified Traditional Methods of Preparation and Mode of Consumption

To prepare it, sorghum/rice/ragi flour is made into a thick batter with addition of water and allowed to ferment for 14 h in an earthen pot. The fermented sorghum/

Fig. 13.6 Photograph of *Ambil*

rice/ragi batter is then cooked with water (if required), stirred to avoid lump formation during cooking. After cooking it is allowed to cool. Then mixed with sour buttermilk and spices (ginger, garlic, salt, green chilli, coriander, and cumin) before consumption. The lactic acid bacteria present in buttermilk degrade the starch of sorghum/rice/ragi into simple sugars, resulting in a decrease of pH from 6.0 to 5.4 in 5–6 h, while the volume increases by about 20%, indicating CO_2 production. The fact that CO_2 is produced may be due to the hetero-fermentative lactic acid bacteria.

Microorganisms LAB (Lactic acid bacteria) viz. *Lactobacillus bulgaricus* and *Lactobacillus acidophilus* (Shinde 2011).

Nutritional Composition (Table 13.3)

Health Benefits *Ambil* is known to be very nutritious for the body as it is rich in minerals. *Ambil* is good for satiating hunger and has a cooling effect in the stomach (Fig. 13.7).

Table 13.3 Nutritive composition of *Ambil*

Nutrients	Content[a]
Moisture content (%)	91.5
Crude protein (%)	1
Crude fat (%)	0.32
Ash (%)	1.22
Carbohydrate (%)	3.42
Calcium (mg)	58
Sodium (mg)	23
Potassium (mg)	06

Source: Shinde (2011)
[a]Per 100 g

Sorghum/rice/ragi
↓
Cleaning and washing
↓
Surface drying
↓
Milling to obtained flour
↓
Addition of water to the flour
↓
Addition of starter culture (curd) to slurry
↓
Fermentation (14-16 hours, 28°C)
↓
Cooking of fermented slurry (15 min)
↓
Churning of cooked slurry with addition of buttermilk and spices
↓
Ambil

Fig. 13.7 Flowsheet of *Ambil* preparation

13.3.2 *Amboli*

Amboli is the specialty of *Malavan* area of *Konkan* region. *Amboli* is a special type of fermented flattened rice bread. It resembles *Dosa* or *Uttapam*. However, as it is fluffy and spongy, it is consumed along with spicy curries (Fig. 13.8).

13.3.2.1 Traditional Methods of Preparation and Mode of Consumption

Amboli is prepared using rice, black gram splits, and chickpea splits (2:1:0.5). The clean rice, black gram, split chickpea, and fenugreek seeds are washed and soaked

Fig. 13.8 Photograph of *Amboli*

Wash rice, black gram splits and chickpea splits

↓

Soaking separately for 6 hours

↓

Coarse grinding of rice

↓

Fine grinding of black gram splits and chickpea splits

↓

Mix and allow to ferment for 10-12 hours (preferably overnight)

↓

Spread a scoop of batter on pre-heated frying pan (1-1.5cm thick)

↓

Cook till it turns golden brown in colour

↓

Amboli

Fig. 13.9 Flowsheet of *Amboli* preparation

for 6–8 h separately. The rice is then drained and coarsely ground separately in stone grinder. The black gram, split chickpea, and fenugreek are drained and finely ground together in a stone grinder. During grinding coriander and cumin seeds are also added. Both the slurries are combined to form a thick batter with continuous stirring. Salt is added as per taste. The batter is then allowed to ferment overnight (10–12 h) in warm and humid condition. The fermentation time must be sufficient to allow a definite leavening of the batter and to develop a pleasant acidic flavour. The fermented batter is then spread over (approximate 20 cm diameter and 1–1.5 cm thickness) pre-greased hot frying pan and allowed to cook till it gets golden brown colour. It may be served with chicken, mutton, or black peas curry. It can be served in breakfast or as evening snack (Fig. 13.9).

13.3.3 *Olya Fenya*

Olya fenya is mainly prepared in *Konkan* region. It is prepared using rice. The process for making this product is highly time consuming. It is also known as '*Papdya*'. These foods are an important source of calories in the diet and nutrition of Konkani people (Fig. 13.10).

13.3.3.1 Traditional Methods of Preparation and Mode of Consumption

Olya fenya is mainly prepared from rice. Cleaned and washed rice is soaked in water (ratio of rice:water, 1:2) and allowed to ferment for 3 days without changing the water. Normally, fermented smell is perceived on the third day. After fermentation, on the fourth day the rice is washed with fresh water and ground finely using

Fig. 13.10 Photograph of *Olya fenya*

stone grinder (traditional mill) to obtain a thick batter. Poppy seeds and salt are added in the batter. The batter is spread (1–2 mm thickness) over the metal plates with thin coating of curd. A set of 4–6 plates is placed in *feni* stand (special structure made to hold 4–6 plates) and allowed for steam cooking for 5–10 min. The *fenya* is removed from plate and served hot with curd. *Olya fenya* are consumed for breakfast (Fig. 13.11).

Rice

↓

Cleaning and soaking for 72 hours at room temperature for fermentation

↓

Washing with fresh water and ground finely to obtain batter

↓

Add poppy seeds and salt in batter

↓

Thin coating of metal plates with curd

↓

Spread the batter on curd coated plates

↓

Steaming for 5-10 minutes

↓

Removing the *fenya* from plate

↓

Olya fenya

Fig. 13.11 Flowsheet of *Olya fenya* preparation

13.3.4 *Sandan*

Sandan is a snack or breakfast food of Maharashtra. It is prepared by fermented rice batter with addition of jackfruit, jaggery, grated coconut, and then steaming like *idli*. It is specially prepared by Konkani people.

13.3.4.1 Traditional Methods of Preparation and Mode of Consumption

The rice grains are soaked overnight (7–8 h) at ambient temperature for fermentation. The water is then drained and soaked rice is spread over a clean cloth and kept under shade (1–2 h) for surface drying. The rice is ground into a coarse powder without using water. The fully ripened jackfruit (*Artocarpus heterophyllus*) is selected for preparation of *Sandan* and chopped into small pieces. A mixture of coarse ground rice, chopped jackfruit, grated coconut, salt and powdered jaggery was prepared and pulverized to obtain thick batter. Transfer a ladle full of batter into a *Sandan* mould/plate which is greased with ghee and allow for steam cooking for 15–20 min in modak (Maharashtrian sweet dumpling) steaming vessel. *Sandan* are consumed for breakfast or as a tea time snack (Fig. 13.12).

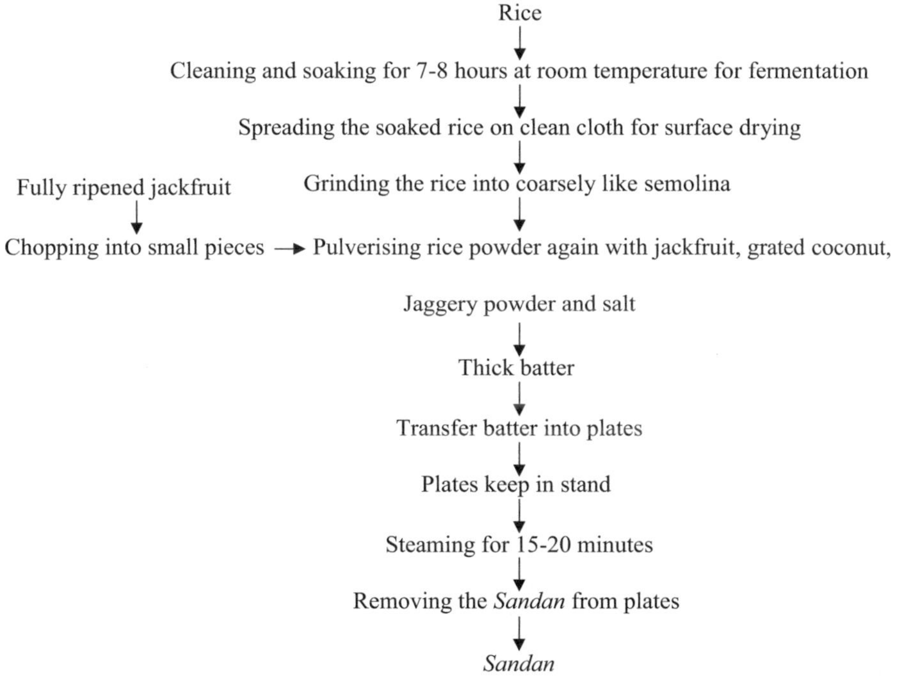

Fig. 13.12 Flowsheet of *Sandan* preparation

Fig. 13.13 Photograph of *Anarasa*

13.3.5 *Anarasa*

Anarasa is a fermented product made from rice, jaggery, poppy seeds, and ghee. It is an Indian pastry-like snack usually associated with the Hindu festival of Diwali in Maharashtra and Bihar. It is also a popular snack in Pakistan and known as *Andrasa*. Special skills are required to prepare a good-quality *Anarasa* (Fig. 13.13).

13.3.5.1 Modified Traditional Methods of Preparation and Mode of Consumption

During *Anarasa* preparation (Fig. 13.14), the rice is cleaned, washed, and soaked in enough water for 3–4 days during winter season (at ambient temperature 12–34 °C) and 2–3 days during summer season (at ambient temperature 22–43 °C) for partial fermentation. The soaking water is replaced with fresh water everyday to control the fermentation. The water is then drained and soaked rice is spread over a clean cloth and kept under shade (1–2 h) for surface drying. It is then pulverized into a fine flour using a mortar and pestle followed by sieving. The finely ground rice flour is mixed with jaggery in the proportion of 2:1 and kneaded to form a hard dough without addition of water and allow to ferment overnight. This dough can be stored under refrigerated conditions till further use. The fresh dough is divided into small balls (approximately 15–20 g) and flattened by hand into a circular disc with 5–6 cm in diameter and 7–8 mm thickness. A small amount of poppy seeds are sprinkled over the surface of flattened dough and subjected to deep fat frying at 170 °C for 5–7 min using low flame until it turns to golden brown in colour. After completing the frying *Anarasa* is removed from the frying oil and allow to drain and cool for some time. A porous textured of *Anarasa* is developed during frying due to release of gases produced during the process of fermentation. The traditionally accepted quality criterion for *Anarasa* are porous structure, light in weight, fluffy and crispy texture with golden brown colour.

Rice
↓
Soaking for 3 to 4 days at ambient temperature
↓
Changing the soaked water with fresh water everyday
↓
Soaked water draining
↓
Surface drying of rice in shade for 1-2 hrs
↓
Grinding into fine flour
↓
Mixing rice flour and ground sugar/ jagggery (2:1)
↓
To form a hard dough
↓
Kept for overnight at ambient temperature (for fermentation)
↓
Small balls (approximately15-20 gm) and flattened by hand into a circular disc
↓
Sprinkling the poppy seed on surface of flattened dough
↓
Deep fat frying at low flame till golden brown colour
↓
Anarasa

Fig. 13.14 Flowsheet of *Anarasa* preparation

13.3.6 *Salpapadya*

Salpapadya are mainly prepared from rice. *Salpapadya* is also known as '*Vaphevarche tandalache papad*' in Maharashtra. It is mainly prepared in summer season and stored throughout the year. Consumption of *Salpapadya* is quite high as it is consumed at many occasions and during festivals in the Maharashtra (Fig. 13.15).

13.3.6.1 Traditional Methods of Preparation and Mode of Consumption

Salpapadya is mainly prepared from rice. Rice is washed and soaked for 4 days at ambient temperature. The soaking water is replaced with fresh water everyday to control the fermentation. On 5th day, soaked water is drained and rice is spread over a clean cloth for surface drying (1–2 h). After that, soaked rice is finally ground into flour using a stone mill. The water is added to rice flour along with salt and cumin and sesame seeds (1–2%) to obtain a batter with thin consistency. Then the batter is spread (2–3 mm thickness and 10–15 cm diameter) over the metal plates. A set of

Fig. 13.15 Photograph of *Salpapadya*

4–6 plates is placed in *Salpapadya* stand (special structure made to hold 4–6 plates) and subjected to steam cooking for 5–10 min and then allowed to cool for 5 min. After cooling, *Papadya* were removed from plates and transferred to a plastic sheet for sun drying (for 3–4 days). Store in dry and airtight container (Fig. 13.16).

Rice
↓
Soaking the rice in water for 4 days (change the water everyday)
↓
Draining the soaked water and spread rice on cloth for surface drying
↓
Grinding rice into fine flour
↓
Mix the rice flour with salt and cumin & sesame seeds, water to obtain thin batter.
↓
Spreading the batter on plates (2-3 mm thick and 10-15 cm in diameter)
↓
Plates kept for steaming (5-10 min)
↓
Cooling the plates for 5 min
↓
Removing *Papadya* from plates transfer into a plastic sheet
↓
Sun drying for 3-4 days
↓
Salpapadya

Fig. 13.16 Flowsheet of *Salpapadya* preparation

13.3.7 *Bibdya*

Bibdya is the speciality of *Khandesh* (northern Maharashtra). It is made from sorghum, which is a staple crop of the region. The traditional method ensures that the original taste of sorghum remains the same in the product, *Bibdya.* Spices, chilli powder, and garlic are also added to enhance the taste and flavour.

Bibdya is very popular among the people in *Khandesh.* Traditional recipe and method of preparation are being handed over from generation to generation. Even today for preparation of *Bibdya* many nearby families in the society come together to help each other. *Bibdya* is a light and healthy snack since it is roasted and not fried (Fig. 13.17).

13.3.7.1 Traditional Method of Preparation of *Bibdya*

Bibdya prepared from sorghum. Sorghum is washed and soaked for 3–4 days during winter season (at ambient temperature 12–34 °C) and 2–3 days during summer season (at ambient temperature 22–43 °C) for fermentation. The sorghum is again washed on third day and drained. The washed sorghum is coarsely ground in stone grinder and crushed garlic, cumin seeds, sesame seeds, and red chilli powder are added to this mass. Then this mass was added to water in the proportion of 1:2 and cooked for 20–30 min. Appearance of bubbles on top of the batter indicates that the batter is cooked. This thick batter is spread on a cloth with hand in circular disc having thickness of 8–10 mm and 7–10 in. diameter and sun dried for 3–5 days. They are carefully staked one above the other and stored in containers and are used throughout the year as snack food (Fig. 13.18).

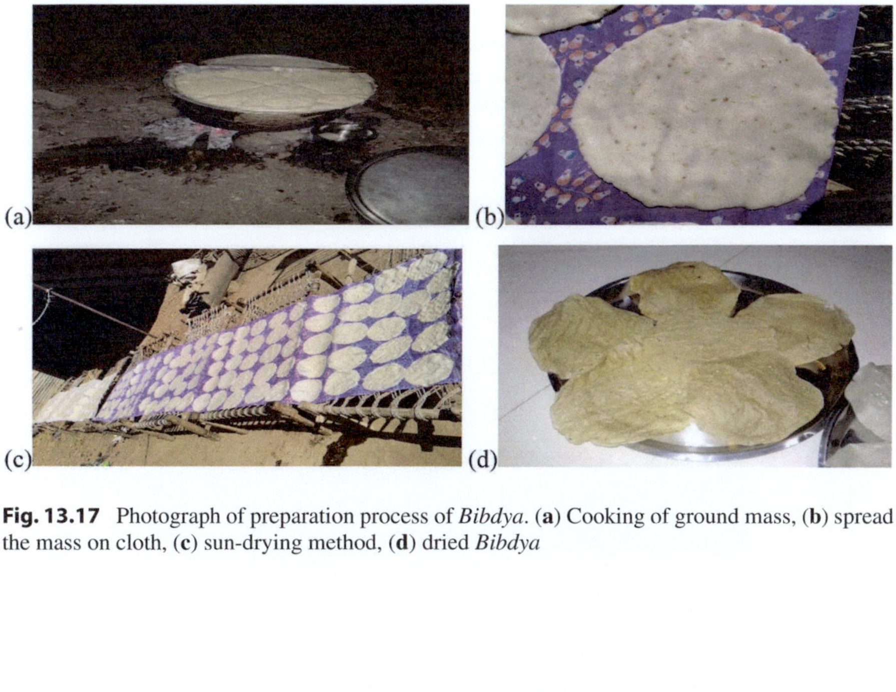

Fig. 13.17 Photograph of preparation process of *Bibdya*. (**a**) Cooking of ground mass, (**b**) spread the mass on cloth, (**c**) sun-drying method, (**d**) dried *Bibdya*

Sorghum
↓
Soaking at ambient temperature
↓
Washing and grinding (sorghum crushed mass)
↓
Add salt and seasonings into the crushed mass
↓
Add crushed mass into boiling water
↓
Cooking for 20-30 mins
↓
Thick batter is spread on a cloth with hand in circular disc (thickness of 8-10 mm and 7-10 inch diameter)
↓
Sun drying the *Bibdya* for 3-5 days
↓
Bibdya

Fig. 13.18 Flowsheet of *Bibdya* preparation

13.3.8 *Kurdai*

Kurdai is wheat-based fermented product prepared and consumed in Maharashtra. It is generally deep fat fried before consumption. It is also used for making curry when fresh vegetables are not available, particularly in summer season (Fig. 13.19).

13.3.8.1 Modified Traditional Method of Preparation

Kurdai is prepared from fermented and cooked wheat gruel. Wheat is soaked in water for 3–4 days at an ambient temperature (30–35 °C). The soaking water is replaced with fresh water everyday to control the fermentation. On the 5th day, the soaked wheat is then pulverized either in a crusher or traditionally by using stone grinder (locally known as a *pata-warwanta*). The crushed wheat is then washed

Fig. 13.19 Photograph of *Kurdai*

with fresh water to separate the wheat endosperm in the form of milky extract. The extract obtained is allowed to settle the colloidal matter for 2 days. Then the supernatant is separated. The settled residue is mixed with fresh water and cooked to obtain a thick translucent semi-solid mass (45 min). The cooked mass is then transferred to the hand extruder having a piston and die arrangement from which it is extruded in the form of strands in circular motion on wet cloth and sun dried for 4–5 days. *Kurdai* is carefully stacked one above the other and stored in containers and are used throughout the year as snack food. It is consumed after deep fat frying.

Microorganisms *P. pentosaceus, P. acidilactici, Pedicoccus* sp., *Lactobacillus* sp., or/and fungi like *R. oligosporus, A. oryzae, S. rouxii, S. cerevisiae* (Surve et al. 2014) (Fig. 13.20).

13.3.9 *Dhapode*

Dhapode are produced primarily in Khandesh in Maharashtra. It is made from sorghum flour. It is just like a *Bibdya*. It is thicker and small in size as compare with the *Bibdya*. *Dhapode* is the important source of the fibre and calories in the diet and nutrition of Maharashtrian (Fig. 13.21).

13.3.9.1 Traditional Method of Preparation and Mode of Consumption

Dhapode are prepared from fermented and cooked sorghum flour. The sorghum flour is soaked in water for 10–12 h at an ambient temperature (30–35 °C) for fermentation. On the next day the fermented batter is cooked with water. During this, make sure there is no lump form in the cooked mass with continuous stirring by wooden ladle (*Chatu*) and then add the seasonings during cooking (chilli powder, turmeric powder, salt, sesame seeds and carom seeds) and cook the batter for 30–45 min. The cooked batter is taken in small quantity and spread on a clean and wet cloth in circular shape, approximately 5–6 cm in diameter and 4–5 mm in thickness, and sun dried. After 7–8 h wet cloth again and remove *dhapode* from cloth and keep in tray for sun drying for 1–2 days during summer. Dried *Dhapode* are cautiously staked one above the other and stored in airtight container, and is used throughout the year as snack food. These are served in fried form with jaggery or peanuts, and are popular in rural areas. They are prepared in summer season and used throughout the year, but are more popular during rainy season.

Nutritional Value The nutrient composition of sorghum grain indicates that it is a good source of energy, protein, vitamins, minerals, and phytochemicals. The sorghum grain content is 2.1% of fibre, 1.6% of mineral content, 72.6% of carbohydrate, and 21.2% of amylose. It is a good source of energy and provides about 349 kcal/100 g. Starch is the major carbohydrate of the sorghum grain. About

Wheat
↓
Soaking in water for 3-4 days at ambient temperature
↓
Draining the soaked water
↓
Grinding/crushing soaked wheat
↓
Washing the crushed wheat with fresh water
↓
Separating the wheat endosperm from crushed wheat → Bran
↓
Milky extract
↓
Settlement of colloids particles and fermentation (2 days)
↓
Separating settled residue → Supernatant
↓
Mixing with water (1:1)
↓
Cooking for 45 min
↓
Addition of salt and mixing continuously with wooden ladle (locally known as a *Chatu*)
↓
Thick translucent semi-solid mass
↓
Take little mass and press with hand to make thin circular shape
↓
Hand extruder with circular die at the bottom with small holes
↓
Folded in circular fashion on clean wet cloth
↓
Sun drying for 4-5 days
↓
Kurdai

Fig. 13.20 Flowsheet of *Kurdai* preparation (Kumari et al. 2016)

32–79% of sorghum grain weight is due to starch. The other carbohydrates present are simple sugars, cellulose, and hemicellulose (Dudhate et al. 2017) (Fig. 13.22).

13.3.10 *Kharodya*

Kharodya is prepared from the pearl millet. It is prepared in the summer season in rural areas and stored throughout the year (Fig. 13.23).

Fig. 13.21 Photograph of *Dhapode*

Sorghum flour
↓
Soaking 10-12 hrs at ambient temperature
↓
Add the fermented batter in boiling water and mix well
↓
Add Seasoning in the batter
↓
Cooking the batter for 30-45 mins
↓
Spread the batter on wet cotton cloth in circular shape approximate 4-5mm thickness and 5-6 cm diameter
↓
Sun drying for 2-3 days
↓
Keep in airtight container
↓
Dhapode

Fig. 13.22 Preparation of *Dhapode*

13.3.10.1 Traditional Method of Preparation and Mode of Consumption

Kharodya is prepared from pearl millet. The pearl millet is washed with fresh water for 2–3 times to remove dirt and impurities. Cleaned pearl millet is soaked in water for 5–6 h and then ground to coarse particles using stone grinder. The ground pearl millet is steeped in buttermilk for 8–12 h for fermentation. Then the fermented mixture is added into boiling water and cooked for 20–25 min to obtain the semi-solid thick mass to which sesame seeds and salt are added. In addition to this, sometimes seasonings such as garlic paste, cumin seeds, red chili powder, or green chilli can

Fig. 13.23 Photograph of *Kharodya*

also be added as per the requirement. This semi-solid mass is divided into small portions and dropped by hand on a plastic sheet and sun dried for 3–4 days in summer season.

Kharodya is a popular snack. It is consumed as such or after frying as snacks along with meal.

Microorganisms *L. mesenteroides, L. fermentum,* and *S. faecalis* microorganisms and *Saccharomyces* sp., *Eleusine coracana L. Gaertn.* (Steinkraus 1983) (Fig. 13.24).

Pearl millet
↓
Washing with fresh water
↓
Soaking in water for 5-6 hours
↓
Coarsely grinding in stone grinder
↓
Steeping the coarsely ground mass in buttermilk for 8-12 hours
↓
Add fermented batter in boiling water
↓
Add Seasonings and mix well
↓
Cooking for 20-25 mins
↓
Adding sesame seeds to semi solid thick mass and mix well
↓
The semi-solid mass is divided into small portions and dropped by hand on a plastic sheet
↓
Sun drying for 3-4 days
↓
Kharodya

Fig. 13.24 Flowsheet of *kharodya* preparation

Fig. 13.25 Photograph of *Kohalache Sandge*

13.3.11 *Kohalache Sandge*

Sandge are prepared from the coarse ground black gram splits after fermentation. *Kohalache Sandge* are made from grated *Kohala* (wax guard) and fermented batter of legume. Wax guard is the good source of calcium and phosphorus. The wax gourd (*Benincasa hispida*) is also called white gourd and winter melon. In Marathi it is called as '*Kohala*' which is easily available in Maharashtra region (Gopalan and Balasubramanian 2004) (Fig. 13.25).

13.3.11.1 Traditional Method of Preparation and Mode of Consumption

The black gram splits are washed and soaked for 8–12 h at ambient temperature. The water is drained and the soaked black gram splits coarsely ground in stone mortar (locally known as *Pata-warwanta*). The semi-solid ground mass is then allowed to ferment for 6–8 h at room temperature. To this fermented semi-solid mass of black gram, grated wax guard was added and mixed together to form semi-solid thick mass. Salt, chilli powder, fenugreek powder, and asafoetida are added to this mass as per requirement. This semi-solid mass is then divided into small portions and dropped by hand on a plastic sheet and sun dried for 3–4 days in summer season. After optimum drying is achieved, the product *Sandge* is stored in an air tight container. *Sandge* is used throughout the year as snack food. The *Sandge* can be a substitute for vegetable dish when there is a shortage of vegetable in rural areas. *Kohalache Sandge* are cooked in coconut curry or consumed as fried *Sandge* with onion and coconut as a salad. It is mainly consumed in rainy season.

Nutritional Value The wax guard contains protein—0.4%, fat—0.1%, carbohydrates—3.2%, mineral matter—0.3%, and vitamin B1—21 I.U./100 g. It is a source of vitamin B1. The fruit is laxative, diuretic, tonic, aphrodisiac; it cures strangury, urinary diseases. In Ayurveda, it is said to be a heart tonic (De and Parikh 1985) (Fig. 13.26).

Black gram splits
↓
Soaking in water for 8-12 hours at ambient temperature
↓
Grinding coarsely soaked black gram splits
↓
Kept ground mass for fermentation at ambient temperature (6-8 hrs)
↓
Mixing with the grated pumpkin, seasoning and fermented mass to form semi solid thick mass
↓
Semi-solid thick mass is divided into small portions and dropped by hand on a plastic sheet
↓
Sun drying for 3-4 days
↓
Keep *Sandge* in air tight container
↓
Kohalyache sandge

Fig. 13.26 Preparation *kohalyache sandge*

13.4 Dried Fish in Maharashtra

Introduction Dried fish is mainly consumed in coastal area of Maharashtra (Konkan and Mumbai). Drying fish consist of different steps. After catching fish it is sorted and cleaned by the koli fisherwomen. Once the fish is cleaned, it is ready to be either dried or salted. Some commonly sold fish are dried shrimp and prawns. There is *Jawla* (tiny shrimp), *Kardi* (prawn), *Tengli* (a kind of anchovy), *Sode* (black or red shrimp), *Bombil* (Bombay duck), and *Mandeli* (golden anchovy), as well as salted, dehydrated *Bangda* (mackerel), *Surmai* (kingfish), *Halwa* (black pomfret), *Vaktya* (ribbon fish), and *Sukke Makul* (dried squid). Usually big fish and some particular species are salted. Most of the fish is dried in the sun. Fish species like shrimp and Bombay duck are very well known among the dry fish lovers. Fish species like mackerel and kingfish are mainly salted and is consumed at a large scale by the fish eaters.

The traditional technique for fish preservation in the Maharashtra regions involves dried and salted dried fish. The purpose of preservation of fish is that fresh fish is not available in rainy season (Fig. 13.27).

Microorganisms Unknown.

Fig. 13.27 Photograph dried fish market of Alibaug (Raigad, Maharashtra, India)

13.4.1 *Jawla* (Tiny Shrimp)

Jawla is an ethnic dried fish product commonly prepared in coastal part of Maharashtra. It is also known as '*Sukat*', '*Sukha jawla*', and '*Sukke kolim*' (Fig. 13.28).

During preparation of *Jawla*, the shrimps are collected, washed with fresh water, and sun dried for 4–7 days. The sun-dried fish product can be stored at room temperature for 5–6 months and used for consumption. *Jawla* is consumed as salad, curry, chutney, and *Jawla* khichadi or pulav. During curry preparation, *Jawla* is roasted, soaked in water for 10–15 min, fried, and mixed with dry chilli, turmeric, coconut–onion paste, and salt. It is usually eaten with flat bread (rice bhakri) or

Fig. 13.28 Photograph of *Jawla*

Fig. 13.29 Flowsheet of *Jawla* preparation

Shrimp
↓
Washing wish fresh water
↓
Sun drying 4-7 days
↓
Jawla

boiled rice. Sometimes, it is cooked with vegetables and eaten with bhakri (Fig. 13.29).

13.4.2 *Karndi* (Prawns)

Karndi is a sun-dried fish product commonly prepared in coastal part of Maharashtra. It is also known as '*Sukke Ambadi*' and '*Sukke Karndi*' (Fig. 13.30).

13.4.2.1 Traditional Method of Preparation and Mode of Consumption

During preparation of *Karndi*, the small prawns are collected, washed with fresh water, and sun dried for 7–10 days. The sun-dried fish product can be stored at room temperature for 5–6 months for consumption. Before using the *karndi*, head, tail, and legs of prawns are removed. Then it is soaked in water for 20–30 min and is used for cooking preparation. *Karndi* is fried and mixed with dry chilli, turmeric, coconut–onion paste, and salt to make curry. It is usually eaten with flat bread (rice bhakri) or boiled rice. Sometimes, it is cooked with vegetables and eaten with *bhakri* (Fig. 13.31).

Fig. 13.30 Photograph of *Karndi*

Fig. 13.31 Flowsheet of *Karndi* preparation

Small prawns
↓
Wash with fresh water
↓
Sun drying (7-10 days)
↓
Krandi

Fig. 13.32 Photograph of *Tengli*

13.4.3 *Tengli* (A Kind of Anchovy)

Tengli is a sun-dried fish product commonly prepared in coastal part of Maharashtra. It is also known as '*khade*' (Fig. 13.32).

13.4.3.1 Traditional Method of Preparation and Mode of Consumption

During preparation of *Tengli*, the fish are collected, washed, and dried in the sun for 7–10 days. The sun-dried fish product is stored at room temperature for 5–6 months for consumption. Before using the *Tengli*, it is soaked in water for 20–30 min. Then it is fried and mixed with dry chilli, turmeric, coconut–onion paste, and salt to make *Kalvan*. It is usually eaten with rice flat bread (rice *bhakri*) or boiled rice (Fig. 13.33).

Fig. 13.33 Flowsheet of *Tengli* preparation

Fish
↓
Wash with fresh water
↓
Sun drying for 7-10 days
↓
Tengli

Fig. 13.34 Photograph of *Sode*

13.4.4 *Sode* (Black or Red Shrimp)

Sode is a sun-dried fish product commonly prepared in coastal part of Maharashtra. It is most expensive dried fish in Maharashtra because of its cleaning and drying process method (Fig. 13.34).

13.4.4.1 Traditional Method of Preparation and Mode of Consumption

During preparation of *Sode*, the shrimps are collected and washed and outer shell, head, tail, and legs are removed. Then it is dried on bamboo basket or wooden board for 7–10 days. The sun-dried *Sode* can be stored at room temperature for 5–6 months for consumption. It is sold at local markets of coastline of Maharashtra. Before using the *Sode*, it is soaked in water for 20–30 min. After that it is used for cooking preparation. *Sode* is shallow fried in oil and mixed with dry chilli, turmeric, coconut–onion paste, and salt to make curry. It is usually eaten with rice flat bread (*Bhakri*) or boiled rice. Sometimes, it is cooked with vegetables and eaten with *Bhakri* (Fig. 13.35).

Fig. 13.35 Flowsheet of *Sode* preparation

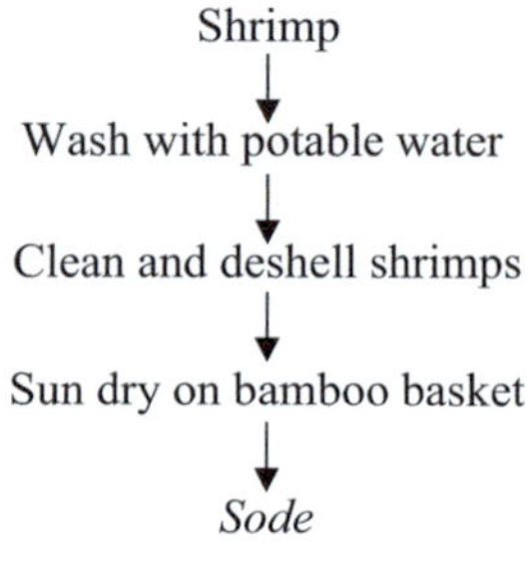

Fig. 13.36 Photograph of *Sukhe Bombil*

13.4.5 *Sukhe Bombil* (Dry Bombay Duck)

Bombay duck (*Harpodon neherus*) is popularly known as 'Bombil' in Maharashtra. Only a small quantity of the total landing is consumed in fresh condition, and the remaining is being converted into dried form for national as well as international market. *Sukke Bombil* is very popular sun-dried fish in all over Maharashtra (Fig. 13.36).

13.4.5.1 Traditional Method of Preparation and Mode of Consumption

During preparation of *Sukke Bombil*, the fish are collected, washed, rubbed with salt, and dried in the sun for 4–7 days. The sun-dried fish product is stored at room temperature for 5–6 months for consumption. It is sold at local markets in Maharashtra. *Sukke Bombil* is consumed as fried *Bombil,* curry, and chutney. During curry preparation, it is soaked in water for 10–15 min, fried, and mixed with dry chilli, turmeric, coconut–onion paste, and salt. It is usually eaten with flat bread (rice bhakri) or boiled rice (Fig. 13.37).

Microorganisms Lactic acid bacteria: *Lactococcus lactis* subps. *Cremoris, Lc. Lactis* subsp. *lactis, Lc. Plantarum, Leuconostoc mesenteroides, Enterococcus*

Fig. 13.37 Preparation of *Sukke Bombil*

Fish (Bombay duck)
↓
Washing with fresh water
↓
Rub with salt
↓
Sun drying
↓
Sukke bombil

Fig. 13.38 Photograph of *Sukke Mandeli*

faecium, E. faecalis, Pediococcus; Candila chiropterorum, C. bombicola, and Saccharomycopsis spp. (Tamang 2010).

13.4.6 *Sukke Mandeli* (Golden Anchovy)

Sukke Mandeli is very popular sun-dried fish in Maharashtra (Fig. 13.38).

13.4.6.1 Traditional Method of Preparation and Mode of Consumption

During preparation of *Mandeli*, the fish are collected, washed, and dried in the sun for 4–7 days. The sun-dried fish product is stored at room temperature for 5–6 months for consumption. It is sold at local markets in Maharashtra. *Mandeli* is consumed as curry. During curry preparation, it is soaked in water for 10–15 min, fried, and mixed with dry chilli, turmeric, coconut–onion paste, and salt. It is usually eaten with flat bread (rice bhakri) or boiled rice (Fig. 13.39).

13.4.7 *Khara Masa* (*Bangda, Surmai, Halwa*)

Khara masa is very popular sun-dried fish in Maharashtra. The word Khara masa means salted fish in Marathi. It is mainly prepared from mackerel (*Bangda*), king-fish (*Surmai*), and black pomfret (*Halwa*) (Fig. 13.40).

Fig. 13.39 Flowsheet of *Sukke Mandeli* preparation

Fish
↓
Washing with clean water
↓
Sun drying (10-15 days)
↓
Sukke Mandeli

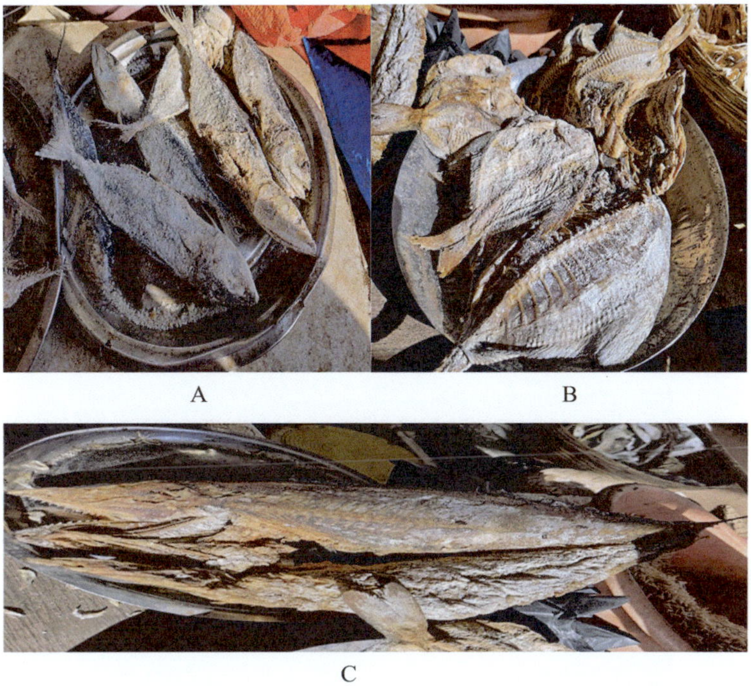

Fig. 13.40 Photograph of *Khara masa.* (**a**) Mackerel (*Bangda*). (**b**) Black pomfret (*Halwa*). (**c**) Kingfish (*Surmai*)

13.4.7.1 Traditional Method of Preparation and Mode of Consumption

During preparation of *Khara masa*, the fish are collected, washed, rubbed with salt, and dried in the sun for 4–7 days. The sun-dried fish product is stored at room temperature for 5–6 months for consumption. It is sold at local markets in Maharashtra. *Khara masa* is consumed as fried *Khara masa* and *Kalvan* (curry). During curry preparation, it is soaked in water for 30–40 min, fried, and mixed with dry chilli, turmeric, coconut–onion paste, and salt. It is usually eaten with rice flat bread (*Tandlachi* bhakri) or boiled rice (Fig. 13.41).

13.4.8 *Sukke Makul*

Sukke Makul is sun-dried fish in Maharashtra. It mainly prepared from squid (Makul) and *sukke* means dried (Fig. 13.42).

Fig. 13.41 Flowsheet of *khara masa* preparation

Fish
↓
Cleaning and washing
↓
Rub with Salt
↓
Sun drying
↓
Khara masa

Fig. 13.42 Photograph of *Sukke Makul*

13.4.8.1 Traditional Method of Preparation and Mode of Consumption

During preparation of *Sukke Makul*, the fish are collected, washed, and dried in the sun for 4–7 days. The sun-dried fish product is stored at room temperature for 5–6 months for consumption. It is sold at local markets in Maharashtra. *Sukke Makul* is consumed as *Kalvan* (curry). During curry preparation, it is soaked in lukewarm water for 30–40 min, fried, and mixed with dry chilli, turmeric, coconut–onion paste, and salt. It is usually eaten with rice flat bread (*Tandlachi bhakri*) or boiled rice (Fig. 13.43).

Fig. 13.43 Flowsheet of *Sukke Makul* preparation

Fish
↓
Washing with fresh water
↓
Sun drying
↓
Sukke Makul

Fig. 13.44 Photograph of *Vaktya*

13.4.9 *Vaktya*

Vaktya is sun-dried fish in Maharashtra. It is mainly prepared from silver ribbon fish (Fig. 13.44).

13.4.9.1 Traditional Method of Preparation and Mode of Consumption

During preparation of *Vaktya*, the fish are collected, washed, and dried in the sun for 4–7 days. The sun-dried fish product is stored at room temperature for 5–6 months for consumption. It is sold at local markets in Maharashtra.

Vaktya is consumed as *Kalvan* (curry). In Koli community the curry is called as *Vaktyachi chichvani* which is made up of the tamarind (locally known as *chinch*). During curry preparation, *Vaktya* is soaked in water for 30–40 min, fried in oil, and mixed with dry chilli, turmeric, coconut–onion paste, and salt. It is usually consumed with rice flat bread (*Tandlachi bhakri*) or boiled rice (Fig. 13.45).

13.5 Pickles (*Lonche*)

Introduction The most common fruit and vegetable pickles are made by simple brining, which results in lactic acid bacterial fermentation. Pickle products add spice to meals and snacks. The skilful blending of spice, sugar, and oil with fruit and vegetable gives crisp, firm texture and pungent, sweet-sour flavour. Pickles serve as

Fig. 13.45 Flowsheet of *Vaktya* preparation

appetizers and help in digestion by aiding flow of gastric juices. *Lonche* is the Marathi word for pickles.

Influence of salt in pickles: Salt withdraws water and nutrients from the fruits and vegetable tissue, and these nutrients become the substrate for the growth of lactic acid bacteria. Salt, in combination with the acids produced by fermentation, prevents the growth of undesirable bacteria and interruptions enzymatic softening of the fruits and vegetables. Inadequate amount of salt results in softening and lacking flavour of fruits and vegetables. An adequate amount of salt is added in pickle, which flavour with proper balance of salt to acid. In Maharashtra, pickles are stored in porcelain jar as well as glass jar with airtight lid and a muslin cloth is tightly placed over the jar at ambient temperature.

Microorganisms Unknown.

Health Benefits The health benefits of pickles include a good supply of essential vitamins, minerals, and antioxidants, which contribute to diabetes control, improved digestion, liver protection, a supply of probiotics, and the ability to heal ulcers. Fermented pickles also have beneficial bacteria that can control harmful intestinal microbes. Culinary socio-economy and ethnical value: Pickles are available in local markets of Maharashtra. All fruit and vegetable pickles are sold all over Maharashtra. Prawn pickle is mainly prepared in the coastal region of Maharashtra, for preservation of prawns. In Vidarbha region, the Savji community preserves mutton in pickle form.

13.5.1 Lemon Pickle (*Limbuch Lonche*)

Lemon pickle is prepared from the fresh and fully grown lemon (*Citrus limon*). Lemon is easily available in market of Maharashtra. Lemon is the good source of vitamin C.

13.5.1.1 Traditional Method of Preparation and Mode of Consumption

Fully ripened lemon is selected for pickle preparation. Washing with potable water is performed to sanitize the fruit surface and avoid microbiological contaminants to survive in the brine solution. The fruit is cut into small pieces and then a little salt and turmeric powder are applied to the lemon pieces and kept aside for fermentation. A muslin cloth is tightly placed over the container and kept at a very sunny place to initiate brine formation as the salt application on fresh lemon helps to ooze out water from the fruit. It takes around 24 h for brine formation and fermentation to initiate that is usually indicated in the form of bubbles of CO_2 on the surface of brine. Jar is further incubated in the same conditions for a period of 1 week to continue the fermentation process. Incubation for lactic acid fermentation as desired in pickle development is recommended at 21 °C, and the process is continued until no

Fig. 13.46 Photograph of lemon pickle

further bubbles appear in the jar. Generally, the composition of the ingredients depends on the local taste and included split mustard seeds, asafoetida, turmeric powder, fenugreek powder, and red chilli powder. A few spoons of oil is heated in a pan; mustard seeds are added and cooked until the seeds pop up, and then heating is stopped; then the spices are added in oil. All the spices are added to the fermented lemon and the ingredients are homogeneously mixed in a glass or porcelain jar. This jar is kept for 15–20 days. After 2 days, agitate the pickle in correct manner. This jar is closed with airtight lid and allowed to ageing for 2–3 weeks. It is consumed after 2–3 weeks (Figs. 13.46 and 13.47).

13.5.2 Sweet Lemon Pickle (*Limbuch Goad Lonche*)

13.5.2.1 Traditional Method of Preparation

Fully ripened lemon is selected for pickle preparation which is then washed with potable water to sanitize the fruit surface and avoid microbiological contaminants to survive in the brine solution and soaked in water overnight. The fruit is cut into small pieces and then salt and sugar are applied and set aside for 4 days. Remove the

Fig. 13.47 Flowsheet of lemon pickle preparation

Yellow, mature lemon
↓
Sorting
↓
Washing
↓
Cutting/ slicing into pieces
↓
Brining (1 day)
↓
Spices addition
↓
Stirring
↓
Pickle store in glass or porcelain jar

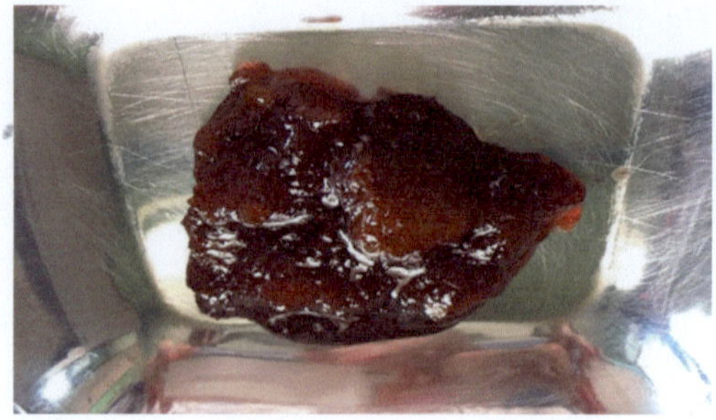

Fig. 13.48 Photograph of sweet lemon pickle

juice from the lemon left out after 4 days. Oil is heated in a pan; mustard seeds, crushed garlic, asafoetida, turmeric powder, and chilli powder are added and fried for 1–2 min on low flame. This spices mixture and jaggery syrup are added to fermented lemon pieces and pickle is homogenously mixed. Sweet lime pickle is stored in a glass or porcelain jar at room temperature. This jar is closed with airtight lid and allowed to ageing for 2–3 weeks and consumed after 3–4 weeks (Figs. 13.48 and 13.49).

13.5.3 Raw Mango Pickle (*Ambaych Lonche*)

The mango is the national fruit of India. In India, harvest and sale of mangoes is during March–May. Mango is a low-calorie fruit that is high in fibre, and is an excellent source of vitamins A and C. It also contains folate, B6, iron, and a little calcium, zinc, and vitamin E.

Fig. 13.49 Flowsheet of sweet lemon pickle preparation

Yellow, mature lemon
↓
Sorting
↓
Washing
↓
Draining
↓
Cutting/ slicing
↓
Apply Salt and sugar for lemon pieces (4 day)
↓
Add jaggery syrup and spices
↓
Store in Jar

Fig. 13.50 Photograph of raw mango pickle

13.5.3.1 Traditional Method of Preparation

Matured raw mango is selected for pickle preparation which is then washed with water and cut into medium pieces. Turmeric and salt are applied on the raw mango pieces and kept for 2–3 days for fermentation at ambient temperature. The mustard seeds are ground to obtain split mustard seed. The oil is heated on medium flame and split mustard seeds, fenugreek seed, asafoetida, fenugreek powder, chilli powder, and turmeric powder are added. This spices mixture is added to fermented mango pieces. Raw mango pickle is stored in a glass jar or traditional ceramic jar (pickle barni) at room temperature and consumed after 3–4 weeks (Figs. 13.50 and 13.51).

13.5.4 Dry Mango Pickle (*Amboshich Lonche*)

Amboshi is prepared with raw mango and then it is sun dried for 7–8 days. Amboshi is used in many curries, vegetables, fish recipes to enhance the recipe by the sour

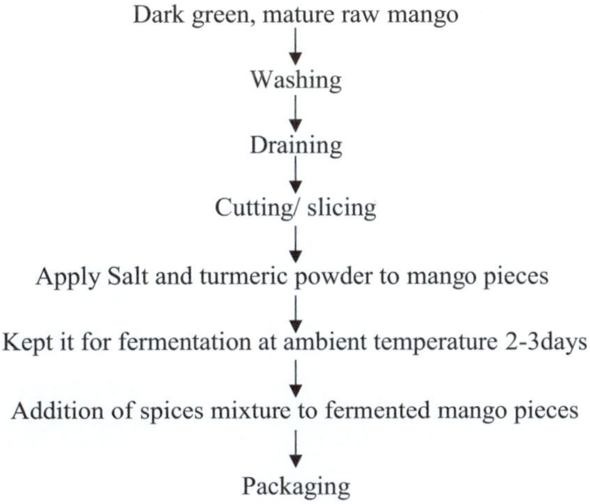

Fig. 13.51 Flowsheet of raw mango pickle preparation

taste. It is main ingredient of *amboshich lonche.* Amboshi is mainly consumed in the Konkan region of the Maharashtra.

13.5.4.1 Traditional Method of Preparation

Matured raw mango is selected for pickle preparation which is then washed with potable water and removed peel of raw mango. It is cut into medium cubes. Little salt is applied. Dry these pieces under sunlight for 7–8 days, until they become completely dry. Blanching of these dried pieces of mango for removing dust and impurities. After that mango pieces are soaked in water for rehydration for 7–8 h. The soaked water is drained and rubbed with salt again. Mustard seed powder and water are added to blender and blend until the mixture becomes whitish paste. Oil is heated in a small pan. It is tempered with mustard seeds, asafoetida, and turmeric powder and cooled. A small amount of water is added to jaggery and a syrup of jaggery is made. The following is mixed together—blended mustard mixture, fenugreek powder, jaggery syrup, and prepared tempering and mango pieces. Pickle is stored in a glass jar/porcelain jar at ambient temperature. This jar is closed with airtight lid and allowed to ageing for 3–4 weeks and consumed after 3–4 weeks (Fig. 13.52).

13.5.5 Anola Pickle (*Aavlyach Lonche*)

Anola pickle is mainly prepared from Anola (Indian gooseberry) which is hard and light green colour in nature. Gooseberries are available in winter season of Maharashtra. The fruit is the richest source of vitamin C. It cures insomnia and is healthy for hair. It is used as the cardio protective, useful in haemorrhage, menorrhagia, leucorrhoea, and discharge of blood from uterus.

13.5.5.1 Traditional Method of Preparation

Fresh matured anola is selected for pickle preparation which is then washed with potable water. Anola is cut into uniform pieces, salt, turmeric, and sugar are added and homogenously mixed and set aside till the sugar melts. The oil is heated on a low flame; asafoetida, split mustard seeds, and red chilli powder are added; when it is cooled it is mixed with anola pieces. Pickle is stored in a glass jar at ambient temperature for ageing and consumed after 4–5 weeks (Fig. 13.53).

Fig. 13.52 Photograph of dry mango pickle

Fig. 13.53 Photograph of anola pickle

13.5.6 Karonda Pickle (*Karvandch Lonch*)

Karonda pickle is made from karonda (*Carissa carandas*) which is a type of berry which is green and sour when unripe and turns red and sweet as it ripens. Karonda fruit is a good source of iron and contains a fair amount of vitamin C. Mature fruit contains high amount of pectin. It is available from March to August in the local markets of Maharashtra.

13.5.6.1 Traditional Method of Preparation

Fresh matured green coloured karonda is selected for pickle preparation which is then washed with potable water. Washed karonda is added into the brine solution for 4–6 h. After that the brine solution was drained off and Karonda was surface dried using a cloth. A few spoons of oil is heated in a pan; split mustard seeds are added and cooked until the seeds pop up and then heating is stopped; then the spices are added in oil. All the spices are added to the fermented karonda and the ingredients are homogeneously mixed in a glass or porcelain jar. This jar is kept for 15–20 days. After 2 days, the pickle is agitated in correct manner. This jar is closed with airtight lid and allowed to ageing for 2–3 weeks at ambient temperature. It is consumed after 2–3 weeks (Fig. 13.54).

13.5.7 Mainmula Pickle (*Mainmula Lonche*)

Coleus Root (*Plectranthus barbatus*) is a tropical plant also known as **Indian Coleus or Mainmula.** The pickle is supposed to help strengthen the heart muscles.

Fig. 13.54 Photograph of Karonda pickle

Fig. 13.55 Photograph of *Mainmula* pickle

Coleus is native to India, grows in tropical climate, and is related to the lavender and mint family. Mainmula pickle is mainly prepared and consumed in Kolhapur region of Maharashtra.

13.5.7.1 Traditional Method of Preparation

Matured roots of mainmula is chosen for pickle preparation which is then washed with water for removing dirt and impurities. The roots are peeled and dried with cloth and then cut into long slices; salt and turmeric are added and kept for fermentation for 1 day. A few spoons of oil is heated in a pan; mustard seeds, fenugreek seeds, and asafoetida are added and cooked until the seeds pop up, and the heating is stopped; then the split mustard seed, red chilli powder, and turmeric powder are added in oil. The spices mixture and lemon juice are added to the brined mainmula and the ingredients are homogeneously mixed in container. The pickle is transferred to a glass jar or porcelain jar. This jar is kept for 15–20 days. After 2 days, the pickle is agitated in right manner. This jar is closed with airtight lid and allowed to ageing for 2–3 weeks. It is consumed after 2–3 weeks (Fig. 13.55).

13.5.8 Bitter Guard Pickle (*Karlyach Lonch*)

Bitter guard is mostly consumed in all over Maharashtra. In Marathi it known as '*karle*'. It is used in curry such as '*karlyachi bhaji*' or stuffed with spices and then cooked in oil. Bitter guard fruit has a lots of medicinal uses including cancer prevention, treatment of diabetes, fever, HIV, AIDS, and infection.

13.5.8.1 Traditional Method of Preparation

Fresh matured green coloured bitter guard (*Momordica charantia*) is selected for pickle preparation which is then washed with potable water. Bitter gourd is cut into round slices, salt and turmeric are added and kept aside for 1 day. A few spoons of oil is heated in a pan; mustard seeds, fenugreek seeds, and asafoetida are added and cooked until the seeds pop up, and then heating is stopped; then the split mustard seed, red chilli powder, and turmeric powder are added in oil. The spices mixture and lemon juice are added to the brined bitter guard and the ingredients are homogeneously mixed in a container and transferred into a glass or porcelain jar. This jar

Fig. 13.56 Bitter guard pickle (*Karlyach Lonch*)

is kept for 20–30 days. This jar is closed with airtight lid and allowed to ageing for 3–4 weeks at ambient temperature. It is consumed after 3–4 weeks (Fig. 13.56).

13.5.9 Turmeric Rhizomes Pickle (*Olya Haldich Lonch*) (Bulla 2010)

Turmeric pickle is mainly prepared from rhizomes (roots) of turmeric. Turmeric roots is bright orange flesh and is earthy, peppery, and slightly bitter in taste. It is easily available in winter season of Maharashtra.

13.5.9.1 Traditional Method of Preparation

Turmeric rhizomes are washed with potable water to remove mud and other impurities. Roots are peeled and cut into uniform pieces or the rhizomes are grated. The salt is added in grated turmeric roots and kept aside for 1 day. A few spoons of oil is heated in a pan; mustard seeds, fenugreek seeds, and asafoetida are added and cooked until the mustard seeds pop up, and then heating is stopped; then the split mustard seed, red chilli powder, and turmeric powder are added in oil. The spices mixture and lemon juice are added to the grated turmeric and the ingredients are homogeneously mixed in a glass or porcelain jar. This jar is kept for 15–20 days. This jar is closed with airtight lid and allowed to ageing for 2–3 weeks at ambient temperature. It is consumed after 2–3 weeks. It is also consumed sometimes as fresh pickle.

13.5.10 Garlic Pickle (*Lasunch Lonch*) (Bulla 2010)

Garlic (*Allium sativum*) is a species in the onion genus, Allium. Garlic is very common seasoning used in Maharashtrian cuisine. It is easily available in local markets. It is excellent source of manganese, vitamin B6, vitamin C, copper, and selenium.

13.5.10.1 Traditional Method of Preparation

The garlic cloves are peeled and salt and turmeric and red chilli powder are added and kept for 1 day. Oil is heated in a pan; fenugreek seeds and asafoetida are added and cooked for 2–3 s, and then heating is stopped; then the split mustard seed, red

Fig. 13.57 Photograph of garlic pickle

chilli powder, and turmeric powder are added in oil. The spices mixture and lemon juice are added to the garlic and the ingredients are homogeneously mixed in a glass or porcelain jar. This jar is kept for 15–20 days. This jar is closed with airtight lid and allowed to ageing for 2–3 weeks at ambient temperature. It is consumed after 2–3 weeks (Fig. 13.57).

13.5.11 Green Chilli Pickle

Green chillies are those peppers that don't grow fully and are harvested before they mature. It is excellent source of vitamin B6, vitamin A, iron, copper, potassium, and a small amount of protein and carbohydrates. It is available in local markets throughout the year.

13.5.11.1 Traditional Method of Preparation

Fresh green chilli is selected for preparation of green chilli pickle which is then washed with potable water. Green chillis are cut into 1–2 in. pieces; salt and turmeric are added and kept for 1 day. Oil is heated in a pan; fenugreek seeds and asafoetida are added and cooked for 2–3 s, and the heating is stopped; then the split mustard seed and turmeric powder are added in oil. The spices mixture and lemon juice are added to the green chilli and the ingredients are homogeneously mixed in a glass or porcelain jar. This jar is kept for 15–20 days. This jar is closed with airtight lid and allowed to ageing for 2–3 weeks at ambient temperature. It is consumed after 2–3 weeks (Fig. 13.58).

Fig. 13.58 Photograph of Green Chilli pickle

13.5.12 Prawns Pickle (*Kolambich Lonch*)

Prawn pickle is prepared from the fresh prawns. It is mainly consumed in Konkan region of the Maharashtra. Prawn pickle is now available in local markets.

13.5.12.1 Traditional Method of Preparation

The prawns are marinated with the garlic paste, salt, and turmeric and kept for about 15–20 min. The oil is heated in pan and the prawns are fried until prawns are cooked (al dente). The oil is drained and kept aside. Then the same oil is heated on slow flame; the fenugreek seeds and the fenugreek powder, chilli powder, turmeric powder, and lemon juice are added and all the ingredients are mixed homogenously. The prawns are added in spices mixture and cooked for another 3–4 min, on a very low heat, and the heating is stopped. It is cooled and filled in a glass jar and stored at room temperature for 2–3 weeks.

13.5.13 Goat Meat Pickle (*Muttonach Lonch*)

It is mainly prepared from the goat meat which is fresh meat. It is mainly consumed in Nagpur region of Maharashtra. Savaji community preserves the meat as a pickle form.

13.5.13.1 Traditional Method of Preparation

Oil is heated in a vessel roughly crushed dried red chillies, mustard seeds, garlic and ginger paste, red chillies, and turmeric powder are added and mixed homogenously. Heating is stopped, and the mixture is cooled down to room temperature and cumin powder, mustard powder, salt, and red chilli powder are added and mixed well. The mutton pieces are deep-fried in hot oil till crisp and the oil is drained. Then the mutton pieces are added to the spices mixture and mixed well. The mixture is cooled down to room temperature. Then lemon juice and vinegar are added to the mutton pickle and mixed well. The mutton pickle is transferred into a sterilized bottle and another 1–2 ladles full of hot oil is poured from the vessel on top. It is fixed with the lid and stored for 2–3 weeks.

References

Bulla RS (2010) Documentation of pickles and development. University of Agricultural Sciences, Dharwad

Census of India (2011) Population by religion community—2011. The Registrar General & Census Commissioner, India

De S, Parikh KM (1985) Medicinal importance of some common indian vegetables part I. Ancient Sci Life 4:232–237

Dudhate AK, More DR, Syed IH (2017) Studies on process standardization and nutritional value of Indian heritage Food-Kharodi. J Pharmacogn Phytochem 6:590–593

Gopalan C, Rama Sastri BV, Balasubramanian SC (2004) Nutritive value of Indian foods. National Institute of Nutrition, ICMR, Hyderabad

Kumari A, Pandey A, Raj A et al (2016) Cereal-based non-alcoholic indigenous fermented foods. In: Indigenous Fermented Foods of South Asia. CRC Press, Boca Raton

Lonkar SP, Mahajan AP, Ranveer RC, Sahoo AK (2011) Development of instant mattha mix. World J Dairy Food Sci 6:125–129

Shinde SV (2011) Development and quality characterization of traditional sorghum (*Sorghum bicolor* L.) fermented'Ambil' bevarage. Vasantrao Naik Marathwada Krishi Vidyapeeth, Parbhani

Singh KS (1992) People of India: an introduction, vol I. Anthropological Survey of India, Calcutta, 234 pp

Steinkraus KH (1983) Handbook of indigenous fermented foods. Marcel Dekker, New York. 671 pp

Surve VD, Kadam PG, Mhaske ST, Annapure US (2014) Studies in rheological properties of wheat batter prepared from wheat grains fermented at different temperatures used to prepare Kurdi—a traditional Indian food. Int Food Res J 21:1877–1885

Swapna G, Chavannavar SV (2013) Shrikhand—value added traditional dairy product. Int J food Nutr Sci 2:45–51

Tamang J (2010) Himalayan fermented foods. CRC Press, Boca Raton. https://doi.org/10.1201/9781420093254

Web Resources

https://www.maharashtra.gov.in/1128/Districts (n.d.)
https://www.maharashtratourism.gov.in (n.d.)
https://mahades.maharashtra.gov.in/files/publication/ESM_17_18_mar.pdf (n.d.)
https://www.maayboli.com/hitguj/marathi-recipes (2019)

Ethnic Fermented Foods and Alcoholic Beverages of Manipur

14

Romi Wahengbam, Anand Singh Thangjam,
Santosh Keisam, Ibemhal D. Asem,
Debananda S. Ningthoujam, and Kumaraswamy Jeyaram

Abstract

Manipur is characterized by its rich culture, indigenous food habits, and diverse population of over 30 different tribal and non-tribal ethnic communities. The major non-tribal Meitei community settles in the valley area, while the Naga and Kuki-Chin-Mizo tribes inhabit the hilly regions. Different ethnic communities have their unique food habits. Rice is their staple food and is consumed with fresh or boiled vegetables, fish, and meat with a variety of fermented foods. Ethnic fermented foods form an intrinsic part of the traditional diet system and the cultural integrity of all ethnic communities. These fermented products are unique in terms of the substrate used, the traditional indigenous knowledge of preparation, and the culinary practices. Ethnic fermented foods of Manipur are fermented bamboo shoots (*soidon, soidon mahi/soijin/soijim, soibum, thunkheng, thunbin, thunkhengkang, thunbinkang*), fermented fish (*ngari, hentak, ithiitongka*), fermented soybean (*hawaijar, khuichang, bethu, bekanthu, theisui*), fermented mustard leaves (*inziangsang, ziangdui, ziangsang, ankamthu, ganang tamdui*), fermented sesame seed (*sithu*), fermented mustard/rapeseed seed

R. Wahengbam (✉)
Life Sciences Division, Institute of Advanced Study in Science and Technology (IASST), Vigyan Path, Paschim Boragaon, Garchuk, Guwahati, Assam, India

A. S. Thangjam
AICRP on Post-Harvest Engineering Technology, Directorate of Research, Central Agricultural University, Iroishemba , Imphal, Manipur, India

S. Keisam
Advanced Level Institutional Biotech Hub, Modern College, Imphal, Manipur, India

I. D. Asem · D. S. Ningthoujam
Department of Biochemistry, Manipur University, Canchipur, Imphal, Manipur, India

K. Jeyaram
Institute of Bioresources and Sustainable Development (IBSD), Nursery Veng, Khatla, Aizawl, Mizoram, India

J. P. Tamang (ed.), *Ethnic Fermented Foods and Beverages of India: Science History and Culture*, https://doi.org/10.1007/978-981-15-1486-9_14

(*hangammaru hawaichar*), fermented kenaf seed (*gankhiangkhui*), fermented milk (*sangom aphamba*), fermented meat (*sathu*, *saayung*, *guaighi kang*, *sahro*, *bongkarot*, *aithu*), and fermented alcoholic rice beverages (ethnic starters—*hamei*, *chamri*, *khai*; non-distilled beverages—*yu angouba*, *khor*, *zoungao*, *atingba*, *patso*, *timpui*, *waiyu*, *pheijou*; distilled liquors—*yu*, *acham*, *zouju/zouzu*). This chapter discusses the history and culture of different ethnic groups of Manipur and focuses on the traditional indigenous knowledge, socioeconomic and ethical values, microbiology, and health and nutritional benefits of different traditional ethnic fermented foods and beverages of Manipur.

Keywords
Ethnic food culture · Fermented foods · Fermented alcoholic beverages · Traditional indigenous knowledge · Starter culture · Microbiology · Health benefits · Food safety

14.1 Introduction

Manipur is a place of splendid natural beauty and serene climate, a culturally rich state lying at the 23.83°N–25.68°N latitude and 93.03°E–94.78°E longitude on the northeastern corner of India. It is flanked by Myanmar in the south and the east, Nagaland in the north, Assam in the west, and Mizoram in the southwest (Map 1). Manipur has a total geographical area of 22,327 km^2 and is rectangular in shape having a little saucer-shaped centrally located valley surrounded by hill and mountain ranges on all sides (Singh and Baghel 2003; Asem et al. 2017). It falls under the regime of the Eastern Himalayan and the Indo-Burma Biodiversity hotspots. Manipur has a salubrious climate and can be broadly divided into temperate, subtropical, and tropical, which are influenced by the topography of the hilly region. The average temperature ranges from 15.4 to 25 °C with an annual rainfall of 161.62 cm and an altitude of about 750 m in the valley to about 3000 m in the hills and mountain peaks. The soils of Manipur are broadly divided into two types, viz., red ferruginous soil in hills and alluvium in the valley (Asem et al. 2017). Manipur has a total population of 2,855,794 with 1,438,586 males and 1,417,208 females distributed across its 16 districts (Census of India 2011). Imphal East, Imphal West, Bishnupur, Thoubal, and Kakching form the lowland valley districts, while the remaining 11 districts constitute the upland hilly areas. About 67.5% of the total population lives in rural areas, whereas 32.5% inhabits the urban areas. It is home to many religions with Hinduism and Christianity being the major religions, followed by Islam, Sanamahism, Jainism, Sikhism, and Buddhism. The official language spoken here is Manipuri language (called *Meiteilon*), which is included in the eighth Schedule of the Indian Constitution and is spoken by the majority of the population although different tribal ethnic groups speak their languages. There are more than 30 ethnic groups, with varying responses to the changing sociocultural environment, forming a heterogeneous population including both tribal and non-tribal communities (Census of India 2011). The women population play a

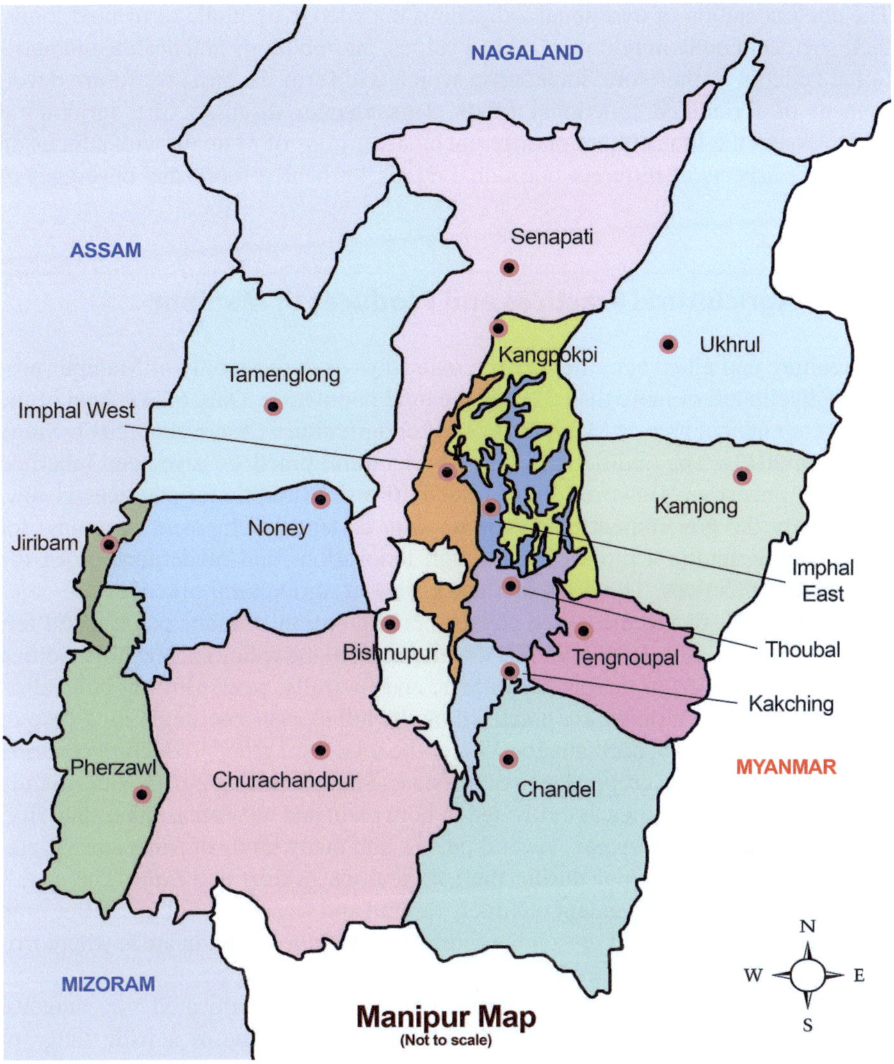

Map 1 Map of Manipur

significant role in the sociocultural and economic fronts of the state, as they are the major stakeholders in livelihood generation in agriculture and allied activities and the traditional food sectors (Jain et al. 2011). Though rice is the staple food of the people of Manipur, the indigenous fermented foods and beverages occupy an indispensable position in the traditional diet system and the cultural integrity of all the ethnic communities and contribute to a large proportion of the daily food consumption in the state (Jeyaram et al. 2009). However, adaption to new food habits and lifestyle changes accompanied by decreased intake of traditional foods has resulted in an increased incidence of lifestyle-associated diseases (Devi and Kumar 2012).

The documentation of traditional indigenous knowledge of ethnic fermented foods, their socioeconomic importance, ethical values, microbiology and health, and nutritional benefits is, therefore, imperative, which will form the basis for future development of fermented functional foods. This chapter discusses the agricultural practices and the food culture of different ethnic groups of Manipur with a focus on the above aspects of different traditional ethnic fermented foods and beverages of the state.

14.2 Agricultural Practices and Produces of Manipur

Agriculture and allied activities are the mainstays of the economy of Manipur providing livelihood of more than 52% of the total population. Only 6.74% land of the total geographical area of Manipur is used for agriculture (Asem et al. 2017; Singh and Bera 2017). The traditional ways of agricultural practices have been inherited from the ancestors. However, the productivity using traditional practices is low; thus, lately the government has come up with certain development programs for enhancing agricultural production through innovations and modernization of the agricultural practices. Three distinct categories of agricultural practices are common in Manipur due to the varied physical configuration, climatic pattern, soil fertility, and technological and scientific development in the valley and the hills. Settled farming is practiced in the plains, valleys, and foothills, while shifting cultivation (*jhum*) and terrace farming are practiced on the hill slopes. The staple food crop of Manipur is rice, which accounts for 95% of the total food grain production covering about 72% of the total cropped area of the state (Singh and Bera 2017). Rice, mainly of local glutinous varieties, is cultivated in both plain and hilly area. Other than rice, cereals like maize, wheat, etc., several pulses, and many kinds of fruits and vegetables are grown in Manipur during the two seasons, *Kharif* and *Rabi*. The state's agriculture is mainly dependent on timely rainfall and weather conditions. Rice crop is usually grown in *Kharif* season but rarely during *Rabi* season in areas where irrigation system could be afforded.

In Manipur, three types of horticultural crops are widely cultivated, viz., annuals, biennials, and perennials (Devi 2014). Horticulture consists of a wide range of crops, vegetables, fruits, flowers, spices, medicinal and aromatic plants, and plantation crops. Based on the geo-climatic conditions and ethnic preferences, cropping pattern, i.e., growing of horticultural crops with field crops in the same field, may increase the crop production, thereby meeting the need of the increasing population. Due to the difference in soil type and favorable climatic conditions, the valley and the hill areas are suitable for growing different varieties of vegetables and fruits. Horticultural crops grown in Manipur are *Kharif* vegetables, viz., French bean, cucurbit, bitter guard, bottle guard, tomato, brinjal, ladyfinger, *Colocasia*, *Alocasia*, etc., and *Rabi* vegetables, viz., cabbage, cauliflower, mustard, potato, pea, broad bean, radish, carrot, broccoli, lettuce, and capsicum. Spices grown in Manipur are onion, garlic, chili, ginger, turmeric, etc. and many more indigenous aromatic herbs. Fruits and plantation crops include tree bean, litchi, areca nut, cashew nut, walnut, orange, lemon, banana, pineapple, passion fruit, pomegranate, apricot, peach, plum, pear, papaya, sugarcane, mango, jackfruit, guava, palm, and kiwi (Devi and Kumar 2012). Major forest products include oak, teak, pine, cane bamboo, *leihao* (*Michelia* spp.), *uningthou* (*Phoebe hainesiana*), etc. Bamboos are also widely grown for bamboo shoots and other usage. Cultivation of low-volume cash crops like cardamom, black pepper, clove, nutmeg, cinnamon, etc. increases socioeconomic conditions of poor farmers particularly in hill areas. All ethnic communities of Manipur follow the tradition of collecting and selling wild edible plant resources for livelihood and to meet the day-to-day food demands. They are traditionally consumed as an essential constituent of the daily local meal thereby fulfilling the nutritional requirement of the people. Over 70 wild and semidomesticated edible plant species belonging to 42 families are consumed in Manipur for their shoots, aerial parts, rhizome, roots, corms, flowers, inflorescence, leaves, and fruits (Jain et al. 2011; Konsam et al. 2016). The trade of edible wild plants is an important alternative source of income and is mainly done by women folks. In the valley alone, over 110 tons of most commonly consumed wild edible plants are sold annually, which accounts for a business of more than ₹ 4 lakhs. Some of the exotic, commonly consumed, and economically important wild and semidomesticated edible plants are *Allium hookeri* (*maroi napakpi*), *Allium odorum* (*maroi nakuppi*), *Alocasia cucullata* (*palukabi/singju-paan*), *Alpinia nigra* (*pullei*), *Amomum aromaticum* (*namra*), *Centella asiatica* (*peruk*), *Chenopodium album* (*monsaobi*), *Chimonobambusa callosa* (*laiwa*), *Colocasia esculenta* (*paan*), *Curcuma zedoaria* (*yaipal*), *Eryngium foetidum* (*awaphadigom*), *Euryale ferox* (*thangjing*), *Hedychium coronarium* (*loklei*), *Hibiscus cannabinus* (*sougri*), *Houttuynia cordata* (*tokningkhok*), *Ipomoea aquatica* (*kolamni*), *Lemanea australis* (*nung-sam*), *Nelumbo nucifera* (*thambal mana*), *Neptunia oleracea* (*esing ekaithabi*), *Oenanthe javanica* (*komprek*), *Parkia javanica* (*yongchak*), *Passiflora edulis* (*sitaphal mana*), *Persicaria posumba* (*ken-goi*), *Polygonum barbatum* (*yelang*), *Sagittaria sagittifolia* (*koukha*), and *Zizania latifolia* (*esing kambong*). Livestock, which are reared in Manipur, are poultry, cattle, buffalo, sheep, goat, pig, etc. Pig and poultry farming generates good income.

Manipur has 155 wetlands (locally known as *pat*) covering an area of 529 km^2 that comprise 21 lakes, 2 oxbows, 2 reservoirs, and 130 waterlogged sites (Jain et al. 2011). Loktak Lake (*Loktak Pat*), located in Moirang in Bishnupur district of Manipur, is the largest of all (289 km^2 area) and is one of the "Wetlands of International Importance" designated by the Ramsar Convention. It is famous for the floating heterogeneous mass of vegetation, soil, and organic matter, locally known as *phumdi*. Wetlands are very important in the socioeconomic and cultural life of the people and are considered as a lifeline of Manipur. They serve as the natural sources for a variety of wild edible and medicinal plants and edible insects and supply substantial amount of freshwater fishes for household consumption and livelihood earning (Jain et al. 2011; Bharati et al. 2017). Fish is the most important item of Manipuri, the only animal protein acceptable by all, and it is an easily digestible protein food. Rearing of both indigenous and exotic fishes is practiced in the state. Manipur has the largest number of endemic fishes (110 species) in Northeast India, which are mainly of food and ornamental values (Anal 2014; Sarma et al. 2018). Loktak Lake accounts for more than 50% of the fish-producing area of Manipur and is the largest fishery resource of the state (Bharati et al. 2017). Some of the indigenous fishes of food value in Manipur are *mukanga* (*Amblypharyngodon mola*), *ngasang* (*Esomus danricus*), *ukabi* (*Anabas testudineus*), *sareng khoibi* (*Botia histrionica*), *ngamhai* (*Chanda nama*, *Parambassis ranga*, *Parambassis waikhomi*), *ngamu bogra* (*Channa punctata*), *nailon ngamu* (*Glossogobius giuris*), *ngachik* (*Heteropneustes fossilis*), *ngathi* (*Labeo calbasu*), *ngaton* (*Labeo bata*, *Labeo dero*, *Labeo dyocheilus*), *ngakijou* (*Lepidocephalichthys guntea*), *ngaril* (*Mastacembelus armatus*), *ngasep* (*Mystus* spp.), *ngapai* (*Notopterus notopterus*), *nganap* (*Pangio pangia*), *phabounga* (*Puntius chola*, *Puntius sophore*, *Pethia* spp.), *ngakha* (*Pethia manipurensis*), *pengba* (*Osteobrama belangeri*), etc. (Anal 2014; Devi et al. 2017; Sarma et al. 2018).

14.3 Ethnicity and Food History of Manipur

The different ethnic communities inhabiting Manipur can be broadly grouped into two main categories, i.e., the non-tribal group, comprising Meitei/Meetei, Meitei-Pangal, and Mayang communities, and the tribal group, comprising Naga and Kuki-Chin-Mizo tribes. The Meiteis have been identified as the core ethnic community of Manipur. They are the dominant non-tribal community settled throughout the Manipur valley (covering 92% of the total valley population) and make up about 61.7% of the total population of Manipur (Census of India 2011; Jain et al. 2011). The ethnic Muslim community of Manipur is called Meitei-Pangal. They claim divergence from other Muslims of India as they have a different history and a unique process of ethnic evolution, thus, claiming that they are naturalized people of Manipur who speak *Meiteilon* and profess Islam. They settled in the valley area with most of the population inhabiting Thoubal district. Those who do not fall into any of the ethnic groups of Manipur are included under Mayang (Devi 2012). They belong to several groups who originally came from other states of India and are, now, distinct inhabitants of Manipur. According to the Census of India (2011), there are 33 recognized tribes in Manipur that are classified under the two major tribes, viz., Naga and Kuki-Chin-Mizo. They are Aimol, Anal, Angami, Chiru, Chothe, Gangte, Hmar, Kabui (Rongmei, Puimei), Kacha Naga (Liangmei and Zemei, collectively called Zeliangrong), Kharam, Koirao, Koireng, Kom, Lamkang, Lushai (any Mizo tribes), Mao, Maram, Maring, Monsang, Moyon, Paite, Poumai Naga, Purum, Ralte, Sema, Simte, Suhte (Sahlte), Tangkhul, Tarao, Thadou (Kuki), Vaiphei, Zou, and any Kuki tribe. They inhabit the hilly regions of the state and have different languages and cultural practices. Certain tribes such as Kabui, Kacha Naga, Mao, Maram, Maring, and Tangkhul, occupy northern and northwestern parts of Manipur. The other tribes occupy southwestern and southeastern hills of Manipur, and most of them usually live in Churachandpur, Tengnoupal, and Sadar Hills districts. They migrated to Manipur in various periods of history and have been given another exogenous identity called Kuki (Devi and Kumar 2012).

The inhabitants of Manipur dates back as far as 40,000 BP in the Paleolithic period of the prehistoric Stone Age. According to the archaeological findings from caves and prehistoric hilltop villages in Ukhrul, Chandel, and Senapati districts, the pre-Neolithic people of Manipur were hill-dwellers settled in caves and hilltops. They were hunter-gatherers, foraging wild animals (deer, bovid, wild fowls) for meat, who knew growing plants for food and started pottery since the Mesolithic period (Devi 1988). Pottery in Manipur is known as *chaphu* (*cha* meaning "food" and *phu* meaning "vessel/pot") that literally translates into a vessel engaged in food preparation and consumption. According to oral history and *Chakparol Khuntaba Puya* (a sacred scripture of Meitei), *chaphu* originated from the potter community of *Chakpa Loi* (an aborigine of Manipur) who consumes pork meat and traditional fermented liquor and is presently settled at villages such as Andro, Thongjao, Sekta, Sekmai, Phayeng, Chairel, etc. (Samurailatpam 2014). They practice production of fermented alcoholic beverages as the main occupation apart from agriculture and rearing of animals like pork, duck, cock, etc. for their domestic consumption (Devi 2012). The settlement of the people in the valley of Manipur started during the proto-historic period with the foundation of Ningthouja/Meitei principality by the Meitei Lord Nongdalairen Pakhangba (33–154 AD). The main tribes who had settled in the historical days in Manipur are Kabui, Chothe, Koireng, etc. on the western hills, Mao, Thangkhul, Thangal, and Maram on the eastern hills, and Anal, Kom, Kabui, Lamgang, Mayon, Monsang, Aimol, etc. on the southern and south-eastern hills (Samurailatpam 2014). The people of Manipur have been using fermented items such as drinks and fish in their rites and rituals since time immemorial. The uses of fermented items are mentioned even in the folktales of Manipur from the very early period and monarchical rule. The folktales of *Hanuba Hanubi Pan Thaba*, *Houdong Lamboiba amasung Pebet*, and *Sanarembi Cheishra* clearly depict that people had begun eating *hentak* (fermented fish) (Devi 2017). Migration of people carrying religion, food culture, and habits influences the settlers leading to cultural consolidation or fusion. After the advent of Hinduism some 300 years ago during the reign of King Pamheiba (latter known as King Garib Niwaz after adopting Hinduism in 1717 AD), the traditional fervor and enthusiasm of the indigenous people of Manipur shifted with an enormous impact on the food culture of the Meiteis inhabiting the valley. For example, in the past, a community hunt was performed during festivals and ritual ceremonies, and the animal killed or captured was sacrificed to the Father God and Mother Goddess along with the traditional liquor. Such a community hunt is no longer performed, and the tradition of taking liquor and meat curry has now been replaced with water, fish curry, and vegetarian diets (Devi 2012). The present food culture of the people in the valley region is largely a fusion of the Hindu and Meitei cultures, with modifications based on ethnic preferences and social ethos. Ever since the aboriginal people inhabited Manipur—ranging from the hilltops to the valley region—hunting, gathering, and utilization of plants and animals for consumption started, followed by their rearing and cultivation that gradually emerged as the ethnic food culture of the present day (Tamang 2010).

14.4 Food Culture of Manipur

Manipuri cuisines are simple, tasty, and healthy. Cuisines of Manipur are very seasonal, prepared from the seasonal vegetables. Thus, every season has its own specialty cuisines. Various aromatic vegetables and herbs that are wild and semidomesticated edible plants are regular constituents of the traditional Manipuri cuisine. Usually all the ethnic groups in Manipur eat two meals a day, one in the morning and another in the evening/night, and take rice as the main course with vegetables, meat, fish, and various fermented items. They prefer boiled, smoked, roasted, and fermented fish, meat, and vegetables. The principal food items of the non-tribal Meitei community in the valley region are fish, poultry, pork, and vegetables, and fermented fish remains as the essential ingredient of the boiled vegetable cuisines (Tuikhar et al. 2019). Many other cultures have influenced most of the Meitei diet due to various sociopolitical reasons. Sanskritization is the most important factor that influences diet and has led to the emergence of vegetarian population among the community (Devi and Kumar 2012). Breakfast is very light. *Chabon* (boiled rice water) was taken earlier, but now it is replaced with milk, tea, coffee, and snacks. In the *Loi* villages of the valley, the tradition of drinking fermented alcoholic rice beverages is still practiced among the Meitei community. The major traditional cuisines unique to the Meitei community are *eromba*, *morokmetpa*, *singju*, *kangsu*, *paknam*, *ooti*, *chagempomba*, *kangsoi/chamthong*, and *champhut* (Konsam et al. 2016). Manipuri are fond of *eromba*, an ethnic delicacy prepared by smashing boiled potato or pea and vegetables such as brinjal, tomato, banana pseudostem, banana flower, tree bean (*P. javanica*), broad bean (*Vicia faba*), rhizome of *H. coronarium*, corm of *C. esculenta*, seeds of *E. ferox*, stem and rhizome of *A. aromaticum*, etc. and mixing with abundant amount of chilies and steamed or roasted ethnic fermented fish. Apart from vegetables, fermented bamboo shoots are also used to prepare delicious *eromba*. *Morokmetpa* is a spicy chutney prepared with a good number of chilies and fermented fish and garnished with fresh aromatic herbs. *Singju* is a finely shredded fresh vegetables salad mixed with chili, fermented fish, and crushed powder of roasted pea or roasted gram powder. Depending on individual choice, some people may use dried or fermented meat or soybean in *morokmetpa* and *singju*. *Kangsu* is a dry preparation of finely shredded boiled wild vegetables or herbs (such as *C. asiatica*, bamboo shoots, etc.) mixed with fermented fish and chili. *Paknam* is a pancake style cuisine prepared from a mixture of gram flour, herbs, vegetables (mainly *A. hookeri*, *A. odorum*, onion leaves), green chili, and roasted fermented fish by steaming and baking in turmeric leaves. It is also prepared with wild edible split gill mushroom (*kanglayen*) (*Schizophyllum commune*) or fresh small fishes without gram flour. *Ootti* is prepared by boiling pea with green leafy vegetable mixture along with some rice and a pinch of sodium bicarbonate. *Kangsoi* is a stew prepared by boiling vegetables and potato with chili, salt, fermented fish, and small smoked or roasted dried fish.

The tribal communities in the hill districts consume heavy meat (pork, mithun, beef), boiled vegetables, and plentiful rice-based fermented alcoholic beverages with other fermented or dried food items and no milk products (Tuikhar et al. 2019).

The Anal tribe eats fish, egg, beef, pork, and other kinds of meat as well as fruits and vegetables. They prepare and drink traditional rice beer, known as *zupar* or *zuhrin* (Bareh 2007). The Liangmai Naga tribes, believed to have been migrated from Mongolia during the seventeenth century, eat rice, fresh meat, green vegetable, etc. along with other ethnic food items such as fermented juice of mustard leaves (*ziang-dui*), potted bamboo shoot (*tasun*), and fermented soybean (*tasang*). During family worshipping and death feasts, the Chiru tribes offer rice and meat (pork, mithun, beef) to the neighbors and village elders along with the traditional wine served in dry, hollow bottle gourds (Maremmei 2007). Hmar tribes eat rice with simple meat and vegetable dishes with a few spices and a good amount of chili or pepper (Hmar 2013). Some special dishes exclusively identifiable as of *Hmar* origin are *chartang* (boiled meat curry with ginger, salt, a good amount of chili or pepper, and the aromatic herb *E. foetidum* locally called *bachikhawm*), *hmepawk* (porridge prepared with rice, meat, and vegetables), and *chi-al hme* (boiled vegetable curry with salt, hot chili or pepper, salt, ashes of burnt wood or straw and fermented fish, or fermented pork or dried beef, pork or fish). The locally brewed rice beer called *zu* is essentially consumed in almost all festivals and celebrations. The Rongmei tribes consume a variety of alcoholic beverages, commonly known as *jou*, since ancient times as an essential item in their everyday life and also during festivals and feast days (Thaimei 2002). The regular food items of the Marings are rice, meat, and fish with a variety of vegetables preferably in boiled and roasted forms. Due to some historical significance, mithun is the only accepted animal for important rituals and ceremonies. Dried and fermented fish and fermented soybeans are delicacies, and they are fond of rice beer, distilled liquor, and puffing pipe (Meitei 2015). The Thadous drink plentiful rice-based fermented alcoholic beverages called *ju* and consume heavy meats of mithun, pigs, and ox. (Siltou 2010). It is a customary to drink ethnic alcoholic rice beer and eat pig and mithun meats among the Vaiphei and Zou tribes during traditional celebrations such as post-harvest feasting (Sinha 2007).

Edible insects are also an essential constituent of the traditional foods among the ethnic communities of Manipur and play important role in their nutrition, as they are a rich source of protein, minerals, and energy (Shantibala et al. 2014). Among the edible insects, aquatic species are one of the most favorable groups due to their taste and high availability in inland wetlands such as lakes, rivers, ponds, puddles, etc. Meitei, Kabui, Chothe, Tarao, Tangkhul, and Thadou are the highest consumers of edible insects. Some of the commonly consumed edible insects are *Lethocerus indicus* (*naosek*), *Laccotrephes maculatus* (*haonaosek*), *Hydrophilus olivaceus* (*thara-ikokpi*), *Cybister* spp. (*tengbi*), *Crocothemis servilia* (*maikhumbi*), *Diplonychus rusticus* (*kongjeng kokphai*), *Oxya hyla hyla* (*loubuk koujeng*), *Apis cerana indica* (*haying khoi*), *Apis dorsata* (*khoiren*), *Omphisa fuscidentalis* (*watin*), and *Udonga montana* (*usingsa*). They are consumed at the larval, pupa, and adult stages in roasted or fried form with oil, salt, chili, spices, and traditional aromatic vegetables and herbs (Sangma et al. 2016).

The traditional food habits of different ethnic groups are simple and exhibit cultural ethnicity. The peculiarity of the ethnic foods is the dominant preparation and consumption of many traditional fermented items. Different ethnic communities

Table 14.1 Ethnic fermented foods and beverages of Manipur

Fermented food/ beverage	Substrate	Sensory property	Major ethnic consumers	Region/district
Fermented bamboo shoot				
Soibum	Succulent, young bamboo shoot slices	Wet, solid, sour, acidic, non-salted	Meitei	Imphal, Andro, Kwatha, Tengnoupal, Kakching, Noney, Bishnupur
Soidon	Young, tender apical meristem exclusively of *Schizostachyum munroi*, *soidon mahi/soijin/soijim/*, leaves/juice of *Garcinia pedunculata*, rice-washed water	Wet, solid, sour, acidic, non-salted	Meitei	Bishnupur, Imphal
Soidon mahi/soijin/soijim	Juice extract of fruits of *Garcinia pedunculata*, rice-washed water	Liquid, sour, acidic, non-salted	Meitei	Bishnupur, Imphal
Thunkheng	Whole succulent, young bamboo shoot	Wet, solid, sour, acidic, non-salted	Zeliangrong, Kuki	Tamenglong, Noney
Thunbin	Succulent, young bamboo shoot slices	Wet, solid, sour, acidic, non-salted	Zeliangrong, Kuki	Tamenglong, Noney
Thunkhengkang	*Thunkheng*	Dried, solid, sour, acidic, non-salted	Zeliangrong, Kuki	Tamenglong, Noney
Thunbinkang	*Thunbin*	Dried, solid, sour, acidic, non-salted	Zeliangrong, Kuki	Tamenglong, Noney
Fermented leafy vegetable				
Inziangsang	Mustard leaf	Wet or dry, solid, sour, acidic, non-salted	Naga	Naga inhabited hill areas
Ziangsang	Mustard leaf	Liquid, sour, acidic, non-salted	Naga	Naga inhabited hill areas
Ziangdui	Mustard leaf	Concentrated liquid, sour, acidic, non-salted	Naga	Naga inhabited hill areas
Ankamthu	Mustard leaf	Concentrated liquid, non-salted	Kuki-Mizo	Churachandpur

(continued)

Table 14.1 (continued)

Fermented food/ beverage	Substrate	Sensory property	Major ethnic consumers	Region/district
Ganang tamdui	Mustard leaf	Concentrated liquid, non-salted	Zeliangrong, Kuki	Tamenglong, Noney
Fermented seed				
Hawaijar	Soybean	Wet, solid, alkaline, pungent, mucilaginous, sticky, non-salted	Meitei	Meitei inhabited valley areas (Imphal, Thoubal, Bishnupur, Kakching)
Khuichang	Soybean	Wet, solid, alkaline, pungent, mucilaginous, sticky, non-salted	Zeliangrong	Tamenglong, Noney
Bethu	Soybean	Wet, solid, alkaline, pungent, mucilaginous, sticky, non-salted	Kuki, Thadou	Churachandpur
Bekanthu	Soybean	Wet, solid, alkaline, pungent, mucilaginous, sticky, non-salted	Hmar	Churachandpur
Theisui	Soybean	Wet, solid, alkaline, pungent, mucilaginous, sticky, non-salted	Tangkhul	Eastern hill areas such as Ukhrul
Sithu	Sesame seed	Wet, colloid, thick, non-salted	Kuki-Chin-Mizo	Churachandpur
Gankhiangkhui	Kenaf seed	Wet, solid, alkaline, pungent, sticky, non-salted	Zeliangrong, Kuki	Tamenglong, Noney
Hangammaru hawaichar	Mustard/rapeseed seed	Thick paste, pungent, non-salted	Meitei (Chakpa clan)	Chakpa Meitei inhabited valley areas
Fermented fish				

(continued)

Table 14.1 (continued)

Fermented food/ beverage	Substrate	Sensory property	Major ethnic consumers	Region/district
Ngari	*Puntius sophore*	Solid, pungent, non-salted	Meitei	Meitei inhabited valley areas (Imphal, Thoubal, Bishnupur, Kakching)
Hentak	*Esomus danricus*, petioles of *Alocasia macrorrhiza*	Thick paste, pungent, non-salted	Meitei	Meitei inhabited valley areas (Imphal, Thoubal, Bishnupur, Kakching)
Ithiitongka	Small freshwater fishes	Solid, non-salted	Moyon	Chandel
Fermented milk				
Sangom aphamba	Unboiled, raw cow milk	Curd, colloid	Meitei	Imphal
Fermented meat				
Sathu	Pork fat	Thick paste, non-salted, strong smell, oily	Zeliangrong, Kuki	Western hills
Gwag ruum	Pork fat	Thick paste, non-salted	Rongmei	Western hills
Bongthu	Cow or buffalo fat	Thick paste, non-salted	Kuki, Hmar, Paite, Vaiphei	Hill areas
Saayung	Pork meat	Solid, hard, salted	Tangkhul	Tangkhul inhabited hill areas
Guaighi Kang	Cow or buffalo skin	Solid, dried, hard	Rongmei	Western hills
Sahro	Any meat, usually pork	Solid, dried, hard	Hmar	Hmar inhabited hill areas
Bongkarot	Cow leg meat, *sathu*	Liquid with solid meat pieces, sticky	Vaiphei, Paite, Hmar	Hill areas
Aithu	Crab	Solid, alkaline, pungent	Kuki-Chin-Mizo	Hill areas
Amylolytic starter culture for fermented alcoholic beverage production				
Hamei	Rice, bark of *Albizia myriophylla*	Dry, solid, alcoholic flavor	Meitei, Kabui	Imphal, Bishnupur, Tengnoupal, Jiribam

(continued)

Table 14.1 (continued)

Fermented food/ beverage	Substrate	Sensory property	Major ethnic consumers	Region/district
Chamri	Rice, bark of *Albizia myriophylla*	Dry, solid, alcoholic flavor	Tangkhul	Tangkhul inhabited hill areas
Khai	Rice, bark of wild tree locally known as *khaipuroi*	Dry, solid, alcoholic flavor	Zeliangrong, Kuki	Tamenglong, Noney
Fermented alcoholic beverage				
Yu angouba	Rice, germinated paddy	Non-distilled, mild alcoholic flavor	Meitei, Kabui	Imphal
Khor	Germinated paddy	Non-distilled, mild alcoholic flavor	Tangkhul	Tangkhul inhabited hill areas
Zoungao	Rice, germinated paddy	Non-distilled, mild alcoholic flavor	Zeliangrong, Kuki	Tamenglong, Noney
Atingba	Rice (glutinous variety), *hamei*	Non-distilled, filtered, sweet-bitter, mild alcoholic flavor	Meitei, Kabui	Imphal
Patso	Rice, *chamri*	Non-distilled, mild alcoholic flavor	Tangkhul	Tangkhul inhabited hill areas
Timpui	Rice, *khai*	Non-distilled, mild alcoholic flavor	Zeliangrong, Kuki	Tamenglong, Noney
Waiyu	Rice, husk, *hamei*	Non-distilled, mild alcoholic flavor	Meitei, Kabui	Imphal
Pheijou	Rice, husk, *khai*	Non-distilled, mild alcoholic flavor	Zeliangrong, Kuki	Tamenglong, Noney
Yu	Rice (glutinous variety), *hamei*	Distilled liquor, strong alcoholic flavor	Meitei, Kabui	Imphal
Acham	Rice, *chamri*	Distilled liquor, strong alcoholic flavor	Tangkhul	Tangkhul inhabited hill areas
Zouju/zouzu	Rice, *khai*	Distilled liquor, strong alcoholic flavor	Zeliangrong, Kuki	Tamenglong, Noney

have their own traditional fermented food and beverages that are unique in the traditional knowledge of preparation and the organoleptic characteristics of the fermented products (Table 14.1).

14.5 Ethnic Fermented Foods and Beverages of Manipur

The ethnic communities of Manipur prepare and consume various kinds of fermented foods and beverages as an indispensable part of their traditional diet. Based on the raw material used and nature of the fermented product, the ethnic fermented foods can be broadly classified as (1) fermented bamboo shoots, (2) fermented leafy vegetables (fermented mustard leaves), (3) fermented seeds (fermented soybean, fermented kenaf seed, fermented sesame seed, fermented mustard/rapeseed seed), (4) fermented fish, (5) fermented meat (fermented pork fat, fermented pork meat, fermented cow or buffalo fat, fermented cow or buffalo skin, fermented crab, etc.), (6) fermented milk, and (7) fermented alcoholic beverages (Jeyaram et al. 2009; Tamang 2010; Devi and Kumar 2012; Singh et al. 2018c).

14.6 Ethnic Fermented Bamboo Shoots

Bamboo shoots are the young, tender saplings of bamboo plant, the giant grass belonging to subfamily Bambusoideae of family Poaceae. Globally, over 1250 species of bamboo across 75 genera are reported, to which India contributes 136 species belonging to 23 genera (Singh et al. 2003; Forest Survey of India 2017). Beyond its utility in paper industries and constructions, bamboo has a great potential as medicinal food. Nearly 100 species of bamboo are commonly known to utilize for consumption of bamboo shoots as food (Chongtham et al. 2011; Waikhom et al. 2013; Nongdam and Tikendra 2014). Northeast region of India is considered as a treasure house of edible bamboo shoots as it harbors more than 50% of India's bamboo species (Nongdam and Tikendra 2014; Forest Survey of India 2017). Manipur has a vast bamboo resource with 54 species under 11 genera that provides an almost unimaginable amount of edible bamboo shoots. Bhatt et al. (2004) reported that 5078 tons of fresh bamboo shoots were harvested annually for consumption in the region, with Manipur taking the highest consumption of 2188 tons per year. The edible bamboo shoots in Manipur are from *Bambusa balcooa* (locally called *leewa/ching saneibi*), *Bambusa cacharensis* (*moriang wa*), *Bambusa kingiana* (*watangkhoi*), *Bambusa manipureana* (*uma*), *Bambusa nana* (*khokwa*), *Bambusa nutans* (*saneibi*), *Bambusa oliveriana* (*khokwa*), *Bambusa tulda* (*utang/watang*), *Bambusa tuldoides*, *C. callosa* (*laiwa*), *Dendrocalamus brandisii* (*wamu*), *Dendrocalamus flagellifer* (*longwa*), *Dendrocalamus giganteus* (*maribob*), *Dendrocalamus hamiltonii* (*wanap/unap*), *Dendrocalamus hookeri* (*watangkhoi/utangkhoi*), *Dendrocalamus latiflorus* (*wui*), *Dendrocalamus longispathus* (*ui/unap manbi/chingwa*), *Dendrocalamus manipureanus* (*ooe/oei/ui*), *Dendrocalamus strictus* (*unan*), *Melocanna baccifera* (*moubi wa*), *Schizostachyum dullooa* (*phisautong/tolluwa*), *Schizostachyum munroi* (*naat*), *Schizostachyum pergracile* (*pungsang/pongshang*), and *Thyrsostachys oliveri* (*kabo wa*). The total import and export economy of edible bamboo shoots in the global market is about US$ 1.2 billion, while the Indian edible bamboo shoot economy is estimated to be US$ 0.8 million indicating its great potential for growth and share in the international market. In Northeast India, a net annual income of US$ 0.7 million is estimated from bamboo shoot trade (Nongdam and Tikendrajit 2014).

Consumption of young succulent bamboo shoots is becoming popular in the modern world because of their exotic taste, flavor, nutritive value, and numerous health benefits. Bamboo shoots are rich in dietary fibers, carbohydrates, minerals (potassium, phosphorus, magnesium, calcium, and iron), vitamins, and phytochemicals (polyphenols and phytosterols) and possess antioxidant, anti-inflammatory, neuroprotective, antiapoptotic, antimicrobial, antifungal, anticancer, antihypertensive, and cholesterol-lowering properties (Chongtham et al. 2011; Singhal et al. 2013; Nirmala et al. 2014). The bulk seasonal availability only during the monsoon months (May–September) and the highly perishable nature (shelf life of 3–4 days) of fresh bamboo shoots necessitate their preservation to ensure availability throughout the year. Age-old indigenous practices of traditional fermentation have been widely practiced in Asian countries to preserve these edible bamboo shoots. They are traditionally processed and used as a delicacy in Northeast India. In Manipur, fermented bamboo shoot is an important part of the traditional cuisine, and various ethnic communities produced varieties of indigenous fermented bamboo shoots that differ in terms of the raw materials used, preparation process, texture, and flavor. Traditionally, fermentation of bamboo shoot is done in specially designed earthen pots or bamboo baskets/chambers in solid or submerged states via spontaneous fermentation or backslopping approach. Fresh shoots of *B. balcooa*, *B. tulda*, *D. hamiltonii*, *D. giganteus*, *M. baccifera*, and *S. munroi* are exclusively used for preparing the ethnic fermented foods (Jeyaram et al. 2009, 2010; Romi et al. 2015). They are mainly produced by the Meitei community and commonly known as *soibum*, *soidon*, and *soidon mahi/soijin/soijim*. Other lesser known fermented bamboo shoots are *thunkheng*, *thunbin*, *thunkhengkang*, and *thunbinkang* of the Naga and Kuki communities. Based on the duration of fermentation, fermented bamboo shoots can be broadly categorized into two types: short-duration (2–7 days) fermented bamboo shoots, which comprised *soidon* and *soidon mahi/soijin/soijim*, and long-duration (6–12 months) fermented bamboo shoots, which comprised *soibum*, *thunkheng*, and *thunbin*. Based on the difference in the traditional preparation methods and their common practice in specific region of Manipur, the fermented bamboo shoots are further classified into five types: Kwatha-type; Noney-type; Kakching-type; Andro-type, which comprise *soibum*, *thunkheng*, and *thunbin*; and Lamangdong-type, which includes only *soidon*.

14.6.1 *Soidon*

Soidon is a non-salted, sour, short-duration fermented bamboo shoot product of the Meitei community of Manipur. It is exclusively produced in the Bishnupur (old name: Lamangdong), which is known as the "place of *soidon* origin." In earlier times, *soidon* was known as *Lamangdong soidon*. The main production of *soidon* is from the small-scale traditional production centers of Bishnupur. Unlike all other fermented bamboo shoots of Northeast India and other Asian countries, *soidon* is unique in the sense that it is prepared exclusively from the apical shoot meristem of *S. munroi*, locally known as *naat*. *S. munroi* is endemic to the Ngariyan hill range of Bishnupur district. The commercial production of *soidon* stills depends on the age-old practice of indigenous traditional fermentation (Romi et al. 2015).

14.6.1.1 Traditional Preparation Method

The local tribal people of Ngariyan hill range harvest the succulent tender apical meristems, which are available irrespective of season, from the wild by shaking the bamboo plant. They are then brought to the Bishnupur market to sell to the traditional *soidon* producers. The tender meristems are defoliated to remove the outer casing, and unwanted hard portions are cut off before cutting the shoots transversely into small pieces (Fig. 14.1). The traditional fermentation of *soidon* is done by backslopping using the milky, sour fermentation liquid (*soidon mahi/soijin/soijim*) of the previous batch as the starter for a subsequent batch. The transversely cut shoot meristems (5–10 cm length) are submerged in *soidon mahi/soijin/soijim* in 1:1 dilution in an earthen pot (locally known *chaphu*) or a bamboo basket layered with polythene sheet. In recent times, plastic containers are used for the fermentation. The preparation is allowed for submerged batch fermentation for 5–7 days in winter or 2–4 days in summer at the ambient environment with intermittent stirring. Once the fermentation is complete, *soidon* is removed from the container to drain off the milky fermentation soup before selling in the market. It can be stored for over a year in airtight container. In some cases, acidic juice extract or leaves of *Garcinia pedunculata* Roxb., locally known as *heibung*, and rice-washed water, locally called *chenghi*, are added during the fermentation to impart a unique flavor and enhance quality, texture, and color of the fermented product. Backslopping results in the dominance of the best adapted microbial strains (Jeyaram et al. 2009, 2010; Das and Deka 2012; Soibam and Ayam 2018).

14.6.1.2 Gastronomy, Socioeconomy, and Ethical Values

Soidon is a favorite ethnic food of the Meitei community and is consumed as an indispensable constituent of the traditional diets. It is mainly consumed after boiling as an ethnic cuisine called *eromba* by mixing with mashed, boiled potato, petiole of *Colocasia gigantea* locally called *yendem*, chili, fermented fish *ngari*, and salt. Sometimes, it is fried with potato, vegetables, and fish to prepare the local delicacy *kangou*. *Soidon* is sold in loose form in polythene bags or by wrapping in banana leaves by womenfolk in the local vegetable markets called *Keithel*, mainly in Bishnupur Bazar (earlier known as *Lamangdong Keithel*) in Bishnupur and

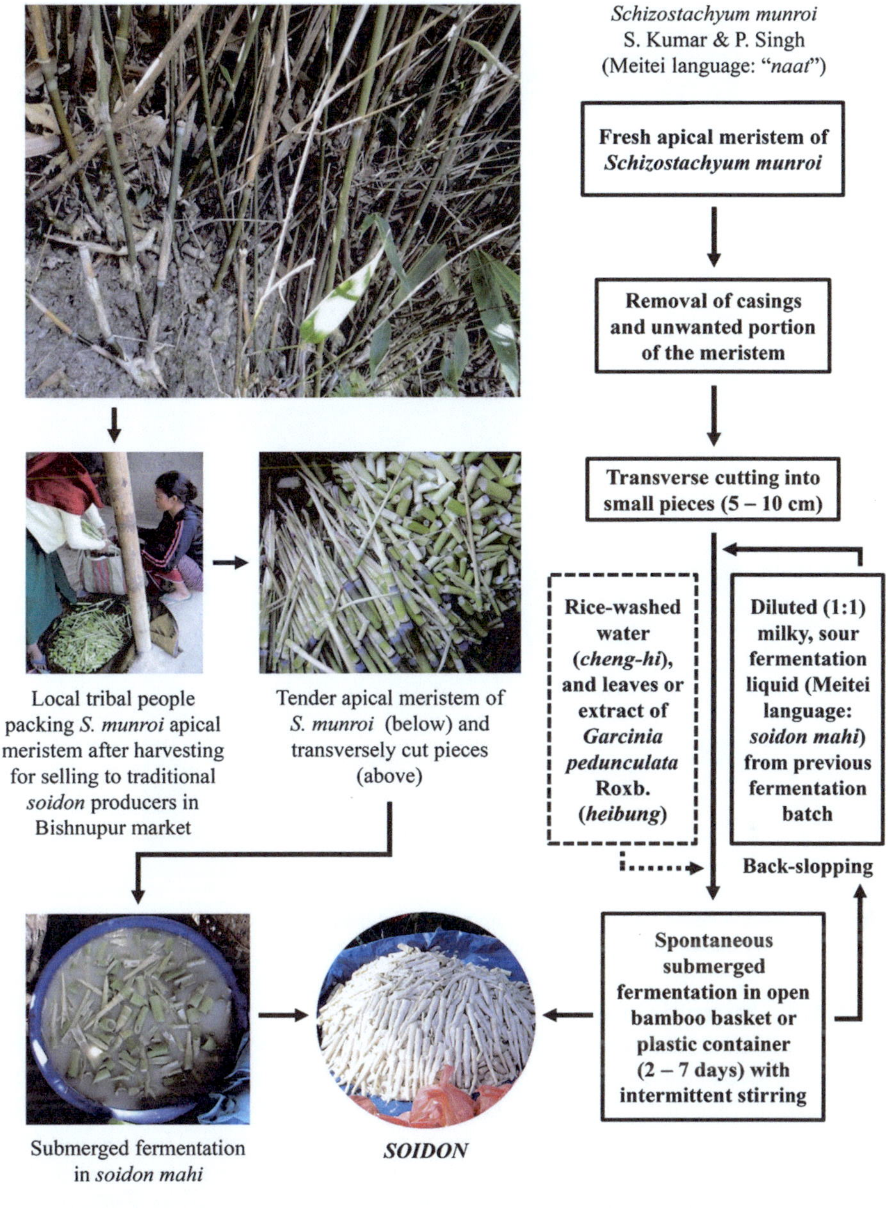

Fig. 14.1 The indigenous traditional preparation process for *soidon* production in Manipur, India

Khwairamband Bazar (popularly known as *Ema Keithel*) in Imphal (Fig. 14.2). It acts as a source of income for the artisanal producers and the tribal people harvesting the shoots. Each kilogram of fresh apical shoot meristem fetches ₹ 20–40, while *soidon* is sold for ₹ 80–100 per kg.

Fig. 14.2 *Soidon* being sold in the Bishnupur market, Manipur by the local traditional producers, while preparing the fresh apical meristem of *naat* for fermentation

14.6.1.3 Microbiology

Soidon is acidic in nature with pH of 3.9–4.3, and the fermentation is dominated by lactic acid bacteria and yeasts. The final fermented and marketed *soidon* has a lactic acid bacterial load of 10^6–10^9 cfu/g and yeast load of 10^4 cfu/g. The culturable lactic acid bacteria associated with *soidon* comprise *Leuconostoc* spp. (*Leuconostoc citreum*, *Leuconostoc fallax*, *Leuconostoc lactis*), *Lactococcus* spp. (*Lactococcus lactis* subsp. *lactis*, *Lactococcus lactis* subsp. *cremoris*, *Lactococcus raffinolactis*), *Lactobacillus* spp. (*Lactobacillus brevis*, *Lactobacillus plantarum*), *Weissella* spp. (*Weissella cibaria*, *Weissella oryzae*), and *Pediococcus pentosaceus*, while the reported yeasts are *Candida humilis*, *Kazachstania* spp., *Hanseniaspora thailandica*, *Bandoniozyma tunnelae*, and *Cryptococcus diffluens*. Traditional fermentation of *soidon* is a complex, multi-species microbial ecosystem, and it occurs through a distinct and rapid three-phase successional dynamics of selective, autochthonous lactic acid bacteria. *Lc. lactis* subsp. *cremoris* and *Weissella* spp. (*W. cibaria*, uncultured *Weissella ghanensis*) predominate the early phase (1–2 days) of fermentation, which is joined by *Leuc. citreum* during the mid-phase (3 days), while in the late phase (5–7 days), *Lb. plantarum* and *Lb. brevis* emerge and become dominant with contemporaneous disappearance of *W. cibaria* and *Lc. lactis* subsp. *cremoris*. Although fresh meristem of *S. munroi* is predominated by *Proteobacteria*, lactic acid bacteria belonging to *Leuconostocaceae*, *Streptococcaceae*, and *Lactobacillaceae* rapidly dominate the fermentation within 1 day. Next-generation sequencing and DGGE-based metagenomic analyses also reveal additional lactic acid bacterial species under the genera *Weissella* (*Weissella* spp., *Weissella confusa*) and *Lactobacillus* (uncultured *Lactobacillus acetotolerans*, *Lactobacillus sakei*, *Lactobacillus johnsonii*, *Lactobacillus intestinalis*, *Lactobacillus zymae*, *Lactobacillus mindensis*) during the fermentation. These dominant lactic acid bacteria are potential targets for designing functional starter culture for industrial production of *soidon* with improved quality, safety, and functionality (Tamang et al. 2008; Romi et al. 2015).

14.6.1.4 Nutritional Composition, Health Benefits, and Functionalities

Soidon is a good source of protein (20.7–37.2% of dry weight), carbohydrate (10.3–46.6%), dietary fiber (4.38–5.01%), and minerals (Ca, 0.19 mg/g; Na, 0.04–2.65 mg/g; K, 2.46–14.35 mg/g; Mg, 3.14 mg/g) (Tamang et al. 2012; Premlata et al. 2015; Sonar et al. 2015). It has high moisture content (50.8–92.2%) and less fat content (3.1–3.7%) with a food value of 363.1 kcal/100 g (Tamang et al. 2012; Sonar et al. 2015). Over 55 volatile organic compounds belonging to alkanes, alkenes, esters, aldehydes, ketones, alcohols, and phenolic aromatic compounds are detected in *soidon* and are responsible for imparting odor, taste, and texture characteristic of the food (Sonar et al. 2015). The most important volatile compounds are p-cresol, 2-methylnaphthalene, 2-heptanol, acetic acid, linalool, and phenyl acetaldehyde. *Soidon* has a considerable amount of phytosterol (2068.80 μg/g). It is rich in polyphenols with the total phenol, flavonoid, and tannin contents of 718.03–9466.70 μg/g, 342.43 μg/g, and 30.81 μg/g, respectively, that contribute to its high antioxidant property (DPPH and H_2O_2 radical scavenging activity of 70.84% and 58.86%, respectively) (Premlata et al. 2015; Sonar et al. 2015). It is reported to possess the lowest α-glucosidase inhibitory activity (IC_{50}: 29.78 ppm) among the fermented bamboo shoots of Manipur. The inhibition of this enzyme delays the intestinal absorption of glucose into the blood, thereby preventing an increase in the level of blood sugar, and thus helpful for the management of diabetes. The most abundant vitamins in *soidon* are ascorbic acid (3.20 mg/g), folic acid (3.23 mg/g), and cyanocobalamin (0.29 mg/g) (Sonar et al. 2015). *Soidon* has the lowest cyanogenic glycoside content (0.7722 ppm) making it to be the safest fermented bamboo shoot for consumption. The functional value of *soidon* is attributed to the lactic acid bacteria mainly lactobacilli that are reported with probiotic properties, hydrolytic enzymes production, anti-nutritive compound-degrading properties, protease activities, and broad-spectrum antibacterial activities against common food-borne pathogens (Sonar and Halami 2014; Romi 2015).

14.6.2 *Soidon Mahi*

Soidon mahi is the traditional fermentation liquid prepared and used by the Meitei community of Bishnupur village of Manipur as starter for the indigenous fermentation of *soidon*. It is also commonly known as *soijin* or *soijim*. It is milky, acidic, sour, and non-salted in nature. Once a freshly prepared batch of *soidon mahi* is utilized for preparation of *soidon*, this fermentation liquid is utilized for backslopping of the next fed-batched fermentations.

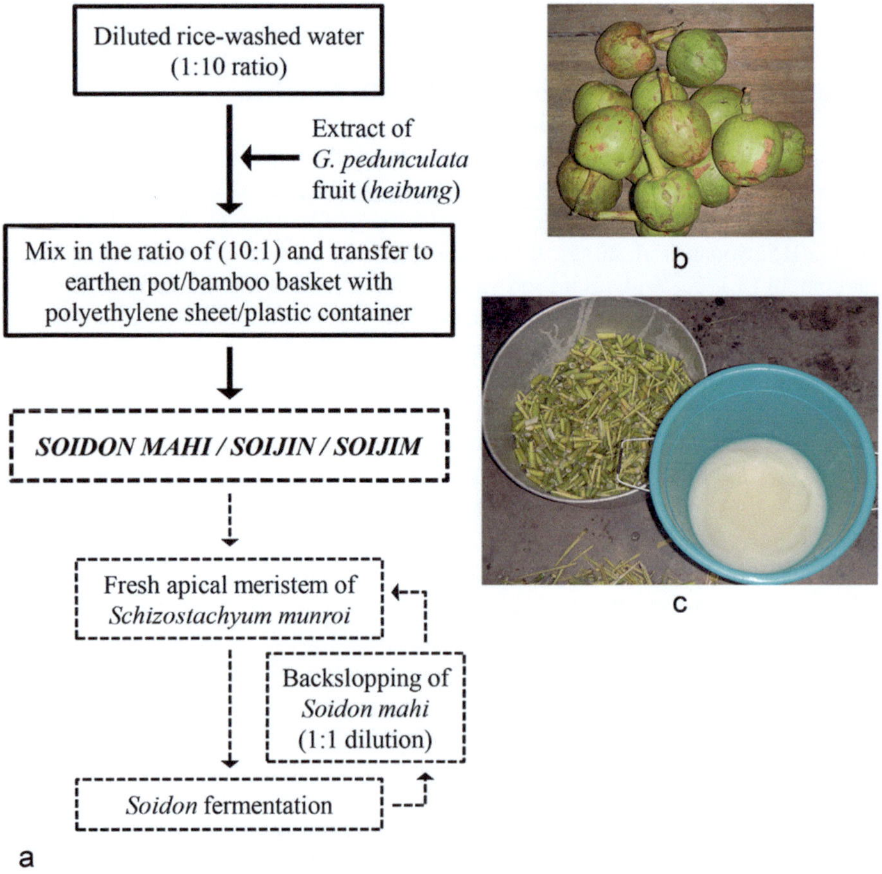

Fig. 14.3 The indigenous traditional process for the preparation of *soidon mahi* in Manipur, India (**a**). The fruits of *G. pedunculata* (**b**), and the fresh apical meristem of *S. munroi* (*naat*) ready for submerging in the *soidon mahi* for its fermentation (**c**)

14.6.2.1 Traditional Preparation Method

For the preparation of a fresh batch of *soidon mahi*, rice-washed water is prepared by washing a local variety of rice with an equal amount of water. The rice-washed water is further diluted at a 1:10 ratio and mixed with the acidic juice extract of *G. pedunculata* fruits (*heibung*) (Fig. 14.3). For every 10–15 L of diluted rice-washed water, the juice extract of 1–1.5 kg of *heibung* is used for the preparation. The preparation is then transferred to an earthen pot (*chaphu*) or a bamboo basket layered with polythene sheet or a plastic container, to which the freshly prepared apical meristems of *S. munroi* (*naat*) are added for the submerged fermentation with intermittent stirring (3–5 times per day) with a bamboo stick for aeration. Once *soidon* fermentation is completed, the milky fermentation liquid developed during this fresh batch of fermentation is kept as the master starter and used by diluting with water at 1:1 ratio for backslopping of the subsequent batches of *soidon* fermentation. The master starter can be kept for a year and used continuously by diluting with water (Jeyaram et al. 2010).

14.6.2.2 Gastronomy, Socioeconomy, and Ethical Values

Soidon mahi can be kept in a bottle and stored at room temperature for a year or more. Mostly, the tribal people of Manipur use the sour acidic liquid as condiment in the preparation of local cuisine of meat, fish, and vegetables. It represents an important cooking ingredient that imparts a special taste highly relished by ethnic tribes. It is similar to the fermented bamboo shoot liquid (*bastengapani*) of Nagaland (Tamang et al. 2008).

14.6.2.3 Microbiology

Soidon mahi is acidic in nature (pH of 4.1–4.5). *Bacillus* spp. and lactic acid bacteria with a culturable population load of 10^5–10^8 cfu/mL dominated the fermentation liquid. Viable yeast and fungi are not detected in *soidon mahi*. The microbial community is dominated *by Bacillus* spp. (*Bacillus subtilis*, *Bacillus cereus*, *Bacillus pumilus*), *Lactobacillus* spp. (*Lb. brevis*, *Lb. plantarum*), *Carnobacterium* sp., *Enterococcus faecium*, *Pseudomonas fluorescens*, *Leuc. fallax*, and *Leuc. lactis*. It is a complex ecosystem driven by highly diverse multi-strain bacterial species belonging to *B. subtilis*, *B. cereus*, and *Lb. brevis*. Next-generation sequencing-based metagenomics study has identified other *Lactobacillus* spp. (uncultured *Lb. acetotolerans*, *Lb. intestinalis*, *Lb. johnsonii*, *Lb. zymae*), *W. ghanensis*, and *Paralactobacillus selangorensis* in *soidon mahi*. However, the presence of alarming population of *B. cereus* load (10^6–10^7 cfu/mL) raises safety concern regarding the consumption of *soidon mahi* (Tamang et al. 2008; Jeyaram et al. 2010; Romi et al. 2015).

14.6.3 *Soibum*

Soibum is one of the most common and sought-after fermented bamboo shoot of Manipur. It is a non-salted, acidic, and long-duration ethnic fermented product produced by the Meitei community. It is prepared exclusively from the young succulent shoot sprouts of *D. hamiltonii* (*wanap/unap*), *D. giganteus* (*maribob*), *B. balcooa* (*leewa/ching saneibi*), *B. tulda* (*utang/watang*), and *M. baccifera* (*moubi wa*). It is consumed as an important constituent of the traditional diet. It is similar to *mesu* of Sikkim (Tamang 2010).

14.6.3.1 Traditional Preparation Method

The preparation of *soibum* starts during the monsoon season from June to September when the young succulent shoots emerge from the soil. In the traditional process, the harvested young shoots are defoliated, chopped into slices, and densely packed into earthen pots or bamboo baskets/chambers or polythene bags for spontaneous natural fermentation. Based on different traditional preparation methods, three types of *soibum* are produced, which are unique to the Meitei population of different geographical regions of Manipur (Jeyaram et al. 2009; Devi and Kumar 2012; Nongdam 2015; Romi 2015).

The preparation of Andro *soibum* is exclusively practiced in Andro village of Imphal East district by submerged fermentation in roasted, leak-proof traditional earthen pots (*andro chaphu*) (Fig. 14.4). The ethnic locals of nearby hilly areas such as Chaningpokpi and Pheitaiching harvest the young succulent bamboo shoot sprouts from the wild, defoliate them, cut in angular position to uniformly thick slices, and bring to the traditional artisan producers in Andro village after wrapping in the leaves of banana plant, *Alocasia macrorrhiza* (locally called *hongu*), etc. After washing using *long* or *polang* (traditional bamboo baskets) and draining excess water by spreading on *yangkok* (traditional bamboo based winnowing tray), the bamboo slices are compactly packed in the earthen pot and kept undisturbed for spontaneous fermentation for 6–12 months at the ambient environment condition. The milky, acidic fermentation liquid (locally called *soibum mahi*) released during fermentation is not drained out. When the volume of the fermenting mass reduces as the fermentation continues, additional fresh slices are added, and this process is repeated until the pot is completely filled.

Kwatha *soibum* is prepared by solid-state fermentation and is practiced in Kwatha and other adjoining villages in the Indo-Burma border of Chandel and Tengnoupal districts of Manipur (Fig. 14.5). The young succulent bamboo shoots are harvested from the nearby hills of Kwatha and Khudengthabi and cut uniformly into thin slices. The shoot slices are densely packed in traditionally designed bamboo chambers having an inner lining of forest leaves such as banana, *Colocasia*, *Alocasia*, etc., or polythene sheet. After filling to its capacity in batches with adequate pressing, the chamber is covered and sealed with polythene sheets, and heavy stones are placed on the top for tight packing. Unlike Andro *soibum*, *soibum mahi* is not retained, and the bottom of the chamber is perforated to facilitate evacuation. After 6–12 months of undisturbed spontaneous fermentation at the ambient environment condition, the seal is opened to remove the fermented product.

The traditional preparation of Kakching *soibum* is similar to that of Kwatha *soibum* but involves two stages. In the first stage, fine bamboo shoot slices are incubated with water in batches in plastic containers for a week, after which they are

Fig. 14.4 The indigenous traditional preparation process of Andro *soibum* in Manipur, India

Fig. 14.5 The indigenous traditional method for the preparation of Kwatha *soibum* in Manipur, India

transferred and compactly packed in airtight plastic bags and kept undisturbed in an inverted position for another 2 weeks. In the second stage, the 3-week-old partially fermented bamboo shoots are tightly packed with adequate pressing in specially designed bamboo and cane basket (width, 3–4 ft; height, 4–5 ft) having an inner lining of polythene sheets with perforations at the bottom designed to drain out *soibum mahi*. After sealing with polythene sheets and heavy stones, the airtight chamber is kept for spontaneous fermentation for 5–6 months. Irrespective of the preparation methods, the longer the incubation, the better is the quality of *soibum*. *Soibum* preserved for 5–6 years is considered to have the best quality and fetch a much higher price than the one fermented for 6–12 months.

14.6.3.2 Gastronomy, Socioeconomy, and Ethical Values

Kwatha and Kaching *soibum* are produced in commercial bulk quantity and commonly sold in the local markets, whereas Andro *soibum* is prepared in small scale for household consumption and religious purposes. Like *soidon*, it is mainly consumed by preparing *eromba*. *Kangou* is prepared by frying with potato, vegetables, and fish. Various kinds of curry commonly known as *athongba* are prepared with vegetables, fish, and meat. *Soibum* is cooked with catfish *Clarias magur* (locally called *ngakra*) to prepare a delicious curry known as *soibum ngakra thongba*. Pork prepared with *soibum* is a favorite dish among the ethnic people of the state. The business of *soibum* production is a flourishing source for income generation for the ethnic communities and the artisan producers. Fresh shoots (locally called *ushoi*) of

Fig. 14.6 *Soibum* being sold in the *Ema Keithel* in Imphal, Manipur, by the local women vendors along with other traditional foods

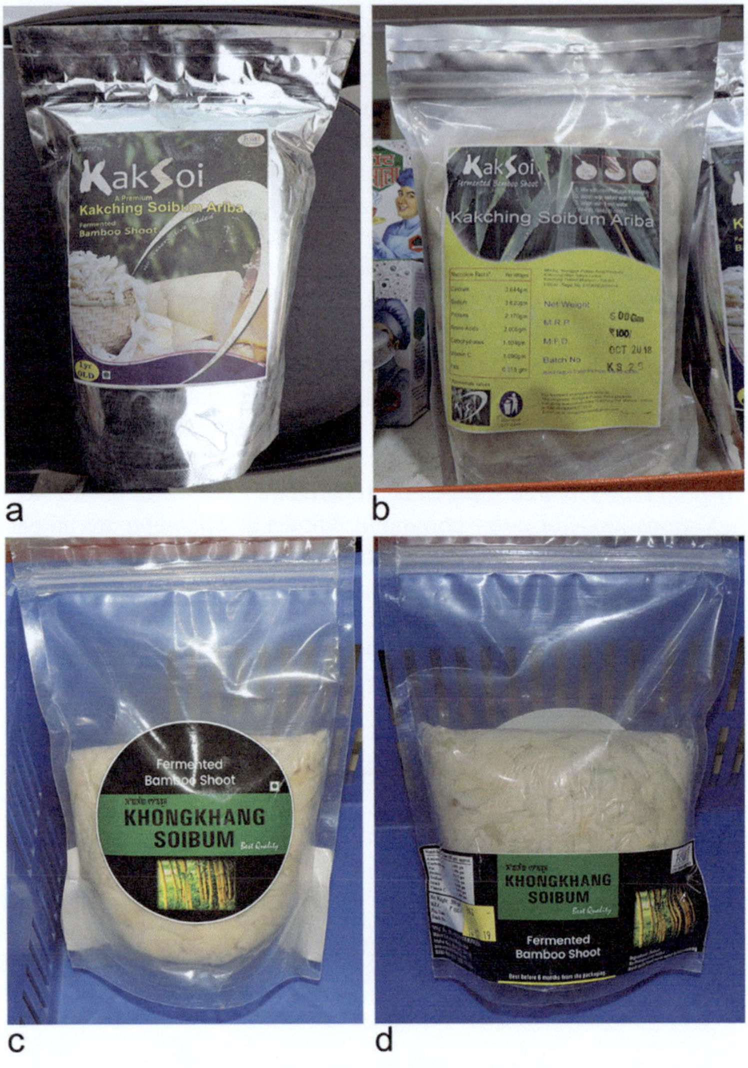

Fig. 14.7 *Soibum* of Manipur marketed in attractive and hygienic packages with nutritional tags

B. tulda are sold at ₹ 10–50 per average-sized piece, while *soibum* costs ₹ 30–50 per kg. It is largely sold in loose form in polythene bags or by wrapping in banana leaves by womenfolk in the local vegetable markets, mainly in Kakching Bazar in Kakching and *Ema Keithel* in Imphal (Fig. 14.6). *Soibum* production has evolved as an organized sector in the processed food industry of the state. Some artisan producers and entrepreneurs are marketing *soibum* with nutritional tags in hygienic and attractive packaging under different brands, which are available in local outlets, departmental stores, grocery shops, and online shopping at the cost of ₹ 100–170 per kg (Fig. 14.7). The fresh shoots and *soibum* are used by the ethnic people in the treatment of various human ailments due to their therapeutic effects. Crushed shoots and juice of *B. tulda* are applied to the injury of nails due to iron sword or arrows and to cure skin problem like dandruff. It is believed that taking a decoction of *D. strictus* shoots aids in parturition. The boiled decoction of *soibum* is used for the treatment of ringworm, tumors, and meningitis. In earlier days, ethnic women used *soibum* along with a fermented fish against the plague disease (Singh et al. 2007, 2010, 2018a).

14.6.3.3 Microbiology

The pH of *soibum* ranges from 3.9 to 4.3. *Soibum* fermentation is dominated by lactic acid bacteria, *Bacillus*, and yeasts and is characterized by population dynamics of bacteria and yeasts ranging from 10^9–10^6 cfu/g to 10^7–$<10^4$ cfu/g, respectively. The final fermented and marketed *soibum* has a lactic acid bacterial load of 10^6–10^7 cfu/g and yeast load of 10^4–10^7 cfu/g, and the *Bacillus* load is below 10^4 cfu/g. The cultivated microbiota associated with *soibum* production can be grouped into four types: lactic acid bacteria (*Lactobacillus* spp. [*Lb. plantarum, Lb. brevis, Lactobacillus hammesii, Lactobacillus paracollinoides, Lactobacillus spicheri, Lb. zymae, Lactobacillus coryniformis, Lactobacillus delbrueckii*], *Leuconostoc* spp. [*Leuc. fallax, Leuc. lactis, Leuconostoc mesenteroides*], *Enterococcus durans, Carnobacterium* sp.), *Bacillus* spp. (*B. subtilis* subsp. *subtilis, Bacillus firmus, Bacillus licheniformis, Bacillus coagulans*), yeasts (*Meyerozyma guilliermondii, Meyerozyma caribbica, Debaryomyces hansenii, Candida zeylanoides, Candida* sp., *Saccharomyces* sp., *Torulopsis* sp.), and other minor bacteria (*Brevibacterium iodinum, Achromobacter xylosoxidans, Elizabethkingia meningoseptica, Microbacterium resistens, Oceanobacillus oncorhynchi, P. fluorescens, Stenotrophomonas maltophilia, Micrococcus* sp.). The traditional fermentations of Andro *soibum* and Kwatha *soibum* have different microbial community composition and structure; however, assemblage of the microbial community is driven by a complex multiphase succession of selective lactic acid bacteria, comprising both cultured and uncultured species under *Lactobacillaceae, Leuconostocaceae*, and *Streptococcaceae*. The uncultured lactic acid bacteria and yeasts in *soibum* detected by metagenomics analyses are *Lactobacillus* spp. (*Lb. acetotolerans, Lactobacillus farciminis, Lactobacillus helveticus, Lactobacillus odoratitofui, Lactobacillus frumenti, Lactobacillus rossiae, Lactobacillus versmoldensis, Lactobacillus similis, Lb. mindensis, Lactobacillus fermentum*), *Lc. lactis, P. pentosaceus, Weissella* spp. (*Weissella beninensis, W. ghanensis, W. confusa*), *Trichosporon laibachii, Geotrichum silvicola, Issatchenkia siamensis, Cryptococcus victoriae Sterigmatomyces halophilus, Candida rugosa, Puccinia triticina, Saccharomyces paradoxus*, and *Mazocraeoides gonialosae* (Giri and Janmejay 2000; Tamang et al. 2008; Sarangthem and Singh 2013; Romi et al. 2014; Romi 2015).

14.6.3.4 Nutritional Composition, Health Benefits, and Functionalities

The nutritional composition of *soibum* comprises carbohydrate (14.6–47.2% of dry weight), dietary fiber (5.92–7.24%), protein (23.6–36.3%), fat (2.6–3.2%), ash (6.6–13.3%), and minerals (Ca, 0.16 mg/g; Na, 0.03–1.75 mg/g; K, 2.12–13.45 mg/g;

Mg, 2.53 mg/g) with a food value of 362.8 kcal/100 g (Tamang et al. 2012; Premlata et al. 2015; Sonar et al. 2015). All the nutrient components except dietary fiber are decreased in *soibum* as compared to the freshly harvested young shoots (Nirmala et al. 2008). Consumption of such fiber-rich food decreases body fat and serum and hepatic lipids, especially cholesterol, and increases peristaltic movement of the intestines thereby enhancing digestion and preventing constipation. Trans-fatty acid is absent in *soibum* indicating its health value (Singh et al. 2011). Alkanes, alkenes, aldehydes, esters, ketones, alcohols, and phenolic aromatic compounds are the volatile compounds detected in *soibum* (Sonar et al. 2015). It is rich in phytosterols (3047.30 μg/g) and polyphenols (total phenol, 854.7–9131.10 μg/g; total flavonoids, 408.16 μg/g) that correlate with the high antioxidant properties (DPPH radical scavenging activity: >90%) (Premlata et al. 2015; Sonar et al. 2015). *Soibum* is a good source of water-soluble vitamins (ascorbic acid, 170.17–174.70 mg/g; riboflavin, 11.23–13.55 mg/g; folic acid, 3.25 mg/g; cyanocobalamin, 0.31 mg/g) (Singh et al. 2011; Sonar et al. 2015). The functional value of *soibum* is attributed to the lactic acid bacteria mainly *Lactobacillus* spp. that are reported with probiotic properties, high riboflavin production, hydrolytic enzyme production, anti-nutritive compound-degrading properties, protease activities, and broad-spectrum antibacterial activities against common food-borne pathogens. *Soibum* fermentation drastically reduces toxic and anti-nutritive factors (cyanogenic glycoside, phytate, and saponin) in edible bamboo shoots enhancing its safety, nutritional quality, and health benefits (Sarangthem and Singh 2013; Sonar and Halami 2014; Romi 2015; Sonar et al. 2015; Thakur and Tomar 2015, 2016).

14.6.4 *Thunkheng* and *Thunbin*

Thunkheng and *thunbin* are minor ethnic fermented bamboo shoots of Manipur, which is prepared and consumed by the Zeliangrong and Kuki tribes of Tamenglong and Noney hill districts. *Thunkheng* is the fermented whole succulent bamboo shoot, while *thunbin* is the fermented bamboo shoot slices. They are exclusively from *D. hamiltonii*, *Dendrocalamus sikkimensis*, *D. giganteus*, *Melocanna bambusoides*, *B. tulda*, and *B. balcooa*. The traditional preparation for the fermentation of bamboo shoots (locally called *thun*) is carried out during May to June when new shoots sprout during the monsoon. The process is similar to that of Kwatha *soibum* production but is unique in the sense that the fermentation is carried out inside a pit in the soil. This preparation method is commonly known as Noney-type fermentation. A bamboo basket (locally called *khoupak*) in the shape of the pit is made inside it, and the inner wall is lined with banana or wild *Colocasia* leaves to keep airtight during fermentation. A hole is made at the bottom for draining the fermentation liquid. The harvested whole tender bamboo shoots are defoliated, cleaned, and packed inside the basket for *thunkheng* preparation, while longitudinally sliced, thin small pieces of the shoots are used for the preparation of *thunbin*. The upper portion of the basket is then covered with banana leaves, and stones are put on it as a weight to initiate the natural spontaneous fermentation. Though the fermentation is carried

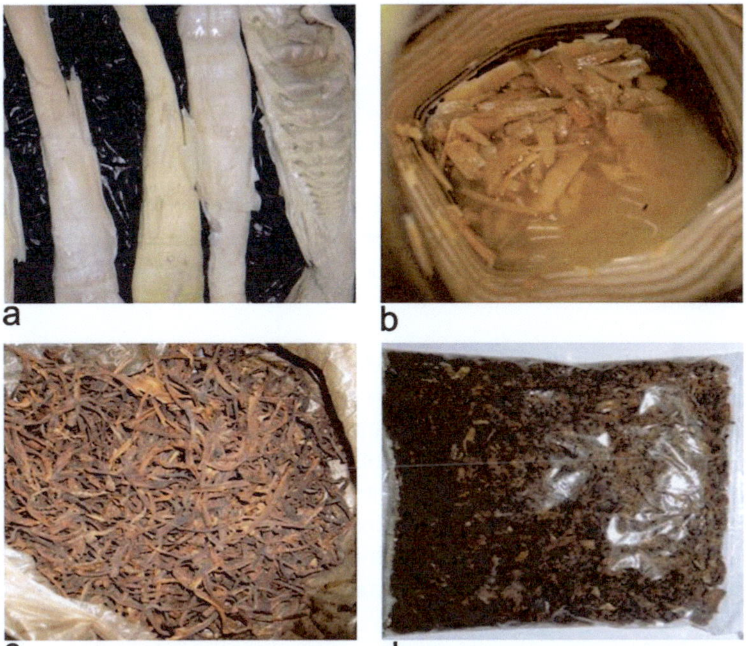

Fig. 14.8 Minor fermented bamboo shoots, *thunkheng* (**a**), *thunbin* (**b**), *thunkhengkang* (**c**), and *thunbinkang* (**d**) of Naga and Kuki tribes of Manipur. (Reproduced from Singh et al. 2018c with permission from Excellent Publishers)

out for months, *thunkheng* and *thunbin* are sold and used for making curry within 5–10 days of fermentation because of the preference of high degree of sourness by the local tribes. For longer storage, *thunkheng* and *thunbin* are sun-dried to produce *thunkhengkang* and *thunbinkang*, respectively (Fig. 14.8). These dried fermented shoots are kept in a basket to use in the off-season. The fermented bamboo shoots are consumed as traditional delicacies by these tribes after boiling with pork, meat, and vegetables. *Thunkhengkang* is graded according to size for preparing different dishes. Microbiology, nutritional composition, and potential health benefits of these products are unknown though lactic acid bacteria are likely to be the functional microorganism (Jeyaram et al. 2009; Singh et al. 2018c). *Thunkheng* and *thunbin* are similar to *ekung* of Nyishing tribes and *hirring* of Apatani tribes of Arunachal Pradesh, respectively.

14.7 Ethnic Fermented Leafy Vegetables

In Manipur, the ethnic people of the Naga and Kuki-Chin-Mizo communities prepare and consume fermented leafy vegetables, mainly mustard greens, which form an integral part of their traditional diets. Depending on the nature of the final fermented product, they are popularly known by various names such as *inziang-sang*, *ziangdui*, *ziangsang*, *ankamthu*, and *ganang tamdui*.

14.7.1 *Inziangsang, Ziangdui,* and *Ziangsang*

Inziangsang, *ziangdui*, and *ziangsang* are different forms of sour, non-salted ethnic fermented mustard (*Brassica* sp.) leaves of the Naga communities of Manipur. Mainly the womenfolk produce them. It is similar to *gundruk* of Nepali communities of Sikkim.

14.7.1.1 Traditional Preparation Method

Inziangsang, *ziangdui*, and *ziangsang* are prepared during the winter season when mustards are grown and available in surplus amount. Withered mustard leaves (locally known as *hanggam*) (2–3 days old) are crushed using traditional wooden mortar and pestle and soaked in warm water (Fig. 14.9). After squeezing to remove excess water, the crushed leaves are packed in an airtight container and kept undisturbed in the ambient environment condition for 7–10 days semisolid spontaneous fermentation. The fermented mustard leaves are called *inziangsang*. The fermented liquid is extracted from *inziangsang* by squeezing with hands to get *ziangdui*. *Ziangdui* is further concentrated by boiling to produce a condensed paste known as *ziangsang*. For future consumption and long-term storage (> 1 year), *inziangsang* (freshly prepared as well as after *ziangdui* extraction) is sun-dried for 4–5 days and stored at room temperature, while *ziangsang* is stored in the traditional bamboo container for more than a year (Jeyaram et al. 2009; Tamang and Tamang 2009).

14.7.1.2 Gastronomy, Socioeconomy, and Ethical Values

Inziangsang is consumed as a soup. *Ziangdui* and *ziangsang* are used as a condiment for flavoring special local dishes and ethnic soup, which are consumed with local boiled rice. It is sold in local tribal vegetable markets of Manipur, mainly in Naga inhabited hill areas.

14.7.1.3 Microbiology

The pH of *inziangsang* ranges from 4.6 to 4.9 and is dominated by lactic acid bacteria (population load: 10^7 cfu/g) with no presence of yeasts and molds. The culturable lactic acid bacteria associated with *inziangsang* are *Lactobacillus* spp. (*Lb. plantarum*, *Lb. brevis*) and *Pediococcus acidilactici* (Tamang et al. 2005).

14.7.1.4 Nutritional Composition, Health Benefits, and Functionalities

The nutritional value of *inziangsang* comes from its protein (38.7% of dry weight), fat (3.2%), carbohydrate (41.2%), ash (16.9%), minerals (Ca, 2.40 mg/g; Na, 1.34 mg/g; K, 6.58 mg/g), and moisture (17.6%) contents (Tamang et al. 2012). It has a food value of 348.4 kcal/100 g. The lactic acid bacteria associated with *inziangsang* contribute to its functional value as these microbes have probiotic

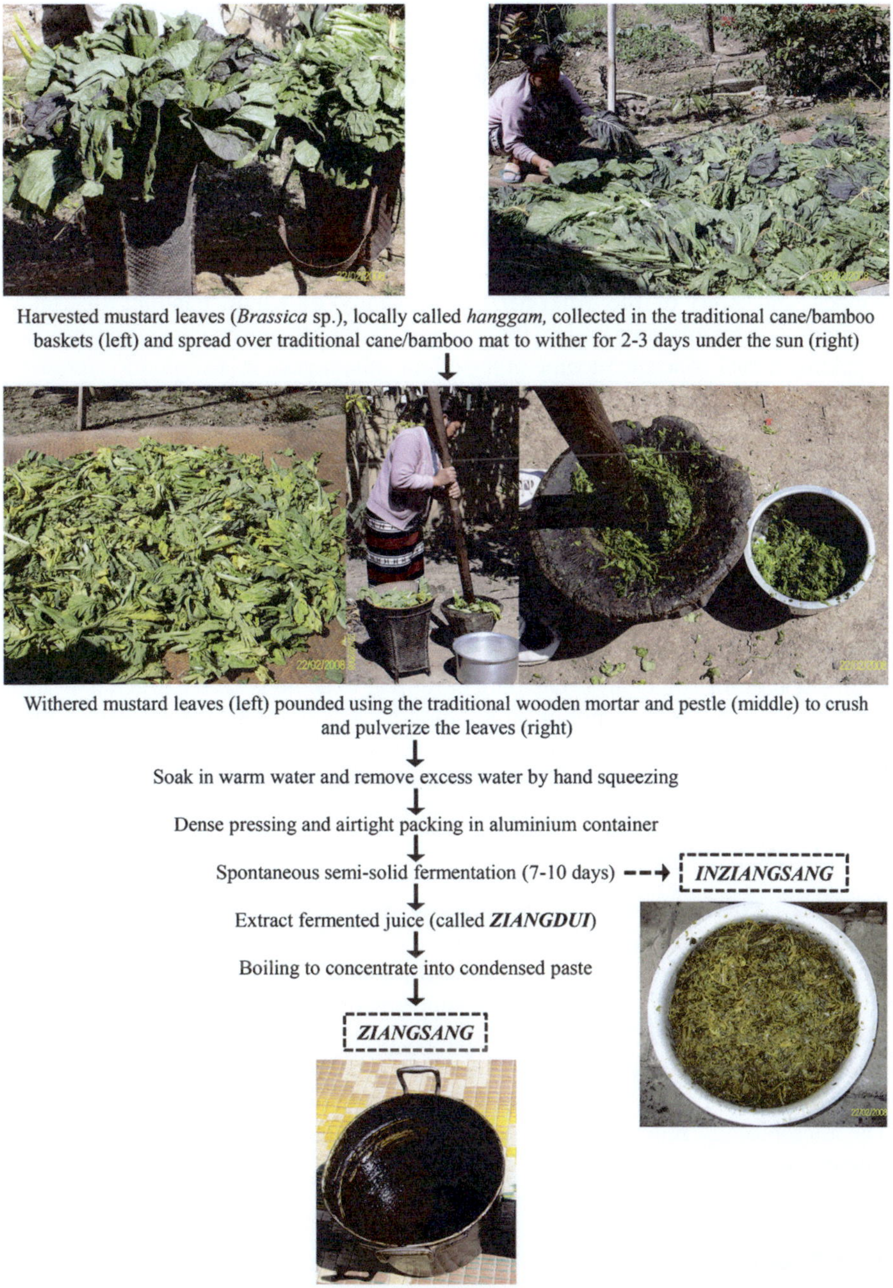

Fig. 14.9 The traditional method for the preparation of *inziangsang*, *ziangdui*, and *ziangsang* in Manipur

properties (hydrophobicity, adherence to intestinal cells), strong acidification properties, degrade anti-nutritive compounds and oligosaccharides, and produce bacteriocin against the food-borne pathogen (Tamang et al. 2009).

14.7.2 *Ankamthu*

Ankamthu is the ethnic fermented mustard green extract prepared and consumed by the tribes (Chiru, Chin, Gangte, Hmar, Paite, Kom, Mate, Lushai, Simte, Thadou, Vaphei, and Zou) of the Kuki-Mizo community of Churachandpur district of Manipur. It is mainly prepared by the womenfolk for off-season consumption and to supplement the economic income of the ethnic inhabitants. It is similar to *ziangsang* of Naga communities of Manipur (Hoikhokim 2017).

14.7.2.1 Traditional Preparation Method

Ankamthu is prepared from the extract of mustard green leaves (mainly *Brassica juncea*, *Brassica campestris*) (Fig. 14.10). In the generations-old preparation process, a bulk amount of mustard green leaves are plucked, washed in tap water, and wilted under the sun by spreading over a traditional bamboo winnowing tray (locally known as *pheh*). After crushing the wilted leaves by pounding using a traditional wooden mortar and pestle, the mesh is transferred to an airtight bamboo basket with a lid and kept for some days. The dark color leaf exudate is then extracted by hand squeezing and boiled in a pot or an aluminum container until the extract becomes condensed. The condensed extract is then filled in batches inside hollow bamboo internodes with one end sealed with the nodes. After sealing the other ends with banana leaves, the bamboo containers are kept near traditional kitchen fire or under the sun for 3–5 days for spontaneous fermentation.

14.7.2.2 Gastronomy, Socioeconomy, and Ethical Values

Ankamthu is a traditional delicacy popular among all tribal communities in the hill district of Churachandpur. It serves as an appetizer. It is consumed on a regular basis either directly without other vegetable supplements or as a salad with green chili, onion, and salt. *Ankamthu* has a good market value within the local inhabitants and contributes substantially in the family income as it is sold in huge amount in different local markets of all hill districts of Manipur. It costs ₹ 100–200 per internode. The microbiology, nutritional values, and health benefits of *ankamthu* have not been studied yet.

14.7.3 *Ganang Tamdui*

Ganang tamdui is the fermented mustard leaf extract of the Zeliangrong and Kuki tribes of Tamenglong and Noney hill districts of Manipur. It is similar to *ankamthu* of Kuki-Mizo community. In the traditional preparation process, mustard leaves harvested from jhum field are wilted until the leaves turn yellow. The leaves are

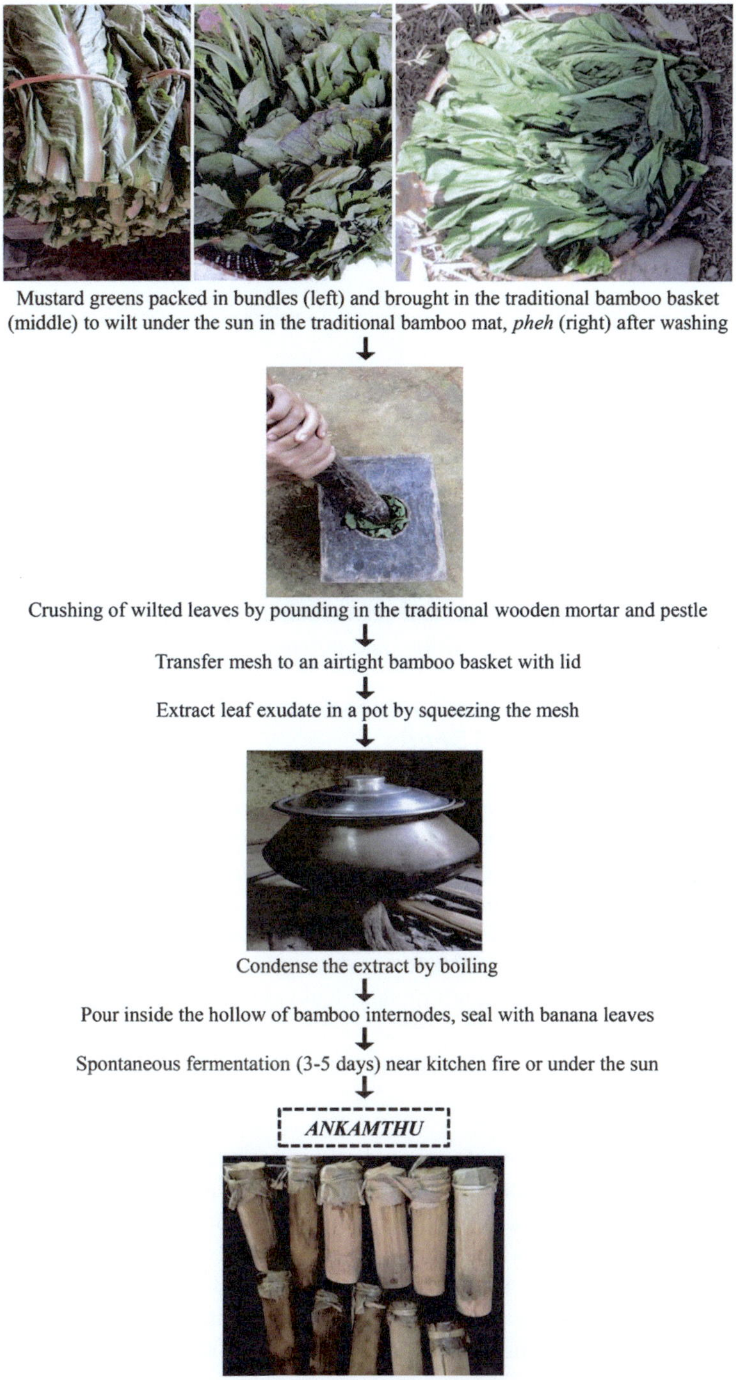

Fig. 14.10 The traditional method for the preparation of *ankamthu* of Kuki-Mizo community of Manipur. (Photographs reproduced from Hoikhokim 2017 with permission from MG Aricent Pvt. Ltd.)

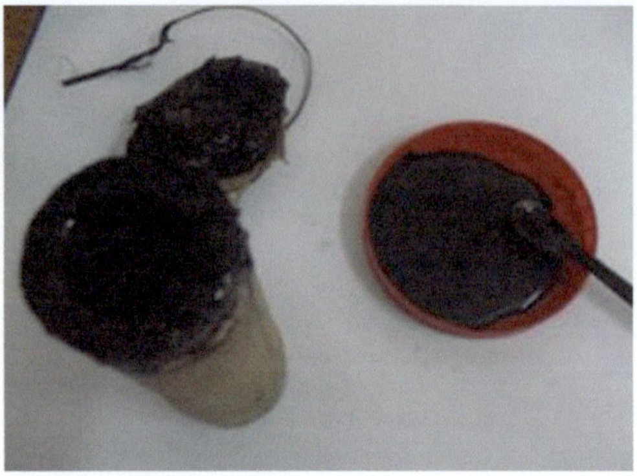

Fig. 14.11 The fermented mustard leaf extract, *ganang tamdui*, of Zeliangrong and Kuki tribes of Manipur. (Reproduced from Singh et al. 2018c with permission from Excellent Publishers)

pressed inside a bamboo culm, which is then sealed with banana leaves and mud. The packed culms are kept undisturbed until a pungent smell is released. The fermented leaves are removed and squeezed with both hands to extract the dark color liquid. The extract is concentrated by boiling in a pot to form a condensed liquid known as *ganang tamdui* (Fig. 14.11). It can be preserved in a tightly closed container up to a year. *Ganang tamdui* is mainly used in the preparation of *tam*, an ethnic soup. It is used as a taste and flavor enhancer in various local dishes. Its microbiology, nutritional composition, and potential health benefits are unknown.

14.8 Ethnic Fermented Seeds

Almost all ethnic people in Manipur produce and consume some types of fermented seed products. The most important among fermented seed product is the fermented soybean. It is locally known as *hawaijar* in Meitei, *khuichang* in Zeliangrong, *bethu* in Thadou and Kuki, *bekanthu* in Hmar, and *theisui* in Tangkhul. Other fermented seed products include *sithu* (fermented sesame seed), *hangammaru hawaichar* (fermented mustard/rapeseed seed), and *gankhiangkhui* (fermented kenaf seed) of Kuki-Chin-Mizo, Meitei, Zeliangrong, and Kuki communities, respectively.

14.8.1 *Hawaijar*

Hawaijar is a sticky, naturally fermented ethnic soybean food of Meitei people in Manipur. It is alkaline and slightly ammoniacal making it slightly pungent. It is non-salted, mucilaginous, and sticky in nature and has a unique flavor, but not very soft in texture. The color of *hawaijar* ranges from light gray-yellow to tan-yellow. It is produced in small quantity in household level and sold in local market by woman vendors. It serves as a cheap source of quality protein supplement to the ethnic people from the ancient times (Jeyaram et al. 2008b, 2009).

Clean and sort soybean

↓

Soak in water overnight (8-10 h)

↓

Boil and cook the soybeans in a pot or pressure cooker until soft (2-3 h)

↓

Wash with hot water, drain excess water

↓

Wrap with *F. hispida* leaves and loosely pack in bamboo basket (*lubak*)

↓

Wrap with gunny bags/old clean cloths

↓

Incubate near fireplace or buried under paddy

↓

Solid-state, spontaneous natural fermentation (2-7 days)

↓

HAWAIJAR

(Mucilaginous, sticky with characteristic ammoniacal smell)

Fig. 14.12 The traditional method for *hawaijar* preparation in Manipur

14.8.1.1 Traditional Preparation Method

Hawaijar is produced and consumed primarily by the Meitei community of Manipur residing in the central valley areas. Traditionally, *hawaijar* is produced by ladies of Meitei Brahmin families; all Meitei ladies know the art of production of *hawaijar*, and they usually prepare it for their household consumption from time to time.

During the traditional production of *hawaijar*, the soybean seeds (*Glycine max* L.) (locally known as *nunghawai*) are washed and soaked in excess water overnight (usually 8–10 h) (Fig. 14.12). Next day, soaked soybeans are properly cooked in a pot or pressure cooker, which is checked by crushing the seeds between fingers and is also indicated by a characteristic aroma released from the broth after boiling several hours (2–3 h). The cooked seeds are washed using boiled water and allowed to drain excess water. The drained seeds are wrapped with the leaves of *Ficus hispida* (locally called *asiheibong*) while they are still warm and kept in a traditional bamboo basket called *lubak*. It is then wrapped with gunny bags or old cloths and placed undisturbed near the fireplace or buried in paddy straw or husk for spontaneous natural fermentation. It takes 2–3 days during summer and 5–7 days during winter for the fermentation to complete. The release of characteristic ammoniacal, pungent odor with sticky mucilage indicated the proper fermentation of *hawaijar*. *Hawaijar* is then packed in banana leaves in small quantities for sales in the local market. A medium to small seed variety of soybean is preferred for *hawaijar* fermentation over large seed variety. *Hawaijar* has a very less shelf life (2–3 days), but it can be stored for about a week in a refrigerator without changing its taste and other characters. For longer storage, it is dried for 2–3 days in sunlight by spreading on a bamboo mat or in the fireplace. The dried *hawaijar* can be stored for several months to even 1 year. Similar products are prepared in hill areas by different ethnic tribal communities. In Tamenglong and Noney districts of the western hills of Manipur, the Zeliangrong people produce fermented soybean known as *khuichang* (Fig. 14.13), which is very similar with *hawaijar* in production and consumption except that the incubation is slightly longer (3–5 days during summer and 7–9 days during winter). In the eastern hills of Manipur, the Tangkhul tribes produce fermented soybean called *theisui*, which is similar to *hawaijar*, but ash is added during incubation like the *kinema* of Sikkim and Darjeeling. In the southwestern hills of Churachandpur district, the Hmar and Kuki ethnic tribes prepare fermented soybean

a b

Fig. 14.13 Other fermented soybean products, *khuichang* of the Zeliangrong community in Tamenglong and Noney districts (**a**) and *bekanthu* of the Hmar community in Churachandpur district (**b**) of Manipur

known as *bekanthu* (Fig. 14.13) and *bethu*, respectively (Jeyaram et al. 2008b, 2009; Singh et al. 2018c).

14.8.1.2 Gastronomy, Socioeconomy, and Ethical Values

Hawaijar is the favorite food item of the Meitei community in particular, but all ethnic communities of Manipur, in general, take the product regularly. It is consumed raw by mixing with green chili or *umorok* (king chilli), salt, and onion as side dish along with other traditional dishes and steamed rice. An exotic dish called *chagempomba* is prepared with *hawaijar*, rice grains, green vegetables, and fish and is a price delicacy in Manipur. Due to Sanskritization, some people change to vegetarianism or avoid non-vegetarian items on certain days during which traditional dish items of *hawaijar* are being used as a substitute of fermented fish (*ngari*), which is ubiquitous in all Manipuri dishes (Devi and Kumar 2012). During *utsav* or traditional feast for offering to deities where fish along with other meats are prohibited and use of only vegetables are allowed, *hawaijar* items are usually incorporated. Thus, regular production and selling of *hawaijar* is a source of additional income to the rural women in Manipur. A woman can easily earn up to ₹ 20,000/month by producing and selling 2–3 kg of *hawaijar* daily without much investment. These immensely help the socioeconomic position of rural households. In the hill areas, the fermented soybean is cooked with pork and meat during festivals. They make a popular dish by cooking the fermented soybean, pork meat, rice grain, and other vegetables, which is similar to *chagempomba*. The ethnic tribal people consume boiled foods in their traditional diet and use very less or no oil in cooking. Fermented soybean along with other fermented fish or fermented meat products is used as taste enhancer or condiments.

14.8.1.3 Microbiology

The pH of *hawaijar* is alkaline ranging from 8.0 to 8.2, which normally supports *Bacillus* growth. *B. subtilis* is the most abundant culturable bacteria present in *hawaijar*. Other predominant culturable bacteria are *Bacillus* spp. (*B. cereus*, *B. licheniformis*, *Bacillus amyloliquefaciens*, *Bacillus tequilensis*), *Staphylococcus* spp. (*Staphylococcus aureus*, *Staphylococcus sciuri*, *Staphylococcus cohnii* subsp. *cohnii*, *Staphylococcus equorum* subsp. *linens*), lactic acid bacteria (*Lc. lactis* subsp. *lactis*, *Weissella thailandensis*, *E. durans*, *Proteus mirabilis*), *Alcaligenes* spp., *Providencia rettgeri*, and *Clostridium* spp. (*Clostridium botulinum*, *Clostridium perfringens*) (Jeyaram et al. 2008b; Singh et al. 2014; Thokchom and Joshi 2015; Keisam et al. 2019). It has an aerobic mesophilic bacterial load of 10^7–10^9 cfu/g with lactic acid bacteria load of 10^6–10^8 cfu/g. The high prevalence of *B. cereus*, *Staph. aureus*, *P. mirabilis*, *C. botulinum*, and *C. perfringens* in traditionally produced and marketed *hawaijar* raises concerns on the microbial safety of *hawaijar* consumption.

14.8.1.4 Nutritional Composition, Health Benefits, and Functionalities

Hawaijar is a good source protein (43.9% of dry weight) with food value of 521.2 kcal/100 g and carbohydrate and fat content of 23.4% and 27.9%, respectively (Tamang et al. 2012). Minerals such as calcium and potassium are present in significant amount, whereas sodium is very less. Fermentation increases the crude protein contents in *hawaijar*, whereas the carbohydrate content is decreased. All the essential amino acids are present in the *hawaijar*. Moreover, the presence of polyglutamate enhances the taste of foods when it is cooked with *hawaijar*. Some of the reported health-promoting effects of fermented soybean includes fibrinolytic and thrombolytic activities, antiviral activity, antioxidant activity, antidiabetic activity, angiotensin-I-converting enzyme (ACE) inhibiting activity, anti-α-glucosidase activity, acetyl-choline esterase (AChE) inhibiting activity, etc., indicating it to be a health food for cardiovascular diseases, hypertension, glucose metabolisms, and Alzheimer's disease (Singh et al. 2014; Rai and Jeyaram, 2015). Among the protein-rich ethnic fermented foods (fermented soybean and fermented fish products), *hawaijar* has the highest specificity of fibrinolytic activity toward human fibrin, and the fibrinolytic activity is of microbial origin primarily contributed by *B. subtilis* and *B. amyloliquefaciens* (Singh et al. 2014).

14.8.2 *Gankhiangkhui*

Gankhiangkhui is an alkaline ethnic fermented kenaf seed (*H. cannabinus*), locally known as *shougri/gankhiang*, prepared and consumed by the Zeliangrong and Kuki tribes of Tamenglong district of Manipur. It is consumed as a flavor enhancer and protein supplement.

14.8.2.1 Traditional Preparation Method

The mature seeds are harvested during August after *H. cannabinus* plant no longer gives edible leaves. The seeds are harvested by trashing the plants and then sun-dried and kept in tin containers for storage until it is required to prepare *gankhiangkhui*. The process of preparation of *gankhiangkhui* is similar to the preparation of *hawaijar*. The seeds were first cleaned, sorted, and soaked overnight (Fig. 14.14). In the morning the seeds were boiled for about 3–4 hours until they are soft and easily crushed between fingers. The cooked seeds are drained, crushed lightly, packed in an earthen pot, and allowed to ferment (primary) for 3–4 days during which the texture of seeds changes and the color of the seeds are darkened. Ashes obtained after burning rice straw are sometimes added during the packing. The fermented seeds are then taken out and wrapped in banana leaves in smaller packs and kept in the fireplace for 2–3 days until the peculiar odor is released from the incubated products. Thus, fermented *shougri* seeds are ready to eat (Singh et al. 2018c).

Clean kenaf (*H. cannabinus*) seeds, soak in water for 8 h

↓

Drain out the water and boil for 3-4 h or until cooked

↓

Drain out the water using bamboo basket, crush the seeds lightly

↓

Add ash from rice straw burning, pack in an earthern pot

↓

Incubate for 3-4 days (solid-state, spontaneous natural fermentation)

↓

Remove and wrap with banana leaves and cover with gunny cloths

↓

Keep in sunlight or near fire place for 2-3 days until peculiar odor develops
(Solid-state, spontaneous natural fermentation)

↓

GANKHIANGKHUI

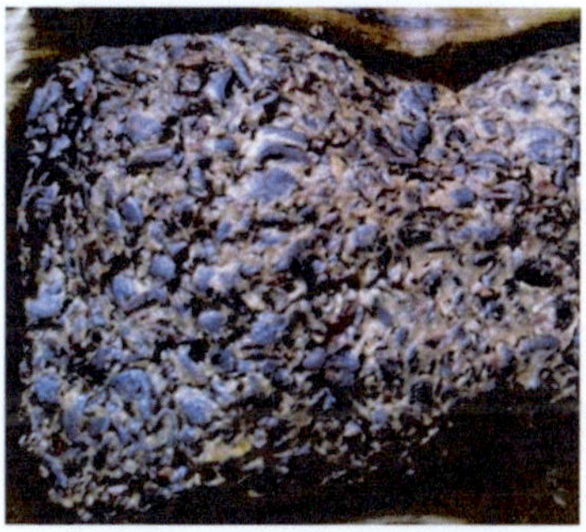

Fig. 14.14 The traditional method for *gankhiangkhui* preparation in Manipur. (Photograph reproduced from Singh et al. 2018c with permission from Excellent Publishers)

14.8.2.2 Gastronomy, Socioeconomy, and Ethical Values

Gankhiangkhui is used in the preparation of local delicacies known as *tam* (Singh et al. 2018c). It is also used as side dish by mixing with salt and chilies. It serves as a taste enhancer in many local dishes, particularly boiled vegetables. For prolonged storage, the product is added with salt and packed in the bamboo culm, which is then closed tightly by banana leaves then by mud. This is usually kept in the traditional kitchen and stored up to 6 months. The local also claimed that products stored for a longer period taste better, and the smell is also softer. The microbiology, nutritional composition, and potential health benefits of *gankhiangkhui* are unknown.

Clean the seeds of *sith* (*S. indicum*)

↓

Roast until characteristic smell is released

↓

Pounded with the addition of some water in a traditional wooden crusher

↓

Sticky thick colloid

↓

Put in an airtight container, keep near fire place

↓

Spontaneous natural fermentation for 4-5 days

↓

SITHU

Fig. 14.15 The traditional method for *sithu* preparation in Manipur

14.8.3 *Sithu*

Sithu is a traditional fermented product made from the seeds of sesame plant (*Sesamum indicum* L.), locally known as *sith*. It is produced and consumed by the Kuki-Chin-Mizo ethnic communities of Churachandpur district of Manipur. It is produced in the household level and regularly consumed as a food item. The product is prepared and kept in every household throughout the year for its addition in almost every vegetable item to make them tenderer and impart a characteristic flavor to the dishes.

14.8.3.1 Traditional Preparation Method

The seeds of *sith* (preferably black variety) are cleaned by removing dirt and unwanted materials after spreading in a bamboo matt (Fig. 14.15). It is roasted until it produces a distinctive smell, which indicates the proper roasting of the seeds. Then, it is pounded by a wooden crusher with the addition of some water to make a thick sticky colloid. This is then transferred to an airtight container and placed near the fireplace for 4–5 days to undergo fermentation.

14.8.3.2 Gastronomy, Socioeconomy, and Ethical Values

Sithu may be taken as raw (uncooked) as a side dish or used as flavoring agent in vegetable salad and vegetable cooked items. The microbiology, nutritional composition, and potential health benefits of *sithu* are unknown.

14.8.4 *Hangammaru Hawaichar*

Hangammaru hawaichar is a fermented mustard/rapeseed seed product, mostly prepared and consumed by the Meitei community belonging to *Chakpa* clan. It is strongly pungent and produces a distinctive taste. This product is produced during

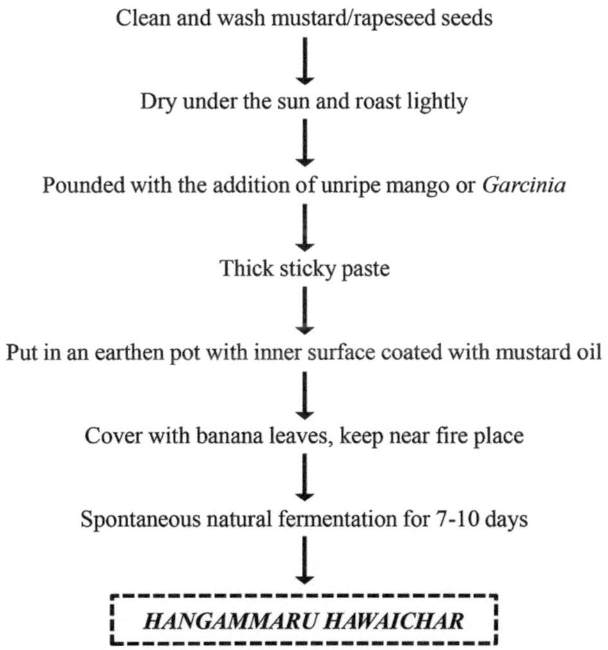

Fig. 14.16 The traditional method for the preparation of *hangammaru hawaichar* in Manipur

the months of March–April when the rapeseeds are harvested. The traditional method of preparation is similar to that of *kahudi* of Assam. The seeds locally known as *hangammaru* are cleaned, washed, and sun-dried (Fig. 14.16). It is then roasted lightly on a pan before pounding with the addition of unripe mango or *Garcinia*. The mixture forms a thick paste, which is transferred in an earthen pot whose inner part is coated with mustard oil. It is covered with banana leaves and allowed to ferment for a week or up to 10 days. This product does not require cooking, and it is consumed directly from the storing container. It is usually taken as a side dish along with other dishes. The microbiology, nutritional composition, and potential health benefits of *hangammaru hawaichar* are unknown.

14.9 Ethnic Fermented Fish

14.9.1 *Ngari*

Ngari is a non-salted fermented fish product intrinsically bound with the diet of the Meitei. It is part of the traditional diet that forms an important part of the cultural identity of Meitei community. It is prepared in the valley region by the Meitei community but relished by almost all communities of Manipur. *Ngari* is traditionally prepared from sun-dried *P. sophore* (locally *phabou*) caught from streams and rivers and fermented in traditional pots. *Utong-ngari* is a similar type of fermented fish, which is prepared in bamboo trunks instead of earthen pot, but this method is rarely used. *Ngari* is generally prepared during October to January as the raw materials are abundantly available during this period. Due to the non-availability of *phabou* in Manipur for large-scale production, it is imported from Assam, Bangladesh, and Gujarat (Jeyaram et al. 2009; Tamang 2010; Das and Deka 2012; Soibam and Ayam 2018).

14.9.1.1 Traditional Preparation Method

The traditional method of *ngari* preparation and its associated indigenous knowledge passed through different generations are properly documented. Briefly, the sun-dried whole fish is washed in a porous bamboo basket, and the water is allowed to drain off for 12–24 h (Fig. 14.17). Next, the washed fish is spread on a polythene sheet or gunny bags and pressed by legs (which are already covered with gunny bags) or mechanical rollers to remove excess water and also to break the heads and bones of the fishes. The pressed fish is then packed inside the *kharung* (earthen pot) in batches of about 5 kg until it is completely filled. *Kharung* is the traditional round-bottom and narrow-neck earthen pots with 45–50 kg capacity used in *ngari* preparation. Before packing, the inner surface of the pot is coated with mustard oil in order to establish anaerobic environment inside the chamber, and it also prevents fish from sticking to its surface. Generally, pre-used *kharungs* are preferred for *ngari* preparation as they need to be oil-coated only once before fermentation, while the new pots have to be coated at least 8–10 times intermittently for efficient fermentation. This packing is done by skilled artisans using legs or wooden rods.

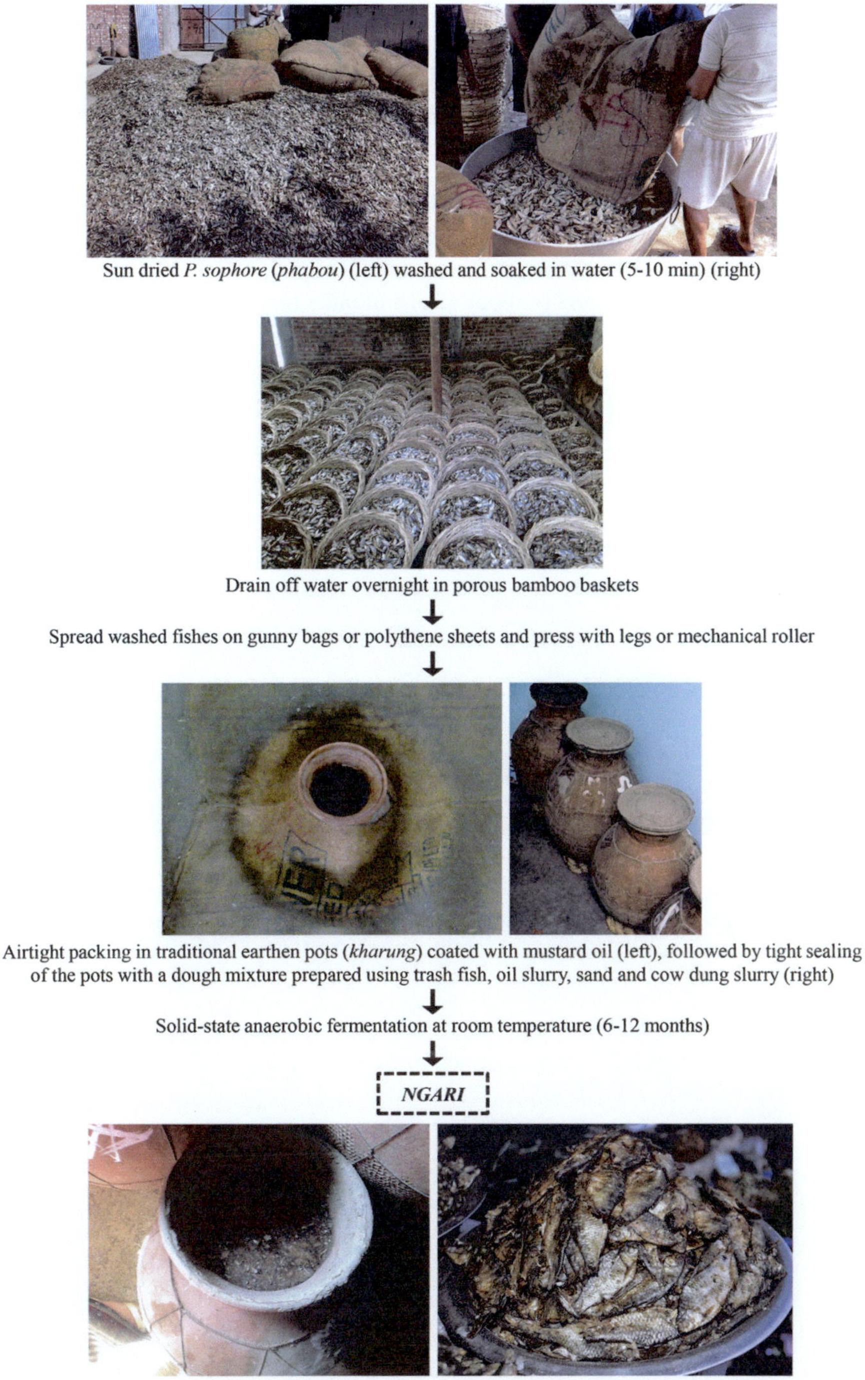

Fig. 14.17 The traditional method of *ngari* preparation in Manipur

During packing, the pots are kept in pits lined with moist gunny bags or straw so that the pots remain stable and withstand the pressure of packing. The pots are then finally sealed with a dough mixture prepared by mixing trash fish, oil slurry, sand, and cow dung slurry. When the sealed layer becomes dry and starts cracking after 3–4 days, the slurry mixture is applied again to maintain anaerobic conditions. These sealed pots are stacked together and incubated in room temperature for 6–12 months. After the fermentation is completed, the sealing layer is removed, and the top layer of the fish known as *phumai* is taken out. *Phumai* is considered to be of low quality and sold with less price compared to *ngari*, which sells for ₹ 500–600 per kg. The shelf life of *ngari* is about 12–18 months (Jeyaram et al. 2009; Tamang

Fig. 14.18 The two best-quality *ngari* of Manipur—Singjamei (**a**, **b**) and Sengmai (**c**, **d**), marketed in attractive and hygienic packages with nutritional tags

2010; Devi and Kumar 2012; Das and Deka 2012; Keishing and Banu 2015; Thapa 2016; Soibam and Ayam 2018).

14.9.1.2 Gastronomy, Socioeconomy, and Ethical Values

As much as *ngari* is ubiquitous in the daily diet of the Meitei community, it is found sold ubiquitously in all major markets of Manipur, including departmental stores and street-corner grocery shops. Traditionally, *ngari* is sold in loose form from the production pots, but nowadays many small-scale industries have started producing *ngari* with attractive industrial packaging making it attractive and hygienic for the consumers (Fig. 14.18). It is a source of income for the producers, shops, and artisans and acts as a common thread that binds the social fabric of the different communities residing in Manipur. *Ngari*, with its appetizing taste and characteristic aroma is an integral part of the daily diet. The roasted *ngari* mixed with salt is consumed as an appetizing side dish. The most common mode of consuming *ngari* is by preparing local delicacies like *eromba*, *singju*, *kangsoi*, *morokmetpa*, *paknam*, etc. *Ngari* is also believed to have therapeutic value as drinking *ngari* homogenate prepared with water in an empty stomach is believed to cure stomach ulcers. As part of an age-old custom, *ngari* along with other items is offered to evil spirits to ward off their evil eye. This traditional ritual known as *Saroi Khaangba* is performed by the elderly womenfolk of the Meitei community in the lunar month of *Lamta* (March–April) (Tamang 2010; Tamang et al. 2012; Thapa 2016; Soibam and Ayam 2018).

14.9.1.3 Microbiology

Ngari is a pH-balanced food with a pH of 6.2–6.7, and the fermentation is dominated by lactic acid bacteria. A bacterial load of 10^7–10^8 cfu/g and yeast load of 10^4–10^7 cfu/g are reported in *ngari*, and the microbial species commonly detected are lactic acid bacteria (*Tetragenococcus halophilus*, *Lc. lactis* subsp. *cremoris*, *Lb. plantarum*, *E. faecium*, and *Lactobacillus fructosus*), *Bacillus* spp. (*B. subtilis*, *B. pumilus*, *Bacillus vallismortis*, *B. licheniformis*, *Bacillus velezensis*, *Bacillus aerius*, *Bacillus sonorensis*), *Staph. cohnii*, *Staphylococcus nepalensis*, *Micrococcus* spp., and yeasts (*Candida* sp. and *Saccharomycopsis* sp.) (Thapa et al. 2004; Devi et al. 2015; Majumdar et al. 2015; Singh et al. 2018b). Although a similar bacterial load is observed in both the raw material (*phabou*) and fermented product (*ngari*), Devi et al. (2015) studied the bacterial dynamics during its yearlong fermentation and revealed the succession of different species at different fermentation stages, especially a drastic microbial change in the sixth month of fermentation. The dominant bacterial flora reported during this study were *Staph. cohnii* subsp. *cohnii*, *T. halophilus* subsp. *flandriensis*, a novel phylotype related to *Lactobacillus pobuzihii*, *E. faecium*, *Bacillus indicus*, and *Staphylococcus carnosus*. The emergence of *T. halophilus* and *Lb. pobuzihii* during the third and sixth month of fermentation, respectively, was also observed during the same investigation. Cultivation-independent analysis further detected additional species (*Kocuria halotolerans*, *Clostridium irregulare*, *Azorhizobium caulinodans*, and *Macrococcus caseolyticus*) during the fermentation. The presence of potentially pathogenic *B. cereus* and

Staph. aureus at a load of 10^3 cfu/g is a cause of concern regarding its microbial risk (Thapa et al. 2004). For the industrial production of *ngari* with assured safety and quality, *T. halophilus*, *Lb. pobuzihii*, *Staph. carnosus*, and *B. indicus* are potential target species to design *ngari* starter culture (Devi et al. 2015).

14.9.1.4 Nutritional Composition, Health Benefits, and Functionalities

Ngari is rich in proteins, peptides, amino acids thereby the nutritive value of the local diet. It has a high content of amino acids (both essential and nonessential), lipids (omega-3, omega-6 fatty acids), and minerals (calcium, sodium, potassium, magnesium) with a food value of 381.6 kcal/100 g (Devi et al. 2015; Majumdar et al. 2015; Thapa 2016). *Ngari* is rich in glycine (4.95% of dry weight), proline (3.15%), aspartic acid (3.64%), and the essential amino acids [phenylalanine (3.23%), leucine (2.46%) and lysine (3%)], while threonine and histidine are found in traces (Majumdar et al. 2015). The essential amino acid content in *ngari* comprises about 39.6% of its total amino acid content. Among the saturated fatty acids, palmitic acid was predominant contributing about 7% of the total fatty acids, while vaccenic acid (29.23%) and oleic acid (23.58%) were the major monoenoic fatty acids present. Linoleic acid (11.68%) and arachidonic acid (0.65%) were detected as major PUFA in *ngari*. The higher amount of volatile nitrogenous compounds in *ngari* compared to other similar fish products is responsible for developing its organoleptic properties during fermentation (Tamang 2010;Majumdar et al. 2015). Several studies have reported the antioxidant, fibrinolytic, and probiotic abilities of *ngari*. Antioxidant properties of *ngari* as exhibited by its DPPH scavenging activity showed that the activity was dependent on the protein concentration and period of fermentation (Phadke et al. 2014). The total antioxidant activity (μg ascorbic acid equivalent per mg protein) was found to be 17.41 for *ngari*, while the DPPH radical scavenging activity was reported to be about 87% (Majumdar et al. 2015). With the increase in fermentation period and protein concentration, its ACE inhibitory activity also increases. The fibrinolytic activity of *ngari* and other fermented *P. sophore* products (*shidal* of Assam, *berma* of Tripura, *ngawum* of Mizoram, and *tungtap* of Meghalaya) is higher than most of the fermented soybean products, including *hawaijar*, and is of non-microbial endogenous origin. The ACE inhibitory activity coupled with the endogenous fibrinolytic activity of *ngari* (43.02 ± 1.67 KPU/g) is believed to protect the regular consumers against hypertension and cardiovascular diseases (Phadke et al. 2014; Singh et al. 2014; Keishing and Banu 2015). *Ngari* also possesses probiotic properties as *E. faecium* BDU7 isolated from *ngari* shows characteristic probiotic properties like bile acid tolerance, autoaggregation, and hydrophobicity (Abdhul et al. 2014).

14.9.2 *Hentak*

Hentak is a non-salted fermented fish paste prepared by mixing freshwater fishes (*E. danricus*, locally called *ngasang*) and *A. macrorrhiza*. It is prepared and consumed

Fig. 14.19 The traditional method of *hentak* preparation in Manipur

by the Meitei community in the valley region of Manipur. It is believed that *hentak* was used commonly before *ngari* became popular (Jeyaram et al. 2009; Tamang 2010; Das and Deka 2012; Devi and Kumar 2012).

14.9.2.1 Traditional Preparation Method

For preparing *hentak*, the sun-dried *E. danricus* is grinded into powder and kept aside. Similarly, fresh and chopped petioles of *A. macrorrhiza* are washed and sun-dried for a few hours. In the next step, equal proportion of the fish powder and the chopped petioles are mixed together and grinded in a traditional mortar and pestle to form a thick paste (Fig. 14.19). This paste is kneaded into small balls and kept in sealed earthen pots or bamboo tubes and kept at room temperature for natural fermentation. After 10–14 days of fermentation, the fish balls attain their unique aroma and desirable texture and are ready for consumption. The fish balls become hard on long storage and need to be kneaded with water and few drops of mustard oil for proper storage and to retain the taste. Although *hentak* can also be prepared with small onions (local variety commonly known as *meitei tilhou*) instead of *A. macrorrhiza*, it results in inferior quality *hentak* with a shorter shelf life. *Hentak* is similar to *utonga-kupsu*, a fermented mixed, small fish paste commonly prepared and consumed by the Meitei people settled in the Cachar district of Assam (Jeyaram et al. 2009; Tamang et al. 2012; Kakati and Goswami 2013; Singh et al. 2018b; Soibam and Ayam 2018).

14.9.2.2 Gastronomy, Socioeconomy, and Ethical Values

Hentak is a good source of protein for the local population. It is used for preparing *kangsoi* with green vegetables and consumed with rice or mixed with chilies to prepare *morokmetpa*, which is a favorite side dish of Manipur. It is also used as a flavoring agent in vegetable soup and stews. Traditionally, *hentak* is consumed as an alternative to *ngari* during the postpartum period (about 3 months) as consumption of *ngari* during this period is believed to cause melasma on the face. *Hentak* is not as commonly available like *ngari*, but it can be still found sold by elder womenfolk in only the major markets of Imphal (Tamang 2010; Tamang et al. 2012; Soibam and Ayam 2018).

14.9.2.3 Microbiology

Similar to *ngari*, *hentak* is also a pH-balanced food (pH 6.2–6.5), and its fermentation is dominated by lactic acid bacteria. A total bacterial count of 10^7–10^8 cfu/g and lactic acid bacterial count of 10^4–10^5 cfu/g were detected. The bacterial species detected are *Lc. lactis*, *Lb. plantarum*, *Lb. fructosus*, *Lactobacillus amylophilus*, *Lb. coryniformis*, *Lb. brevis*, *Lb. pentosus*, *E. faecium*, *B. subtilis*, *B. pumilus*, *Bacillus altitudinis*, *B. cereus*, *B. licheniformis*, *Bacillus methylotrophicus*, *Bacillus safensis*, *Bacillus siamensis*, *B. tequilensis*, *B. vallismortis*, *Bacillus* spp., *Micrococcus* sp., *Staph. cohnii*, *Staph. nepalensis*, and *Staphylococcus* sp. (Thapa et al. 2004; Majumdar et al. 2015; Aarti et al. 2016, 2017; Thapa 2016; Singh et al. 2018b). Although the cultivation-based study by Thapa et al. (2004) did not detect any yeast and fungal flora, *Aspergillus fumigatus*, *Aspergillus niger*, *Cladosporium* sp.,

Geotrichum sp., *Myrothecium striatisporum*, *Penicillium rubrum*, *Penicillium rugulosum*, *Rhizopus nigricans*, and *Debaryomyces fabryi* were dominantly detected (10^5 cfu/g) in other studies (Vishwanath and Sarojnalini 1989; Singh et al. 2014). A detailed study using other types of culture media and cultivation-independent molecular techniques to understand its bacterial and yeast communities including its microbial dynamics during fermentation will help in designing a starter culture for industrial production. The presence of *B. cereus* and *Staph. aureus* (10^2–10^3 cfu/g) in *hentak* samples is a public health concern (Thapa et al. 2004).

14.9.2.4 Nutritional Composition, Health Benefits, and Functionalities

Hentak is a rich source of proteins, lipids, and minerals with a food value of 408 kcal/100 g (Majumdar et al. 2015; Thapa 2016). The low moisture content of *hentak* (35%) is explained by the use of sun-dried fish as its raw material. It is a rich source of glycine (5.72% of dry weight), alanine (4.09%), proline (4.45%), aspartic acid (3.84%), glutamic acid (3.35), and the essential amino acids (comprising 44% of total amino acids) (Majumdar et al. 2015). The major essential amino acids detected were phenylalanine (4.91%), lysine (3.81%), and leucine (4.79%). Among the saturated fatty acids, stearic acid was dominant (25% of the total fatty acids), while palmitoleic acid and oleic acid were the major monoenoic acids (comprising 23% of the total fatty acids). Arachidonic acid was the only PUFA detected in *hentak*. Similar to *ngari*, *hentak* has antioxidant, fibrinolytic, and probiotic properties. The antioxidant property of *hentak* as exhibited by its DPPH scavenging activity was reported to be about 36% (Singh et al. 2018b). The fibrinolytic activity of *hentak* (59.28 ± 5.84 KPU/g) is higher than that of *ngari*, which may be of both microbial and endogenous origin as the yeast *D. fabryi* showed good fibrinolytic activity (Singh et al. 2014). *Lb. brevis*, *Lb. pentosus*, and *B. subtilis* isolated from *hentak* are reported with probiotic properties, strong antioxidant properties, and broad-spectrum antibacterial activities (Aarti et al. 2016, 2017; Singh et al. 2018b).

Fig. 14.20 The fermented fish, *ithiitongka*, of the Moyon tribe of Manipur (Photograph reproduced from Wanglar et al. 2018 with permission from the journal office)

14.9.3 *Ithiitongka*

Ithiitongka is a unique type of fermented fish, prepared and consumed by the Moyon tribe of Chandel district in Manipur. For *ithiitongka* preparation, small fishes caught from local rivers and streams are cleaned and washed thoroughly. These fishes are put into a young bamboo trunk (locally known as *tongka*) and the mouth sealed tightly with turmeric leaves. This sealed bamboo is kept in the fire till its external layer is partially burnt (Fig. 14.20). It is then allowed to cool off and kept above a fireplace to prevent it from spoilage. It can be stored up to 2 weeks by sealing the bamboo trunk tightly and keeping it over the fireplace. This fermented fish with unique flavor and taste is mainly prepared for household consumption when the fish catch is large. It is not available commercially, and no investigation on its microbial profile and nutritional composition is reported (Wanglar et al. 2018).

14.10 Ethnic Fermented Meat

All the ethnic communities in Manipur consume meat traditionally but with strong influence of Hinduism. People, particularly the Meitei, stop eating meat although fish is consumed except for few highly religious people who have converted to vegetarian. However, the majority of tribal populace in the hill districts consumes meat products (Tuikhar et al. 2019). Chicken, pork, and beef are the most common meat consumed by them. The hill people occasionally hunt wild animals for food although it is restricted nowadays by the state government. Every dish has one or other meat item in preparations with pork being the most cherished of the lot. The meats are required to be stored due to their heavy consumption, and therefore, they are usually processed for longer shelf life. Drying in sunlight and fireplace and smoking are commonly practiced in the hills of Manipur, but in fewer cases the meats and other parts of the animal carcass are fermented. The most common fermented meat products are fermented pork fat, cow or buffalo fat, fermented cow or buffalo skin, fermented pork meat, and fermented bone marrow.

14.10.1 *Sathu, Gwag Ruum*, and *Bongthu*

Many tribal ethnic communities in Manipur produced and consumed fermented pork fat as traditional delicacies. Kuki, Hmar, Paite, and Vaiphei communities called the fermented pork fat as *sathu* (Fig. 14.21). It is slightly acidic (pH 6.3–6.5), has a strong odor, and is oily. It is usually used as taste enhancer in many of the traditional dishes. The boiled pork fats are fermented for 3–5 days in an airtight container to get the product. *Sathu* is similar to *sa'um* of Mizoram, India, and similarly it could be infested with *P. mirabilis*, *E. faecium*, *C. botulinum*, and *C. perfringens* (Singh et al. 2014; Keisam et al. 2019). Not much study has been done on this product, and hence the microbiology and nutritional composition are unknown. A similar

Internal pork fat removed, clean and cut into small pieces

↓

Boil for 20 min

↓

Cool and pour in a container or dried hollowed out gourd (*tumba*)

↓

Place in sunlight or near fire place

↓

Spontaneous natural fermentation (3-5 days)

↓

Characteristic desired smell

↓

SATHU

Fig. 14.21 The traditional method for the preparation of *sathu* in Manipur

product called *gwag ruum* is produced and consumed by the Rongmei community in the western hills of Manipur.

Similar products are also prepared and consumed by these communities using cow or buffalo fat, and they are called *bongthu* (Fig. 14.22). The consumption and use of the product are similar to those of *sathu*, but *sathu* is more preferred than the other.

14.10.2 *Saayung*

Saayung is the fermented pork meat commonly produced by the Tangkhul community during the late December, usually after Christmas festival. The excess pork meats during festivals are prepared into *saayung* and stored for months for later

Fig. 14.22 The fermented cow or buffalo fat, *bongthu* of Manipur

consumption. The process of fermentation gives a peculiar taste and also makes the meat tender. The product is prepared by boiling and adding salt as well as spices (Fig. 14.23). *Saayung* is given to pregnant women and lactating mothers due to the traditional belief that it gives more energy and nutrition to such women.

14.10.3 *Guaighi Kang*

Guaighi kang is a dried fermented skin of cow or buffalo prepared and consumed by Rongmei community in western hills of Manipur. This product is specially used for preparation of special dishes called *tam* (Fig. 14.24). It is prepared by stacking the skin of the slaughter animals in an airtight container and keeping for about 7 days to allow fermentation process. The product is taken out from the container, and the hairs are scraped off completely using a knife. Then it is either sun-dried or kept near the fireplace. It can be stored in ambient temperature for 6–8 months. It is soaked in water to soften before preparation and cooking as the dried skin is very hard.

14.10.4 *Sahro*

Sahro is a valued meat product of the Hmar community (Fig. 14.25). During the preparation of *sahro*, any meat but usually pork is half cooked to remove fat from the meat. It is spread in the sun for another 7 days during which the fermentation

Cut pork meat into small pieces

↓

Boil in water with 2% salt for 15-20 min

↓

Drain excess water and mix with turmeric (10 g/kg meat)

↓

Hang in the traditional storage chamber near fire place

↓

Spontaneous natural fermentation (6-9 months)

↓

SAAYUNG

Fig. 14.23 The traditional method for the preparation of *saayung* by Tangkhul community in Manipur

takes place. Then the meat is placed near the fireplace. *Sahro* is usually given to older people and small kids due to its low-fat content.

14.10.5 *Bongkarot*

Bongkarot is a fermented meat product made from cow legs. Vaiphei, Paite, and Hmar communities prepare this product to enhance its shelf life and for the unique taste. It is prepared by boiling a few pieces of cow leg until the meat gets tender. *Sathu* is also added during its cooking to give its aroma. After cooking for half an hour, the meat becomes very sticky. During this time, the broth is cooled down and packed in a bamboo culm, which is then covered tightly using banana leaves. It is

a b

Fig. 14.24 *Guaighi kang* of Rongmei community of Manipur (**a**) and the ethnic food *tam* (**b**) prepared using *guaighi kang*. (The second photograph is reproduced from Singh et al. 2018c with permission from Excellent Publishers)

Fig. 14.25 *Sahro* of Hmar community of Manipur

believed that the broth spoils if touched by bare hand. It is kept for future use and can be stored for 1 month (Fig. 14.26).

14.10.6 *Aithu*

Aithu is a fermented crab, a slightly alkaline and pungent food product of Kuki-Chin-Mizo communities of Manipur. This item is a delicacy and served to special guests. It is not the main course of any meal but a side dish and a taste maker in many vegetarian and non-vegetarian traditional dishes. This product is similar to *japangangnagtsu* of Ao Naga of Nagaland. *Aithu* is prepared by grinding the already washed and cleaned crab and mixing with ground *sith* (Fig. 14.27). The mixture is wrapped tightly using banana leaves and kept in an earthen pot. The pot is kept near

Fig. 14.26 The cow leg meat (**a**) used for the preparation of *bongkarot* (**b**) in Manipur

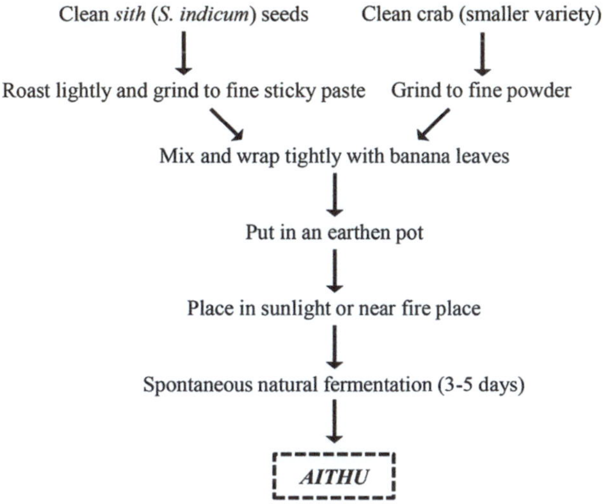

Fig. 14.27 The traditional method for the preparation of *aithu* in Manipur

the fireplace or under the sun until the characteristic smell comes from the product, which usually takes 3–5 days.

14.11 Ethnic Fermented Milk

14.11.1 *Sangom Aphamba*

Traditionally, the ethnic people of Manipur do not prefer milk product. In Manipur, traditionally milk is consumed fresh and seldom processed. The only known fermented milk is curd locally called *sangom aphamba* of Meitei community. The production of *sangom aphamba* is simple and similar to *dahi* preparation. During its preparation, fresh raw milk is kept in a traditional earthen pot, and a spoonful of

a b

Fig. 14.28 Women vendors in *Ema Keithel*, Imphal, Manipur selling *sangom aphamba* (**a**). *Sangom aphamba* in traditional earthen pots (**b**)

curd from the previous batch is added, and the pot is covered with a muslin cloth locally called *sana*. It is allowed to stand for 2–3 days to get the *sangom aphamba* (Fig. 14.28). Traditionally it is not used in any local dish preparations but usually taken as raw. It is also taken with steam rice. It serves as one of the essential requirements for performing rituals in certain traditional religious ceremony such as *nahutpa* (ear-piercing ceremony of children).

14.12 Ethnic Fermented Alcoholic Beverages

The preparation and consumption of ethnic fermented alcoholic beverages are a common traditional practice among many ethnic communities of Manipur since time immemorial. Rice-based alcoholic beverages are the main ethnic beverages of the state and commonly known as *yu* by the Meitei and Kabui communities and *zou* by the Zeliangrong and Kuki tribes. The traditional process of making fermented alcoholic beverages starts with the preparation of the indigenous amylolytic starters, followed by the fermentation of rice using the starter culture. The ethnic amylolytic starters for alcoholic rice beverages are *hamei* of Meitei and Kabui communities, *chamri* of Tangkhul community, and *khai* of Zeliangrong and Kuki tribes. The different fermented alcoholic rice beverages of Manipur are *atingba*, *yu angouba*, *yu*, and *waiyu* of Meitei and Kabui communities; *patso*, *khor*, and *acham* of Tangkhul tribes; and *zoungao*, *timpui*, *zouju/zouzu*, and *pheijou* of Zeliangrong and Kuki tribes.

14.12.1 *Hamei, Chamri,* and *Khai*

Hamei, *chamri*, and *khai* are the traditionally prepared dry, solid, flattened/rounded, rice dough starter cultures produced by the Meitei, Kabui, Tangkhul, Zeliangrong, and Kuki communities of Manipur and are used for the preparation of various indigenous fermented alcoholic rice beverages (Jeyaram et al. 2008a; Devi and Kumar 2012; Singh et al. 2018c). They are similar to *marcha* of Sikkim, *pham* and *phut* of Arunachal Pradesh, *thiat* of Meghalaya, *dowdim* of Mizoram, *chowan* of Tripura, *humao* of Assam, *chu* of China, *naruk* of Korea, *ragi* of Indonesia, *budob* of the Philippines, and *loogpang* of Thailand (Tamang 2010; Sha et al. 2018; Anupma et al. 2018).

14.12.1.1 Traditional Preparation Method

Hamei is prepared by womenfolk in many villages of Manipur such as Andro, Sekmai, Phayeng, Bishnupur, Tengnoupal, and Jiribam. During the traditional preparation, raw rice of local varieties, with or without presoaking in water for half an hour and drying for 15 min to remove excess water, are crushed into powder mass (locally called *yam*) by pounding in a wooden mortar (called *shumban*) with a wooden pestle (called *shuk*). Dried bark of *Albizia myriophylla* (known as *yangli*) is powdered or finely chopped and mixed homogeneously with the rice powder. For every 1 kg of rice, 100–250 g of *yangli* powder is used. A proportionate amount of water is added to the mixture to form a dough. It is then inoculated with a pinch of powdered *hamei* from previous batches and mixed thoroughly. From this dough, elliptical or rounded flattened cake like structures (2–7 cm diameter, 0.6–1.5 cm thickness) called *hamei* are prepared by pressing between the palms. The *hamei* cakes are kept over paddy husk or paddy straw in a bamboo basket or floor, covered with straw and sackcloths and incubated undisturbed for 2–3 days in the ambient environment condition for fermentation (Fig. 14.29). *Hamei* is ready for production of fermented rice beverages when it swells up, produces the alcoholic smell, and becomes yellowish. It is air-dried and can be stored yearlong in a cool, dry place. It is believed that *hamei* prepared from unpolished rice gave better quality alcoholic beverage than using polished rice. *Hamei* preparation is done during the summer months of May–July. The preparation of *khai* and *chamri* is similar to that of *hamei* (Singh and Singh 2006; Jeyaram et al. 2008a, 2009; Singh et al. 2018c).

14.12.1.2 Socioeconomy

Hamei, *chamri*, and *khai* are used to prepare a rice-based fermented alcoholic beverages *atingba*, *patso*, *timpui*, *waiyu*, *pheijou*, *zoungao*, *yu*, and *zouju*. Though *hamei* is prepared for household production of beverages, it is also sold in the local markets of Manipur.

14.12.1.3 Microbiology

Hamei is a mixed starter culture predominated by culturable filamentous molds (*Mucor* spp., *Rhizopus* spp.), yeasts (*Saccharomyces cerevisiae*, *Pichia anomala*, *Pichia guilliermondi*, *Pichia fabianii*, *Pichia kudriavzevii*, *Candida tropicalis*,

Hulled, unpolished rice (local variety)

↓

Soak in water (30 min)

↓

Spread out and air dry to remove excess water (15 min)

↓

Crush to powder using traditional wooden mortar (*shumban*) and wooden pestle (*shuk*)

↓ ← Powdered/finely chopped *A. myriophylla* (*yangli*) bark

Add boiled water and mix thoroughly to form a dough

↓ ← Powdered *hamei* from previous batch

Make elliptical or rounded flattened cakes by pressing between palms

↓

Spread over paddy husk/straw in a bamboo basket/floor and cover with straw and sackcloths

↓

Incubate 2-3 days in the ambient environment (20-30°C)

↓

Characteristic alcoholic smell formation

↓

HAMEI

↓

Spread in open air for drying

↓

Store in a cool dry place

Fig. 14.29 The traditional method for the preparation of *hamei* in Manipur. (The first photograph of *hamei* is reproduced from Jeyaram et al. 2008a with permission from Elsevier)

Candida parapsilosis, *Candida montana*, *Candida glabrata*, *Torulaspora delbrueckii*, *Trichosporon* sp., *Wickerhamomyces anomalus*), and lactic acid bacteria (*Lb. plantarum*, *P. pentosaceus*). It has mold, yeast, and bacterial population of 10^6–10^9 cfu/g, 10^8–10^9 cfu/g, and 10^7–10^9 cfu/g, respectively. The molds are responsible for the production of amylolytic enzymes that drive the saccharification process of the rice starch. PCR-DGGE-based culture-independent study of *hamei* reveals additional yeasts and molds belonging to the phyla *Ascomycota* (*Saccharomycopsis capsularis*, *Saccharomycopsis fibuligera*, *Saccharomycopsis malanga*, *S. paradoxus*, *Aspergillus* sp., *Aspergillus oryzae*, *Candida* sp.,) and *Zygomycota* (*Rhizopus oryzae*). An in-depth next-generation sequencing-based metagenomics study has shown that *hamei* has the highest yeast and fungal operational taxonomic units among the traditional amylolytic starters of Northeast India, and the taxa are mostly limited to the phyla *Ascomycota*, *Zygomycota*, and *Basidiomycota*. It also reveals that *hamei* has various noble or unique yeast and fungal species, which are dominated by *Solicoccozyma terrea* and *Thermomyces lanuginosus*. *Hamei* also harbors numerous rare species belonging to *Ascomycota* (*Acidomyces acidothermus*, *Aspergillus gracilis*, *Auricularia* spp., *Cephaliophora tropica*, *Cladosporium* spp., *Clonostachys rosea*, *Epicoccum nigrum*, *Fusarium* spp., *Ganoderma enigmaticum*, *Geastrum kotlabae*, *Hannaella siamensis*, *Lachnum virgineum*, *Malassezia globosa*, *Metschnikowia pulcherrima*, *Paracamarosporium fagi*, *Penicillium* spp., *Phoma omnivirens*, *Physciella chloantha*, *Pichia kluyveri*, *Pseudocercospora stephanandrae*, *Pseudocercosporella chaenomelis*, *Trichomonascus ciferrii*, *Truncospora macrospora*, *Volvariella murinella*, *Xerochrysium dermatitidis*), Basidiomycota (*Acremonium persicinum*, *Arthrinium* spp., *Aureobasidium* spp., *Camptophora hylomeconis*, *Curvularia hawaiiensis*, *Cutaneotrichosporon terricola*, *Eupenidiella venezuelensis*, *Gilbertella persicaria*, *Kwoniella heveanensis*, *Neoceratosperma yunnanensis*, *Sporobolomyces bannaensis*, *Trichomerium dioscoreae*, *Wallemia* spp., *Wallrothiella subiculosa*, *Yarrowia lipolytica*), and Zygomycota (*Cryptococcus* spp., *Papiliotrema* spp.) (Tamang et al. 2007; Jeyaram et al. 2008a; Mangang et al. 2017a; Sha et al. 2018, 2019).

14.12.2 Fermented Alcoholic Rice Beverages

Cooked or uncooked rice are traditionally fermented with or without the amylolytic starter cultures to prepare different forms of fermented alcoholic beverages by the non-tribal Meitei community as well as by some ethnic communities of the major Naga and Kuki tribes. Fermented alcoholic rice beverages can be broadly classified as non-distilled, which comprises *yu angouba*, *khor*, *zoungao*, *atingba*, *patso*, *timpui*, *waiyu*, and *pheijou*, and distilled liquor, which are *yu*, *acham*, and *zouju/zouzu* (Fig. 14.30).

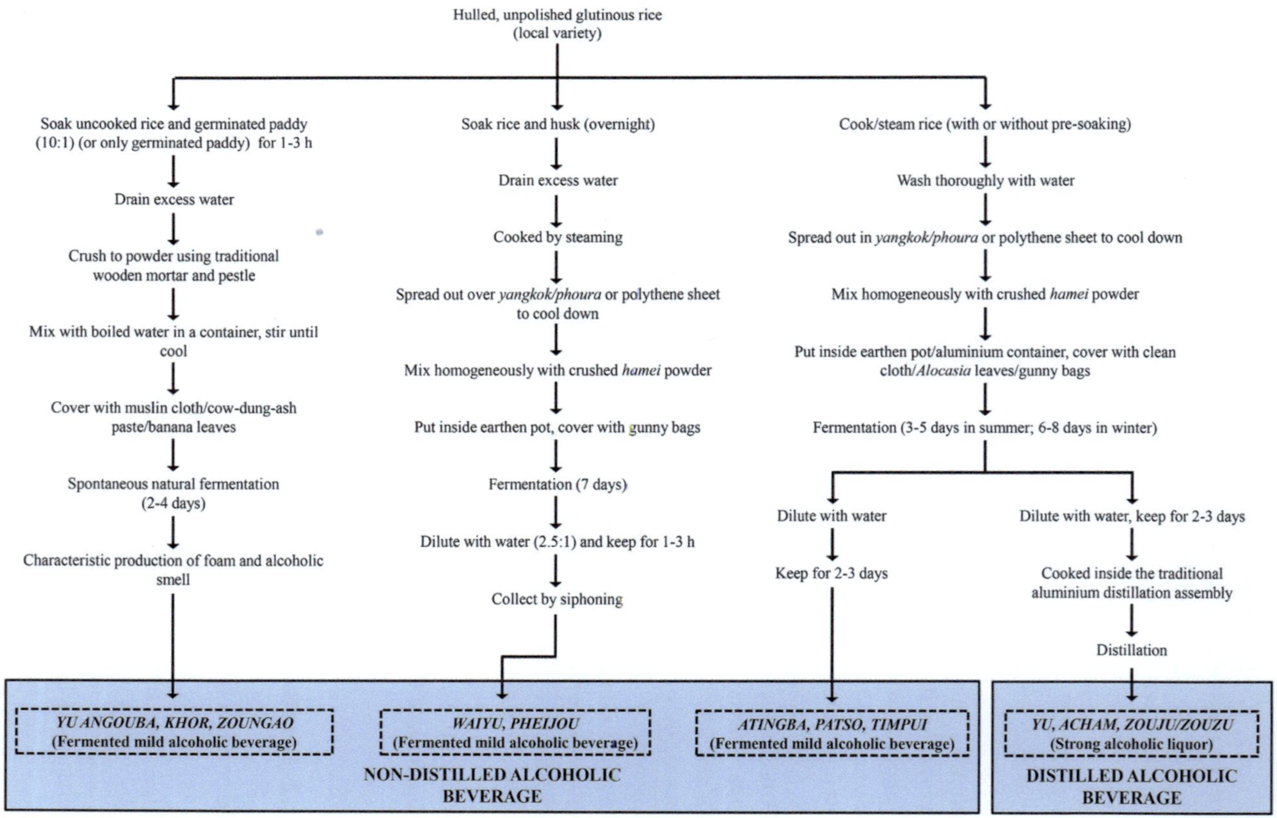

Fig. 14.30 The traditional methods for the preparation of different types of ethnic fermented alcoholic beverages of Manipur

14.12.2.1 *Yu Angouba, Khor,* and *Zoungao*

Traditional Preparation Method

Yu angouba is a non-distilled, mild alcoholic beverage of Meitei and Kabui communities traditionally made from uncooked raw rice without the use of starter culture *hamei*. For its preparation, local variety of rice and some germinated paddy in 10:1 ratio are soaked in water for 1–3 h (Fig. 14.30). The germinated paddy acts as the source of amylolytic enzymes for the saccharification of rice starch. After draining out excess water, the soaked rice paddy mixture is crushed by pounding in traditional wooden mortar and pestle, and the powdered rice is mixed with boiled water in a container with continuous stirring until it cools down completely. The container is then covered with a muslin cloth and kept undisturbed for 2–4 days for spontaneous natural fermentation to produce *yu angouba*. In the fermentation container, the emergence of foams and production of typical alcoholic odor and flavor indicate that the beverage is ready for consumption. The Tangkhul tribes prepare similar kind of alcoholic beverage, which is known as *khor*. The preparation process is same with that of *yu angouba* except that *khor* is prepared from germinated paddy instead of rice and fermentation is carried out in a traditional pot having a narrow slender neck, which is sealed with a mixture of cow dung and ashes. Among the Zeliangrong and Kuki tribes, such beverage is called *zoungao*. During its preparation, the powdered rice paddy mixture is churned with hot water inside a traditional wooden barrel (called *bu*) or earthen pot, which is then covered with banana leaves and allowed for fermentation. *Yu angouba*, *khor*, and *zoungao* have a short shelf life of 7 days. Their consumption is considered not only good for health but also good for fair complexion when used within 3–4 days. The ethnic people in the hill areas regularly drink the beverages instead of tea and milk. It is also used in ritual ceremony (Devi and Kumar 2012; Singh et al. 2018c; Tuikhar et al. 2019).

14.12.2.2 *Atingba, Patso,* and *Timpui*

Traditional Preparation Method

Atingba is the most famous non-distilled, mild alcoholic beverage of Meitei and Kabui communities. The ethnic food beverage is traditionally prepared from cooked glutinous rice using the starter culture *hamei* (Fig. 14.30). For the preparation of *atingba*, properly cooked or steamed local varieties of glutinous rice (*moirang phou*, *kumbi phou*, *KD phou*, etc.) are cooled and air-dried by spreading over traditional bamboo winnowing trays (*yangkok* or *phoura*) and mixed homogenously with crushed *hamei* powder (40–50 g/kg of rice). In Andro village, an ancient village of Meitei community, the rice is soaked in water overnight, and the wet rice is cooked by exposing to hot steam. Before the addition of *hamei*, washing of the cooked/steamed rice is practiced in Andro, Sekmai, and Phayeng villages, which are known to produce best-quality *atingba* in Manipur. The *hamei*-mixed rice is then put inside specially designed traditional earthen pots (15–20 L capacity) already cleaned and dried in the sun and allowed for undisturbed solid-state fermentation after covering with *Alocasia* leaves or clean cloth. The fermentation is carried out for 3–4 days in summer and 6–8 days in winter. In some villages, the fermentation during winter is carried out inside a traditional bamboo basket internally wrapped with leaves of *Tectona grandis*, *F. hispida*, banana, giant taro, etc. and baked under direct sunlight. During the fermentation, heat is released along with pungent and alcoholic smell and water is poured to reduce the temperature to an optimum level and allowed for another 2–3 days of submerged fermentation to get the final fermented beverage called *atingba*. *Atingba* is consumed within 1–2 days after fermentation but can be kept for around 1–2 months to be used for distillation for preparing the distilled liquor called *yu*. *Atingba* is called *patso* among the Tangkhul community and *timpui* among the Zeliangrong and Kuki tribes. Preparation of *patso* and *timpui* is similar to that *atingba* (Singh and Singh 2006; Jeyaram et al. 2009; Devi and Kumar 2012; Kumari et al. 2015; Singh et al. 2018c).

14.12.2.3 *Waiyu* and *Pheijou*

Traditional Preparation Method

Waiyu is a non-distilled, mild alcoholic beverage prepared from cooked rice and husk. It is a delicious alcoholic drink popular among the Meitei, Kabui, Maring, and Zeliangrong communities. In Andro village, hand-pounded rice is mixed with husk and soaked overnight in water (Fig. 14.30). After draining excess water, the rice husk mixture is cooked by steaming and mixed with *hamei* after spreading and cooling over *yangkok* or *phoura*. The mixture is fermented for 7 days inside an earthen pot after covering with gunny bags. The fermented liquid is diluted with water (2.5 L/kg of rice) in another pot and kept undisturbed for another 1–3 h. The yellowish fermented liquid thus obtained is called *waiyu* (*wai* meaning husk and *yu* meaning alcoholic rice beverage). In Langthabal, a Kabui-inhabited village, the husk is mixed with cooked hand-pounded rice in the ratio of 1:4 after adding *hamei*, and fermentation is carried out for 3–4 days in summer and 3–6 days in winter. *Waiyu* is drawn out from the pot using a pipe (locally called *yu thaba*) through the process of siphoning and is served fresh. *Waiyu* is locally known as *pheijou* among the Zeliangrong community (Kumari et al. 2015).

14.12.2.4 *Yu*, *Acham*, and *Zouju/Zouzu*

Traditional Preparation Method

Yu is the ethnic distilled, strong alcoholic liquor of the Meitei and Kabui communities. It is prepared from fermented rice beverages that are made using the indigenous starter cultures. It is the ethnic rice-based alcoholic beverage produced for general consumption and commercial purpose. The preparation of *yu* involves two stages, viz., alcoholic fermentation of rice and distillation (Figs. 14.30 and 14.31). In the first stage, which is similar to the preparation of *atingba*, hulled, unpolished rice is cooked in a pot. In Andro village, the rice is soaked in water overnight before

Fig. 14.31 The traditional method for the preparation of distilled fermented alcoholic rice beverage, *yu*, of Meitei community of Manipur. (**a**) Hulled, unpolished glutinous rice. (**b**) Cooking of rice (*chakngan thongba*). (**c**) Cooked rice (*chakngan*) after washing (*chakngan chamba*) in the traditional bamboo basket. (**d**) *Hamei* before adding to the washed *chakngan*. (**e**) Fermentation of *hamei*-mixed *chakngan* in the traditional earthen pots. (**f**) Partially fermented *chakngan* inside aluminum and plastic containers. (**g**) Allowing fermentation to continue while covered with gunny bags. (**h**) Fermented rice alcoholic liquid mixture (*khajee*) being cooked (*yu thongba*) in the traditional aluminum distillation assembly. (**i**) View of the traditional production center. (**j**) Final distilled liquor *yu*

cooking by steaming. The cooked rice is called *chakngan*, and its preparation process is known as *chakngan thongba. Chakngan* is properly washed with cold water (a process known as *chakngan chamba*) and air-dried by spreading over traditional bamboo winnowing trays or polythene sheets. The thoroughly dried *chakgan* is mixed with powdered *hamei*, put inside an earthen pot, covered with gunny bags, and kept undisturbed in the ambient environment condition for 3–5 days of fermentation. In Sekmai, *hamei*-mixed *chakgan* is fermented inside a bamboo basket lined interiorly with polythene sheets. The process of transferring the *hamei-chakgan* mixture to the basket is known as *luthup chanba*. In earlier times, leaves of giant taro, *Datura metel*, etc. were used for the interior lining of the basket. In recent times, the use of earthen pot and basket has been replaced with wide-mouth aluminum and plastic containers. The mixture of properly fermented rice and the alcoholic liquid released during the fermentation (locally known as *khajee*) is mixed with water in roughly 1:1 ratio to start the distillation process. In Andro and Langthabal, *khajee* is kept for another 1–2 days after mixing with water. In the second stage, the aluminum container with the mixture of *khajee* and water (locally called *yuphu*) is lidded with another similar but smaller size hollow aluminum container of same mouth width (locally called *yukok*). A traditional aluminum flat funnel designed for condensing and collecting the evaporated alcohol is fitted on the inside wall of the *yukok* and connected outside with a pipe through the lid. Between the *yuphu* and *yukok*, a perforated aluminum plate (called *yumai*) is placed to streamline the flow of steam inside *yukok*. Finally, an aluminum pot containing cold water is placed over the hollow end of *yukok* to enhance condensation and restrict evaporation outside. All the connecting points are then sealed with mud. The *yuphu* is then heated by burning firewood to start the distillation process, which is locally known as *yu thongba*. The vapor condenses inside *yukok* and the distillate is collected outside in a container through the funnel and the pipe. This distillate is called *yu*. The distillation continues until all the alcohol present in the boiling mixture is removed. *Yu* is known as *acham* in Tangkhul and *zouju*/*zouzu* among the Zeliangrong and Kuki tribes. They are prepared from *patso* and *timpui*, respectively, with similar distillation process (Devi and Kumar 2012; Kumari et al. 2015; Soibam and Ayam 2018; Singh et al. 2018c).

14.12.2.5 Gastronomy, Socioeconomy, and Ethical Values

The non-distilled fermented rice beverages are mainly meant for use during ritual ceremonies, and as such, they are prepared only on rare occasions and not for general consumption or commercial purposes. Though liquors are prohibited and not openly for sale in Manipur, mainly in the valley districts, the ethnic fermented alcoholic beverages are available and consumed in the villages where they are produced as a part of their tradition. The ethnic people in the hill areas regularly drink the non-distilled rice beverages instead of tea and milk. The *yu* of Sekmai (Awang Sekmai), Andro, and Phayeng villages are acknowledged as the best and commonly popular among the Meitei people. The occupation of alcoholic fermented rice beverage distillation is a lucrative one. Almost every house in Sekmai, Andro, and Phayeng is involved in the business of alcoholic beverage production in one way or

a b

Fig. 14.32 The traditionally produced ethnic distilled liquor, *yu*, of Phayeng (**a**) and Awang Sekmai (**b**) packed in attractive packages and glass bottles

the other. Apart from agriculture, it is the main source of income for the traditional producers and considerably supports other economic enterprises such as piggery. The solid residue of fermented rice obtained after the distillation of *yu* is a very good feed for pigs. The traditional artisan producers have started bottling in attractive glass bottles to increase their sale and income (Fig. 14.32) (Devi and Kumar 2012; Singh et al. 2018c).

14.12.2.6 Microbiology

The microbes of *atingba* are similar to that of *hamei* with the predominant presence of *P. anomala* and *P. pentosaceus*. Molds and *Enterobacteriaceae* are not detected in *atingba*. The fermentation of *atingba* involves microbial population dynamics of mesophilic aerobes, lactic acid bacteria, and yeasts with the population varying in the range of 10^1–10^7 cfu/mL, 10^4–10^9 cfu/mL, and 10^3–10^{10} cfu/mL, respectively. No potential food-borne pathogen is detected in *atingba*. The microbiology of other fermented beverages of Manipur is unknown (Mangang et al. 2017a).

14.12.2.7 Nutritional Composition, Health Benefits, and Functionalities

Atingba prepared from glutinous black rice of Manipur has an alcohol content of 5.71–7.75%, titratable acidity of 4.46% lactic acid equivalent, and anthocyanin content of 0.44–1.17 μg/mL with no biogenic amines (Mangang et al. 2017a, b). The main ingredient of *hamei*, *A. myriophylla* bark (*yangli*), is a good source of polyphenols, soluble sugars, terpenoids, flavonoids, and anthraquinone, has antioxidant and antimicrobial activities, and inhibits lipid peroxidation. The consumption of *yu angouba/khor/zoungao* is considered not only good for health but also good for fair complexion when used within 3–4 days. Drinking of *atingba* is believed to enhance the removal of kidney stone during urination. The initial drops of distillate collected during the distillation of *yu* form a special liquid called *machin*. It has a very high percentage of alcohol and is used for medical purposes. It is believed to have a soothing effect when massaged on the lower surface of the feet, relieves pain during arthritis and muscle soreness, enhances quick healing of wound stiches, etc. Various traditional herbal preparations are made by traditional healers and taken orally or massaged over the body for improving poor health conditions of women such as irregular menstruation, infertility, obesity, loss of appetite, and postpartum health (Singh and Singh 2006; Soibam and Ayam 2018; Singh et al. 2018c).

14.13 Food Safety of Marketed Ethnic Fermented Foods in Manipur

Sporadic incidences of food-borne outbreaks (food poisoning and enteric infection) due to the consumption of ethnic fermented foods, mainly fermented soybean *hawaijar*, are reported in Imphal, Kaching, Bishnupur, Thoubal, and Ukhrul districts of Manipur in the last 10 years (Keisam et al. 2019). However, the mildness in symptoms and lack of infrastructure and awareness make it challenging to go for proper surveillance, and the outbreaks are underreported. Moreover, their scientific investigation or intervention is rarely carried out, except few records of detecting potential pathogens in the food products. The traditionally produced and marketed *hawaijar* has a huge load of WHO prioritized food-borne pathogens, namely, *B. cereus* (10^8–10^9 cfu/g) and *Staph. aureus* (10^7–10^9 cfu/g) (Jeyaram et al. 2008b). Apart from these two pathogens, culture-dependent and metagenomic analyses reveal the presence of high population (>10^7 cells/g) of *P. mirabilis*, *C. botulinum*, and *C. perfringens* as the predominant enteric bacterial pathogens in *hawaijar* (Singh et al. 2014; Keisam et al. 2019). A first-ever baseline study on the molecular surveillance of microbial risk associated with the consumption of marketed ethnic fermented foods of Northeast India by Keisam et al. (2019) highlighted the alarming toxigenic and pathogenic potentials of enteric bacterial pathogens prevalent in these foods, particularly fermented soybean, fish, and pork products. *B. cereus* and *P. mirabilis* originated from *hawaijar* possess various genes for the virulence factors (nonhemolytic enterotoxin, hemolysin, emetic toxin, and urease), form swarming colonies, and show resistance to multiple antibiotics, indicating the potential to cause food poisoning and infection. The prevalence of these pathogens in the fermented soybean products is higher during the colder months, which may explain the increased incidence of *hawaijar*-borne outbreaks during the colder season. The marketed *ngari* and *hentak* have a low population of *B. cereus* and *Staph. aureus* (10^2–10^3 cfu/g), while *soidon mahi* is detected with an alarming population of *B. cereus* (10^6–10^7 cfu/mL) though their toxigenic and pathogenic potentials are unknown (Thapa et al. 2004; Jeyaram et al. 2010; Devi et al. 2015). The unusual presence of *P. mirabilis* in the marketed fermented food with the high incidences of urolithiasis (kidney stone) cases in Manipur is a concern due to its uropathogenic potential leading to urease-induced urolithiasis and multidrug-resistant status (Keisam et al. 2019). A population study has linked the consumption of *hentak* with a high prevalence of urolithiasis, which is attributed to the traditional method of its preparation (Singh et al. 1986). These health hazard potentials emphasize the need of the hour to have a coordinated action to control and prevent the spread of enteric bacterial pathogens through the fermented food chain.

14.14 Conclusion

The ethnic fermented foods and beverages of Manipur are prepared through the age-old indigenous process of natural fermentation. Fermented food is still an unorganized sector in the processed food industry of Manipur though very few artisan producers and entrepreneurs are marketing them with nutritional tags in commercial packages. Changing the small-scale, traditional production into commercial processing units will create profitable businesses and generate gainful employment and income for the workers of many unorganized sector. Although they form an integral part of the diets of the ethnic communities as a nutrient-rich food with numerous health benefits, the uncontrolled fermentation leads to safety issues, health hazards, and highly heterogeneous product quality. This provokes the need to regulate the traditional preparations and marketing of the products and the development of standardized controlled fermentation technology using starter cultures to improve their quality, safety, and functionalities. The traditional beliefs associated with ethnic foods should be scientifically validated to add values to these food products. Overall, a hygienic production, packaging, and value addition by claiming their health benefits will industrialize these traditional food products and support the socioeconomic development in the state.

References

Aarti C, Khusro A, Arasu MV, Agastian P, Al-Dhabi NA (2016) Biological potency and characterization of antibacterial substances produced by *Lactobacillus pentosus* isolated from *Hentak*, a fermented fish product of North-East India. Springerplus 5:1743. https://doi.org/10.1186/s40064-016-3452-2

Aarti C, Khusro A, Varghese R, Arasu MV, Agastian P, Al-Dhabi NA, Ilavenil S, Choi KC (2017) *In vitro* studies on probiotic and antioxidant properties of *Lactobacillus brevis* strain LAP2 isolated from *Hentak*, a fermented fish product of North-East India. LWT Food Sci Technol 86:438–446

Abdhul K, Ganesh M, Shanmughapriya S, Kanagavel M, Anbarasu K, Natarajaseenivasan K (2014) Antioxidant activity of exopolysaccharide from probiotic strain *Enterococcus faecium* (BDU7) from *Ngari*. Int J Biol Macromol 70:450–454

Anal BS (2014) Fishes of the Chindwin River Basin. Ph.D. thesis, Department of Life Science, Manipur University, Imphal, Manipur, India, p 293. http://hdl.handle.net/10603/39727

Anupma A, Pradhan P, Sha SP, Tamang JP (2018) Traditional skill of ethnic people of the Eastern Himalayas and North East India in preserving microbiota as dry amylolytic starters. Indian J Tradit Knowl 17:184–190

Asem ID, Imotomba RK, Mazumder PB (2017) The deep purple color and the scent are two great qualities of the black scented rice (*Chakhao*) of Manipur. In: Li JQ (ed) Advances in international rice research. InechOpen Ltd., Rijeka, pp 125–136. isbn:978-953-51-3010-9

Bareh HM (2007) Encyclopaedia of North-East India- Manipur, vol vol. III. Mittal Publication, New Delhi, p 312. isbn:81-7099-790-9

Bharati H, Landge AT, Sharma A, Singh YJ, Chanu TN (2017) Insight into the socio-economic life of fishers of Loktak Lake, Manipur-A Ramsar Site. Fish Technol 54:215–220

Bhatt BP, Singha LB, Sachan MS, Singh K (2004) Commercial edible bamboo species of the North-Eastern Himalayan region, India. Part II: Young shoot sales. J Bamboo Rattan 3:337–364

Census of India (2011) Last access on 18 Dec 2018. https://www.census2011.co.in/

Chongtham N, Bisht MS, Haorongbam S (2011) Nutritional properties of bamboo shoots: potential and prospects for utilization as a health food. Compr Rev Food Sci Food Saf 10:153–168

Das AJ, Deka SC (2012) Fermented foods and beverages of the North-East India. Int Food Res J 19:377–392

Devi BT (2014) Rural marketing of agricultural produce in Bishnupur and Thoubal districts Manipur. Ph.D. thesis, Department of Commerce, Manipur university, Imphal, Manipur, India, p 224. http://hdl.handle.net/10603/39996

Devi CS (2017) Oral tradition: an anthropological insight on the folktales of Meitei. Natl J Multidiscip Res Dev 2:182–186

Devi KR, Deka M, Jeyaram K (2015) Bacterial dynamics during yearlong spontaneous fermentation for production of *ngari*, a dry fermented fish product of Northeast India. Int J Food Microbiol 99:62–71

Devi LK (1988) A study of archaeology in Manipur since prehistoric times. Ph.D. thesis, Department of History, Manipur University, Imphal, Manipur, India, p 322. http://hdl.handle.net/10603/103663

Devi P, Kumar SP (2012) Traditional, ethnic and fermented foods of different tribes of Manipur. Indian J Tradit Knowl 11:70–77

Devi PS (2012) Social change among the Lois of Manipur. Ph.D. thesis, Department of Sociology, Assam University Silchar, Assam, India, p 283. http://hdl.handle.net/10603/93241

Devi WS, Uma D, Papia D, Sulata K, Devashish K (2017) Limnological study with emphasis on fish diversity of Loktak Lake, Manipur, India. Bull Environ Pharmacol Life Sci 6:75–77

Forest Survey of India (2017) Bamboo resources of the country. In: State of Forest Report 2017, Forest Survey of India, Ministry of Environment & Forests, Government of India, Dehradun, India. Last access on 18 Dec 2018. http://fsi.nic.in/forest-report-2017

Giri SS, Janmejay LS (2000) Effect of bamboo shoot fermentation and ageing on nutritional and sensory qualities of *soibum*. J Food Sci Technol 37:423–426

Hmar D (2013) The Hmars. Akansha Publishing House, New Delhi, p 98. ISBN: 9788183703550

Hoikhokim (2017) Ankamthu: traditional mustard leaves fermented food preparation of Churachandpur district, Manipur. Int J Res Advent Technol 5:24–26

Jain A, Sundriyal M, Roshnibala S, Kotoky R, Kanjilal PB, Singh HB, Sundriyal RC (2011) Dietary use and conservation concern of edible wetland plants at Indo-Burma hotspot: a case study from Northeast India. J Ethnobiol Ethnomed 7:29. https://doi.org/10.1186/1746-4269-7-29

Jeyaram K, Romi W, Singh TA, Devi AR, Devi SS (2010) Bacterial species associated with traditional starter cultures used for fermented bamboo shoot production in Manipur state of India. Int J Food Microbiol 143:1–8

Jeyaram K, Singh TA, Romi W, Devi AR, Singh WM, Dayanidhi H, Singh NR, Tamang JP (2009) Traditional fermented foods of Manipur. Indian J Tradit Knowl 8:115–121

Jeyaram K, Singh WM, Capece A, Romano P (2008a) Molecular identification of yeast species associated with '*Hamei*'—a traditional starter used for rice wine production in Manipur, India. Int J Food Microbiol 124:115–125

Jeyaram K, Singh WM, Premarani T, Devi AR, Chanu KS, Talukdar NC, Singh MR (2008b) Molecular identification of dominant microflora associated with '*Hawaijar*'—a traditional fermented soybean (*Glycine max* (L.)) food of Manipur, India. Int J Food Microbiol 122:259–268
Kakati B, Goswami CU (2013) Microorganisms and the nutritive value of traditional fermented fish products of Northeast India. Global J Biosci Biotechnol 2:124–127
Keisam S, Tuikhar N, Ahmed G, Jeyaram K (2019) Toxigenic and pathogenic potential of enteric bacterial pathogens prevalent in the traditional fermented foods marketed in the Northeast region of India. Int J Food Microbiol 296:21–30
Keishing S, Banu T (2015) Fermented fish (*ngari*) of Manipur–preparation technique and its potential as a functional food ingredient. Elixir Food Sci 85:34502–34507
Konsam S, Thongam B, Handique AK (2016) Assessment of wild leafy vegetables traditionally consumed by the ethnic communities of Manipur, northeast India. J Ethnobiol Ethnomed 12(9):9. https://doi.org/10.1186/s13002-016-0080-4.
Kumari A, Pandey A, Ann A, Raj A, Gupta A, Chauhan A, Sharma A, Das AJ, Kumar A, Attri AL et al (2015) Indigenous alcoholic beverages of South Asia. In: Joshi VK (ed) Indigenous fermented foods of South Asia. CRC Press, Taylor & Francis Group, Boca Raton, pp 501–596. ISBN: 9781439887837
Majumdar RK, Bejjanki SK, Roy D, Shitole S, Saha A, Narayan B (2015) Biochemical and microbial characterization of *Ngari* and *Hentaak*-traditional fermented fish products of India. J Food Sci Technol 52:8284–8291
Mangang KCS, Das AJ, Deka SC (2017a) Comparative shelf life study of two different rice beers prepared using wild-type and established microbial starters. J Inst Brew 123:579–586
Mangang KCS, Das AJ, Deka SC (2017b) Shelf life improvement of rice beer by incorporation of *Albizia myriophylla* extracts. J Food Process Preserv 41:1–13
Maremmei K (2007) The ethno-history of the Chiru Tribe of Manipur. Ph.D. thesis, Department of History, Manipur University, Imphal, Manipur, India, p 130
Meitei NJ (2015) The Maring Tribe of Manipur—an anthropological study. Ruby Press and Co., New Delhi, p 260. ISBN: 9382395784
Nirmala C, Bisht MS, Laishram M (2014) Bioactive compounds in bamboo shoots: health benefits and prospects for developing functional foods. Int J Food Sci Technol 49:1425–1431
Nirmala C, Sharma ML, David E (2008) A comparative study of nutrient components of freshly harvested, fermented and canned bamboo shoots of *Dendrocalamus giganteus* Munro. Bamboo Sci Cult J Am Bamboo Soc 21:41–47
Nongdam P (2015) Traditional fermented bamboo shoot foods of North-East India and their characteristic natural microbial flora. In: Proceedings of the 10th World Bamboo congress, September 17–22, 2015, Damyang, Korea, vol 1, pp 876–888
Nongdam P, Tikendra L (2014) The nutritional facts of bamboo shoots and their usage as important traditional foods of Northeast India. Int Sch Res Notices 2014:1–17
Phadke G, Elavarasan K, Shamasundar B (2014) Angiotensin-I converting enzyme (ACE) inhibitory activity and antioxidant activity of fermented fish product *ngari* as influenced by fermentation period. Int J Pharm Bio Sci 5:134–142
Premlata T, Nirmala C, Bisht MS (2015) Bioactive compounds in traditional fermented bamboo shoots of Manipur, 7th Indo-Global Summit and Expo on Food & Beverages, October 08–10, 2015, New Delhi. J Food Process Technol 6:93. https://doi.org/10.4172/2157-7110.C1.030
Rai AK, Jeyaram K (2015) Health benefits of functional proteins in fermented foods. In: Tamang JP (ed) Health benefits of fermented foods and beverages. CRC Press, Taylor & Francis Group, Boca Raton, pp 455–474. ISBN: 9781466588097
Romi W (2015) Metagenomic and culture-dependent analyses of microbial communities associated with natural fermentation of different indigenous bamboo shoot products in Northeast India. Ph.D. thesis, Department of Biotechnology, Gauhati University, Guwahati, Assam, India, p 347. http://hdl.handle.net/10603/98978
Romi W, Ahmed G, Jeyaram K (2015) Three-phase succession of autochthonous lactic acid bacteria to reach a stable ecosystem within 7 days of natural bamboo shoot fermentation as revealed by different molecular approaches. Mol Ecol 24:3372–3389

Romi W, Keisam S, Ahmed G, Jeyaram K (2014) Reliable differentiation of *Meyerozyma guilliermondii* from *Meyerozyma caribbica* by internal transcribed spacer restriction fingerprinting. BMC Microbiol 14:52. https://doi.org/10.1186/1471-2180-14-52

Samurailatpam SD (2014) Comparative study of late medieval URN-burial sites in Manipur. Ph.D. thesis, Department of Archaeology and Ancient History, Maharaja Sayajirao University of Baroda, Gujarat, India, p 127. http://hdl.handle.net/10603/72884

Sangma RHC, Pal R, Singh DR (2016) Edible insects of Northeast India. In: Purkayastha J (ed) Bioprospecting of indigenous bioresources of North-East India. Springer Nature, Singapore, pp 253–267. isbn:978-981-10-0619-7

Sarangthem K, Singh TN (2013) Fermentation decreases the antinutritional content in bamboo shoots. Int J Curr Microbiol App Sci 2:361–369

Sarma D, Singh AK, Baruah D (2018) Checklist of endemic ichthyofauna of North-East India. Ind J Fisheries 65:1–15

Sha SP, Suryavanshi MS, Jani K, Sharma A, Shouche Y, Tamang JP (2018) Diversity of yeasts and molds by culture-dependent and culture-independent methods for mycobiome surveillance of traditionally prepared dried starters for the production of Indian alcoholic beverages. Front Microbiol 9:2237. https://doi.org/10.3389/fmicb.2018.02237

Sha SP, Suryavanshi MS, Tamang JP (2019) Mycobiome diversity in traditionally prepared starters for alcoholic beverages in India by high-throughput sequencing method. Front Microbiol 10:348. https://doi.org/10.3389/fmicb.2019.00348

Shantibala T, Lokeshwari RK, Debaraj H (2014) Nutritional and antinutritional composition of the five species of aquatic edible insects consumed in Manipur, India. J Insect Sci 14:14. https://doi.org/10.1093/jis/14.1.14

Siltou R (2010) Thadou customs, ceremonies and festivals: an anthro–historical study. Ph.D. thesis, Department of History, Manipur University, Imphal, Manipur, India, p 151

Singh A, Singh RK, Sureja AK (2007) Cultural significance and diversities of ethnic foods of Northeast India. Indian J Tradit Knowl 6:79–94

Singh HB, Kumar B, Singh RS (2003) Bamboo resources of Manipur: an overview for management and conservation. J Bamboo Rattan 2:43–55

Singh HD, Singh AS, Singh NR, Singh NR (2011) Biochemical composition of *soibum*—a fermented bamboo shoot and its dynamics during fermentation in real time model. 2011 international conference on food engineering and biotechnology, May 7–9, 2011, Bangkok, Thailand. Int Proc Chem Biol Environ Eng 9:198–202

Singh MK, Ibrahim M, Bordoloi S, Meena D, Kakoti S (2018a) Bamboo diversity, distribution and utility in forest fringe villages of Manipur (India). Pharma Innov J 7:503–507

Singh PK, Devi SP, Devi KK, Ningombam DS, Athokpam P (2010) *Bambusa tulda* Roxb. in Manipur state, India: exploring the local values and commercial implications. Not Sci Biol 2:35–40

Singh PK, Singh KI (2006) Traditional alcoholic beverage, *Yu* of Meitei communities of Manipur. Ind J Tradit Knowl 5:184–190

Singh PP, Rathore V, Singh LB (1986) A possible role of fish preparation '*Hentak*' in urolithiasis in Manipur—an experimental study. Indian J Exp Biol 24:88–90

Singh RK, Baghel SS (2003) Aromatic rices of Manipur. In: Singh RK, Singh US (eds) A treatise on the scented rices of India. Kalyani Publishers, New Delhi, pp 347–335. ISBM: 9788127210311

Singh SS, Mandal SE, Lalnunmawii E, Kumar NS (2018b) Antimicrobial, antioxidant and probiotics characterization of dominant bacterial isolates from traditional fermented fish of Manipur, North-East India. J Food Sci Technol 55:1870–1879

Singh SY, Bera BK (2017) Present scenario of agricultural sector of Manipur. Econ Aff 62:225–232

Singh TA, Devi KR, Ahmed G, Jeyaram K (2014) Microbial and endogenous origin of fibrinolytic activity in traditional fermented foods of Northeast India. Food Res Int 55:356–362

Singh TA, Sarangi PK, Singh NJ (2018c) Traditional process foods of the ethnic tribes of western hills of Manipur, India. Int J Curr Microbiol App Sci 7:1100–1110

Singhal P, Bal LM, Satya S, Sudhakar P, Naik SN (2013) Bamboo shoots: a novel source of nutrition and medicine. Crit Rev Food Sci Nutr 53:517–534

Sinha SP (2007) Lost opportunities: 50 years of insurgency in the North-East and India's response. Lancer Publishers, New Delhi, p 357. ISBN: 8170621623

Soibam H, Ayam VS (2018) The traditional fermented foods of Meiteis of Manipur, India: a case study. J Pharmacogn Phytochem 7:535–539

Sonar NR, Halami PM (2014) Phenotypic identification and technological attributes of native lactic acid bacteria present in fermented bamboo shoot products from North-East India. J Food Sci Technol 51:4143–4148

Sonar NR, Vijayendra SVN, Prakash M, Saikia M, Tamang JP, Halami PM (2015) Nutritional and functional profile of traditional fermented bamboo shoot based products from Arunachal Pradesh and Manipur states of India. Int Food Res J 22:788–797

Tamang B, Tamang JP (2009) Traditional knowledge of biopreservation of perishable vegetables and bamboo shoots in Northeast India as food resources. Indian J Tradit Knowl 8:89–95

Tamang B, Tamang JP, Schillinger U, Franz CMAP, Gores M, Holzapfel WH (2008) Phenotypic and genotypic identification of lactic acid bacteria isolated from ethnic fermented bamboo tender shoots of North East India. Int J Food Microbiol 121:35–40

Tamang JP (2010) Himalayan fermented foods: microbiology, nutrition and ethnic values. CRC Press, Taylor & Francis Group, New York, p 295. isbn:978-1-4200-9324-7

Tamang JP, Dewan S, Tamang B, Rai A, Schillinger U, Holzapfel WH (2007) Lactic acid bacteria in *Hamei* and *Marcha* of North East India. Indian J Microbiol 47:119–125

Tamang JP, Tamang B, Schillinger U, Franz CMAP, Gores M, Holzapfel WH (2005) Identification of predominant lactic acid bacteria isolated from traditionally fermented vegetable products of the Eastern Himalayas. Int J Food Microbiol 105:347–356

Tamang JP, Tamang B, Schillinger U, Guigas C, Holzapfel WH (2009) Functional properties of lactic acid bacteria isolated from ethnic fermented vegetables of the Himalayas. Int J Food Microbiol 135:28–33

Tamang JP, Tamang N, Thapa S, Dewan S, Tamang B, Yonzan H, Rai AK, Chettri R, Chakrabarty J, Kharel N (2012) Microorganisms and nutritional value of ethnic fermented foods and alcoholic beverages of North East India. Indian J Tradit Knowl 11:7–25

Thaimei L (2002) History and culture of the Kabui Rongmei people of North-East India. Ph.D. thesis, Department of History, Manipur University, Imphal, Manipur, India, p 170. http://hdl.handle.net/10603/103877

Thakur K, Tomar SK (2015) Exploring indigenous *Lactobacillus* species from diverse niches for riboflavin production. J Young Pharm 7:122–127

Thakur K, Tomar SK (2016) *In vitro* study of riboflavin producing lactobacilli as potential probiotic. LWT Food Sci Technol 68:570–578

Thapa N (2016) Ethnic fermented and preserved fish products of India and Nepal. J Ethnic Foods 3:69–77

Thapa N, Pal J, Tamang JP (2004) Microbial diversity in *ngari*, *hentak* and *tungtap*, fermented fish products of North-East India. World J Microbiol Biotechnol 20:599–607

Thokchom S, Joshi SR (2015) Physicochemical analysis of ethnically fermented soybean products of North-East India and molecular characterization of associated lactic acid bacteria. Proc Natl Acad Sci India B Biol Sci 85:527–533

Tuikhar N, Keisam S, Labala RK, Imrat, Ramakrishnan P, Arunkumar MC, Ahmed G, Biagi E, Jeyaram K (2019) Comparative analysis of the gut microbiota in centenarians and young adults shows a common signature across genotypically non-related populations. Mech Age Dev 179:23–35

Vishwanath W, Sarojnalini C (1989) Changes in the total oxalate content in the fermentation of fish paste *Hentak*. J Food Sci 54:754–755

Waikhom SD, Louis B, Sharma CK, Kumari P, Somkuwar BG, Singh WM, Talukdar NC (2013) Grappling the high altitude for safe edible bamboo shoots with rich nutritional attributes and escaping cyanogenic toxicity. Biomed Res Int 2013:1–11

Wanglar WA, Arunkumar L, Tarao SM (2018) Fish preservation techniques practised by the Moyon tribe, Chandel district Manipur. Int J Recent Sci Res 9:28622–28626

Ethnic Fermented Foods and Beverages of Meghalaya

15

Santa Ram Joshi, Welfareson Khongriah, and Koel Biswas

Abstract

Meghalaya is primarily inhabited by the indigenous tribes comprising of Khasi, Jaintia and Garo. The tribal population is known for practicing various traditional ways of preserving and storing foods. Among them, spontaneous fermentation of soybean and fish using indigenous technology are most prevalent. The raw materials for the process are acquired locally and are subjected to fermentation where microflora appearing spontaneously carry out the biochemical process. The resultant products have unique flavour and taste which forms an ingredient of the regular food habits of the tribes. Among various microorganisms that have been described from these foods, lactic acid bacteria and yeasts are primarily involved in the fermentation. The end product of fermentation not only helps in processing but also is an indigenous method of long-term storage of the products. The microflora prevalent in the processing consists of health-promoting bacteria of probiotics category. The fermented soybean is locally called as *tungrymbai*, while the fermented fish product is called *tungtap*. The traditional production of *tungrymbai* uses soybean (*Glycine max* (L.)) where the preparation involves washing, boiling and wrapping in leaves (*Pyrnium pubinerve*) followed by incubation. In case of *tungtap*, fresh fish (*Puntius* spp. or *Danio* spp.) are used as raw materials. Preparation of *tungtap* can be achieved by two methods, i.e. pre-fermentation salting or post-fermentation salting.

Keywords

Ethnic foods · Fermentation · Lactic acid bacteria · Probiotic properties

---

S. R. Joshi (✉) · W. Khongriah · K. Biswas
Microbiology Laboratory, Department of Biotechnology and Bioinformatics, North-Eastern Hill University, Shillong, Meghalaya, India

J. P. Tamang (ed.), *Ethnic Fermented Foods and Beverages of India: Science History and Culture*, https://doi.org/10.1007/978-981-15-1486-9_15

15.1 Introduction

Meghalaya lies between latitude 20°1′ N and 26°5′ N and longitude 85°49′ E and 92°53′ E and covers an area of 22,429 km^2 (Chakraborty et al. 2012). It is home to a unique array of biodiversity of flora, fauna, microbes and even aquaculture ranging from tropical and sub-tropical to temperate or near temperate and is located within the biodiversity hotspot of Indo-Burma region. State is a treasure of ethnic diversity of fermented foods and their related products consumed by the local people. Among many of the traditionally fermented products consumed by the ethnic tribes, non-alcoholic fermented foods such as fish and soybean products (*tungtap* and *tungrymbai*) occupy important place in the dietary habits of the indigenous people.

Tungrymbai and *tungtap* are traditional non-alcoholic fermented foods of Meghalaya prepared from soybean and dried fish, respectively. These are essential source of proteins especially in remote areas where meat is not abundantly available. These are used as food additives by the indigenous people mainly Khasi, Jaintia and Garo tribes of Meghalaya (Sohliya et al. 2009) from among more than 250 different varieties of fermented foods and alcoholic beverages prepared and consumed in the Northeast region (Tamang et al. 2012). Preparation of fermented soybeans and fish dates back since time immemorial. Fermented soybean is known by different names in different regions of Northeast India, viz. *kinema* of Sikkim, *hawaijar* in Manipur, *bekang* in Mizoram, *aakhone* in Nagaland, *peruyyan* in Arunachal Pradesh and *tungrymbai* in Meghalaya. They are similar to *natto* of Japan, *thua-nao* of Thailand, *pe-poke* of Myanmar and *douchi* of China (Tamang 2015).

Similarly, fermented fish is an important food product in East and Southeast Asia and are known by different local names in various parts such as *peda, bekasam, kecap and jambal* in Indonesia; *Patis* in Philippines; *Nuoc-mam* in Cambodia dan Vietnam; *Nam pla* in dan Laos Thailand; *luxia-you dan Jepang* in China; *Sikhe* in liquid part of *Jeotkal* in Korea. *Ngari* of Manipur, *shidal* and *numsing* of Assam and Tripura and *tungtap* of Meghalaya are some of the commonly known fermented fishes of Northeast India.

According to Nout and Motarjemi (1997), fermented foods are typically unique and vary according to regions due to the variation in climate, social patterns, consumption practices and most importantly the availability of raw materials. Availability of raw materials brings about the conversion of these resources to different form of fermented food products in order to increase the food varieties as well as to maintain food security. Even before refrigeration, canning and other modern preservation techniques became available; food fermentation had been an important preservation method. A modern approach of biopreservation has been gaining importance where there is application of lactic acid bacteria (LAB) called probiotic bacteria to the fish and soybean. Probiotics are one of the functional foods that link diet and health. They are health-promoting microbial food ingredients that have a beneficial effect on humans (Chuayana Jr. et al. 2003).

15.2 History of Fermented Soybeans and Fish: Sources of Protein-Rich Foods

Foods are the basic survival needs for human being. Since ancient times, various methods have been used to process and to preserve foods. Various techniques of preserving cereal and seafood products are well-developed in Southeast Asia. Fermented foods are usually prevalent in Southeast Asia to balance the fluctuation in food availability in the area during the stage of monsoonal circulation (Law et al. 2011). Fermentation is the oldest known form of food biotechnology providing a means for producing safe and well-preserved food which is used in households, in small-scale food industries as well as in large enterprise.

Although some studies have been conducted on the microbial diversity of the fermented soybean foods of Northeast India (Singh and Umabati 1995; Tamang 2003; Jeyaram et al. 2008; Sohliya et al. 2009; Thokchom and Joshi 2012a, b), very few studies have evaluated the probiotic properties, antibiotic tolerance, and anti-bacterial activity of the associated microbes (Thokchom and Joshi 2012a, b). There are very few reports on the bile salt and low acid tolerance and cell surface hydrophobicity of LAB isolated from these fermented foods. Normal bacteria and other microbial yeasts responsible for these food fermentations and the percentage of dominant LAB of various species have been analysed in some fermented products of Northeast India (Tamang et al. 2012). The occurrence of *Bacillus subtilis* and *Bacillus cereus* in higher proportion compared to *Staphylococcus* spp. with another LAB are reported in *Hawaijar* from Manipur (Jeyaram et al. 2008).

15.3 Ethnic Fermented Foods of Meghalaya

15.3.1 *Tungrymbai*

Tungrymbai is an ethnically fermented soybean food prepared by the indigenous *Khasi* tribes of Meghalaya. It forms an intricate part of the diet and serves as a cheap

Soybean seeds (*Glycine max* (L.)) boiled for about 45 min to 2 hours until softened

↓

Excess water drained off and left opened for temperature to drop down to 25-30°C

↓

Transferred to bamboo basket aligned with *Pyrnium pubinerve* leaves

↓

Fermentation at 25-30°C for 3-4 days

↓

Fermented soybean seeds crushed lightly in wooden mortar and pestle

↓

Tungrymbai

Fig. 15.1 Flowchart for preparation of *Tungrymbai*

Fig. 15.2 Raw material and product of fermented soybean- *Tungrymbai*. (**a**) Raw soybean (*Glycine max* (L.); (**b**) fermented soybean-Tungrymbai; and (**c**) cooked *Tungrymbai*

source of high protein food. Locals of this region eat *tungrymbai* as a side dish with rice.

15.3.1.1 Traditional Method of Preparation

It is prepared by boiling soybean seeds for about 45 min to 2 h until softened. The excess water is drained off and left opened for temperature to drop down to 25–30 °C. The seeds are transferred to bamboo basket aligned with *Pyrnium pubinerve* leaves and fermented at 25–30 °C for 3–4 days. The fermented soybean seeds are crushed lightly in wooden mortar and pestle to obtain the final product (Sohliya et al. 2009) (Figs. 15.1 and 15.2).

15.3.1.2 Microbiology

Bacillus subtilis, B. pumilus and *B. licheniformis, Enterococcus faecium, E. hirae, E. raffinosus, E. durans, E. cecorum, Lactobacillus brevis* and yeasts like *Saccharomyces cerevisiae*, *Debaryomyces hansenii* and *Pichia burtonii* (Tamang et al. 2012; Chettri and Tamang 2015) are the common microbes found in *Tungrymbai*.

15.3.1.3 Nutritional Value

Moistures 60.0%; pH 7.6; protein 45.9 g/100 g, fat 30.2 g/100 g, fibre 12.8 g/100 g, carotene 212.7 μg/100 g and folic acid 200 μg/100 g. Antioxidant activities such as DPPH scavenging activity 670.9 μg/mL, ABTS radical scavenging activity 190.9 μg/mL and total phenol content 2.6 mg GAE/g fresh weight. Saponin content (Group B) 447.9 mg/100 g (Tamang et al. 2012; Chettri and Tamang 2014).

15.3.2 *Tungtap*

Tungtap is a popular fermented fish (*Puntius* spp. and/or *Danio* spp.) product, commercially prepared and consumed by the *Khasi* and *Jaintia* tribes of Meghalaya. *Tungtap* is consumed as pickle and taste enhancer (Rapsang and Joshi 2012).

Fresh fish (*Puntius* spp. or *Danio* spp.) collected and sundried for 3-4 days

↓

Dried fish kept in Jute bag (30-40Kg batches)

↓

Dried fish from packed jute bag taken out and washed properly, dried in the sun for 3-4 days

(a)

↓

30-40 Kg batches taken and supplemented with some amount of fish fats.

↓

Fish salted in about 1:10 ratio (salt:fish)

↓

Packed in oil smeared earthen pots which sealed using fish scales, clay oil slurry or polythene sheets to make the earthen pots air tight.

↓

The earthen pots are then kept at room temperature (18-28°C) for 3-6 months for fermentation.

↓

Products taken out and sold in retailed market.

(a)

(b)

↓

30 -40 Kg batches taken and supplemented with some amount of fish fats

↓

Packed with oil smeared earthen pots which sealed using fish scales, clay, oil slurry or polythene sheets to make the earthen pots airtight

↓

The earthen pots are then kept at room temperature (18-28°C) for 3-6 months for fermentation.

↓

The fish are taken out and salted about 1:10 ratio (salt:fish)

↓

Products taken out and sold in in retail market.

(b)

Fig. 15.3 Flow chart showing the traditional preparatory steps involved in the fermentation of fish into *Tungtap* through (**a**) pre-fermentation salting and (**b**) post-fermentation salting

15.3.2.1 Traditional Method of Preparation

Traditional preparation of *tungtap* can be done by two methods, the pre-fermentation salting and the post-fermentation salting. In pre-fermentation salting, the salting is done prior to fermentation. After sun-drying, about 30–40 kg batches are supplemented with some amount of fish fats followed by salting in about 1:10 ratio (salt/fish). After salting, the mixture is then packed into oil smeared earthen pots which

Fig. 15.4 Pictorial representation of the steps in preparation of Tungtap. (**a**) Fresh *Puntius* spp.; (**b**) drying; (**c**) collection after sundried; (**d**), (**e**) and (**f**) washing steps; (**g**) dewater; (**h**) salting; (**i**) fermented *tungtap*

are then sealed using fish scales, clay, oil slurry or even polythene sheets to make the earthen pot airtight. The earthen pots are then kept at room temperature (18–28 °C) for 3–6 months for fermentation. The product are taken out after the fermentation and sold in the local retail markets. Whereas in post-fermentation salting, the method of preparation is similar except that the salting of fish was done after fermentation prior to selling in the local market (Rapsang and Joshi 2012) (Figs. 15.3 and 15.4).

15.3.2.2 Microbiology

Microorganisms involved during fermentation of *tungtap* are as follows: *Lactobacillus amylophilus, Lactobacillus coryniformis* subsp. *torques, Lactobacillus plantarum, Lactobacillus fructosus, Lactococcus lactis* subsp. *cremoris, Lactococcus plantarum, Enterococcus faecium, Bacillus subtilis, Bacillus pumilus and Micrococcus* and yeast species of *Candida* and *Saccharomyces* (Thapa et al. 2004; Rapsang and Joshi 2012).

15.3.2.3 Nutritional Value

Moisture 35.4%, pH 6.2, ash 18.9%, protein 32.0%, fat 12.0%, carbohydrate 37.1%, food value 384.4 kcal/100 g, Ca 25.8 mg/100 g, Fe 0.9 mg/100 g, Mg 1.6 mg/100 g, Mn 0.8 mg/100 g and Zn 2.4 mg/100 g (Tamang et al. 2012).

Apart from *tungrymbai* and *tungtap*, *thiat* is also produced in Meghalaya which is an amylolytic starter culture used for the preparation of local fermented alcoholic beverages. For its production, glutinous rice is used as a substrate. The rice is soaked in water overnight after which it is sun-dried to drain off the excess water. Dried leaves and roots of herbs *khaw-iang* (*Amomum aromaticum*) are added to the soaked rice, and then mixture is made into thick dough by adding water. Flat to round balls are made from the dough and fermented for 1–3 days. The freshly prepared *thiat* balls are sun-dried for 3–5 days (Anupma et al. 2018).

15.3.2.4 Importance of Lactic Acid Bacteria Prevalent in Fermented Foods

Lactic acid bacteria (LAB) are the most common microorganisms found in fermented foods (Caplice and Fitzgerald 1999). They are a group of Gram-positive, low GC content, catalase-negative, non-motile bacteria characterized by their capability to ferment sugars to lactic acid. They are devoid of cytochromes and are of anaerobic nature but can tolerate air. Phylogenetically, they are a very diverse group of microorganisms and belong to the clostridial branch of Gram-positive bacteria (Aguirre and Collins 1993; Zhu et al. 2009).

In the concept of functional food, importance of LAB is mainly associated with their safe metabolic activity while growing in foods utilizing available sugar to produce organic acids and other metabolites. Apart from these, there is also an increasing interest for its probiotic properties and antimicrobial activities. The antimicrobial activities and probiotic properties have been attributed to the production of metabolites such as organic acids (lactic and acetic acid), hydrogen peroxide, ethanol, diacetyl, acetaldehyde, other low molecular mass compounds with antimicrobial activity and bacteriocins (Kwon et al. 2001; Welman and Maddox 2003; Burgess et al. 2004; De Vuyst and Leroy 2007; Rapsang et al. 2011).

15.3.2.5 Probiotic Properties of Lactic Acid Bacteria (LAB) from Fermented Foods

Living microorganisms, which get ingested in considerable numbers by host and which exert health benefits apart from the benefits of basic nutrition, are defined as probiotics (Guarner and Schaafsma 1998; Tannock 2002).

There are a number of definitions of the term 'probiotic' used over the years, but the one derived by the World Health Organization/Food and Agriculture Organization(FAO/WHO 2001; Homayouni 2009) and approved by the International Scientific Association for Probiotics and Prebiotics (Reid et al. 2003) best exemplifies the breadth and scope of probiotics as they are known today, 'live microorganisms which, when administered in adequate amounts, exert health benefit on the host'. Use of probiotic term for microorganisms needs criteria such as these microorganism must be capable of being prepared in a viable manner and on a large scale, during use and under storage, the probiotics should remain viable and stable, they should be able to survive in the intestinal ecosystem and also the host animal should gain beneficially from harbouring the probiotic (Ogueke et al. 2010). The first recorded probiotic was fermented milk for human consumption.

Some species of *Lactobacillus*, *Bifidobacterium*, *Saccharomyces boulardii* are known probiotic strains. They have numerous health benefits which take account of regulation of bowel activity and increasing of well-being to more specific, such as neutralizing food mutagens generated in colon; exerting antagonistic effect on the gastroenteric pathogens *Clostridium difficile, Campylobacter jejuni, Helicobacter pylori* and rotavirus; shifting the immune response towards a Th2 response and thereby modulating allergic reactions; and bringing down the level of serum cholesterol (Tannock 2002). However, very little work on controlled studies in humans and only some reports of animal or in vitro studies have been done (Ljungh and Wadstrom 2006).

A matter of debate remains whether the probiotic strains for use should be sourced from human or fermented foods but if the strains can be shown to survive the passage in the human gastrointestinal (GI) tract and grow in the large intestine of human, it is not a matter of concern (Ljungh and Wadstrom 2006). Survival of probiotics in GI tract depends on their growth in the stressful niche of the stomach (acidic pH and bile) with acquiring of new genes that can encode several proteins useful in stressed environments. Even if there is no microbial survival throughout the GIT, there are convincing data on beneficial immunological effects generated from dead cells (Mottet and Michetti 2005). Production of antioxidants by colonic bacteria confers a beneficial effect in free radicals scavenging, since the availability of antioxidants decreases rostrally in the GI tract, which otherwise results in lipid peroxidation in mammalian blood and liver. Lipid peroxidation can be defined as oxidative deterioration of lipids containing several carbon-carbon double bonds (Rice-Evans and Burdon 1993). LAB isolates generally produce antimicrobials metabolite(s) with action against the strain of homologous nature; nonetheless these bacterial strains additionally can also produce microbicidal substances which inhibit pathogens infecting gastric colon and intestine or compete for sites on the cell surface and mucin-binding locations (Ljungh and Wadstrom 2006). This may well be the phenomenon supporting reports that some probiotic strains hinder or reduce translocation of bacteria to the liver and to the gut. A protective consequence against development of cancer can be credited to mutagens binding by bacteria of the intestine, β-glucuronidase activity reduction, β-glucosidase and urease and bile salts deconjugation or just by enhancing the host immune system (Ljungh and Wadstrom 2006). LABs have been tested in several clinical trials in allergic diseases, and the latter has attracted considerable research interest. Probiotic strains are in general strain specific, and their property must be established individually. Endurance of bacterial isolates during production, packing, and storage of a living cell mass must be tested and appropriately declared.

Table 15.1 Probiotic bacteria and their reported effects[a]

Strain	Reported effects
Lactobacillus acidophilus LA1	Immune enhancer, adheres to human intestinal cells, modifies intestinal microflora
Lactobacillus acidophilus NCFB 1748	Decreases faecal mutagenicity, prevents radiotherapy-related diarrhoea, used in treatment of constipation
Lactobacillus GG (ATCC 53013)	Prevents diarrhoea associated with antibiotics, rotavirus and *Clostridium difficile*. Used in the treatment of Crohn's disease
Lactobacillus casei Shirota	Treatment of rotavirus diarrhoea, modifies intestinal microflora, has positive effects in the treatment of superficial bladder cancer, immune enhancer
Lactobacillus bulgaricus	Treatment of rotavirus and viral diarrhoea, modifies intestinal microflora
Bifidobacterium bifidum	Treatment of rotavirus and viral diarrhoea, modifies intestinal microflora
Lactobacillus gasseri	Survives in the gastrointestinal environment
Lactobacillus reuteri	Colonizes the intestinal tract, mainly animal studies, an emerging human probiotic
Lactobacillus plantarum	Safe in immunosuppressed patients, inhibits bacterial translocation and secondary septic responses.
Lactobacillus salivarius UCC118	Reduces faecal coliforms, *Clostridium perfringens* and enterococci levels attenuate gastrointestinal inflammation, prevents tumour formation in an IL-10 knockout mouse model
Bifidobacterium infantis UCC 35624	Reduces clostridia levels and increases lactobacilli and bifidobacteria, increased blood phagocytic activity, reduces inflammation in CD45Rbhi mouse model

[a]Adapted from O'Sullivan et al. (2005)

15.3.2.6 Health Effects of Probiotic Lactic Acid Bacteria from Fermented Foods

While there appears to be special health benefits from the consumption of traditionally fermented soy and fish foods, non-fermented one can still make a very nourishing contribution to the diet. There is no doubt that LAB and some non-LAB are responsible for food fermentation, of which some of them exert probiotic properties. Over the years probiotics have gained considerable attention on their claims that have been made on health benefits which bring in improvement of human health (Table 15.1). A decisive outcome has failed to be reached due to controversies that have been raised on various aspects despite the claimed benefits of LAB from the human clinical studies carried out for the last two decades. For example, probiotics have been proposed for cholesterol removal by assimilation of cholesterol by growing cells, incorporation of cholesterol into the cellular membrane, binding of cholesterol to cellular surface, co-precipitation of deconjugated bile and cholesterol and deconjugation of bile via bile salt hydrolase (Kumar et al. 2012). However, conditions generated under laboratory conditions would not be applicable and impractical in the in vivo systems in all the cases as these mechanisms are strain dependent.

Bile is a yellow-green aqueous solution consisting of bile acids, cholesterol, phospholipids and biliverdin (Carey and Duane 1994; Hofmann 1994). Pericentral hepatocytes of the liver synthesize bile, and it is stored and concentrated in the gall bladder, and the duodenum receives it after intake of food. Bile emulsifies and solubilizes lipids and plays an important and essential role in the digestion of fat. This property of bile also bestows potent antimicrobial activity, basically through the solubilization of bacterial membranes (Begley et al. 2005). It is an established fact that cholesterol and bile salt metabolism are closely associated even though the mechanism of hypocholesterolaemic activity of probiotics has not yet been completely understood. To explain cholesterol-lowering effects of probiotics, 'BSH hypothesis' has been proposed in the recent past (Kumar et al. 2012). The hypocholesterolaemic effects of some probiotics displaying high BSH activities from in vitro trials have been established in human as well as in animal systems (Kumar et al. 2012). On the other hand, the hypocholesterolaemic property of probiotics based on the BSH hypothesis is yet to be adequately elucidated. In addition, more studies are needed to be carried out to establish whether the BSH activity of the probiotic strains is good or harmful to the host considering that several commercial probiotic strains exhibit high BSH activities. Tolerance to bile is primary and fundamentally important in probiotic research, as such property makes bacteria to cope with and remain viable in its journey along the duodenum and ultimately to survive and colonize the epithelia of the gut. This parameter considered primary importance in the selection of probiotic strains. Thus, it is important to recognize the physiological and molecular mechanisms by which microorganisms like bifidobacteria have come through the evolutionary passage to survive the antimicrobial potential of bile in the GI tract. Added research on the genes such as *bsh* with regard to their conserved and variable elements from various species could aid in the development of nontraditional phylogenetic markers. Also, exploring bile salt hydrolase activity on the physiology of the bacterial and mammalian cells will be one of the major future challenges (Kumar et al. 2012).

15.3.2.7 Safety Uses of Lactic Acid Bacteria

Lactic acid bacteria are GRAS (generally recognized as safe) organisms. However, there are several case reports in the literature on systemic infections caused by LAB. The clear majority deals with severely immune-compromised patients. It should also be borne in mind that the taxonomy of several LAB has been reconstructed during the last decade, and the use of modern polyphasic taxonomy has reclassified several probiotic strains (Klein et al. 1998; Hoa et al. 2000; Temmermann et al. 2004). There have been case reports of systemic spread of LAB, including strains of species used as probiotics. Molecular biology techniques, however, have been successful in distinguishing the causative agent, in patient with a liver abscess, which was indistinguishable from a strain used as probiotics, *L. rhamnosus* GG (Rautio et al. 1999). Most probiotic microorganisms belong to LAB, such as *Lactobacillus* sp., *Bifidobacterium* sp. and *Enterococcus* sp. (Klein et al. 1998). In some countries, the use of *Enterococcus* sp. as a probiotic has been questioned because of safety aspects with regard to transfer of genes conferring antibiotic resistance (Lund and Edlund 2001).

Probiotic strains should carry few, if any, mechanisms for antibiotic resistance and preferably no plasmids with antibiotic resistance. Several of the available probiotic *Bacillus* products expressed high levels of antibiotic resistance (Hoa et al. 2000). Generally, probiotic strains carry a very low risk of causing infection. Many probiotic products have been used traditionally over generations and proven to be safe. Since different characteristics of strains are strain specific, reports of systemic infections caused by various species should not make one to exclude these species as probiotics. Each strain should be evaluated in tests for safety, but so far, there is no standard test(s) recommended (Ljungh and Wadstrom 2006).

15.4 Conclusion

Traditional fermentation of Tungrymbai, Tungtap and local fermented beverages in Meghalaya is an old age ethnic practiced method whereby people have adopted the spontaneous fermentation based on the oral knowledge that has been inherited from their forefathers. This tradition is still in practice at present which has not been properly documented scientifically. There is a need for research studies for documentation as well as scientific validation of the process as well as the product. This will result in developing technology to obtain nutritionally enriched fermented products that have benefits from the prospective of both health promoting and commercialization of this traditional technology. The present write up is an attempt to provide the information regarding the traditional method of preparation of fermented foods, identification of culturable microorganisms, nutritive value and beneficial aspects like probiotic properties of traditionally fermented products prepared and consumed by the indigenous tribes of Northeastern part of India. However, metagenomic study on these fermented foods is expected to provide deeper insights into the microbial diversity and dynamics encountered at different stages during the fermentation of these products.

References

Aguirre M, Collins MD (1993) Lactic acid bacteria and human clinical infection. J Appl Bacteriol 75(2):95–107

Anupma A, Pradhan P, Sha SP, Tamang JP (2018) Traditional skill of ethnic people of the Eastern Himalayas and North East India in preserving microbiota as dry amylolytic starters. Indian J Tradit Knowl 17(1):184–190

Begley M, Gahan CGM, Hill C (2005) The interaction between bacteria and bile. FEMS Microbiol Rev 29:625–651

Burgess C, O'Connell-Motherway M, Sybesma W, Hugenholtz J, van Sinderen D (2004) Riboflavin production in *Lactococcus lactis*: potential for in situ production of vitamin-enriched foods. Appl Environ Microbiol 70:5769–5777

Caplice E, Fitzgerald GF (1999) Food fermentations: role of microorganisms in food production and preservation. Int J Food Microbiol 50:131–149

Carey MC, Duane WC (1994) Enterohepatic circulation. In: Arias IM, Boyer N, Fausto N, Jackoby WB, Schachter DA, Shafritz DA (eds) The liver: biology and pathobiology. Raven Press, Ltd., New York

Chakraborty R, Deb B, Devanna N, Sena S (2012) North-East India an ethnic storehouse of unexplored medicinal plants. J Nat Prod Plant Resour 2(1):143–152

Chettri R, Tamang JP (2014) Functional properties of *tungrymbai* and *bekang,* naturally fermented soybean foods of India. Int J Ferment Foods 3:87–103

Chettri R, Tamang JP (2015) *Bacillus* species isolated from *tungrymbai* and *bekang*, naturally fermented soybean foods of India. Int J Food Microbiol 197:72–76

Chuayana EL Jr, Ponce CV, Rivera MRB, Cabrera EC (2003) Antimicrobial activity of probiotics from milk products. Phil J Microb Infect Dis 32:71–74

De Vuyst L, Leroy F (2007) Bacteriocins from lactic acid bacteria: production, purification, and food applications. J Mol Microbiol Biotechnol 13:194–199

FAO/WHO (2001) Health and nutritional properties of probiotics in food including powder milk with live lactic acid bacteria. Food and Agriculture Organization of the United Nations and World Health Organization Expert Consultation Report

Guarner F, Schaafsma GJ (1998) Probiotics. Int J Food Microbiol 39:237–238

Hoa NT, Baccigalupi L, Huxham A et al (2000) Characterization of *Bacillus* species used for oral bacteriotherapy and bacteriprophylaxis of gastrointestinal disorders. Appl Environ Microbiol 66:5241–5247

Hofmann AF (1994) Bile acids. In: Arias IM, Boyer JL, Fausto N, Jackoby WB, Schachter DA, Shafritz DA (eds) The liver: biology and pathobiology. Raven Press, Ltd., New York

Homayouni A (2009) Letter to the editor. Food Chem 114(3):1073

Jeyaram K, Mohendro SW, Premarani T et al (2008) Molecular identification of dominant microflora associated with 'Hawaijar'—a traditional fermented soybean (*Glycine max* (L.)) food of Manipur, India. Int J Food Microb 122:259–268

Klein G, Pack A, Bonaparte C, Reuter G (1998) Taxonomy and physiology of probiotic lactic acid bacteria. Int J Food Microbiol 41:103–125

Kumar M, Nagpal R, Kumar R et al (2012) Cholesterol-lowering probiotics as potential biotherapeutics for metabolic diseases. Exp Diabetes Res. https://doi.org/10.1155/2012/902917

Kwon S, Yoo IK, Lee WG, Chang HN, Chang YK (2001) High-rate continuous production of lactic acid by *Lactobacillus rhamnosus* in a two-stage membrane cell-recycle bioreactor. Biotechnol Bioeng 73:25–34

Law SV, Abu BF, Mat HD, Abdul HA (2011) Popular fermented foods and beverages in Southeast Asia. Int Food Res J 18:475–484

Ljungh A, Wadstrom T (2006) Lactic acid Bacteria as probiotics. Curr Issues Intest Microbiol 7:73–90

Lund B, Edlund C (2001) Probiotic *Enterococcus faecium* strains is a possible recipient of the vanA gene cluster. Clin Infect Dis 32:1384–1385

Mottet C, Michetti P (2005) Probiotics: wanted dead or alive. Dig Liver Dis 37:3–6

Nout MJR, Motarjemi Y (1997) Assessment of fermentation as a household technology for improving food safety: a joint FAO/WHO workshop. Food Control 8(5–6):221–226

O'Sullivan GC, Kelly P, O'Halloran S, Collins C, Collins JK, Dunne C, Shanahan F (2005) Probiotics: an emerging therapy. Curr Pharm Des 11:3–10

Ogueke CC, Owuamanam CI, Ihediohanma NC, Iwouno JO (2010) Probiotics and prebiotics: unfolding prospects for better human health. Pak J Nutr 9:833–843

Rapsang GF, Kumar R, Joshi SR (2011) Identification of *Lactobacillus pobuzihii* from *tungtap*: a traditionally fermented fish food and analysis of its bacteriocinogenic potential. Afr J Biotechnol 10(57):12237–12243

Rapsang GF, Joshi SR (2012) Bacterial diversity associated with *tungtap,* an ethnic traditionally fermented fish product of Meghalaya. Indian J Tradit Knowl 11(1):134–138

Rautio M, Jousimies-Somer H, Kauma H et al (1999) Liver abscess due to a *Lactobacillus rhamnosus* indistinguishable from *L. rhamnosus* strain GG. Clin Infect Dis 28:1159–1160

Reid G, Jass J, Sebulsky MT, McCormick JK (2003) Potential uses of probiotics in clinical practice. Clin Microbiol Rev 16:658–672

Rice-Evans C, Burdon R (1993) Free radical-lipid interactions and their pathological consequences. Prog Lipid Res 32(1):71–110
Singh NI, Umabati DA (1995) Fermentation prospects of two phylloplane bacteria in traditional *hawaijar* made from boiled soybean (*Glycine max* L.). J Food Sci Technol 32:219–220
Sohliya I, Joshi SR, Bhagobaty R, Kumar R (2009) *Tungrymbai*-a traditional fermented soybean food of the ethnic tribes of Meghalaya. Indian J Tradit Knowl 8(4):559–561
Tamang JP (2003) Native microorganisms in fermentation of kinema. Indian J Microbiol 43(2):127–130
Tamang JP, Tamang N, Thapa S et al (2012) Microorganisms and nutritional value of ethnic fermented foods and alcoholic beverages of North East India. Indian J Tradit Knowl 11(1):7–25
Tamang JP (2015) Naturally fermented ethnic soybean foods of India. J Ethnic Foods 2:8–17
Tannock GW (2002) Probiotics and prebiotics. Where are we going? Caister Acad Press, Norfolk
Temmermann R, Huys G, Swings J (2004) Identification of lactic acid bacteria: culture-dependent and culture-independent methods. Trends Food Sci Technol 15:348–359
Thapa N, Pal J, Tamang JP (2004) Microbial diversity in *Ngari, Hentak* and *Tungtap* fermented fish products of North-East India. World J Microbiol Biotechnol 20:599–607
Thokchom S, Joshi SR (2012a) Microbial and chemical changes during preparation in the traditionally fermented soybean product *Tungrymbai* of ethnic tribes of Meghalaya. Indian J Tradit Knowl 11(1):139–142
Thokchom S, Joshi SR (2012b) Antibiotic resistance and probiotic properties of dominant lactic microflora from *Tungrymbai*, an ethnic fermented soybean food of India. J Microbiol 50(3):535–539
Welman AD, Maddox IS (2003) Exopolysaccharides from lactic acid bacteria: perspectives and challenges. Trends Biotechnol 21:269–274
Zhu Y, Zhang Y, Li Y (2009) Understanding the industrial application potential of lactic acid bacteria through genomics. Appl Microbiol Biotechnol 83(4):597–610

Ethnic Fermented Foods and Beverages of Mizoram

16

K. Thanzami and H. Lalhlenmawia

Abstract

A description of Mizoram state and its agricultural status, dietary habits and different ethnic fermented products are briefly discussed in this chapter. Traditional preparation methods of different fermented foods like *bekang-um* (fermented soybean), *sa-um* (fermented pork fat), *chhi-um* (fermented sesame seeds), *ai-um* (fermented crabs with crushed sesame seeds), *aite-um* (fermented whole crab), *tam-um* (fermented mustard leaves) and *tuai-um* (fermented bamboo shoots) and ethnic fermented alcoholic drinks such as *zupui* (main alcoholic beverage), *zufâng* (sweet rice beer), *tin-zu* (strong undistilled rice beer) and *rakzu* (strong distilled rice beer) are discussed in details with flowchart. The role of women and poorer section of the population on socio-economic importance of the fermented products for income generation has also been mentioned. The ethnical importance of traditional alcoholic drinks during religious rituals and festivals of Mizo before embracing Christianity has been discussed. The nutritional contents, microorganisms involved and other biochemical parameters of some of the traditionally fermented food products are highlighted and compared to similar fermented foods and beverages of other regions.

Keywords

Mizoram · Traditional foods · Traditional drinks · Fermentation

K. Thanzami (✉) · H. Lalhlenmawia
Department of Pharmacy, Regional Institute of Paramedical and Nursing Sciences (RIPANS), Aizawl, Mizoram, India

J. P. Tamang (ed.), *Ethnic Fermented Foods and Beverages of India: Science History and Culture*, https://doi.org/10.1007/978-981-15-1486-9_16

16.1 Introduction

Mizoram is one of the eight states of Northeast India situated in the southernmost landlocked state which shares an international boundary with Myanmar in the east and south and Bangladesh to the west and to the northwest and shares an interstate boundary with Tripura to the north and Assam and Manipur to the northeast. The capital of Mizoram is Aizawl and is situated in the north-central part of the state at an altitude of approximately 4000 ft. above the sea level. Mizoram, 'Land of the Mizos', was known as the Lushai Hills District of Assam before it was renamed the Mizo Hills District in 1954 (The Lushai Hill District (Change of Name) Act 1954]. In 1972, it became a centrally administered union territory under the name of Mizoram, and in 1987 it achieved statehood. There are eight districts in the state, namely, Aizawl, Lunglei, Siaha, Champhai, Kolasib, Champhai, Mamit and Lawngtlai, with average square density of 52 per sq. km and literacy rate of 91.85%. The total area of the state is 21,087 sq. km with population of 1,091,014 as per 2011 census (www.mizoram.nic.in). The state has tremendous biodiversity, and climatic conditions vary from sub-tropical humid to temperate (Sati and Rinawma 2014). As per economic classification of workers 2011 census, more than 70% of the total workers are engaged in agricultural and allied sector in Mizoram (Economic Survey Mizoram 2017–2018).

Agriculture is the main source of occupation of the state, and the main crops include cereals like wheat, rice, barley, maize, rye, oats and millets. There is also considerable cultivation of orange, hatkora, passion fruit, pineapple, banana, papaya and ginger (Sati 2017). Since rice is the staple food in Mizoram, the minimum rice requirement of the state is estimated to be around 1,80,000 MT per year. Out of the yearly requirement, only about 58,994 MT of rice is produced which is a meagre 32% of the total requirement, and the remaining has to be imported from outside (Economic Survey Mizoram 2014–2015). Shifting cultivation or 'jhumming' is a characteristic of traditional agriculture in Mizoram that is practised mainly in the highlands till today; it is also the main source of income in the rural area and is called the 'way of life of Mizos' (Anup et al. 2006). However, this type of cultivation is said to lead to enormous forest degradation and soil erosion which is not environmentally and economically sustainable (Sati 2017). In order to solve various problems faced in traditional agricultural methods, crop diversification programme has been introduced which largely focused on promoting the cultivation of cash crops like sugarcane, pulses, ginger, turmeric, Mizo chilli, oil seeds, etc. that have seasonal advantage over other crops (Economic Survey Mizoram 2014–2015).

As mentioned earlier, the staple diet of Mizo people is rice, along with vegetables and meats as side dishes, which is usually boiled and with fewer spices. Earlier, the methods of food processing among the Mizo were very simple and found to be more or less similar throughout the state, even though a slight difference may be observed from place to place. With the onset of Christianity by the British missionaries in this state in the 1890s, the methods of food preparation improved tremendously (Lalthanpuii et al. 2015).

Fermented foods and beverages are a primary part of every traditional heritage and culture. Fermentation not only presents foods which have diverse taste and nutrition but also offers a preservation technique so that fermented foods may be consumed during the off-season. Indigenous fermented foods constitute the essential component of the daily diet of Mizo people. Indigenous inhabitants of Mizoram prepared fermented food such as *sa-um* (fermented pork fats), *bekang-um* (fermented soybean), *tam-um* (fermented mustard), *ai-um* (fermented crabs), etc. (Tungoe 2016). Fermented alcoholic rice-based beverages of different forms have been reported to be consumed by the tribal people of Mizoram since time immemorial. Some fermented alcoholic beverages are *zupui*, *zufâng*, *tin-zû* and *rakzû* (Dokhuma 1992) in which *zû* means alcohol and the prefix or suffix of *zu* indicates the way the beverage is prepared. These alcoholic products have some similarities with *Yu* of Meitei tribe in Manipur; *Vatejnard* of Sikkim; *Zutho* of Angami tribe in Nagaland; *Apong* of Arunachal Pradesh; *Jou* of Bodo tribe, *Chu* of Garo tribe and *Chako* of Rabha tribe in Assam; *Chubitchi* of Garo tribe, *Kyiad* of Khasi tribe and *Sadhiar* of Jaintia tribe in Meghalaya; and *ChuwakBwtwk* and *Chuwarak* of Tripura (Tamang et al. 2012, 2016).

Earlier, the practice of consuming fermented foods and alcoholic beverages has its roots in many cultural and religious practices of the people of Mizoram and practically supplements health-promoting benefits to consumers (Table 16.1). Nowadays, many of the Mizo youths are reluctant to consume fermented products and especially fermented alcoholic beverages as they are convinced that the foods and drinks may cause harm to their body and spiritual life.

16.2 Ethnic Fermented Foods of Mizoram

16.2.1 *Bekang-um* (Fermented Soybean)

Bekang-um (Picture 16.1) is an ethnic fermented food prepared from soybean seeds by the indigenous Mizo people.

It is similar to other sticky fermented soybean foods of northeast such as *kinema* of Sikkim, *hawaijar* of Manipur, *aaxone* of Nagaland and *tungrymbai* of Meghalaya (Tamang et al. 2012). *Bekang-um* is a fermented sticky soybean which serves as a cheap source of high protein in local diet. It is consumed as it is or made into curry with the addition of salt, green chillies and tomatoes. It is commonly used as a side dish and often used as a seasoning food. During the preparation, the seeds of soybean are cleaned, washed and soaked in water for 10–12 h. After removing the excess water, the beans are cooked/boiled for about 1 h until they become soft. Excess water is drained off; placed on a traditional tray made of bamboo called *kho* lined with previously cleansed leaves of *Callicarpa arborea*, locally called *hnahkiah*; sprinkled with ashes; and again covered with *hnahkiah* leaves. The tray is near the earthen oven or in warm place and is allowed to ferment naturally for 3–4 days (Chettri and Tamang 2015). The method of preparation of *bekang-um* is highlighted in Fig. 16.1.

Table 16.1 Ethnic fermented foods and beverages of Mizoram

Food	Substrate	Sensory property of product	Major ethnic consumers	Region/district
Bekang-um	Soybean	Sticky, ammonical smell, alkaline taste	Whole Mizo tribes	Whole Mizoram and other states/regions where Mizo tribes reside
Sa-um	Pork fat	Creamy to pale yellow in colour, semi-solid with characteristic rancid smell	Whole Mizo tribes	Whole Mizoram and other states/regions where Mizo tribes reside
Chhi-um	Sesame seeds	Coarse paste with nutty taste with slight cheesy note, ammonical smell	Whole Mizo tribes	Whole Mizoram and other states/regions where Mizo tribes reside
Ai-um	Fresh water crab and sesame seeds	Nutty and alkaline taste with strong ammonical smell	Whole Mizo tribes	Whole Mizoram and other states/regions where Mizo tribes reside
Aite-um	Juvenile fresh water crab	Alkaline taste with strong ammonical smell	Whole Mizo tribes	Whole Mizoram mainly in the southern part
Tuai-um	Bamboo shoot	Slightly acidic and pungent smell	Whole Mizo tribes	In Mizoram, mainly in the northern belt
Dawidim	Rice, wild herbs	Dry white round cake-like mass with no characteristic smell	Whole Mizo tribes	Traditional *dawidim* is very difficult to find nowadays; maybe found in remote villages
Zupui	Rice + starter (*Dawidim*)	High alcohol content with bitter taste, whitish or pale yellow in colour, consumed with undissolved rice	Whole Mizo tribes	Not popular anymore but may be found in some remote villages
Zufâng	Sticky rice + *dawidim*	Low-alcoholic, sweet beverage, whitish or pale yellow in colour, consumed with undissolved rice	Whole Mizo tribes	Whole Mizoram, but prepared only for one's own consumption
Tin-zu	Rice + *dawidim*	High alcohol content with bitter taste, whitish or pale yellow in colour, consumed after filtering off the undissolved rice	Whole Mizo tribes	Not usually consumed anymore

(continued)

Table 16.1 (continued)

Food	Substrate	Sensory property of product	Major ethnic consumers	Region/district
Rakzu	Rice + *dawidim*	High alcohol content with bitter taste, clear liquid, consumed after distilling the filtered liquor	Whole Mizo tribes	Most popular among all the alcoholic beverages; but manufacturing and consumption is totally prohibited in the state

Picture 16.1 '*Bekang-um*', a Mizo traditional fermented soybean

16.2.2 *Sa-um* (Fermented Pork Fat)

Sa-um is an ethnic fermented pork fat common to every Mizo family. The preparation method of *sa-um* is highlighted in Fig. 16.2. Pork fat is mainly collected from the inner abdominal portion (sometimes fats from other parts of the pig's body were also used) (Lalthanpuii et al. 2015), cut into little pieces and boiled in water for an hour or 2, till most of the water evaporates out. Then, the fat is transferred into the already cleansed and sun-dried *sa-um bur* (Picture 16.2a) which is prepared from the dried fruit of the plant *ûm* (*Lagenaria siceraria*) bottle gourd (Sawmliana 2013). The container was closed tightly with a wooden lid and then kept over the fireplace or in the sun for 3–5 days. The anaerobic fermentation process could have eliminated the growth of aerobic spoilage bacteria and hence results in perfectly fermented *sa-um* (Picture 16.2b). However, no inoculation of starter culture is being done as the microbe responsible for the production of *sa-um* has not been identified yet. The fermented fat is then used to make *bai*, a typical Mizo vegetable stew, by emulsification with cooking soda. It is also used to make *bâwl*, typical Mizo chutney, again with cooking soda.

Sa-um is considered to be one of the food courses in typical Mizo's kitchen, but the risk of stomach cancer caused by *sa-um* was reported by Malakar et al. (2014).

Soybean

↓

Cleaned and washed with water

↓

Soaked overnight

↓

Cooked for 45 min-1 h with minimum water

↓

Cooled at room temperature (normal temperature) and drained off the excess water

↓

Put in a bamboo tray (kho) properly lined with clean 'hnahkiah' leaves

↓

Sprinkled clean fire ashes over the beans uniformly

↓

Covered the beans with 'hnahkiah' leaves in 2-3 Layers

↓

Keep the 'kho' in a section over kiln (rapchung) for 3-4 days.

↓

BEKANG-UM

Fig. 16.1 Steps involved in traditional method of preparation of *bekang-um*

Pork fat (lard), minced

↓ Added water

Cooked for an hour till water evaporates out

↓

Cooked lard transferred to cleansed and dried 'Ûm'

↓ Fireplace/Sun for 3-5 days

SA-UM

Fig. 16.2 Flowchart of preparation of *sa-um*

(a) (b)

Picture 16.2 (**a**) '*Sa-um*', fermented pork fat in traditional container '*sa-um bur*'. (**b**) Final product of '*sa-um*' to be stored in refrigerator for longer shelf life (Photo credit: Zawna Pachuau)

Sesame seeds lightly roasted and ground using traditional pestle and mortar (*Suk & Sum*) to make coarse paste

↓

Wrap the paste in clean 'hnahthial leaves and tie it properly

↓

Put the wrapped paste over kiln (*rapchung*) for 3-4 days

↓

CHHI-UM

Fig. 16.3 Steps involved in traditional method of preparation of *chhi-um*

It is also suggested that *H. pylori* infection interacted with consumption of *sa-um* plays a role in promoting stomach cancer in Mizoram (Phukan et al. 2006).

16.2.3 *Chhi-um* (Fermented Sesame Seeds)

Sesamum indicum, locally known as *chhibung*, is used to prepare this fermented food item.

Traditional method of *chhi-um* preparation is depicted in Fig. 16.3. Sesame seed is lightly roasted and crushed with wooden mortar and wooden pestle, which is then tightly wrapped with *hnahthial* leaves. The tightly wrapped crushed '*chhibung*' is placed over the fireplace for fermentation. After 3 days, *chhi-um* (Picture 16.3) is ready to be used as taste enhancer for other food items or consumed as chutney.

Picture 16.3 '*Chhi-um*', a Mizo traditional fermented sesame seeds (Photo credit: Zothansanga)

Sesame seeds lightly roasted and ground using traditional pestle and mortar (*Suk & Sum*)

↓

Added cleansed crabs to the sesame in *sum* and crush with ground sesame seeds to make coarse paste

↓

Wrap the paste in clean *hnahthial* leaves and tie it properly

↓

Put the wrapped paste over kiln (*rapchung*) for 3-4 days

↓

Fry the paste in oil for about 5-10 min in medium flame

↓

AI-UM

Fig. 16.4 Steps involved in traditional method of preparation of *ai-um*

16.2.4 *Ai-um* (Fermented Crabs with Sesame Seeds)

Ai-um is a traditional fermented Mizo food which is usually consumed as side dish or used as taste enhancer. Sesame seeds (*chhibung*) are lightly roasted in medium heat and pounded with *sum* and *suk*, a traditional mortar and pestle made from wood. Juvenile fresh water crabs mainly about the size of a thumb, and sometimes matured crabs, captured from the rivers and streams are killed with hot water and pounded with the ground sesame seeds to make coarse paste. Bigger crabs with thick shells may also be used to prepare *ai-um*. The steps involved in the preparation of *ai-um* are given in Fig. 16.4.

(a) (b)

Picture 16.4 (**a**) Incubation of '*ai-um*' over '*thuk*', a Mizo traditional fireplace. (**b**) '*Ai-um*', fermented crab with sesame seeds (Photo credit: Zosangliana)

Kill small crabs with hot water and wash them properly

↓

Put the crabs in a clean and dry bamboo cavity and plug with banana leaves

↓

Put the bamboo near fireplace (*tapchhak*) for 3-4 days

↓

AITE-UM

Fig. 16.5 Steps involved in traditional method of preparation of *aite-um*

The coarse paste is wrapped with *hnahthial* leaves (*Phrynium capitatum* Wild.) and placed above the fireplace (Lalthanpuii et al. 2015) (Picture 16.4a). After about 3 days, the coarse paste has been fermented to *ai-um*. This fermented product is usually served as side dish or may be consumed as it is, after frying with oil. *Ai-um* (Picture 16.4b) is sometimes mixed with roasted chilli and tomatoes to make tasty chutney. It may also be added to *bai*, a Mizo traditional vegetable stew, to enhance the flavour and the aroma.

16.2.5 *Aite-um* (Fermented Small Crabs)

This fermented food is prepared only with whole juvenile fresh water crabs which are small with soft carapace, and steps involved are highlighted in Fig. 16.5. The crabs are first killed with hot boiling water and washed properly. The crabs are then

Picture 16.5 '*Aite-um*', fermented freshwater crab (Photo credit: Zosangliana)

placed in a bamboo cavity with the opening plugged with banana leaves or wrapped with *hnahthial* leaves.

The bamboo or the wrapped leaves containing the small crabs are kept near the fireplace for 3–4 days. The crabs are fermented in the process which may be mixed with chilli and salt to make chutney or added to vegetable stew called *bai* for flavour enhancement (Picture 16.5).

16.2.6 *Ṭam-um* (Fermented Mustard Leaves)

Ṭam-um—fermented mustard leaves—is one of the most popular foods of the Mizo. For this, a variety of mustard locally known as *Ṭampui* (*Brassica rapa/B. juncea*) is used. *Ṭampui* is washed meticulously and is pounded with a wooden mortar and pestle called *sum* and *suk* and is wrapped with the leaf of *P. capitatum* locally known as *hnahthial* and kept over the fireplace for 2–3 days. After this the water is squeezed, which is locally known as *ṭam-er*, which can also be taken with food as side dish. Fresh fermented *ṭam-um* is either sun-dried or dried over the fire so that it can be stored for a longer period of time. The dried *ṭam-um* (Picture 16.6) may be lightly roasted and grinded with salt, chilli, and small amount of hot water, which is then ready to be served as side dish (Lalthanpuii et al. 2015). The method of preparation of tam-um is given in Fig. 16.6.

16.2.7 *Tuai-um* (Fermented Bamboo Shoot)

Fermented bamboo shoot or *tuai-um* is not as popular as the other traditional fermented foods of Mizo but is mainly prepared and eaten in the northern parts of Mizoram. *Mautuai* or bamboo shoot is washed, cut into smaller pieces, and pounded

Picture 16.6 '*Tam-um*', fermented mustard leaves

Clean and wash the mustard leaves and cut into smaller pieces

↓

Crush the leaves into coarse paste and wrap with clean *hnahthial* leaves

↓

Put the wrapped paste over kiln (*rapchung*) for 3-4 days

↓

ṬAM-UM

↓

Sundry the fermented product for longer shelf-life

Fig. 16.6 Steps involved in traditional method of preparation of *tam-um*

with a wooden mortar and pestle called *sum* and *suk*. The pounded *mautuai* is then wrapped with the leaf of *P. capitatum* locally known as *hnahthial* and kept over the fireplace for about 3 days. The steps involved in the traditional preparation of *tuai-um* are given in Fig. 16.7.

The pounded and fermented bamboo shoot is either sun-dried or dried over the fire so that it can be stored for a longer period of time. The dried *tuai-um* may be soaked with hot water, mixed with salt and chilli, and is then ready to be served as a very tasty side dish.

16.3 Ethnic Fermented Beverages of Mizoram

All the fermented beverages of Mizoram are alcoholic drinks and are prepared from different varieties of rice, and the processes discussed are based on the traditional methods of Mizo tribes. Most surprisingly, without any scientific knowledge, the

Clean and wash the bamboo shoots and cut into smaller pieces

↓

Crush the shoots into coarse paste and wrap with clean *hnahthial* leaves

↓

Put the wrapped paste over kiln (*rapchung*) for 3-4 days

↓

TUAI-UM

↓

Sundry the fermented product for storage

Fig. 16.7 Steps involved in traditional method of preparation of *tuai-um*

forefathers knew that starter culture is required to make alcoholic drinks, and they even developed yeast starter culture from scratch. This yeast starter culture is called *chawl* or *dawdim* which is used to make all the different kinds of traditional alcoholic beverages of Mizo tribe.

16.3.1 *Dawdim/Chawl* (Dried Starter Culture)

Dawdim or *chawl* is a traditionally prepared dried starter to produce various alcoholic beverages in Mizoram (Anupma et al. 2018). The starter culture used in all Mizo traditional alcoholic drinks is made from a bark of climber called *zangzu*, *nilengthlum* or *hawhmathlum*. Rice is first soaked in water overnight and dried for about 15 min to remove excess water. The soaked rice is made into powder by pounding it and made into homogenous paste by adding small amount of water. The paste mass is made into a small elliptical cake-like mass, about 2 in. diameter. The powdered *zangzu* is sprinkled evenly on the surface of the rice cake which is kept in a container with paddy husk. The container is incubated in a hearth for 4–5 days after which the rice ball has swollen a bit which is now called *chawl* (Picture 16.8a). Then, *chawl* is taken out and air-dried for storage. The preparation of *dawdim/chawl* is similar to other starters of Northeast India such as *marcha* of Sikkim, *hamei* of Manipur, *humao* of Assam, *thiat* of Meghalaya, etc. (Singh and Singh 2006; Jeyaram et al. 2009; Tamang 2010; Anupma et al. 2018).

The flowchart in Fig. 16.8 shows the steps involved in the making of *chawl/dawidim*.

(a) (b)

Picture 16.8 (**a**) '*Chawl*', Mizo traditional yeast starter culture and (**b**) '*zufâng*', a mild Mizo traditional rice beer (Photo credit: Zawna Pachuau)

Rice soaked in water overnight

↓

Drained water and let it sit for about 15 min

↓

Pounded with water into homogenous paste to make an elliptical mass

↓

Powdered *Zangzu* sprinkled over the elliptical mass

↓

Incubated in a warm place for 4-5 days

↓

CHAWL

↓

Air-dried for storage

Fig. 16.8 Steps involved in preparation of *chawl*

16.4 Alcoholic Beverages and Drinks of Mizoram

16.4.1 *Zupui*

Zupui is one of the most popular fermented rice-based beverages of Mizo (Dokhuma 1992). During preparation of '*Zupui*', rice is milled manually into fine powder, removing its husk using the traditional wooden mortar and pestle (*sum* and *suk*). The rice is processed specially to be made into *Zupui* from the beginning and hence is sometimes called *Hranden Zû*. *Bellâm* is a vessel dedicated specially to cook/steam the milled rice and consists of two pots. The lower pot is a bigger one where water is kept, and the smaller upper pot has some holes punched in its bottom. A

(a) (b)

Picture 16.7 (**a**) '*Nganbel*', a vessel dedicated to ferment '*Zu-pui*', a Mizo traditional rice beer. (**b**) A picture showing '*Nganbel*' and '*Ngansap*'

sieve made from bamboo called *ngânsap* is kept to cover the holes so that the milled rice stays on the upper pot without falling through the holes. The upper pot is covered as tight as possible, and the space between the upper and lower pot is also sealed using clay. Then, the rice is steamed till it is cooked properly. After the rice is cooked, it is transferred to another vessel, and a starter yeast culture called *chawl* is sprinkled over the cooked rice to start fermentation.

A special earthen pot called *ngânbel* (Picture 16.7a) is used for fermentation of the beverage. After fermentation, a small pipe made of iron called *dawnkâwn* (Picture 16.7b) is immersed into the fermenting mixture, and the rice beer is sucked through the pipe. The steps involved in traditional method of preparation of *Zupui* are given in Fig. 16.9. *Zupui* is the main beverage usually served during festive occasions and has a special place in the Mizo religion and culture.

16.4.2 *Zufâng*

Zufâng is the alcoholic beverage prepared from sticky rice. As the name suggests, *zufâng* is not a concentrated alcohol beverage as *zû* means 'alcoholic drink' and *fâng* means 'not real' and is considered to be a lesser alcoholic drink (Dokhuma 1992).

The steps involved in traditional method of preparation of *zufâng* are given in Fig. 16.10. Mizo sticky rice called *buhban*, usually *kâwnglâwng* variety, is used to prepare *zufâng*. Sticky rice is cooked and fermented in *zufângbêl*, a special vessel

Rice with husk

↓ Put in sunlight for a day for easier milling

Hand-milled using *sum* and *suk*

↓

Steam rice powder using *Bellam*

↓

Steam-cooked rice transferred to *Ngânbel*

↓

Added *chawl* to start fermentation

↓

Ferment till it becomes sweet

↓

Add water to the fermented rice beer in the vessel to increase the volume

↓

Suck liquid using a special small pipe called *dawnkâwn*

↓

ZUPUI

Fig. 16.9 Steps involved in traditional method of preparation of *Zupui*

Clean sticky rice *buhban*

↓

Cook *buhban*

↓

Transfer cooked *buhban* to *zufâng bêl*

↓

Sprinkle *chawl* over *buhban* to start fermentation

↓

Keep in warm place for 3 days & 3 nights

↓

ZUFÂNG

Fig. 16.10 Steps involved in traditional method of preparation of *zufâng*

dedicated only to make *zufâng* after a starter yeast culture called *chawl* is sprinkled over the cooked rice to start fermentation. *Zufâng* (Picture 16.8b) is a mild-alcoholic beverage with sweet taste and is consumed anytime along with the undissolved rice.

Clean rice *faisa*

↓

Cook *faisa*

↓

Transfer cooked *faisa* to any vessel for fermentation, usually a tin can

↓

Sprinkle *chawl* evenly over cooked *faisa* to start fermentation

↓

Keep in warm place for 3 days & 3 nights

↓

FAISA ZÛ/TIN ZÛ

Fig. 16.11 Steps involved in traditional method of preparation of *faisazû/tin-zû*

16.4.3 *Tin-zû*

Tin-zu also called as *faisazû* is one of the alcoholic beverages made from rice grain after the husk is removed. It does not have a vessel dedicated only for its fermentation; hence any type of vessel is used (Dokhuma 1992). It is prepared exactly similar to *zufâng* except that sticky rice is not used. Moreover, the undissolved rice is filtered out, and only the liquid part is consumed. Most of the time, a big tin can is used as a fermenting vessel and hence the name *tin-zû*. This alcoholic drink may be consumed anytime with no occasion. The method of preparation of *tin-zû* is given in Fig. 16.11.

Tin-zû was introduced by the non-Mizo soldiers during the 1890s, and many do not consider it as a Mizo traditional beverage (Lalthangliana 2005).

16.4.4 *Rakzû*

This is an alcoholic drink developed after non-Mizos entered from outside the Mizo Hills, and steam distillation principle is utilised for its preparation. The fermentation procedure is similar to *tin-zû*, but the liquefied fermentation product is again distilled using traditional steam distillation unit to obtain a clear and strong *rakzû*. The distillation unit of *rakzû* has three main components and has three different sizes of vessels which are kept over one another. The lowest vessel is the biggest one where the fermented mash of rice mixed with water is kept. The middle vessel has a big hole bored at the bottom, and a small vessel may be kept in the middle to collect the condensed alcohol. The uppermost and the smallest vessel contains cold water, the bottom of which acts as a condenser. All the vessels, after arranging them properly, are sealed with clay of wood ash so that steam cannot escape from the sides (Pictures 16.9a, b). The vessel containing the fermented rice is heated up where the steam rises up to be condensed by the topmost vessel containing cold water. The condensed alcohol is then collected by the small vessel below it, which

(a) (b)

Picture 16.9 Two ways of distillation to make '*rak-zu*'. (**a**) The lowermost container holds the fermenting mixture which is being heated; the middle container has collector for condensed alcohol fitted with pipe and is collected outside; the uppermost vessel holds cold water and acts as condenser. (**b**) The lowermost container holds the fermenting mixture which is being heated; the middle container has collector fitted inside (probably by hanging a small vessel) to collect condensed alcohol; the uppermost vessel holds cold water and acts as condenser Photo credit: (**a**) Zosangliana and (**b**) Zawna Pachuau

Tin-Zu/Faisa Zû

↓

Steam distillation using three-tiered distillation unit

↓

RAK-ZÛ

Fig. 16.12 Steps involved in method of preparation of *rakzû*

may be sampled out using the small pipe connected to the collector. The water at the topmost vessel needs to be replaced with cold water after it becomes warm.

The flowchart in Fig. 16.12 depicts the method of preparation of *rakzu*. *Rakzû* is as clear as water, but since the alcohol content is very high, one needs to be careful when consuming it. And since its concentration is high, it fetches a good price in the market and is still popular in the present days (Lalthangliana 2005). This type of alcoholic drink is also made by Chakma tribe of Mizoram, but double distilled, and goes by the name *Duichuani*.

16.5 Socio-economic and Ethnical Value

Mizoram holds lots of indigenous knowledge when it comes to medicine, food, agriculture and natural resources management. The people of Mizoram have been utilising both produces of the forest and of shifting cultivation that provides vast arrays of ethnic nutritious foods. Many wild plants and animals had been selectively consumed, and their conservation had been done for food and nutritional security (Singh et al. 2007). Fermentation plays very important role in food conservation as fermented foods and beverages are significant components of the daily dietary culture of the Mizos. Mostly, women are engaged and play very important role in the production of the fermented products as well as in preserving the ethnic knowledge (Advani 2005). Among all the traditional fermented products, alcoholic beverages had a strong importance among ethnic people of Mizoram, where most of the olden traditional festivals and many important rituals were involved with the consumption of appreciable quantities of alcohol. Traditional alcoholic beverages were offered to perform ethnic religious practices to please god, or sometimes the devil. *Zupui* was usually consumed during *Chapchar Kut*, the biggest Mizo festival which usually lasted for 6 whole days. The alcoholic beverage was served from the third day, called as *Zupui Ni*, the day which all the youths in the village come out in the open field to perform a dance called *Chai*. The fourth and the fifth days of *Chapchar Kut* were called as *Zuthing Ni* and *Thing Leh Ni*, respectively, in which the remains of alcoholic beverage *Zupui* were consumed and *Chai* dance was performed again by the youths in the open field. One can say that the last 2 days were simply repetition of *Zupui Ni*. *Zufâng*, as mentioned earlier, was not considered a real alcoholic drink and was usually consumed after doing any heavy work or offered to guests visiting their home, as how any other non-alcoholic drink is offered nowadays (Dokhuma 1992). However, with the onset of Christianity, the Mizos have been taught that consumption of alcohol is an offence and a vessel of the devil by the early Christians, and hence, consumption of alcoholic beverages was prohibited. This mentality still remains large in the minds of Mizos, and hence, much of the traditional knowledge of alcoholic beverage preparation methods is lost with time.

16.6 Microbiology and Nutritional Composition

Very few scientific studies are available on the microbiology and nutritional components of the fermented foods and beverages of Mizoram. Even the processes of preparation of those fermented products are not usually well-documented and were passed on verbally from generation to generation. It is a well-known fact that microorganisms change the nutritional constituents of substrates, be it of plant or animal origins, during fermentation. The fermentation process is usually brought about by microorganisms, mainly lactic acid bacteria, yeast and filamentous moulds. It may be believed that the same microbial reactions are involved in the production of various traditionally fermented foods and beverages of Mizoram.

Among the fermented products of Mizoram, only *bekang-um* (fermented soybean food) has been studied, and its microbiology and functionality have been reported (Omizu et al. 2011; Chettri and Tamang 2014, 2015; Tungoe 2016). It has been observed that the nutritional value of *bekang-um* (carbohydrate 267.5 mg/100 g, fat 9.952 mg/100 g, protein 5.09 mg/100 g, moisture content 54.99%, ash value 7.744%, 1.375 calories per g) slightly varies when compared with the cooked non-fermented soybean (carbohydrate 339 mg/100 g, fat 12.688 mg/100 g, protein 4.4125 mg/100 g, moisture content 55.04%, ash value 9.52%, 1.64 calories per g) (Vanlalruati and Das 2012). The pH of *bekang-um* is usually 7.2–8.0 signifying its alkaline character with ammonical flavour (Chhetri and Tamang 2014). It has also been found that *bekang-um*, after the pH crosses 8.5, is considered to be non-palatable as its ammonic smell and taste become too strong. Moreover, with prolonged fermentation time, the microbial load has increased which resulted in the high alkalinity of the fermented product (Tungoe 2016). The microorganisms involved in *bekang-um* are *Bacillus subtilis*, *B. pumilus*, *B. licheniformis*, *B. sphaericus*, *B. brevis*, *B. coagulans*, *B. circulans*, *Enterococcus faecium*, *E. hirae*, *E. raffinosus*, *E. durans* and *E. cecorum* and yeasts *Saccharomyces cerevisiae*, *Debaryomyces hansenii* and *Pichia burtonii* (Chettri and Tamang 2015), and *Bacillus* species are the predominant organisms reported in similar non-salted sticky fermented soybean foods of Asia (Tamang 2015). *Bekang-um* prepared in the laboratory of Department of Pharmacy, Regional Institute of Paramedical and Nursing Sciences (RIPANS), Aizawl, Mizoram, using Mizo traditional method showed the presence of *Bacillus sphaericus*, *B. macernas* and *B. brevis* (Tungoe 2016). With increase in fermentation time, the amount of reducing sugar present decreases, while the amount of free amino acids, protease and amylase activity increases in *bekang-um* which is also observed with other traditionally fermented soybean (Omafuvbe et al. 2000). Other studies on *bekang-um* had revealed that *Bacillus* sp. isolated shows high proteolytic activities which explains the increase in free amino acids as fermentation proceeds and improves the digestibility of proteins (Chettri and Tamang 2014). The strains of lactic acid bacteria (LAB) from *bekang-um* do not show amylolytic and proteolytic activity, but many were found to have probiotic properties (Chettri and Tamang 2014). 49% of LAB strains isolated from *bekang-um* could cause coagulation of milk at 30 °C with significant drop in pH that shows their potential as starters or adjunct cultures in the production of fermented products (Chettri and Tamang 2014). LAB isolated could not produce antimicrobial bacteriocin and biogenic amines under the applied study condition, while 45% of LAB strains degraded phytic acid, 35% degraded raffinose and 20% degraded both phytic acid and raffinose that are antinutritive factors, hence making soybean fit for consumption (Chettri and Tamang 2014). *Bekang-um* is also found to contain phenolic compounds, considered to be one of the most important antioxidant compounds, ranging from 2.6 to 4.2 mg GAE/g fresh weight depending upon the place of sampling (Chettri and Tamang 2014). Moreover, *bekang-um* has been shown to exhibit certain amount of DPPH scavenging (IC50 values ranging from 456.7 to 493.3 μg/mL) and ABTS scavenging (IC50 values ranging from 145 to 170 μg/mL) activities (Chhetri and Tamang 2014).

Biochemical analysis of *sa-um* showed that the moisture content is 6.21% by weight and fat content accounted for 91%; protein, 0.7%; and carbohydrates, 2% with a total ash value of 0.1 (Mandal et al. 2018). The high fat content is reflected in its calorific value, i.e. 830 kcal/100 g, and the pH of *sa-um* is found to be slightly acidic (6.60). It is also seen that *sa-um* contains traces of iron, zinc and magnesium, while sodium, potassium and calcium were present in higher quantity. Microbial analysis of *sa-um* showed that the dominant microbes present are from the phyla *Firmicutes*, *Proteobacteria*, *Bacteroides*, *Actinobacteria*, *Chloroflexi*, *Planctomycetes*, *Synergistetes* and *Acidobacteria* and at the genus level dominated by *Clostridium*, *Bacteroides*, *Oscillospira*, *Corynebacterium*, *Megamonas*, *Faecalibacterium*, *Proteus*, *Ruminococcus* and *Prevotella* (Mandal et al. 2018).

Nutritional value determination and biochemical and microbiological analysis studies on *chhi-um*, *ai-um*, *aite-um*, *tam-um* and *tuai-um* are not yet available but are discussed in the light of similar fermentation products of other regions.

Chhi-um or fermented sesame could be compared with *ogiri-saro* of Nigeria even though the preparation is slightly different; the substrate for both the fermentation products is the same. Studies have shown that fermentation of such sesame seeds resulted in significant increase in mineral concentrations which could be due to conversion of insoluble reserve foods during fermentation (Makinde and Akinoso 2014).

Crabs are excellent source of proteins while they are fair source of minerals like calcium, potassium, magnesium, sodium and iron (Varadharajan and Soundarapandian 2014). The Indian freshwater crabs have high contents of monounsaturated fatty acids (MUFA) and polyunsaturated fatty acids (PUFA), exhibiting a higher percentage of essential fatty acids out of total fatty acids (Das et al. 2015). However, consumption of fermented crabs is mainly localised in Northeast India, mainly in Mizoram and other areas where Mizo tribes reside. Fermentation of crabs is also practised by Nigerians in which juvenile crabs are fermented wholly with carapace, or for mature crabs, carapaces are crushed and wrapped in layers of banana leaves after carefully washing the crabs. The crabs are allowed to ferment by keeping them in warm place for 3–4 days to produce *ogiri-nsiko* (Achi et al. 2007). The method of preparation of *ogiri-nsiko* is quite similar to the Mizos' preparation of *aite-um* and hence may have similarities in many aspects. Fermentation of crabs with sesame seeds appears to be practised by Naga communities whose method of preparation is similar with that of *ai-um* (Mao and Odyuo 2007). However, a scientific study for the fermented crabs of both Mizo and Naga tribes is not available.

Wickerhamomyces anomalus, *Candida glabrata*, *Pichia anomala* and *Saccharomycopsis fibuligera* were reported from *dawidim*, traditionally prepared dried starter for alcohol production in Mizoram (Sha et al. 2018). *Zufâng*, an alcoholic beverage that utilise *dawidim* for its production, has an average alcohol content of 6.3% and pH of 4.32 with total acidity of 0.59 g per 100 mL and total antioxidant capacity of about 11% (Zonunmawii et al. 2017). It should be kept in mind that *zufâng* is the mildest drink among all the Mizo alcoholic beverages and is not considered as real alcoholic drink. The *rakzû* sample collected from Tuirial area, Aizawl district, is found to contain 20.6% (v/v) of ethanol with 0.826% (w/v) of residue on evaporation and 0.198% (w/v) ash content (Devi 2012). The ethanol content of *rakzû* is quite high as compared to other rice beer from Northeast India whose ethanol content ranges from 12 to 13% but is low when compared to international standard.

16.7 Conclusion

Preservation methods for many of the traditional food items are not available mainly due to unavailability of scientific in-depth studies. Most of the indigenous foods are seasonal and not available throughout the year which makes development of food preservation process necessary. Traditional food processing techniques of Mizo, mainly fermentation, are developed not only to improve the taste of food but also to preserve it. It is always the case that, when the season for certain vegetable/food is over, the demand for such vegetable/food becomes very high as is with the price. Fermentation allows such vegetable/food to be preserved for certain amount of time prolonging their availability. It is wonderful to learn that the Mizo forefathers somehow utilised the minimal available resources such as fire and sun and devised their own way of food preservation. It is indeed quite intriguing that our forefathers practised the science of biotechnology without knowing it. However, most of the fermented products are still in the dark without proper scientific studies. A study on the nutritive values, microorganisms involved, physical and chemical changes brought about by fermentation, etc. of the traditionally fermented products of Mizo would be highly significant. It is also of utmost importance to study the possible effect, whether beneficial or harmful, of the fermented products on the health and well-being of consumers. Scientific study on the traditional fermentation process is a must in order to improve the technique which will have a positive impact on the quality and safety of the final fermented products. By utilising the current scientific knowledge and technologies, one can be innovative about many of the foods produced using fermentation and indigenous knowledge systems, and they must be used to add value to such products, such as increased shelf life, improved flavour and appealing packaging and labelling. This will definitely help in promoting ethnic fermented foods and beverages that will result in improving the economy of the state. While applying innovative techniques in producing such fermented products, documentation of the traditional indigenous technologies must be done for future generations. It will also help in creating a reference database to be utilized by future generations for the benefit of society in the long run.

Acknowledgements The authors acknowledge kind support from Director, RIPANS, Institutional Level Biotech Hub (IBThub, RIPANS), and Prof. N. Senthil Kumar, Dept. of Biotechnology, Mizoram University. The authors also acknowledge the photographs of fermented foods and beverages of Mizoram contributed by Bawihi Fanai, Thangtea Photographix, Zawna Pachuau, Zosangliana and Zothansanga Ralte.

References

Achi OK, Anokwuru IC, Ogbo FC (2007) Microbiological and chemical changes during fermentation of crabs for *ogiri-nsiko* production. Am J Food Technol 2(4):301–306. https://doi.org/10.3923/ajft.2007.301.306

Advani P (2005) Agriculture sector in India. A report on the impact of WTO on women in agriculture, new.nic.in/pdfreports/impact of WTO women in agriculture.pdf

Anup D, Ghosh PK, Choudhury BU, Patel DP, Munda GC, Ngachan SV, Chowdhury P (2006) Climate change in northeast India: recent facts and events-worry for agricultural management. In: ISPRS archives XXXVIII-8/W3 workshop proceedings: impact of climate change on agriculture

Anupma A, Pradhan P, Sha SP, Tamang JP (2018) Traditional skill of ethnic people of the Eastern Himalayas and North East India in preserving microbiota as dry amylolytic starters. Indian J Tradit Knowl 17(1):184–190

Chhetri R, Tamang JP (2014) Functional properties of tungrymbai and bekang, naturally fermented soybean foods of India. Int J Ferment Foods 3:87–103. https://doi.org/10.5958/2321-712X.2014.01311.8

Chhetri R, Tamang JP (2015) *Bacillus* species isolated from *tungrymbai* and *bekang*, naturally fermented soybean foods of India. Int J Food Microbiol 197:72–76. https://doi.org/10.1016/j.ijfoodmicro.2014.12.021

Das M, Kundu JK, Misra KK (2015) Nutritional aspect of crustaceans especially freshwater crabs of India. Int J Adv Res Biol Sci 2(12):7–19

Devi CN (2012) Quantitative analysis of local alcohol in Mizoram. B. Pharm Dissertation, Department of Pharmacy, Regional Institute of Paramedical & Nursing Sciences (RIPANS), Aizawl, pp 12–14

Dokhuma J (1992) Khawtlang Dan. Hmanlai Mizo pi pute kalphung (Ed: Padma Shri James Dokhuma), Published by James Dokhuma at JD Press, Mizoram, pp. 177–230

Economic Survey Mizoram (2017–2018) Planning & Programme Implementation Department (Research & Development Branch), Govt of Mizoram, pp 195–217

'Employment & Labour Welfare' in Mizoram Economic Survey (2014–2015) Government of Mizoram Planning & Programme Implementation Department (Research & Development Branch), pp 22–26

Jeyaram K, Singh TA, Romi W, Devi AR, Singh WM, Dayanidhi H, Singh NR, Tamang JP (2009) Traditional fermented foods of Manipur. Indian J Tradit Knowl 8(1):115–121

Lalthangliana B (2005) Zu-intoxicating dinks. In: Palkhiwala K (ed) Culture and folklore of Mizoram. Publications Division, New Delhi, pp 248–256. ISBN 10: 8123013094/ISBN 13: 9788123013091

Lalthanpuii PB, Lalruatfela B, Zoramdinthara, Lalthanzara H (2015) Traditional food processing techniques of the Mizo people of Northeast India. Sci Vis 15(1):39–45

Makinde FM, Akinoso R (2014) Comparison between the nutritional quality of flour obtained from raw, roasted and fermented sesame (*Sesamum indicum* L.) seed grown in Nigeria. Acta Sci Pol Technol Aliment 13(3):309–319. https://doi.org/10.17306/J.AFS.2014.3.9

Malakar M, Devi KR, Phukan RK, Kaur T, Deka M, Puia LH, Sailo L, Lalhmangaihi T, Barua D, Rajguru SK, Mahanta J, Narain K (2014) p53 Codon 72 polymorphism interactions with dietary and tobacco related habits and risk of stomach cancer in Mizoram, India. Asian Pac J Cancer Prev 15(2):717–723. https://doi.org/10.7314/APJCP.2014.15.2.717

Mandal SD, Singh SS, Muthukumaran RB, Thanzami K, Kumar V, Kumar NS (2018) Metagenomic analysis and the functional profiles of traditional fermented pork fat 'sa-um' of Northeast India. Appl Ind Microb Biotechnol Express 8:163. https://doi.org/10.1186/s13568-018-0695-z

Mao AA, Odyuo N (2007) Traditional fermented foods of the Naga tribes of Northeastern, India. Ind J Tradit Knowl 6(1):37–41

Omafuvbe BO, Shonukan OO, Abiose SH (2000) Microbiological and biochemical changes in the traditional fermentation of soybean for soy-daddawa—Nigeria food condiments. Food Microbiol 17:469–474. https://doi.org/10.1006/fmic.1999.0332

Omizu Y, Tsukamoto C, Chettri R, Tamang JP (2011) Determination of saponin contents in raw soybean and fermented soybean foods of India. J Sci Ind Res 70:533–538

Phukan RK, Narain K, Zomawia E, Hazarika NC, Mahanta J (2006) Dietary habits and stomach cancer in Mizoram, India. J Gastroenterol 41(5):418–424. https://doi.org/10.1007/s00535-006-1761-x

Sati VP (2017) Changing agriculture and cropping pattern in Mizoram, Northeast India. https://www.researchgate.net/publication/322100703

Sati VP, Rinawma P (2014) Practices of shifting cultivation and its implications in Mizoram, North-East: a review of existing research. Nat Environ 19(2):179–187

Sawmliana M (2013) In: Sawmliana M, editor. The book of mizoram plants (includes wild animals, birds, etc.). P. Zakhuma, Chanmari West, Aizawl, pp 1–526

Sha SP, Suryavanshi MS, Jani K, Sharma A, Shouche Y, Tamang JP (2018) Diversity of yeasts and molds by culture-dependent and culture-independent methods for mycobiome surveillance of traditionally prepared dried starters for the production of Indian alcoholic beverages. Front Microbiol 9:2237. https://doi.org/10.3389/fmicb.2018.02237

Singh PK, Singh KI (2006) Traditional alcoholic beverage, *Yu* of *Meitei* communities of Manipur. Indian J Tradit Knowl 5(2):184–190

Singh A, Singh RK, Sureja AK (2007) Cultural significance and diversities of ethnic foods of Northeast India. Indian J Tradit Knowl 6(1):79–94

Tamang JP (2010) Diversity of fermented foods. In: Tamang JP, Kailasapathy K (eds) Fermented foods and beverages of the world. CRC Press, pp 41–84. isbn:978-1-4200-9496-1

Tamang JP (2015) Naturally fermented ethnic soybean foods of India. J Ethnic Foods 2:8–17. https://doi.org/10.1007/s12275-012-1409-x

Tamang JP, Tamang N, Thapa S, Dewan S, Tamang B, Yonzan H, Rai AK, Chettri R, Chakrabarty J, Kharel N (2012) Microorganisms and nutritional value of ethnic fermented foods and alcoholic beverages of North East India. Indian J Tradit Knowl 11(1):7–25

Tamang JP, Thapa N, Bhalla TC, Savitri (2016) Ethnic fermented foods and beverages of India. In: Tamang JP (ed) Ethnic fermented foods and alcoholic beverages of Asia. Springer Nature, New Delhi, pp 17–72. 9788132227984

The Lushai Hill District (Change of Name) Act (1954) Dated 29th April, 1954

Tungoe PC (2016) Isolation and characterization of bacteria from selected traditional fermented foods of Mizoram. M.Pharm Thesis, Department of Pharmacy, Regional Institute of Paramedical & Nursing Sciences (RIPANS), Aizawl, pp 46–47

Vanlalruati, Das S (2012) Comparison between nutrient content of fermented and nonfermented soyabean. B. Pharm dissertation, Department of Pharmacy, Regional Institute of Paramedical & Nursing Sciences (RIPANS), Aizawl, pp 14–22

Varadharajan D, Soundarapandian P (2014) Proximate composition and mineral contents of freshwater crab *Spiralothelphusa hydrodroma* (Herbst, 1794) from Parangipettai, South East Coast of India. J Aquacult Res Dev 5(2):1–6. https://doi.org/10.4172/2155-9546.1000217

Zonunmawii, Lalduhsaki P, Laltlanzovi PC (2017) Comparison of rice beer fermented by 'Chawl', a traditional Mizo yeast and other commercial yeasts. B.Pharm dissertation, Department of Pharmacy, Regional Institute of Paramedical & Nursing Sciences (RIPANS), Aizawl, pp 9–12

Some Ethnic Fermented Foods and Beverages of Nagaland

17

T. Ajungla, Lydia Yeptho, Asangla Kichu, and Gloria Nyenthang

Abstract

Ethnic fermented foods and beverages are important components of the dietary culture of the indigenous people of Nagaland. The raw materials used during fermentation are either domesticated or collected from the jungle. The use of wild herbs and forest products is a common practice employed during preparation of the fermented food products. The dietary culture of Naga also expresses their love for food and its variety as these foods help in shaping the economy of the people, but till date only limited research have been made to study the fermented foods and beverages in detail although many foods have been reported. Literatures on the microorganisms of the finished fermented products and also the microorganisms involved during the fermentation process are scarce or insufficient so attempt have been made to describe some of the common ethnic food products, their method of preparation, the substrates used along with their sensory characteristics. It also includes information on the most popular fermented foods of Nagaland and the tribes and regions associated with it. Microorganisms identified are only for *axone* (fermented soybean) and *zutho* (fermented local rice beverage) and the nutritional value of these two foods have also been mentioned. All the fermented foods and beverages are naturally fermented except for *zutho* which uses an amylolytic starter (*khrei*). Most of the fruit beverages are low alcoholic and are prepared locally at home during the seasons and used for various purposes.

Keywords

Ethnic · Fermented foods and beverages · Nagaland

T. Ajungla (✉) · L. Yeptho · A. Kichu · G. Nyenthang
Department of Botany, Nagaland University, Zunheboto, Nagaland, India

J. P. Tamang (ed.), *Ethnic Fermented Foods and Beverages of India: Science History and Culture*, https://doi.org/10.1007/978-981-15-1486-9_17

17.1 Introduction

Nagaland, the 16th state of the Indian Union, was established on December 1, 1963, and lies between 26.6° and 27.4° North latitude and 98° to 96° East longitude. It is a land of agriculture and beautiful terraces predominated by mountains and is an abode to the Naga's, people with strong warrior background. It is exquisitely rich in flora and fauna, and the world's tallest *Rhododendron* is found in Mt. Japfu, and *Cymbidium tigrinum*, a species of orchid, was first discovered in Nagaland.

Nagaland supports a history of tribes where 16 tribes are recognised as major tribes, and except the difference in languages, all tribes have similar yet unique traditions and practices. While irrigation is the main occupation of the Naga people, dexterity is especially seen among women in the society where they are regarded with respect and honour. A central feature of Naga life is the essence of celebration of series of festivals throughout the year where each tribe has its own folklore and song that expresses all the exuberant concern for life.

Most Naga live in small villages, and its land supports a considerable amount of crops like corn, pulses, fibres, potatoes, tobacco, oilseeds, sugarcane, millets, and rice besides the abundance of wild fruits and vegetables. Food is reflected as a part of their rich culture, and the range of fermented food products produced and consumed serves as a gateway to understand their ethnic and unique food habits prepared at its best. Distinct cuisines separate a certain tribe from others, while most food overlaps between the various tribes. Every household in Nagaland gives out delicious smell of foods prepared, and the traditional and culinary knowledge are passed down to younger generations mostly by mothers to their daughters by cooking together. The method of preparation also varies according to individual perceptions which increases a diversity of traditional knowledge. Besides food, preparation of varieties of indigenous drinks is also very popular among the Naga. *Zutho*, a fermented rice beverage prepared by the Angami Naga, is used widely during festivals. During ancient times, this drink was served to all, irrespective of gender and status; however with the blend of religion and western culture, it has now being replaced by tea and is only used by a few people. Drinking of fermented fruit beverage is also very common among the Naga tribes. Wild as well as domesticated fruits are used for preparation of these beverages; some of the fruits commonly used are *Vitis vinifera* (grapes), *Passiflora edulis* (passion fruit), *Ananas comosus* (pineapple), *Musa paradisiaca* (banana), *Docynia indica* (Naga wild apple), *Phyllanthus emblica* (gooseberry), *Prunus persica* (peach), and *Artocarpus heterophyllus* (jackfruit).

Nagaland supports unique and numerous fermented food products and beverages, many of which are yet to be documented and explored, but some of the ethnic fermented foods and beverages are discussed in Table 17.1.

Table 17.1 Ethnic fermented foods and beverages of Nagaland

Food	Substrate	Sensory property of product	Major ethnic consumers	Region/district
Axone/ Akhonii	Soybean	Sticky, flavoured; curry	Sumi	Zunheboto
Anishi	Taro leaves	Dry, flavoured; curry	Ao	Mokokchung
Ashikumna/ Thevochie	Pork fats	Semi-solid, flavoured; Curry	Angami and Sumi	Kohima and Pughoboto
Bastenga	Bamboo shoot	Wet and dry, flavoured; curry	Lotha	Wokha
Bell dzu/dzii	Passion fruit	Low-alcoholic, sweet beverage	Angami, Aos, and Sumis	Kohima, Mokokchung, and Zunheboto
Cutocie	Cucumber fruits and leaves	Liquid, flavoured; curry	Angami	Jotsoma
Kothal dzu	Jackfruit	Low-alcoholic, sweet beverage	Sumi	Zunheboto
Khulushi dzii	Gooseberry	Low-alcoholic, sweet beverage	Angami	Jotsoma
Tsugu/ Tsiingen ngashi	Crab and sesame seeds	Semi-solid, flavoured; curry	Aos and Rengma	Changki and Tsuminyu
Zutho	Rice + starter (*Khrei*)	Alcoholic; sweet beverage	Angami	Kohima
Khrei	Unhulled rice grains	Dried starter to produce alcoholic beverages, round cake-like	Angami	Kohima

Soybean

↓

Washed with water

↓

Boiled for 1-2 h

↓

Excess water drained off and cooled

↓

Cooked soybean placed in container lined with banana leaves and covered loosely

↓

Container kept near earthen- oven in kitchen

↓

Fermented (20-35 C, 6-8 d)

↓

Axone

Fig. 17.1 Traditional method of preparation of *Axone* by Sumi Naga at Lumami village

17.2 *Axone/Aakhone*

It is a fermented soybean product mostly prepared by Sumi Naga but also popular among other Naga tribes especially Angami Naga. The method of preparation of *axone* differs among tribes and also within individuals resulting into its unique flavour and taste.

Traditional method of preparation (details).

Flow sheet as follows (Fig. 17.1):

17.2.1 Culinary and Mode of Consumption

Axone/aakhone is best cooked with smoked pork which is a delicacy among the Nagas. The Sumi Naga use it in preparation of almost all vegetable curries. It is also used to prepare chutneys, and sometimes boiled egg is added in curry preparation. It is also deep fried with oil which is a common preparation by mothers for their children studying outside.

17.2.2 Socio-economy and Ethical or Religious Values

Axone/aakhone is sold at the local markets and also at homes which supports the income of a local home. The Naga women also use the money generated by selling *axone* to contribute and support the local churches. Also during religious ceremonies and even during weddings, *axone* is served as one of the main dishes.

Fresh and matured leaves collected

↓

Washed and stacked together

↓

Wrapped with banana leaf and sundried

↓

Leaves turn yellow (5-8 days)

↓

Grounded into paste (salt, chilly and ginger may be added if desired)

↓

Made into cakes

↓

Dried over earthen- oven in the kitchen till it becomes hard

↓

Anishi

Fig. 17.2 Traditional method of preparation of *anishi* by Ao Naga at Khensa village

Microorganisms: *Bacillus subtilis*, *B. licheniformis*, and *B. cereus* (Jamir and Deb 2018). *Aeromonas hydrophila*, *A. eucranophila*, *A. salmonicida*, *B. coagulans*, *B. pantothenticus*, *B. lentus and B. stearothermophilus*, *Erwinia ananas*, *Enterobacter* sp., *Klebsiella oxytoca*, *K. pneumoniae*, *Hafnia alvei*, *Salmonella enterica ser Typhimurium*, *Morganella morganii*, *Pseudomonas* sp., *Providencia rettgeri*, and *Proteus* sp. (Singh et al. 2014)

17.2.3 Nutritional Value

Moisture (%), 50; pH, 8; protein (g/100 g), 42.1; crude fibre (g/100 g), 1.61; reducing sugars (%), 29.7 (Jamir and Deb. 2018)

17.3 *Anishi*

It is a fermented food product made from the leaves of *Colocasia* sp. usually mixed with spices such as ginger, chilli powder, and salt. It is mostly prepared by Ao Nagas and is stored in airtight containers to prevent contamination and can be consumed yearlong when stored properly.

Traditional method of preparation (details).

Flow sheet as follows (Fig. 17.2):

17.3.1 Culinary and Mode of Consumption

Anishi is a delicacy of Ao Nagas which is prepared best with smoked pork. Before cooking, the *Anishi* cakes are roasted and lightly crushed which enhances the flavour and taste of the curry. It is also prepared along with other vegetables as a side dish. Another common preparation of *Anishi* includes addition of smoked and dried eels.

17.3.2 Socio-economy and Ethical and Religious Value

Anishi is prepared during festive seasons and sometimes during weddings as a side dish. It is also sold at a very good price in the market. *Anishi* with smoked pork is a popular menu at many local hotels in Mokokchung.

Microorganisms: unknown.

17.4 *Ashikumna/Thevocie*

It is a delicacy prepared by fermenting pork fats (belly portion) used mainly during vegetable curry preparation in handful amounts.

Traditional method of preparation (details).

Fig. 17.3 Traditional method of preparation of *thevocie* by Angamis in Jotsoma and *ashikumna* by Lazamis in Pughoboto

1. Preparation by Angami's (*Thevochie)*

Pork fats collected and washed
↓
Cut into small pieces and put inside a bamboo
↓
Left to ferment (2-3 months)
↓
A strong smell is emitted
↓
Thevochie

2. Preparation by Lazami's (*Ashikumna*)

Pork fats collected and washed
↓
Cut into small pieces
↓
Pour in an airtight container
↓
Container kept near earthen- oven in kitchen
↓
Fermented (7-10 d)
↓
Ashikumna

Flow sheet as follows (Fig. 17.3):

17.4.1 Culinary and Mode of Consumption

It is added in small quantities in almost every vegetable curry preparation which enhances the flavour and taste of the curry. It is mostly used as a substitute for meat and is used as a side dish.

Socio-economy and ethical values. *Thevocie* is rarely prepared these days, and only a few people living in remote villages are known to prepare it. *Ashikumna* on the other hand is a common condiment found in almost all local households of Lazami. It can be stored almost yearlong in an airtight container.

Microorganisms: unknown.

17.5 *Bastenga*

It is a fermented bamboo shoot product, is acidic in nature, and is of two types—dry and wet.

Fresh, young and tender bamboo shoots collected and washed
↓
Sliced into small pieces
↓
Kept in airtight containers
↓
Left to ferment (2-3 months)
↓
Wet *bastenga*

Bamboo shoot are mashed and dried in open sunlight
↓
Dried *bastenga*

Fig. 17.4 Traditional method of preparation of dried and wet *bastenga* prepared by Lotha Naga at Wokha

Traditional method of preparation (details).

Flow sheet as follows (Fig. 17.4):

17.5.1 Culinary and Mode of Consumption

Dried *bastenga* is best prepared with pork and fish. It is also deep fried with salt and chilly. Wet *bastenga* is mostly added in vegetable curry preparation and more commonly with *Colocasia* tubers which is a delicacy among the Nagas used as a side dish. The brine of fermented bamboo shoot is also used as a substitute of tomato which enhances the flavour of the curry.

17.5.2 Socio-economy and Ethical Values

Bastenga is a very popular fermented food product of Nagaland and is widely used by the Nagas. It is well known that the tastiest *bastengas* are prepared by the Lothas found at the local markets of Wokha town, and *bastenga* is particularly associated with Lothas. It has high market value and can be stored and kept up to a year in airtight containers, but long-term storage may affect the quality and taste of the product.

Microorganisms: unknown.

17.6 *Bell Dzu/Dzii*

It is a sweet low-alcoholic fermented beverage made from the fruits of *Passiflora edulis*.

Traditional method of preparation (details).

Flow sheet as follows (Fig. 17.5):

Fig. 17.5 Traditional method of preparation of *bell dzu/dzii* by Angami Naga at Kohima

Matured and ripped fruit collected

↓

Rind removed and put into an airtight container

↓

Sugar is added

↓

Container filled with water

↓

Fermented (10-12 d)

↓

Bell dzu/ dzii

Fig. 17.6 Traditional method of preparation of *cutocie* by Angami Naga at Jotsoma

Collect cucumber and cucumber leaves

↓

Remove the skin of the fruit

↓

Half fill a glass bottle with sliced cucumber

↓

Mash the leaves and put on top of it

↓

Airtight the glass bottle

↓

Fermented (2-3 months)

↓

Cutocie

17.6.1 Culinary and Mode of Consumption

Sugar is added according to individual taste, and sometimes water is added to dilute the drink.

17.6.2 Socio-economy and Ethical Values

It is enjoyed best in community gatherings among the youngsters. It is also served to guests and is used as a substitute for packaged fruit juices. It has high market value and is one of the most expensive fruit beverages in Nagaland.

Microorganisms: unknown.

17.7 *Cutocie*

It is a fermented liquid product prepared by fermenting the fruits and leaves of cucumber. It is unique to the Angami Nagas and is used to add flavour and taste.

Traditional method of preparation (details).

Flow sheet as follows (Fig. 17.6):

17.7.1 Mode of Consumption and Culinary

It is mostly used for preparing vegetable curries used as a side dish. It is also used for making chutneys and mostly used as a seasoning.

17.7.2 Socio-economy and Ethical Values

Cutocie has not been recorded to be commercialised at local markets, and today it is only prepared by few families in rural areas.

Microorganisms: unknown.

17.8 *Kothal Dzu*

It is a sweet alcoholic beverage prepared by the Nagas by fermenting the fruits of *Artocarpus heterophyllus*.

Traditional method of preparation (details).

Flow sheet as follows (Fig. 17.7):

Fig. 17.7 Traditional method of preparation of *kothal dzu* by Sumi Naga at Lumami village

Matured fruit collected

↓

Outer skin peeled off

↓

Seeds put inside an airtight container

↓

Sugar and water is added

↓

Fermented (7-9 d)

↓

Kothal dzu

17.8.1 Culinary and Mode of Consumption

Fermented jackfruit juice is diluted with water and sugar is added according to individual taste.

17.8.2 Socio-economy and Ethical Values

Kothal dzu is served to guests in community gatherings. Young people and sometimes elders gather together after dinner and share the simple joys of life over a cup of this fermented drink.

Microorganisms: unknown.

17.9 *Khulushi Dzii*

It is a fermented juice prepared from gooseberry (Fig. 17.8).

17.9.1 Mode of Consumption and Culinary

Sugar is added in *khulushi* and drank in small amounts after dinner which is believed to help in indigestion.

17.9.2 Socio-economy and Ethical Issues

This fruit juice is more common than any other fermented fruit juice as the tree grows wild in many parts of Nagaland. It is mostly used by the young people in community gatherings or used to serve visitors at home.

Microorganisms: unknown.

Fig. 17.8 Traditional method of preparation of *khulushi* by Angami Naga at Jotsoma village

Fresh gooseberries collected

↓

Half filled inside an airtight container

↓

Sugar and water is added

↓

Fermented (10-15 d)

↓

Khulushi

Crab collected and washed

↓

The exoskeleton removed along with the pair of claws and legs

↓

Sesame seeds roasted and crushed

↓

Crab and sesame seeds grinded into paste

↓

Paste covered with banana leaves

↓

Kept over earthen –oven (2-3 d)

↓

Tsugu / Tsiingen ngashi

Fig. 17.9 Traditional method of preparation of *tsugu/tsiingen ngashi* by Ao Naga at Changki village

17.10 *Tsugu/Tsiingen Ngashi*

It is a fermented food product produced by fermenting crab and roasted sesame seeds. It is unique to Ao and Rengma tribes of Nagaland.

Traditional method of preparation (details).

Flow sheet as follows (Fig. 17.9):

17.10.1 Mode of Consumption and Culinary

It is cooked along with other vegetables during curry preparation. It adds a distinct flavour and aroma to the curry. It is usually added in small quantities to prepare chutneys.

17.10.2 Socio-economy and Ethical Issues

Tsugu/tsiingen ngashi is sold commercially at the local markets for a decent price. So, the locals prefer to buy it from the market rather than preparing it at home as collecting crabs from the rivers and streams is laborious and tedious.

Microorganisms: unknown.

Fig. 17.10 Traditional method of preparation of *zutho* by Angami Naga at Kohima

Rice soaked overnight
↓
Water drained out and sundried
↓
Grinded into powder
↓
Add a small amount of *Khrei*
↓
Add boiled water
↓
Store in earthen or aluminium pots (3-4 d)
↓
Formation of white bubbles
↓
Discard the bubbles and filter it
↓
Zutho

Preparation of *Khrei*

Unhulled rice soaked (2-3 d)
↓
Sundried
↓
Allowed to germinate
↓
Grinded into powder
↓
Khrei

17.11 *Zutho*

Zutho, a sweet alcoholic beverage, whitish in colour is made from rice and 207 an amylolytic starter, *Khrei*. It is prepared by the Angami Nagas and is the most popular alcoholic beverage of Nagaland.

Traditional method of preparation (details).

Flow sheet as follows (Fig. 17.10):

17.11.1 Culinary and Mode of Consumption

It is especially prepared during festivals like *Sekrenyi* where they are served in bamboo mugs for consumption. It is also sold commercially as an alcoholic beverage.

17.11.2 Socio-economy and Ethical Issues

Some women in Naga society commercially sells *zutho*, but their moral value is questioned by the society, and so only few women practice the art of preparing *zutho* as a source of income. Some households in the village act as hubs for selling *zutho* at wholesale rates to urban areas. Nevertheless it is the most common drink in all local wine stores *(zuki)* and is mostly consumed by the male gender in Naga society.

17.11.3 Microorganisms

Zutho: *Saccharomyces cerevisiae* (Teramoto et al. 2002 Tamang et al. 2012)

Khrei: *Wickerhamomyces anomalus, Saccharomycopsis fibuligera, Pichia anomala, P. terricola, P. kudriavzevii, and Candida glabrata* (Sha et al. 2018)

Nutritional value: pH 3.6, acidity 5.1%, alcohol 5% (v/v) (Teramoto et al. 2002 Tamang et al. 2012)

Fig. 17.11 *Bell dzii/dzu*

Fig. 17.12 *Bastenga*

Fig. 17.13 *Cutocie*

Fig. 17.14 *Ashikumna/Thevocie*

Fig. 17.15 *Axone* wrapped with *Colocasia esculenta* leaves

17.12 Conclusion

Extensive research is needed to understand the microbiota of fermented food products, the microorganisms involved, and the scope of these foods to be developed as standard products. Besides, documenting the methods of preparation is needed to preserve the indigenous knowledge of these foods requiring the need for trained and knowledgeable microbiologist to apply the basic microbiological skills acting as buffer to promote the fermented food products at the global level without damaging

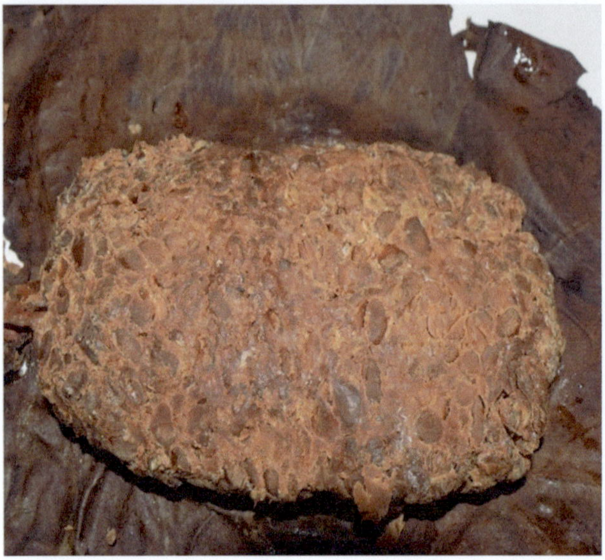

Fig. 17.16 *Axonelaakhone*

Fig. 17.17 Bastenga (dried)

the dietary culture (Figs. 17.11, 17.12, 17.13, 17.14, 17.15, 17.16, 17.17, 17.18, 17.19, 17.20, 17.21, and 17.22).

Fig. 17.18 *Tsugultsiingen ngashi*

Fig. 17.19 *Zutho*

Fig. 17.20 *Khulushi dzii*

Fig. 17.21 *Anishi*

Fig. 17.22 Fermented jackfruit juice

References

Jamir B, Deb CR (2018) Nutritional assessment and molecular identification of microorganisms from *Akhuni/Axone*: a soybean based fermented food of Nagaland, India. J Adv Biol 11:2347–6893

Sha SP, Suryavanshi MV, Jani K, Sharma A, Shouche Y, Tamang JP (2018) Diversity of yeasts and molds by culture-dependent and culture-independent methods for Mycobiome surveillance of traditionally prepared dried starters for the production of Indian alcoholic beverages. Front Microbiol 9:2237. https://doi.org/10.3389/fmicb.2018.02237

Singh BR, Karan R, Singh V (2014) Microbial quality and safety of Axone-Akhuni, a fermented soybean food of Nagaland. Noto-are 15185525: Medicine. 2014-03-14

Tamang JP, Tamang N, Thapa S, Dewan S, Tamang B, Yonzan H, Rai AK, Chettri R, Chakrabarty J, Kharel N (2012) Microorganisms and nutritional value of ethnic fermented foods and alcoholic beverages of north East India. Ind J Tradit Knowl 11:7–25

Teramoto Y, Yoshida S, Ueda S (2002) Characteristics of a rice beer (*zutho*) and a yeast isolated from the fermented product in Nagaland, India. Int J Food Microbiol 18:813–816

Ethnic Fermented Foods and Beverages of Sikkim and Darjeeling Hills (Gorkhaland Territorial Administration)

18

Namrata Thapa and Jyoti Prakash Tamang

Abstract

More than 48 types of major and minor ethnic fermented foods and alcoholic beverages are produced using indigenous knowledge of food fermentation in Sikkim and Darjeeling hills. There are eight different categories of ethnic fermented foods and beverages based on substrates which are prepared and consumed in these regions: fermented vegetable products, fermented legumes, fermented cereals, fermented dairy products, traditionally preserved fish products, traditionally preserved meat products, amylolytic starter cultures and alcoholic beverages. Almost all categories of fermented foods are prepared by natural fermentation without starter cultures, except in alcoholic fermentation which is traditionally prepared by 'back-sloping' dry amylolytic starters. Among the ethnicity, the Gorkha is the largest stakeholder as well as consumers of 80% of the fermented foods and beverages of Darjeeling hills and Sikkim. Functional microorganisms play important roles in the traditional fermentation processes by their functionality and impart health-promoting benefits. Extensive researches have been carried on almost all ethnic fermented foods and beverages of Sikkim and Darjeeling hills by Tamang and team for the last 33 years documenting and profiling microbial community, functionality, nutritional aspects and some health benefits. Our finding proves that microbial diversity in ethnic fermented foods contributes significant genetic resources due to diverse food cultures of the multi-ethnic groups of people in Sikkim and Darjeeling hills.

N. Thapa
Biotech Hub, Department of Zoology, Nar Bahadur Bhandari Degree College, Gangtok, Sikkim, India

J. P. Tamang (✉)
Department of Microbiology, DAICENTER (DBT-AIST International Centre for Translational and Environmental Research) and Bioinformatics Centre, School of Life Sciences, Sikkim University (Central University), Gangtok, Sikkim, India

J. P. Tamang (ed.), *Ethnic Fermented Foods and Beverages of India: Science History and Culture*, https://doi.org/10.1007/978-981-15-1486-9_18

Keywords
Sikkim · GTA · Darjeeling · Fermented foods · Beverages

18.1 Introduction

The State of Sikkim (www.sikkimtourism.gov.in) and Darjeeling hills, administratively known as Gorkhaland Territorial Administration (GTA), in the state of West Bengal (www.tourismdarjeeling.com) is located in the Eastern Himalayan regions of India (Fig. 18.1). Sikkim is a mountainous state of India with an area of 7096 km^2 and altitudes ranging from 300 to 8500 m. Sikkim borders with China in the north and northeast, Bhutan in the east, Nepal in the west and Darjeeling hills (GTA) in the south. The state comprises four districts: North, East, South and West. The total population of Sikkim is 610,577 (www.census2011.co.in). The State of Sikkim and Darjeeling hills (GTA) is also mountainous and one of the best tourist spots in the world in the background of Mt. Kanchenjunga (the third highest mountain in the world) with an area of 3303.98 km^2 and a population of 878,002 (www.census2011.co.in). The altitudinal variation of Darjeeling hills ranges from 150 m in foothills in

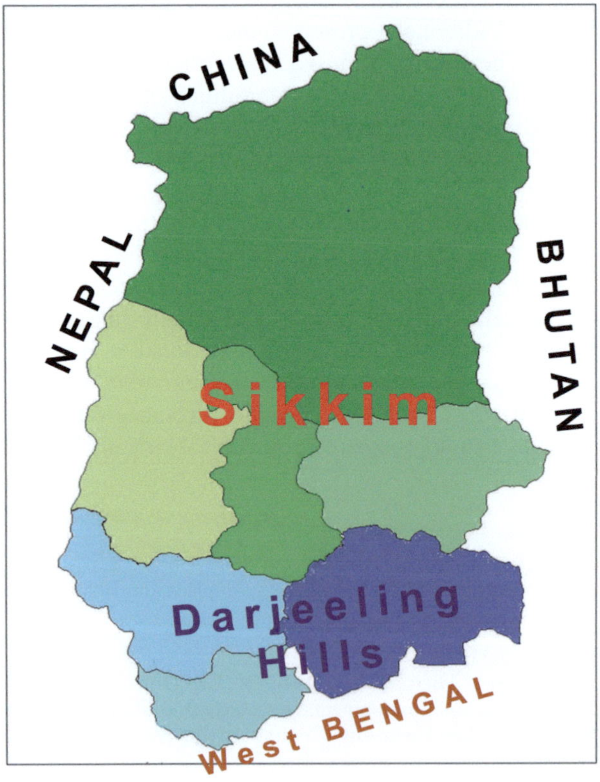

Fig. 18.1 Location map of Sikkim and Darjeeling hills in India

north-westwards up to Sandakphu with location of Darjeeling town at an elevation of 6700 ft (2042.2 m). Darjeeling borders with Sikkim in the north, Nepal in the west, Bhutan in the east and Siliguri plain areas of West Bengal in the south.

18.2 History of Sikkim and Darjeeling

Sikkim Kingdom was established by the fifth-generation descendant of Guru Tashi, *Phuntsog Namgyal*, as the first *Denjong Gyalpo* or the *Chogyal* (means King) belonging to Bhutia community in 1642 CE (Risley 1928; Bareh 2001). However the history of pre-Namgyal dynasty of Sikkim before 1641 recorded that the Lepcha and the Gorkha/Nepali communities mostly the Limboo and the Mangar were ruling in different villages independently (Subba 1999; Chumlung 2014). Sikkim was ruled by 12 Kings or *Chogyal* from 1642 CE to 1975 CE, and the last King or *Chogyal* of Independent Sikkim was Palden Thondup Namgyal (Bhanja 1993). Sikkim was declared as 22nd state of India on 16 May 1975.

Nepal invaded Darjeeling in 1780 which was originally a part of Sikkim Kingdom (Bhanja 1993). After the Anglo-Gorkha War, Nepal ceded one-third of its territories including Darjeeling hills to the British under the Treaty of Sugauli signed on 2 December 1815 and ratified on 4 March 1816 CE between the East India Company and King of Nepal following the Anglo-Nepalese War of 1814–1816 CE (Bareh 2001). On 10 February 1817, the Treaty of Titalia was signed between the *Chogyal* of Sikkim and the British East India Company under which the British returned Darjeeling annexed by Nepal to Sikkim. Darjeeling was gifted to British India by the *Chogyal* of Sikkim as per the Deed of Grant signed on 1 February 1835 (Dozey 1922). On 11 November 1865, the King of Bhutan and the British signed the Treaty of Sinchula that ceded Kalimpong hills to the British India (Bhanja 1993). By 1866 CE, Darjeeling had assumed its current shape and size, covering an area of 1234 square miles (3200 km^2) in British India (O'Malley 1907). After the independence of India in 1947, Darjeeling was made to merge with the state of West Bengal as a district (Bhanja 1993). However, new autonomous administrative body called Gorkhaland Territorial Administration (GTA) comprising hills of Darjeeling, Kurseong and Kalimpong and some plain areas of Siliguri was created on 12 March 2012 (www.gta-darjeeling.org).

18.3 Ethnicity

Gorkha or Nepali is the major dominant ethnic communities in Darjeeling hills with several castes within this community such as Rai, Limboo, Tamang, Gurung, Chettri, Magar, Bahun, Pradhan/Newar, Dewan, Sunwar, Bhujel, Khagatey, Sherpa, Sanyasi or Giri, Kami, Damai, Sarki and Maji, along with other minority ethnic groups of Lepcha and Tibetan. Similarly in Sikkim 80% ethnic community is Gorkha or Nepali followed by Bhutia and Lepcha. The Gorkha or Nepali community has several castes, including Limboo, Tamang, Rai, Chettri, Sanyasi/Giri,

Bahun, Magar, Pradhan/Newar, Gurung, Bhujel, Dewan, Sunwar, Khagatey, Sherpa, Kami, Damai, Sarki and Maji, and 20% of ethnic communities of Lepcha and Bhutia (Tamang 2010). Ethnicity, language, culture, rituals and topography of both Sikkim and Darjeeling hills are same.

18.4 Climates and Edible Bio-resources

The wide altitudinal variation from 230 to above 8000 m and adverse climatic conditions from cold in the north to extremely wet conditions provide uniqueness to Sikkim and Darjeeling hills and make these regions rich in both floral and faunal diversity. Climate ranges from sub-tropical to temperate with temperatures seldom exceeding 28 °C in summer and ranging from −40 to 10 °C in winter (Hooker 1854). Mixed farming system for production of edible agro-resources is common agricultural practice in mountainous regions of Sikkim and Darjeeling hills which reflects the food culture of the community (Tamang 2005). Depending on the agro-climatic conditions, various food crops are cultivated in these regions which include rice, maize, finger millet, wheat, buckwheat and barley; pulse crops such as black gram, soybeans, green gram and garden peas; vegetable such as cabbage, cauliflower, leafy mustard (*rayo sag*) and young tendrils; fruits of squash (*iskus*), brinjal, chilli, cucumber, young tendrils and fruits of pumpkin, tomato, tree tomato, sponge gourd, etc.; tubers and rhizome crops such as potato, sweet potato, cassava, *Colocasia*, greater yam, ginger, turmeric and large cardamom; and root crops like radish, carrot, etc. (Subba 1984). Seasonal fruits such as orange, banana, mango, papaya, guava, pear, peach, apple, fig, avocado, etc. are cultivated and eaten (Tamang 2005). Various wild edible plants including young edible bamboo shoots, ferns, mushrooms and stinging nettles and their parts such as seeds, fruits, roots, leaves and flowers in local diet are important components of local cuisine (Rai et al. 2005; Pradhan and Tamang 2015). Tea is one of the most common cash crop in Sikkim and Darjeeling hills. Livestock mostly plays a subsidiary role in the mixed farming system in Sikkim and Darjeeling hills which includes cattle, sheep, goats, pigs, yaks, poultry, etc., which is mainly used for meat, milk and milk products. Yaks (*Bos grunniens*) are reared mostly on extensive alpine and subalpine scrublands between 2100 and 4500 m altitude for milk products and meat (Balaraman and Golay 1991; Sharma et al. 2006).

18.5 Dietary Culture

Drinking of full mug of milk tea (with or without sugar) at early morning is the traditional practice in Sikkim and Darjeeling hills. Frequency of daily intake of staple diet is two times. The food in these regions is less spicy, nonoily or lesser oily and semi-boiled. The first meal, the traditional Nepali *thali*, starts in the morning around 8:00 to 9:00 a.m. with cooked rice, *dal* (legume soup); vegetable, mostly leafy green vegetables; or wild edible herbs, mixed with potatoes, some fermented

Fig. 18.2 Gorkha *thali*

food items, meat or milk products (*dahi*, *mohi*, *chhurpi*) and pickles commonly called *bhat-dal-tharkari-achar* (Fig. 18.2). It is followed by light refreshment with mostly traditional snacks and tea with/without milk/salt/sugar in the afternoon. The second meal is dinner around early evening, which consists of the same *bhat-dal-tharkari-achar*. Bhutia and Lepcha usually eat *thug-pa*, noodles in soup. Traditionally the people of these regions prefer cooked rice as staple diet; however, wheat-based baked bread called *roti* or *chapatti* is also a popular staple food mostly among urban population. Cooked grinded maize is also eaten as staple food mostly in rural areas (Tamang 2005). A traditional ethnic food of Gorkha called *dheroh*, boiled maize/finger millet/buckwheat, is a popular staple diet in villages which is consumed with *mohi*, buttermilk and *gundruk*, fermented vegetable. Ethnic people of North Sikkim in high mountains drink *pheuja*, butter tea prepared from yak milk.

Besides consumption, some of these ethnic foods have social values especially in festivals and occasions. Celebration of festivals with *sel roti*, fermented cereal-based fried donut-like confectionery, is a custom of the Gorkha. *Dahi*, fermented milk product, is consumed as savoury in daily diets in these regions and is also used by the Gorkha as an adhesive for rice grains and colour to make *tika* which is applied on the foreheads of the younger members of the family by their elders during festivals and marriages. Besides entertainment for drinking, alcoholic beverages and distilled liquor are offered to pray ancestors and gods, and also in spirit possession practiced by some ethnic communities.

18.6 Consumption Habits

In Sikkim about 88.3% are non-vegetarians, whereas 11.7% are vegetarians (Tamang et al. 2007a). The Bhutia and the Lepcha are non-vegetarians and prefer beef and pork and even yak meat in North Sikkim. Some Brahmin Gorkha are vegetarians. Non-vegetarians eat chicken, mutton, buffalo, pork and beef. Beef is taboo to a majority of the Gorkha except Tamang and Sherpa. Newar prefers buffalo meat. About 67.7% of people prepare ethnic fermented foods at home for consumption (Tamang et al. 2007a). In Sikkim, per capita consumption of ethnic fermented foods and beverages is 163.8 g/day representing 12.6% of fermented foods in daily intake of meal (Tamang et al. 2007a).

Cooking is usually done by women, sometimes assisted by men members. Traditionally in rural areas, male members in the family are served with the meals first, and women eat afterwards in the kitchen. Traditionally members of the family sit together in bamboo-made mats on clean floor in the kitchen, and meals are served by women members of the family and then usually eaten by hands. Bamboo-made chopstick is commonly used by the Bhutia and Tibetans. Plates made up of brass or lining with brass called *kasa ko thal* are traditionally used to serve the foods. Locally available tree leaves, locally called *nevera* or fig (*Ficus* spp.) plant, are also used occasionally in festivals and marriages for serving foods. Nowadays even in rural areas, people prefer to eat in dining tables with different cutleries. Change from traditional feeding practice to modern cutlery systems in culinary cultures has been observed in these regions.

'Cooked rice-soybean-fish-alcoholic beverage' diet is the characteristic dietary culture of the Far East and South East Asia, whereas 'wheat/barely-milk and milk products-meat-wine' is the food culture of the Western world (Tamang and Samuel 2010). The dietary culture of the ethnic people in the Sikkim Himalayas shows a 'combination' of dietary cultures of both Eastern and Western world probably due to the geotechnical location of and diverse ethnicity in a mixed society. Culinary and typical cuisine of different ethnic people is unique and unparalleled since Sikkim and Darjeeling hills are wedged in between Nepal and Bhutan with vast expanse of the Tibetan Plateau in the north. Probably the food culture of these regions is a fusion of the Gorkha and Tibetan cuisines with modifications based on preference, acceptability, adaptability and other social ethos.

18.7 History of Fermented Foods and Alcoholic Beverages

Various types of ethnic fermented foods and beverages in these regions have been prepared and consumed for many centuries and have become a part and parcel of each ethnic community originated or settled or migrated within the Himalayan regions. Traditional knowledge of food fermentation might have been passed from generation to generation by elders, mostly grandmothers/grandfathers and village elders, and also by self-practice, family tradition and community knowledge based on migration, settlement pattern and political history. Each community dwelling in

Sikkim and Darjeeling hills might have brought with them their original culinary skills and cuisine of ethnic foods and alcoholic beverages and shared within the different communities in a mixed society which resulted into unique fermented foods and beverages of Sikkim and Darjeeling hills.

The origin of soybean in Northeast India and Nepal has not been historically recorded. However, domestication of soybeans {'भटमास' (*Bhaṭamāsa*) in Nepali language} was recorded in one of the myths of the Limboo, ethnic community of Gorkha, known as 'mundhuns' (Tamang 2010), which was written in between 2500 and 100 BC (Subba 2008). According to Shurtleff and Aoyagi (2010), soybean was probably introduced to the Eastern Himalayan regions of India, Nepal and Bhutan from Yunnan province of China. Traditionally two local native varieties of small-sized (5–6 mm) soybeans of 'yellow cultivar' and 'brown cultivar' are cultivated in summer in lower altitudes of Sikkim and Darjeeling hills (Tamang 2010). *Kinema*, a fermented soybean food of the Eastern Himalayan regions of Nepal, India (Sikkim and Darjeeling hills) and Bhutan, might have originated in eastern part of Nepal around 600 BC to 100 AD during the Kirat (Limboo belongs to Kirat race) dynasty (Tamang 2001). The word *kinema* was originated from the word *kinamba* of the Limboo language: *ki* means fermented and *namba* means flavour (Tamang 2010). Hypothetically the Limboo community was the first to start production and consumption of *kinema*, though no historical records on the origin of *kinema* is available. The time of invasion of Nepal to territories of Sikkim including present Darjeeling hills in 17th AD (Bhanja 1993) resulted in the movement and settlement of Gorkha from one place to another, exchanging culture and food habits within modern Sikkim and Darjeeling hills, and subsequently kinema was also adopted as a highly accepted delicacy in food culture of other castes of Gorkha. The Lepcha calls it *satlyangser*, and the Tibetan and Bhutia call it *bari*.

Historical descriptions of *chyang* or *chee/chi*, mild-alcoholic fermented finger millet beverages, were recorded in Sikkim and Darjeeling hills as early as 1800 by Hooker (1854), Risley (1928) and Gorer (1938). In *The Gazetteer of Sikkim*, Risley (1928) wrote '*marwa*, *chang*, is a kind of beer brewed by everyone in Sikkim, and might be called their staple food and drink', which interprets the long history of alcoholic beverages in Sikkim.

18.8 Ethnic Fermented Foods

Diverse ethnic people of Sikkim and Darjeeling hills consume more than 48 different types of ethnic fermented foods (40 varieties) and beverages (8 varieties) prepared from available plant and animal sources (Table 18.1). Some of these fermented foods and alcoholic beverages are common, whereas some are uncommon confined to particular community and regions. Type of ethnic fermented foods, their methods of preparation, mode of consumption, economy, functional microorganisms and nutritive value have been mentioned below.

Table 18.1 Ethnic fermented foods of Sikkim and Darjeeling hills

Food products	Plant/animal sources	Sensory property and edibles	Major ethnic consumer	Region/district
Fermented vegetable products				
Gundruk	Leafy vegetable	Dried, sour; soup and pickle	All	All
Sinki	Radish taproot	Dried, sour; soup and pickle	All	All
Khalpi	Cucumber	Sour; pickle	Bahun-Chettri	East, South and West District of Sikkim, Kalimpong
Goyang	Leafy wild plant	Sour; soup	Sherpa	East Sikkim
Mesu	Bamboo shoots	Sour; pickle	All	All
Fermented legume products				
Kinema	Soybean	Sticky soybeans; curry	Non-Brahmin Gorkha	All
Maseura	Black lentil	Dry, ball-like; curry and soup	Newar (Pradhan)	East Sikkim, Rimbik of Darjeeling hills
Fermented cereal products				
Sel roti	Rice-wheat flour	Round, deep-fried bread	Gorkha	All
Fermented dairy products				
Chhurpi (soft)	Cow/yak milk	Soft, cheese-like; curry and pickle	All	All
Chhurpi (hard)	Cow/yak milk	Hard mass; masticator	All	All
Chhu/ Sheden	Cow/yak milk	Soft, strong-flavoured; dish	Bhutia, Sherpa, Lepcha	North and East Sikkim
Philu	Cow/yak milk	Cream; fried curry with butter	Bhutia, Sherpa	North and East Sikkim
Somar	Cow/yak milk	Paste, flavoured; condiment	Sherpa	West Sikkim, Rimbik of Darjeeling
Dahi	Cow/yak milk	Curd; savoury	All	All
Shyow	Yak milk	Curd; savoury	Bhutia, Lepcha	North and West Sikkim
Mohi	Cow milk	Buttermilk	All	All
Gheu	Cow milk	Butter	All	All
Maa	Yak milk	Butter	Bhutia	North and East Sikkim
Chur	Cow milk	Wet cheese-like; curry	Bhutia	North Sikkim
Sun-dried fish products				
Sidra	Fish	Dried fish; side dish curry and pickle	Non-vegetarian Gorkha	All

(continued)

Table 18.1 (continued)

Food products	Plant/animal sources	Sensory property and edibles	Major ethnic consumer	Region/district
Sukuti	Fish	Dried, salted; side dish curry and pickle	Non-vegetarian Gorkha	All
Suka ko maacha	River fish	Dried, smoked; side dish curry and pickle	Non-vegetarian Gorkha	All
Gnuchi	River fish	Smoked fish; side dish curry	Lepcha	North Sikkim and Kalimpong
Fermented and sun-dried meat products				
Shakampo	Beef/yak/pork	Smoked meat; curry	Bhutia, Lepcha	North Sikkim
Sukula	Buffalo meat	Dried meat; curry	Newar	All
Suka ko masu	Mutton/pork	Smoked meat; curry	Non-vegetarian Gorkha	All
Lang kargyong	Beef, intestine (fatty), chopped meat, fat, garlic, ginger, salt	Soft or hard, brownish; sausage, curry	Bhutia, Lepcha, Sherpa	North and East Sikkim
Yak kargyong	Yak, intestine (fatty), chopped meat, fat, garlic, ginger, salt	Soft or hard, brownish; sausage, curry	Bhutia	North Sikkim
Faak kargyong	Pork, intestine, boiled rice, blood, chopped meat, salt, garlic, ginger	Soft or hard, brownish; sausage, curry	Bhutia, Lepcha, Sherpa	North and East Sikkim
Lang satchu	Beef, long strand-like dried meat	Hard, brownish; dried meat, side dish curry	Bhutia, Tibetan, Lepcha, Tamang, Sherpa	All
Yak satchu	Yak, long strand-like dried meat	Hard, brownish; dried meat, side dish curry	Bhutia, Lepcha, Sherpa	North Sikkim
Suka ko masu	Mutton, long strand-like dried meat	Hard, brownish-chocolate; side dish curry	Non-vegetarian Gorkha	All
Sheakua	Buffalo, long strand-like dried meat	Hard, brownish-chocolate; grilled; smoked; snack	Newar	All
Yak chilu	Yak fat	Hard, used as substitute of an edible oil	Bhutia	North Sikkim
Lang chilu	Beef fat	Hard, used as an edible oil	Bhutia, Lepcha, Sherpa	North and East Sikkim
Luk chilu	Sheep fat	Hard, used as an edible oil	Bhutia, Lepcha, Sherpa	North Sikkim

(continued)

Table 18.1 (continued)

Food products	Plant/animal sources	Sensory property and edibles	Major ethnic consumer	Region/district
Yak kheuri	Yak, chopped intestine (fatty), meat, fat, abdomen, salt	Curry	Bhutia	North Sikkim
Lang kheuri	Beef, chopped intestine (fatty), meat, fat, abdomen, salt	Curry	Bhutia, Lepcha, Sherpa	Sikkim
Khyopeh	Meat mixtures stuffed into the rumen (stomach) of yak	Soft or hard and brownish; cooked with nettle leaves as side dish	Bhutia mostly Lachungpa	North Sikkim

Fig. 18.3 Traditional method of preparation of *gundruk*

18.9 Fermented Vegetable Products

18.9.1 *Gundruk*

Gundruk is an ethnic fermented leafy vegetable product of Sikkim and Darjeeling hills commonly prepared during winter when leafy vegetables are plenty. The word *gundruk* has been derived from Newar/Pradhan dialect *gunnu* meaning dried taro (*Colocasia*) stalk. *Gundruk* is similar to fermented vegetable products of other countries such as *kimchi* of Korea, *pao cai/suan-cai* of China, *sunki* of Japan, and *sauerkraut* of Germany (Tamang et al. 2016a).

18.9.1.1 Traditional Method of Preparation

During preparation of *gundruk* (Fig. 18.3), fresh leaves of *rayo-sag* [*Brassica rapa* L. ssp. *campestris* (L.) Clapham variety *cumifolia* Roxb.], leaves of mustard [*Brassica juncea* (L.) Czern], leaves of radish (*Raphanus sativus* L.), leaves of cauliflower (*Brassica oleracea* L. variety *botrytis* L.) and leaves of cabbages (*Brassica oleracea* L. variety *capitata*) are wilted and shredded using a sickle or knife for 1–2 days. The shredded leaves are crushed mildly and pressed into a container or earthen jar, without adding any salt or spices, made air-tight, kept in a warm place for natural fermentation till 7–10 days. After 7–10 days, typical flavour with acidic taste predicts the completion of the fermentation, and *gundruk* is removed from jar and sun-dried for 2–4 days. Only dried *gundruk* (15% moisture) is traditionally consumed. Dried *gundruk* can be preserved for 2 years or more at room temperature for consumption.

18.9.1.2 Mode of Consumption

Per capita consumption of *gundruk* is 1.4 g/day with approximate annual production of 3.2 kg/house in Sikkim (Tamang et al. 2007a). *Gundruk* is eaten as soup or pickle. Soup is made by soaking *gundruk* in water for 10 min, squeezing and frying in edible oil with chopped onions, tomatoes, chillies, turmeric powder and salt. It is then boiled for 10–15 min and served hot with steamed rice. *Gundruk* soup is a good appetizer in a bland and starchy diet. *Gundruk* is sold in all local markets mostly by women.

18.9.1.3 Microorganisms

The microorganisms found are *Lactobacillus fermentum*, *Lb. plantarum*, *Lb. casei*, *Lb. casei* subsp. *pseudoplantarum* and *Pediococcus pentosaceus* (Tamang et al. 2005). During *gundruk* fermentation, *Lactobacillus fermentum* initiates the fermentation and is followed by *Pediococcus pentosaceus*, *Lactobacillus casei* and finally *Lactobacillus plantarum* dominating the fermentation (Tamang and Tamang 2010). These bacteria produce lactic acid and acetic acid which lower the pH of the substrates making the products more acidic in nature.

18.9.1.4 Nutritional Value and Health Benefits

Nutritional values of *gundruk* are protein (38.7% DM (dry matter)), fat (2.1% DM), carbohydrate (38.3% DM), calorie value (321.9 kcal/100 g DM), ash (22.2% DM) and minerals such as calcium (234.6 mg/100 g), sodium (142.2 mg/100 g) and potassium (677.6 mg/100 g) (Tamang et al. 2012). Due to low pH (3.3–3.8) and high acid content (1.0–1.3%), *gundruk* can be preserved for longer periods without refrigeration, unique type of biopreservation of perishable vegetable in the mountainous regions. Many functional properties were exhibited by same species of lactic acid bacteria isolated from *gundruk* such as strong acidification, coagulation activities, antimicrobial activities, non-producers of biogenic amines and probiotics candidates promoting the health benefits to the consumers (Tamang et al. 2009).

18.9.1.5 Upgradation of Traditional Process

Traditional methods for preparation of *gundruk* vary from place to place causing inconsistency in the final product. An attempt is made to upgrade the traditional processing of perishable vegetables using pure strains of lactic acid bacteria isolated from naturally fermented *gundruk* (Tamang and Tamang 2010). On the basis of superior technological properties of LAB strains such as acidification ability, antimicrobial activities, non-production of biogenic amines, ability to degrade antinutritive factors and even high degree of hydrophobicity (Tamang et al. 2009), *Lb. plantarum* GLn:R1 (MTCC 9483) and *P. pentosaceus* GLn:R2 (MTCC 9484) were selected as a starter culture for production of *gundruk* (Tamang and Tamang 2010). Leaves of local vegetable 'rayo saag' (*Brassica rapa* L. subsp. *campestris* (L.) Clapham variety *cumifolia* Roxb.) were washed in sterile distilled water, wilted in oven (~30 °C) for 6 h, crushed, put into sterile warm water (about 90 °C) for 5 min and transferred into another sterile glass container. Excess water in the leaves was removed by squeezing, and about 400 g of crushed leaves are distributed aseptically into each sterile 500 mL capped bottles. Each bottle was inoculated by a mixture of actively grown culture strains of *Lactobacillus plantarum* GLn: R1 (MTCC 9483) and *Pediococcus pentosaceus* GLn:R2 (MTCC 9484), previously isolated from market samples of *gundruk* (Tamang et al. 2005), at the ratio of 10^7 cfu/g, and bottles were tightly capped and incubated at 20, 25 and 30 °C, respectively, for 6 days. *Gundruk* prepared by the starter culture was evaluated organoleptically, and it was found that the sixth day fermented 'rayo' leaves at 20 °C had the highest score of general acceptability among consumers with better aroma and acidic taste typical of *gundruk* (Tamang and Tamang 2010).

18.9.2 *Sinki*

Sinki is an ethnic fermented radish taproot product which is unique to these regions. *Sinki* is prepared by pit fermentation technique during winter when radish taproots are available in plenty.

18.9.2.1 Traditional Method of Preparation

Sinki is prepared by 'pit fermentation technique' which is unique to Sikkim and Darjeeling hills using the native skill of ethnic people. Pit fermentation of *sinki* preparation is one of the traditional methods of biopreservation of perishable radish taproot by lactic acid fermentation. A pit of 2.0–3.0 ft deep with diameter of 2.0–2.5 ft is dug in a dry place nearby the house in a village. Then the pit is cleared, plastered with fine mud and dried up by burning with dry leaves. After removing the ashes, the pit is lined with bamboo sheaths and dry paddy straw. Taproots of radish (*Raphanus sativus* L.) are wilted for 2–3 days, crushed, dipped in lukewarm water, squeezed and pressed tightly into the pit, then covered with dry leaves and weighted down by heavy planks or stones. The surface of the pit is plastered with mud to make it anaerobic and left to ferment naturally for 22–30 days. After completion of fermentation, fresh *sinki* is removed from pit, cut into small pieces, sun-dried for

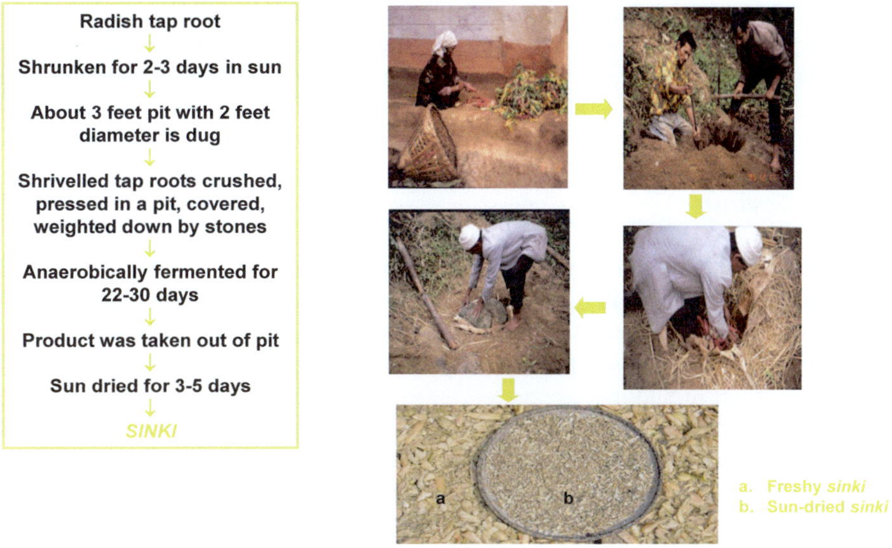

Fig. 18.4 Traditional method of preparation of *sinki*

3–5 days (Fig. 18.4) and stored at room temperature for future consumption. Dry *sinki* can be kept for 2 years or more at room temperature.

18.9.2.2 Mode of Consumption

Sinki has acidic flavour and is consumed as soup and pickle. The soup is made by soaking *sinki* in water for about 10 min and squeezing and frying it along with chopped onions, tomatoes and green chillies and salt. Soup is served hot along with meals. It is said to be a good appetizer, and people use it as a remedy for indigestion. The pickle is prepared by soaking *sinki* in water, squeezing it dry and mixing it with salt, mustard oil and chillies. *Sinki* is also sold in all local markets mostly by the rural women for their livelihood.

18.9.2.3 Microorganisms

Sinki is a naturally fermented product and is mostly dominated by species of lactic acid bacteria including *Lactobacillus plantarum*, *Lb. brevis*, *Lb. casei* and *Leuconostoc fallax* (Tamang and Sarkar 1993; Tamang et al. 2005).

18.9.2.4 Nutritional Value and Health Benefits

Nutritional value of *sinki* shows protein content (14.6 g DM), fat (2.5 g DM), carbohydrate (68.0% DM), calorie value (344.2 kcal/100 g DM), ash (11.5 g DM), calcium (223.9 mg/100 g, sodium (737.3 mg/100 g) and potassium (2320.4 mg/100 g) (Tamang et al. 2012). During fermentation pH dropped from 6.7 to 3.3 with increase in acidity of 1.1% (Tamang and Sarkar 1993). *Sinki* contains high amount of organic acids which are good appetizers. Due to low pH and high acids, *sinki* can be preserved for several months for consumption. *Lactobacillus plantarum*, *Lb. brevis*, *Lb.*

casei and *Leuconostoc fallax* isolated from *sinki* samples showed many functional properties including antimicrobial activities, non-production of biogenic amines and even few probiotic properties, which may have health-promoting benefits to the consumers (Tamang et al. 2009).

18.9.3 *Goyang*

Goyang is a fermented wild leafy vegetable product with acidic flavour (pH 6.5 and acidity 0.13%), traditionally consumed by the Sherpas of Sikkim. *Goyang* is prepared from the leaves of wild plant locally called 'magane-saag' (*Cardamine macrophylla* Willd.) that belongs to the family Brassicaceae which are available in high altitude >6000 ft in Sikkim.

18.9.3.1 Traditional Method of Preparation

Leaves of *Cardamine macrophylla* Willd. are collected from natural resources, washed and cut into pieces. Leaves are then squeezed to drain off excess water and are tightly pressed into the bamboo baskets lined with 2–3 layers of leaves of fig plants. The top of the baskets is then covered with fresh fig plant leaves, fermented naturally at room temperature (~15–25 °C) for a month (Fig. 18.5). Poly bags or glass jars are sometimes replaced by bamboo baskets which are easy to maintain anaerobic condition during preparation of *goyang*. Freshly prepared *goyang* is transferred into an air-tight container which can be stored for 2–3 months. However,

Fresh leafy wild herbs
↓
Cut into pieces; Squeezed
↓
Put into container, made anaerobic condition
↓
Naturally fermented for 25-30 days
↓
Products are made into balls
↓
Dry in sun light for 2-3 days
↓
GOYANG

Fig. 18.5 Traditional method of preparation of *goyong*

shelf life can be prolonged, by making *goyang* into balls and sun drying it for 3–5 days, and be stored for several months at the room temperature.

18.9.3.2 Mode of Consumption

Goyang is an important dietary item for the Sherpas who live in high altitudes. It is boiled in a soup along with yak or beef meat and noodles to make a thick 'thukpa', a common staple food of the Sherpas.

18.9.3.3 Microorganisms

Lactobacillus plantarum, *Lb. brevis*, *Lactococcus lactis*, *Enterococcus faecium* and *Pediococcus pentosaceus* are present in *goyang* which makes it acidic (Tamang and Tamang 2007).

18.9.3.4 Nutritional Value

Composition of nutritional value of *goyang* is protein (35.9% DM), fat (2.1% DM), carbohydrate (48.9% DM), calorie value (357.2 kcal/100 g), ash (12.9% DM), calcium (92.2 mg/100 g), sodium (6.7 mg/100 g) and potassium (268.4 mg/100 g) (Tamang and Tamang 2007).

Fig. 18.6. Traditional method of preparation of *khalpi*

18.9.4 *Khalpi*

In Sikkim and Darjeeling hills, fresh cucumber is commonly eaten as salad. Meanwhile, ripened cucumber is fermented into the product called *khalpi* or *khaipi* which is mostly consumed by the Brahmin Gorkha as pickle.

18.9.4.1 Traditional Method of Preparation

Fully ripened cucumber is usually preferred for *khalpi* preparation. During preparation, selected ripened cucumber is cut into pieces, sun-dried for 2 days and then put into a bamboo vessel, locally called 'dhungroo', made air-tight and kept at room temperature for natural fermentation for 4–7 days (Fig. 18.6.). *Khalpi* is acidic product with pH of 3.9 and acidity of 0.95% (Tamang and Tamang 2010).

18.9.4.2 Mode of Consumption

Khalpi is consumed as pickle, by adding mustard oil, salt and powdered chillies, in meal with the boiled rice.

18.9.4.3 Microorganisms

Lactobacillus plantarum, *Lb. brevis* and *Leuconostoc fallax* are reported from *khalpi* (Tamang et al. 2005).

18.9.4.4 Nutritional Value

Nutritional value of *khalpi* shows the protein content of 12.3% DM, fat 2.6% DM, carbohydrate 70.9% DM, calorie value 356.2 kcal/100 g DM, ash 14.2% DM, calcium 6.4 mg/100 g, sodium 2.2 mg/100 g and potassium 125.1 mg/100 g (Tamang et al. 2012).

18.9.4.5 Upgradation of Traditional Process

Lactobacillus plantarum KG:B1 (MTCC 9485), *Lb. brevis* KG:B2 (MTCC 9486) and *Leuc. fallax* KB:C1 (MTCC 9487) previously isolated from *khalpi* (Tamang et al. 2005) were selected on the basis of superior technological properties such as acidifying capacity, antimicrobial activities, non-production of biogenic amines and ability to degrade antinutritive factors of the raw materials (Tamang et al. 2009). Cucumber was inoculated with abovementioned consortia of bacteria (10^7 cfu/g) kept in tightly capped bottles and incubated at 20 °C for 72 h which resulted into *khalpi*. Organoleptically *khalpi* produced by the optimized process was highly acceptable to consumers (Tamang and Tamang 2010).

18.9.5 *Mesu*

There are varieties of edible bamboo shoots which are commonly eaten as fresh vegetable in Sikkim and Darjeeling hills during rainy seasons. Practice of fermenting fresh edible bamboo shoots is also popular in these regions, and the product is called *mesu* with sour-acidic taste (pH 3.9 and acidity 0.88%). The word *mesu* was derived from the Limboo language: *me* means young bamboo shoot and *su* means sour (Tamang 2010). The Lepcha calls it *satit*. *Mesu* is similar to *lung-siej* of Meghalaya, *soibum* and *soidon* of Manipur and *naw-mai-dong* of Thailand (Tamang 2010).

18.9.5.1 Traditional Method of Preparation

Young and tender sprouting shoots of edible varieties of bamboo (*Dendrocalamus sikkimensis* Gamble, *Dendrocalamus hamiltonii* Nees et Arn. ex Munro and *Bambusa tulda* Roxb.) are collected from forest, and outer hard casings are removed, and inner portion is then chopped into small pieces with a knife. The chopped pieces are washed thoroughly with clean water, drained well, pressed tightly into hollow bamboo shoot locally called as 'dhungro'—made air-tight with a lid, placed in an upside-down position to drain out any liquid—and fermented under natural anaerobic condition for 7–15 days (Fig. 18.7). After desired fermentation, *mesu* is kept in an air-tight container for a year at ambient temperature. It is also preserved mixing with mustard oil and mustard powder for consumption as pickle. This is a good example of biopreservation of edible bamboo shoots. *Mesu* contained in the same fermenting bamboo vessel covered by bamboo leaves is seen in the local market for sale by rural women.

Fig. 18.7 Traditional method of preparation of *mesu*

18.9.5.2 Mode of Consumption

Mesu is consumed as pickle. It is mixed with oil, chillies and salt in a closed jar and is kept for several months.

18.9.5.3 Microorganisms

Species of LAB dominate *mesu* fermentation. *Lactobacillus plantarum*, *Lb. brevis*, *Lb. curvatus*, *Leuconostoc citreum* and *Pediococcus pentosaceus* were isolated from samples of *mesu* (Tamang and Sarkar 1996; Tamang et al. 2008).

18.9.5.4 Nutritional Value

Nutritional value of *mesu* shows the protein content of 27.0% DM, fat 2.6% DM, carbohydrate 55.6% DM, calorie value 352.4 kcal/100 g DM, ash 15.0% DM, calcium 7.9 mg/100 g, sodium 2.8 mg/100 g and potassium 282.6 mg/100 g (Tamang and Sarkar 1996).

18.10 Fermented Legumes

18.10.1 *Kinema*

Two local varieties of soybeans (*Glycine max*), 'yellow cultivar' and 'dark brown cultivar' are cultivated in summer as mix-crop with rice and maize in Sikkim and Darjeeling hills up to 1500 m, and are harvested in early winter. Dry soybean seeds, locally known as '*bhatmas*' are commonly consumed as both non-fermented (roasted) and fermented recipes in Sikkim and Darjeeling hills. *Kinema* is an ethnic fermented soybean food of Limboo community of Gorkha, which is sticky, flavorsome with slightly ammoniacal flavor. *Kinema* is a typical example alkaline fermented food with pH of 7.9 produced by natural fermentation. However, delicacy of *kinema* is now shared among other castes of the Gorkha. The Lepcha calls it *satlyangser* and the Bhutia calls *bari* in Sikkim. Similar to kinema are *tungrymbai* of Meghalaya, *hawaijar* of Manipur, *bekang* of Mizoram, *aakhuni* of Nagaland, *peron naming/peruyann/peha* of Arunachal Pradesh, *natto* of Japan, *chungkokjang* of Korea, *thua nao* of Thailand, and *sieng* of Campodia (Tamang 2015; Tamang et al. 2016a).

18.10.1.1 Traditional Method of Preparation

Local yellow variety of small-sized (~6 mm) soybeans is usually preferred for *kinema* preparation in these regions. Soybean seeds are soaked overnight in water and cooked by boiling until they are soft. Excess water is drained off, and cooked soybean seeds are cracked lightly by a wooden pestle (locally called *muslo*) in a wooden mortar (locally called *okhli*) to split the cotyledons, probably to accelerate the fermentation. About 1% of firewood ash collected from earthen oven in kitchen is added to cracked soybeans to maintain the alkaline condition of the product. Practice of cracking cooked seeds of soybeans and addition of firewood ash during *kinema* production is unique to Sikkim and Darjeeling hills unlike other similar

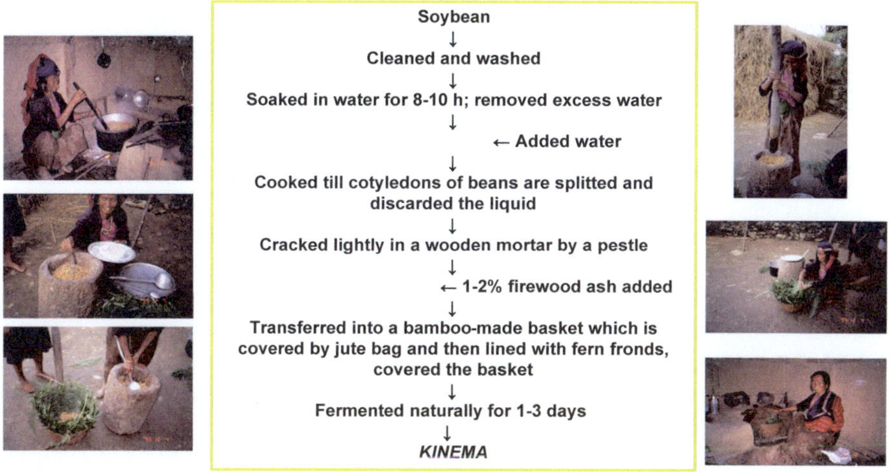

Fig. 18.8 Traditional method of preparation of *kinema*

Fig. 18.9 Freshly fermented *kinema* and *kinema* curry

fermented soybean foods of Asia and Northeast India probably to increase the surface areas for speed fermentation by aerobic spore-forming *Bacillus* spp. Soybean grits are then placed into a bamboo basket lined with locally grown fresh fern (*Glaphylopteriolopsis erubescens*) leaves, covered in a jute bag and left to ferment naturally at ambient temperatures (25–40 °C) for 1–2 days above earthen oven in kitchen. Appearance of white viscous and sticky mass on the surface of fermenting soybeans with 'umami' flavour generated during proteolysis of soya proteins (Kawamura and Kara 1987) indicates completion of fermentation (Fig. 18.8). Quality of *kinema* depends on the stickiness when fresh fermented *kinema* is touched/stirred by spoon or wooden paddle, when threads are pooled and if the threads become longer and more unbroken that *kinema* is considered as the good quality. Stickiness in *kinema* is presumed as Poly-γ-glutamic acid (PGA), an amino acid polymer, one of the glutamic acids (Chettri et al. 2016).

Shelf life of fresh *kinema* is very short about 2–3 days during summer and a maximum of 1 week in winter without refrigeration. By sun drying for 2–3 days, the shelf life of *kinema* can be prolonged, and dry *kinema* is stored for several months at room temperature. Preparation of *kinema* mostly by women varies from place to place and is still restricted to household level. This unique traditional knowledge of *kinema*-making has been protected as hereditary right and passes from mother to daughter.

18.10.1.2 Mode of Consumption

Kinema is eaten as side dish curry (Fig. 18.9) with cooked rice. Fresh *kinema* is fried in vegetable oil, and chopped onions, tomatoes and turmeric powder are added and fried for 2 min. Then salt and sliced green chillies are added and fried for 3–5 min. A little water is added to make thick gravy, and it is cooked for 5–7 min. *Kinema* curry is ready for serving with boiled rice (Tamang and Tamang 1998).

18.10.1.3 Socio-economy

Kinema is sold in local markets commonly called 'haats' in different places of Sikkim and Darjeeling hills. It is sold in amount equivalent to weights of 150–200 g packages of *kinema*, usually packed in *nevara* (fig plant) leaves and tied loosely by straw. The 100 g of *kinema* costs about Rs.10 to Rs.20.

Survey on *kinema* production and its impacts on socio-economy was conducted at Aho village of East Sikkim, which is considered as the centre of kinema supply to nearby local markets in and around Gangtok and adjoining areas. The outcome of survey showed that about 40% of profit is made out of selling *kinema* by Limboo women of Aho who sell the products at Gangtok vegetable markets and the rest 60% is spent on cost of raw soybeans, fuel for cooking, transportation from village to town, etc. (Tamang 2000). It is also observed that the little profit the women sellers make is spent on children's education and procuring of essential commodities not locally available in the village and also on domestic expenses. *Kinema* production is an income generation for some families. *Kinema*-making is a home-based cottage-industry for marginal income generation practiced by some rural women, however, such food processing has not been included in small-cottage industry scheme of the government and or financial institutions.

18.10.1.4 Microorganisms

Bacillus subtilis, *B. licheniformis* and *B. sonorensis*, *Enterococcus faecium*, *Candida parapsilosis* and *Geotrichum candidum* have been reported in *kinema* (Sarkar et al. 1994; Tamang et al. 2002; Tamang 2003; Chettri et al. 2016). However, *Bacillus subtilis* is the dominant functional bacterium in *kinema* fermentation (Sarkar and Tamang 1994, 1995; Tamang and Nikkuni, 1996; Sarkar et al. 2002). Sources of inoculum in natural fermentation of *kinema* might be surrounding environmental conditions including soybean, used equipment and leaves as wrapping materials that may supplement microorganisms for spontaneous fermentation of *kinema* without using starter cultures (Tamang 2003). The culinary use of traditionally prepared *kinema* as curry is safe for consumption (Nout et al. 1998).

Recently, we did the metagenomics of *kinema* samples collected from various places of Sikkim and Darjeeling hills using 16S rRNA gene amplicon sequencing which generated taxonomic reads of eight major phyla representing *Firmicutes* (84.5%), *Proteobacteria* (11.1%), *Actinobacteria* (3.4%) and others (1.2%). Taxonomic reads also showed the wide diversity of bacteria at genera level mostly dominated by *Bacillus* (74.9%) with species *Bacillus subtilis*, *B. pumilus*, *B. licheniformis* and *B. clausii*, followed by *Wohlfahrtiimonas* (5.9%), *Corynebacterium* (3.0%), *Lachnoclostridium* (2.8%), *Acinetobacter* (1.7%), *Enterobacteriaceae* (1.4%), *Staphylococcus* (1.4%), *Enterobacter* (1.3%), *Lysinibacillus* (1.2%), *Vagococcus* (1.2%), *Brevibacillus* (1.1%) and other genera (4.1%). The unknown and uncultured bacteria were found to be abundant in kinema samples (Tamang et al. 2019 unpublished).

18.10.1.5 Nutritional Value

Kinema is a high plant protein-based food in the local diet. Nutritional value of *kinema* shows protein content of 47.7% DM, fat 17.0% DM, carbohydrate 28.1% DM, calorie value 454 kcal/100 g DM, total amino acids 42618.0 mg/100 g, free amino acids 5129.0 mg/100 g, ash 7.2% DM, calcium 432.0 mg/100 g, sodium 27.7 mg/100 g, iron 17.7 mg/100 g, manganese 5.4 mg/100 g and zinc 4.5 mg/100 g (Sarkar et al. 1994). *Kinema* is the cheapest source of plant protein as compared to milk and animal products on the basis of protein cost per kg (Tamang 2001). Total amino acids, free amino acids and mineral contents are increased during *kinema* fermentation (Sarkar and Tamang 1995; Tamang and Nikkuni 1998). Proteolytic enzymes are produced by *Bacillus subtilis* during fermentation into peptides and amino acids enhancing digestibility (Tamang and Nikkuni 1998) and hydrolyse soy proteins, which may be denatured by cooking process. Remarkable increase in water-soluble nitrogen and trichloroacetic acid (TCA)-soluble nitrogen contents was observed during fermentation (Sarkar and Tamang 1995).

18.10.1.6 Health Benefits

Kinema may be considered as functional foods due to several health-promoting benefits to consumers (Tamang 2015; Tamang et al. 2016b). *Kinema* is rich in an essential fatty acid mostly linoleic acid (Sarkar et al. 1996) and contains all essential amino acids (Sarkar et al. 1997). During fermentation, cholesterol-lowering phytosterols are increased in *kinema* (Sarkar et al. 1996). Vitamins mostly riboflavin and niacin increase in *kinema* during fermentation (Sarkar et al. 1998). *Kinema* has antioxidant activities (Moktan et al. 2008). Due to presence of a large amount of Group B saponins, kinema claims to have health-promoting benefits (Omizu et al. 2011).

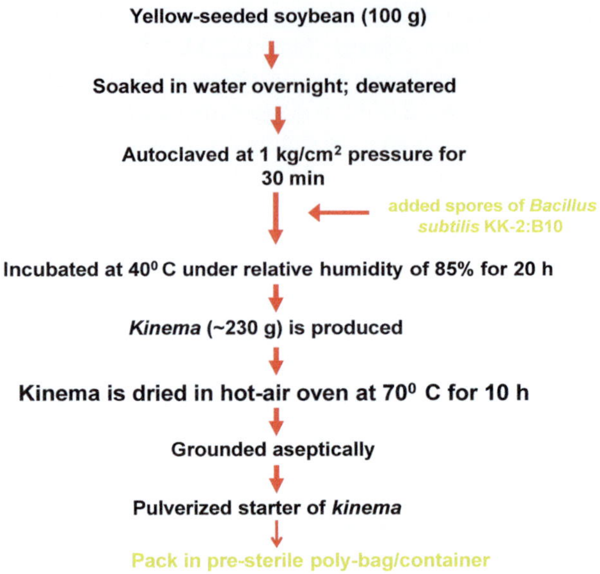

Fig. 18.10 Production method of pulverized *kinema* (Tamang 1999)

18.10.1.7 Improved Method for *Kinema* Production

Optimization of traditional process for *kinema* production was aimed to maintain the standard quality. Pure strains of *Bacillus subtilis* KK-2:B10 (MTCC, Microbial Type Culture Collection, Chandigarh 2756) and GK-2:B10 (Microbial Type Culture Collection, Chandigarh 2757), previously isolated from naturally fermented *kinema*, were selected as starters, inoculated on boiled soybeans, and fermented at 40 °C for 20 h followed by maturation at 5^0 C for 1 day to get the high-quality *kinema* (Tamang and Nikkuni 1998). Ready-to-use pulverized and dry starter culture of *Bacillus subtilis* was developed for *kinema* production (Tamang 1999). *Kinema* prepared by using *B. subtilis* KK2:B10 strain which was harvested in soybean extract broth was dried in an oven at 70 °C for 10 h and grounded aseptically to make pulverized starter. The 1% of pulverized starter was added aseptically to autoclaved soybeans and fermented to get *kinema* (Fig. 18.10). *Kinema* prepared by using pulverized starter had more advantages over traditional method due to shorter fermentation time that eliminates the chance of growth of contaminants and hygienic conditions, maintaining consistency with better quality and flavour (Tamang 1999). Ready-to-use pulverized starter may be appropriate and economic for *kinema* producers in Sikkim and Darjeeling hills due to it being cost-effective and easy to prepare. Indian Patent was granted to Prof. Jyoti Prakash Tamang on "A process for production of kinema, fermented soybean food, using a pure starter culture" (Patent No: 25346) in 2012. This may foresee the commercialization of *kinema* as low-cost, high plant protein food with functionality and health benefits in the country who is not familiar with traditional *kinema*.

18.10.1.8 HACCP

We formulated an application of Hazard Analysis Critical Control Points (HACCP) system in both traditionally prepared and laboratory-prepared *kinema* to identify specific hazards to ensure the safety of the product. HACCP model for optimized production of *kinema* was proposed which may help to reduce the pathogenic load to an acceptable level (Rai et al. 2014).

18.10.2 *Maseura*

Maseura, a cone-shaped hollow, dry (moisture, 8–10%; pH, 5.6–6.3), brittle and friable product, is one of the lesser-known fermented black lentil (*Vigna mungo*) foods consumed by Newar/Pradhan of Gorkha as condiment or an adjunct in cooking vegetable.

18.10.2.1 Traditional Method of Preparation

Beans of black gram (*Vigna mungo*) or black lentil (*Phaseolus mungo*) or rice bean (*Phaseolus calcaratus*) are cleaned, washed and soaked overnight and dehulled by pressing through hands, and the hulls are flown off, ground into thick paste using mortar and pestle. Water is carefully added while grinding, until paste becomes sticky, which is then made hand-moulded into small balls or cones. If rice bean is used, then boiled potato or squash or yam is mixed with the paste to make it sticky. The mixture is placed on a bamboo mat and fermented naturally in open kitchen for 2–3 days. After fermentation the product is sun-dried for 3–5 days depending upon the weather condition (Fig. 18.11). It is similar to Punjabi *wari or dal bodi.*

18.10.2.2 Mode of Consumption

Maseura is used as condiment or adjunct to vegetable in main meal, mostly consumed by Newar caste of Gorkha. Preparation and consumption of *maseura* is declining in Sikkim and Darjeeling hills. The new generation hardly knows the product.

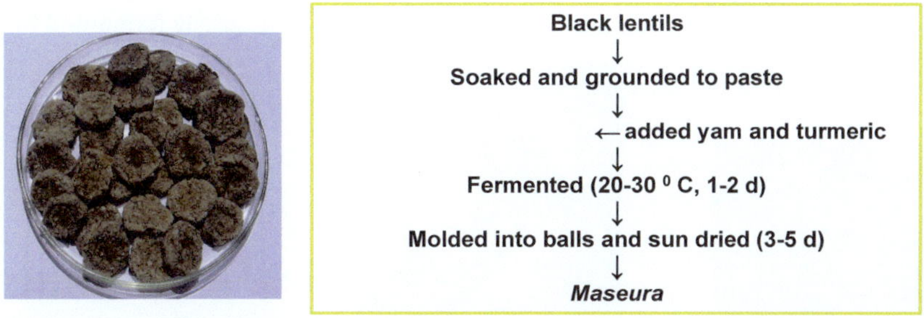

Fig. 18.11 Traditional method of preparation of *maseura*

18.10.2.3 Microorganisms

Bacillus subtilis, *B. mycoides*, *B. pumilus*, *B. laterosporus*, *Pediococcus acidilactici*, *P. pentosaceus*, *Enterococcus durans*, *Lb. fermentum*, *Lb. salivarius* and yeasts *Saccharomyces cerevisiae*, *Pichia burtonii* and *Candida castellii* were reported from samples of *maseura* collected from different places of Sikkim and Darjeeling hills (Chettri and Tamang 2008).

18.10.2.4 Nutritional Value

Protein content in *maseura* is 8–10%, and carbohydrate is 67–70% (Chettri and Tamang 2008).

18.11 Fermented Cereal Products

18.11.1 *Sel roti*

Sel roti or *Sel* is deep-fried, ring-shaped, spongy, pretzel-like fermented food product prepared from rice batter. *Sel roti* is one of the important dietary items of Gorkha commonly eaten as confectionery bread in festivals and special occasions (Yonzan and Tamang 2010a).

18.11.1.1 Traditional Method of Preparation

During *sel roti* preparation, rice (*Oryza sativa* L.) local variety 'attey' is sorted, washed and soaked in cold water overnight or 4–8 h at ambient temperature. The

Fig. 18.12 Traditional method of preparation of *sel roti*

use of milled rice in *Sel roti* preparation is also commonly practiced in these regions. Water is then decanted from the rice by using bamboo-made sieve called 'chalni' and spread over a woven tray made up of bamboo, locally called 'naanglo', and dried for 1 h. Soaked rice is pounded into coarse powder in a wooden mortar and pestle known as 'okhali' and 'mushli', respectively. Larger particles of pounded rice flour are separated from the rest by winnowing in a bamboo tray. Then, the rice flour is mixed with nearly 25% refined wheat (*Triticum aestivum* L.) flour, 25% sugar, 10% butter or fresh cream and 2.5% spices/condiments containing large cardamom (*Amomum subulatum* Roxb.), cloves (*Syzygium aromaticum* Merr.), coconut (*Cocos nucifera* L.), fennel (*Foeniculum vulgare* Mill.), nutmeg (*Myristica fragrans* Houtt.), cinnamon (*Cinnamomum zeylanicum* Bl.) and small cardamom (*Elettaria cardamomum* Maton.) (added to the rice flour and mixed thoroughly). Milk (boiled/unboiled) or water is added, kneaded into a soft dough and finally into batter with easy flow. Batter is left to ferment naturally at ambient temperature (20–28 °C) for 4–6 h during summer and at 10–18 °C for 6–8 h during winter. The leavened batter (pH 5.8) is ready for frying to get deep-fried *sel roti* (Fig. 18.12).

18.11.1.2 Mode of Consumption

Leavened batter is squeezed by hand or 'daaru' (metallic serving spoon), deposited as continuous ring circles like a doughnut onto hot edible oil in a cast-iron frying pan locally called 'tawa' and fried until golden brown and crisp on both sides. Oil is drained out by poker locally called 'jheer' or 'suiro' or also by a spatula locally called 'jharna' for few seconds. Deep-fried *sel roti* is served as confectionery bread with *aalu dam* (boiled potato curry) and *simi ko acchar* (pickle prepared from string beans) and meat curry. It can be served hot or cold. *Sel roti* is kept at room temperature for a month or more. Per capita consumption of *sel roti* in Sikkim is 8 g/day with average annual production of 18.5 kg per household (Yonzan and Tamang 2010a). *Sel roti* is sold in local food stalls and restaurants. Some people are economically dependent upon these products.

18.11.1.3 Microorganisms

Sel roti batters contain both lactic acid bacteria and yeasts. Species of bacteria are *Leuconostoc mesenteroides*, *Enterococcus faecium*, *Pediococcus pentosaceus* and *Lactobacillus curvatus*; and yeasts are *Saccharomyces cerevisiae*, *Saccharomyces kluyveri*, *Debaryomyces hansenii*, *Pichia burtonii* and *Zygosaccharomyces rouxii* (Yonzan and Tamang 2010b).

18.11.1.4 Nutritional Value

Nutritional value of *sel roti* batter shows protein (5.7% DM), fat (2.7% DM), carbohydrate (91.3% DM), calorie value (410.3 kcal/100 g DM), ash (0.8% DM), sodium (8.9 mg/100 g), potassium (29.7 mg/100 g) and calcium (23.8 mg/100 g) (Yonzan and Tamang 2010b).

18.11.1.5 Optimization of Traditional Process

Mixtures of a bacterium *Leuconostoc mesenteroides* BS1:B1 and a yeast *Saccharomyces cerevisiae* BA1:Y2, previously isolated from naturally fermented *sel roti* batters (Yonzan and Tamang 2010a, b), were selected based on their functional properties (Yonzan and Tamang 2013) and were mixed with rice batters and fermented at 28° C for 4 h to get *sel roti* batters. Batters were then deep-fried and served to consumers for sensory evaluation and consumers' preference trial test. The result showed that majority of consumers preferred *sel roti* prepared by consortium of starter cultures than traditionally prepared *sel roti* (Yonzan and Tamang 2013).

18.12 Fermented Milk Products

18.12.1 *Dahi*

Dahi is the most popular traditional curd prepared from cow milk by 'back-sloping method' in Sikkim and Darjeeling hills. Varieties of fermented dairy products are obtained from *dahi* such as *gheu*, *mohi*, soft *chhurpi*, *chhu*, etc. *Dahi* is the Nepali word; the Bhutia and Lepcha call it *shyow* (Rai et al. 2016).

18.12.1.1 Traditional Method of Preparation

Fresh milk obtained from cow or yak is boiled in a vessel. After boiling, the milk is cooled to room temperature, and often, a little amount of previously fermented *dahi* is mixed to the fresh milk to speed up the fermentation process. This is left for 1–2 days in summer or for 2–4 days in winter at room temperature for fermentation (Fig. 18.13). The duration of fermentation depends on the season as well as on the geographical location of the place in these regions.

Fig. 18.13 Traditional method of preparation of *dahi*

18.12.1.2 Mode of Consumption

Dahi (pH 4.2, acidity 0.73%) is consumed directly as a refreshing non-alcoholic beverage in meal with cooked rice or maize/buckwheat/finger millet porridge called *dheoroh*. It is also consumed after mixing with rice or *chiura* (beaten rice). Apart from its dietary properties, *dahi* plays an important role in the social, religious and ritualistic activities.

18.12.1.3 Microorganisms

Many species of LAB were reported, based on culture-dependent methods, from *dahi* samples of these regions including *Lactobacillus bifermentans*, *Lactobacillus alimentarius*, *Lactobacillus casei* subsp. *pseudoplantarum*, *Lactococcus lactis* subsp. *lactis*, *Lactococcus lactis* subsp. *cremoris* and also few yeasts *Saccharomycopsis* and *Candida* (Dewan and Tamang 2007). We extracted DNA directly from *dahi* samples and performed 16S rRNA amplicon sequencing using high-throughput sequencing technique (Shangpliang et al. 2018). We found that *Firmicutes* (61.2%) were predominant members of bacterial communities followed by *Proteobacteria* (35.1%) in *dahi* samples of Sikkim (Shangpliang et al. 2018). Wide microbial diversity was observed in *dahi* samples including *Lactococcus lactis* (17.1%), *Acetobacter pasteurianus* (15.2%), *Clostridium tyrobutyricum* (14.0%), *Lactococcus raffinolactis* (8.1%), *Leuconostoc mesenteroides* (7.5%), *Pseudomonas fluorescens* (6.2%), *Acetobacter syzygii* (4.8%), *Leuconostoc pseudomesenteroides* (4.0%), *Lactobacillus helveticus* (3.8%), *Lactobacillus gasseri* (2.9%), *Acetobacter lovaniensis* (2.3%), *Gluconobacter cerinus* (1.3%) and *Leuconostoc lactis* (1.1%) (Shangpliang et al. 2018). Microorganisms were also isolated from *dahi* samples collected from different places of Sikkim and identified by culture-dependent method using 16S rRNA Sanger sequencing technique, and it was found that *Lactococcus lactis* (60%) is predominant bacterium followed by *Leuconostoc mesenteroides* (30%) and *Enterococcus italicus* (10%) (Tamang et al. 2019 unpublished data).

Fig. 18.14 Traditional method of preparation of *mohi*

Milk
↓
Dahi
↓
Churned in bamboo or wooden vessel (30-45 min)
↓
Warm or cold water added
↓ → ***Gheu*(butter)**
↓
Mohi

18.12.1.4 Nutritional Value

Nutritional composition of *dahi* shows protein content of 22.5% DM, fat 24.5% DM, carbohydrate 48.2% DM, ash 4.7% and calorie value 503.6 k cal/100 g DM (Dewan and Tamang 2007).

18.12.2 *Mohi*

Mohi is produced as by-product liquid during the preparation of butter from *dahi*. *Mohi* is a Nepali name. The Bhutia and Lepchas call it *kachhu*.

18.12.2.1 Traditional Method of Preparation

Dahi is churned to produce butter, locally called *gheu*, and liquid portion is *mohi* (Fig. 18.14).

18.12.2.2 Mode of Consumption

Mohi is usually mixed with cooked rice or maize of *dheroh* by Gorkha. It is drunk as a cooling beverage to overcome tiredness. It is also processed further to produce other fermented milk products like soft *chhurpi*, *chhu*, *dudh chhurpi*, etc.

18.12.2.3 Microorganisms

Lactococcus lactis (67%) is predominant bacterium in *mohi*, followed by *Leuconostoc mesenteroides* (33%) (Tamang et al. 2019 unpublished data).

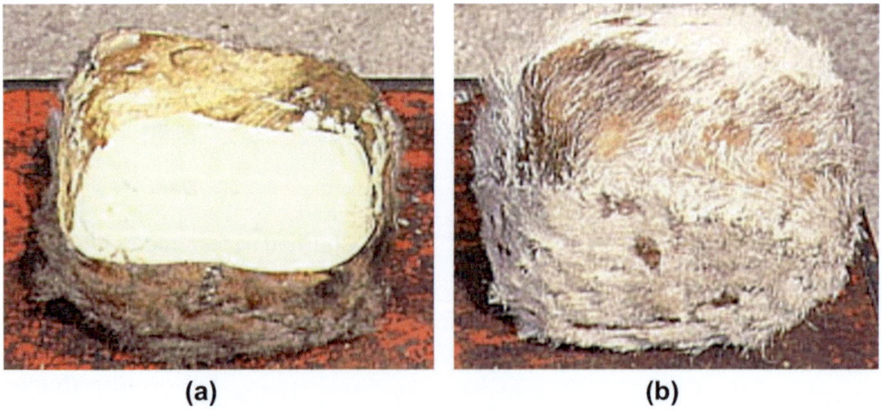

Fig. 18.15 (**a**) *Maa* and (**b**) Maata

18.12.3 *Gheu*

Gheu is a butter prepared by churning *dahi* from cow milk with a typical flavour and aroma. It is a popular milk product of Sikkim and Darjeeling hills used for cooking purposes. *Gheu* is a Nepali word; Bhutia calls it *maa*; Lepcha calls it *mor*.

18.12.3.1 Traditional Method of Preparation

Dahi is poured into the bamboo vessel locally called as 'tolung', or 'somg', or 'padung' and is churned by lifting and lowering of the 'madani' made up of wooden long stick 'shar' with a circular or star-shaped flat wooden disc 'pangra' at one end inside the bamboo vessel. The churning is done for 15–30 min with the addition of either cold or warm water as the weather demands to facilitate better separation of the *gheu* from the liquid *mohi*. After a big lump of soft *gheu* is formed and seen floating on the *mohi*, it is carefully lifted out with the hand and transferred to another vessel. Another method of churning is by using a 'theki', a hollow wooden vessel, and a 'madani' consisting of the 'ghurra' and the pulling string called 'neti'. *Dahi* is kept inside the 'theki' and churned by pulling the 'neti' with either hand so that the 'madani' rotates in alternating clockwise and anticlockwise direction in the 'theki' to precipitate *gheu*.

Butter is churned from yak milk in North Sikkim locally called *maa* (Fig. 18.15a) which is stored by wrapping in a cubical-shaped dried skin of sheep by stitching on all the edges from the inside called as *maata* (Fig. 18.15b). *Maata* can be stored for several months or years for later consumption of the *maa* inside.

Dahi
↓
Churned in bamboo-made vessel or wooden-made vessel
↓
→ produced *ghee*, extracted and collected
↓
Buttermilk (*mohi*) produced
↓
Cooked for 15 -20 min till whitish soft mass is formed
↓
Mass sieved out and put inside a muslin cloth
↓
Hung by a string to drain out the remaining whey
↓
CHHURPI

a

b

Fig. 18.16 Traditional method of preparation of chhurpi (**a**) fresh *chhurpi* and (**b**) *Chhurpi* curry

18.12.3.2 Mode of Consumption

Freshly prepared *gheu* (or *nauni gheu* in Nepali) is also consumed as it is. It is clarified further by boiling till the oily liquid separates from the unwanted dark brown precipitate 'khar'. The clarified *gheu* is then consumed in a variety of ways. *Gheu* is mixed with cooked rice and eaten. It is also spread over *roti* and bread. *Gheu* is used to prepare a large number of dishes like *sel roti*, *khapjay*, varieties of sweets, etc.

18.12.3.3 Economy

Gheu is a highly priced milk product and serves as a major source of income for farmers in Sikkim and Darjeeling hills. It is sold in the local markets all the year round. The price of freshly prepared *gheu* is quite high ranging from Rs.400 to Rs.600 per kg.

18.12.4 *Chhurpi*

Chhurpi (pH 4.2, acidity 0.61%) is a cottage cheese-like fermented milk product prepared from cow milk. It has a rubbery texture with slightly sour taste and excellent aroma when it is fresh. It is used to prepare various dishes with delicacy in local diets.

18.12.4.1 Traditional Method of Preparation

Mohi is continuously cooked for about 15 min till whitish soft mass is formed. This mass is sieved out and put inside a muslin cloth, which is hung by a string to drain out the remaining whey. The product is called *chhurpi* which is soft and white in colour (Fig. 18.16a, b).

18.12.4.2 Mode of Consumption

Soft *chhurpi* is prepared as curry by cooking it in oil along with onions, tomato and chillies. *Chhurpi* curry with edible ferns 'sauney ningro' (*Diplazium polypodioides*) and 'kali ningro' (*Diplazium* sp.) makes the delicious cuisine in local diet which is eaten as side dish with cooked rice. It is also used to prepare 'achar' or pickle by mixing it with chopped cucumber, radish and green chillies (Tamang and Tamang 1998). Soup prepared from soft *chhurpi* can be consumed as a substitute for *dal* along with cooked rice.

18.12.4.3 Economy

Chhurpi is sold in all local periodical markets by rural women. It is packed in the leaves of fig plant and then tied loosely by straw. At local market *chhurpi* is sold at Rs.120 to Rs. 180 per kg.

18.12.4.4 Microorganisms

Chhurpi is characterized by lactic acid bacteria fermentation. The predominant lactic acid bacteria are *Lactobacillus plantarum*, *Lactobacillus curvatus*, *Lactobacillus fermentum*, *Lactobacillus paracasei* subsp. *pseudoplantarum*, *Lactobacillus alimentarius*, *Lactobacillus kefir*, *Lactobacillus hilgardii*, *Enterococcus faecium* and *Leuconostoc mesenteroides* (Tamang et al. 2000; Dewan and Tamang 2007). Huge diversity of bacterial community was observed in *chhurpi* samples of Sikkim analysed by high-throughput sequencing which showed that phylum *Firmicutes* (62.6%) was dominant followed by *Proteobacteria* (30.0%). Species level in *chhurpi* samples revealed the presence of *Lactococcus lactis* (25.3%), *Lactobacillus helveticus* (10.1%), *Acetobacter syzygii* (6.5%), *Acetobacter lovaniensis* (5.8%), *Acetobacter pasteurianus* (6.1%), *Leuconostoc mesenteroides* (3.2%), *Staphylococcus cohnii* (3.3%), *Gluconobacter oxydans* (3.0%), *Lactibacllus delbrueckii* (3.1%), *Leuconostoc pseudomesenteroides* (2.9%), *Pseudomonas fluorescens* (2.5%), *Lactococcus raffinolactis* (1.9%), *Acetobacter tropicalis* (1.4%) and *Lactobacillus gasseri* (1.2%) (Shangpliang et al. 2018). Soft *chhurpi* is dominated by *Leuconostoc mesenteroides* (70%), *Lactococcus lactis* subsp. *cremoris* (20%) and *Lactococcus lactis* subsp. *hordniae* (10%) (Tamang et al. 2019 unpublished data).

18.12.4.5 Nutritional Value

Nutritional value of chhurpi is estimated: moisture 73.8%, protein 65.3% DM, fat 11.8% DM, carbohydrate 16.3% DM, ash 6.6% DM, calcium 44.1 mg/100 g, iron 1.2 mg/100 g, magnesium 16.7 mg/100 g, manganese 0.6 mg/100 g and zinc 25.1 mg/100 g (Dewan and Tamang 2007; Tamang et al. 2012).

18.12.4.6 Health Benefits

Chhurpi is consumed as an excellent source of protein and as a substitute for vegetables. The lactic acid bacteria present in *chhurpi* showed the probiotic properties advocating the *chhurpi* as functional or health food (Tamang et al. 2000).

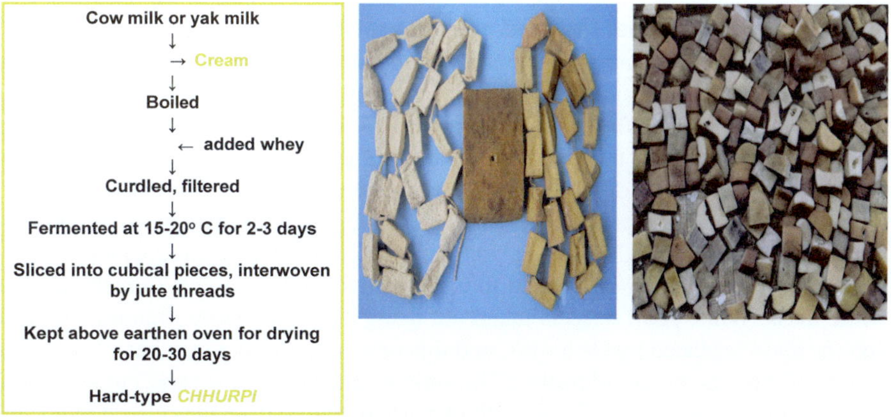

Fig. 18.17 Traditional method of preparation of hard-*chhurpi*

Boiled cow milk or yak milk
↓
***Dahi* churned**
↓
***Mohi* (butter-milk)**
↓
Cooked continuously till white mass is formed
↓
Solid sieved, pressed in a jute sack with stones or heavy materials
↓
Desirable cubical pieces made
↓ ← **Added *thake* paste on product**
↓
Woven in a thread and dry above earthen oven for 3-4 days
↓
Dudh Chhurpi

Dudh chhurpi

Fig. 18.18 Traditional method of preparation of *dudh chhurpi*

18.12.5 Hard Variety of *Chhurpi*

Hard-variety *chhurpi* is mostly prepared from yak milk in North Sikkim and West Sikkim (2100–4500 m) and is used as masticator due to its characteristic gumminess and chewiness. Two types of hard *chhurpi* are prepared: one is common *chhurpi* (hard), and the other is *dudh chhurpi* (slightly softer than the previous). The Bhutia and Lepcha call it *khamu*.

18.12.5.1 Traditional Method of Preparation

During preparation, cream is separated from yak milk, and the skimmed milk is boiled, curdled by adding whey, and after straining, the coagulum is cooked until the free water dries up. The highly stringy mass is wrapped in a cloth and fermented under pressure at room temperature for about 2 days. After pressing, the mass is sliced and allowed to dry by keeping above earthen oven for about a month (Fig. 18.17). During preparation for *dudh chhurpi*, *dahi* is churned in a bamboo vessel locally called 'tolung' or 'somg' for about 1–2 h to produce *gheu* which is separated from the liquid *mohi*, cooked in a vessel over fire for 1–2 h to make a solid white mass. White solid mass is taken out using a bamboo sieve, liquid is drained and the mass is placed inside a sack, and this mass is pressed with a heavy stone for 2–4 h to drain out the liquid further. The solid is taken out of the sack, cut to cubical pieces (about 6.0 g weight). *Thake*, a thick paste prepared from the cooked mohi and milk, is applied on the surface of these pieces, which, then, are threaded and hung in the open air for sun drying for 3–4 days (Fig. 18.18). After drying, this unique dairy product is consumed as masticator. A sweet, dried skimmed milk powder covers the hard texture of *chhurpi*; hence it is called *dudh* (milk)-*chhurpi*.

18.12.5.2 Mode of Consumption

Both varieties of hard *chhurpi* are consumed as a nutritious masticator like a chewing gum.

18.12.5.3 Economy

Hard-variety *chhurpi* is sold in all local periodical markets, seen as long threads of cubical-shaped *chhurpi* being hung outside the shop for sale. One kg of *chhurpi* costs about Rs.200 to Rs. 350.

18.12.5.4 Microorganisms

Bacterial community in hard variety of *chhurpi* prepared from yak milk of North Sikkim, based on 16S rRNA sequencing, shows the presence of *Leuconostoc mesenteroides* subsp. *jonggajibkimchii* (50%), *Leuconostoc mesenteroides* (40%) and *Lactococcus lactis* subsp. *cremoris* (10%) (Tamang et al. 2019 unpublished).

18.12.5.5 Nutritional Value

Nutritional value of hard-variety chhurpi is moisture 3.9–13%, ash 6.6–7.7%, fat 7.7–12.3%, protein 53.4–68.5% and carbohydrate 20.4–23.2% (Katiyar et al. 1991; Pal et al. 1993, 1996; Hossain et al. 1996).

Boiled or raw milk of cow or yak
↓
Shyow* or *dahi
↓
Churned
↓
***Kachhu* (buttermilk)**
↓
Cooked till solid, white mass formed, sieved
↓
Whey portion squeezed by muslin cloth
↓
Chhu
↓
Further fermentation till 30 days
↓
CHHU

Fig. 18.19 Traditional method of preparation of *chhu*

18.12.6 *Chhu*

Chhu (pH 6.0) is a traditional fermented milk product consumed mostly by the Bhutia in Sikkim. It is prepared from boiled or raw milk of cow or yak. Like *chhurpi*, initially, it has a rubbery texture with slightly sour taste when it is fresh; after further fermentation it becomes more liquid and creamish to pale yellow in colour and develops a strong flavour.

18.12.6.1 Traditional Method of Preparation

Shyow or *dahi* s prepared from boiled or unboiled milk (which helps in the formation of a highly sour and rapidly fermenting *chhu* later on). *Shyow* is churned in a bamboo or wooden vessel, with addition of warm or cold water to produce butter *maa* and *kachhu*. *Kachhu* is cooked for 15 min till a soft, whitish mass is formed. This mass is sieved out and put inside a muslin cloth, which is hung by a string to drain out the remaining whey, and the final product is called *chhu* (Fig. 18.19). *Chhu* is placed in a closed vessel and kept for several days to months.

18.12.6.2 Mode of Consumption

Chhu is prepared into a curry by cooking it in *maa* (butter) along with onions, tomato and chillies and is eaten as side dish with cooked rice in meal. Soup is also prepared from strong-flavoured *chhu* (kept for several months). It has a sour taste with strong aroma and is used as appetizer.

18.12.6.3 Microorganisms

Lactic acid bacteria dominate the microflora of *chhu* which include *Lactobacillus farciminis*, *Lb. brevis*, *Lb. alimentarius*, *Lb. salivarius* and *Leuconostoc lactis*

Boiled milk of cow or yak
↓
Shyow
↓
Churned → *Mor* (butter)
↓
***Thara* (buttermilk); boiled till coagulated**
↓
***Shergem* (soft *chhurpi*)**
↓
Dewatered the excess liquid
↓
Fermented in a closed container for 15 days
↓
Cooked with fresh boiled milk, *mor*, turmeric to form a thick and brown paste for 30 min
↓
Stored for 4-7 months
↓
SOMAR

Fig. 18.20 Traditional method of preparation of *somar*

subsp. *cremoris* and also some yeasts *Saccharomycopsis* and *Candida* (Dewan and Tamang 2006).

18.12.6.4 Nutritional Value

Nutritional value of *chhu* shows moisture 75.5%, protein 58.4% DM, fat 5.8% DM, carbohydrate 33.9% DM, ash 1.9% DM, calcium 111.1 mg/100 g, iron 4.5 mg/100 g, Mg 64.3 mg/100 g, manganese 3.1 mg/100 g and zinc 87.6 mg/100 g (Dewan and Tamang 2006).

18.12.7 *Somar*

Somar (pH 6.0), a brownish soft paste with strong flavour, is an ethnic fermented milk product of the Sherpa community in Sikkim and Darjeeling hills. It is prepared from cow or yak milk.

18.12.7.1 Traditional Method of Preparation

Shyow (*dahi*) is prepared from boiled and cooled cow's milk and is churned in a bamboo or wooden vessel to produce butter (*mor*) and buttermilk (*thara*). *Thara* is cooked till a soft, whitish mass is formed (Fig. 18.20). This mass is sieved out with a cloth or plastic sieve to get the product called *shergem* (same as fresh soft *chhurpi*) and is kept in a closed vessel for fermentation for 15–20 days. The end product is called *somar* and is cooked with milk, *mor* and turmeric to form a thick brown paste, which forms another type of *somar*. This type of *somar* can be stored up to 10–12 years for future consumption.

18.12.7.2 Mode of Consumption

Somar soup is prepared by frying garlic or Szechwan pepper *(Litsaea citrate)* locally called *timbur* and salt in little oil. Water is added to produce a thick soup. *Somar* soup is consumed with rice or *dheroh*. *Somar* is kept for a long period of time by cooking it with *gheu* and turmeric to produce a soft brown paste to form another type of *somar*. This is consumed as soup along with boiled rice or *dheroh*.

18.12.7.3 Microorganisms

Lactobacillus paracasei subsp. *pseudoplantarum* and *Lactococcus lactis* subsp. *cremoris* have been reported from *somar* (Dewan and Tamang 2007).

18.12.7.4 Nutritional Value

Nutritional value of *somar* is protein 35% DM, fat 15.4% DM, carbohydrate 46.9% DM, ash 2.7% DM, calcium 31.2 mg/100 g, iron 0.4 mg/100 g, magnesium 13.7 mg/100 g, manganese 0.5 mg/100 g and zinc 17.2 mg/100 g (Tamang et al. 2012).

Raw milk of cow or yak
↓
Poured in a hollow bamboo vessel or wooden vessel
↓
Continuously rotated and swirled in the vessel till milk particles stuck inner walls of vessels
↓
Milk poured off, vessel upturned
↓
Process repeated till thick creamy layers formed in inner walls of vessel
↓
Allowed creamy mass for fermented for till days
↓
Fermented mass is scrapped off
↓
PHILU

Fig. 18.21 Traditional method of preparation of *philu;* (**a**) swirling of cream and (**b**) *philu*

18.12.8 *Philu*

Philu (pH 4.3) is a typical indigenous fermented butter-like milk product obtained from cow or yak milk, with an inconsistent semi-solid texture. It is commonly eaten by the Bhutia and also by Sherpa who calls it *philuk* of these regions.

18.12.8.1 Traditional Method of Preparation

Fresh milk of cow or yak is collected in cylindrical bamboo vessels (locally called 'dzydung') or in wooden vessels (called 'yadung') and is slowly swirled around the walls of these vessels by rotating the vessels for a few minutes. Sometimes a thick mesh of dried creeper is kept inside the vessel to increase the surface area for cream to stick inside the vessel, and creamy mass sticks to the walls of the vessels and also around the creeper. The milk is then poured off from vessel and utilized elsewhere, and the vessel is then kept in an upside-down position to drain out the remaining milk. This process is repeated daily for about 6–7 days until a thick, white cream layer is formed on the vessel walls and the creeper surface (Fig. 18.21a, b). This soft mass cream is called *philu* which is scraped off and stored in dry place for later consumption. *Philu* obtained from yak milk has a cream-white colour with an inconsistent semi-solid texture.

18.12.8.2 Mode of Consumption

Philu is cooked with butter and little salt is added. Rich gravy is prepared and consumed as side dish along with cooked rice.

18.12.8.3 Economy

Philu is an expensive traditional milk product sold in local markets in Sikkim costing more than Rs. 300 per kg. In North Sikkim, *philu* is produced mostly from yak milk and is produced and consumed at the household level by Bhutia.

18.12.8.4 Microorganisms

Lactobacillus paracasei subsp. *paracasei*, *Lb. bifermentans* and *Enterococcus faecium* are reported in *philu* (Dewan and Tamang 2007). The predominant bacterium in yak milk *philu* is *Lactococcus lactis* subsp. *cremoris* (90%) followed by *Lactococcus lactis* subsp. *tructae* (10%) (Tamang et al. 2019 unpublished).

18.12.8.5 Nutritional Value

Nutritional value of *philu* is protein 52% DM, fat 32% DM, carbohydrate 12.5% DM, ash 3.6% DM, calcium 34.9 mg/100 g, iron 0.8 mg/100 g, magnesium 16.9 mg/100 g, manganese 0.9 mg/100 g and zinc 27.1 mg/100 g (Tamang et al. 2012).

18.13 Traditionally Preserved Fish Products

18.13.1 *Suka Ko Maacha*

Suka ko maacha is an ethnic smoked fish product prepared in the river-site villages in Sikkim and Darjeeling hills. *Suka* means dry and *maacha* means fish in Nepali language (Thapa 2016a).

18.13.1.1 Method of Preparation

Two types of river fishes mostly 'dothay asala' (*Schizothorax richardsoni*) and 'chuchay asala' (*Schizothorax progastus*) are preferred for the preparation of *suka ko maacha* by the people residing near streams or river. Fish is caught from

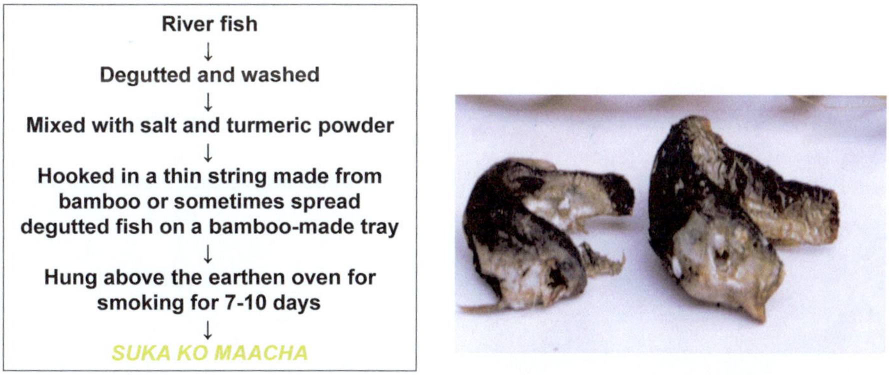

Fig. 18.22 Traditional method of preparation of *suka ko maacha*

mountain river of streams and collected in a bamboo basket, locally called 'bhukh', and is gutted, washed and mixed with salt and turmeric powder. Gutted fishes are hooked in a bamboo-made string and are hung above the earthen oven in kitchen for 7–10 days (Fig. 18.22). The smoked fish is called *suka ko maacha*. *Suka ko maacha* can be kept up to 3–4 months at room temperature (Thapa 2016b).

18.13.1.2 Mode of Consumption

Suka ko maacha is prepared as curry mixed with tomato, chillies and salt. It is also cooked with vegetable. It is eaten as side dish in main meal with boiled rice by Gorkha.

18.13.1.3 Microorganisms

Lactococcus lactis subsp. *cremoris*, *Lc. lactis* subsp. *lactis*, *Lc. plantarum*, *Leuconostoc mesenteroides*, *Enterococcus faecium*, *E. faecalis*, *Pediococcus pentosaceus* and yeasts *Candida chiropterorum*, *C. bombicola* and *Saccharomycopsis* spp. were isolated from *suka ko maccha* (Thapa et al. 2006).

18.13.1.4 Nutritional Value

Nutritional value of *suka ko maccha* is estimated: protein 35.0%, fat 12.0%, carbohydrate 36.8%, calorie value 395.2 kcal/100 g, calcium 38.7 mg/100 g, iron 0.8 mg/100 g, magnesium 5.0 mg/100 g, manganese 1.0 mg/100 g and zinc 5.2 mg/100 g (Thapa and Pal 2007).

Fig. 18.23 Traditional method of preparation of *gnuchi*

Mountain river fish
↓
Collected and degutted
↓
Washed
↓
Mixed with salt and turmeric powder thoroughly
↓
Sieved in a bamboo-made tray and spread
↓
Smoked above the kitchen for 10-14 days
↓
GNUCHI

18.13.2 *Gnuchi*

Gnuchi is a typical smoked and dried fish product common to the Lepcha. *Gnuchi* means smoked fish in the Lepcha language (Thapa 2016a).

18.13.2.1 Method of Preparation

River fish (*Schizothorax richardsonii* Gray, *Labeo dero* Hamilton, *Acrossocheilus* spp., *Channa* sp.) is caught early in the morning from the river because the fishermen believe that during this hour the fishes migrate near the banks which make it easier for them to catch the fishes using fishing net locally called 'sangli'. Fishes are collected in a bamboo basket locally called 'tamfyok', which is woven by bamboo strips and is properly tied around the waist of the fisherman while fishing. Fishes are kept on a big bamboo tray called 'sarhang' to drain off water, gutted and mixed with salt and turmeric powder. Fish is separated according to their size. The bigger-sized fish is spread upside-down on 'sarhang', and the small-sized fish is hung one after the other in a bamboo strip above the earthen oven in kitchen and is smoked for 10–14 days to get *gnuchi* (Fig. 18.23). *Gnuchi* can be kept at room temperature for 2–3 months.

18.13.2.2 Mode of Consumption

Gnuchi is eaten as curry with cooked rice. It is also cooked with vegetable.

18.13.2.3 Microorganisms

Some species of lactic acid bacteria were isolated from *gnuchi*: *Lactobacillus plantarum*, *Lactococcus lactis* subsp. *cremoris*, *Lc. lactis* subsp. *lactis*, *Leuconostoc mesenteroides*, *Enterococcus faecium*, *E. faecalis*, *Pediococcus pentosaceus* and some species of yeasts – *Candida chiropterorum*, *C. bombicola* and *Saccharomycopsis* spp. (Thapa et al. 2006).

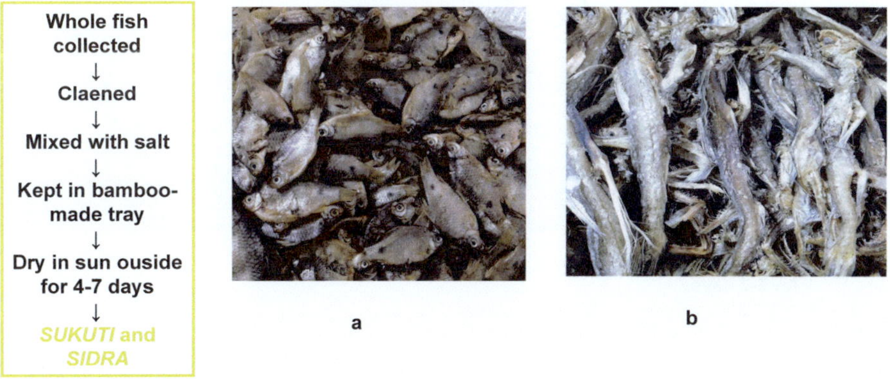

Fig. 18.24 Traditional method of preparation of (**a**) *sidra* and (**b**) *sukuti*

18.13.2.4 Nutritional Value

Nutritional value of *gnuchi* shows protein 21.3%, fat 14.5%, carbohydrate 47.3%, calorie value 404.9 kcal/100 g, ash 16.9%, calcium 37.0 mg/100 g, iron 1.1 mg/100 g, magnesium 8.8 mg/100 g, manganese 1.1 mg/100 g and zinc 7.5 mg/100 g (Thapa and Pal 2007).

18.13.3 *Sidra*

Sidra is sun-dried fish product commonly consumed by Gorkha of Sikkim and Darjeeling hills.

18.13.3.1 Method of Preparation

During its preparation, the whole fish (*Puntius sarana* Hamilton) is collected, washed, placed in bamboo-made mats and dried outside in the sun for 4–7 days (Fig. 18.24a). *Sidra* can be stored at room temperature for 3–4 months for consumption.

18.13.3.2 Mode of Consumption

Sidra is consumed as *achar* or pickle. *Sidra* is roasted and is mixed with dry chillies, boiled tomato and salt to make a thick pickle paste. *Sidra ko achar* or pickle is the traditional cuisine of the Gorkha eaten with cooked rice and 'khalo dal' (black gram soup). *Sidra* is commonly sold in local markets of Sikkim and Darjeeling hills.

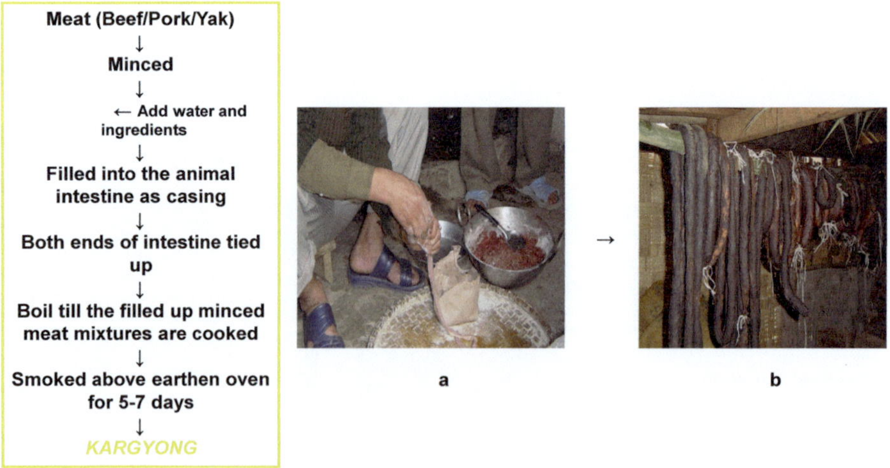

Fig. 18.25 Traditional method of preparation of *kargyong*: stuffing meat mixtures into animal intestine and (**b**) *kargyong* being smoked above earthen-oven

18.13.3.3 Microorganisms

Lactococcus lactis subsp. *cremoris*, *Lc. lactis* subsp. *lactis*, *Lc. plantarum*, *Leuconostoc mesenteroides*, *Enterococcus faecium*, *E. faecalis*, *Pediococcus pentosaceus*, *Weissella confuse* and few yeasts *Candida chiropterorum*, *C. bombicola* and *Saccharomycopsis* spp. were reported from *sidra* (Thapa et al. 2006). We have identified bacteria isolated from *sidra* of Sikkim on the basis of 16S rRNA gene using Sanger sequencing method which included *Enterobacter cloacae*, *Klebsiella pneumonia*, *Salmonella enterica*, *Escherichia fergusonii*, *Escherichia coli*, *Providencia* spp., *Bacillus cereus*, *Staphylococcus aureus*, *Staphylococcus nepalensis* and *Staphylococcus sciuri* (Tamang et al. 2019 unpublished).

18.13.3.4 Nutritional Value

Nutritional value of *sidra* shows protein content of 25.5%, fat 12.2%, carbohydrate 45.7%, calorie value 394.6 kcal/100 g, ash 16.6%, calcium 25.8 mg/100 g, iron 0.9 mg/100 g, magnesium 1.6 mg/100 g, manganese 0.8 mg/100 g and zinc 2.4 mg/100 g (Thapa and Pal 2007).

18.13.4 *Sukuti*

Sukuti is also very popular sun-dried fish product among the Gorkha.

18.13.4.1 Method of Preparation

During preparation of *sukuti*, fish (*Harpodon nehereus* Hamilton) is collected, washed, rubbed with salt and dried in an open air in the sun for 4–7 days (Fig. 18.25b). *Sukuti* is stored at room temperature for 3–4 months for consumption.

18.13.4.2 Mode of Consumption

Sukuti is consumed as pickle, soup and curry. During curry preparation, *sukuti* is fried, mixed with dry chillies, onion and salt to make a pickle. It is usually eaten with cooked rice and 'kalo dal' or black gram soup. *Sukuti* is sold in local markets.

18.13.4.3 Microorganisms

Lactococcus lactis subsp. *cremoris*, *Lc. lactis* subsp. *lactis*, *Lc. plantarum*, *Leuc. mesenteroides*, *Enterococcus faecium*, *E. faecalis*, *Pediococcus pentosaceus* and yeasts *Candida chiropterorum*, *C. bombicola* and *Saccharomycopsis* spp. are microbiota of *sukuti* (Thapa et al. 2006). Based on 16S rRNA sequencing method, diverse species of bacteria were identified from *sukuti* which included *Enterobacter hormaechei*, *Enterobacter cancerogenus*, *Klebsiella pneumonia Salmonella enterica*, *Acinetobacter radioresistens*, *Staphylococcus vitulinus*, *Pseudomonas plecoglossicida*, *Staphylococcus aureus* and *Providencia vermicola* (Tamang et al. 2019 unpublished).

18.13.4.4 Nutritional Value

Nutritional value of *sukuti* shows the protein content of 36.8%, fat 11.4%, carbohydrate 38.2%, calorie value 402.6 kcal/100 g, calcium 17.7 mg/100 g, iron 0.3 mg/100 g, magnesium 1.4 mg/100 g, manganese 0.2 mg/100 g and zinc 1.3 mg/100 g (Thapa and Pal 2007).

18.14 Traditionally Preserved Meat Products

18.14.1 *Kargyong*

In Sikkim mostly in North Sikkim, the ethnic people prepare and consume traditionally prepared sausages for centuries. *Kargyong* is an ethnic fermented sausage with soft or hard texture prepared from beef/pork/yak meats and is mostly consumed by the Bhutia, Tibetan, Drukpa, Lepcha and Sherpa. The Lepcha calls it *tiklee* (Rai et al. 2009).

18.14.1.1 Traditional Method of Preparation

According to meat preference (beef/pork/yak) of the consumers, lean meat is usually selected for *kargyong* preparation. Fatty portions of meat are chopped finely; mixed with crushed ginger and garlic; and added with desirable amount of salt and water. The mixture is then stuffed into the segment of cleaned gastrointestinal tract of animal locally called *gyuma*, which is used as natural casings with a diameter of 3–4 cm and length of 40–60 cm. One end of the casing is tied up with rope, and other end is sealed after stuffing the meat mixture, and then it is boiled for 20–30 min. Cooked sausages are drained out and hung in the bamboo strips above the earthen oven in kitchen for smoking and drying for 10–15 days to produce the final product *kargyong* (Fig. 18.25). Depending on use of meats, *kargyong* may be *lang* (means beef) *kargyong*, *faak* (means pork) *kargyong* and yak *kargyong*. And unlike European sausages, no sugar, nitrite and nitrate are added during *kargyong* preparation. *Kargyong* shows natural curve shape due to use of natural casings. *Kargyong* can be kept at room temperature till 1 month in high altitudes in North Sikkim.

18.14.1.2 Mode of Consumption

Kargyong is boiled for 10–15 min, sliced and fried in edible oil or *maa* (yak butter) in pan by mixing with onion, tomato, chillies and salt to make curry. It is also consumed as fried sausage with *raksi*, distilled liquor, or *chyang/kodo ko jaanr*, mild-alcoholic finger millet-based beverage by Bhutia, Lepcha and some non-vegetarian Gorkha.

18.14.1.3 Microorganisms

Microorganisms present in *kargyong* are *Lactobacillus sake*, *Lb. divergens*, *Lb. carnis*, *Lb. sanfransisco*, *Lb. curvatus*, *Leuc. mesenteroides*, *Enterococcus faecium*, *Bacillus subtilis*, *B. mycoides*, *B. thuringiensis*, *Staphylococcus aureus*, *Micrococcus* and few yeasts *Debaryomyces hansenii* and *Pichia anomala* (Rai et al. 2010a; Rai

Fig. 18.26 Traditional method of preparation of *satchu*

and Tamang 2017). Based on 16S rRNA gene amplicon sequence by high-throughput sequencing technique, we have recently profiled the bacterial community in *lang* (beef) *kargyong*—*Psychrobacter pulmonis* (66.3%), *Pseudomonas veronii* (10.6%), *Kurthia zopfii* (7.1%), *Psychrobacter sanguinis* (5.0%), *Staphylococcus sciuri* (4.2%), *Myroides odoratimimus* (1.0%) and others (5.9%)—and in *faak* (pork) *kargyong*: *Lactobacillus sakei* (56.6%), *Psychrobacter pulmonis* (23.7%), *Lactococcus garvieae* (9.4%), *Weissella cibaria* (4.7%), *Macrococcus caseolyticus* (2.9%), *Weissella ceti* (1.5%) and others (1.2%) (Tamang et al. 2019 unpublished).

18.14.1.4 Nutritional Value

Nutritional value of *kargyong* shows protein content of 16.0% DM, fat 49.1% DM, carbohydrate 32.0% DM, ash 2.8% DM and calorie value 634.5 kcal/100 g (Rai et al. 2010b).

18.14.2 *Satchu*

Satchu is a dried and smoked meat (beef/pork/yak) products commonly consumed by Tibetans, Drukpa, Bhutia, Lepcha, Tamang and Sherpa (Rai et al. 2009). Preparation of *satchu* reflects the natural preservation of perishable fresh raw meats in absence of refrigeration or cold storage in these regions.

18.14.2.1 Traditional Method of Preparation

As per the food preference, any animal flesh mostly red meat of beef or pork or yak is sliced into several strands (60–90 cm) and is mixed thoroughly with turmeric powder, edible oil or butter and salt. The mixtures of meat strands are hung in the bamboo strips or wooden stick and are kept in an open air in corridor of the house or are smoked above the kitchen oven for 10–15 days as per the convenience of the producers (Fig. 18.26). Based on types of meat used, *satchu* may be *lang* (beef) *satchu*, *faak* (pork) *satchu* and yak *satchu* (prepared and consumed in North Sikkim only). *Satchu* is kept at room temperature for several weeks.

18.14.2.2 Mode of Consumption

Satchu is made into curry by washing and soaking in water briefly, then squeezed and fried with chopped garlic, ginger, chilli and salt in pan. Thick gravy is made and consumed with *thukpa* (noodles in soup) or boiled/baked potato by Bhutia. Deep-fried *satchu* is popular side dish in the local cuisine served with traditional alcoholic beverages/drink in Sikkim and also in Darjeeling hills.

18.14.2.3 Microorganisms

Pediococcus pentosaceus, *Lb. casei*, *Lb. carnis*, *E. faecium*, *B. subtilis*, *B. mycoides*, *B. lentus*, *S. aureus*, *Micrococcus* and few yeasts *D. hansenii* and *Pichia anomala* were reported from satchu (Rai et al. 2010a). We profiled the bacterial community in yak *satchu* based on 16S rRNA gene amplicon sequence by high-throughput sequencing technique which included *Staphylococcus succinus* (23.5%), *Bdellovibrio bacteriovorus* (23.5%), *Staphylococcus equorum* (17.7%), *Psychrobacter pulmonis* (11.8%), *Ketogulonicigenium vulgare* (11.8%) and *Pseudomonas stutzeri* (11.8%) (Tamang et al. 2019 unpublished).

18.14.2.4 Nutritional Value

Nutritional value of *satchu* shows the protein content of 51.0% DM, fat 4.7% DM, carbohydrate 37.0% DM, calorie value 405.8 kcal/100 g and ash 7.3% DM (Rai et al. 2010b).

18.14.3 *Suka Ko Masu*

Suka ko masu is a dried or smoked mutton meat product prepared and consumed by non-vegetarian Gorkha, which is similar to *satchu*. The Newar and the Nepali call it *sheakua*, prepared from buffalo meat.

18.14.3.1 Traditional Method of Preparation

Suka ko masu is prepared by cutting the red meat of buffalo or chevon (goat meat) into 25–30-cm-long stripes and mixed with turmeric powder, mustard oil and salt. Mixed meat stripes are hung on bamboo and are kept above the earthen kitchen oven and smoked for 7–10 days to get smoked and dried product *suka ko masu* which can be stored at room temperature for several weeks.

18.14.3.2 Mode of Consumption

Suka ko masu is washed and soaked in lukewarm water for 10 min—excess water is squeezed out—and fried in hot edible oil, with chopped onion, ginger, chilli powder and salt. Coriander leaves are sprinkled over the curry and are eaten with boiled rice. *Suka ko masu* or *sheakua* is usually grilled in charcoal and is served as a popular side dish in the region.

Liver, lungs, fats, intestines and other innards of yak chopped
↓ ←Added salt
Chopped meat filled and stuffed inside the removed rumen/stomach previously from dead yak
↓
Opne portion of rumen closed and tied with rope
↓
Hang in a bamboo stripes above kitchen oven or at attic for drying
↓
Ferment for 4-6 months
↓
Khyopeh

Fig. 18.27 Traditional method of preparation of *khyopeh*: (a) 6-months fermented *khyopeh*

18.14.3.3 Microorganisms

Lactobacillus carnis, *Lb. plantarum*, *Enterococcus faecium*, *Bacillus subtilis*, *B. mycoides*, *B. thuringiensis*, *Staphylococcus aureus*, *Micrococcus* spp. and few yeasts *Debaryomyces hansenii* and *Pichia burtonii* are present in samples of *suka ko masu* (Rai et al. 2010a; Rai and Tamang 2017).

18.14.3.4 Nutritional Value

Nutritional composition of *suka ko masu* is as follows: ash 1.8% DM, protein 44.8% DM, fat 2.0% DM, carbohydrate 51.4% DM and calorie value 403.1 kcal/100 g (Rai et al. 2010b).

18.14.4 *Khyopeh*

Khyopeh is a unique meat product of North Sikkim which is naturally fermented yak meat mixtures in the removed part of the rumen or stomach of yak. The ethnic *Lachungpa* community of North Sikkim mostly practices the preparation and consumption of *khyopeh* during the month of December when yaks are slaughtered for the Buddhist festivals during early winter for fresh meats, and remaining will be preserved as *khyopeh*, *satchu*, etc.

18.14.4.1 Traditional Method of Preparation

Ingredient used for preparation of *khyopeh* includes parts of liver, lungs, fats, intestines and innards of yak, which are chopped finely and mixed with required amount of salt. The meat mixtures are stuffed into the rumen (stomach) of yak and are tied up with rope. It is then hanged in a bamboo strips above the kitchen oven or at the attic for smoking and drying for 4–6 months or even for a year to make *khyopeh* (Fig. 18.27). *Khyopeh* is soft or hard and brownish in colour.

Table 18.2 Ethnic alcoholic beverages of Sikkim and Darjeeling hills

Alcoholic beverage	Substrate	Use and sensory property of the product	Major ethnic consumer	Region/ district
Starter culture				
Marcha *Khesung* *Phab* *Buth*	Rice, wild herbs, spices	Sun-dried, mixed amylolytic starter culture to ferment alcoholic beverages	Limboo, few Rai, Lepcha women	All (in rural areas only)
Kodo ko jaanr/ Chyang/ Chee	Finger millet fermented by *marcha*	Mild-alcoholic, slightly sweet-acidic; beverage	Non-Brahmin Gorkha, Bhutia, Lepcha	All
Bhaati jaanr	Rice	Mild-alcoholic, sweet-sour, food beverage; paste	Non-Brahmin Gorkha, Bhutia, Lepcha	
Makai ko jaanr	Maize	Mild-alcoholic, sweet-sour, food beverage; paste	Non-Brahmin Gorkha, Bhutia, Lepcha	
Gahoon ko jaanr	Wheat	Mild-alcoholic, slightly acidic, beverage	Non-Brahmin Gorkha, Bhutia, Lepcha	
Simal tarul ko jaanr	Cassava tuber	Mild-alcoholic, sweet-sour, food beverage; paste	Non-Brahmin Gorkha, Bhutia, Lepcha	
Jao ko jaanr	Barley	Mild-alcoholic, slightly acidic, beverage	Non-Brahmin Gorkha, Bhutia, Lepcha	
Faapar ko jaanr	Buck wheat	Mild-alcoholic, slightly acidic, beverage	Non-Brahmin Gorkha, Bhutia, Lepcha	
Raksi/Arak	Cereals	Clear distilled liquor; high alcohol content drink	Non-Brahmin Gorkha, Bhutia, Lepcha	

18.14.4.2 Mode of Consumption

Khyopeh is eaten as raw or cooked with nettle leaves locally called *sishnu* (*Urtica dioica* L.) in the main meal with cooked rice in North Sikkim.

18.14.4.3 Microorganisms

Phenotypic characterization of isolates from *khyopeh* samples revealed the presence of species of *Staphylococcus*, *Escherichia* and *Enterococcus* (Bhutia et al. 2019). We have profiled the bacterial community in *khyopeh* based on 16S rRNA gene amplicon sequence by high-throughput sequencing technique which included *Lactobacillus sakei* (48.9%), *Carnobacterium divergens* (45.6%) and *Staphylococcus equorum* (5.6%) (Tamang et al. 2019 unpublished).

18.15 Alcoholic Beverages

Wine culture and malting process are not traditiona of Indian dietary culture including Sikkim and Darjeeling hill, instead alcoholic beverages are prepared from cereals-based as substrates by using dry starter cultures which contain consortium of essential microorganisms for alcoholic fermentation. Drinking of alcoholic beverage and its distilled liquor is provision for pleasure as well as for ritualistic purposes in Sikkim and Darjeeling hills (Tamang et al. 1996). Alcoholic beverages (Table 18.2) are exclusively prepared from locally cultivated cereal grains such as finger millets, rice, maize, etc. using traditionally prepared mixed inocula or starter called *marcha* at household level mostly by rural women. Varieties of alcoholic beverages, their traditional methods of preparation, mode of consumption, microorganisms and nutritive value have been mentioned below.

18.15.1 *Marcha*

Dry, round to flat, creamy white to dusty white, solid ball-like cake with a dimeter of 1.9–11.8 cm is prepared traditionally as nonedible starter commonly called *marcha* for production of alcoholic beverages in Sikkim and Darjeeling hills. *Marcha* is a Nepali word. Limboo calls it *khesung*, Tamang calls it *bharama*, Rai calls it *bopkha* or *khabed*, Bhutia calls it *phab*, and the Lepcha calls it *buth/thanbum* (Tamang et al. 1996). *Marcha*-making technique is very unique to these regions due to application of 'back-sloping' methods using native knowledge of

Fig. 18.28 Traditional method of preparation of *marcha*

'ethno-microbiology'. *Marcha* is similar to other amylolytic starters such as *emao/humao*, *xaaz pitha* and *modor pitha* of Assam; *hamei* of Manipur; *thiat* of Meghalaya; *chowan* of Tripura; *kherie/khekhrii* of Nagaland; *pee*, *paa*, *phut* and *phab* of Arunachal Pradesh and *dawdim* of Mizoram (Anupma et al. 2018); *pho* of Bhutan; *chiu/chu/daque* of China; *nuruk* of Korea; *ragi* of Indonesia; *loogpang* of Thailand; *benh men* of Vietnam; *bubod* of the Philippines; and *dombea* of Cambodia (Tamang et al. 2016a; Sha et al. 2019).

18.15.1.1 Traditional Method of Preparation

During *marcha* preparation, rice (*Oryza sativa* L.) grains are soaked in water for 6–8 h—excess water is drained out—and soaked rice grains are crushed in a foot-driven heavy wooden mortar by a pestle. In 1 kg of ground rice, several ingredients are added which include 2.5 g of roots of 'guliyo jara' or 'chitu' (*Plumbago zeylanica* L.), 1.2 g of leaves of 'bheemsen paate' (*Buddleja asiatica* Lour), 1.2 g of flowers of 'sengreknna' {*Vernonia cinerea* (L.) Less}, 5 g of ginger, few pieces of dry red chilli and 2–3% of previously prepared powered *marcha* as a mother culture (back-sloping). The mixture is then made into paste by adding water and kneaded into flat cakes of desirable sizes and shapes and placed individually on the ceiling floor made up of bamboo strips above the kitchen, bedded with fresh fronds of ferns, locally called 'pire uneu' {*Glaphylopteriolopsis erubescens* (Wall ex Hook.)}, and covered with dry ferns and jute bags. These are left to ferment for 1–2 days in summer and 2–4 days in winter. Completion of fermentation is indicated by distinct alcoholic and ester aroma and puffy/swollen appearance of *marcha*. These are immediately sun-dried in bamboo-made trays outside for 2–3 days to get dry *marcha* (Fig. 18.28). *Marcha* retains its potency in situ for over a year or more.

18.15.1.2 Economy

Marcha is produced at home in few villages in Sikkim exclusively by the rural women belonging to the Limboo, Rai and the Lepcha. These *marcha*-making villages in Sikkim are Jhosing and Tibuk in North Sikkim, Chhejo and Lingchom in West Sikkim, Salghari and Barnyak in South Sikkim and Aho and Kopchey in East Sikkim, Nor busty in Darjeeling and Kashyong and Mangzing in Kalimpong. Men help women in collecting wild herbs and pounding the herbs during *marcha* preparation. This art of technology is protected as hereditary secret and passes from mother to daughters. Some people are economically dependent upon this product. *Marcha*-making villages have linkages to nearby markets where *marcha*-makers can sell the products once or twice in a week. The cost of *marcha* depends on its size. Small-sized *marcha* (~2.0 cm) costs about Re.1.0 paisa each, while the large-sized (~12 cm) *marcha* is sold at Rs.10 per piece. Earnings out of selling *marcha* supplement the domestic expenses.

18.15.1.3 Microorganisms

Marcha-making method complements a traditional knowledge and wisdom of ethnic Himalayan people on subculturing of desirable inocula (microorganisms consisting of filamentous moulds, amylolytic and alcohol-producing yeasts and bacteria) from previous batch to a new culture using rice or wheat as a starchy base or medium. This indigenous technique of 'ethno-microbiology' preserves microbial diversity for fermentation of starchy substrates to alcoholic beverages in the mountainous regions applying 'back-slopping' practices. The first report of microbial composition of *marcha* of Sikkim was published by Kobayashi et al. (1961) in 1961 reporting *Rhizopus oryzae*, *Mucor praini* and *Absidia lichtheimi* in *marcha*. Batra and Millner (1974) reported *Hansenula anomala* var. *schneggii* (=*Pichia anomala)* in *marcha* collected from Kalimpong of Darjeeling hills. Species of filamentous moulds present in *marcha* are *Mucor circinelloides* forma *circinelloides*, *Mucor* sp. close to *M. hiemalis*, *R. chinensis* and *R. stolonifer* variety *lyococcus* (Tamang et al. 1988). Yeasts present in *marcha* are *Saccharomycopsis fibuligera*, *Sm. capsularis*, *Saccharomyces cerevisiae*, *S. bayanus*, *Pichia anomala*, *P. burtonii* and *Candida glabrata* (Tsuyoshi et al. 2005). *Sm. fibuligera*, *Sm. capsularis* and *P. burtonii* have high amylolytic activities suggesting the amylolytic yeasts, whereas *S. bayanus*, *C. glabrata* and *P. anomala* are alcohol-producing yeasts (Tsuyoshi et al. 2005). *Sm. fibuligera* is the most dominant yeasts in *marcha* (Tamang and Sarkar 1995). Saccharifying activities are mostly shown by *Rhizopus* spp. and *Sm. fibuligera*, whereas liquefying activities are shown by *Sm. fibuligera* and *S. cerevisiae* (Thapa and Tamang 2004). *Rhizopus* spp. and *Sm. fibuligera* degrade cereal starch and produce glucose, and then alcohol-producing yeasts species of *Saccharomyces* and *Pichia* rapidly grow on the resultant glucose to produce ethanol. Among LAB *P. pentosaceus*, *Lb. bifermentans* and *Lb. brevis* are present in *marcha* (Tamang and Sarkar 1995; Tamang et al. 2007b). Recently Pradhan and Tamang (2019) reported several species of bacteria in *marcha* of Darjeeling hills and Sikkim based on 16S rRNA sequence method which included LAB: *Enterococcus durans, E. faecium, E. faecalis, Pediococcus acidilactici, P. pentosaceus, Leuconostoc mesenteroides,*

Fig. 18.29 Traditional method of preparation of *kodo ko jaanr*

Weissella cibaria; and non-LAB: *Staphylococcus hominis subsp. hominis, Lysinibacillus boronitolerans, Micrococcus yunnanensis, Bacillus subtilis subsp. inaquosorum , B. pseudomycoides, B. nakamurai.* LAB present in *marcha* play a role in imparting flavour, antagonism and acidification of the substrates (Tamang et al. 2007b).

Based on PCR-DGGE technique, Sha et al. (2018) reported yeasts species in *marcha* of Sikkim: *Saccharomyces cerevisiae* (16.5%), *Saccharomycopsis fibuligera* (15.3%), *Wickerhamomyces anomalus* (11.3%), *Sm. malanga* (11.7%), *Kluyveromyces marxianus* (5.3%), *Meyerozyma* sp. (2.7%), *Candida glabrata* (2.7%) and many strains below 2%. Microbial community in *marcha* of Sikkim analysed by high-throughput sequencing technique was observed which showed bacterial genera *Acetobacter* (52.6%), *Fructobacillus* (21.1%), *Lactococcus* (10.3%), *Lactobacillus* (8.4%), *Leuconostoc* (4.0%), *Burkholderia* (2.1%) and *Gluconacetobacter* (1.4%) and yeasts species *Wickerhamomyces anomalus*, *Candida quercitrus* and *Kazachstania exigua*, followed by *Saccharomyces* and *Pestalotiopsis* and filamentous moulds *Aspergillus penicillioides*, *Aureobasidium pullulans* and *Mucor circinelloides* (Sha et al. 2017).

18.15.1.4 Composition

Marcha is slightly acidic in nature containing pH 5.6 with 0.1% acidity with 14% moisture and 1.4% ash (Sha et al. 2016).

18.15.2 *Kodo Ko Jaanr or Chyang*

Most popular mild-alcoholic beverage in Sikkim and Darjeeling hills is *kodo ko jaanr* or *chyang* which is fermented from dry seeds of finger millet [*Eleusine coracana* (L.) Gaertn.], locally called ‘kodo’, by *marcha. Kodo* (means finger millet*) jaanr* is common name for alcoholic beverage in Nepali. Each caste has its own vernacular name for *kodo ko jaanr* such as *mandokpenaa thee* by Limboo, *sampicha ummaak* by Rai, *naarr paa* by Gurung, *saangla chi* by Tamang, *chirs shyaabu* by Sunwar, *paadaare haan* by Magar, *gyaar chyang* by Sherpa, *minchaa chyang* by Bhutia and *mong chee* by Lepcha (Tamang 2010).

18.15.2.1 Traditional Method of Preparation

Seeds of finger millets are collected, cleaned, washed and cooked for about 30 min in an open cooker. Excess water is drained off, and cooked millets are spread on a bamboo mat called ‘mandro’ for cooling. About 1–2% of powdered *marcha* is sprinkled over cooked seeds; mixed thoroughly and packed in a bamboo basket lined with fresh fern, locally called ‘thadre unioon’ (*Thelypteris erubescens* Well ex Hook.), or banana leaves; then covered with sack clothes; and kept for 2–4 days at room temperature for saccharification. During saccharification sweet aroma is emitted out, and the saccharified mass is transferred into an earthen pot or into specially made bamboo basket called ‘septu’ made air-tight for alcoholic fermentation for 3–4 days during summer and 5–7 days in winter at room temperature (Fig. 18.29).

Good quality of *jaanr* has sweet taste with mild-alcoholic flavour. Prolonged fermentation makes the product sour in taste and more alcoholic and to have unpleasant flavour which is unacceptable to consumers.

18.15.2.2 Mode of Consumption

Kodo ko jaanr is consumed in a unique way in Sikkim and Darjeeling hills. About 200–500 g of *jaanr* is put into a vessel called *toongbaa*, and lukewarm water is added up to the edge of it. After 10–15 min, milky white extract of *jaanr* is sipped through a narrow bamboo straw called *pipsing* having a hole in a side near the bottom to avoid passing of grits. Water can be added 2–3 times after sipping up the extract. Guests are served with *toongbaa* along with fried meat or pickles. Alternately, thick milky white liquid pressed from the *kodo ko jaanr* is filtered under pressure using a filter called 'chhapani'. Such liquor is believed to be good tonic for ailing persons and post-natal women. After consumption, grits of *kodo ko jaanr* are used as fodder for pigs and cattle. This is a good example of total utilization of substrate as food and fodder.

18.15.2.3 Microorganisms

Microorganisms for fermentation of finger millets into *kodo ko jaanr* are mostly supplemented by *marcha* used as starters during preparation of *kodo ko jaanr* (Thapa and Tamang 2004). Two main species of fungi *Mucor circinelloides* and *Rhizopus chinensis*; amylolytic and alcohol-producing yeasts *Saccharomycopsis fibuligera*, *Pichia anomala*, *S. cerevisiae* and *Candida glabrata*; and few species of bacteria *Pediococcus pentosaceus* and *Lb. bifermentans* were detected in *kodo ko jaanr* (Thapa and Tamang 2004, 2006).

18.15.2.4 Nutritional Value

Moisture, 69.7%; pH, 4.1; alcohol, 4.8%; ash, 5.1% DM; protein, 9.3% DM; fat, 2.0% DM; crude fibre, 4.7% DM; carbohydrate, 83.7% DM; calorie value, 389.6 kcal/100 g; calcium, 281.0 mg/100 g; potassium, 398.0 mg/100 g; phosphorous, 326.0 mg/100 g; iron, 24.0 mg/100 g; magnesium, 118.0 mg/100 g; manganese, 9.0 mg/100 g; and zinc, 1.2 mg/100 g (Thapa and Tamang 2004).

18.15.2.5 Health Benefits

Kodo ko jaanr is high-calorie food beverage with mild-alcoholic content of 4%. Remarkable increase in mineral contents such as calcium, magnesium, manganese, iron, potassium and phosphorous has been observed in *jaanr* during fermentation (Thapa and Tamang 2004, 2006); *Kodo ko jaanr*, thus, contributes to the mineral intake in daily diet of the local people (Thapa and Tamang 2004). Because of high calorie, ailing persons and post-natal women consume the extract of *kodo ko jaanr* to regain the strength. Vitamin cyanocobalamin, which is not present in finger millet, is synthesized by the fermenting microorganisms (Basappa et al. 1997). The essential amino acids like valine, threonine, leucine and isoleucine are in higher concentration in *kodo ko jaanr* or *chyang* (Basappa et al. 1997).

18.15.2.6 Upgradation of Traditional Process

Each selected strain (filamentous mould and yeast, previously isolated from *marcha*) was tested for its ability to produce the final product *kodo ko jaanr* either by monoculture or mixed culture (Thapa and Tamang 2006). *Jaanr* produced by selected combination of strains was evaluated for reducing sugar and alcohol contents and also subjected to sensory evaluation. *Kodo ko jaanr* prepared by a combination of *Rhizopus chinensis* MJ:R3 and *Saccharomyces cerevisiae* MJ:YS2 scored highest in general acceptability, due to its mild-alcoholic-sweet flavour, significantly ($P < 0.05$) acceptable to consumers (Thapa and Tamang 2007). Consortium of a selected strain of mould *(Rhizopus)* and amylolytic and alcohol-producing yeast (*Saccharomyces cerevisiae*) is recommended to produce a good-quality *kodo ko jaanr* which also maintains consistency and maximizes utilization of substrates.

18.15.3 *Bhaati Jaanr*

Bhaati jaanr is a mild-alcoholic and juicy soft product with distinct sweet aroma, prepared from steamed glutinous rice. *Bhaati jaanr* is Nepali word for fermented rice beverage. Different ethnic people call it by their own dialect such as *tak thee* (Limboo), *kok umaak* (Rai), *kaiyan paa* (Gurung), *kaan chi* (Tamang), *kameshyaabu* (Sunwar), *chho haan* (Magar), *ja thon* (Newar), *dacchhang* (Sherpa), *laayakaa chhyaang* (Bhutia) and *jo chee* (Lepcha) (Tamang 2005). *Bhaati jaanr* is similar to other fermented rice beverages of Asia such as *makgeolli* of Korea, *tapé ketan* of Indonesia and *lao-chao* of China (Tamang et al. 2016a).

18.15.3.1 Traditional Method of Preparation

During production of *bhaati jaanr*, rice, mainly glutinous, is cooked for about 15 min in an open cooker. Excess water is drained off and spread on a bamboo mat called *mandro* for cooling (~ 40 °C). Powdered *marcha* (1–2%) is sprinkled over cooked rice, mixed well and kept in a vessel or an earthen pot for 1–2 days at room temperature for saccharification. During saccharification sweet aroma is emitted out. After saccharification, the vessel is made air-tight; *bhaati jaanr* is fermented for 2–3 days in summer and 7–8 days in winter, and after fermentation (Fig. 18.30), it is collected in earthenware crock.

18.15.3.2 Mode of Consumption

Bhaati jaanr is made into a thick paste by stirring fermented mass with the help of a hand-driven wooden or bamboo-made stirrer. It is consumed directly as food beverage. Sometimes, *bhaati jaanr* is stored in an earthenware crock for a week or more to make yellowish-white supernatant liquor called *nigaar*, collected at the bottom of the earthenware crock. *Nigaar* is drunk directly with or without water.

18.15.3.3 Microorganisms

Marcha supplements the essential microbiota for *bhaati jaanr fermentation* (Tamang and Thapa 2006). *Rhizopus chinensis*; yeasts *Saccharomycopsis fibuligera*, *Pichia*

Fig. 18.30 Traditional method of preparation of *bhaati jaanr*

anomala, *Saccharomyces cerevisiae* and *Candida glabrata*; and few lactic acid bacteria *Pediococcus pentosaceus* and *Lb. bifermentans* were isolated from *bhaati jaanr* (Tamang and Thapa 2006).

18.15.3.4 Nutritional Value

Alcohol, 5.9%; ash, 1.7% DM; protein, 9.5% DM; fat, 2.0% DM; crude fibre, 1.5% DM; carbohydrate, 86.9% DM; calorie value, 404.1 kcal/100 g; calcium, 12.8 mg/100 g; potassium, 146.0 mg/100 g; phosphorous, 595.0 mg/100 g; iron, 7.7 mg/100 g; magnesium, 50.0 mg/100 g; manganese, 1.4 mg/100 g; and zinc, 2.7 mg/100 g (Tamang and Thapa 2006).

18.15.3.5 Health Benefits

Bhaati jaanr is a traditional diet for new mothers in villages who believe that it helps them to regain their strength. *Bhaati jaanr* is high-calorie food beverage. Considerable increase in calcium, manganese, iron, zinc, sodium, potassium and phosphorous has been observed in *bhaati jaanr* (Tamang and Thapa 2006).

18.15.4 *Raksi*

Raksi is a clear distilled liquor obtained from continuous distillation of fermented cereal beverages such as *kodo ko jaanr*, *bhaati jaanr*, *makai ko jaanr*, *gahoon ko jaanr*, etc. with characteristic aroma (Kozaki et al. 2000). Fermented masses of buckwheat, potato and cassava roots are also distilled to get *raksi*. *Raksi* is a common term in Nepali meaning alcoholic drink. *Rakshi* has many vernacular names among ethnic groups of Gorkha such as *sijongwaa aara* in Limboo, *aarakha/hemma*

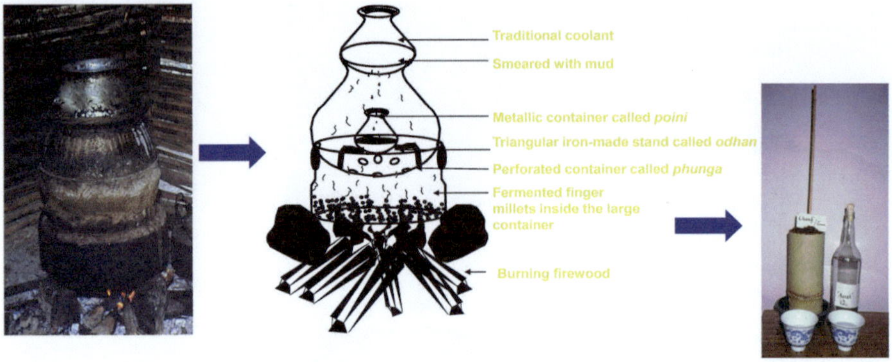

Fig. 18.31 Traditional distillation apparatus for *raksi* production: (**a**) traditional distillation apparatus and (**b**) anatomical sketch of apparatus showing the inside views

in Rai, *paa* in Gurung, *aaerak* in Tamang, *rindho* in Sunwar, *aayala* in Newar, *dhise* in Magar, *aarak* in Bhutia and Sherpa and *aarok* in Lepcha (Tamang 2005).

18.15.4.1 Method of Preparation

Kodo jo jaanr, *bhaati jaanr* and other fermented cereal beverages are distilled in a big cylindrical metallic vessel continuously for 2–3 h in an earthen oven over firewood. At the top of the distilling vessel, cold water is kept in a metallic container as condenser; water is replaced 3–5 times after it gets heated. Condensed *raksi* is collected in a small collecting metallic vessel called *poini* (Fig. 18.31). *Raksi* prepared after replacing condensing water for three times is known as *theen pani raksi* which contains high alcohol and traditionally prepared for religious purposes. *Raksi* prepared after replacing the condensing water for five times is known as *panch pani raksi* which is a common alcoholic drink. *Raksi* is usually stored in bottle capped with piece of dry corncob.

18.15.4.2 Mode of Consumption

Raksi is drunk directly without addition of water along with fried meat or side dish. *Raksi* is traditionally drunk by the 'matwali' (alcohol drinkers) including non-Brahmin Gorkha, Bhutias and Lepchas. *Jhane ko raksi* is common in rural areas. *Raksi* is sizzled in a hot vessel containing butter, and vessel is covered with a lid as long as it emits a sizzling sound. *Jhane ko raksi* is drunk hot to get relief from cough and cold.

18.15.4.3 Traditional Apparatus for *Raksi* Distillation

It is worth to document the traditional methods of distillation for *raksi* production by ethnic people of these regions using their native skill of ethnic people of Sikkim and Darjeeling hills. *Raksi* distillation apparatus is made up of a metallic vessel (Fig. 18.31). In main cylindrical metallic vessel measuring 40 cm × 30 cm × 25 cm, fermented grits (*kodo ko jaanr or bhaati jaanr or gahoon ko jaanr*) are steamed continuously for 2–3 h over firewood. Above main cylindrical vessel, a perforated container called *phunga* is placed. Inside *phunga*, a small metallic collector called *poini* is placed on iron-made tripod called *odhan* to collect distillate (*raksi*). Above *phunga*, metallic vessel with cold water used as condenser is placed. The bottom of the condenser vessel is plastered by mud with the tip of *phunga* to prevent excess ventilation during distillation. This apparatus can distil 2–4 kg of *jaanr* to get 1–2 L of *raksi* after replacing condensing water three times.

18.15.4.4 Proximate Composition

Alcohol content of *raksi* is 22.9%. *Raksi* distilled from *bhaati jaanr* mixed with few petals of *Rhododendron* showed highest alcohol content (~27%) comparable to *raksi* prepared from other fermented cereals (Kozaki et al. 2000; Tamang 2010).

18.16 Conclusion

Ethnic fermented foods of Sikkim and Darjeeling hills cover all types of locally cultivated or grown substrates ranging from milk to alcohol, soybeans to cereals, vegetables to bamboo, meat to fish, etc. Among the ethnicity, the Gorkha is the largest stakeholder as well as consumer of 80% of the fermented foods and beverages of Darjeeling hills and Sikkim. This is mainly due to ethnic diversity within the Gorkha community from Brahmin to Kirat, Aryan to Mongoloid, and projection of indispensable food culture of the single Gorkha. Ethnic food reflects the dietary culture, history, socio-economy of a particular community, and also promotes health benefits to consumers with functional microorganisms. Diversity of common and uncommon ethnic fermented foods and alcoholic beverages is unknown outside the regions, but no doubt these ethnic foods and alcoholic drinks have been consumed by the Himalayan people for more than 2500 years old as per the historical records. These ethnic fermented foods and beverages are, no doubt, considered as the heritage foods.

References

Anupma A, Pradhan P, Sha SP, Tamang JP (2018) Traditional skill of ethnic people of the Eastern Himalayas and North East India in preserving microbiota as dry amylolytic starters. Indian J Tradit Knowl 17(1):184–190

Balaraman N, Golay MM (1991) Livestock production in Sikkim. Sikkim Science Society, Gangtok

Bareh H (2001) Encyclopaedia of North-East India: Sikkim. Mittal Publications, New Delhi. isbn:978-81-7099-794-8

Basappa SC, Somashekar D, Agrawal R, Suma K, Bharathi K (1997) Nutritional composition of fermented ragi (*chhang*) by *phab* and defined starter cultures as compared to unfermented ragi (*Eleucine coracana* G.). Int J Food Sc Nutr 48:313–319

Batra LR, Millner PD (1974) Some Asian fermented foods and beverages and associated fungi. Mycologia 66:942–950

Bhanja KC (1993) History of Darjeeling and the Sikkim Himalaya. Gyan Publishing House, New Delhi

Bhutia, M., Thapa, N. and Tamang, J.P. (2019). *Khyopeh*, a traditional fermented yak meat product of Sikkim. Ind J Tradit Knowl (in press)

Chettri R, Bhutia M, Tamang JP (2016) Poly-γ-glutamic acid (PGA)-producing *Bacillus* species isolated from *Kinema*, Indian fermented soybean food. Front Microbiol 7:971. https://doi.org/10.3389/fmicb.2016.00971

Chettri R, Tamang JP (2008) Microbiological evaluation of *maseura*, an ethnic fermented legume-based condiment of Sikkim. J Hill Res 21(1):1–7

Chumlung Y (2014) Kirat history and culture: all about South Asian Monoglians. Kindle Edition, p 189

Dewan S, Tamang JP (2006) Microbial and analytical characterization of Chhu, a traditional fermented milk product of the Sikkim Himalayas. J Sci Ind Res 65:747–752

Dewan S, Tamang JP (2007) Dominant lactic acid bacteria and their technological properties isolated from the Himalayan ethnic fermented milk products. Anton Leeuw Int J Gen Mol Microbiol 92(3):343–352

Dozey EC (1922) A concise history of Darjeeling District since 1835. University of Michigan Library, Ann Arbor, p 412

Gorer G (1938) The Lepchas of Sikkim. Gian Publishing House, Delhi

Hooker JD (1854) Himalayan journals: notes of a naturalist in Bengal, the Sikkim and Nepal Himalayas, the Khasia Mountains. John Murray, London

Hossain SA, Pal PK, Sarkar PK, Patil GR (1996) Sensory characteristics, manufacturing methods and cost of producing of milk churpi. J Hill Res 9(1):121–127

Katiyar SK, Bhasin AK, Bhatia AK (1991) Traditionally processed and preserved milk products of Sikkimese Tribes. Sci Cult 57(10 and 11):256–258

Kawamura Y, Kara MR (1987) Umami: a basic taste. Marcel Dekker, New York

Kobayashi Y, Tubaki K, Soneda M (1961) Several moulds and a yeast used for brewing native beer (Kodok Jar) among the Sikkimese of India. J Japanese Bot 36:321–331

Kozaki M, Tamang JP, Kataoka J, Yamanaka S, Yoshida S (2000) Cereal wine *(jaanr)* and distilled wine *(raksi)* in Sikkim. J Brew Soc Jpn 95(2):115–122

Moktan B, Saha J, Sarkar PK (2008) Antioxidant activities of soybean as affected by *Bacillus*-fermentation to Kinema. Food Res Int 4(6):586–593

Nout MJR, Bakshi D, Sarkar PK (1998) Microbiology safety of kinema, a fermented soya bean food. Food Control 9(6):357–362

O' Malley LSS (1907) Darjeeling District Gazetteer. Gyan Publishing House, New Delhi

Omizu Y, Tsukamoto C, Chettri R, Tamang JP (2011) Determination of saponin contents in raw soybean and fermented soybean foods of India. J Sci Ind Res 70:533–538

Pal PK, Hossain SA, Sarkar PK (1993) An assessment of manufacturing methods and sensory characteristics of market chhurpi. J Hill Res 6(2):73–76

Pal PK, Hossain SA, Sarkar PK (1996) Optimisation of process parameters in the manufacture of chhurpi. J Food Sci Technol 33:219–223

Pradhan S, Tamang JP (2015) Ethnobiology of wild leafy vegetables of Sikkim. Indian J Tradit Knowl 14(2):290–297

Pradhan P, Tamang JP (2019) Phenotypic and genotypic identification of bacteria isolated from traditionally prepared dry starters of the Eastern Himalayas. Front Microbiol 10:2526. https://doi.org/10.3389/fmicb.2019.02526

Rai AK, Palni U, Tamang JP (2009) Traditional knowledge of the Himalayan people on production of indigenous meat products. Indian J Tradit Knowl 8(1):104–109

Rai AK, Sharma RM, Tamang JP (2005) Food value of common edible plants of Sikkim. J Hill Res 18(2):99–103

Rai AK, Tamang JP (2017) Prevalence of *Staphylococcus* and *Micrococcus* in traditionally prepared meat products. J Sci Ind Res 76:351–354

Rai AK, Tamang JP, Palni U (2010a) Microbiological studies of ethnic meat products of the Eastern Himalayas. Meat Sci 85:560–567

Rai AK, Tamang JP, Palni U (2010b) Nutritional value of lesser-known ethnic meat products of the Himalayas. J Hill Res 23(1 and 2):22–25

Rai R, Kharel N, Tamang JP (2014) HACCP model of *kinema*, a fermented soybean food. J Sci Ind Res 73:588–592

Rai R, Shangpliang HNJ, Tamang JP (2016) Naturally fermented milk products of the Eastern Himalayas. J Ethnic Foods 3:270–275

Risley HH (1928) The Gazetteer of Sikkim. D.K. Publishing Distributors (P) Ltd., New Delhi

Sarkar PK, Hasenack B, Nout MJR (2002) Diversity and functionality of *Bacillus* and related genera isolated from spontaneously fermented soybeans (Indian Kinema) and locust beans (African Soumbala). Int J Food Microbiol 77:175–186

Sarkar PK, Jones LJ, Craven GS, Somerset SM, Palmer C (1997) Amino acid profiles of kinema, a soybean-fermented food. Food Chem 59(1):69–75

Sarkar PK, Jones LJ, Gore W, Craven GS (1996) Changes in soya bean lipid profiles during kinema production. J Sci Food Agric 71:321–328

Sarkar PK, Morrison E, Tingii U, Somerset SM, Craven GS (1998) B-group vitamin and mineral contents of soybeans during kinema production. J Sci Food Agric 78:498–502

Sarkar PK, Tamang JP (1994) The influence of process variables and inoculum composition on the sensory quality of kinema. Food Microbiol 11:317–325

Sarkar PK, Tamang JP (1995) Changes in the microbial profile and proximate composition during natural and controlled fermentations of soybeans to produce kinema. Food Microbiol 12:317–325

Sarkar PK, Tamang JP, Cook PE, Owens JD (1994) Kinema—a traditional soybean fermented food: proximate composition and microflora. Food Microbiol 11:47–55

Sha SP, Anupma A, Pradhan P, Prasad GS, Tamang JP (2016) Identification of yeasts by PCR-mediated DGGE in *marcha*, an ethnic amylolytic starter of India. J Ethnic Foods (Elsevier) 3:292–296

Sha SP, Jani K, Sharma A, Anupma A, Pradhan P, Shouche Y, Tamang JP (2017) Analysis of bacterial and fungal communities in *Marcha* and *Thiat*, traditionally prepared amylolytic starters of India. Sci Rep 7:10967. https://doi.org/10.1038/s41598-017-11609-y

Sha SP, Suryavanshi MS, Jani K, Sharma A, Shouche Y, Tamang JP (2018) Diversity of yeasts and molds by culture-dependent and culture-independent methods for mycobiome surveillance of traditionally prepared dried starters for the production of Indian alcoholic beverages. Front Microbiol 9:2237. https://doi.org/10.3389/fmicb.2018.02237

Sha SP, Suryavanshi MS, Tamang JP (2019) Mycobiome diversity in traditionally prepared starters for alcoholic beverages in India by high-throughput sequencing method. Front Microbiol 10:348. https://doi.org/10.3389/fmicb.2019.003482237

Shangpliang HNK, Rai R, Keisam S, Jeyaram K, Tamang JP (2018) Bacterial community in naturally fermented milk products of Arunachal Pradesh and Sikkim of India analysed by high-throughput amplicon sequencing. Sci Rep 8:1532. https://doi.org/10.1038/s41598-018-19524-6

Sharma DK, Ghosh K, Raquib M, Bhattacharya M (2006) Yak products' profile: an overview. J Food Sci Technol 43:447

Shurtleff W, Aoyagi A (2010) History of soybeans and soyfoods in South Asia/Indian subcontinent (1656–2010): Extensively annotated bibliography and sourcebook. Soyinfo Center, Lafayette

Subba JR (1984) Agriculture in the Hills of Sikkim. Sikkim Science Society, Gangtok, p 268

Subba JR (1999) The Limboos of the Eastern Himalayas with special reference to Sikkim. Sikkim Yakthung Mundhum Saplopa, Gangtok, p 779

Subba JR (2008) History, culture and customs of Sikkim. Gyan Publishin House, New Delhi

Tamang B, Tamang JP (2007) Role of lactic acid bacteria and their functional properties in *Goyang*, a fermented leafy vegetable product of the Sherpas. J Hill Res 20(20):53–61

Tamang B, Tamang JP (2010) *In situ* fermentation dynamics during production of *gundruk* and *khalpi*, ethnic fermented vegetables products of the Himalayas. Indian J Microbiol 50(Suppl 1):S93–S98

Tamang B, Tamang JP, Schillinger U, Franz CMAP, Gores M, Holzapfel WH (2008) Phenotypic and genotypic identification of lactic acid bacteria isolated from ethnic fermented tender bamboo shoots of North East India. Int J Food Microbiol 121:35–40

Tamang JP (1999) Development of pulverised starter for *kinema* production. J Food Sci Technol 36(5):475–478

Tamang JP (2000) Case study on socio-economical prospective of kinema, a traditional fermented soybean food. In: Director, CFTRI, editor. The proceeding of the 1997 international conference on traditional foods, March 6–8, 1997., Central Food Technological Research Institute, Mysore, India, pp 180–185

Tamang JP (2001) Kinema. Food Cult 3:11–14

Tamang JP (2003) Native microorganisms in the fermentation of *kinema*. Indian J Microbiol 43(2):1–4

Tamang JP (2005) Food culture of Sikkim, Sikkim Study Series volume IV. Information and Public Relations Department, Government of Sikkim, Gangtok, p 120

Tamang JP (2010) Himalayan fermented foods: microbiology, nutrition, and ethnic values. CRC Press, Taylor & Francis Group, New York, p 295. 9781420093247

Tamang JP (2015) Naturally fermented ethnic soybean foods of India. J Ethnic Foods 2:8–17

Tamang JP, Dewan S, Tamang B, Rai A, Schillinger U, Holzapfel WH (2007b) Lactic acid bacteria in *Hamei* and *Marcha* of North East India. Indian J Microbiol 47(2):119–125

Tamang JP, Dewan S, Thapa S, Olasupo NA, Schillinger U, Holzapfel WH (2000) Identification and enzymatic profiles of predominant lactic acid bacteria isolated from soft-variety *chhurpi*, a traditional cheese typical of the Sikkim Himalayas. Food Biotechnol 14(1&2):99–112

Tamang JP, Holzapfel WH, Watanabe K (2016a) Diversity of microorganisms in global fermented foods and beverages. Front Microbiol 7:377. https://doi.org/10.3389/fmicb.2016.00377

Tamang JP, Nikkuni S (1996) Selection of starter culture for production of kinema, a fermented soybean food of the Himalaya. World J Microbiol Biotechnol 12:629–635

Tamang JP, Nikkuni S (1998) Effect of temperatures during pure culture fermentation of kinema. Int J Microbiol Biotechnol 14:847–850

Tamang JP, Samuel D (2010) Dietary culture and antiquity of fermented foods and beverages. In: Tamang JP, Kailasapathy K (eds) Fermented foods and beverages of the world. CRC Press, Taylor & Francis Group, New York, pp 1–40. 97814-20094954

Tamang JP, Sarkar PK (1993) *Sinki*: a traditional lactic acid fermented radish tap root product. J Gen Appl Microbiol 39:395–408

Tamang JP, Sarkar PK (1995) Microbiology of murcha—an amylolytic fermentation starter. Microbios 81:115–122

Tamang JP, Sarkar PK (1996) Microbiology of *mesu*, a traditional fermented bamboo shoot product. Int J Food Microbiol 29:49–58

Tamang JP, Sarkar PK, Hesseltine CW (1988) Traditional fermented foods and beverages of Darjeeling hills and Sikkim—a review. J Sci Food Agric 44:375–385

Tamang JP, Shin DH, Jung SJ, Chae SW (2016b) Functional properties of microorganisms in fermented foods. Front Microbiol 7:578. https://doi.org/10.3389/fmicb.2016.00578

Tamang JP, Tamang B, Schillinger U, Franz CMAP, Gores M, Holzapfel WH (2005) Identification of predominant lactic acid bacteria isolated from traditional fermented vegetable products of the Eastern Himalayas. Int J Food Microbiol 105(3):347–356

Tamang JP, Tamang B, Schillinger U, Guigas C, Holzapfel WH (2009) Functional properties of lactic acid bacteria isolated from ethnic fermented vegetables of the Himalayas. Int J Food Microbiol 135:28–33

Tamang JP, Tamang N (1998). Traditional food recipes of the Sikkim Himalayas. Report, Sikkim biodiversity and eco-tourism project, GBPIHED, Tadong, Sikkim, India, p 31

Tamang JP, Tamang N, Thapa S, Dewan S, Tamang BM, Yonzan H, Rai AK, Chettri R, Chakrabarty J, Kharel N (2012) Microorganisms and nutritional value of ethnic fermented foods and alcoholic beverages of North East India. Indian J Tradit Knowl 11(1):7–25

Tamang JP, Thapa N, Rai B, Thapa S, Yonzan H, Dewan S, Tamang B, Sharma RM, Rai AK, Chettri R, Mukhopadhyay B, Pal B (2007a) Food consumption in Sikkim with special reference to traditional fermented foods and beverages: a micro-level survey. J Hill Res 20(1):1–37

Tamang JP, Thapa S (2006) Fermentation dynamics during production of bhaati jaanr, a traditional fermented rice beverage of the Eastern Himalayas. Food Biotechnol 20(3):251–261

Tamang JP, Thapa S, Dewan S, Jojima Y, Fudou R, Yamanaka S (2002) Phylogenetic analysis of *Bacillus* strains isolated from fermented soybean foods of Asia: *kinema*, *chungkokjang* and *natto*. J Hill Res 15(2):56–62

Tamang JP, Thapa S, Tamang N, Rai B (1996) Indigenous fermented food beverages of Darjeeling hills and Sikkim: process and product characterization. J Hill Res 9(2):401–411

Thapa N (2016a) Microbiology and nutrition of ethnic fermented and preserved fish products of the eastern Himalayas. Today and Tomorrow's Printers & Publishers, New Delhi. 116 p. ISBN: 81-7019-538-8

Thapa N (2016b) Ethnic fermented and preserved fish products of India and Nepal. J Ethnic Foods (Elsevier) 3:69–77

Thapa N, Pal J (2007) Proximate composition of traditionally processed fish products of the Eastern Himalayas. J Hill Res 20(2):75–77

Thapa N, Pal J, Tamang JP (2006) Phenotypic identification and technological properties of lactic acid bacteria isolated from traditionally processed fish products of the Eastern Himalayas. Int J Food Microbiol 107(1):33–38

Thapa S, Tamang JP (2004) Product characterization of kodo ko jaanr: fermented finger millet beverage of the Himalayas. Food Microbiol 21:617–622

Thapa S, Tamang JP (2006) Microbiological and physico-chemical changes during fermentation of kodo ko jaanr, a traditional alcoholic beverage of the Darjeeling hills and Sikkim. Indian J Microbiol 46(4):333–341

Tsuyoshi N, Fudou R, Yamanaka S, Kozaki M, Thapa N, Thapa S, Tamang JP (2005) Identification and characterisation of yeast strains isolated from marcha, a microbial starter for amylolytic fermentation. Int J Food Microbiol 99(2):135–146

Yonzan H, Tamang JP (2010a) Indigenous knowledge of traditional processing of *Selroti,* a cereal-based ethnic fermented food of the Nepalis. Indian J Tradit Knowl 9(2):271–274

Yonzan H, Tamang JP (2010b) Microbiology and nutritional value of *selroti*, an ethnic fermented cereal food of the Himalayas. Food Biotechnol 24(3):227–247

Yonzan H, Tamang JP (2013) Optimization of traditional processing of *Selroti*, a popular cereal-based fermented food. J Sci Ind Res 72:43–47

Ethnic Fermented Foods and Beverages of Tamil Nadu

19

Usha Antony, Shankar Ilango, Ramachandran Chelliah, Sudha Rani Ramakrishnan, and Kavitha Ravichandran

Abstract

Traditional foods and beverages derived from the natural fermentation of different food substrates have contributed to the dietary intake of the people of Tamil Nadu including Sri Lankan Tamils and states bordering Tamil Nadu since the early times. These can be broadly grouped as cereal, largely rice- and or millet-based foods and beverages produced by mixed natural fermentation which are nonalcoholic and palm sap-derived alcoholic beverages. The ethnicity and cultural traditions associated with these foods are discussed in addition to the research studies that have documented the science and the newer technologies that have been explored. While data on many foods is minimal or lacking such as *pathaneer/neera* and toddy (*kallu*) from palm, extensive studies have been done on *idli* especially over the last 10 to 15 years. *Idli* has moved from pan-Indian to global, and the batter production has been successfully commercialized both within and outside India. However, others like *pazhaya sadham*, which was

U. Antony (✉)
College of Fish Nutrition and Food Technology, Tamilnadu Dr J Jayalalithaa Fisheries University, Madhavaram, Chennai, Tamilnadu, India

S. Ilango · K. Ravichandran
Department of Biotechnology, Centre for Food Technology, Anna University, Chennai, Tamilnadu, India

R. Chelliah
Department of Food Science and Biotechnology, Kangwon National University, Chuncheon, Republic of Korea

S. R. Ramakrishnan
School of Food Science and Biotechnology, Kyungpook National University, Daegu, Republic of Korea

J. P. Tamang (ed.), *Ethnic Fermented Foods and Beverages of India: Science History and Culture*, https://doi.org/10.1007/978-981-15-1486-9_19

considered everyday food, are mainly used only among the economically poor and rural households. Millet-based *koozh* considered a health food is widely consumed by the majority during specific festivals and remains a popular street food even today. Typical dairy fermented foods include curd rice/*dahi bath* called *thayir sadham*, spiced buttermilk called *moru* and the gravy made from curd or buttermilk called *mor kuzhambu* which are commonly consumed. This chapter focusses on some of these foods based on the data available on them. Recently, a few niche restaurants are introducing many traditional foods for the migrant consumer who is nostalgic for ethnic and traditional foods indicating a trend for revival. The need to study and document these foods is emphasized especially in the current context of the role of microbes in human and animal health as well as ecological sustainability.

Keywords
Fermented foods of Tamil Nadu · *Pazhya sadham* · *Idli* · *Koozh* · *Thayir* · *Kallu*

19.1 Introduction

Tamil Nadu is one of the states of India, located in the southern end with an area of 130,058 km^2, a population of about 72 million spread over 32 districts, with Tamil as its official language. The state has two mountain ranges, the Western and Eastern Ghats which both meet at the Nilgiri Hills. The Western Ghats dominate the entire western border with Kerala. The western, southern and the northwestern parts are hilly and rich in vegetation, and in the eastern parts are fertile coastal plains. While the northern parts are a mix of hills and plains, the central and the south-central regions consist of arid plains. Tamil Nadu has the second largest coastline in the country. The state therefore has seven agroclimatic zones from dry semi-humid to semi-arid zones conducive to varied flora and fauna and crop cultivation (http://tnhorticulture.tn.gov.in/horti/agro-climatic-zones).

The state has an ancient past, recorded in Sangam literature, ruled by the Cholas, the Cheras, the Pandyas and the Pallavas. In the fourteenth century, it came under the Vijayanagar Kingdom. In the sixteenth century, European commercial interest brought the Portuguese, the Dutch and the French; the English followed, and Tamil Nadu was one of the first British settlements in India. The neighbouring states are Andhra Pradesh in the north, Karnataka in the northwest and Kerala in the west, with the Bay of Bengal on the east and the Indian Ocean in the south. Cape Comorin (Kanyakumari) the southernmost tip of the Indian peninsula is part of Tamil Nadu.

Tamil Nadu is the largest contributor to India's GDP, with 70% of the people engaged in agriculture and allied activities which contribute significantly to the state's economy. The principal food crops grown include paddy, millets and pulses. Commercial crops grown are sugarcane, cotton, sunflower, coconut, cashew, chillies, gingelly and groundnut. Plantation crops are tea, coffee, cardamom and rubber.

Archaeological excavations and anthropological studies have revealed that as early as the third millennium BCE, the valleys of Godavari, Krishna and Kaveri and other rivers were settled by people who spoke Dravidian languages. By 2000 BCE, they had spread over a wide area including Maharashtra, Andhra Pradesh, Kerala and Tamil Nadu of the present day (Sen 2015). There are different ethnic people within Tamil Nadu, which is one of oldest civilizations within India and believed to be 1.6 million years old. But people of Tamil origin are distributed in native Tamil Nadu, Puducherry and Sri Lanka. Historical records show the early cultivation of some of the cereals/millets and legumes, which have their origin in southern India. For example, black gram/*urud* (*Vigna mungo*) and green gram/*mung* (*Vigna radiata*) were said to be grown in South India in early to mid-third millennium BCE together with two local millets (Fuller 2011). India was home to several millets along with varieties imported from sub-Saharan Africa and China in the second millennium BCE. In the Neolithic period (2800–1200 BCE), the dietary staples of people in South India were two legumes and millets (Southworth 2014). Cereals were ground into flour and mixed with legume flour to make probably ancient versions of such typical South Indian foods such as *idli*, *vada* and *dosa*. The use of large open bowls and pots discovered at archaeological sites suggests the boiling of flour-based gruels (Fuller and Rowlands 2009). Today, India leads in millet production which stood at 16.14 million tonnes in 2016–2017 and Tamil Nadu with 36.4 metric tonnes and a very high productivity of 1500 tonnes per hectare (http://www.thedailyrecords.com).

The southern states of India, where the consumption of rice is more prominent compared to wheat or maize, have distinct foods that are biotransformed by fermentation (Tamang et al. 2016). Rice- and millet-based fermented foods singly or in combination or with legumes are very popular, and some have even became global. Research attention on some of these is more recent. However, the demand for ethnic foods and increasing health and environment consciousness of the consumer is driving the revival of many fermented foods in the market. This chapter focusses on such traditional fermented foods (processing, nutritional significance and health benefits) of Tamil Nadu. While the preparation and consumption of many such foods are restricted to specific cultural events/festivals/worship, some are everyday foods being consumed frequently (Akmal and Suneetha 2016) as seen from Table 19.1.

19.2 Fermented Foods from Rice

India is the second largest producer of rice in the world. Rice is the most common cereal for approximately half of the global population, and contains 80% carbohydrates, 7–8% protein, 3% fat and 3% fibre. Different varieties, both traditional and modern are cultivated, milled and used with or without parboiling. Multiple food products are made with rice or its flour. Some of the fermented foods made from rice in Tamil Nadu are discussed below.

Table 19.1 Ethnic fermented foods and beverages of Tamil Nadu

Food	Substrate	Sensory property of product	Major ethnic consumers	State
Fermented rice (pazhaya sadham)	Cooked rice + water	The mixture of rice and water has a fermented acidic odour and mild sour taste with distinct flavour and is consumed with salt/ buttermilk and chillies or vegetable pickle	Rural households and lower socio-economic groups	Tamil Nadu, Kerala
Idli	Parboiled rice + black gram dhal	The steamed product, which is circular, disc shaped and white in colour, is soft and spongy with mild fermented acidic odour and sour taste with distinct flavour and is consumed with coconut chutney, vegetable sambar and chilli chutney powder mixed with sesame oil/ghee/coconut oil. May also be served with mutton *paya* on special occasions	Consumed by all communities across Tamil Nadu, by all age groups; a typical breakfast food also consumed as evening snack or main course of the evening meal	Tamil Nadu, Kerala, Karnataka, Andhra Pradesh

(continued)

Table 19.1 (continued)

Food	Substrate	Sensory property of product	Major ethnic consumers	State
Dosa/Dosai	Parboiled rice + black gram dhal	Cooked on a hot *tawa* (iron or non-stick griddle) in the shape of pancakes. Golden brown, on one side, white on the other side; thickness and texture vary from crisp, thin, paper-like, to thick, crisp on one side and soft on the other; has unique mild fermented flavour and aroma and sour taste. May be served with various types of fillings, usually potato masala, coconut chutney and sambar. May also be served with chicken/mutton curry	Consumed by all communities across Tamil Nadu, by all age groups; a typical breakfast food also consumed as evening snack or main course of the evening meal. It is so popular nationally and internationally, that it is synonymous with South Indian vegetarian food	Tamil Nadu, Kerala, Karnataka, Andhra Pradesh
Uthappam	Rice + black gram dhal	Made from the same batter of *dosai* but is a thicker, soft and spongy version made with onion or other toppings	Consumed by all communities across Tamil Nadu, by all age groups	Tamil Nadu, Karnataka
Aappam	Raw rice fermented with yeast	The batter is cooked on a special metal or earthen *chatti*/*wok*. Has mild sweet fermented flavour. Spongy at the centre with lacy edges. Eaten with sweetened coconut milk or vegetable/ chicken/mutton stew/*kurma*/gravy	Consumed by all communities across Tamil Nadu	Tamil Nadu, Kerala

(continued)

Table 19.1 (continued)

Food	Substrate	Sensory property of product	Major ethnic consumers	State
Koozh	Finger/pearl millet flour + broken rice	A fermented porridge, reddish brown to brown in colour, with unique sour flavour and taste and texture. Served with raw onion, green chillies, fried fermented chilly (*mor milagai*), lemon or mango pickle	A meal replacer that may be consumed at breakfast or lunch, popular among labourers. Is a special offering at temples during devotional festivals	Tamil Nadu, called *ambali* in Andhra Pradesh
Dahi/thayir	Fermented cow or buffalo milk set like yoghurt	A white semi-solid product of jelly-like consistency, mild sour taste and unique fermented flavour	Consumed by all communities across Tamil Nadu, by all age groups; served with rice as the last course of a traditional meal, used as salad dressing for vegetable salad called *pachadi*	Tamil Nadu, Kerala, Karnataka, Andhra Pradesh
Dahi rice/ thayir sadham	Mixture of rice and curd	Served plain or with garnishing of green chilli, ginger, curry leaves, coriander leaves. Has a unique bland, mildly sour taste	Consumed by all communities across Tamil Nadu, by all age groups; served as the last course of a traditional vegetarian meal	Tamil Nadu
Moru or buttermilk	Curd with salt, garnish with spices/herbs is optional	A traditional beverage served to quench thirst in summer. Has mild sour flavour and taste of fermented milk. Addition of salt, chilli, ginger, curry leaves and coriander leaves modifies the flavour	Consumed by all communities across Tamil Nadu, by all age groups. Distributed at festivals and in the streets during summer	Tamil Nadu, Kerala, Karnataka, Andhra Pradesh

(continued)

Table 19.1 (continued)

Food	Substrate	Sensory property of product	Major ethnic consumers	State
Kallu/toddy	Sap of inflorescence of the coconut or palm tree	A traditional alcoholic white cloudy beverage with sweet-sour taste, often produced and consumed fresh; has very low shelf life	Consumed largely by men, often produced and sold locally	Tamil Nadu, Kerala, Karnataka, Andhra Pradesh

Davidar (2001), Wickramasinghe and Rajah (2002), Bali (2011), Chandrasekhar et al. (2012), Akmal and Suneetha (2016)

Fig. 19.1 Traditional method of preparation of *pazhaya sadham*/fermented rice in Tamil Nadu

Left-over cooked rice from the day

↓

Soaked in excess water, covered in a metal or earthen vessel

↓

Fermented overnight for 12-16 h

↓

PAZHAYA SADHAM

19.2.1 *Pazhaya Sadham*

Fermented rice is one of the traditional staple foods in Tamil Nadu. Traditionally the leftover cooked rice is allowed to ferment overnight and consumed the following day to preserve the food resources. The 1-day-old, cooked, water-soaked and fermented rice is known as '*pazhaya sadham*' or '*palam soru*' (fermented rice) as seen from Fig. 19.1. Fermented rice water known as *pazhaya sadham kanchi* or *Neeragaram* is also used as healthy water in natural medicine in folklore practice (Fig. 19.1).

This traditional staple food (Fig. 19.2) is usually consumed with salt and buttermilk accompanied by lemon/mango pickle, raw onions and green chillies.

However, in recent days, fermented rice is one of the neglected foods in countries both in terms of research and consumption. Before the use of refrigeration technology around 30–40 years ago, a vast majority of the farming community in Asia especially South India had a habit of consuming fermented rice for breakfast which provides adequate nutrition (Varnkulendran et al. 2016). More recently, this practice is seen only among economically weaker sections, as most families/communities refrigerate the rice and use it after steaming or boiling again.

The traditional fermented rice is found to contain beneficial microflora which provides health-promoting macro- and micronutrients, phytochemicals, and other

Fig. 19.2 Fermented rice (*pazhaya sadham*)

functional components as an effect of fermentation. Analysis of samples from two local houses in Thanjavur, a southern district of Tamil Nadu, revealed presence of minerals including selenium (0.02 ± 0.01 and 0.03 ± 0.01 mg/100 g, respectively) which may have a preventive role against cancer (Arjun et al. 2014).

The Tamil population in Sri Lanka share many cultural practices and traditional foods of Tamil Nadu. Studies done with the Sri Lankan fermented red rice showed the presence of the lactic acid bacteria like *Leuconostoc* (*Leuc.*) sp. and *Lactobacillus* (*Lb.*) sp. which have potential as natural preservatives due to their properties to combat the growth of pathogens (Varnkulendran et al. 2016). Cooked and uncooked raw white and red rice from Sri Lanka were fermented by soaking in sterile water (rice/water, 1:3) overnight (12–16 h) in clay pots at ambient temperature (27 °C). Microbial studies on 122 bacterial isolates revealed coccus (48), diplococcus (30) and bacillus (30), of which 7 were *Lb* sp. and could therefore be potential probiotics (Jeyagowri et al. 2015).

According to traditional systems of medicine like Siddha and Ayurveda, fermented rice provides energy and also helps in controlling stomach issues like bloating, constipation, ulcer and diarrhoea. It prevents dehydration by acting as an effective electrolyte solution. In addition, it regulates the body temperature and also protects the skin by its cooling effect and cures acne and red blisters of the face. The optimal pH upon application helps in providing shiny, long hair, improves skin elasticity, reduces surface friction and prevents grey hair (Varnakulendran et al. 2016).

19.2.2 Traditional and Novel Products from Fermented Rice

Traditional snacks and novel products from fermented rice may be used. Fermented rice (leftover)s are made into *vadam* by mashing and addition of spices followed by drying in the sun and then are later stored. It is a traditional ready-to-cook food, which is deep-fried in vegetable oil for use as a snack or accompaniment to the main course of a meal.

Novel product formulations from fermented rice have been reported. Micronutrient-enriched spread using fermented cooked red rice and other ingredients such as tomato, chick pea, spices, corn flour, salt, sugar, coconut oil and bee's honey was developed in Sri Lanka. The product was enriched with 2.3 μg of vitamin B_{12} and 40 mg of ascorbic acid per 100 g (Jayawardena and Wansapala 2015). It was microbiologically stable and safe with shelf life of 1.5 months under refrigerated conditions. The solids and liquid from traditional fermented rice are yet to be studied in depth and commercially leveraged for new functional foods that have ethnic appeal.

19.2.3 *Idli*

Among several Indian traditional fermented foods, *idli* is widely consumed throughout India and is also becoming popular in other countries, and market studies indicate that the humble *idli* has become part of the global cuisine. *Idli* is a cereal-pulse-based fermented breakfast food consumed mostly in the southern part of India as well as Sri Lanka. Koh and Singh (2009) reported that locally sourced traditional foods represent a major portion that is more than 50% of the diet. The respondents in Chennai and Bangalore (49.7% and 32.8%, respectively) consumed *idli* as their breakfast item on a daily basis when compared to Pondicherry (29.7%). A majority of the respondents in Pondicherry consumed *idli* twice a week. The other traditional foods consumed frequently included rice, *aappam*, *adai* and *pittu* (Fig. 19.3).

19.2.3.1 Characteristics and Fermentation of *Idli* Batter

Idli is a white, fermented (acid-leavened), soft, spongy textured, steamed cake made from a mixture of rice (*Oryza sativa*) and dehulled black gram (*Phaseolus mungo*). It is preferred for its mild pleasant flavour and aroma, easy digestibility and known health and nutritional benefits, apart from its unique textural properties. *Idli* fermentation has been the subject of many research investigations, covering aspects such as methods of preparation, microbiology and nutritive value (Rao et al. 2005).

Traditionally, for *idli* preparation, parboiled rice is preferred over raw rice, with rice and black gram dhal usually in 3:1 (v/v) ratio. The rice and black gram dhal are soaked separately in water. After draining the water, while rice is coarsely ground, dhal is finely ground separately, with occasional addition of water during the process. Then both batters are mixed together with a little salt and allowed to ferment overnight at room temperature (about 30 °C). The fermented batter is placed in

Parboiled *Idli* rice and Black gram dhal (3:1 v/v)

↓

Washed and soaked separately in excess water for 4 to 6 h

↓

Ground separately in a traditional stone grinder to form a thick batter of pouring consistency

↓

Rice and Dhal batters mixed well with salt (2%)

↓

Fermented overnight for 12-18 h

↓

Batter poured into *idli* moulds and steamed for 7 to 10 min

↓

IDLI

Fig. 19.3 Traditional method of preparation of *idli* in Tamil Nadu

Fig. 19.4 *Idli*

special *idli* pans and steamed for 5–8 min (Deshmukh and Pawar 2016). It is generally served with accompaniments such as coconut chutney, sambar and chilli dhal powder with sesame/coconut oil or *ghee* (clarified butter) (Fig. 19.4). The total time required for its preparation is about 12–18 h that involves washing, soaking, grinding, fermentation and steaming processes.

Black gram dhal (*Phaseolus mungo*), the leguminous component of *idli* batter, serves not only as an effective substrate but also provides the maximum number of microorganisms for fermentation (Balasubramanian and Viswanathan 2007). Black gram is mucilaginous which makes it a valuable ingredient in *idli* preparation (Susheelamma and Rao 1979). The chief proteins present in black gram are albumins, globulins and glutelins. A noteworthy feature of batters containing this legume is their high viscosity ascribed to the presence of polysaccharide arabinogalactan, and this is also responsible for the gas-holding and batter-raising qualities. Other legumes lack this polysaccharide and hence are unsuitable for such food preparations.

The natural flora in the batter causes rapid acidification of the raw material through the production of organic acids, mainly lactic acid as well as acetic acid. The microbes involved in many fermented foods enhance shelf life and microbial safety, improve texture and contribute to the pleasant sensory profile of the end product by production of ethanol, aroma compounds, bacteriocins, exopolysaccharides and several enzymes (Sridevi et al. 2010). Two significant changes occurring in *idli* fermentation are leavening and acidification of the batter (Das et al. 2012). *Idli* batter prepared from rice varieties with higher amylose content had higher acidity, gretaer volume and non-protein nitrogen than low and non-amylose varieties (Kaw and Mabesa 1987).

Nagaraju and Manohar (2000) reported that rheology and bulk density showed no relation to fermentation time and rice to black gram ratio, but there was a steep increase in volume after 4 h of fermentation. Balasubramanian and Viswanathan (2007) showed that the flow behaviour index indicated strong non-Newtonian fluid behaviour (pseudoplastic) of *idli* batter at different fermentation times and blend ratios of *idli* batter from soaked polished parboiled rice and decorticated black gram.

The studies on isolation and identification of microorganisms responsible for the characteristic changes in the batter were done by Mukherjee et al. as early as 1965. Although there is a sequential change in the bacterial flora, the predominant microorganism responsible for souring and gas production was found to be *Leuc. mesenteroides*. In the later stages of fermentation, *Streptococcus (S.) faecalis* and *Pediococcus (Ped.) cerevisiae* became significant. They concluded that fermentation of *idli* was caused by the leavening action of the hetero-fermentative lactic acid bacterium *Leuc. mesenteroides*.

Soni and Sandhu (1989) analysed different *idli* batters, both commercial and laboratory-prepared. They observed six species of bacteria (*Leuc. mesenteroides*, *Str. faecalis*, *Lb. fermentum*, *Ped. cerevisiae*, *Lb. delbrueckii*, *B. amyloliquefaciens*) and yeast (*S. cerevisiae*, *D. hansenii*, *H. anomala*, *Trich. beigelii*, *Torul. candida*, *Trich. pullulans*), in the range of 10^6–10^9 and 10^6 g^{-1}, respectively. *Leuc. mesenteroides* and *Str. faecalis* were found to be essential for leavening of batter and acid production in *idli* (Rhee et al. 2011). A large number of LAB have been identified to be a part of the microflora responsible for fermentation of *idli* batter, and they include *Leuc. mesenteroides*, *Lb. coryneformis*, *Lb. delbrueckii*, *Lb. fermentum*, *Lc. lactis*, *Str. faecalis* and *Ped. cerevisiae*. Even though the common bacteria involved in fermentation are species of *Leuconostoc*, *Lactobacillus*, *Streptococcus*,

Pediococcus, *Micrococcus* and *Bacillus*, in certain products, the fungal species, viz. *Aspergillus*, *Paecilomyces*, *Cladosporium*, *Fusarium* and *Saccharomyces* (yeast), are also found (Blandino et al. 2003).

A study by Kannan et al. (2015) conclusively showed that in natural *idli* fermentation, LAB and yeasts play a major role and affect the physicochemical and the textural properties of the batter, thereby enhancing the textural and organoleptic properties of the steamed product. The traditional method of natural fermentation for a period of 12–16 h at 37 °C to an end point of pH 4.5 yielded *idli* with required textural and sensory properties.

Experiments carried out by Chelliah et al. (2014) with *idli* prepared from heat-treated rice showed lower fermentation time than untreated rice based on the microbiological profile, pH and titratable acidity. The heat treatment gave *idli* with better appearance, colour, taste, flavour, texture and mouthfeel, along with reduced fermentation time.

19.2.3.2 Probiotic Microbes in *Idli* Batter Fermentation

Further, *idli* batter can be used as a source for isolating probiotic strains mainly lactic acid bacteria (LAB). Agaliya and Jeevaratnam (2012) identified eight potential probiotic *Lactobacillus plantarum* strains from fermented *idli* batter using in vitro assays, and the isolates showed resistance towards antibiotics like gentamycin, ciprofloxacin, nalidixic acid and norfloxacin. All the isolates showed bile salt hydrolase activity with cholesterol-lowering capacity, the highest being 73% by *Lb. plantarum* JJ18. The isolates possessed galactosidase activity; but none showed haemolytic activity.

Iyer et al. (2013) isolated 15 bacterial strains from *idli* batter based on their catalase activity and gram nature. One *Lactobacillus* strain and one *Lactococcus* strain showed antimicrobial activity. They were identified to be *Lb. plantarum* and *Lc. lactis*. Both isolates showed good tolerance against high concentration of bile salts and acidic pH, were able to adhere on the mucosal surfaces and non-pathogenic bacteria and are generally regarded as safe.

The antimicrobial activity and probiotic properties of a wild strain of *Pichia (P.) kudriavzevii* (Chelliah et al. 2016a) as well as the safety, antimicrobial activity and probiotic properties of *Escherichia coli Nissle 1917* isolated from frozen *idli* batter (Chelliah et al. 2016b) have been documented.

19.2.3.3 Nutritional Quality of *Idli*

Additionally, apart from its organoleptic properties, *idli* is highly nutritive and a significant source of calories, proteins and micronutrients. *Idli* is an important dietary source of vitamins, especially B complex vitamins, compared to the raw unfermented ingredients. Fermentation of the batter causes an increase in the protein efficiency ratio, essential amino acids and vitamins such as niacin, riboflavin and thiamine, together with a decrease in antinutrient content (such as phytic acid), enzyme inhibitors and flatus sugars, over the unfermented counterpart (Nout 2009; Hamad and Fields 1979; Reddy et al. 1982; Sharma et al. 2016). The fermented batter of *idli* and *dosa* contains higher amount of available lysine, cysteine and methionine. After processing, maximum retention of lysine, methionine and cysteine was observed in steamed *idli* (Riat and Sadana 2009). From the nutritional and health point, *idli* appears to be an ideal human food for all ages and at all times. Moreover, for infants, it is used as a weaning food owing to its predigested components from fermentation with additional benefit of sterilization during steam cooking.

19.2.3.4 Stabilization and Commercialization of *Idli* Batter Production

Idli in general has immense scope for commercialization as a food with improved nutritional value as well as functional properties. Even with rapid urbanization, *idli* still remains the choice of breakfast for the population either at home or away from home. Until about 10 years ago, *idli* production was confined to the domestic level and catering establishments. Over the last decade, there has been a significant growth in the commercial production and marketing of *idli* batter contributed by the changing lifestyles, urbanization and the growing number of women joining the work force. In spite of the demand, batter production although commercialized to some extent, the issues with standardized quality parameters and shelf life of the product remain.

Observation of industry practices revealed that *idli* batter manufacturing was performed by washing and grinding the raw materials such as rice, black gram dhal and fenugreek separately. Rice and black gram were mixed in 4:1 proportion with further addition of fenugreek paste (0.3%) and sodium chloride (3%). The standardized batter was packed in different-sized pouches based on the customer needs (Chelliah et al. 2014, unpublished data).

Idli preparation is a time-consuming process; and although instant *idli* premixes as powder or batter are available in the market, they do not have the distinctive taste and aroma similar to the *idli* prepared at home (Unika and Jaffar 2014). The freshness of product is typically defined by the qualities of taste, texture and appearance. Because of the immense size of the produce industry, it would be highly advantageous from an economic standpoint to preserve *idli* batter, for a longer period of time. Hence, research has focused on different methods for prolonging shelf life (drying), with varying degrees of success because the product loses its textural and organoleptic properties (Rajendran and Chander 2014).

Nisha et al. (2005) stabilized the *idli* batter at room temperature (28–30 °C) as well as refrigerated storage (4–8 °C) using hydrocolloids and surface-active agents. Hydrocolloids gave good stabilization, whereas surface-active agents failed to stabilize the batter. Among the various hydrocolloids, 0.1% guar gum gave the best batter stabilization, and *idli* made after 10 days of room temperature (30 ± 2 °C) storage and 30 days of refrigerated (4 °C) storage of batter were found to be of acceptable quality. Chelliah (2016) reported that the shelf life of *idli* batter was extended up to 30 days at 4 °C by process modification, wherein the batter was prefrozen to −20 °C and maintained at that temperature for 1 h (Figs. 19.5 and 19.6).

Mulmule et al. (2017) prepared ready-to-eat (RTE) *idli* with shelf life of up to 2 months at ambient temperature using combination of electron beam irradiation

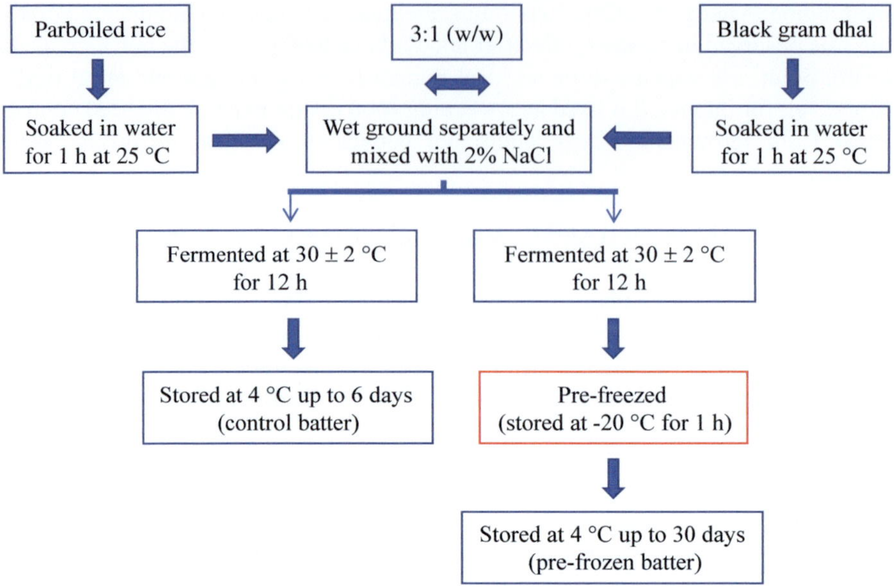

Fig. 19.5 Shelf life extension of *idli* batter by process modification

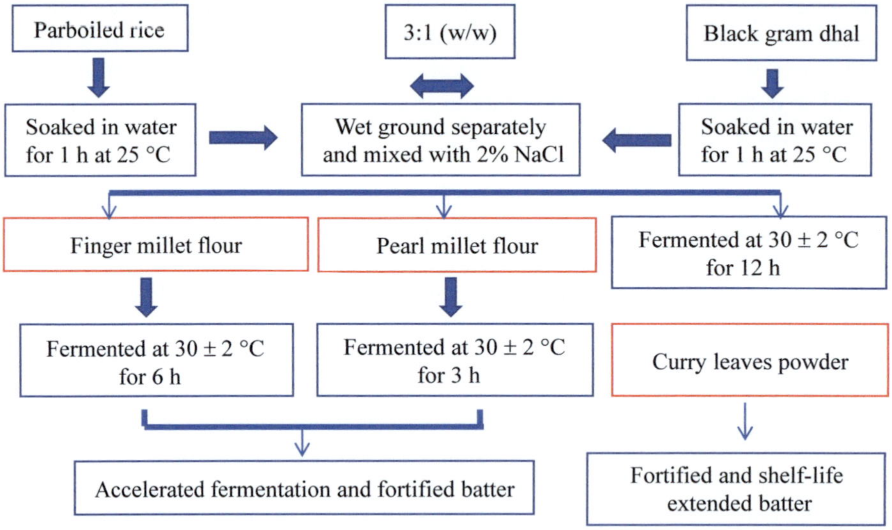

Fig. 19.6 Reduction in fermentation time and shelf life extension of *idli* batter by biofortification (Chelliah 2016)

(EBI) at dosage 2.5 kGy and thermal treatment at 80 °C for 20 min. It was inferred that only combination process could be useful to achieve extended shelf life without impairing the organoleptic quality of RTE *idli*.

19.2.3.5 Use of Other Ingredients in *Idli*

Novel *idli* batters with soybeans and mung beans replacing the conventional black gram revealed differences in the levels of biochemical constituents. Soybean batters contained the highest levels of soluble solids, reducing sugars, total nitrogen, proteins, amylases, proteinases and group B vitamins, which are possibly due to high nitrogen, amylase and proteinase contents. The organoleptic evaluation of *idli* revealed wide acceptance of the products due to their attractive aroma and taste (Sathe and Mandal 2016).

The study done by Iyer and Ananthanarayan (2008) was to explore the possibility of expediting the process of *idli* batter fermentation by adding an exogenous source of amylase enzyme. Amylase (5, 15 and 25 U per 100 g) was added to the *idli* batter and allowed to ferment. The fermentation time was reduced to 8 h, and the sensory attributes of the final product were also successfully maintained.

Nazni and Shalini (2010) prepared three types of *idli*, namely, standard *idli*, mixed *idli* (black gram + sorghum) and sorghum *idli* using different combinations of ingredients such as rice (parboiled), black gram and sorghum. The overall score of standard *idli* was found to be slightly higher (8.58) than mixed *idli* (7.78) and sorghum *idli* (7.00). Complete or partial replacing of rice with sorghum had good impact on the nutritive value by increasing the protein, fat, fibre, calcium and iron content in the *idli*. The sorghum *idli* was found to be acceptable with respect to both sensory and nutritional quality. Rekha and Vijayalakshmi (2011) reduced the fermentation time of *idli* batter from the conventional 14 to 10 h by partial substitution of underutilized *okara* (soy residue) instead of black gram in the ratio of 1:1.

Use of xylooligosaccharides (XOS) in *idli* was investigated by Aachary et al. (2011). Batter with different concentrations of XOS (0, 0.2, 0.4 and 0.6% w/v) was allowed to ferment for 4–18 h. Addition of XOS (0.4%) increased lactic acid bacteria number (9.88 ± 0.08 log cfu g^{-1}) with rapid reduction in pH after 6 h of fermentation compared to conventional batter without XOS (9.46 ± 0.06 log cfu g^{-1}). *Idli* with XOS had higher moisture content and softer texture. Addition of XOS benefited both fermentation and *idli* quality.

The incorporation of curry leaf (*Murraya koenigii*) powder (5%) in the *idli* batter increased the shelf life at room temperature (30 °C) up to 5 days with retention of quality. The sensory qualities of the fortified *idli* were preferred by panelists, and nutritional enhancement of dietary fibre and calcium content was an added benefit (Chelliah et al. 2016c). According to Chelliah et al. (2017), addition of finger millet and pearl millet flour (10% w/w) to the batter decreased the fermentation time from 12 h to 6 h and 8 h, respectively (Fig. 19.6).

19.3 Fermented Beverages

19.3.1 *Koozh*

In Tamil Nadu, especially in rural areas, the traditional fermented millet porridge *koozh* is widely consumed. *Koozh* is made with pearl or finger millet flour and rice (Ilango and Antony 2014) and consumed extensively during festival seasons and religious function and ceremonies.

Research on *koozh* sold by street vendors has been done in localities of Madurai, Salem and Chennai and Pondicherry (Kumar et al. 2010; Patel et al. 2013). People from lower-income groups preferred *koozh* as a meal instead of snack (Patel et al. 2013). The 65% who consumed *koozh* were vendors and hawkers. *Koozh* is a food for working-class people and infants due to the fact that it is considered a nutritional food/beverage and energy source. Farmers, daily wage labourers and lorry drivers consume it due its satiety and economic value. Also, *koozh* has a unique tangy flavour that is preferred by the consumer, in addition to health, religious and festive reasons. This fermented gruel *koozh* acts as an alternative nourishment source. The preference of pearl and finger millet *koozh* differs according to locality. Pearl millet porridge is consumed in Madurai and finger millet in Chennai and Salem (Patel et al. 2013).

Consumers have stated they feel satiated and energized after consumption and perceive it as giving stamina over an extended time span. It is also believed to help manage diabetes mellitus (Patel et al. 2013). The food is also said to have prebiotic effect because of the presence of oligosaccharides and non-starch polysaccharides (dietary fibre).

19.3.1.1 Preparation of *Koozh*

The methodology of preparation slightly varies according to geographical location within Tamil Nadu. In general, the preparation of *koozh* involves a two-step natural fermentation process—first fermentation with the raw millet slurry and second fermentation after cooking the slurry (Figs. 19.7 and 19.8). The finger/pearl millet flour is mixed with water in the ratio of 2:1. Primary fermentation is carried out at ambient temperature overnight (12–15 h). The next day the fermented millet slurry is mixed into a pot of boiling broken parboiled rice (*noyee*) in an earthen pot. Later this cooled mixture is hand crushed and kept in a closed container overnight. The

Fig. 19.7 Preparation and consumption of *koozh*

Finger /Pearl millet is cleaned and ground into flour

↓

Millet flour slurry is made with drinking water (1:1w/v)

↓

Millet slurry is fermented over night at ambient temperature

↓

Fermented slurry is cooked along with rice grits into a porridge

↓

Cooked mixture is fermented over night at ambient temperature

↓

The thick porridge is shaped into spheres called *kali* and stored at ambient temperature

↓

Kali is mixed with drinking water (1:6 w/v) and salt and served with accompaniments

↓

KOOZH

Fig. 19.8 Steps in the preparation of *koozh* (Ilango and Antony 2014)

semi-solid product is made into a hand-size ball called *kali*, which can be stored for a week at ambient temperature. A single ball is then used to prepare 2–3 L of *koozh* by diluting with water, stored at ambient temperature and usually consumed within 12 h (Ilango and Antony 2014). The nutrition density of *koozh* increases when the *noyee* and millet undergo spontaneous naturally fermentation, decreasing the viscosity of millet gruel. The product is a double fermented nonalcoholic one.

Koozh is served from a 12 to 15 L plastic/stainless steel vessel. In some rural areas, the *koozh*-containing clay pot is laid over wet sand and placed on a table. This table is kept in market stalls made of thatched roofing. It is served from morning to noon in 500–600 mL wide-mouthed stainless steel containers (*sembhu*). Accompaniments like sliced onion (*Allium cepa*), green coriander (*Coriandrum sativum*), chutney and pickles made from mango (*Mangifera indica*) or lime (*Citrus aurantifolia*) are served with *koozh*. The others are sun-dried vegetable and spices like chilli pepper (*Capsicum annuum*), turkey berry (*Solanum torvum*) and cluster beans (*Cyamopsis tetragonoloba*) which are fried in oil and served. The preparative handling and service of the product varies, altering the taste, probably contributed by difference in the microbial flora.

19.3.1.2 Microbes in *Koozh*

Studies on *Koozh* are limited. The microflora of *koozh* prepared from finger millet alone under laboratory conditions with fermentation for 2 days showed the presence of *Weissella paramesenteroides* with probiotic properties and *Lb. fermentum* with antibacterial activity towards *Salmonella typhi*, *Vibrio parahaemolyticus* and *Listeria monocytogenes* (Kumar et al. 2010). Geetha and Kalaichelvan (2013) have reported the predominance of lactic acid bacteria, in their study where finger millet, pearl millet, sorghum and maize were made separately into *koozh* by fermenting for 20 h in the laboratory.

A study of six market samples of *koozh* revealed that the pH of *koozh* ranged from 4.3 to 4.9 with acidity ranging from 0.16 to 0.35%, in all samples. Microbial enumeration indicated the predominance of LAB followed by yeasts; moulds were absent (Ilango and Antony 2014). Further, the evaluation of their microbial quality and safety showed the presence of coliforms in all samples indicating unhygienic handling of the product. The probiotic potential of LAB isolates from *koozh* samples has been documented (Ilango et al. 2016).

19.3.2 *Kallu* (Toddy/Palm Wine)

Toddy is the traditional fermented alcoholic beverage made by fermenting the sap from the coconut, palmyra and toddy palm tree. The tip of an unopened flower is sliced, causing sap to ooze out. This is then collected in an earthenware container tied underneath the flower. In South India the commonly tapped palm species are *Borassus flabellifer* (palmyra palm) and *Cocos nucifera* (coconut palm). Many palm species have been tapped for centuries to produce fresh sap, fermented drinks, syrup, jaggery, sugar and vinegar. The fresh, unfermented palm sap, rich in sugar (about 10–18%), is called *neera* or *pathaneer*. It is sweet-tasting and usually refrigerated, stored and sold. It is one of the cheapest sources of vitamin B and also contains ascorbic acid (Anal 2019). The amounts of thiamine, riboflavin, and pyridoxine increase with fermentation. Fermented coconut toddy contains approximately 1.8–7.9 g alcohol, 0.29 g sucrose, 0.9–3.0 g invert sugar and 3.72 g/dL total solids. It contains nitrogen, phosphorus, potassium, calcium and magnesium <0.5 g/100 mL (Ekanayake 2016). The palm sap also contains natural yeasts leading to fermentation of the sugar into ethanol, and the traditional alcoholic beverage obtained is called *kallu* or *toddy*. The alcohol content varies from 4 to 9%. Fermentation time is usually 1 day, when the beverage is allowed to ferment in the collection pots. After this it becomes increasingly sour due to conversion of alcohol to acetic acid.

Shamala and Sreekantia (1988) have documented the microorganisms in the fresh sap and fermented palm wine or toddy. These include *Saccharomyces cerevisiae, Schizosaccharomyces pombe*, *Acetobacter aceti*, *Acetobacter rancens*, *Acetobacter suboxydans*, *Leuconostoc dextranicum*, *Micrococcus* sp., *Pediococcus* sp., *Bacillus* sp. and *Sarcina* sp.

19.3.3 *Thayir* (Curd/Dahi)

Thayir is the Tamil term for the traditionally lactic acid fermented milk in almost every household and is also known by other terms as *curd* or *dahi*. Generally, it is

Fig. 19.9 Steps in the traditional preparation of *thayir*

Cow/Buffalo milk

↓

Boiled for few minutes

↓

Cooled to about 40 °C

↓

Previously fermented curd added and mixed well (2 – 5%)

↓

Fermented at room temperature (25 to 35 °C) for 6 to 8 h

↓

THAYIR / CURD

made from cow or buffalo milk and known for its palatability, digestibility, nutritional quality and therapeutic role (Kanade 2008). Typical bacterial strains are variable as the product is prepared by back-slopping or using the fermented material as a starter for the next batch and is usually a mixed consortium (Fig. 19.9).

Typical strains include *Lb. bulgaricus*, *Str. thermophilus*, *Lac. lactis*, *Lb. cremoris*, *Lb. helveticus*, *Lb. casei* and *Lb. acidophilus* (Harun-ur-Rashid (2007). The total viable count ranges from 10^6 to 10^7 per millilitre. The traditional product is prepared fresh every day providing live microbes and is a potential source of probiotics in the diet. The prepared curd is consumed fresh or stored in refrigerator for about 24 h before use (Sakore et al. 2007). Bharadwaj et al. (2012) who analysed 40 *dahi* samples from local markets of rural and urban areas in North India have also reported the presence of *Lb. casei*, *Lb. brevis*, *Lb. fermentum*, *Lb. plantarum*, *Lb. helveticus*, *Lb. rhamnosus*, *Lb. viridiscence*, *Lac. lactis* and *Lb. acidophilus*, where *L. acidophilus* was the most prevalent.

19.4 Conclusion

Several other fermented foods of Tamil Nadu that are popular locally and enjoyed in countries where Indians have migrated like *dosai*, *adai*, *athirasam* and *kuzhi paniyaram* have not received research attention. The science behind their unique texture, flavour, and taste remain unexplored. That they form a part of the regular diet is well recognized; however, their nutrient contribution and health benefits though perceived need scientific substantiation. The scrutiny of the functional role of microbes involved in their production can throw light on their possible use in preventive and therapeutic nutrition and thus worthy of in-depth study in the near future.

References

Aachary AA, Gobinath D, Prapulla SG (2011) Short chain xylooligosaccharides: a potential prebiotic used to improve batter fermentation and its effect on the quality attributes of *idli*, a cereal legume- based Indian traditional food. Int J Food Sci Technol 46(7):1346–1355

Agaliya PJ, Jeevaratnam K (2012) Screening of *Lactobacillus plantarum* isolated from fermented *idli* batter for probiotic properties. Afr J Biotechnol 11(65):12856–12864

Akmal S, Suneetha V (2016) Food culture in Tamilnadu—a study. Int J Pharm Technol 8(4):22246–22253

Anal AK (2019) Quality ingredients and safety concerns for traditional fermented foods and beverages from Asia: a review. Fermentation 5(1). https://doi.org/10.3390/fermentation5010008

Arjun D, Kumar R, Singh C (2014) Nutrient content of '*Pazhaya sadham*': a traditionally fermented food of Tamil Nadu, India. Adv Food Sci Technol 2(9):269–270

Balasubramanian S, Viswanathan R (2007) Properties of *idli* batter during its fermentation time. J Food Process Preserv 31(1):32–40

Bali PS (2011) Quantity food production operations and Indian cuisine. Oxford University Press, New Delhi, pp 182–193

Bhardwaj A, Puniya M, Sangu KPS, Kumar S, Dhewa T (2012) Isolation and biochemical characterization of Lactobacillus species isolated from *Dahi*. Res Rev J Dairy Sci Technol 1(2):12

Blandino A, Al-Aseeri ME, Pandiella SS, Cantero D, Webb C (2003) Cereal-based fermented foods and beverages. Food Res Int 36(6):527–543

Chandrasekhar K, Sreevani S, Seshapani P, Pramodhakumari J (2012) A review on palm wine. Int J Res Biol Sci 2(1):33–38

Chelliah R (2016) Process modification to ensure shelf-life extension of *Idli* batter and characterization of probiotic yeast and bacteria. Ph.D. thesis, Anna University, Chennai, India

Chelliah R, Ramakrishnan SR, Anbarasan S, Antony U (2014) Reduction in fermentation time of *idli* batter by process modification. J Biochem Sci 1(5):106–117

Chelliah R, Ramakrishnan SR, Antony U (2016b) Evaluation of safety, antimicrobial activity and probiotic properties of *Escherichia coli* Nissle isolated from *idli* batter. Res J Biotechnol 11(7):42–49

Chelliah R, Ramakrishnan SR, Prabhu PR, Antony U (2016a) Evaluation of antimicrobial activity and probiotic properties of wild strain *Pichia kudriavzevii* isolated from frozen *idli* batter. Yeast 33:385–401. https://doi.org/10.1002/yea.3181

Chelliah R, Ramakrishnan SR, Premkumar D, Antony U (2016c) Bio-fortification and self-life extension of *idli* batter using curry leaves (*Murraya koenigii*). J Food Sci Technol 53(6):2851–2862

Chelliah R, Ramakrishnan SR, Premkumar D, Antony U (2017) Accelerated fermentation of *idli* batter using *Eleusine coracana* and *Pennisetum glaucum*. J Food Sci Technol 54(9):2626–2637

Das A, Raychaudhur U, Chakraborty R (2012) Cereal based functional food of Indian subcontinent: a review. J Food Sci Technol 49(6):665–672

Davidar RN (2001) Indian food sense—a health and nutrition guide to traditional recipes. East West Books (Madras) Pvt Ltd., Chennai, pp 85–241

Deshmukh GP, Pawar PP (2016) Optimization of formulation and development of carrot fortified *idli* and its physico-chemical characterization. Int J Eng Sci Res Technol 5(8):783–786

Ekanayake S (2016) Ethnic fermented foods and beverages of Sri Lanka. In: Tamang JP (ed) Ethnic fermented foods and alcoholic beverages of Asia. Springer, New Delhi, pp 139–150

Fuller DQ (2011) Finding plant domestication in the Indian sub-continent. Curr Anthropol LII/S4:S347–D362

Fuller DQ, Rowlands M (2009) Towards a long-term macro-geography of cultural substances; food and sacrifice in East, West and South Asia. Chinese Rev Anthropol 12:32–33

Geetha T, Kalaichelvan G (2013) A study on the fermentation pattern of common millets in *Koozh* preparation—a traditional South Indian food. Indian J Tradit Knowl 12(3):512–517

Hamad AM, Fields ML (1979) Evaluation of the protein quality and available lysine of germinated and fermented cereals. J Food Sci 44(2):456–459

Harun-ur-Rashid M, Togo K, Ueda M, Miyamoto T (2007) Probiotic characteristics of lactic acid bacteria isolated from traditional fermented milk '*Dahi*' in Bangladesh. Pak J Nutr 6(6):647–652

Ilango S, Antony U (2014) Assessment of the microbiological quality of *koozh*, a fermented millet beverage. Afr J Microbiol Res 8(3):308–312

Ilango S, Pandey R, Antony U (2016) Functional characterization and microencapsulation of probiotic bacteria from *koozh*. J Food Sci Technol 53(2):977–989

Iyer BK, Ananthanarayan L (2008) Effect of α-amylase addition on fermentation of *idli*—a popular South Indian cereal-legume-based snack food. LWT-Food Sci Technol 41(6):1053–1059

Iyer BK, Singhal RS, Ananthanarayan L (2013) Characterization and *in vitro* probiotic evaluation of lactic acid bacteria isolated from *idli* batter. J Food Sci Technol 50(6):1114–1121

Jayawardena JAEC, Wansapala MAJ (2015) Study on developing micronutrients enriched spread using fermented cooked rice. Int J Innov Res Technol 1(12):1603–1605

Jeygowri N, Parahitiyawa N, Jeyatilake S, Ranadheera S, Madhujith T (2015) Study on isolation of potentially probiotic *Lactobacillus* species from fermented rice. Trop Agric Res 26(3):428–440

Kanade P (2008) Probiotic dairy foods: present status and future potential in India. Indian Dairyman 60(3):86–92

Kannan D, Chelliah R, Rajamanickam EV, Venkatraman RS, Antony U (2015) Fermented batter characteristics in relation with the sensory properties of *idli*. Croatian J Food Technol Biotechnol Nutr 10(1–2):37–43

Kaw V, Mabesa LB (1987) Physico-chemical properties of *idli* batters from rice with varied amylose content. Phil J Crop Sci 12(2):106–109

Koh BK, Singh V (2009) Cooking behavior of rice and black gram in the preparation of *idli*, a traditional fermented product of Indian origin, by viscography. J Texture Stud 40(1):36–50

Kumar PP, Begum HV, Kumaravel S (2013) Mineral nutrients of '*pazhaya sadham*': a traditional fermented food of Tamil Nadu, India. Int J Nutr Metab 4(11):51–152

Kumar RS, Varman DR, Kanmani P, Yuvaraj N, Paari KA, Pattukumar V, Arul V (2010) Isolation, characterization and identification of a potential probiont from South Indian fermented foods (*Kallappam*, *Koozh* and *Mor Kuzhambu*) and its use as biopreservative. Probiot Antimicrob Proteins 2:145–151

Mulmule MD, Shimmy SM, Bambole V, Jamdar SN, Rawat KP, Sarma KSS (2017) Combination of electron beam irradiation and thermal treatment to enhance the shelf-life of traditional Indian fermented food (*Idli*). Radiat Phys Chem 131:95–99

Nagaraju VD, Manohar B (2000) Rheology and particle size changes during *idli* fermentation. J Food Eng 43(3):167–171

Nazni P, Shalini S (2010) Physical and nutritional evaluation of *idli* prepared from sorghum (*Sorghum bicolor I. moench*). Asian J Sci Technol 2:44–48

Nisha P, Ananthanarayan L, Singhal RS (2005) Effect of stabilizers on stabilization of *idli* (traditional South Indian food) batter during storage. Food Hydrocoll 19(2):179–186

Nout MR (2009) Rich nutrition from the poorest—cereal fermentations in Africa and Asia. Food Microbiol 26(7):685–692

Patel K, Guenther D, Wiebe K, Seburn RA (2013) Marginalized street food vendors promoting consumption of millets among the urban poor: a case study of millet porridge vendors in Madurai, Tamil Nadu, India. In: Food sovereignty: a critical dialogue, international conference at Yale University, pp 1–37

Rajendran C, Chander S (2014) Processing and microbiology of *idli*, Indian cereal-legume fermented food. Ph.D. thesis, University of Helsinki

Rao RE, Varadaraj MC, Vijayendra SVN (2005) Fermentation biotechnology of traditional foods of the Indian subcontinent. CRC Press, New Delhi

Reddy RNR, Sathe SK, Salunkhe DK (1982) Phytates in legumes and cereals. Adv Food Res 28:1–92

Rekha CR, Vijayalakshmi G (2011) Accelerated fermentation of *idli* batter using soy residue *okara*. J Food Sci Technol 48(3):329–334

Rhee SJ, Lee JE, Lee CH (2011) Importance of lactic acid bacteria in Asian fermented foods. Microb Cell Factories 10(1):1–13

Riat P, Sadana B (2009) Effect of fermentation on amino acid composition of cereal and pulse-based foods. J Food Sci Technol 46(3):247–250

Sakore DB, Dhole PT, Chavan KD, Pawar BK (2007) Role and viability of probiotic cultures in cow milk *dahi*. J Dairy Foods Home Sci 26:63–68

Sathe GB, Mandal S (2016) Fermented products of India and its implication: a review. Asian J Dairy Food Res 35(1):1–9

Sen CT (2015) Climate, crops and prehistory, in feasts and fasts—a history of food in India. Speaking Tiger Publishing Pvt. Ltd., New Delhi, pp 10–33

Shamala TR, Sreekantiah KR (1988) Microbiological and biochemical studies on traditional Indian palm wine fermentation. Food Microbiol 5(3):157–162

Sharma A, Kumari S, Nout MJR, Sarkar PK (2016) Minimization of antinutrients in *idli* by using response surface process optimization. J Food Process Preserv 41(5). https://doi.org/10.1111/jfpp.13099

Soni SK, Sandhu DK (1989) Fermentation of *idli*: Effects of changes in raw material and physico-chemical conditions. J Cereal Sci 10(3):227–238

Southworth F C (2014). Proto-Dravidian Agriculture. www.upenn.edu

Sridevi J, Halami PM, Vijayendra SVN (2010) Selection of starter cultures for *idli* batter fermentation and their effect on quality of *idlis*. J Food Sci Technol 47(5):557–563

Susheelamma NS, Rao MVL (1979) Functional role of the arabinogalactan QF black gram (*Phaseolus mungo*) in the texture of leavened foods (steamed puddings). J Food Sci 44(5):1310–1313

Tamang JP, Thapa N, Bhalla TC, Savitri (2016) Ethnic fermented foods of India. In: Tamang JP (ed) Ethnic fermented foods and alcoholic beverages of Asia. Springer, New Delhi, pp 17–72. 978-82-322-2798-4

Top Ten millet producing states of Tamilnadu, the Daily Record 2019. http://www.thedailyrecords.com/2018-2019-2020-2021/world-famous-top-10-list/world/largest-millets-producing-countries-world/12787/

Unika D, Jaffar AM (2014) A study on consumer behavior towards instant food products in Tamilnadu. Int J Business Admin Res Rev 3(5):65–69

Varnakulendran N, Ravimannan N, Navaneetha S, Shiyamala V (2016) Study on the fermented boiled Srilankan Red rice—scientific and traditional view. Int J Adv Ayurveda Naturopathy 1(1):1–5

Wickramasinghe P, Raja CS (2002) The Food of India. Murdoch Books Pvt. Ltd., London, p 296

Ethnic Fermented Foods and Beverages of Telangana and Andhra Pradesh

20

Ravindranadh Palika, Teena Dasi, Bharati Kulkarni, and Raghu Pullakhandam

Abstract

Food habits of different regions across the world are driven by agricultural systems. Cereal-legumes and milk are integral part of the routine meal in the states of Telangana and Andhra Pradesh. Traditionally fermented foods in these states are almost always prepared with fermented rice and black gram batter, and subtle differences in their preparation give rise to multiple recipes such as *idli*, *dosa*, *uttapam*, *vada*, etc. The millet flours (finger millet) or excess water collected after boiling the rice are also fermented and consumed, but their preparation and consumption are almost declining. In addition the fermented milk products such as dahi (locally referred as *perugu*) or curds and butter milk (locally referred as *majjiga* or *salla*) are consumed almost every day, particularly at the end of the meal. Toddy (locally referred as kallu), a fermented alcoholic beverage, made by fermentation of palm sap, is also popular, particularly among rural folks. A description of traditional methods of preparation, fermentation chemistry, nutritional value, and associated microbiome of fermented foods of Telangana and Andhra Pradesh was provided in the current chapter along with recent developments in their commercial use and value addition. Finally, a brief account of health benefits of fermented foods and underlying mechanisms has been presented.

Keywords

Fermented foods · Fermented alcoholic beverages · Probiotics · Telangana · Andhra Pradesh · Idli · Dosa · Vada

R. Palika · R. Pullakhandam (✉)
Biochemistry Division, National Institute of Nutrition, Indian Council of Medical Research, Jamai Osmania, Hyderabad, Telangana, India

T. Dasi · B. Kulkarni
Clinical Division, National Institute of Nutrition, Indian Council of Medical Research, Jamai Osmania, Hyderabad, Telangana, India

J. P. Tamang (ed.), *Ethnic Fermented Foods and Beverages of India: Science History and Culture*, https://doi.org/10.1007/978-981-15-1486-9_20

20.1 Introduction

Andhra Pradesh and Telangana are the constituent states of South India. Earlier Telangana is a part of Andhra Pradesh for almost six decades, but on 2 June 2014, it was separated through the Andhra Pradesh Reorganisation Act (Kumar et al. 2016). Hyderabad remains the common capital for both states for a period not less than 10 years from the date of bifurcation, and Amaravathi is being developed as a new capital city of Andhra Pradesh. In both states more than 85% are Hindus followed by Muslims and others. Geographical areas of the Andhra Pradesh and Telangana states are 160,205 and 112,077 km^2, respectively, with a total population of 4.97 and 3.52 crores, respectively (Census of India 2011). Climate in Telangana is usually hot, but in Andhra Pradesh it varies based on geographical region, with humid conditions among coastal regions.

The people of Telangana and Andhra Pradesh belong to a Dravidian ethnic group who speak Telugu as their mother tongue (Krishnamurti 1978; Majumder 2001). The economy of these two South Indian states is mainly supported by agriculture. Two important rivers of India, the Godavari and Krishna, flow through the Telangana state, while four rivers, the Godavari, Krishna, Penna, and Tungabhadra, flow through the Andhra Pradesh state, providing irrigation (Jain et al. 2009). Rice is the major staple food in both of these states and is also the major crop (Reddy 2011). Other important local crops are maize, peanuts, cotton, sugarcane, chili, mango, and tobacco. The pulses such as toor/arhar dal (pigeon pea), moong dal (green gram), and urad dal (black gram) are also sowed in large quantities and are widely used in routine cuisine as a source of protein apart from milk and its products *dahi*, butter (local name *venna*), and clarified butter (local name *neyyi* or *ghee*). In recent years crops used for vegetable oil production, such as sunflower and soybean, are being grown increasingly due to their profitability. Though there are subtle but unique differences among regional cuisines of different regions of Telangana and Andhra Pradesh, rice is the most predominant staple food, while millet-based breads or other preparations are also consumed widely. The routine meal in general in these two states consists of early morning breakfast (predominantly the fermented recipes as described below), and the typical lunch and dinner consists of steamed rice along with lentils, a variety of green leafy vegetables or vegetable curries (including chili, turmeric, garlic, ginger, onion, and spices), and *rasam* or *charu* (tamarind juice in water tampered with spices, tomatoes, onion, green chilies, and mustard) followed by fermented milk products such as *perugu* (*dahi*) or *majjiga* or *salla* (*butter milk*). There are also wide varieties of pickles prepared from seasonal produce such as *raw mango*, *tamarind*, *gongura*, etc. *Egg*, *chicken*, *fish*, and *meat* are also consumed widely but occasionally along with rice in the form of curries. The liberal use of *chili* and *tamarind* along with spices in curries gives them unique hot, spicy, and tangy taste which is now savored across Indian restaurants.

20.2 Ethnic Fermented Foods of Telangana and Andhra Pradesh

Andhra Pradesh and Telangana are the leading food grain-producing states as well as among the top three rice-producing states in India (Tiwari 2002). Possibly due to abundant availability, rice and lentils/dals are consumed routinely across all sections of population. Traditionally fermented foods in these states are almost always prepared with combination of rice/millets and/or de-hulled grains of black gram, but minor differences in their preparation and added condiments give rise to multiple recipes such as idli, *dosa*, *uttapam*, *vada*, etc. (DasMohapatra et al. 2017; Ray et al. 2016; Reddy et al. 1982). The millet flours or excess water collected after boiling the rice are also fermented and consumed, but their preparation and consumption are almost declining (DasMohapatra et al. 2017; Ray et al. 2016). In addition milk is fermented by back slopping (with previous day's curd) to make *dahi*, *butter milk*, etc. and consumed almost every day in many households. Due to tropical weather, palmyra trees (*Borassus flabellifer*) and silver date palms (*Phoenix sylvestris*) are found abundantly in these two states; the sap collected from these palms is fermented and consumed as an alcoholic beverage or palm wine (Kumari et al. 2016; Shamala and Sreekantiah 1988; Theivendirarajah and Chrystopher 1987). A description of traditional methods of their preparation is described below along with an account on fermentation chemistry, nutritional aspects, and microbiome involved.

20.3 Fermented Foods

20.3.1 *Idli*

Idli also known as "rice cake" is a popular traditional cereal-/legume-based fermented food of Telangana and Andhra Pradesh savored across South India. *Idli* is a favorite breakfast food, which has a white appearance, spongy texture, and appetizing taste and flavor (Table 20.1). The earliest references to *idli* were made in the Kannada writing of Shivakotiacharya (920 AD), Lokopakara (1025 AD), and in subsequent Sanskrit *Manasollasa* (1130 AD). It is postulated that modern *idli* making that includes the use of rice grits and urad dal and fermentation of the batter followed by steaming could have originated from Indonesia (Prasad 2017).

Traditional Method of Preparation: Traditionally, *idli* is prepared from the fermented batter of rice (*Oryza sativa*) and de-hulled black gram (*Phaseolus mungo*) (Ghosh and Chattopadhyay 2011; Mukherjee et al. 1965; Nisha et al. 2005; Purushothaman et al. 1993). The key ingredients used for preparing *idli* are black gram cotyledons (de-hulled beans) and rice typically at the ratio of 1:2 (Fig. 20.1). Both the raw materials are washed several times with running tap water to remove adhering dirt and soaked in water for 6 h, separately. After soaking, the rice is ground to a coarse batter, while the black gram beans are ground to a fine gelatinous paste with required amount of water (1.5–2 volumes of water/dry weight) in a kitchen mixer grinder or stone grinder (electric or manual). The batters are then

Table 20.1 Fermented foods of Telangana and Andhra Pradesh

Food	Substrate	Sensory property	Microbiome involved
Idli	Black gram and rice Rice may be replaced with idli rava or semolina	Mild-acidic, soft, moist, spongy; breakfast food	*L. mesenteroides, L. fermentum, L. lactis, S. faecalis, L. delbrueckii, and P. cerevisiae.* (Blandino et al. 2003; Ramakrishnan 1980; Ray et al. 2016; Saravanan et al. 2015; Tamang et al. 2016) *S. cerevisiae, D. hansenii, H. anomala, G. candidum, T pullulans, and Torulopsis* (Aidoo et al. 2006; Tamang and Fleet 2009)
Dosa	Black gram and rice	Thin, crispy pancake; shallow-fried, staple	*L. mesenteroides, L. fermentum, L. lactis, S. faecalis, L. delbrueckii, and P. cerevisiae.*(Blandino et al. 2003; Ray et al. 2016; Tamang et al. 2016) *S. cerevisiae, D. hansenii, H. anomala, G. candidum, T. pullulans, and Torulopsis* (Tamang 2012; Tamang and Fleet 2009)
Uttapam	Black gram and rice	Thick and spongy, mild sour taste	*Lactobacillus pentosus, L. plantarum, L. plantarum* sp., *and L. plantarum* sp. *from uttapam* (Ray et al. 2016, Saraniya and Jeevaratnam 2014)
Vada	Black gram and rice	Fried donut-shaped patties; crispy and spongy	*Pediococcus* sp., *Streptococcus* sp., *and Leuconostoc* sp. (Blandino et al. 2003; Ray et al. 2016)
Ambali	Finger millet flour	Acidic porridge	*L. mesenteroides, L. fermentum, and S. faecalis* (Blandino et al. 2003, Ray et al. 2016)
Taravani or kali	Rice	Acidic porridge	*Lactic acid bacteria like Lactobacillus bulgaricus, Lactobacillus casei, Pediococcus acidilactici, S. faecalis, Streptococcus thermophilus, Microbacterium flavum, and Saccharomyces* sp. (Ray et al. 2016; Tamang 2012)
Dahi	Cow/buffalo milk	Semisolid, sweet and sour taste	*L. bifermentans, L. alimentarius, L. paracasei, L. lactis, S. cremoris, S. lactis, S. thermophilus, L. bulgaricus, L. acidophilus, L. helveticus, L. cremoris, P. pentosaceus, P. acidilactici, W. cibaria, W. paramesenteroides, L. fermentum, L. delbrueckii* subsp. *indicus, Saccharomycopsis* sp., *Candida* sp. (Shangpliang et al. 2018a, b; Tamang et al. 2016)
Buttermilk	Butter extracted dahi	Acidic liquid	*L. bifermentans, L. alimentarius, L. paracasei, L. lactis, S. cremoris, S. lactis, S. thermophilus, L. bulgaricus, L. acidophilus, L. helveticus, L. cremoris, P. pentosaceus, P. acidilactici, W. cibaria, W. paramesenteroides, L. fermentum, L. delbrueckii* subsp. *indicus, Saccharomycopsis* sp., *Candida* sp. (Shangpliang et al. 2018a, b; Tamang et al. 2016)

(continued)

Table 20.1 (continued)

Food	Substrate	Sensory property	Microbiome involved
Toddy or kallu	Palm sap	White-colored liquid with sweet and sour taste	*S. cerevisiae*, *S. pombe*, *A. aceti*, *A. rancens*, *A. suboxydans*, *Leuconostoc dextranicum*, *Micrococcus* sp., *Pediococcus* sp., *Bacillus* sp., *and Sarcina* sp. (Shamala and Sreekantiah 1988; Tamang 2012)

De-hulled black gram cotyledons and rice (1 :2 ratio)

↓

Washed and soaked in water for 6 h, separately

↓

Ground to a course texture in stone grinder

↓

Mix the batters and ferment overnight

↓

Place the batter in idli pans and steam for 5–8 min

Fig. 20.1 Traditional method of idli preparation

combined and allowed to ferment overnight or longer depending on the season or temperature. After fermentation, salt and other ingredients like coriander and carrot were also added to the fermented batter to enhance flavor. The batter is then placed in *idli* pans, steamed for 5–8 min, and served hot. The key consumer preferences in idli are its coarseness and spongy texture. The black gram paste due to its gelatinous nature holds air generated during fermentation and thus increases the batter volume. Many households use different ratios of legume to cereal from 4:1 to 1:1, to attain the desired textural preference.

Culinary and mode of consumption: Traditionally steamed hot *idlies* are consumed as breakfast and served along with tampered ground nut or coconut chutney and *sambar*. *Sambar* is a soup of lentils in tamarind pulp seasoned with spices and vegetables.

Fermentation chemistry and nutritional value: Acidification and leavening are the two important changes that occur during fermentation of cereal/black gram blend used in preparation of *idli* (Balasubramanian and Viswanathan 2007; Desikachar et al. 1960; Mukherjee et al. 1965; Nisha et al. 2005; Reddy et al. 1982; Sandhu and Soni 1988; Soni and Sandhu 1989b). The sour taste of the fermented cereal/legume foods is mainly due to the production of lactic acid (Mukherjee et al. 1965; Reddy et al. 1982). The soft spongy texture and flavor observed in the leavened steamed *idli* is due to gas and acid produced by the microorganisms during fermentation. The fluffiness of the product depends on many factors like quality of raw ingredients, soaking time, grinding conditions, proportion of blending, fermentation time, and temperature (Durgadevi and Shetty 2012; Koh and Singh 2009; Nisha et al. 2005; Radhakrishnamurty et al. 1961; Reddy et al. 1982). Proteins and carbohydrates present in the black gram have the capacity to retain CO_2 released during fermentation, which results in soft and fluffy texture (Nisha et al. 2005). The porous structure is stabilized during steaming by arabinogalactan present in black gram (Susheelamma and Rao 1978). In addition, the organic compounds such as ketones, diols, and acids released during fermentation impart flavor to the food (Agrawal et al. 2000; Tamang 2012). Black gram has a major role in fermentation as it serves as a source of both microorganisms and fermenting substrate (Radhakrishnamurty et al. 1961). In most cases the microflora required for fermentation is present naturally on the surface of the raw ingredients, the black gram and rice. *L. mesenteroides* initiates fermentation at a wide range of salinity and temperatures, which results in the formation of lactic acid and CO_2 through anaerobic glycolysis, thus reducing the pH of the batter and providing optimal conditions for the growth of other high acid-producing bacteria and yeast (Mukherjee et al. 1965).

Early studies identified an increase in free sugar, nonprotein nitrogen, thiamine, folic acid, and riboflavin and a reduction in phytic acid content during fermentation of rice-black gram batter (Blandino et al. 2003; Ghosh and Chattopadhyay 2011; Reddy et al. 1982; Soni and Sandhu 1989b). In agreement with these results, the vitamin and protein status of experimental rats fed with fermented *idli* was also higher compared to unfermented *idli* (Joseph and Swanson 1994; Rajalakshmi and Vanaja 1967). The phytic acid and tannins are potent inhibitors of mineral absorption in cereal based staple diets, and their higher abundance in foods is the major etiological factor for the wide spread prevalence of anemia in India (Nair and Iyengar 2009). Indeed, reduction of phytate and tannin content has been reported during fermentation of *idli* batter (Reddy et al. 1982). Interestingly, bioavailability of iron and zinc was also found to be high from these batters (Hemalatha et al. 2007). Similarly, fermentation of finger millet flour was reported to reduce antinutrients and enhance the availability of protein and minerals (Antony and Chandra 1998). The combination of cereals/legumes during fermentation also appears to aid in improving overall protein quality (Campbell-Platt 1994). Cereals are a rich source of sulfur-containing amino acids, while black gram is a source of lysine; therefore their combination partly balances the protein quality. In addition, the digestibility of protein as measured by in vitro methods was also reported to increase due to fermentation (Padhye and Salunkhe 1978). Together these studies suggest that both combination of cereals/legumes and their fermentation improve the balance of nutrients; fermentation further aids in digestion and absorption of minerals. Though lactic acid bacteria are known to synthesize cobalamines (Burgess et al. 2009), the B_{12} content of these traditionally fermented foods has not been addressed as yet, which is particularly relevant in populations with high vegetarianism.

Microbiome: Though multiple bacteria and yeast are identified in rice-black gram batters, *Leuconostoc mesenteroides, L. fermentum, L. lactis, Streptococcus faecalis, L. delbrueckii, and Pediococcus cerevisiae* appear to be the predominant bacteria (Agaliya and Jeevaratnam 2013; Saravanan et al. 2015). Particularly, *Leuconostoc mesenteroides* and *Streptococcus faecalis* are identified as essential microbes involved in the leavening action of *idli* batter (Blandino et al. 2003; Mukherjee et al. 1965; Ramakrishnan 1980). In addition multiple yeast species (*Saccharomyces cerevisiae, Pediococcus cerevisiae, Debaryomyces hansenii, Hansenula anomala, Geotrichum candidum, Trichosporon pullulans, and Torulopsis*) were also identified in these batters (Aidoo et al. 2006; Tamang and Fleet 2009). Back slopping with previous day's batter and addition of *butter milk* or yeast are generally practiced to accelerate fermentation particularly during winter seasons.

Optimization and commercialization: A systematic study on textural characteristics of *idli* with different ratios of black gram and cereal blends showed that a ratio of 3:1.57 gives the desirable textural characteristics (Durgadevi and Shetty 2012). The coarseness of the rice batter gives idli a grainy texture with a good mouth feel; thus consistency in grinding is critical. To overcome this, commercially available *idli* rava (semolina of wheat or rice) is now being used instead of rice batter. In this case the soaked *idli* rava is mixed with black gram batter and fermented. In addition the unpolished rice has been shown to ferment better with higher microbial load compared to polished rice, but the sensory score of the latter was better (Kumar et al. 2005). Replacement of milled polished rice with either parboiled or extruded rice has also been demonstrated to improve the overall quality of *idli* (Mukherjee et al. 1965; Singh et al. 1995).

Instant *idli* mixes are widely available in the commercial market either in the dry form or in ready-to-cook batter form, but only the latter is preferred due to its better sensory and textural characteristics. To assist in industrialization, fermentation by defined starter cultures is also being researched that aid in accelerated and controlled fermentation times and consistency in textural and flavor profiles of the product (Agrawal et al. 2000; Sridevi et al. 2010). One of the challenges with *idli* batter from an industrial perspective is its stability; prolonged fermentation leads to collapse of batter with whey separation resulting in unacceptable changes in texture of the final product. The addition of 0.1% mustard oil, 7.5 ppm nicin/2000 ppm potassium sorbate, or 0.1% guar gum has been shown to improve the shelf life of *idli* batter for up to 5–10 days at room temperature and to a month under refrigeration (Nisha et al. 2005; Regubalan and Ananthanarayan 2018).

20.3.2 *Dosa* and *Uttapam*

Dosa is a crispy pancake and is popular in Telangana and Andhra Pradesh but now savored across India (Table 20.1). *Uttapam* is a variation of *dosa* prepared with the same batter except that it is a thick and spongy pancake as opposed to crispy *dosa*. The earliest references to *dosa* are mentioned in Sangam literature in 600 AD (Prasad 2017), which is made with only rice. But current-day traditional *dosa* is prepared with rice and urad dal batter, whose fermentation ingredients are very similar to *idli* batter.

Traditional method of preparation: *Dosa* batter preparation is similar to idli, except that the soaked rice and black gram are ground finely to get a smooth consistency (Fig. 20.2). The raw rice and black gram beans are washed, soaked, and ground separately with added water (1.5–2 v/v) to get a smooth consistency. The batters are then combined at 1:1 ratio and fermented overnight at room temperature (Sandhu and Soni 1988; Soni et al. 1985; Soni and Sandhu 1989a, b). The batter is then spread in thin layer over preheated hot plate spread with oil or clarified butter (ghee), resulting in a circular semisoft *dosa* or pancake. Multiple ingredients such as herbs, shredded carrots, boiled and tampered potatoes, tomato, onions, etc. are seasoned on top of the *dosa* and served hot along with either ground nut or coconut chutney or sambar (Fig. 20.2). The key consumer preference in *dosa* is its color and

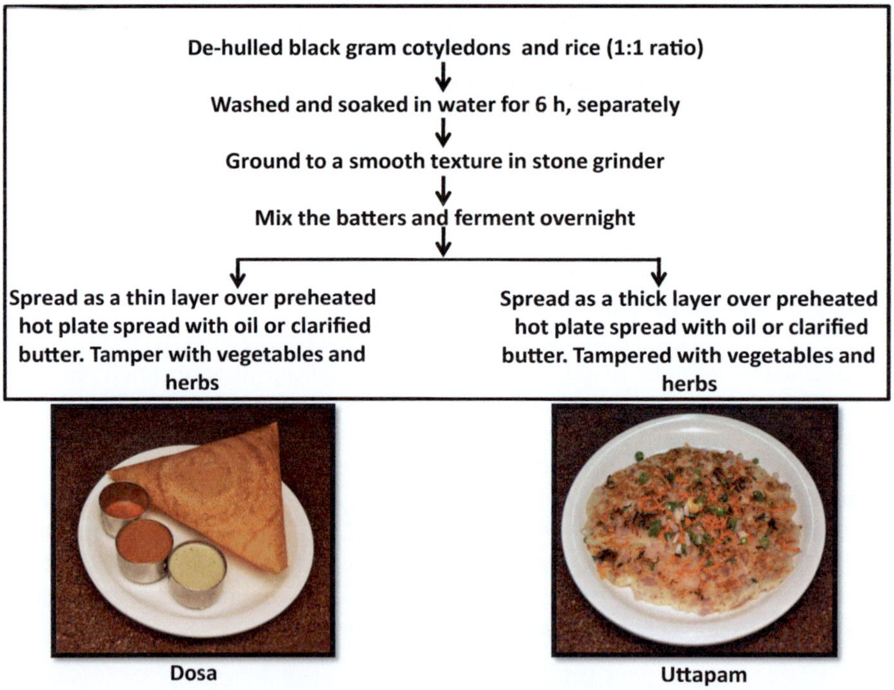

Fig. 20.2 Traditional method of *Dosa* and *Utappam* preparation

crispiness. Small quantities of fenugreek seeds or soaked chick pea dal are often added to the black gram dal during soaking that gives a characteristic yellow-brown color when it is roasted on a hot plate. Either flattened rice or steamed rice is often added during the batter preparation that gives the *dosa* additional crispiness. In addition, the particle size of the *dosa* batter is thinner compared to that of *idli* batter, which might potentially impact the overall fermentation process. Indeed, different grinding methods, and thus differences in particle size, have been demonstrated to modulate the viscosity of batter, which has a bearing the texture of the product (Koh and Singh 2009).

Culinary and mode of consumption: Traditionally hot *dosa or uttapam* is consumed as breakfast and occasionally as dinner. *Dosa* or *uttapam* is served along with tampered ground nut or coconut chutney or sambar. In addition, *dosa* is also tampered with a variety of vegetables including onions (onion *dosa*), upma (boiled semolina; upma *dosa*), or spiced potatoes (*masala dosa*); cottage cheese (*paneer dosa*) and eggs (*omelet dosa*) were also added, thus giving rise to many of its variations. *Dosa* is also made with battered whole green gram or mung bean, known as *pesarattu*, and served with chutney or *sambar*.

Fermentation chemistry and nutrition values: Since the ingredients and mode of *dosa* batter preparation are similar to that of *idli*, the fermentation chemistry and nutritive values are expected to remain similar to that of *idli* batter.

Microbiome: Microbiome associated with *idli* and *dosa* batters is similar, except that the prevalence of yeast in *dosa* batters is reported to be low (Sandhu and Soni 1988; Soni and Sandhu 1989a). The microbes associated with fermentation are LAB and yeast. *Lactobacillus pentosus*, *L. plantarum*, *L. plantarum* sp., and *L. plantarum* sp. are isolated from *uttapam* batter (Ray et al. 2016; Saraniya and Jeevaratnam 2014).

Optimization and commercialization: A systematic study indicated that the best combination of parameters for *dosa* batter fermentation in improving the nutritional quality is incubation at 28 °C, initial pH of 4.5, and supplementation of the batter with 1–2% sugar (Sandhu and Soni 1988). Similar to the ready-to-cook *idli* batters, *dosa* batters are also increasingly popular across India. The same preservation methods and starter cultures may also be used to improve the shelf life of *dosa* batter (Nisha et al. 2005; Regubalan and Ananthanarayan 2018).

20.3.3 *Vada*

Vada is a fried donut-shaped food that is crispy on the outside and soft on the inside prepared with a variety of lentils such as black gram, chick peas, mung bean, etc. (Table 20.1).

Traditional method of preparation: Traditionally black gram dal is soaked and ground to a course and thick paste (Fig. 20.3). The batter is used either immediately or after fermentation for a period of 5–6 h at room temperature. The batter made into a fine ball is flattened with a hole in the middle and fried by dropping it in hot

Black gram cotyledons
↓
Washed and soaked in water for 6 h
↓
Ground to a thick paste in stone grinder
↓
Ferment 6-8h
↓
Add rice flour or semolina, salt, herbs
↓
Deep fry in vegetable oil

Fig. 20.3 Traditional method of *Vada* preparation

vegetable oil. A small amount of rice flour or semolina is added to improve the texture and crispiness of *vada*.

Culinary and mode of consumption: *Vada* is typically served hot, along with chutneys. Alternatively, *vadas* are immersed in curds or *dahi* for 2–3 h (referred as *dahi vada*), tampered with spices and herbs, and served cool. *Dahi vada* is particularly preferred during hot summers.

Fermentation chemistry and nutritional values: Since the ingredients and mode of *dosa* batter preparation is similar to that of *Idli*, the fermentation chemistry and nutritive values are expected to remain similar. However, since *vadas* are deep fried, their fat content could be higher compared to *idli* and *dosa*.

Microbiome: The microbiome associated could be very similar to that of *idli* and *dosa*, and perhaps the microbial load could lower owing to its short fermentation time. The predominant microorganisms identified are *Pediococcus*, *Leuconostoc*, and *Streptococcus* (Das et al. 2012; Ray et al. 2016; Tamang 2012).

Optimization and commercialization: Ready-to-cook *vada* batters are becoming popular across India. Due to similar cereal/legume blends, fermentation of *vada* batter is very similar to that of *idli*; therefore, the same preservation methods and starter cultures may also apply to *vada* batter (Nisha et al. 2005; Regubalan and Ananthanarayan 2018).

20.3.4 *Ambali*

Ambali (referred locally as java) is a fermented beverage made with ragi flour (finger millet) and water, typically consumed as energy-rich drink (Table 20.1).

Traditional method of preparation: *Ragi* flour is mixed with water to obtain a smooth paste, cooked, and subjected to fermentation overnight. The next day the product is diluted with butter milk or water, seasoned with salt and herbs (Fig. 20.4). In some households, the fermented ragi flour is added to cooked rice and subjected to second fermentation.

Culinary and mode of consumption: The fermented beverage is consumed after seasoning with salt and other spices. In some households *butter milk* is added to the fermented beverage before consumption.

Fermentation chemistry and nutritional value: The fermentation of ragi flour is associated with gradual acidification over a period of 6–72 h (Antony and Chandra 1997). The total free sugars tended to decrease from 6 to 24 h, followed by an increase. It is postulated that increased microbial load during the same time could consume free sugars. However, free amino acid concentration increased gradually throughout the fermentation process, particularly after 24 h (Antony and Chandra 1997; Usha et al. 1996). Further increase in volume due to fermentation also suggests production of CO_2 (Ramakrishnan 1980). The increased biological value and net protein utilization and reduced long-chain fatty acid content have also been reported during fermentation of finger millet flour (Aliya and Geervani 1981; Usha et al. 1996). Fermentation is also reported to reduce the phytic acid content in finger

Mix Ragi flour and water in to thick slurry

↓

Cook on low flame until boiling

↓

Ferment overnight

↓

Add salt and dilute with butter milk or water

Fig. 20.4 Traditional method of *Ambali* preparation

millet flour along with the increase in acid extractable minerals which could enhance their bioavailability (Sripriya et al. 1997).

Microbiome: Microbiome is dominated by lactobacilli; *Leuconostoc mesenteroides*, *Lactobacillus fermentum*, and *Streptococcus faecalis* have been isolated from fermented ragi flour (Antony and Chandra 1997; Das et al. 2012).

Optimization and commercialization: There are no systematic optimization studies or commercial preparations of *Ambali* available to the best of the author's knowledge.

20.3.5 *Taravani* or *Kali*

Taravani or *kali* (also referred to as *Ganji* in Telangana) is a traditionally fermented rice beverage prepared in parts of Telangana and Andhra Pradesh but is almost extinct (Table 20.1).

Traditional method of preparation: Traditionally, the excess starchy water drained after boiling rice is collected into a mud pot and fermented for a couple of days.

Culinary and mode of consumption: The fermented beverage is consumed after seasoning with salt and other spices. Interestingly, portions of this fermented beverage are also added to the rice while cooking, which adds slight sour taste.

Fermentation chemistry and nutritional value: Acidification is the major change during fermentation of *taravani* or *kali.*

Microbiome: Microbiome associated with fermentation of rice-based fermented foods are lactic acid bacteria like *Lactobacillus bulgaricus*, *Lactobacillus casei*, *Pediococcus acidilactici*, *S. faecalis*, *Streptococcus thermophilus*, *Microbacterium flavum*, and *Saccharomyces* sp. (Ray et al. 2016; Tamang 2012).

Optimization and commercialization: This recipe is almost extinct now, and there are no equivalent commercial products available.

20.3.6 *Dahi* and *Butter Milk*

Dahi (*local name Perugu*) is a traditionally fermented milk product which is routinely consumed in both Telangana and Andhra Pradesh (Shangpliang et al. 2018a; Tamang et al. 2012) and is very similar to yogurt (Table 20.1). *Dahi* is prepared with cow or buffalo milk. *Butter milk* (*local name majjiga or salla*) is a by-product generated during the churning of butter from *dahi.*

Traditional method of preparation: Traditionally the boiled milk of cattle (cow or buffalo) is cooled and supplemented with a pinch of previous day's *dahi*, as a starter culture (back slopping), and left at room temperature for fermentation for 6–8 h, depending on the temperature (Fig. 20.5). The fermentation results in conversion of liquid milk into semisolid mass, which is then consumed directly or processed further.

Fig. 20.5 Traditional method of *Dahi and butter milk* preparation

Culinary and mode of consumption: Traditionally, *dahi* is consumed along with steamed rice or wheat-based flat bread (*chapatti*). The liquid fraction that is left over when fat/cream is churned out of *dahi* is known as *butter milk* and is consumed directly or seasoned with a variety of herbs (Table 20.1), salt and spices, or sugar, particularly during summer season (Ray et al. 2016; Tamang et al. 2012). The lower fat content and fewer calories compared with *dahi* make it a preferred beverage among health-conscious individuals.

Fermentation chemistry and nutritional values: The major components of milk are water, fat, proteins, lactose, and minerals apart from vitamins and enzymes (Longvah et al. 2017). Also milk is probably the sole source of animal protein among typical vegetarian populations in these states. The traditional fermentation of milk at households is predominantly achieved by back slopping with previous day's *dahi* (Shangpliang et al. 2018a). The biochemical transformation during milk fermentation is coagulation of proteins. The conversion of lactose to lactic acid by lactic acid bacteria (LAB) reduces the pH, leading to coagulation of milk protein, casein, into a semisolid mass. Apart from lactose digestion, a myriad of proteolytic enzymes also digest the milk proteins during fermentation leading to the release of peptides and free amino acids. Although there is no definitive evidence for improved protein quality of *dahi* compared to natural milk as yet, it has been reported that free amino acid content in fermented milk is higher than raw milk and denaturation itself might improve digestibility of proteins (Gurr 1987). In addition natural decrease in lactose due to fermentation renders fermented milk suitable for persons with lactose intolerance, but caution should be taken on the amount of lactose present (Gurr 1987). Studies in postmenopausal women demonstrated that calcium absorption was higher from fermented milk compared to natural milk (Narva et al. 2004). Though mechanisms of increased calcium absorption are not clear, redistribution of most of the minerals including calcium from sedimentable phospho-casein fraction to the soluble whey fraction has been reported in fermented milk and could account for improved absorption of calcium. Although there are no changes in overall fat-soluble vitamin content during fermentation of milk, B vitamins are reported to be reduced or unchanged in fermented milk (Alm 1982; Gurr 1987). The fact that heat treatment of milk may reduce the B vitamins, and few strains of bacteria is capable of de novo synthesis of these vitamins (Burgess et al. 2009), caution must be ensured in interpreting these results. The conjugated linolenic acid (CLA; isomers of C18:2 with conjugated double), a known anticarcinogenic and atheroprotective fatty acid, is exclusively present in dairy products (Ferlay et al. 2017). Interestingly, fermentation of milk with lactic acid bacteria is reported to increase the CLA content (Jiang et al. 1998; Kim and Liu 2002), which could promote health.

Microbiology: The predominant microorganism identified in traditionally fermented milk products belongs to lactobacilli (Agaliya and Jeevaratnam 2013; Patil et al. 2010; Shangpliang et al. 2018a; Tamang et al. 2016). A recent meta-genomic study (Shangpliang et al. 2018b) in fermented milk products identified microbes belonging to six families, *Streptococcaceae* (24.2%), *Lactobacillaceae* (16.8%), *Leuconostocaceae* (8.0%), *Staphylococcaceae* (6.8%), *Bacillaceae* (1.6%), and *Clostridiaceae* (1.3%), wherein *Lactococcus lactis* (19.7%), *Lactobacillus helveticus* (9.6%), and *Leuconostoc mesenteroides* (4.5%) were found to be the predominant microbes.

Optimization and commercialization: As *dahi* is regularly consumed across households, it is widely available commercially under different brands. It is typically packed airtight and sold under refrigeration. Attempts are also being made to supplement *dahi* with probiotic strains of lactobacilli, such as *Lactobacillus acidophilus*, *Lactobacillus casei*, and *Bifidobacterium bifidum* which are demonstrated to have antidiabetic and immunomodulatory activities in animal models (Shandilya et al. 2016; Sonal and Kansal 2009; Yadav et al. 2007, 2008).

20.4 Fermented Alcoholic Beverages

20.4.1 *Toddy* or *Kallu*

Toddy or *kallu* is a traditional alcoholic beverage of Telangana and Andhra Pradesh and is popular among rural folks (Table 20.1). There are two types of *toddy*, *thadi kallu* obtained from palmyra trees (*Borassus flabellifer*) and the *eetha kallu* extracted from silver date palms (*Phoenix sylvestris*). Both of these palm trees are considered as tree of life, as almost all parts of the tree are used for a variety of applications. For instance, their leaves are used for making huts and used as fuel, tender fruits (munjalu) are eaten, and ripened fruits with sweet and sour taste are consumed directly or used in making sweets, etc. (Chaurasiya et al. 2014; Das and Das 2003; Mohanty et al. 2018). In fact palm *toddy* collection is done by local men referred to as *toddy* tappers, for whom it is a traditional community occupation and a source of income (Latha 2016).

Traditional method of preparation: In traditional method, the *toddy* men climb the tree and make an angular cut at the edge of budding flower (spadix); the sap draining from the flower is then collected in mud clay pots hanged to the spadix with a rope (Fig. 20.6). The freshly collected oyster white-colored and translucent sap/*toddy* is called neera, which is sweet in taste. Neera when fermented is converted to effervescent, milky opaque color solution (Shamala and Sreekantiah 1988; Theivendirarajah and Chrystopher 1987). Fermentation is initiated traditionally by back slopping with a small amount of previous day's *toddy* left in clay pots (Shamala and Sreekantiah 1988). Sap that dripped into the pot is collected typically in the morning and evening and should be consumed within 5–10 h; otherwise it will develop a sour taste and unacceptable odor (Shamala and Sreekantiah 1988; Steinkraus 1983b).

Fermentation chemistry and nutritional value: The palm tree sap is rich in carbohydrates, particularly fructose and macro- and micronutrients, thus acting as an excellent media for microbial growth (Barh et al. 2008; Longvah et al. 2017). Natural fermentation begins as soon as the sap drip into the clay pot at ambient temperature by natural and residual yeasts in the collecting pot (Shamala and Sreekantiah 1988; Steinkraus 1983a, 1995). A small amount of fermented *toddy* was always left in the clay pot to initiate fermentation. Fermentation of sap was reported to occur in different stages (Chandrasekhar et al. 2012; Shamala and

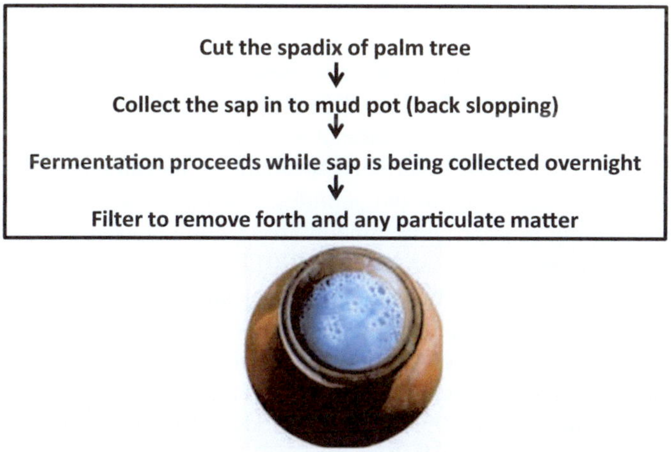

Fig. 20.6 Traditional method of *toddy* preparation

Sreekantiah 1988; Tamang 2012). During the first phase, acid-producing bacteria grow rapidly by utilizing carbohydrates present in the sap, leading to acidification. The acidic pH (4.5) favors the growth of yeast, leading to alcohol production. The third phase is dominated by acetic acid-producing bacteria. Within 24 h 50% of the total sugars is utilized for fermentation, and the alcohol concentration reaches 5–6% (v/v). If allowed further, the acetic acid concentration increases, which is used as vinegar (Shamala and Sreekantiah 1988; Theivendirarajah and Chrystopher 1987). The specific aroma and flavor of palm *toddy* is due to the presence of 3-isobutyl-2-methoxypyrazine (earthy), acetoin (buttery), ethyl hexanoate (fruity), and 2-acetyl-1-pyrroline (popcorn-like) (Lasekan and Abbas 2010). The difference in odor between sap and *toddy* is due to the absence of esters and alcohols in sap (Uzochukwu et al. 1994). The sugars present in the *toddy* appear to be almost in the fructose (5.7 g/100 g) form, and it has a balance of saturated (30 mg/100 g), mono-unsaturated (34 mg/100 g), and polyunsaturated (65 g/100 g) fatty acids (Longvah et al. 2017). Interestingly, *toddy* also provides alpha-linolenic acid (ALA; 20 mg/100 g), an essential omega-3 fatty acid, which has been reported to reduce the risk of cardiovascular disease (Longvah et al. 2017). The metabolic products of ALA, EPA and DHA, are also known to reduce inflammation and reduce depression (Lin and Su 2007; Simopoulos 2002). *Toddy* is also a source of vitamins such as B2 (0.27 mg/100 g), B3 (0.35 mg/100 g), B7 (1.96 μg/100 g), folic acid (0.73 μg/100 g), and vitamin C (0.92/100 g) (Longvah et al. 2017). The sodium (1.6 mg/100 g) content in *toddy* was also much less compared to its potassium (83 mg/100 g) content; indeed the ratio of K/Na is much higher in *toddy* water (51) compared to coconut water (7.6).

Microbiology: *Leuconostoc*, *Lactobacilli*, *Streptococci*, *Bacilli*, and *Enterobacter* are the acid-producing bacteria, and yeasts such as *Saccharomyces*, *Pichia*, *and Candida* are involved in alcoholic fermentation (Shamala and Sreekantiah 1988; Tamang 2012).

Optimization and commercialization: Since *toddy* is an alcoholic beverage, it is transported to government-notified centers for selling under the control of excise department.

Health benefits of fermented foods and underlying mechanisms: As described above the nutritive quality of fermented foods particularly that of legume /cereal-based food is increased by shear combination of nutrients from either source, and fermentation appears to assist in further improving the nutritional quality by partial digestion, release of amino acids and peptides, and reduction of anti-nutrient content and thus helping in improving mineral absorption (Chavan and Kadam 1989; Rajalakshmi and Vanaja 1967; Tamang et al. 2012). Therefore, consumption of fermented foods per se might improve nutritional status and thus general health. Although the effect of these typical cereal–/legume-based fermented foods have not been explored systematically, consumption of fermented milk products was identified to reduce the risk of breast cancer (van't Veer et al. 1989), modulate immune response (Link-Amster et al. 1994), reduce blood cholesterol levels (St-Onge et al. 2000), control hypertension (Beltran-Barrientos et al. 2018), aid in weight management, improve cardiovascular health, and contribute to a myriad of other health benefits (Marco et al. 2017) among human subjects.

It is now increasingly recognized that the health benefits of fermented foods are beyond nutritional and may involve multiple components contributed by both food and fermenting microbiome. The gut microbiome, which refers to the resident bacteria in the human gastrointestinal tract, has been recognized as an important mediator of individual's health and is subject to extensive research. The gut microbiome mainly consists of anaerobic bacteria, dominated by 2 phyla, *Bacteroidetes* and *Firmicutes*, with more than 1000 members at the species level. The general function of gut microbiome is to synthesize essential vitamins and amino acids and help in the digestion of nondigestible carbohydrates and thus supply energy to the host. In addition they also compete against invading harmful microorganisms by producing antimicrobial compounds, thus inhibiting its growth and colonization. Imbalance in normal gut microbiota has been linked with gastrointestinal diseases such as inflammatory bowel disease (IBD) and irritable bowel syndrome (IBS) and with systemic diseases such as obesity and type 2 diabetes, which are typically associated with inflammation.

It is now widely recognized that organic acids, prebiotics (exopolysaccharides), and probiotics (live bacteria) supplied from fermented foods mediate the beneficial effects of fermented foods via improving gut health (Bull and Plummer 2014; Ehrlich 2016; Marco et al. 2017). For instance, lactic acid and other organic acids are reported to act as anti-inflammatory compounds and antioxidants. Exopolysaccharides and proteins in fermented foods are known to inhibit the adhesion of pathogenic bacteria to the intestinal cells and contribute to the anti-inflammatory actions (Bull and Plummer 2014). Exopolysaccharides also act as prebiotics, giving rise to short-chain fatty acids during fermentation. In addition, the live bacteria, predominant lactobacilli, are known to survive through the digestion process to reach the colon (Ehrlich 2016; Liang et al. 2018; Marco et al. 2017), and consumption of these bacteria regularly might help maintaining the microbiome at balance and thus protecting health.

20.5 Conclusions

It is clear from the above description that fermented foods add variety, improves the digestibility, protein quality and essential vitamins and also reduces the potential antinutrients leading to improved nutritional quality. It is increasingly becoming clear that organic acids produced during fermentation also have health benefits. It is a fact that human gut microbiome is an integral part in sustaining health; hence, the microbiome associated with fermented foods could have a positive effect in protecting and balancing the gut microbiome. Also the exopolysaccharides present in fermented foods appear to have anti-inflammatory properties. Therefore, the cumulative benefits of fermented foods could be beyond their nutritional improvement, which needs to be researched systematically. Also it is interesting to note that commercialization of these foods with the use of defined starter cultures could reduce the microbial diversity and associated health benefits and thus should be dealt with caution.

References

Agaliya PJ, Jeevaratnam K (2013) Molecular characterization of lactobacilli isolated from fermented idli batter. Braz J Microbiol 44:1199–1206. https://doi.org/10.1590/S1517-83822013000400025

Agrawal R, Rati ER, Vijayendra SVN, Varadaraj MC, Prasad MS, Nand K (2000) Flavour profile of idli batter prepared from defined microbial starter cultures. World J Microbiol Biotechnol 16:687–690. https://doi.org/10.1023/A:1008939807778

Aidoo KE, Nout MJ, Sarkar PK (2006) Occurrence and function of yeasts in Asian indigenous fermented foods. FEMS Yeast Res 6:30–39. https://doi.org/10.1111/j.1567-1364.2005.00015.x

Aliya S, Geervani P (1981) An assessment of the protein quality and vitamin b content of commonly used fermented products of legumes and millets. J Sci Food Agric 32:837–842. https://doi.org/10.1002/jsfa.2740320814

Alm L (1982) Effect of fermentation on B-vitamin content of Milk in Sweden. J Dairy Sci 65:353–359. https://doi.org/10.3168/jds.S0022-0302(82)82199-1

Antony U, Chandra TS (1997) Microbial population and biochemical changes in fermenting finger millet (*Eleusine coracana*). World J Microbiol Biotechnol 13:533–537. https://doi.org/10.1023/A:1018561224777

Antony U, Chandra TS (1998) Antinutrient reduction and enhancement in protein, starch, and mineral availability in fermented flour of finger millet (*Eleusine coracana*). J Agric Food Chem 46:2578–2582

Balasubramanian S, Viswanathan R (2007) Properties of idli batter during its fermentation time. J Food Process Preserv 31:32–40. https://doi.org/10.1111/j.1745-4549.2007.00104.x

Barh D, Mazumdar BJRJ o M, Sciences M (2008) Comparative nutritive values of palm saps before and after their partial fermentation and effective use of wild date (Phoenix sylvestris Roxb.) sap in treatment of anemia. Res J Med Med Sci 3:173–176

Beltran-Barrientos LM, Gonzalez-Cordova AF, Hernandez-Mendoza A, Torres-Inguanzo EH, Astiazaran-Garcia H, Esparza-Romero J, Vallejo-Cordoba B (2018) Randomized double-blind controlled clinical trial of the blood pressure-lowering effect of fermented milk with Lactococcus lactis: a pilot study. J Dairy Sci 101:2819–2825. https://doi.org/10.3168/jds.2017-13189

Blandino A, Al-Aseeri ME, Pandiella SS, Cantero D, Webb C (2003) Cereal-based fermented foods and beverages. Food Res Int 36:527–543. https://doi.org/10.1016/S0963-9969(03)00009-7

Bull MJ, Plummer NT (2014) Part 1: The human gut microbiome in health and disease. Integr Med (Encinitas) 13:17–22

Burgess CM, Smid EJ, van Sinderen D (2009) Bacterial vitamin B2, B11 and B12 overproduction: an overview. Int J Food Microbiol 133:1–7. https://doi.org/10.1016/j.ijfoodmicro.2009.04.012

Campbell-Platt GJFRI (1994) Fermented foods—a world perspective. Food Res Int 27:253–257

Census of India (2011) Provisional population totals. Office of the Registrar General and Census Commissioner New Delhi

Chandrasekhar K, Sreevani S, Seshapani P, Pramodhakumari J (2012) A review on palm wine. Int J Res Biol Sci 2:33–38

Chaurasiya AK, Chakraborty I, Saha J (2014) Value addition of Palmyra palm and studies on the storage life. J Food Sci Technol 51:768–773. https://doi.org/10.1007/s13197-011-0561-3

Chavan JK, Kadam SS (1989) Nutritional improvement of cereals by fermentation. Crit Rev Food Sci Nutr 28:349–400. https://doi.org/10.1080/10408398909527507

Das BC, Das SN (2003) Cultivation of minor fruits. Kalyani Publishers, Ludhiana

Das A, Raychaudhuri U, Chakraborty R (2012) Cereal based functional food of Indian subcontinent: a review. J Food Sci Technol 49:665–672. https://doi.org/10.1007/s13197-011-0474-1

DasMohapatra PK, Parua S, Mitra B, Ghosh K, Mondal KC (2017) Rice-based Fermented Foods and Beverages Functional and Nutraceutical Properties. Fermented Foods, Part II. CRC Press, Boca Raton, pp 150–176

Desikachar H, RadhakrishnaMurty R, Rama Rao G, Kadkol SB, Srinivasan M, Subrahmanyan V (1960) Studies on idli fermentation: Part I—some accompanying changes in the batter. J Sci Ind Res 19:168–172

Durgadevi M, Shetty PH (2012) Effect of ingredients on texture profile of fermented food, idli. APCBEE Proc 2:190–198. https://doi.org/10.1016/j.apcbee.2012.06.034

Ehrlich SD (2016) The human gut microbiome impacts health and disease. C R Biol 339:319–323. https://doi.org/10.1016/j.crvi.2016.04.008

Ferlay A, Bernard L, Meynadier A, Malpuech-Brugere C (2017) Production of trans and conjugated fatty acids in dairy ruminants and their putative effects on human health: a review. Biochimie 141:107–120. https://doi.org/10.1016/j.biochi.2017.08.006

Ghosh D, Chattopadhyay P (2011) Preparation of idli batter, its properties and nutritional improvement during fermentation. J Food Sci Technol 48:610–615. https://doi.org/10.1007/s13197-010-0148-4

Gurr MI (1987) Nutritional aspects of fermented milk products. FEMS Microbiol Lett 46:337–342. https://doi.org/10.1111/j.1574-6968.1987.tb02470.x

Hemalatha S, Platel K, Srinivasan K (2007) Influence of germination and fermentation on bioaccessibility of zinc and iron from food grains. Eur J Clin Nutr 61:342–348. https://doi.org/10.1038/sj.ejcn.1602524

Jain AK, Muralikrishna Rao BM, Rama Mohan Rao MS, Venkata Swamy M (2009) Groundwater scenario in Andhra Pradesh. WASHCost—CESS Working Paper No. 3 WASHCost (India) Project Centre for Economic and Social Studies N. O. Campus, Begumpet, Hyderabad

Jiang J, Bjorck L, Fonden R (1998) Production of conjugated linoleic acid by dairy starter cultures. J Appl Microbiol 85:95–102

Joseph E, Swanson BG (1994) Protein quality of "idli", fermented steamed cakes prepared from beans (*Phaseolus vulgaris*) and rice. Nutr Res 14:553–568. https://doi.org/10.1016/s0271-5317(05)80219-2

Kim YJ, Liu RH (2002) Increase of conjugated linoleic acid content in milk by fermentation with lactic acid bacteria. J Food Sci 67:1731–1737. https://doi.org/10.1111/j.1365-2621.2002.tb08714.x

Koh BK, Singh V (2009) Cooking behavior of rice and black gram in the preparation of idli, a traditional fermented product of indian origin, by viscography. J Texture Stud 40:36–50. https://doi.org/10.1111/j.1745-4603.2008.00168.x

Krishnamurti BH (1978) Language planning and development: the case of Telugu. Contrib Asian Stud XI:37–56

Kumar RP, Prabhakaran P, George JK, Parambath SG (2016) Mapping regional disparities in human development—the case of erstwhile Andhra Pradesh. Proc Technol 24:1843–1850. https://doi.org/10.1016/j.protcy.2016.05.233

Kumar CS, Rao ER, Prakash J (2005) Effect of varietal differences and polishing of rice on quality parameters of idli. J Sensory Stud 20:397–409. https://doi.org/10.1111/j.1745-459X.2005.00035.x

Kumari A, Pandey A, Ann A, Raj A, Gupta A, Chauhan A, Sharma A, Das AJ, Kumar A, Attri B (2016) In: Indigenous alcoholic beverages of South Asia. CRC Press, New York. pp 501–66

Lasekan O, Abbas KA (2010) Flavour chemistry of palm toddy and palm juice: a review. Trends Food Sci Technol 21:494–501. https://doi.org/10.1016/j.tifs.2010.07.007

Latha TM (2016) Role of Bommagani Dharma Biksham in Toddy Tappers movement. Int J Res Econ Soc Sci 6:336–349

Liang D, Leung RK, Guan W, Au WW (2018) Involvement of gut microbiome in human health and disease: brief overview, knowledge gaps and research opportunities. Gut Pathog 10:3. https://doi.org/10.1186/s13099-018-0230-4

Lin PY, Su KP (2007) A meta-analytic review of double-blind, placebo-controlled trials of antidepressant efficacy of omega-3 fatty acids. J Clin Psychiatry 68:1056–1061

Link-Amster H, Rochat F, Saudan K, Mignot O, Aeschlimann JJFI, Microbiology M (1994) Modulation of a specific humoral immune response and changes in intestinal flora mediated through fermented milk intake. FEMS Immunol Med Microbiol 10:55–63

Longvah T, Anantan I, Bhaskarachary K, Venkaiah K (2017) Indian food composition tables. National Institute of Nutrition, Indian Council of Medical Research, Hyderabad

Majumder PP (2001) Ethnic populations of India as seen from an evolutionary perspective. J Biosci 26:533–545. https://doi.org/10.1007/Bf02704750

Marco ML, Heeney D, Binda S, Cifelli CJ, Cotter PD, Foligne B, Ganzle M, Kort R, Pasin G, Pihlanto A, Smid EJ, Hutkins R (2017) Health benefits of fermented foods: microbiota and beyond. Curr Opin Biotechnol 44:94–102. https://doi.org/10.1016/j.copbio.2016.11.010

Mohanty S, Mishra S, Pradhan R (2018) Optimization and storage studies of palm (*Borassus flabellifer*) ready-to-serve (RTS) juice. In: Second international conference on food quality, safety and security (FOOD QUALSS 2018), Colombo, SriLanka, 25–26 October, 2018

Mukherjee S, Albury M, Pederson C, Van Veen A, Steinkraus KJAM (1965) Role of Leuconostoc mesenteroides in leavening the batter of idli, a fermented food of India. Appl Microbiol 13:227–231

Mukherjee SK, Albury MN, Pederson CS, Vanveen AG, Steinkraus KH (1965) Role of leuconostoc mesenteroides in leavening the batter of idli, a fermented food of India. Appl Microbiol 13:227–231

Nair KM, Iyengar V (2009) Iron content, bioavailability & factors affecting iron status of Indians. Indian J Med Res 130:634–645

Narva M, Nevala R, Poussa T, Korpela R (2004) The effect of Lactobacillus helveticus fermented milk on acute changes in calcium metabolism in postmenopausal women. Eur J Nutr 43:61–68. https://doi.org/10.1007/s00394-004-0441-y

Nisha P, Ananthanarayan L, Rekha RS (2005) Effect of stabilizers on stabilization of idli (traditional south Indian food) batter during storage. Food Hydrocolloids 19:179–186

Nisha P, Ananthanarayan L, Singhal RS (2005) Effect of stabilizers on stabilization of idli (traditional south Indian food) batter during storage. Food Hydrocolloids 19:179–186. https://doi.org/10.1016/j.foodhyd.2004.03.007

Padhye V, Salunkhe DK (1978) Biochemical studies on black gram (*Phaseolus mungo* L.) 111. Fermentation of the black gram and rice blend and its influence on the in vitro digestibility of the proteins. J Food Biochem 2:327–347

Patil MM, Pal A, Anand T, Ramana KV (2010) Isolation and characterization of lactic acid bacteria from curd and cucumber. Indian J Biotechnol 9:166–172

Prasad G (2017) Idli, dosai, sambar, coffee: consuming tamil identity. The English paradigm in India. Springer, Singapore, pp 91–100

Purushothaman, D., Dhanapal, N., Rangaswami, G (1993) Indian idli, dosa, dhokla, khaman, and related fermentations. In: Steinkraus KH (ed) Handbook of indigenous fermented foods. Marcel Dekker, New York, pp 149–165

Radhakrishnamurty R, Desikachar HS, Srinivasan M, Subrahmanyan V (1961) Studies on Idli fermentation. II. Relative participation of black gram flour & rice semolina in the fermentation. J Sci Ind Res 20C:342–345

Rajalakshmi R, Vanaja K (1967) Chemical and biological evaluation of the effects of fermentation on the nutritive value of foods prepared from rice and grams. Br J Nutr 21:467–473

Ramakrishnan C (1980) Studies on Indian fermented foods. Baroda J Nutr 6:57

Ray M, Ghosh K, Singh S, Chandra Mondal K (2016) Folk to functional: an explorative overview of rice-based fermented foods and beverages in India. J Ethnic Foods 3:5–18. https://doi.org/10.1016/j.jef.2016.02.002

Reddy AA (2011) Dynamics of the agricultural economy of Andhra Pradesh, India since the last five decades. J Dev Agric Econ 3:394–410

Reddy NR, Pierson MD, Sathe SK, Salunkhe DK (1982) Legume-based fermented foods: their preparation and nutritional quality. Crit Rev. Food Sci Nutr 17:335–370. https://doi.org/10.1080/10408398209527353

Reddy NR, Sathe SK, Pierson MD, Salunkhe DK (1982) Idli, an indian fermented food: a review. J Food Qual 5:89–101. https://doi.org/10.1111/j.1745-4557.1982.tb00736.x

Regubalan B, Ananthanarayan L (2018) Shelf life improvement of idli batter by addition of mustard essential oil as bio-preservative. J Food Sci Technol Mysore 55:3417–3426. https://doi.org/10.1007/s13197-018-3247-2

Sandhu DK, Soni SK (1988) Optimization of physicochemical parameters for Indian dosa batter fermentation. Biotechnol Lett 10:277–282. https://doi.org/10.1007/bf01024419

Saraniya A, Jeevaratnam K (2014) Purification and mode of action of antilisterial bacteriocin produced by lactobacillus pentosus SJ65 isolated from uttapam batter. J Food Biochem 38:612–619

Saravanan C, Gopu V, Shetty PH (2015) Diversity and functional characterization of microflora isolated from traditional fermented food idli. J Food Sci Technol Mysore 52:7425–7432. https://doi.org/10.1007/s13197-015-1791-6

Shamala T, Sreekantiah KJFM (1988) Microbiological and biochemical studies on traditional Indian palm wine fermentation. Food Microbiol 5:157–162

Shandilya UK, Sharma A, Kapila R, Kansal VK (2016) Probiotic Dahi containing Lactobacillus acidophilus and Bifidobacterium bifidum modulates immunoglobulin levels and cytokines expression in whey proteins sensitised mice. J Sci Food Agric 96:3180–3187. https://doi.org/10.1002/jsfa.7497

Shangpliang HNJ, Rai R, Keisam S, Jeyaram K, Tamang JP (2018a) Bacterial community in naturally fermented milk products of Arunachal Pradesh and Sikkim of India analysed by high-throughput amplicon sequencing. Sci Rep 8:ARTN 1532. https://doi.org/10.1038/s41598-018-19524-6

Shangpliang HNJ, Rai R, Keisam S, Jeyaram K, Tamang JP (2018b) Bacterial community in naturally fermented milk products of Arunachal Pradesh and Sikkim of India analysed by high-throughput amplicon sequencing. Sci Rep 8:1532. https://doi.org/10.1038/s41598-018-19524-6

Simopoulos AP (2002) Omega-3 fatty acids in inflammation and autoimmune diseases. J Am Coll Nutr 21:495–505

Singh N, Bawa AS, Sekhon KS (1995) Quality Improvement of Idli Using Extruded Rice Flour. J Food Qual 18:193–202. https://doi.org/10.1111/j.1745-4557.1995.tb00374.x

Sonal R, Kansal VKJM (2009) Probiotic Dahi containing Lactobacillus acidophilus and Bifidobacterium bifidum stimulates immune system in mice. Milchwissenschaft 64:147–150

Soni SK, Sandhu DK (1989a) Nutritional improvement of Indian dosa batters by yeast enrichment and black gram replacement. J Ferment Bioeng 68:52–55. https://doi.org/10.1016/0922-338x(89)90214-6

Soni SK, Sandhu DK (1989b) Fermentation of idli: effects of changes in raw material and physico-chemical conditions. J Cereal Sci 10:227–238. https://doi.org/10.1016/s0733-5210(89)80052-9

Soni S, Sandhu D, Vilkhu KJFM (1985) Studies on dosa—an indigenous Indian fermented food: some biochemical changes accompanying fermentation. Food Microbiol 2:175–181

Sridevi J, Halami PM, Vijayendra SV (2010) Selection of starter cultures for idli batter fermentation and their effect on quality of idlis. J Food Sci Technol 47:557–563. https://doi.org/10.1007/s13197-010-0101-6

Sripriya G, Antony U, Chandra TS (1997) Changes in carbohydrate, free amino acids, organic acids, phytate and HCl extractability of minerals during germination and fermentation of finger millet (*Eleusine coracana*). Food Chem 58:345–350. https://doi.org/10.1016/S0308-8146(96)00206-3

Steinkraus KH (1983a) Fermented foods, feeds and beverages. Biotechnol Adv 1:31–46

Steinkraus KHMS (1983b) Handbook of indegenous fermented foods. Microbiology series. Marcel Dekker, New York, pp 315–328

Steinkraus K (1995) Handbook of indigenous fermented foods, revised and expanded. CRC Press, Boca Raton

St-Onge MP, Farnworth ER, Jones PJ (2000) Consumption of fermented and nonfermented dairy products: effects on cholesterol concentrations and metabolism. Am J Clin Nutr 71:674–681. https://doi.org/10.1093/ajcn/71.3.674

Susheelamma NS, Rao MVL (1978) Isolation and characterization of arabino-galactan from black gram (*Phaseolus mungo*). J Agric Food Chem 26:1434–1437. https://doi.org/10.1021/jf60220a048

Tamang JP (2012) Plant-based fermented foods and beverages of Asia. Handbook of plant-based fermented food and beverage technology. CRC Press, Boca Raton, pp 49–92

Tamang JP, Fleet GH (2009) Yeasts diversity in fermented foods and beverages. Yeast biotechnology: diversity and applications. Springer, Dordrecht, pp 169–198

Tamang JP, Tamang N, Thapa S, Dewan S, Tamang B, Yonzan H, Rai AK, Chettri R, Chakrabarty J, Kharel N (2012) Microorganisms and Nutritional value of Ethnic fermented foods and alcoholic beverages of North East India. Indian J Tradit Knowl 11:7–25

Tamang JP, Watanabe K, Holzapfel WH (2016) Review: diversity of microorganisms in global fermented foods and beverages. Front Microbiol 7:377. https://doi.org/10.3389/fmicb.2016.00377

Theivendirarajah K, Chrystopher RK (1987) Microflora and microbial activity in palmyrah (*Borassus flabellifer*) palm wine in Sri Lanka. MIRCEN J Appl Microbiol Biotechnol 3:23–31

Tiwari KJBCI (2002) Rice production and nutrient management in India. Better Crops Int 16:18–22

Usha A, Sripriya G, Chandra TS (1996) Effect of fermentation on the primary nutrients in finger millet (*Eleusine coracana*). J Agric Food Chem 44:2616–2618. https://doi.org/10.1021/jf950787q

Uzochukwu SVA, Balogh E, Tucknot OG, Lewis MJ, Ngoddy PO (1994) Volatile constituents of palm wine and palm sap. J Sci Food Agric 64:405–411. https://doi.org/10.1002/jsfa.2740640403

van't Veer P, Dekker JM, Lamers JW, Kok FJ, Schouten EG, Brants HA, Sturmans F, Hermus RJ (1989) Consumption of fermented milk products and breast cancer: a case-control study in The Netherlands. Cancer Res 49:4020–4023

Yadav H, Jain S, Sinha PR (2007) Antidiabetic effect of probiotic dahi containing Lactobacillus acidophilus and Lactobacillus casei in high fructose fed rats. Nutrition 23:62–68. https://doi.org/10.1016/j.nut.2006.09.002

Yadav H, Jain S, Sinha PR (2008) Oral administration of dahi containing probiotic Lactobacillus acidophilus and Lactobacillus casei delayed the progression of streptozotocin-induced diabetes in rats. J Dairy Res 75:189–195. https://doi.org/10.1017/S0022029908003129

21 Ethnic Fermented Foods and Beverages of Tripura

Ranendra Kumar Majumdar

Abstract

Tripura, the third smallest state of the country, landlocked and characterized by hill ranges, valleys and plains. The state is a homeland of 19 different ethnic groups like *Tripuri*, *Jamatia*, *Reang*, *Noatia*, *Lusai*, *Uchoi*, *Chaimal*, *Halam*, *Kukis*, *Garos*, *Mog* and *Chakma*. Other tribes such as *Bill*, *Munda*, *Orang*, *Santhal*, *Lepcha*, *Khasias* and *Bhutias* are the immigrant tribes came and settled here for economic reasons. Several indigenous technologies especially in the fields of agriculture, food, medicine as well as natural resources management are prevailing in this region from ancient time. Since time immemorial fish are preserved in a very unique method to cater the need during the period when harvesting is not possible. Since *Sheedal* is a salt-free product, it is assumed that it originated long before people knew about the use of salt. Six microbial strains from each of coagulase-negative *Staphylococcus* and lactic acid bacteria (LAB) were found to be the dominant genus in *Sheedal*. There is a salt-fermented fish product known as *Lona ilish*, originated long before partition of India. The traditional knowledge in making of alcoholic beverages uses rice as fermenting medium and local herbs as a source of amylolytic and alcohol-producing yeasts. Using vegetables, namely, bamboo shoot, amla, elephant foot yam, etc. for producing fermented and semi-fermented foods like *Moiya Koshak*, *Melye Amiley*, *Midukeye* and *Moiya Pangsung*, etc. is considered to be an age-old practice by the rural women of the state. The method of production of ethnic fermented foods and beverages of the Tripura is very crude and unhygienic, and no Good Manufacturing Practices are followed at all.

Keywords

Tripura · Fermented foods · Tripura tribals, *Sheedal* · Chuwak bwtwk · Chuwarak

R. K. Majumdar (✉)
College of Fisheries, Central Agricultural University (Imphal), Agartala, Tripura, India

J. P. Tamang (ed.), *Ethnic Fermented Foods and Beverages of India: Science History and Culture*, https://doi.org/10.1007/978-981-15-1486-9_21

21.1 Introduction

Tripura, the third smallest state of the country, was a princely state during British rule in India. Historically, the region was under the rule of the Twipra Kingdom for centuries. The earliest historical records about the Twipra kingdom are apprehended to be in the thirteenth century, when it first came under pressure from the Islamic conquests in India (https://en.wikipedia.org/wiki/ Twipra_Kingdom). The boundaries of the kingdom changed over the centuries, and it reached to south to the Sundarban area on the Bay of Bengal, east to Burma and north up to the Kamarupa kingdom in Assam. From thirteenth century onward, several Muslim invasions occurred in the plains of the kingdom, although their rule never extended to the hill regions. But the Mughals had influence over the appointment of the *Tripuri* kings. In British India, the state had a plain area (Tippera district or Chakla Roshnabad), presently the Comilla district of Bangladesh and a hilly area known as Hill Tippera, the present-day state of Tripura. Following the independence of India in 1947, the Tippera district became a part of erstwhile East Pakistan and Hill Tippera district remained under a Regency Council until 1949. The Tripura Merger Agreement was signed on 9 September 1949 between Union of India and the Maharani Regent of Tripura. In November 1956, the Tripura became a Union Territory without a legislature and finally got statehood in 21 January 1972. During partition of India and especially after 1949, Hindu Bengalis migrated to Tripura to escape religious persecution in Muslim majority East Pakistan.

Tripura is a landlocked state spread over 10,491.69 km^2 and extends from 22°56′N to 24°32′N and 91°09′E to 92°20′E (https://en.wikipedia.org/wiki/ Tripura). The state is bordered by the country of Bangladesh to the west, north and south; the state of Assam to the northeast; and Mizoram to the east. Topographically, the state is characterized by hill ranges, valleys and plains. There are isolated hillocks interspersed throughout the state, which are known as *Tillahs* and the narrow fertile alluvial valleys are called *Lungas*. Soil in the hilly regions are of red laterite type and porous. The valleys and plains that constitute most of the agriculture land are overlaid by alluvial soil. A number of rivers originate in the hills of Tripura and flow into Bangladesh. The state's climate is 'tropical savanna' type with a lengthy wet season and a relatively short dry season. The four main seasons are winter, from December to February; pre-monsoon or summer, from March to April; monsoon, from May to September; and post-monsoon, from October to November. During the monsoon season, the southwest monsoon brings heavy rains, which cause frequent floods. The average annual rainfall is around 2500 mm. During winter, temperature range from 11 to 27 °C, while in summer they fall between 24 and 37 °C.

According to the Census 2011, the population of Tripura was 36.74 lakhs (https://www.census2011.co.in/census/state/tripura.html) and projected to be 41.29 lakhs in 2018 (showing a 12% increase), which constitute about 0.3% of the India's population. Tripura is an agrarian state with more than half of the population dependent on agriculture and allied activities. However, due to hilly terrain and forest cover, only 27% of the land is available for cultivation. Rice, the major crop of the state, is cultivated in 91% of the cropped area. Freshwater fish and prawn culture have made significant advances in the state. A unique method of cultivation, known as *Jhum* or shifting cultivation, is very much popular in which indigenous people are adapted to live and survive with the forest. Such primitive method of cultivation ensures production of a range of ethnic foods which are nutritious and compatible to the civilization. However, the dependency of people on the *Jhum* type of cultivation has declined over the years.

Tripura, a land of vast natural resources, inhabited by people of different religion and cultures. The state is a homeland of 19 different ethnic groups like *Tripuri*, *Jamatia*, *Reang*, *Noatia*, *Lusai*, *Uchoi*, *Chaimal*, *Halam*, *Kukis*, *Garos*, *Mog* and *Chakma*. Other tribes such as *Bill*, *Munda*, *Orang*, *Santhal*, *Lepcha*, *Khasias* and *Bhutias* are the immigrant tribes came and settled here for economic reasons (Uchoi et al. 2015). The *Kokborok*-speaking people of Tripura are the major group amongst 19 tribes and many subtribes, whereas the Bengali-speaking people form the ethnolinguistic majority in Tripura (Debbarma 2002).

The own distinct culture of each ethnolinguistic group confers them a unique cultural identity. Rich culture and tradition of the state is manifested by numerous art forms and festivals, and these are considered to be the integral part of each ethnic group. Historically, the northeast Indian tribals are largely associated with different ethnic groups such as Indo-Mongoloids, Tibeto-Burmese and Proto-Australoids, representing the Asio-Austric culture on the Indian mountainous regions. Although the state is having its own food culture, it is different from rest of the country especially in its taste and flavours. In general, original population of the state is non-vegetarian and having a liking for fermented fish. The rural ethnic population depends substantially on nature for their survival. They hold a strong sense of traditional knowledge for natural resources and environment.

There are many age-old indigenous knowledge-based technologies available for survival of the population in the forest-dominated Tripura. In ancient times, rural women of this region selected several wild plants for food and medicinal purposes simply through trial and error method. To accomplish the demand of food for nutritional security, various fermented and non-fermented foods have been used in different combinations along with traditional vegetables grown in this region. The traditional foods which are usually consumed by tribes of the state are intimately associated with their sociocultural, spiritual life and health. Nevertheless, over the past 30 years, the new generation of each community is going to be accustomed with the modern diets due to their higher education and income leading towards a materialistic life. Intervention of modern crop varieties which are found to be more economic than that of traditional ones and current trends towards use of various processed foods of convenient in nature are also considered as factors responsible for switching over to the modern diets.

21.2 Ethnic Fish Products of Tripura

Since time immemorial fish are preserved in a very unique method to cater the need during the period when harvesting is not possible and also to store this in-house for long time. In ancient times, the tribal preferred to build their homes at hills to avoid water logging and flood situation in the plains. Therefore, most of the hill dwellers used to come down to the plain once or twice in a week (especially in market days) to sell their produce and to buy food and other domestic materials. The details of the ethnic fish products of Tripura are discussed below.

21.2.1 *Sheedal*

Sheedal is a fermented fish product indigenous to the Tripura as well as northeast sector of India. It has several local names, viz. *seedal*, *sepaa*, *hidal* and *verma*. Although Assam also produce *Sheedal*, Tripura is considered as one of the major producers of *Sheedal* and caters the need of almost all the northeast states. The traditional *Sheedal* technology is very old and assumed to have originated in the former undivided India (now Bangladesh). Unlike other salt-fermented fish products available in Southeast Asia, the salt-free *Sheedal* is believed to have developed long before the British Era in northeastern states of India, i.e. before 1824. According to the Indian history, the people of this region were not aware of the use of salt before the British Era. And even after that, the salt was treated to be a highly valued and scarce commodity. In that time, people were accustomed to use ash from banana or papaya plant, known as *khar*, as an alternative of salt. Thus, it was not possible by the poor fishermen of this region to afford spending salt in fish preservation, and this might be one of the reasons for development of *Sheedal* technology where fish is preserved without salt. This is a unique way of fish preservation and completely different from the methods followed in Southeast Asian countries.

Sheedal, the unique technology of preservation of fish without salt, assumed to be developed due to some social compulsion in the past. Once upon a time, the plains of the region (a part is now in Bangladesh) and adjoining northeastern part of India were famous for a kind of shallow water bodies known as *beel fisheries*, which were usually weed-infested and get fully or partially dried up during winter. *Beel fisheries* provide an excellent habitat for variety of weed fish especially *Puntius* spp., 'Darkina' (*Esomus danricus*) and 'Mola' (*Amblypharyngodon mola*). Such weed fish breed naturally with the starting of rainy season and form a good fishery during declining phase of the water level in post-monsoon. Therefore, huge harvesting of such weed fishes during post-monsoon period used to pose a great problem for their marketing, as due to lack of icing facility and good road communications, such perishable commodity could not be possible to be transported to the neighbouring markets in fresh condition where these could fetch a higher price compared to the local markets. Those circumstances compelled the fishermen for alternative method of their preservation, so that they could consume and sell these fish in the lean period (November to April), especially during scarcity of fresh fish in the markets. Drying of fish under sun was the only method known to preserve fish. Northeast India, being a highest rainfall zone, does not provide a congenial environment for simple sun drying of fish. There was frequent rainfall and high humidity of the atmosphere during the peak-harvesting seasons (i.e. from May to September). Therefore, the prevailing situations compelled the fishermen to ferment their produce (mostly *Puntius* spp.), and it was found ideal to cater their needs. Moreover, the taste and food habits of the population of the region those who are basically rice eaters prefer to add some strong flavour-bearing products with their tasteless rice. In traditional technique, usually, no additives/ preservatives or any starter culture are added during the processing steps.

The popularity of *Sheedal* is its strong and attractive flavour. By appearance, the best quality product is a whole raw material fish, bilaterally compressed with little disintegration near belly and caudal portion and with slight sticky surface. But in inferior quality product, the shape of the fish is completely or partially lost. The best quality product is dull white (very often glossy) in colour that gradually changes towards slight brownish to deep brownish/dark on continuous exposure to air. The strong odour permeates the air in and around the storage and gives the area a characteristic smell of *Sheedal*. The quality deteriorates very fast after breaking of seal of the container and exposure to air. Until few years back, *Sheedal* used to be prepared exclusively from *Puntius* sp. only (usually from *Puntius sophore*). But the high retail price of *Punti Sheedal* (due to higher cost of dry or wet Punti fish)

Raw materials (dry *Puntius* spp. or *S. Phasa*)

⇩

Sorting by hands

⇩

Sun drying in open space

⇩

Water washing and overnight partial drying at room temperature

⇩

Packing of oil smeared matkas with partially dried fish and filled up to neck portion

⇩

Sealing of mouth portion with cover paste

⇩

Covering of the paste with paper or banana leaves and keep it undisturbed for 3-4 days

⇩

Removal of the cover leaf and application of thick layer of mud on the mouth

⇩

Keeping the matkas undisturbed for 3-4 months for fermentation at ambient temperature

⇩

Final product *Sheedal* after 3-5 months

Fig. 21.1 Traditional method of production of *Sheedal* in Tripura

Plate 21.1 *Punti Sheedal*

Plate 21.2 *Phasa Sheedal*

necessitated searching for alternative. Therefore, since last 10 years, Phasa fish (*Setipinna Phasa*) has been used as raw material fish in place of *Puntius* spp. to produce low-cost *Sheedal* (known as *Phasa Sheedal*, *Telesch*, *Baspati Sheedal*, etc.). *Phasa Sheedal* is popularly known as *Sheedal for the poor*, as its retail price is almost half of the price best quality *Punti Sheedal*.

In traditional practice, November to February is considered as the ideal time for production of *Sheedal*. Presently, due to lack of sufficient shallow water bodies, dried salt-free *Puntius* species or salted/unsalted dried Phasa fish (*Setipinna Phasa*) are usually collected from other states, viz. West Bengal, Gujarat, UP, Bihar, etc. for production of *Sheedal*. The different steps in *Sheedal* technology and their scientific explanations are discussed below (Fig. 21.1, Plates 21.1, and 21.2).

21.2.1.1 A Brief Description of Traditional Method of *Sheedal* Production

Matka (fermenting container): The earthen container which is used for fermentation of fish is locally known as matka. This is a pear-shaped container with the thickness gradually increased from belly towards bottom, in order to withstand vertical pressure during filling with fish. Although matkas of different capacity are

Plate 21.3 Oil processed matka

Plate 21.4 Sorting of dried fish

available, the neck diameter is fixed (8 in.) in all sizes. Most commonly used matka's dimension is 24 in. (middle-expanded part) × 36 in. (height) having capacity to hold 40 kg of dry fish. Although, small-sized matka of 10 or 20 kg are also used occasionally (not for commercial purpose). Good-quality matka can be used for several times until they break or broken during retailing or transportation. Matkas made from fine black soil are considered to be the best as they are hard, provide very less air permeability and absorb less oil during oil processing.

Oil processing of matka: Smearing of matka with oil is an important step in the *Sheedal* technology, as it makes the matka almost nonpermeable to air and vapour by closing the micropores present in its wall (Plate 21.3). Oil extracted from *Puntius* spp., known as 'Punti oil', is preferably used by commercial producers, if it is available in plenty. Recently the large-scale *Sheedal* producers use low cost vegetable oil especially mustard oil in place of 'Punti oil'. Both inner and outer walls of the matka are properly smeared with oil followed by drying in the sun. Repetition of the process is required 7–10 times in case of new matka and 2–5 times when matka is reused. Oil coating and drying is continued till the matka is fully saturated with oil and absorb no more oil. Regarding reuse of matkas, the producers' perception is that older matkas yield better product with reduced cost of production.

Plate 21.5 Washing of dried fish in river

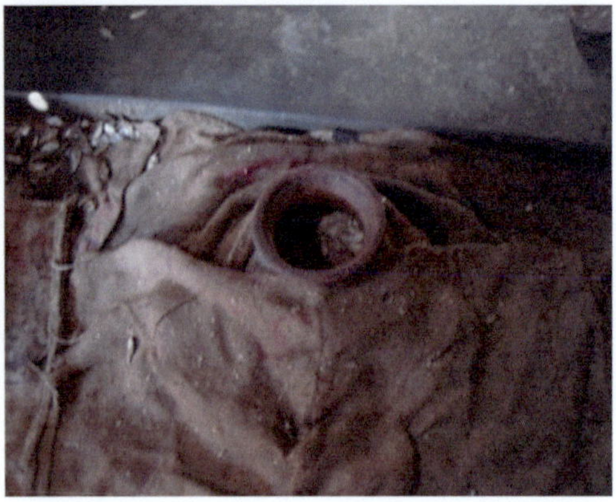

Plate 21.6 Filling of matka

Preprocessing of dried fish: First of all, dried *Puntius* or Phasa fish are cleaned and sorted to remove the broken parts, insect-infested fish and other dirts followed by sun drying for 3–5 days to remove moisture from the fish to maximum possible extent and also to drive away the maggots, if any (Plate 21.4). Next step is water cum water soaking of dried fish in porous bamboo baskets (locally called tukri) (Plate 21.5). Simultaneous water cum water soaking of dried fish is usually done for 3–10 min preferably in the running water. Previous drying of fish enhances the rate of water absorption. Traditionally, this step is done in the evening hours. Water-soaked fish are partially air-dried by spreading over bamboo mattress or cemented floor under shade and left overnight. Such practice of overnight drying for about 8–10 h also avoids the nuisance activities from flies and birds.

This step, i.e. water cum water soaking of dried fish, is very crucial in *Sheedal* technology as it determines the quality of the final product as well as the total fermentation period. However, in traditional practice the duration of water cum water soaking of dried fish is determined by previous experience of producer, and this is mainly based on the requirements, such as quality and expected shelf life of the end product and period of fermentation.

As a thumb rule, the commercial producers believe that more the predetermined fermentation period for good-quality final product lesser the duration of washing cum water soaking. The water soaking and partial drying make the fish soft-textured with dry surface and ready for filling in the matka.

Filling of matka: This step requires setting of the matka in ground to keep it steady and vertical while filling, and this is accomplished by digging a hole in such a way that one-third of the belly remains buried in the ground (Plate 21.6). Steady and vertical position of matka is ensured by putting dug-out soil surrounding the belly of the matka with compression. This helps the matkas to withstand the physical pressure applied by hand or feet during filling of fish with much compaction. Clean gunny bags are spread over the ground to avoid contamination of any spilled out raw material with the soil underneath while filling. Filling is accomplished by putting partially dried fish in the matka in layers with frequent application of uniform pressure with hand or feet so as to avoid any void spaces inside the matka. Sometimes wooden bar with wide circular head is also used for compaction of fish. Filling continued in this manner till the fish layer reach near to the neck part (Plate 21.7). For matka of 40 kg capacity, about 35–37 kg of dried *Puntius* or Phasa fish is required for filling.

Cover paste: Cover paste is prepared from broken dry fish usually collected during sorting. These are ground and made to a thick paste with the help of little oil and water. Some traditional producers also add small quantity of old *Sheedal* during preparation of cover paste. This may possibly be with an idea to provide fermenting microbes for their rapid propagation and activity in the newly filled *matka*.

Sealing of filled matka: Once the matka is filled with fish up to the neck part, it is first sealed with cover paste (Plate 21.8). The thickness of the layer of cover paste is about 2–2.5 in., and it is covered by any broad leaf (preferably banana leaf). *Matka* is finally sealed air-tight by a layer of wet mud (preferably clay soil). This mud layer is checked daily for about a week for appearance of any crack and is

Plate 21.7 Filled matka

Plate 21.8 Sealed matka

Plate 21.9 Fermentation shed

repaired immediately by wet mud. Once the mud seal is dried fully and no further crack appears, then it is covered by a polythene sheet to prevent entry of insects. The filled and sealed matkas are lifted to the surface and shifted carefully to the fermentation shed where it is left undisturbed for maturation.

21.2.1.2 Fermentation Shed

The fermentation shed (Plate 21.9) is a large hall in the ground whose roof and sides are made with bamboo and is made in such a way that entry of direct sunlight to the stored matkas is partially allowed, whereas rainwater is partially prevented. Moreover, air circulation across the fermentation shed remains normal.

The usual period of maturation is 3–5 months till the product gains a characteristic odour, texture and appearance. About 40–42 kg *Sheedal* is obtained from each

Plate 21.10 *Godhak*

Plate 21.11 *Sheedal Chutney*

matka. In wholesale market, intact *matkas* are sold, whereas in retail markets *Sheedal* is sold by taking out from the *matka*.

21.2.1.3 Culinary and Mode of Consumption

Sheedal is consumed basically in three different ways by making three dishes, namely, *Godhak*, *Sheedal Chutney* and *Sheedal Bharta*.

Godhak**:** *Godhak* (Plate 21.10) is an oilless preparation and very appetizing at the beginning of a meal. In earlier days, *Godhak* used to be prepared in bamboo cylinder (single internode of immature bamboo), but presently, it is prepared in small aluminium hundi (Dekchi). The recipe includes *Sheedal*, green chilli, onion and any one or two vegetables such as bamboo shoot (most preferred), banana stem (actually the flower stalk of the banana plant), banana flower (mocha), bitter gourd (uchhe, karela), tree bean seed (*Parkia timoriana*) or any other seasonal ones including potato. All the vegetable ingredients are chopped and washed properly including *Sheedal*, boiled in an aluminium dekchi with addition of salt and water at moderate flame. When the vegetables are boiled, then the mixture is stirred with spoon to break them for making a semi-solid liquid.

***Sheedal Chutney*:** *Sheedal Chutney* (Plate 21.11) is a strong appetizer. The recipe includes *Sheedal*, oil, onion, garlic, green chilli, turmeric power, red chilli powder (optional), tomato (optional) and coriander leaves (optional). First *Sheedal* is lightly fried in hot oil, and green spices are added with salt (up to taste) and tomato and/or coriander leaves and frying continued at moderate heat with constant stirring to make it a thick semi-solid paste.

***Sheedal Bharta*:** This preparation is made by burning *Sheedal* in flame followed by washing and making a hand-made paste with onion and green chilli. This is for immediate consumption of *Sheedal*.

21.2.1.4 Socio-Economy and Ethnical or Religious Values

Sheedal is associated with the socio-economic life of the original inhabitants of the state (both tribals and nontribals) as it serves as a main source of animal protein in their daily diet. Usually hill dwellers come down to the plane during 2 market days in a week and procure *Sheedal* for consumption throughout the week. Sheedal and other dried fish can be stored at ambient temperature in bamboo-made cylinders which are hung over the traditional kitchen. Economically *Sheedal* is comparative cheaper than fresh fish in respect of quantity required for its culinary preparation, as most of the poor men cannot afford for fresh fish.

In traditional custom of ethnic population of the state, *Godhak* is served in any social occasion irrespective of their economic and social status. Although no scientific study has yet been initiated, it is strongly believed by the tribal population of the state that regular consumption of *Sheedal* protects them from malarial infection and also boosts up their immunity against seasonal illnesses. *Godhak* is believed to be useful for heart as it does not require any oil in its preparation. Moreover, *Sheedal Chutney* acts as an appetizer and also believed as a remedy for seasonal mild illnesses.

21.2.1.5 Microbiology of *Sheedal*

The total plate count was found to be 6.87 ± 0.11 and 6.36 ± 0.01 log cfu/g in Punti and Phasa *Sheedal*, respectively. Coagulase-negative *Staphylococcus* and lactic acid bacteria (LAB) were found to be the dominant genus in *Sheedal*. Six species of *Staphylococcus* (*Staph. piscifermentans*, *Staph. condimenti*, *Staph. arlettae*, *Staph. sciuri*, *Staph. warneri* and *Staph. hominis*) and six species of lactic acid bacteria (*Lactobacillus plantarum*, *Enterococcus faecium*, *Enterococcus faecalis*, *Pediococcus pentosaceus*, *Pediococcus acidilactici* and *Pediococcus lolii*) were recognized (Majumdar and Bhaskar 2016). Amongst staphylococci, *Staph. piscifermentans* was found to be most abundant in the *Sheedal* samples analysed. All the six *Staphylococcus* showed their NaCl tolerance from 2–8%. No species was able to grow at 55 °C. Except *S. arlettae* and *S. sciuri*, all the isolated staphylococcal species exhibited growth at pH 4–8. All the six species were found to be sensitive to the antibiotics, namely, erythromycin, norfloxacin, ampicillin, streptomycin and vancomycin, whereas all were resistant to co-trimoxazole. Staphylococcal species, *Staph.*

Table 21.1 Nutritional quality of *Sheedal* and *Lona ilish*

	Punti Sheedal	*Phasa Sheedal*	*Lona ilish*
Biochemical and microbial qualities			
pH	5.86 ± 0.11	6.62 ± 0.07	5.66 ± 0.06
TTA (g %)	0.115 ± 0.01	0.092 ± 0.01	0.98 ± 0.04
Moisture (%)	38.26 ± 0.89	43.48 ± 1.58	54.35 ± 5.06
Crude protein (%)	36.84 ± 1.42	36.75 ± 1.75	20.94 ± 2.62
Total lipid (%)	14.30 ± 2.24	7.85 ± 1.26	9.41 ± 0.74
Salt (%)	–	–	15.75 ± 1.16
TVBN (mg %)	62.53 ± 1.61	120.27 ± 1.24	48.0 ± 6.08
TBA (mg mld/kg)	0.99 ± 0.06	1.10 ± 0.14	1.05 ± 0.21
TPC (log cfu/g)	6.87 ± 0.11	6.36 ± 0.01	3.34 ± 0.07
LAB (log cfu/g)	4.5 ± 0.08	4.8 ± 0.14	3.25 ± 0.09
Minerals			
Calcium	298.09 ± 43.25	329.63 ± 53.89	90.04 ± 14.35
Potassium	118.75 ± 25.85	53.9 ± 9.42	6.77 ± 0.64
Iron	0.012 ± 0.012	0.005 ± 0.03	0.017 ± 0.51
Sodium	180.94 ± 56.3	115.75 ± 45.9	198.22 ± 28.26
Manganese	0.87 ± 0.032	0.10 ± 0.012	0.07 ± 0.002
Cupper	0.02 ± 0.003	0.01 ± 0.001	0.01 ± 0.04
Zinc	0.7 ± 0.034	0.53 ± 0.09	0.31 ± 0.18
Magnesium	12.95 ± 1.23	24.13 ± 1.53	12.54 ± 4.68
Antioxidative qualities			
Superoxide scav. Activity (%)	12.11 ± 0.70	55.02 ± 1.64	4.88 ± 0.18
DPPH assay (%)	79.05 ± 0.39	88.01 ± 0.45	67.45 ± 0.22
ABTS assay (%)	83.87 ± 0.85	78.82 ± 2.25	50.95 ± 0.47
Total antioxidant activity	241.6 ± 5.4	382.5 ± 6.25	111.66 ± 3.71
Amino acids			
Alanine	4.56	3.84	2.16
Glycine	5.78	4.54	2.66
Valine	2.35	2.18	1.46
Leucine	5.16	4.69	3.68
Isoleucine	2.02	1.76	1.23
Threonine	0.60	0.65	0.53
Serine	0.67	1.04	0.60
Proline	4.38	3.53	2.29
Aspartate	3.36	3.04	2.28
Methionine	1.90	1.66	1.25
Hydroxyproline	0.82	0.71	0.35
Glutamic acid	3.35	2.75	1.58
Phenylalanine	4.48	3.59	2.59
Glutamine	0.70	–	–
Lysine	3.07	3.04	2.52
Histidine	0.92	0.71	0.85
Tyrosine	2.24	1.81	1.68
Tryptophan	0.26	0.20	0.13

(continued)

Table 21.1 (continued)

	Punti Sheedal	*Phasa Sheedal*	*Lona ilish*
	Fatty acids		
Lauric acid	1.10 ± 0.12	0.74 ± 0.06	0.22 ± 0.01
Tridecanoic acid	0.19 ± 0.04	0.17 ± 0.01	0.13 ± 0.01
Myristic acid	3.87 ± 0.1	1.37 ± 0.09	6.23 ± 1.6
Myristoleic acid	–	–	0.24 ± 0.01
Pentadecylic acid	1.69 ± 0.07	1.62 ± 0.13	–
Pentadecanoic acid	–	–	1.93 ± 1.08
Palmitic acid	10.25 ± 0.07	10.76 ± 0.36	18.23 ± 9.03
Margaric acid	0.57 ± 0.06	2.21 ± 0.25	3.0 ± 1.3
Stearic acid	0.62 ± 0.19	0.29 ± 0.05	0.67 ± 3.66
Nonadecylic acid	1.23 ± 0.11	0.50 ± 0.11	
Arachidic acid	0.87 ± 0.16	0.18 ± 0.03	2.98 ± 1.5
C16:1(*n*-5)	34.57 ± 1.55	–	–
Palmitoleic acid	–	21.75 ± 0.85	36.84 ± 1.79
Oleic acid	21.71 ± 0.39	44.96 ± 1.65	10.63 ± 6.8
Eicosanoic acid	0.78 ± 0.21	1.24 ± 0.09	2.29 ± 0.07
Linoleic acid	2.08 ± 0.17	9.71 ± 0.55	0.59 ± 0.45
Linolenic acid	0.34 ± 0.05	–	0.36 ± 0.01
Eicosadienoic acid	0.48 ± 0.14	0.18 ± 0.05	0.61 ± 0.03
C20:3(*n*-9)	–	0.11 ± 0.05	–
Arachidonic acid	2.58 ± 0.13	–	2.84 ± 0.06
EPA	1.77 ± 0.17	2.77 ± 0.13	3.13 ± 0.05
DHA	15.29 ± 0.75	1.26 ± 0.13	6.62 ± 1.35

piscifermentans, *Staph. condimenti*, *Staph. warneri* and Staph. *hominis*, exhibited hydrophobicity from medium to high per cent. All the species exhibited moderate to maximum proteolytic and lipolytic activities.

Amongst lactic acid bacteria (LAB), *Lb. plantarum* was found to be most abundant in the *Sheedal* samples analysed. All the LAB isolates were found to be homofermentative nature and low to moderate proteolytic. All the LAB species demonstrated their sensitiveness to erythromycin and norfloxacin, whereas resistances were shown towards co-trimoxazole and vancomycin. Strong resistance against pathogenic indicator bacteria such as *Bacillus subtilis*, *Staph. aureus*, *Salmonella enterica* and *E. coli* by showing wide inhibition zone in agar plate by the entire LAB isolates has been reported (Majumdar and Bhaskar 2016).

21.2.1.6 Nutritional Value of *Sheedal*

The proximate composition and biochemical and microbiological quality of *Sheedal* were reported by Majumdar et al. (2015) (Table 21.1). Low moisture content of *Sheedal* was due to use of sun-dried fish (moisture content <10%). The pH seems to be slightly higher in respect of other fermented fish products, but this may be due to higher amounts of volatile nitrogenous compounds (TVBN) produced during fermentation that accumulate in the product. Absence of salt (a potential prooxidant) and metals (as the fermentation is carried out in earthen container) in the system may be attributed for such moderate values of TBA. Antioxidant potential of *Sheedal* such as superoxide radical scavenging activity, DPPH and ABTS assay as well as total antioxidant activity has been reported by Majumdar and Bhaskar (2016). It has been reported that food-derived peptides exhibit blood-pressure lowering effects, cholesterol-lowering ability and antithrombotic and antioxidant activities (Hartmann and Meisel 2007).

Majumdar et al. (2009) reported about the amino acid contents of *Sheedal* (Table 21.1). Amino acids like glutamic acid, aspartic acid, leucine, alanine and lysine have been found in the higher proportion. Some amino acids such as tyrosine, histidine, arginine and tryptophan have been found very low in amount, and proline was not detected. There is a possibility of formation of derivatives of amino acids such as amines and gluconeogenic substances during fermentation.

Fatty acid profiles of Punti and Phasa *Sheedal* have been estimated by Majumdar et al. (2015). Amongst saturated fatty acids, palmitic acid (C16:0) was found to be dominant in both the fermented fish and contributed about 10% of the total fatty acids. Amongst the monoenoic fatty acids, one unknown with symbol C16:1(*n*-5) contributed about one-third (34.57%) of the total fatty acids followed by oleic acid (21.71%) in Punti *Sheedal*. In case of Phasa *Sheedal*, monoenoic fatty acids were dominated by oleic acid (44.96%) followed by palmitoleic acid (21.75%). The *n*-6 PUFA in Punti *Sheedal* was composed only of arachidonic acid (2.58%) and linoleic acid (2.08%), whereas in Phasa *Sheedal*, it was only linoleic acid (9.71%). Contribution of EPA and DHA in both the *Sheedal*s was 1.77%, 15.29% and 2.77% and 1.26% in Punti and Phasa *Sheedal*, respectively. Presence of EPA, DHA, arachidonic, linolenic and linoleic acid indicates the nutritional significance of *Sheedal*. The *n*-3/*n*-6 ratio was found to be 3.73 and 0.41 in Punti and Phasa *Sheedal*, respectively, which further emphasize the nutritional richness of Punti *Sheedal*. Although, not much work on the fatty acid profile of such salt-free fermented fish products has been reported earlier, but possibilities of loss of PUFAs during fermentation as well as post fermentation exposure of the product out of fermenting containers cannot be ignored.

21.2.1.7 Optimization and Commercialization

Traditional *Sheedal* technology has been optimized in the laboratory. Based on this study, following Good Manufacturing Practices (GMPs) for production of Sheedal has been developed.

Good Manufacturing Practices (GMPs) for Sheedal production:

- Quality raw material (not insect-infested, properly dried and stored).
- Redrying of fish before fermentation (optional, depending on the dryness of the fish).
- Proper screening and sorting to remove insect-infested fish and broken pieces.
- Proper oil processing of matka till it does not absorb oil anymore.
- In controlled system, washing/water soaking of fish should be done in cement tank with spraying of chlorinated water.

- Wash water of each batch of fish should be allowed to move out of the tank.
- Sanitization of the floor of the matka filling room with antimicrobial sanitizer.
- Drying of water-soaked fish in clean and adequately dried bamboo mat or sanitized floor.
- Filling of matka in closed room.
- Persons involved in filling operation should not have any scours/lesions in body especially in hands or legs.
- Spitting, smoking, chewing tobacco, etc. are not allowed in matka filling room.
- Cover paste seal is to be covered with clean leaf or polythene sheet before putting final seal of clay soil.
- Clay used for final sealing should be clean without any humus or debris.
- After redressing the clay seal for a week for crack, seal should be covered with polythene sheet to prevent insect infestation.
- The fermentation shed should be cleaned, well ventilated, mud floor, lightly roofed and fenced with bamboo.
- Entry of rodents, dogs, cats, etc. in the fermentation shed should be protected.

A great deal of malpractices has been reported in the production of *Sheedal* due to its huge demand in all the rural and urban markets of the Northeast India. The principal intension behind such malpractices is to utilize unsold dry fish for *Sheedal* production and early fermentation as well as to produce *Sheedal* which could be sold at very low prices in the rural belt. Some of the malpractices as observed during this study are given below:

- Use of insect-infested/old raw material, i.e. dried *Puntius* spp. and *S. Phasa*.
- Mixing of old and insect-infested and cheap variety of dried marine fish (cutting similar to the size of Punti fish) along with dried *Puntius* spp. and *S. Phasa*.
- Use of synthetic colourants during fermentation.
- Use of vegetable oil during fermentation.

Plate 21.12 *T. ilisha*

Plate 21.13 *Lona ilish*

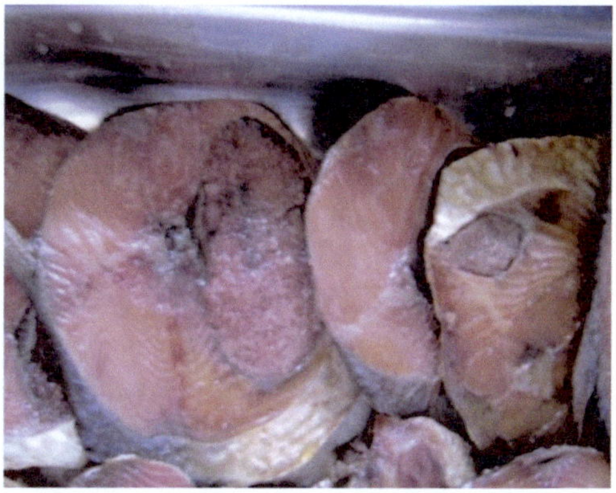

Presently *Sheedal* production is mainly confined to the West Tripura district of the state. Although *Sheedal* production needs a moderate capital and space, however, small scale entrepreneurs are coming forward for this venture owing to its constant demand in all the markets of the state as well as export potential to other northeastern states. Young entrepreneurs are encouraged to improve the marketing of *Sheedal* through packaging and commercialization of its 'ready-to-eat' value added products such as *Godhak*, *Sheedal Chutney*, etc.

21.2.2 *Lona Ilish*

A salt-fermented fish product, popularly known as *Lona ilish*, is prepared exclusively from Indian shad (*Tenualosa ilisha*) (Plate 21.12), a high-fat fish (fat content ranges from 14 to 25%). *Lona ilish* (Plate 21.13) is a very popular product and is widely consumed in Tripura and entire Northeast India and adjoining country Bangladesh, mainly due to its typical flavour, aroma and texture. Tripura is the major producer of *Lona ilish* amongst the NE states, and the product is exported to other neighbouring states. The product is hilsa steaks, about 1.50–2.00 cm in thickness. The distinctive feature of *Lona ilish* is its uniform pink colour with a glossy manifestation immediately after taking the product out of the brine. The texture of muscle remains firm but soft and does not easily separate from the bone. The strong characteristic aroma of *Lona ilish* is some sort of mixed notes of sweetness, fruity, acidic and saltiness. The strong odour spreads in the air during storage and gives the area a characteristic smell of *Lona ilish*. It is kept immersed in saturated brine until consumption. The uniqueness of this product remains in the fact that despite the presence of salt and metals leaching out from the container used for fermentation, the rancidity of this highly unsaturated fatty acid containing fish is kept under control and is not manifested as long as the fish steaks are kept immersed in the fermenting brine.

The *Lona ilish* technology assumed to have originated about 100 years ago on the bank of river 'Padma' and 'Meghna' under Noakhali district in the erstwhile undivided India (now Bangladesh) (Personal communication). It is also presumed that the technology evolved during the period of its overabundance when no other preservation techniques except sun-drying and salting was available. Sun-drying was not considered appropriate for hilsa like high-fat fishes due to oxidation of fat leading to rancidity on being exposed to the sun. Moreover, continuous sun-drying was also not possible during July to August as this period corresponds to the rainy season and also the main harvesting period.

This unique processing technology might have evolved for quick preservation of large quantity of fish in an inexpensive way. But the age-old technology has not changed much with time, and the practice is still considered as one of the major means of preservation of hilsa. Although with time several fish preservation technologies have come into being, but salt drying and to some extent canning are still in practice for preservation of hilsa like high-fat fish. This technology got entry in to Tripura state of India through the immigrants during partition of India in 1947. Although the product is very popular, fetching a good demand in the fish-eating communities in other areas, its localized consumption may be due to the practical difficulties in its transport.

21.2.2.1 A Brief Description of Traditional Method of *Lona Ilish* Production

Collection and preprocessing of fish: *Lona ilish* is exclusively made from hilsa (*Tenualosa ilisha*). The period June to September is the peak time of preparation of *Lona ilish*, when hilsa is caught in large quantity during its upward migration for breeding. During this time the fish has a very high-fat content to the extent of 19–22%. The quality of the final product is entirely dependent on the freshness quality of the raw material fish. Fish is collected from the landing centre or from the wholesale market where fish is imported in ice box within 24 h. The fish are first washed properly with potable water followed by descaling and deheading leaving gut inside. The dressed fish are obliquely cut into steaks (1.5–2.0 cm deep). Fish steaks are not washed again. A slanting cut appears to be helpful in providing more surface area of fish flesh exposed to salt as well as halotolerant bacterial action.

Salt curing of hilsa steaks: Unwashed hilsa steaks are dry salted by rolling thoroughly in salt (fish to salt ratio is 4:1) and placed in a bamboo-made basket (locally called as tukri) layer after layer with little pressure by hand. Some salt is put in the bottom of the basket before keeping fish. Salt is spread between each layer and above the top layer.

Prematuration of hilsa steaks: The dry salted hilsa steaks in bamboo basket are covered with black polythene sheet to prevent oxidation by avoiding entry of light and left undisturbed in a dark place for 48 h. The self-brine formed during store is allowed to drain. This step reduces a considerable amount of moisture content from the hilsa steaks, and pink colour of the flesh changes to dull white. The texture of steaks becomes tough compared to the fresh fish. Loss of a significant amount of water from the fish as self-brine during dry salting prevents dilution of saturated brine during fermentation stage.

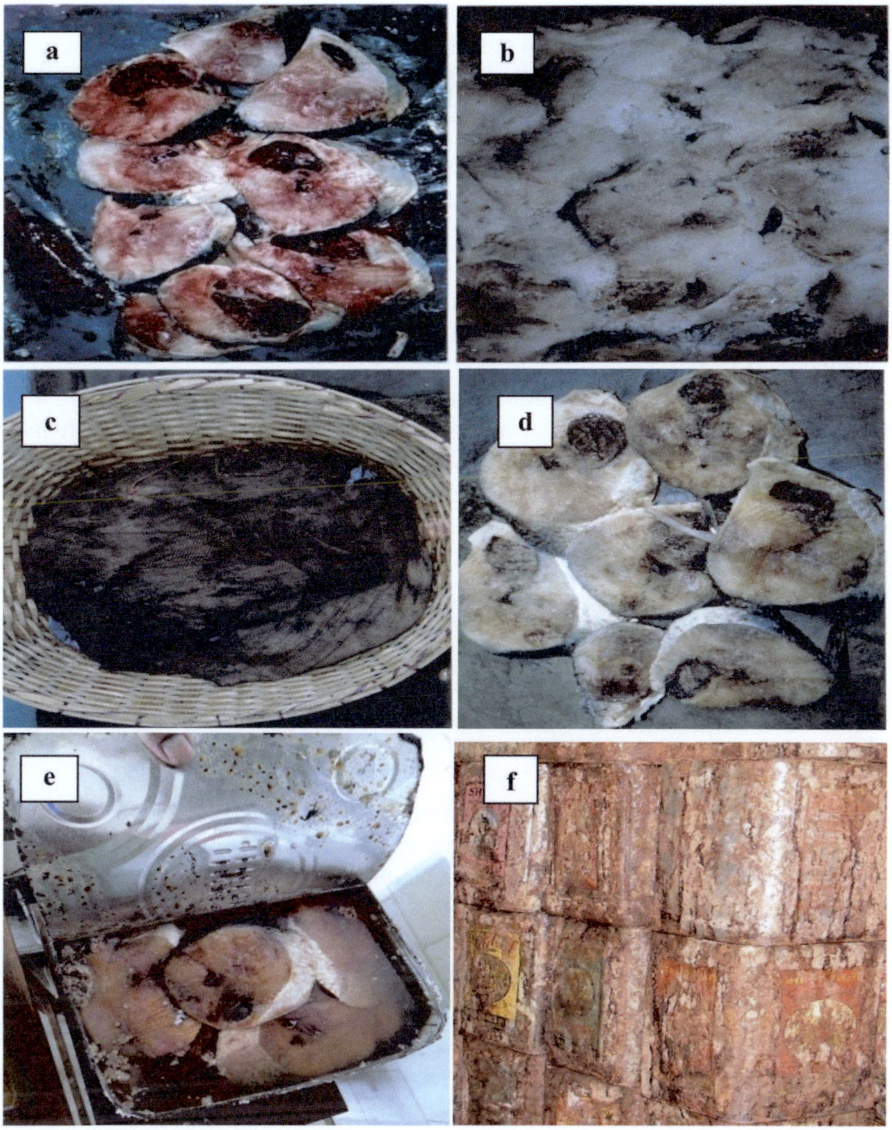

Plate 21.14 Traditional method of *Lona ilish* production. Hilsa steaks (**a**), dry salted hilsa steaks (**b**), prematuration of hilsa steaks (**c**), premature hilsa steaks (**d**), filling of fermenting container (**e**) and fermentation of hilsa steaks (**f**)

Fermenting container and media: Fermentation is done in new empty tin-made containers (cap. 18 L) used for cooking oil. The fermentation media is saturated brine. Saturated brine is prepared and boiled properly followed by cooling overnight. Boiling of brine is done firstly for preparing a saturated solution and secondly to destroy undesirable microorganisms present in the salt that might contaminate the product. The cooled saturated brine (avoiding the salt settled at the bottom) is sieved with nylon net to prevent entry of any undesirable materials in the fermenting container. This is prepared 1–2 days before filling the container with hilsa steaks.

Filling of hilsa steaks: The salt-cured hilsa steaks are taken out from the tukri and made free of adherent salt and dirts, followed by placing in the empty fermenting containers in layers with uniform compaction by hand after each subsequent layer. Filling of fermenting container is continued till the layer of steaks reach at least 4–6 cm below the upper margin. Then the container is filled with cold saturated brine by pouring over the steaks slowly to fill the empty spaces between the steaks till the level of brine reaches 2–3 cm above the fish. The container is then closed with the lid (Plate 21.14).

Fermentation: All the filled and closed containers are kept in a dark room for 4 to 6 months for fermentation. During maturation, the texture of the fish meat changes, and the final product (*Lona ilish*) acquires a texture of cooked meat. Fat is also enzymatically broken down to flavour-bearing substances. Completion of fermentation is perceived by the odour spread in the area and also the colour and texture of the hilsa steaks. The *Lona ilish* tin (fermenting container) can be kept beyond 1 year, but the marketing starts after few months of maturation.

Marketing of *lona ilish*: In wholesale markets, closed intact *Lona ilish* tin is sold. During retail marketing, the product remains submerged until sold. Once a container is opened, the materials are sold part by part. Upon continuous exposure to air and light for few hours, the pink red colour of the products changes to greyish black which is not always acceptable.

21.2.2.2 Culinary and Mode of Consumption

Usually *Lona ilish* is soaked in water for about 10–15 min before cooking for desalting. Most popular culinary preparation of *Lona ilish* is its bada wrapped in ash gourd (Chalkumroh or winter melon) leaves or leaves of pui plant (*Basella alba*). Following desalting, *Lona ilish* is cut into small pieces and mixed with turmeric powder, green chilli, red chilli powder and onion, and this mixture containing 2–3 pieces of fish is wrapped in oil-coated leaves (2–3 layers). These are heated over mild flame in lidded frying pan with little oil till the wrapping leaves turns to black in colour. Other culinary preparation is curry of *Lona ilish* with different vegetables.

21.2.2.3 Socio-economy and Ethnical or Religious Values

As such there is no ethnical or religious values of *Lona ilish* but it's a delicacy to the populace of the state since long time as Hilsa fish was easily available in this region from the erstwhile adjoining country East Pakistan (now Bangladesh). From socio-economic point of view, *Lona ilish* is preferred by all fish-eating population due to its very uncommon flavour and taste, which is not available in kind of fresh water fish and even in marine fish.

21.2.2.4 Microbiology of *Lona Ilish*

Based on the molecular tools, six species of coagulase-negative *Staphylococcus*, namely, *Staph. piscifermentans*, *Staph. condimenti*, *Staph. arlettae*, *Staph. nepalensis*, *Staph. warneri* and *Staph. hominis* were recognized (Gupta 2016). Four lactic acid bacteria, namely, *Lb. plantarum*, *Pediococcus pentosaceus*, *P. acidilactici* and *P. lolii* were recognized (Gupta 2016). Amongst these, *Lb. plantarum* was found to be most abundant in the *Lona ilish* samples analysed. All the LAB isolates were found to be homofermentative nature and low to moderate proteolytic.

21.2.2.5 Nutritional Value of *Lona Ilish*

Evaluation of the biochemical quality of *Lona ilish* sample from the markets of Tripura state was done by Majumdar and Basu (2010) and is presented in Table 21.1. Antioxidant potential of *Lona ilish* such as superoxide radical scavenging activity, DPPH and ABTS assay as well as total antioxidant is given in Table 21.1. Majumdar and Bhaskar (2016) investigated the amino acid fatty acid profiles of *Lona ilish* and is given in Table 21.1. Amino acid analysis indicated the presence of all essential amino acids in the product. *Lona ilish* was found to be a good source of EPA and DHA as well as oleic acid (Table 21.1). The major saturated fatty acid was found to be palmitic acid followed by myristic acid.

21.2.2.6 Optimization and Commercialization

Quality of the product largely depends on the freshness of the raw fish, removal of water during dry salting, period of maturation, concentration of brine, etc. Since, there is no standard method of its preparation, different grades of *Lona ilish* are available in the market. Some of them are of very poor quality especially due to improper salting, use of poor quality fish and lack of proper ripening period, etc.

The *Lona ilish* is different from other fermented fish products. Some distinctive features of *Lona ilish* are given below.

1. It does not come under the category of fish paste or fish sauce.
2. Exclusively hilsa, *Tenualosa ilisha* (Ham-Buch 1822), a very high-fat fish, is used for its preparation.
3. Whole fish is cut in to steaks which remain intact after maturation, whereas, in the case of other salt-fermented fish products, whole fish is either crushed or dressed as whole, and most of them do not retain original form after maturation.
4. Dry salting is followed by brining, whereas either dry salting or simply brining is done in case of most of the similar type of products.
5. Once taken out from the brine and exposed to air, deterioration is very fast, but this is not happened in case of most of the products. This is due to high-fat content of the *Lona ilish* that undergo autoxidation by atmospheric oxygen and the rate of which is further accelerated due to presence of prooxidant, i.e. NaCl.
6. Unlike some of the fermented fish products, *Lona ilish* must be cooked before consumption.

Traditionally, the fishermen themselves process the fish for *Lona ilish*. Hygienic conditions are far from satisfactory in the manufacturing units. A general tendency has been observed amongst most of the *Lona ilish* producers for its preparation either when the market price of raw hilsa goes down or when ice is not available in plenty for re-icing of unsold fish. Besides this, the other common malpractices are as follows:

- The fish are selected irrespective of freshness and size.
- Sometimes, fish are not even washed before being dressed.
- Crude salt used is of cheap and unknown quality.
- Fish to salt ratio is also on a thumb rule basis only.
- Salt-cured fish steaks are often not covered properly, which sometimes results in infestation of flies.
- Insufficient salting often causes dilution of brine during fermentation stage.
- Due to over packing, the container sometimes bursts, the brine drains out, and the top layers remain exposed and get spoiled.
- New and sound tin containers are seldom used.
- The produce is often sold prior to proper maturation to get quick return of the investment.
- Use of colour, mustard oil, vanaspati, etc. during fermenting process with an idea to sell the product after a short maturation period.

On the basis of laboratory production of *Lona ilish* as per traditional practices, following Good Manufacturing Practices (GMPs) for *Lona ilish* production has been developed.

- Hilsa fish should be of best quality.
- Fish should be washed with potable water and descaled before dressing.
- Gut contents should remain adhered to the diagonally cut steaks.
- Saturated brine should be prepared in boiling water and cooled well.
- Hilsa steaks should be properly dry salted with thorough rubbing over salt.
- Fermentation container, i.e. empty mustard oil tin should not be rusted and no oil should be inside.
- Hilsa steak-filling room should be cleaned and sanitized before use.
- Filled tin should be kept in clean, dry and closed room and protected from entry of rodents, dogs and cats.
- *Lona ilish* should be kept immersed in brine during retailing.

The unique taste of *Lona ilish* with very delicious aroma creates a good demand for this product. Market price of the *Lona ilish* depends upon the price of fresh Hilsa fish. Presently, market price of *Lona ilish* is Rs. 1000–1200 per kg. Presently due to price enhancement of fresh Hilsa fish, traditional practice of *Lona ilish* production has been reduced. Since the product is acceptable to any fish-eating communities, therefore a good packaging would add value to it for its export to other places, and maintaining quality of the final product and sanitation in the production area *Lona ilish* technology would regain its earlier status.

21.3 Traditional Beverages of Tripura

The exact period when the preparation and consumption of alcohol had been started by human being is unknown, but it's consumption is mentioned in the oldest books like Veda, Bible, etc. Similarly, the preparation and consumption of alcohol were known amongst Tripuri community since the prehistoric times. There are many varieties of alcohol prepared by Tripuri people, and the women folk are engaged for this. Some of the women are so expert that many consumers would wait for their turn to get chance. The traditional knowledge in making of alcoholic beverages uses rice as fermenting medium and local herbs as a source of amylolytic and alcohol-producing yeasts. It also plays an important role in the sociocultural life of the tribal people as it is found to be associated with many occasions like merrymaking, ritual ceremonies, festivals, marriages and even death ceremonies. It is said that preparation and consumption of this type of indigenously made liquor emerged mainly due to the climatic conditions and to fight against the health problems like insomnia, headache, body ache, inflammation of body parts, diarrhoea and urinary problems, expelling worms and as a treatment of cholera, etc. due to lack of any medical facilities in this jungle-infested region. Moreover, in traditional ethnic culture, alcohol is also used for offering to their deities. All of the tribes prepare their indigenous alcoholic beverages at home using round to flattened solid ball-like mixed dough inocula or starter, and these contain amylolytic and alcohol-producing yeasts, starch-degrading moulds and lactic acid bacteria.

Plate 21.15 Chuwan

Table 21.2 Plants, plant parts and other ingredients used by different tribes for the preparation of rice beer

	Plant details				
Tribes	Local names and common names	Scientific name	Family	Parts used	Other common ingredients
Kalai	Sajnabukur (drumstick)	*Moringa oleifera* Lam.	Moringaceae	Bark	Atop mairom (rice), Rishum (garlic), Mosokwthang (green chilli)
	Chuwanthwichwlabukur and bwlai (rosewood)	*Dysoxylum* Blume.	Meliaceae	Bark and leaves	
	Chuwanbwlai (khae)	*Markhamia stipulata* (Wall.) Seem.	Bignoniaceae	Leaf	
	Theipungbwlai (jackfruit)	*Artocarpus heterophyllus* Lam.	Moraceae	Leaf	
	Kurukbwlai (sugarcane)	*Saccharum officinarum* L.	Poaceae	Leaf	
	Barmajalbwlai (meda)	*Litsea monopetala* (Roxb.) Pers.	Lauraceae	Leaf	
Jamatia	Chuwanbwlai (khae)	*Markhamia stipulata* (Wall.) Seem.	Bignoniaceae	Leaf	Atop mairom (rice), Rishum (garlic), Mosokwthai (red chilli)
	Chuwanthwichwlabukur & bwlai (rosewood)	*Dysoxylum* Blume.	Meliaceae	Bark and leaves	
	Andrasabwlai (pineapple)	*Ananas comosus* Mill.	Bromeliaceae	Leaf	
	Auraibwlai (Rabo De Ranton)	*Casearia aculeate* Jacq.	Saliaceae	Leaf	

Debbarma	Kamala bukur (orange)	*Citrus sinensis* (L.) Osbeck.	Rutaceae	Peel	Mairomrukjak (processed rice), Mosokwthang (green chilli)
	Borbwlai (meda)	*Litsea monopetala* (Roxb.) Pers.	Lauraceae	Leaf	
	Chuwanbwlai (khae)	*Markhamia stipulata* (Wall.) Seem.	Bignoniaceae	Leaf	
	Chindrema (Indian allophylus)	*Allophylus serratus* Kurz.	Liliaceae	Leaf	
	Amang (common aporosa)	*Aporusa diocia* (Roxb.) Muell.	Euphorbiaceae	Leaf	
	Madhumalati (rangoon creeper)	*Combretum indicum* (L.) DeFilipps.	Combretaceae	Leaf	
	Andrasabwlai (pineapple)	*Ananas comosus* Mill.	Bromeliaceae	Leaf	
Molsom	Chuwanbwlai (khae)	*Markhamia stipulata* (Wall.) Seem.	Bignoniaceae	Leaf	Atop mairom (rice), Rishum (garlic), Mosokwthai (red chilli)
	Borbwlai (meda)	*Litsea monopetala* (Roxb.) Pers.	Lauraceae	Leaf	
	Theipungbwlai (jackfruit)	*Artocarpus heterophyllus* Lam.	Moraceae	Leaf	
	Hengra (night Jasmine)	*Nyctanthes arbor-tristis* L.	Oleaceae	Leaf	

Source: Ghosh et al. (2016)

21.3.1 Starter Culture or Chuwan

The methodology of fermentation carried out by different tribes is almost the same, except that the difference comes from the different types of plant species used in starter culture preparation. The taste of alcohol is dependent on this Chuwan (Plate 21.15), apart from the skill and experiences of the concerned brewer. It is like a dry cake made up of many herbal products and rice. Chuwan has to be added in proportion in the cooked rice, and then only the Chuwk bwtwk (rice beer) would be of best quality. There are two kinds of Chuwan, one is Chuwan chwla meaning male Chuwan, and it is elongated elliptical shaped, whereas circular one is known as female Chuwan. Normally there is an umbilical type dimple in the centre of the female Chuwan, but in male Chuwan there are three umbilical dimples instead of one. The preparation of Chuwan is very tedious job, and it takes almost a day for preparation. One has to be very sincere and attentive in preparing the Chuwan. The woman should not have her menstrual cycle at time of preparation of Chuwan. Women are also debarred to prepare in postnatal period of 1 month. There are also many restrictions that have to be observed.

The raw materials used for the preparation of starter culture vary with the tribes, except the basic ingredient, i.e. rice flour. Raw materials required for preparation of starter culture by different tribes are listed in Table 21.2 (Ghosh et al. 2016). Initially, rice is soaked for 2 h and then grinded into a fine paste. Plant materials such as leaves or barks are first washed to remove dirts, made into small pieces, sun-dried and finely powdered and sieved. Next, all the ingredients are mixed, and small amount of water is added to make the dough. Balls of about 100 g each are made out of this dough and pressed by both hands gently so that it becomes circular and flattened to about 2–3 cm thick (Chuwan). With gentle pressure by both hands, some circular flattened cakes are slightly elongated (Chuwan chwla). One dimple is made in the centre of circular Chuwan and three in case of Chuwan chwla. The newly make cakes (Chuwans) are kept covered by gunny bags and left undisturbed for 3 days. Then all the Chuwans are dried under sun for 5–7 days and stored in a closed container and kept in a cool and dry place. Effective storage life of Chuwans is

Plate 21.16 Rice beer

Plate 21.17 Mixing of Chuwan with rice

Plate 21.18 Filling of Chuwan-rice mix in earthen pitcher

3–4 months. Customarily, the first cake (Chuwan chwla) is made in the name of God.

There are two main varieties of alcohol beverages prepared by Tripuri women, viz. rice beer (Chuwk bwtwk) and alcohol (Chuwarak).

21.3.2 Rice Beer

Traditional rice beer (Plate 21.16) or popularly known Langi is an indispensible part of their life, attached economically, culturally and spiritually with them, and plays a very significant role in their sociocultural activities. The production and use of rice beer have a long historical background and are associated with their festivals, marriages, rituals and funerals. Chuwk bwtwk is classified according to the rice used to prepare *Mami ni bwtwk*, *Guria ni bwtwk*, *Khasa ni bwtwk and Maisa ni bwtwk.* According to size of earthen pitcher, it is divided as Langi, Gola, Gora, etc. These alcoholic beverages of Tripura have some similarities with few traditional drinks of the world, such as *Shaosingiju* and *Lao-chao* of China, *Tapuy*, *Brem Bali* and

Tape-Ketan of Indonesia, *Khaomak* of Thailand, *Tapaipulul* of Malaysia, *Chongju* and *Takju* of Korea and *Sake* of Japan, with similarities in basic fermentation process also.

21.3.2.1 Method of Preparation

Rice beer is prepared at a household level for regular consumption in almost all the tribes involving a common process.

Ingredients: Chuwan, Unboiled Rice (Mami Rice)

Rice is first cooked in a big cooking pan with optimum water and spread over a perforated bamboo mat for 1–2 h with frequent stirring for rapid cooling. The Chuwans at two pieces per 5 kg rice are crushed into powder using a hand-operated wooden grinder. The Chuwan powder (Plate 21.17) is then added to the cooled rice slowly with proper mixing by hand. Now the Chuwan mixed rice is covered with banana leaves first and then with some rugs or old clean cloths and left undisturbed for overnight. This is actually done to prevent escape of vapour from the Chuwan-rice mixture and also to maintain warmness inside facilitating better fermentation. In the next day morning, the Chuwan-rice mixture is again evenly spread over bamboo mat, allowed it to cool for 5–10 min followed by pouring inside small earthen pitcher (Gola) with gentle pressure while filling up to the neck region (Plate 21.18). Then, two to three layers of banana leaves are placed over the filled pitcher and tied around the neck of the pitcher so that the cover is firmly tied. Sometimes, banana leave cover is further covered with some rug or old clean cloth and left undisturbed for 3–4 days in summer and about 5–6 days in winter season. The filled pitcher is kept in a dry and warm place, preferably over traditional shelf made oven known as Baka. During this period, maturation of rice beer takes place in presence of amylolytic yeasts.

Drinking of Chuwk bwtwk: After 3–5 days, Chuwk bwtwk or Langi is ready for consumption. First the pitcher is taken out from the shelf, cleaned outside with a soaked piece of cloth. All the covering materials are removed, and the rice is gently pressed inside. Cold drinking water is poured in the pitcher till the edge. A bamboo pipe called Tengi (something like T-shaped) is inserted in to the pitcher to measure the water level (Plate 21.19). Langi is drunk by sucking with Chungi (a

Plate 21.19 Tengi fitted in gola

Table 21.3 Quantitative biochemical analysis of rice beer (tribe wise)

	Tribal communities			
Parameter	Debbarma	Kalai	Molsom	Jamatia
pH	3.64 ± 0.003	3.7 ± 0.004	3.52 ± 0.004	3.61 ± 0.003
Moisture (%)	89.33 ± 0.56	85.5 ± 0.62	86.5 ± 0.43	86.0 ± 0.52
Carbohydrate (mg/mL)	0.425 ± 0.007	0.505 ± 0.008	0.83 ± 0.009	0.51 ± 0.010
Reducing sugar (mg/mL)	0.055 ± 0.004	0.046 ± 0.002	0.068 ± 0.001	1.09 ± 0.026
Non-reducing sugar (mg/mL)	0.355 ± 0.006	0.483 ± 0.008	0.784 ± 0.002	0.392 ± 0.009
Protein (mg/mL)	9.63 ± 0.088	11.45 ± 0.085	12.42 ± 0.125	11.43 ± 0.117
Total acidity (g/100 mL)	0.193 ± 0.006	0.059 ± 0.001	0.217 ± 0.009	0.127 ± 0.010
Volatile acidity (g/100 mL)	0.352 ± 0.005	0.020 ± 0.001	0.020 ± 0.001	0.026 ± 0.001
Alcohol % (% v/v)	6.13 ± 0.072	10.06 ± 0.018	7.38 ± 0.128	9.48 ± 0.021
Alcohol % of distillate (% v/v)	28.14 ± 0.23	35.18 ± 0.14	26.38 ± 0.23	30.41 ± 0.32

Source: Ghosh et al. (2016)

Plate 21.20 Chuwarak

bamboo straw) put up to the bottom of the pitcher till the water level goes down to the bottom of the Tengi. Customarily, the eldest one by age and by relation will consume the rice beer first by saying Khulumkha (i.e. Namaskar) folding both hands together to all. Then the next junior will get his/her turn, and he/she will say Khulumkha by touching feet of their seniors and only folding both hands together to all others. This part is done while consuming in the first round only. It is mandatory to maintain such courtesy by all in the party. When water level goes down to the bottom of the Tengi, water should be poured up to edge every time a person sucks the alcohol. Wahan mosdeng (chutney made from pork), roasted chicken, fried fish and fried tadpole (*ampru*) are usually consumed as Chakhinik along with it as food. The biochemical facts of rice beer have been reported by Ghosh et al. (2016) and are presented in Table 21.3.

Plate 21.21 Traditional distiller

21.3.3 *Chuwarak*

Chuwarak (locally known as Bangla by Bengalis) is a distilled variety of alcohol of Tripuri people (Plate 21.20). It is like scotch or champagne; if it would have been promoted and innovated like these brands, it could have taken the status of Vodka, Feni, etc. in the worldwide, but because of political and bureaucratic bottle neck, it could not be taken to world-class standard. The Tripuri alcohol had been on the culture of Tripuri people since time immoral, but till date not a single case of death due to poisoning by consumption had been reported. That is why it is said that the Tripuri whisky is amongst the safest in the world. There are many varieties of Chuwarak, viz. made of Mami rice, pineapple, jackfruits, Guria rice, etc.

Deshi wine or Bangla is made from the Langi. After proper maturation of Langi for 3–5 days, water is added and kept it for 1–2 days. For wine preparation, all the contents of pot are taken in a big container, and almost equal quantity of water is added. The container is covered with clean cloth and kept undisturbed for 2–3 days till the complete fermentation occurs, and this can be known by the hiss sound along with air bubbles produced in it. Following this, half of the contents are taken in a traditional Indian cooking pot (Dekchi), and a special distiller head made of earthen pot and bamboo pipe line (Batisabasa) is fitted over it (Plate 21.21). The distiller head is made in such a way that vapour of alcohol comes in contact with another container filled with cold water (Patini) and get condensed and stored in glass bottle via bamboo pipe line. No vapour is allowed to exit from the system. Cooking pot is placed over traditional furnace and heated gradually by burning fuelwood from low to medium flame. Care is taken that the content is not boiled; otherwise it would spoil the whole process.

Usually from 5 kg of rice, the output of alcohol is 1.5 L strong wine (known as spirit), 1.5 L medium and 1.5 L light alcohol. The strong wine is diluted 4–5 times during selling. Some traditional producers also use jaggery (Gurh) or sugar during fermentation in order to increase the output of alcohol, but these practices reduce the characteristic flavour of the Chuwarak, and also not good for health. After distillation, Chuwarak is cooled and preserved in air tight bottle. The Chuwarak is divided according to the concentration of alcohol present in it like Arak, Brandy, Johr, etc. It can be preserved for months together or years. Old Chuwarak of 2–5 years are useful for medicinal purposes. Sometimes, used Langi (already consumed) is also converted to wine in the similar process, but, in this case, the alcohol obtained is of medium to light variety only. The slurry of rice after fermentation or used Langi is usually used as feed of pigs.

21.4 Ethnic Fermented Vegetables of Tripura

Rural women of this state selected many wild vegetables to develop foods through trial and error since time immemorial. Accordingly, the state has some traditional fermented and semi-fermented foods made from different vegetables especially bamboo shoot, amla and elephant foot yam. Preparation of such foods of vegetable origin differs amongst the different ethnic groups.

21.4.1 Fermented Products from Bamboo Shoot

Bamboo shoot is one of the popular raw materials for preparation of traditional food items in different processed forms. It is consumed in different cured forms like dried, smoked and fermented by the tribes of Tripura. Some popular fermented bamboo shoot products of Northeast India are *Soibum*, *Soidon* and *Soijin* in Manipur (Jeyaram et al. 2009) and *Bamboo Tenga* in Arunachal Pradesh (Tiwary and Mahanta 2007). *Moiya Koshak*, *Melye Amiley*, *Midukeye* and *Moiya Pangsung* are fermented bamboo shoot products indigenous to the *Debbarma*, *Chakma* and *Uchoi* tribes of Tripura, respectively (Uchoi et al. 2015).

Plate 21.22 *Moiya Koshak*

Plate 21.23 *Moiya Pangsung*

Plate 21.24 *Batema*

21.4.1.1 *Moiya Koshak* and *Midukeye*

Moiya Koshak (Plate 21.22) is an indigenous fermented vegetable exclusively produced from non-bitter variety of bamboo shoot (*Melocanna bambusoides*, Trin) (locally known as *Warthwi Moiya*) by *Debbarma* and *Uchoi* tribes of Tripura. The same product is prepared by *Chakma* tribe, and it is traditionally known as *Midukeye*. The bamboo shoot is cut into pieces, washed and wrapped in banana leaf for providing a good condition for fermentation and also for protecting the product from contaminants. It is then tied with bamboo strip. The product is placed over a raised platform which is locally known as *Baka* (bamboo rack) in order to keep it in undisturbed condition. The products are left to ferment in this condition for 2–3 days.

21.4.1.2 *Melye Amiley*

Melye Amiley is a bamboo shoot (*Melocanna bambusoides*, Trin)-based indigenous fermented vegetable product made exclusively by the *Chakma* tribes. The preparation is almost similar to *Moiya Koshak* except additional step of water soaking. The cut pieces of bamboo shoot are soaked in water for 2 days in a traditional earthen container, and it is kept closed for subsequent fermentation.

21.4.1.3 *Moiya Pangsung*

The *Uchoi* tribe, which belongs to southern parts of Tripura, ferments bamboo shoot (*Melocanna bambusoides*, Trin) in a different manner and is traditionally known as *Moiya Pangsung* (Plate 21.23). Cleaned and washed bamboo shoots (*Melocanna bambusoides*, Trin) are first sliced into medium to large pieces. The sliced bamboo shoots are submerged into plastic container filled with water at a ratio of 1:3. The container is covered airtight and kept for two nights for fermentation. After draining out the water, the sliced bamboo shoots are further chopped into smaller pieces, which can be used immediately for the preparation of dishes or can be stored in airtight container for few more days.

21.4.1.4 Nutritional Value

Fresh bamboo shoot contains approximately 2.65% protein on average. The bamboo protein produces eight essential and two semi-essential amino acids. Besides that it also possesses moderate fibre, selenium, a potent antioxidant, and potassium, a healthy heart mineral (Singhal et al. 2013). It has been reported that bamboo shoot improves appetite, digestion and weight loss and also has potential to alleviate cardiovascular diseases and cancer. The antioxidant capacity of bamboo shoot is due to the presence of phenolic compounds (Chongtham et al. 2011).

21.4.2 Fermented Elephant Foot Yam (*Batema*)

Both domestic and wild varieties of elephant foot yam (*Amorphophallus paeoniifolius*, Nicolson) are used for preparation of traditional vegetable products known as *Batema* (Plate 21.24). The leaf part of domestic elephant foot yam is used for wrapping while preparing different fermented food items like *Midukeye*, *Moiya Koshak*, etc. *Batema* is a very popular indigenous semi-fermented vegetable product of the *Debbarma* tribes of Tripura. The skin of the tuber part of elephant foot yam is

Plate 21.25 *Chakhoi-Kho*

removed, made into smaller pieces and boiled for 30 min. The boiled and cooled tuber pieces are mashed by hand and mixed with traditionally prepared liquid soda (locally known as *Chakhoi*). *Chakhoi* is prepared by mixing ashes in water followed by filtration using traditional bamboo-made very small meshed sieve (locally known as *Chakhoi-Kho*) (Plate 21.25). The paste thus obtained is given a round compressed ball shape with the help of traditional dices. The round balls are then spread over a bamboo-made sieve and sun dried continuously for 2–3 days. Usually the sun drying is continued till the colour of the balls turns in to red. The red-coloured product or *Batema* is then stored in traditional bamboo container and allowed fermentation for 5–6 days before marketing.

21.4.2.1 Nutritional Value

Fresh elephant foot yam is rich in omega 3 fatty acids and key minerals like copper, iron, etc. (http://www.diethealthclub.com/articles/360/diet-and-wellness/elephant-yam-and-health- benefits. html). They are traditionally used in arthralgia, cough, bronchitis, anaemia and general debility (Madhurima et al. 2012).

21.4.3 *Amlai Ntoi*

Amlai Ntoi is fermented Indian gooseberry or amla (*Phyllanthus emblica*, Carl Linnaeus) and is indigenous to the *Uchoi* tribe. This is prepared by boiling green amla in water followed by packing in earthen container (matka) with alternative layer of molasses. The first and final layer should be of molasses. The filled matka is buried under ground and allowed to ferment for 5–6 months.

21.4.3.1 Nutritional Value

The raw amla fruit contains high amount of Vitamin C and adequate levels of minerals. Amla exhibits number of health-promoting activities as it improves digestion,

Plate 21.26 *Kosoi*

Plate 21.27 *Bochu-mba*

reduces cholesterol level from body and lowers blood pressure (http://foodtofitness.com/health-benefits-of-amla).

21.4.4 Fermented *Kosoi* and *Bikang*

One of the ethnic groups *Jamatia* prepares this traditional fermented vegetable product from the Lima bean (*Phaseolus lunatus*, L.) which is locally known as *Kosoi* (Plate 21.26). Another variety of these beans, i.e. Sword bean (*Canavalia gladiata*) is known as *Bikang* by the *Uchoi* tribe. These beans are properly dried under sun for 3–4 days and stored. Before consumption, these dried products are soaked in water and kept overnight for partial fermentation.

21.4.4.1 Nutritional Value

Lima beans have antioxidant and anticancer properties and also reduce cholesterol level (http://www.nutrition-and-you.com/lima-beans.html).

21.4.5 *Bochu-mba*

Bochu-mba (Plate 21.27) is a traditional fermented product and made exclusively from the flower of *Bombax ceiba* (L.), commonly known as 'cotton tree'. This is indigenous to the *Uchoi* tribe. The petal and centre petiole part are removed first from the cotton tree flower and completely dried under sunlight. The dried flowers are soaked in water for 24 h followed by wrapping with banana leaf and left for fermentation for 2 days.

21.4.5.1 Nutritional Value

Each *Bombax ceiba* flower contains 0.4 g protein and 1 g dietary fibre (https://au.nutrihand.com/Nutrihand/pctools/showFoodFacts1.do). This product is used against ulceration of bladder and kidney and for healing of wounds.

21.5 Conclusion

The food fermentation technology practiced by the people of Tripura since time immemorial reveals the strong co-existence of the people with nature and its utilization for survival advantages. Fermented foods of both animal and plant origin are a daily component of the diet of the tribals as well as nontribal rural population of the state. The fermented food products are wholesome in their original state and perform a beneficial role in human nutrition. Fermentation, in addition to an ancient method of food preservation, also imparts additional benefits such as enhancing flavour, increasing digestibility and improving nutritional value and provides pharmaceuticals. These have been undoubtedly contributing to the survival, good health and nutritional security of entire generations that had only a limited quantity of rice and vegetables to eat.

Fermentation of food and beverages involve mixed cultures of microbes that grow either simultaneously or in succession. This can easily be said that indigenous people have been getting the benefit from microorganisms unknowingly for varied purposes. It appears that exploitation of microorganisms for food and beverages purposes seems to be the only contribution of indigenous people. Microflora amplifies the levels of proteins, vitamins, essential amino acids and fatty acids in the fermented foods. Malnutrition associated with protein and vitamin deficiencies are the major problems of the hill-dwelling tribal population of the state. In these circumstances, ethnic fermented foods had a great contribution on the physical wellbeing of the inhabitants of the forest-dominated state Tripura.

References

Chongtham N, Bisht MS, Haorongbam S (2011) Nutritional properties of bamboo shoots: potential and prospects for utilization as a health food. Compr Rev Food Sci 10(3):153–168

Debbarma B (2002) The borok people of Twipra, India and their history and languages. In: Debbarma BR (ed) Yakpai. Kokborok Sahitya Sabha, Tripura, pp 40–57

Ghosh S, Rahaman L, Kaipeng DL, Dipankar Deb D, Nath N, Tribedi P, Sharma BK (2016) Community-wise evaluation of rice beer prepared by some ethnic tribes of Tripura. J Ethnic Foods (Elsevier) 3:251–256

Gupta S (2016) Isolation and characterization of predominant bacteria associated with few fermented fish products of Northeast India. Thesis submitted to the Central Agricultural University, Imphal to fulfil the degree of Master in Fisheries Science (Fish Processing Technology)

Hartmann R, Meisel H (2007) Food-derived peptides with biological activity: from research to food applications. Curr Opin Biotechnol 1:163–169

Jeyaram K, Singh TH, Romi W, Devi RA, Singh WM, Dayanidhi H, Singh NR, Tamang JP (2009) Traditional fermented foods of Manipur. Indian J Tradit Knowl 8(1):115–121

Madhurima P, Kuppast IJ, Mankani KL (2012) A review on *Amorphophallus paeoniifolius*. Int J Adv Sci Res Technol 2(2):99–111
Majumdar RK, Basu S (2010) Characterization of a traditional fermented fish product *Lona ilish* of Northeast India. Indian J Tradit Knowl 9(3):453–458
Majumdar RK, Basu S, Nayak BB (2009) Assessment of nutritional quality of '*Shidal*' a fermented fish product of Northeast India. J Indian Fisheries Assoc 36:25–34
Majumdar RK, Bhaskar N (2016) Fermented fish products of north East India: characterization of microbes and their healthful metabolites for application and scale up. Project Report Submitted to the DBT, GoI
Majumdar RK, Roy D, Bejjanki S, Bhaskar N (2015) Chemical and microbial properties of shidal, a traditional fermented fish of Northeast India. J Food Sci Technol 53(1):401–410
Singhal P, Bal LM, Satya S, Sudhaka P, Naik SN (2013) Bamboo shoots: a novel source of nutrition and medicine. Crit Rev Food Sci 53(5):517–534
Tiwary SC, Mahanta D (2007) Ethnological observations on fermented food products of certain tribes of Arunachal Pradesh. Indian J Tradit Knowl 6(1):106–110
Uchoi D, Roy D, Majumdar RK, Debbarma P (2015) Diversified traditional cured food products of certain indigenous tribes of Tripura, India. Indian J Tradit Knowl 14(3):440–446

Websites

http://foodtofitness.com/health-benefits-of-amla
http://www.diethealthclub.com/articles/360/diet-and-wellness/elephant-yam-and-health-benefits.html
http://www.nutrition-and-you.com/lima-beans.html
https://au.nutrihand.com/Nutrihand/pctools/showFoodFacts1.do
https://en.wikipedia.org/wiki/Twipra_Kingdom
https://en.wikipedia.org/wiki/Tripura
https://www.census2011.co.in/census/state/tripura.html

Ethnic Fermented Foods and Beverages of Uttarakhand, Uttar Pradesh, Haryana, and Punjab

22

Arun Beniwal, Tamoghna Ghosh, Kailash N. Bhardwaj, and Naveen Kumar Navani

Abstract

Fermentation of the food has been the most important method for preservation of food throughout human history. In addition to preservation, fermentation introduces beneficial microbes to the gut of the consumers, removes anti-nutrients components, and strengthens the immune system. Ethnic foods have their origin from the culture and heritage of indigenous tribal communities. Lactic acid bacteria have been used to ferment milk, mostly during the period of Indus Valley Civilization. The fermented products are prepared traditionally through the practice of preservation without knowledge of beneficial microbial strains responsible for carrying out these fermentations. Uttarakhand, Haryana, Punjab, and Uttar Pradesh (UP), states located at the Himalayan foothills, comprise of diverse communities with distinctive cultures. Due to the prevalence of diverse topography and climate, these states are privileged with a myriad of fermented foods. The present chapter discusses the ethnic fermented foods; beverages; meat products indigenous to states of Uttarakhand, UP, Haryana, and Punjab; and their microbial diversity.

Keywords

Ethnic · Fermentation · Food · Microbial · Lactic acid bacteria · Microbiology

---

A. Beniwal · T. Ghosh · N. K. Navani (✉)
Chemical Biology Lab, Department of Biotechnology, Indian Institute of Technology Roorkee, Roorkee, Uttarakhand, India

K. N. Bhardwaj
Uttarakhand State Council for Science and Technology (UCOST), Vigyan Dham, Jhajra, Dehradun, Uttarakhand, India

J. P. Tamang (ed.), *Ethnic Fermented Foods and Beverages of India: Science History and Culture*, https://doi.org/10.1007/978-981-15-1486-9_22

22.1 Introduction

Fermented foods are functional foods which are being consumed by ethnic groups, with a historical indication, and have been found to be essential for the welfare of the people. People belonging to states of Uttarakhand, Punjab, Haryana, and UP have extremely diverse demographic characteristics, including different communities, religions, castes, and tribes. Ethnic origins, languages, types of food, food habits, climatic conditions, and religions are the major sources of cultural diversity of UP, Punjab, and Haryana. The authentic cuisine of Punjab, Haryana, and UP offers a similar delicacy category of fermented food such as *rabadi*, *curd*, and *lassi*. The states such as Punjab and Haryana have transcended on a vast journey and have preserved the footprints of their traditional foods. People of Punjab and Haryana give priority to milk products which they regularly incorporate in their regular diet. However, fermented products found in Uttarakhand possess even greater diversity due to the presence of a diverse geographical environment prevalent in the state (Joshi et al. 2018). These fermented products are prepared by small tribes through the indigenous practice of preservation and processing without knowledge of specific microbes responsible for carrying out these fermentations. The living microorganisms present in these fermented foods have the potential to act as bio-therapeutic agent (Hesseltine 1965). All these states are rich in human resources, history, heritage, and flora and fauna. However, due to the recent moderation of the society and influence from western civilization, there is a further need to preserve the fermented ethnic food of these states. In this chapter we will discuss about fermented ethnic food of Uttarakhand, a Himalayan state in northern part of India, and its surrounding states such as UP, Haryana, and Punjab (Singh 2000).

22.2 Agroclimatic Condition of Uttarakhand, UP, Punjab, and Haryana

Uttarakhand state is the 27th state of our country, carved out of Uttar Pradesh northwestern districts in the year 2000. The state spread over an area of 53,483 km^2 is known for its beauty and comprised of the central Himalayan region. The state is also given the name Devbhoomi due to the presence of holy temples, shrines, and pilgrimages (Maheshwari and Sharma 2018). Only a small area (about 14%) is available for cultivation in the state as the major area of Uttarakhand is covered by wasteland and forest area. Most of the state region includes a hilly region (about 93%) with districts like Dehradun, Tehri, Rudraprayag, Uttarkashi, Pauri, Almora, Chamoli, Bageshwar, Pithoragarh, and Nainital. The plain region (Terai-Bhabar) comprises a smaller proportion and covers districts such as Haridwar, Udham Singh Nagar, and few parts of Nainital and Dehradun districts (Tuteja 2013; Farooquee and Maikhuri 2009). The state of Uttarakhand is administratively divided into two regions: Kumaun and Garhwal. The Garhwal region covers an area of 32,450 km^2, whereas Kumaun has an area of 21,035 km^2. Kumaun region consists of six districts, whereas Garhwal region comprises of seven districts (Chauhan and Bhatt

2000; Tuteja 2013). The livelihood security of the people depends on farming, forest resources, and animal husbandry (Tolia 2010).

The main backbone of the Uttarakhand state is its agriculture which is practiced by a major section of the society. Due to the presence of different agroclimatic endowments and vivid topography in Uttarakhand state, the farmers of the state practice subsistence type of farming (Bodapti and Chander 2013; Parihaar et al. 2014). The soil present in the plain is fertile and supports farming of diverse group of crops, whereas soil of the hills is less fertile as it is constantly eroded due to frequent soil erosion. The area covered by the hills are cultivated for farming of mixed type of crops, whereas in plain areas of the state, single crops such as wheat, rice, oilseeds, soya bean, and pulses are grown (Rais et al. 2009). In Uttarakhand and Uttar Pradesh, farmers practice two modes of agriculture, which include irrigated and rainfed cultivation. However, in the case of Uttarakhand, most of the agricultural land is rainfed. As the state is covered by hilly landscapes, most of the irrigation takes place in plains and valleys. The state of Uttarakhand is also recognized for its economically important horticultural crops and aromatic and medicinal plants. Due to predominance of the temperate climate zone in the hills, only kharif crop is mostly harvested. Therefore, subsistence type of farming is predominant in the state. The principal crops of the state are wheat (30.91%), rice (25.51%), and ragi (12.32%). Presently, to a smaller extent, small millets and sugarcane are also grown in the state. The crop pattern of the state depicts that state is dominated by food grains. Traditional crops like ragi and maize have now lost its share in terms of crop production, while soya bean have gained its share in the state crop production. The mountain terrain and variation in climate offer natural advantage for the production of diverse crops. Therefore, the diversity found in climatic conditions of the state provides a distinctive benefit as well as a competitive advantage over the neighboring states such as UP, Haryana, and Punjab in terms of production of off-season fruits and vegetables and fetches a higher value in the food market.

States such as Haryana and Punjab always stand tall in terms of achievement in agriculture and its cultural heritage. Both these states represent the modern face of India. Both these states have bountiful heritage, culture, vibrant landscape, folklores, and festivals. The state of Haryana was carved from Punjab in 1966 and is popularly known as cradle of ancient culture and was home of the “Bharata” dynasty. Haryana is bound by UP, Rajasthan, Punjab, and Himachal in east, south, west, and north, respectively. Haryana has been observed to contain alluvial plain and Siwalik range. Parts of the Aravalli ranges are also evident in the southern part of Haryana. The cultural life of the Haryana reflects both the agriculture economy and treasure of legends with deep roots in ancient India. A number of festivals in honor of the deities are an important factor that depicts the state’s cultural heritage. Traditional houses of Haryana (commonly known as Haveli) have unique architecture. They have elaborate gates and have huge podiums which are mostly used for religious and social functions. These homes represent the socioeconomic status of the people.

Haryana and Punjab are prosperous in the agriculture sector of India, and both of the states contribute a major amount of rice and wheat crops to the national repository system of our country. Both of the states and parts of western UP are also involved in the production of cotton, mustard and rapeseed, pearl millet, sugarcane, chickpea, maize, sorghum, and potatoes. Haryana has been observed with a variation in the level of precipitation which ranges from 300 to 1100 mm. Punjab, the prosperous state in the northwestern region of India, originally imparted its name on the basis of five rivers, Sutlej, Beas, Ravi, Chenab, and Jhelum, which flow through the state. It stretches from the latitudes 29.30° North to 32.32° North and longitudes 73.55° East to 76.50° East, with a total area of 50,362 km^2. The state has amalgamation of heat in summer, rain in monsoon, and cold in winter to their extreme, which makes it very fertile for the cultivation of staple crops. It contributes two-thirds of the total production of food crops in the country; thus it is also called "India's Bread Basket" or "Granary of India." Dairy buffaloes and cattle are the major livestock in both Punjab and Haryana. Both bullocks and modern agriculture equipments are used for plowing the land.

Recent migration of the young people in search of better job opportunities has reduced the total workforce required for the laborious agricultural operations performed in these states. Due to changing lifestyle and food habits, the residents of Uttarakhand, Punjab, and Haryana have abandoned the cultivation of traditional crops like millets and have mostly shifted to rice and wheat cultivation. These are responsible for the disappearance of the traditional millets and other minor crops of these states. Millets, other ancient grains cultivated in these states, and their food products need to be promoted using modern biotechnological tools to preserve these unique heritages (Chandra et al. 2010).

22.3 Ethnic People

Historically, the northern states of our country, especially Punjab, were the gateway to India which throughout the history have witnessed people from different communities, racial and cultural traits, and inhabiting parts of Punjab, Haryana, UP, and Uttarakhand. According to the reports presented by the Ministry of Tribal Affairs (2003–2004), GOI, there is no tribal population in the state of Haryana and Punjab (Singh 2008). However, most of the tribes are from Uttarakhand state. The most prominent ethnic groups of the Uttarakhand are the Jaunsaris, Bhotiyas, Rajis, Bokshas, and Tharus (Dixit and Singh 2017). In Uttar Pradesh, along with these ethnic communities, additional tribes such as Baiga, Agariya, Bhuiya, Chero, and Parahiya were also found that are mostly located in Sonbhadra district of the state. Jaunsari community comprises of small tribe which has a peculiar style of dressing and is also divided into caste. Bhotiya have their homes at high altitude of Western Himalayas. They are further divided into various subgroups such as Marcha, Jadh, and Shaukas. The community has Mongol-Tibetan-like appearance, but the subgroups have their specific and diverse culture. Rajis are also better known by their other name Van Rawats, and the tribe generally dwells in the forest area of southern Pithoragarh district of state. The Rajis are also known for their matriarch society. The main source of their livelihood is using natural resources for making beautiful wooden utensils and performing shifting cultivation. Buksha are the tribes of Rajput descendants located in the Terai district of Uttarakhand. Tharu are tribal people in the eastern region of Terai. They were considered as original migrant from that of Tibetan tribes. All these tribes have contributed toward the preservation of beneficial microorganisms in the form of various fermented ethnic products which are delineated further (Chandra et al. 2010; Tolia 2010).

22.4 Ethnic Fermented Food of Uttarakhand, UP, Punjab, and Haryana

Ethnic tribes have conserved the knowledge about the procedure of preparation and their associated local ingredients. Such fermented foods are results of natural process which people observed and further developed the method for their cultivation. This method of cultivation differs according to place and surplus food available in that particular region. The process of spontaneous fermentation also varies according to the geographical conditions prevalent in the area (Table 22.1). In upper Himalayan regions of Uttarakhand where cold climate is prevalent throughout the year, fermentation process is completely essential for survival of tribal people. In summer season, people use to ferment local available food for months and utilize them in winter season. People staying in tropical climates of India are not driven by such seasonal requirements, yet fermentation is also an important part of these people in products like dahi (Tamang and Fleet 2009; Shah 2006).

Bacterial species found in fermented food play crucial roles in our body functioning and can further diversify and replenish the microbial ecology of human microbiome. Fermentation process enables the food to remain stable with longer shelf life at ambient temperature present in the region and also has advantage of saving fuel as fermentation can digest complex components which normally require a longer period while being cooked. Moreover, by promoting the health of the intestine, fermented foods can help in controlling disease such as diarrhea. These fermented foods might also improve the survival rate of infants in tribal population (Rawat et al. 2018; Tamang and Kailasapathy 2010).

Haryana and Punjab are always known throughout India for their agriculture and livestock wealth. There is an abundance of ghee, milk, lassi, butter, rabadi, dahi, thandai, and vegetables. Haryana is also rich in eating and cultural habits and used to be known as "Land of Rotis." Punjab on the other hand has been an agrarian society since the ancient civilizations, but it is also popular for its delectable and exotic cuisine. Punjabi cuisine offers a vast variety of dishes, both vegetarian and nonvegetarian, with varied traditional culinary style, like tandoors and chulhas. The cuisine involves the excess use of milk and milk products in most of dishes with lots of traditional Indian spices and fresh vegetables. Fermented foods, like dahi and lassi, are also an important part of Punjabi cuisine. The traditional utensils used by Punjabis are made up of brass and copper. The acclaimed warmth and hospitality of Punjabis assure that no one visiting their land should return with empty stomach. Indeed, the third Guru of Sikhs, Guru Amar Das Ji, also expanded the tradition of

Table 22.1 Ethnic fermented foods and beverages of Uttarakhand, UP, Punjab, and Haryana

Food	Substrate	Sensory property of product	Major ethnic consumers	State
Keem	Barley flour, roots, leaves and bulbils, Jayaras	Sticky, flavored curry	Jaunsaris	Uttarakhand
Soor	Cereals flour, fruits	Beverage	Parvatis and Jaunsaris	Uttarakhand
Balam	Roasted flour of wheat, spices, and herbs	Starters	Bhotiyas	Uttarakhand
Chhang	Boiled barley	Beverage	Bhotiyas	Uttarakhand
Jann	Boiled rice, fruits	Beverage	Bhotiyas	Uttarakhand
Chartayshya	Goat meat	Sausage	Bhotiyas of Kumaun Himalayas	Uttarakhand
Jamma	Goat meat, finger millet, timbur, small intestine	Sausage	Bhotiyas of Pithoragarh	Uttarakhand
Arjia	Chopped lungs of goat, goat meat, timbur, large intestine	Sausage	People of Kumaun	Uttarakhand
Wari	Black gram, asafetida, cardamom, fenugreek, red pepper, clove, ginger	Dried, spicy, and brittle balls	People of Amritsar, Uttarakhand, UP	Uttarakhand, Punjab
Jalebi	Maida, Dahi, water	Crispy, deep-fried	All	Uttarakhand, Punjab, Haryana, and UP
Sez	Rice		Bhotiyas	Uttarakhand, Punjab, and Haryana
Dahi and *Rabadi*	Milk	Savory	All community	Uttarakhand Punjab, Haryana, and UP
Bhallae	Black gram	Deep-fried patties	Punjabi community	Punjab and Haryana
Lassi	Dahi	Savory	All community of Punjab and Haryana	Uttarakhand, Punjab, Haryana, and UP
Kanji	Beet and carrot roots	Red-colored strong alcoholic beverage	People of Haryana and Punjab	Punjab, Haryana
Doli Ki Roti	Whole wheat flour, spices	Golden brown like puri	Multani Punjabi migrated from Pakistan	Punjab and Haryana

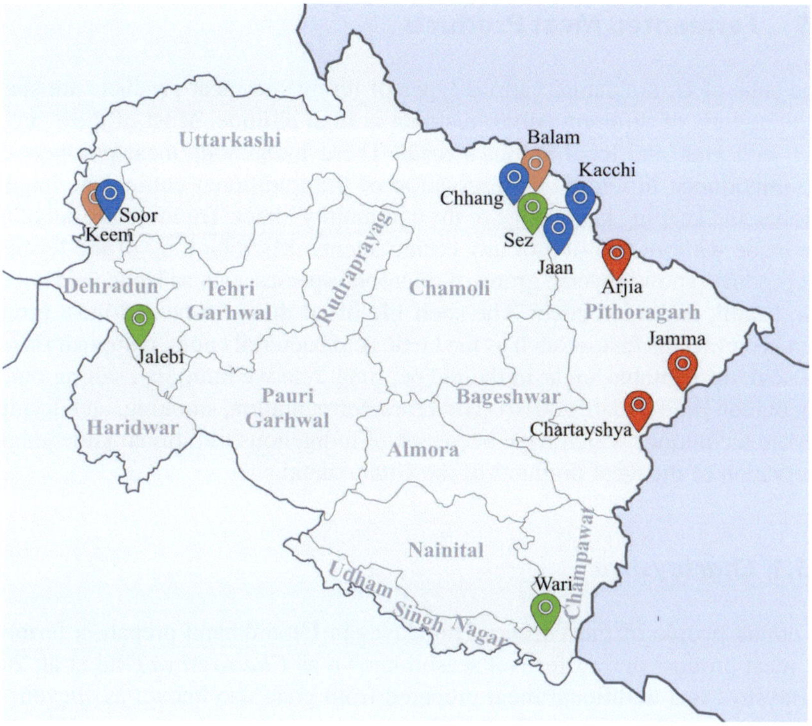

Fig 22.1 Map of Uttarakhand localizing the residing areas of tribes producing different ethnic fermented products

Langar (community kitchen) with a noble thought that "food is central to communal bonding."

Unfortunately, some of the ethnic fermented products are on the brink of extinction as some have already disappeared from the diet of native people, due to the effect of changing lifestyle and urbanization. A diet rich in fermented food protects against numerous diseases and further aids in maintaining the homeostasis during digestion (Fig. 22.1). A loss of interest in these traditional foods has hastened the industrialization of products which have their origin in foreign countries (Roy et al. 2004). Meat products, milk, and vegetables are highly perishable items, and since the dawn of civilization, our lineages have used different techniques which they discovered or maintained to preserve foods for the season when plenty of food is not available. The native people have magnificent skills to prepare the traditional food and ability to identify the palatable products. The high degree of diversity among fermented foods, meat products, and beverages is a major specialty of these states.

22.5 Fermented Meat Products

In the state of Uttarakhand, various types of fermented meat products are specific for the regions of different tribes, residing at high altitude. Most of these are prepared with goat and local animal's meat. These indigenous meat products carry huge importance in terms of preservation of the traditional cultural heritage and microbes and keeping knowledge of the community intact. Traditionally, all of these were made without addition of any curing agents. Microbiology of the fermented meat products shows diverse group of microbial species such as lactic acid bacteria, yeast, bacilli, and micrococci. The shelf life of all these fermented meat products depends on various factors such as the lactic acid bacterial count, composition of the fatty acid, fermentable sugar in the recipe, time, relative humidity, salting out, and fermentation time (Adams 2010). Therefore, fermentation, smoking, and drying are the main techniques which have been part of indigenous traditional knowledge for preservation of the meat products of the Uttarakhand.

22.5.1 *Chartayshya*

The ethnic people of the Western Himalayas in Uttarakhand prepare a fermented goat meat product during festival season known as *Chartayshya* (Rai et al. 2009). Chartayshya is a traditional meat prepared from goat also known as chevon. This product is generally consumed by the Bhotiya community of the Kumaun Himalayas. The product is also consumed by people of Byan and Darma Valley of Dharchula. Chartayshya is also prepared during a religious festival where the ancestors are worshipped known as "Kolatch."

22.5.1.1 Traditional Method of Preparation

Red meat of the goat is cut into various small pieces (3-4 cm)

↓

Mixed with the salt (approximately 1%), sewed in a thread and then suspended on sticks or bamboo strips for a period of 15-20 days outside the house in an open air

↓

The natural microflora of air allows fermentation of the meat

↓

The product can then be stored at ambient temperature for a longer period for further consumption.

22.5.1.2 Culinary and Mode of Consumption

This is the most delicious fermented meat product consumed by the Bhotiya community. It is prepared as curry along with ingredients such as fried garlic, onion, and tomato with a proper amount of salt.

22.5.1.3 Socioeconomic Values

Chartayshya also carries a huge socio-ethnical importance as the ethnic people of Kumaun region prepare the curry of *Chartayshya* in every household and offer it to the ancestor as a part of the festivals.

22.5.1.4 Microbiology

The predominant microbial species found in *Chartayshya* are *Lactobacillus divergens*, *Bacillus subtilis*, *Pediococcus pentosaceus*, *Enterococcus faecium*, *Bacillus mycoides*, *B. thuringiensis*, *Micrococcus* sp., *Debaryomyces hansenii*, *Staphylococcus aureus*, and *Candida famata* (Oki et al. 2011).

22.5.2 *Jamma/Gemma*

Jamma also known as *Gemma* is an ethnic fermented sausage prepared from meat of chevon which is consumed by the people in Kumaun. *Jamma* is also part of the traditional food of Bhotiyas of Pithoragarh district (Oki et al. 2011).

22.5.2.1 Preparation of the *Jamma*

Meat of chevon is chopped into very fine pieces followed by addition of various ingredients such as finger millet, timbur, salt and chilli powder

↓

These ingredients are properly mixed along with small amount of animal blood

↓

The meat is made in semi-liquid form with the addition of water and then the semi-liquid material is stuffed inside the small intestine of the goat

↓

The resulting intestine (100-120 cm length) is then tied at both the ends and then pricked randomly at different positions to avoid rupture during boiling

↓

After boiling for a period of 20 min, the intestine is smoked for a duration of 15-20 days

↓

Natural microflora of intestine, millets and spices serves as starter for fermentation

22.5.2.2 Culinary and Mode of Consumption

This fermented sausage is not available in the market and is prepared generally at individual homes, mostly at the time of festivals and rituals. *Jamma* is consumed in the form of curry along with addition of different ingredients for taste such as tomato, garlic, and salt. Sometimes the sausage is eaten in cooked form. However, this is often taken along with ethnic fermented alcoholic beverage.

22.5.2.3 Microbiology

The predominant microbial species found in *Jamma* are *Lactobacillus sanfrancisco*, *P. pentosaceus*, *E. faecium*, *Leuconostoc mesenteroides*, *Lb. divergens*, *Bacillus subtilis*, *Micrococcus* sp., *S. aureus*, *C. albicans*, and *D. hansenii* (Oki et al. 2011).

22.5.3 *Arjia*

Arjia is one of the ethnic fermented sausages prepared in tribal homes using meat of goat. This is an important diet component of the people of Kumaun (Rai et al. 2009; Oki et al. 2011).

22.5.3.1 Traditional Method of Preparation

Chopped lungs of the goat are chopped into fine pieces

↓

Black pepper known as timbur, chilli powder and salt are added along with blood of the goat.

↓

The prepared mixture is then stuffed in the large intestine of the animal

↓

Prickling was also performed before boiling to prevent the rupture of the large intestine

↓

The boiled intestine was then dried or smoked in the kitchen oven for a period of approximately 20 days

22.5.3.2 Culinary and Mode of Consumption

This fermented meat is consumed along with regular meals in the form of deep fried sausage or in the form of curry.

22.5.3.3 Microbiology

The predominant microbial species found in *Arjia* are *Bacillus subtilis*, *B. thuringiensis*, *E. faecium*, *P. pentosaceus*, *B. mycoides*, *Micrococcus* spp., *Cryptococcus humicola*, *S. aureus*, and *Debaryomyces hansenii* (Oki et al. 2011).

22.5.3.4 Nutritional Value

There are very few studies depicting the nutritional property of *Chartayshya*, *Jamma*, and *Arjia*. All of these fermented meat products have pH range between 5.3 and 6.9 due to the presence of the lactic acid produced by the lactic acid bacteria. Due to the procedure of drying and smoking after boiling, moisture content decreases where *Chartayshya* was found to contain a much lower percentage of moisture content of 17%, whereas the remaining two, *Jamma* and *Arjia*, both were reported to have a higher moisture content of approximately 60%. Therefore, due to the presence of two factors, i.e., lower pH and moisture content, the product can be stored safely for longer duration of period. Titratable acidity of these products was also in range from 0.3 to 2.6%. Most of these fermented meat products are rich in protein content and found to be in range between 6 and 36%. Among the three, *Chartayshya* has the highest nutritional content with a concentration of more than 36% of both protein and carbohydrate. Moreover, food value of these meat products was also found to be very high and ranged from 413 to 454 kcal/100 g of dry matter (Oki et al. 2011; Rai et al. 2010).

22.6 Fermented Beverages

The predilection of human being for the fermented beverage dates back to ancient scriptures in mythology of India. Examples of fermented beverages are present in all the states of India and consumed as a part of culture and diet. These beverages are made from fermentable sugars with some flavoring materials distinctively native to the surrounding area. The use of a microbial consortium in the preparation of beverage is an ancient technique. Beneficial bacteria and yeast have been found to produce ethanol from various carbohydrate substrates which include glucose, lactose, galactose, etc. (Saini et al. 2017, 2018; Tamang 2012). Although fermentation processes have been developed to much advanced stages in commercial level (yoghurt, cheese, ethanol from whey), ethnic beverages of Uttarakhand, Haryana, and Punjab have not been explored at commercial level so far (Anand et al. 2016; Kokkiligadda et al. 2016).

Method of preparation: Fermented beverages can be kept for a longer period of time without adding any type of preservatives due to the production of acid and alcohol, which act as natural preservatives.

22.6.1 *Balam*

Balam is a starter culture used to prepare fermented alcoholic beverage. It was also reported in some studies that Bhotiya prefer *Triticum aestivum* (wheat) as the main cereal for preparation of starter over other cereals such *as Fagopyrum tataricum* (buckwheat), *F. esculentum* (kothu), and *Hordeum vulgare* (barley). It is prepared from wheat with the addition of both spices and herbs. The grains are first washed with water and dried, and its flour is roasted till its brown color appears. The flour is then mixed with the *Piper longum* Linn. (long pepper), *Amomum subulatum* Roxb (black cardamom), *Cinnamomum zeylanicum* Breyn (cinnamon), and seeds of *Ficus religiosa* Linn. (Das and Pandey 2007). The resulting mixture is then mixed with water and prepared as semidried thick paste which is molded into balls called as *balam* balls (Fig. 22.2). These balls are then incubated under bed of *Pinus roxburghii* (chir pine), *Cannabis sativa*, and *Cupressus torulosa* (Himalayan cypress), for a period of 2 weeks or until these balls get dried and turned into white color after colonization of desired microbes. These are dried and used as a starter culture for

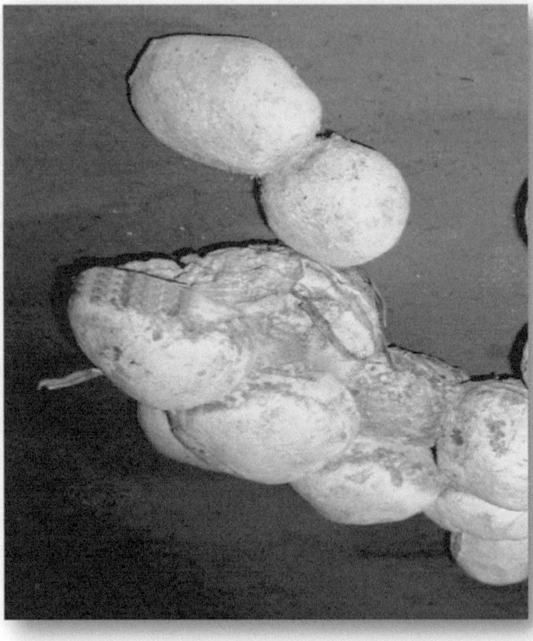

Fig 22.2 Pictorial representation of balam balls used as starter culture for the preparation of fermented beverages Chhang and Jann

important alcoholic beverages *Chhang* and *Jann*. Recently it has been observed that Bhotiya tribes have shifted to back-slopping method using previous balam as starter and have stopped the use of herbs (Bhardwaj et al. 2016). Moreover, balam is also used for the treatment of cholera and weakness of the cattle. It has been found that infusion of herbs allows the beneficial bacteria and yeast flora present in it to colonize the balam balls and also offer antibacterial effects against spoilage-causing bacteria during storage period. Moreover, natural conditions also offer growth advantage to particular species of bacteria, fungi, and yeasts such as lactic acid bacteria that are found associated with the plants.

It is prepared from wheat with the addition of spices and herbs

↓

The grains are first washed with water, dried and its flour is roasted till brown color appears

↓

The flour is then mixed with the *Piper longum* Linn. (Long pepper), *Ammomum subulatum* Roxb (Black cardamom), *Cinnamomum zeylanicum* Breyn (Cinnamon), and seeds of *Ficus religiosa* Linn

↓

The resulting mixture is then mixed with water and prepared as semidried thick paste which is molded into balls called as balam balls

↓

These balls are then incubated under bed of *Pinus roxburghii* (Chir pine), *Cannabis sativa, Cupressus torulosa* (Himalayan cypress), for a period of 2 weeks or until these balls get dried and turned into white color after colonization of desired microbes

↓

These are dried and used as a starter culture for important alcoholic beverages *Chhang* and *Jann*

22.6.1.1 Microbiology

The predominant group of microorganisms found in balam are yeast and lactic acid bacteria, which include *Saccharomycopsis fibuligera*, *R. stolonifera* var. *lyococcus*, *Sm. capsularis*, *M. hiemalis*, *C. glabrata*, *Rhizopus chinensis*, *Mucor circinelloides*, *S. bayanus*, *Lb. bifermentans*, and *Lb. brevis* (Tamang 2010; Kumari et al. 2016).

22.6.2 *Soor*

This alcoholic beverage is a part of culture of Parvati and Jaunsari communities which are inhabitant of Tons Valley in Uttarakhand. This beverage carries importance due to consumption in homes during functions such as birth ceremony, marriage, and some other ritual feast and during festival seasons by these traditional people of Himalayan. The tribal community believes that soor provides relief in urinary tract-related problems, kills intestinal worms, and also is responsible for purification of the blood. Earlier, it was observed that each house of the community carries its own household-level distillation unit. But modernization of the society has also influenced these communities, and the practice of preparing the fermented beverage has been restricted to few households only (Rana et al. 2004; Tamang et al. 2016).

22.6.2.1 Traditional Method of Preparation

The beverage is prepared using a starter known as *Keem*

↓

Different fruits such as pears, apple and apricot are also used as raw ingredients for the preparation of beverage *Soor*

↓

Dough is prepared using traditional cereals present locally with beneficial herbs belonging to Himalayas

↓

Roots, leaves and bulbils of these traditional herbs for preparation of Keem

↓

The prevalent local species of *Cannabis, Sapindus mukorossi* (Indian Soapberry) and other plant species are dried and mixed with barley flour

↓

The resulting mixture is infused with tender and leaves called as Jayaras and doughed into cake like structure

↓

These cakes are then allowed to ferment for a period of 24 days on the bed of plants known as Sathar in a closed room

↓

The cakes are then inverted for proper fermentation and kept for a period of additional 12 days

22.6.2.2 Fermentation and Distillation

The distillation of soor involves cereals and fruits as its main substrate for carrying out fermentation

↓

The cakes are kept in earthen vessel and closed with woolen clothes for maintaining the ideal temperature for fermentation

↓

The fermentation is carried out for a period of about 10 days and in between checked for the generation of gas

↓

The resulting fermented mixture is then transferred to a vessel and put on fire for carrying out distillation

↓

After heating, distillate is slowly carried out in a pot

↓

The first distillate of initial hour is of best quality soor and contain ~ 40% ethanol

22.6.2.3 Microbiology

The predominant genera found in *soor* belong to *C. famata*, *K. marxianus*, *S. cerevisiae*, *C. valida*, and *Sm. fibuligera* (Thakur and Bhalla 2004; Das and Pandey 2007). Starter *keem* is also rich in microbial flora with higher load of yeast such as

Fig 22.3 Picture showing the earthen pot used for storing and fermenting traditional alcoholic beverage Chhang, along with the fermented product

Fig. 22.4 Picture showing the arrangement used by Bhotiya community for distillation of Kacchi, an alcoholic beverage

C. glabrata, *C. tropicalis*, and *Wickerhamomyces* sp. The lactic acid bacteria are also found with a concentration of about 10^7 cfu/mL representing *Enterococcus casseliflavus*, *Weissella paramesenteroides*, and *Enterococcus durans*. Amylolytic bacillus has been also observed in *keem*.

22.6.3 *Jann (Jaanr)* and *Chhang*

Jann or *jaanr* is a fermented finger millet or barley beverage with mild alcoholic content and commonly consumed by the Bhotiya communities of Uttarakhand (Sekar and Mariappan 2007).

22.6.3.1 Traditional Method of Preparation (Figs. 22.3 and 22.4)

Rice is boiled in water till it becomes soft

↓

The excess water is drained off and spread on a container to cool it down

↓

This is then mixed with the Balam and kept in airtight container at a warm place for fermentation

↓

The fermentation is carried out best in the absence of oxygen and at relatively lower temperature in earthen pots

↓

The fermentation takes place for a period of one week

↓

Usually for improving the flavor, it is then allowed to ferment upto a period of one year

Chhang

The beverage is prepared from barley seeds which are first soaked in water and then boiled and allowed to cool

↓

After cooling, these seeds are mixed with the Balam balls in an appropriate ratio and then kept in earthen pot for a period of 4-7 days

↓

The resulting product is then transferred to distillation pot for distillation (Figure.3 ←)

↓

Another similar distilled beverage called Kacchi is prepared by bhotiya community, by carrying out fermentation of barley grains using *balam* balls as starter culture (Figure.4 ←).

22.6.3.2 Microbiology

The predominant microbial species found in *Jann* and *Chhang* are *Rhizopus chinensis*, *S. cerevisiae*, *M. circinelloides*, *Lb. bifermentans*, *S. fibuligera*, *C. glabrata*, *P. anomala*, *K. marxianus*, and *P. pentosaceus* (Tamang and Fleet 2009; Bhardwaj et al. 2016).

22.6.4 *Kanji*

Kanji is a red-colored strong alcoholic beverage which is prepared from the beet root and the carrot in north of India. This is prepared from their juices and is consumed in the summer.

22.6.4.1 Traditional Method of Preparation

Kanji

It is prepared from raw roots of carrot and the beet

↓

These roots are washed and shredded. Salt and mustard seeds along with water are added and left to ferment at 26-34 °C for a period of 4-7 days

↓

Pink color liquid obtained were drained off and used for drinking

↓

Portion of the previously prepared Kanji can be used as the starter culture of the freshly cut carrot and beet

↓

These can be stored in bottles and consumed within a period of 2 weeks

22.6.4.2 Microbiology of the *Kanji*

The yeast is the active strains responsible for fermentation, converting the sugars present in the carrot and beet into ethanol. *H. anomala*, *Geo. candidum*, and *C. tropicalis* are the predominant flora present in the *Kanji* (Campbell-Platt 1987).

22.7 Fermented Cereal and Milk Products

Cereal foods which have always been an important part of daily meal in our country have also been found to provide all the essential nutrients such as proteins, carbohydrates, dietary fibers, and vitamins. Cereals are also sources of phytic acid, phytochemicals, and sterols. Fermentation of these cereals for the production of bread and butter has been associated since the advent of the human civilization (Steinkraus 2002; Poutanen et al. 2009). Sourdough fermentation has been reported to improve the nutritional properties of wheat, oat, and rye baked goods by increasing the concentration of the bioactive molecules discussed above in cereals (Katina et al. 2005). These health-promoting phytochemicals include some phytosterols, vitamins, and folic acid that are found to be present in the outer layer and germ layer of kernel (Glitsø and Knudsen 1999). For example, yeast-based fermentation of wheat has been found to increase the level of folate in baked wheat (Kariluoto et al. 2004). Furthermore, these cereals act as effective substrates for increasing the production of probiotics incorporated as functional food which are responsible for the growth of lactic acid bacteria and *Bifidobacterium* (Swennen et al. 2006). The fermentation carried out by these bacteria yields useful metabolites such as short-chain fatty acid (SCFA). The SCFA are known for stimulating the growth of beneficial microbiota in the large intestine (Macfarlane and Macfarlane 2006; Roopashri and Varadaraj 2009). Different milk and cereal-based ethnic fermented foods of Uttarakhand, Punjab, Haryana, and UP are discussed in details in next section.

22.7.1 *Sez*

Sez is a traditional semi-fermented food consumed in the form of snack mostly by tribal community Bhotiya. This fermented food is prepared from rice (*Oryza sativa*) (Kumar et al. 2013). The *Sez* is prepared as an intermediate step during preparation of *Jann*. However, generally for the preparation of the *Sez*, process of fermentation is slowed down for the extraction of the *Sez*, and lower amount of balam is added as a starter for substrate fermentation. After extraction of the *Sez*, fermentation is increased by the extra addition of balam for the formation of *Jann* (Roy et al. 2004).

This fermented food is prepared from rice (*Oryza sativa*) (Satish Kumar et al. 2013)

↓

The *Sez* is prepared as an intermediate step during preparation of *Jann*

↓

However, generally for the preparation of the *Sez*, process of fermentation is slowed down for the extraction of the *Sez* and lower amount of Balam is added as a starter for substrate fermentation

↓

After extraction of the *Sez*, fermentation is increased by the extra addition of Balam for the formation of *Jann*.

22.7.1.1 Microbiology

The predominant species reported in the *Sez* are *Saccharomyces* sp., *Bacillus* sp., *K. marxianus*, and *S. fibuligera* (Ray et al. 2016)

22.7.2 *Jalebi*

Jalebi is one of the doughnut-shaped, sweet, crispy snacks made from wheat flour. It is consumed in most of the northern state of India. However, it is especially served in Uttarakhand, Uttar Pradesh, and Haryana (Chitale 2000; Sharma and Sarkar 2015; Tamang 1998).

22.7.2.1 Traditional Method for Preparation

Jalebi is prepared by mixing local yoghurt (Dahi) and white wheat flour (maida), followed by addition of water in it and leaving it at room temperature for overnight

↓

This overnight leavened thick batter is then squeezed and allowed to pass through an embroidered hole of about 4 mm diameter using cotton cloth

↓

The resulting batter is passed through the hole into hot edible oil in the form of continuous spirals

↓

The spiral shaped snack is fried from both the sides until a slight golden color crispness appears

↓

When the *jalebi* turns golden it is removed from the oil and dipped for several seconds in a sugar syrup

↓

The resulting spiral shape snack is termed as *jalebi* and is served hot

22.7.2.2 Microbiology

Jalebi batter has been found to contain predominantly lactic acid bacteria (LAB) such as *Lb. fermentum*, *Lb. bulgaricus*, *Lb. buchneri*, *S. thermophilus*, *S. lactis*, and *E. faecalis*. Yeast strains such as *S. cerevisiae*, *S. bayanus*, and *Hansenula anomala* have also been reported from *jalebi* (Soni and Sandhu 1990; Sharma and Sarkar 2015).

22.7.3 *Dahi* and *Rabadi*

Dahi is the oldest traditional fermented milk product found in our country which is alive with all the important nutrients and flavor. The flavors of traditional *dahi* tend to be more distinct as compared with product like yoghurt which is nowadays supplied by the organized dairy sector. People of Uttarakhand, Uttar Pradesh, Haryana, and UP have appreciated the unique flavors of *dahi* resulting from the lactic acid bacteria-based transformation. Eating of indigenous fermented foods like *dahi* is an extremely healthy practice, which directly infuses our digestive tract with live healthy probiotic bacteria essential for breaking down milk proteins and further assimilating nutrients present in milk. *Rabadi* is a Cereal-based fermented dairy product prepared from maize flour and buttermilk. pH of *rabadi* varies from 6.4 to 6.7, whereas for *dahi*, it is from pH 4.2 to 4.5 (Sekar and Mariappan 2007).

Milk is boiled and kept for 30 min for cooling

↓

In a separate bowl two spoons of curd (back slopped starter culture) are added and smoothened

↓

When the milk turns lukewarm, it is mixed with culture by constant stirring with a clean spoon or a spatula

↓

Once the milk and the starter curd culture are mixed well, the bowl is set aside in a dark warm place to allow fermentation for overnight

22.7.3.1 Microbiology

The predominant species reported found in local *yoghurt* or *dahi* are *Lb. acidophilus, Lb. paracasei, Lb. kefir, Lb. alimentarius, S. thermophilus, Lb. paracasei* subsp. *paracasei, Lb. plantarum, Lactococcus lactis* subsp. *lactis, Lc. lactis* subsp. *cremoris*, and *E. faecium* (Dewan and Tamang 2007). *Micrococcus, Bacillus*, and *P. acidilactici* are predominant species found in rabadi. *Dahi* and *rabadi* are nutrition-rich foods where *S. lactis* and *S. cremoris* have been found to raise the concentration of vitamin thiamine from 2 to 20% as compared with that of milk (Sankaran 1998; Sekar and Mariappan 2007; Sarkar 2008).

Lassi: *Lassi* is one of the most popular fermented milk beverages consumed in both Haryana, UP, and Punjab.

22.7.3.2 Traditional Method of Preparation

The traditional *dahi* prepared in every household is broken and mixed with the water

↓

Salt or sugar are added for the flavor according to the requirement

↓

The mixed components were homogenized and poured in glasses and are ready to drink

Microbiology: The micro flora present in the *lassi* are same as that of the *dahi.*

Microbiology: The microflora present in *lassi* is the same as that of *dahi*.

22.7.4 *Wari*

Wari is an endogenous fermented food prepared from black gram product in the states of Punjab, Haryana, Uttar Pradesh, and Uttarakhand. In Punjab it is commonly called Punjabi *wari*. It is dried, spicy, hollow, and brittle ball of 3–8 cm and 15–40 g (Tewary and Muller 1989). These *warries* are somewhat similar to the Japanese product miso. These are highly popular in the Amritsar area of the Punjab region as these Punjabi *warries* are used as condiments while being cooked with legumes, vegetables, and rice.

22.7.4.1 Traditional Method of Preparation

Black gram seeds are submerged in water for 5-10 h followed by grinding the seeds in mortar pestle

↓

The paste formed is inoculated with starter culture from previous batch and spices like asafetida, cardamom, fenugreek, red pepper, caraway, clove and ginger

↓

The mix is allowed to ferment for 1-3 days and balls are hand molded

↓

After 2-8 days of air-drying in bamboo mats, the dish is served with vegetables as side dish (Batra and Millner, 1976; Sandhu and Soni, 1989)

22.7.4.2 Microbiology

The predominant microflora of *wari*, as reported by Rahi and Soni (2007), *are B. subtilis*, *Candida vartiovaarai*, *C. farmata*, *C. curvata*, *Cryptococcus humicolus*, *Debaryomyces tamari*, *Hansenula anomala*, *S. cerevisiae*, *Rhizopus sp.*, *G. candidum*, *C. famata*, and *Trichosporon beigelii* (Rahi and Soni 2007).

22.7.5 *Bhallae*

Bhallae are evening snacks of Punjab and Haryana which are normally prepared from black gram. These are generally soaked in water and fermented in paste form for a period of 5–6 h. In some northern India states, these are also consumed with curd or by soaking in tamarind water.

Microbiology: The predominant flora of the *Bhallae* comprise of yeast such as *C. tamata*, *C. curvata*, *R. polymarina*, *Trichosporon pullulans*, *H. polymorpha*, and *C. parapsilosis*. Bacterial flora is also present in the *Bhallae* and includes *L. fermentum* and *Leuc. mesenteroides*. The microflora present in the bhallae also varies along with the season as bacterial flora predominates in the summer season, while the winter season promotes more of yeast strains (Sankaran 1998; Soni et al. 2013).

22.7.6 *Doli Ki Roti*

This is indigenous fermented cereal product consumed by the ethnic Punjabi community who migrated from Pakistan to India. This Roti (hand tossed round flatbread) is prepared by fermenting wheat flour and spices (Bhatia and Khetarpaul 2012).

Flow chart of preparation of *Doli Ki Roti*

Inoculum:

Poppy seeds with other spices boiled with water

↓

Incubated for 18-24 h at 35°- 40°C

↓

Whole wheat flour added after fermentation, Incubated for 2 hours

Dough:

Inoculum for the Roti and wheat flour mixed and kneaded to make dough

↓

Incubated for a period of 1 hour for fermentation at approx. 40°C for 24 hours

Roti:

Fermented dough casted into equal sized balls which were rolled in form of 'puris'

↓

Filled with legume stuffing's

↓

Closed by hand and fried till it turns golden brown

Microbiology: Unknown microflora

22.7.6.1 Culinary and Mode of Consumption

Traditionally, *Doli Ki Roti* inoculums were prepared in the temples and distributed to the people for the preparation of end product during fast season especially during the festival of Janmashtami.

22.8 Conclusion

The geographical region of Uttarakhand, UP, Haryana, and Punjab states is rich in agricultural resource biodiversity. The variations in the level of rainfall climate found in these states can be utilized for improving the agriculture economy of the states along with the indigenous traditional knowledge. Ethnic fermented beverages and foods have been the integral part of food practices of native people, and fermentation is one of the processes which depicts first human-associated microbial association. It would have been difficult for the tribal people to survive without ethnic fermented foods over the millennia in the harsh environment of the Himalayan region as fermentations are part of natural phenomena which preserve food, enhance digestibility, and incorporate essential amino acids, proteins, vitamins, and other nutrients. However, the available diverse nature of ethnic food available is now at the brim of extinction due to the loss of interest of younger generation of tribes in preparation of fermented food and due to urbanization of society. Ethnic food in the form of standardized product can offer greater prospects having major health and organoleptic, nutritional, and storage attributes. A deeper understanding of such traditionally preserved food using modern molecular biology tools like metagenomics, in addition to studies on the mechanism of probiotic properties of lactic acid bacteria, is the need of the hour. This will lead to development of new food technologies which are likely to have an effect on preservation, production, scale-up, and popularity of ethnic foods.

Acknowledgments This work was supported by Uttarakhand State Council for Science and Technology (UCOST), Dehradun Uttarakhand, and NKN. A.B. and T.G. would like to acknowledge the Ministry of Human Resource Development (MHRD), Government of India, for the financial support.

References

Adams M (2010) Fermented meat products. In: Fermented foods and beverages of the world. CRC Press/Taylor and Francis, Boca Raton, pp 321–334

Anand S, Grover CR, Beniwal A (2016) Evaluation of *Ocimum sanctum* essential oil as potential preservative for fermented dairy products. J Pure Appl Microbiol 10(4):2763–2772

Bhardwaj KN, Jain KK, Kumar S, Kuhad RC (2016) Microbiological analyses of traditional alcoholic beverage (Chhang) and its starter (Balma) prepared by Bhotiya Tribe of Uttarakhand, India. Indian J Microbiol 56(1):28–34. https://doi.org/10.1007/s12088-015-0560-6

Bhatia A, Khetarpaul N (2012) 'Doli Ki Roti'—an indigenously fermented Indian bread: cumulative effect of germination and fermentation on bioavailability of minerals. Indian J Tradit Knowl 11(1):109–113

Bodapti S, Chander M (2013) Integrating indigenous knowledge of farmers for sustainable organic farming: an assessment in Uttarakhand state of India. Indian J Tradit Knowl 12(2):259–264

Campbell-Platt G (1987) Fermented foods of the world, A dictionary and guide. Butterworth's. (Eds. Campbell-Platt and Geoffrey). Wiley-VCH Verlag GmbH & Co. KGaA, Weinheim

Chandra A, Kandari LS, Payal KC, Maikhuri RK, Rao KS, Saxena KG (2010) Conservation and sustainable management of traditional ecosystems in Garhwal Himalaya, India. New York Sci J 3(2):71–77

Chauhan VS, Bhatt JC (2000) Agriculture in Uttarakhand: from subsistence towards self-sufficiency. In: Uttarakhand statehood: dimensions of development. vol 1, pp 168–180

Chitale SR (2000) Commercialization of Indian traditional foods: Jeelebi, Ladoo and Bakervadi. In: The proceedings of the international conference on traditional foods. Central Food Technological Research Institute, Central Food Technological Research India, Mysore, p 331

Das CP, Pandey A (2007) Fermentation of traditional beverages prepared by Bhotiya community of Uttaranchal Himalaya. Indian J Tradit Knowl 6(1):136–140

Dewan S, Tamang JP (2007) Dominant lactic acid bacteria and their technological properties isolated from the Himalayan ethnic fermented milk products. Antonie Van Leeuwenhoek 92(3):343–352

Dixit V, Singh H (2017) Ethnomedicinal plants of Subalpine and Alpine Region of Uttarakhand Himalaya. In: Ethnobotany of India, vol 4. Apple Academic Press, pp 135–168

Farooquee NA, Maikhuri RK (2009) Communities and their agrobiodiversity: priorities for agriculture in Uttarakhand Himalaya, India. Outlook Agric 38(4):383–392

Glitsø LV, Knudsen KB (1999) Milling of whole grain rye to obtain fractions with different dietary fibre characteristics. J Cereal Sci 29(1):89–97

Hesseltine CW (1965) A millennium of fungi, food, and fermentation. Mycologia 57(2):149–197

Joshi SK, Ballabh B, Negi PS, Dwivedi SK (2018) Diversity, distribution, use pattern and evaluation of wild edible plants of Uttarakhand, India. Defence Life Sci J 3(2):126–135

Kariluoto S, Vahteristo L, Salovaara H, Katina K, Liukkonen KH, Piironen V (2004) Effect of baking method and fermentation on folate content of rye and wheat breads. Cereal Chem 81(1):134–139

Katina K, Arendt E, Liukkonen KH, Autio K, Flander L, Poutanen K (2005) Potential of sourdough for healthier cereal products. Trends Food Sci Technol 16(1-3):104–112

Kokkiligadda A, Beniwal A, Saini P, Vij S (2016) Utilization of cheese whey using synergistic immobilization of β-galactosidase and *Saccharomyces cerevisiae* cells in dual matrices. Appl Biochem Biotechnol 179(8):1469–1484

Kumar SR, Kanmani P, Yuvaraj N, Paari KA, Pattukumar V, Arul V (2013) Traditional Indian fermented foods: a rich source of lactic acid bacteria. Int J Food Sci Nutr 64(4):415–428

Kumari A, Pandey A, Ann A, Raj A, Gupta A, Chauhan A, Neopany B (2016) Indigenous alcoholic beverages of South Asia. Taylor and Francis, New York, pp 501–596

Macfarlane S, Macfarlane GT (2006) Composition and metabolic activities of bacterial biofilms colonizing food residues in the human gut. Appl Environ Microbiol 72(9):6204–6211

Maheshwari MA, Sharma P (2018) Community based tourism: a tool for sustainable development in Garhwal Region, Uttarakhand. Int J Sci Res 6(4):557–560

Oki K, Rai AK, Sato S, Watanabe K, Tamang JP (2011) Lactic acid bacteria isolated from ethnic preserved meat products of the Western Himalayas. Food Microbiol 28(7):1308–1315

Parihaar RS, Bargali K, Bargali SS (2014) Diversity and uses of Ethno-medicinal plants associated with traditional agroforestry systems in Kumaun Himalaya. Indian J Agric Sci 84(12):1470–1476

Poutanen K, Flander L, Katina K (2009) Sourdough and cereal fermentation in a nutritional perspective. Food Microbiol 26(7):693–699

Rahi DK, Soni SK (2007) Applications and commercial uses of microorganisms, Microbes: A source of energy in 21st century. New India Publishing House, New Delhi, pp 69–126

Rai AK, Palni U, Tamang JP (2009) Traditional knowledge of the Himalayan people on production of indigenous meat products. Indian J Tradit Knowl 8(4):586–559

Rai AK, Tamang JP, Palni U (2010) Microbiological studies of ethnic meat products of the Eastern Himalayas. Meat Sci 85(3):560–567

Rais M, Pazderka B, VanLoon GW (2009) Agriculture in Uttarakhand, India—Biodiversity, nutrition, and livelihoods. J Sustain Agric 33(3):319–335

Rana TS, Datt B, Rao RR (2004) Soor: a traditional alcoholic beverage in Tons valley, Garhwal Himalaya. Indian J Tradit Knowl 3(1):59–65

Rawat K, Kumari A, Kumar S, Kumar R, Gehlot R (2018) Traditional fermented products of India. Int J Curr Microbiol Appl Sci 7(4):1873–1883

Ray S, Bagyaraj DJ, Thilagar G, Tamang JP (2016) Preparation of Chyang, an ethnic fermented beverage of the Himalayas, using different raw cereals. J Ethnic Foods 3(4):297–299

Roopashri AN, Varadaraj MC (2009) Molecular characterization of native isolates of lactic acid bacteria, *bifidobacteria* and yeasts for beneficial attributes. Appl Microbiol Biotechnol 83(6):1115–1126

Roy B, Kala CP, Farooquee NA, Majila BS (2004) Indigenous fermented food and beverages: a potential for economic development of the high altitude societies in Uttaranchal. J Hum Ecol 15(1):45–49

Saini P, Beniwal A, Malik RK, Vij S (2017) Comparative physiology of *Kluyveromyces marxianus* and *Saccharomyces cerevisiae* during batch cultivation on glucose as a sole carbon source. Indian J Dairy Sci 70(4):427–433
Saini P, Beniwal A, Kokkiligadda A, Vij S (2018) Response and tolerance of yeast to changing environmental stress during ethanol fermentation. Process Biochem 17:1–12
Sankaran R (1998) Fermented foods of the Indian subcontinent. In: Wood BJB (ed) Microbiology of Fermented Foods. Springer, Boston
Sarkar S (2008) Innovations in Indian fermented milk products—a review. Food Biotechnol 22(1):78–97
Sekar S, Mariappan S (2007) Usage of traditional fermented products by Indian rural folks and IPR. Indian J Tradit Knowl 6(1):111–112
Shah NP (2006) Health benefits of yogurt and fermented milks. In: Chandan RC (ed) Manufacturing yogurt and fermented milks. Blackwell Publishing, pp 327–351. isbn:780813823041
Sharma A, Sarkar PK (2015) Microbial diversity in Ethno-fermented foods of Indian Himalayan Region. ENVIS Bull Himalayan Ecol 23:85–91
Singh RB (2000) Environmental consequences of agricultural development: a case study from the Green Revolution state of Haryana, India. Agric Ecosyst Environ 82(1–3):97–103
Singh BP (2008) Ex-criminal tribes of Punjab. Econ Polit Wkly 43:58–65
Soni SK, Sandhu DK (1990) Indian fermented foods: microbiological and biochemical aspects. Indian J Microbiol 30(2):135–157
Soni SK, Soni R, Janveja C (2013) Production of fermented foods. In: Panesar PS, Marwaha SS, editors. Biotechnology in agriculture and food processing opportunities and challenges, pp 219–278. isbn: 9780429111846
Steinkraus KH (2002) Fermentations in world food processing. Compr Rev Food Sci Food Saf 1(1):23–32
Swennen K, Courtin CM, Delcour JA (2006) Non-digestible oligosaccharides with prebiotic properties. Crit Rev Food Sci Nutr 46(6):459–471
Tamang JP (1998) Role of microorganisms in traditional fermented foods. Indian Food Ind 17(3):162–167
Tamang JP (2010) In: Tamang JP (ed) Himalayan fermented foods: microbiology, nutrition, and ethnic values. CRC Press/Taylor and Francis, New York, pp 175–183
Tamang JP (2012) Plant-based fermented foods and beverages of Asia. In: Hui YH, Özgül Evranuzpp E (eds) Handbook of plant-based fermented food and beverage technology. CRC Press, Boca Raton, pp 49–92
Tamang JP, Fleet GH (2009) Yeasts diversity in fermented foods and beverages. In: Satyanarayana T, Kunze G (eds) Yeast biotechnology: diversity and applications. Springer, Dordrecht, pp 169–198
Tamang JP, Kailasapathy K (eds) (2010) Fermented foods and beverages of the world. CRC Press/Taylor and Francis, New York, pp 85–127
Tamang JP, Thapa N, Bhalla TC (2016) Ethnic fermented foods and beverages of India. In: Tamang JP (ed) Ethnic fermented foods and alcoholic beverages of Asia. Springer, New Delhi, pp 17–72
Tewary HK, Muller HG (1989) Some chemical and physical characteristics of wari, an Indian fermented food. Food Chem 32(4):269–276
Thakur N, Bhalla TC (2004) Characterization of some traditional fermented foods and beverages of Himachal Pradesh. Indian J Tradit Knowl 3(3):325–335
Tolia RS (2010) Great tribal diversity of Uttarakhand. Amtikar [magazine distributed by the Rang welfare association, Pithoragarh], pp 49–56
Tuteja U (2013) Agriculture Profile of Uttarakhand. Agricultural Economics Research Centre University of Delhi, pp 1–12

Ethnic Fermented Foods and Beverages of West Bengal and Odisha

23

Kuntal Ghosh, Saswati Parua Mondal, and Keshab Chandra Mondal

Abstract

West Bengal and Odisha are the two neighbouring states in India. Due to the climatic variations, a large number of crops, vegetables and fishes are produced in this region. Therefore, diverse numbers of fermented foods are prepared which can be categorized into four different groups (cereal- and pulse-based, plant-based, dairy-based and fish-based). In this chapter, we focused on the traditional preparation process, microbiology, nutritional value, optimization and commercialization of the fermented foods prepared in West Bengal and Odisha. The native people usually practise their indigenous knowledge for preparation of the fermented foods. They unknowingly use the microbes for fermented food preparation. Among the microbes, yeast, mould, lactic acid bacteria and *Bacillus* (*B.*) sp. majorly took part in the fermentation process. However, most of the fermented foods did not achieve significant scientific attention. This fact may lead to the disappearance of the traditional knowledge. Therefore, it is essential to explore the nutritional content and health benefits, detect critical control point, optimize preparation process and eventually commercialize the traditional fermented foods.

Keywords

Traditional fermented foods · Yeast · Lactic acid bacteria · Chenna poda · Haria

---

K. Ghosh
Department of Biological Sciences, Midnapore City College, Paschim Medinipur, West Bengal, India

S. P. Mondal
Department of Physiology, Bajkul Milani Mahavidyalaya, Purba Medinipur, West Bengal, India

K. C. Mondal (✉)
Department of Microbiology, Vidyasagar University, Midnapore, West Bengal, India

J. P. Tamang (ed.), *Ethnic Fermented Foods and Beverages of India: Science History and Culture*, https://doi.org/10.1007/978-981-15-1486-9_23

23.1 Introduction

Odisha (formerly Orissa) and West Bengal are two neighbouring states in Eastern India. These states were united in the Bengal Presidency of British India. Odisha became a separate province (state) on 1 April 1936, whereas West Bengal separated from East Bengal during India's Independence in 1947. The Bengali people possess some Austro-Asiatic ancestry via gene flow from East and Northeast India (Khan 2017), but Odisha history predates Aryan-Indian history (Choudhury 2013). Due to close proximity and similar geo-climatic environment, West Bengal and Odisha have identical tradition and civilization. Hindu festivals celebrated by the people are very common. Both have enriched cultural foundation. Bengali and Odisha languages are very close in some standards especially the dialects that are neighbouring. Sambalpuri dialect is in some ways sound closer to Calcutta dialect. It is thought that both of the languages are directly descended from Magadhi Prakrit.

Rice is the staple food in both states. The food culture particularly uses mustard oil as a cooking medium; vegetables and spices are very common in both states' cuisine. The diversity of ethnic foods in this region is related to the diversity of ethnicity and abundant bioresources. Rice, pulses, vegetables, milk and fishes are the major foodstuff, and different varieties of non-fermented and fermented foods are prepared. The scarceness information of traditional food culture in these states leads to difficulty to make a portrait of food history in this area. However, this is the inherited food culture of Indian civilization. The culture of preparing food with proper methods was introduced by the two ancient Indian civilizations – the Harappa and the Mohenjo-daro (https://www.dosamatic.com/history-of-indian-food/). The first preparation of food included a number of cereals and pulses. Gradually, the diet consisted of vegetables, fruits, grains, meat, honey, dairy products, beverages and special kind of spices as better forms of cooking were innovated and the civilization moved towards perfection. The Vedic food culture was further enriched due to Aryans influences that travelled from Central Asia with a number of new cuisines. Food components went through the massive changes during the rule of Maurya and Gupta Empires, who forced to incorporate vegetarian diet due to some sacred beliefs (https://www.dosamatic.com/history-of-indian-food/). The food habits underwent revolutionized with the introduction of the most popular Mughal cuisine during the Mediaeval period (https://www.dosamatic.com/history-of-indian-food/). They brought different kinds of meats, fruits and flatbread in the diet of Indians. Sumptuous dishes were prepared during the rule of Shah Jahan and Jahangir. Next, the Nizams and the Portuguese, Chinese, British, and Anglo-Indian influence on Indian food (https://www.dosamatic.com/history-of-indian-food/). This continued in modern days as well.

The techniques of ethnic fermented food and beverage preparation by the rural people are somehow untouched or unmodified with the upgradation of the culinary culture of Indian civilization. The native knowledge and skillful technology for preparing varieties of the fermented products using carefully selected raw ingredients and other materials are unparalleled to the other communities of the world. Moreover, they unknowingly used the beneficial microorganisms (natural or starter added) to maintain the standard of fermented foods and beverages. They employed trial and error methods for improving the quality of the foods. The gathered indigenous knowledge is still maintained and modernized when passed on the generations. However, traditional knowledge is not sufficiently documented scientifically. If it is ignored continuously, the indigenous traditional knowledge will disappear soon without any scientific improvement (Tamang et al. 2016). Hence, there is a need to explore the ethnic foods and beverages scientifically to find out the benefits of those. Finally, modern civilization can achieve health benefits from traditional foods.

The traditional or indigenous fermented foods and beverages are an integral part of the dietary culture in India from ancient times. It was documented that the food fermentation appeared in India in the time between 6000 and 4000 BC (Tamang et al. 2016). The history of the Indian cereal-based fermented foods like *idli*, *dosa* and *dhokla* was reported around 1100 AD (Gode 1943; Nair and Prajapati 2008). *Kinema*, a soybean-based fermented food in Sikkim, India, originated around 600 BC to 100 AD (Tamang et al. 2016). Dairy-based fermented foods, viz., *dahi* and *butter*, are popular in India since the Lord Krishna's time about 3000 BC. However, *dahi* was first mentioned in the Veda and Upanishad during 6000–4000 BC (Yegna Narayan Aiyar 1953).

Most of the traditional fermented foods are scientifically unexplored considering their hidden constitutes, technological upgradation and commercially exploitation. Besides, the majority of minor or restricted community-based traditional foods are undocumented. However, these are ancestrally experienced as safe and nutritious, have ameliorative to the certain ailment and have strong ritual values. Considering the exponential uprising of the market of functional foods, there is enough scope to explore the regional traditional foods in India for the sake of native people and to attract the outsiders. Based on this perspective and rationale, the goal of this article is to document the preparation procedure, the composition of nutrients, microbial composition and other health benefits of major and minor traditional fermented foods of West Bengal and Odisha on the basis of available data and our own observation.

23.2 Fermented Foods and Beverages in West Bengal and Odisha

A total of 5,296,953 and 9,590,756 schedule tribes inhabit in West Bengal and Odisha, respectively, as per 2011 census (http://censusindia.gov.in/2011-Common/CensusData2011.html). Among the schedule tribes, there are *Santal*, *Oraon*, *Munda*, *Bhumij*, *Kora*, *Lodha*, *Mahali*, *Bhutia*, *Bedia* and *Savar*. These vast numbers of ethnic people use their indigenous knowledge to prepare diverse types of foods and beverages. However, the non-ethnic group also followed the methods developed by their ancestors in preparation of the fermented products. This region has a large diversity of fermented food products made from both animal and plant sources. The high diversity of fermented products is probably dependent upon the food habits of ethnic community and the availability of food ingredient (Tiwari and Mahanta 2007). Moreover, the diversity of the natural microorganism also has an impact on the large variation of the indigenous fermented foods (Tamang 2009). These types

Table 23.1 Nutraceuticals and functional properties of popular fermented products in West Bengal and Odisha

Types of fermented foods	Name of the fermented food	Associated microbes	Nutritional composition	Health benefits	References
Cereal- and pulse-based fermented products	*Chakuli pitha*	Not reported	Phytic acid is degraded and thus increased the availability of minerals (zinc and iron) and vitamins (vitamin B and C)	Not reported	Ray et al. (2016), Sahoo et al. (2017)
	Enduri pitha	*Lb. fermentum*	Not reported	It boosts the immune system and acts as a protective agent in different seasonal infections	Ray and Swain (2013), Ray et al. (2016), Roy et al. (2007)
	Podo pitha	Not reported	Abundant carbohydrates, free Sugars and fibres	Not reported	Ray and Swain (2013)
	Sour rice	*Lb. casei*, *Lb. bulgaricus*, *Pd. acidilactici*, *Streptococcus thermophilus and Saccharomyces* sp.	pH (4.89), phytic acid (0.313 mg/g), vitamin B complex and K, minerals [iron (73.91 mg/100 g), sodium (303 mg/100 g), potassium (839 mg/100 g), calcium (850 mg/100 g), phosphorus (80.4 mg/100 g), magnesium (532.3 mg/100 g), zinc (32.8 mg/100 g), etc.]	It is an energy-rich food and helps to restore the gastrointestinal flora during intestinal ailments such as irritable bowel syndrome, duodenal ulcers, Crohn's disease, etc. It prevents the constipation	Choi et al. (2014), Goswami et al. (2016), Ray and Swain (2013), Ray et al. (2016)
	Jalebi[a]	*Lb. fermentum*, *Lb. buchneri*, *Streptococcus lactis*, *Streptococcus faecalis* and *S. cerevisiae*	pH (4.4–3.3), protein (24.1 mg/g), carbohydrate (13.9 mg/g), reducing sugar (29.9 mg/g), fat (4.31 g/100 g), lactic acid	It provides instant energy	Sekar and Mariappan (2007), Sharma et al. (2013)

	Bari[a]	No such data is available for *bari* prepared in this region. Considering the similar products in India, the following bacteria might be involved. *B. subtilis*, *Candida famata*, *Candida curvata*, *Candida parapsilosis*, *Candida krusei*, *Candida vartiovaarai*, *Cryptococcus humicolus*, *Debaryomyces tamarii*, *Debaryomyces hansenii*, *Hansenula anomala*, *Geotrichum candidum*, *S. cerevisiae*, *Kluyveromyces marxianus*, *Enterococcus faecalis*, *Trichosporon beigelii* and *Wingea robertsii*	Protein (16–23.8 mg/g), carbohydrate (9.4–13.9 mg/g), reducing sugar (28–34.1 mg/ g)	It reduces flatulence-causing Oligosaccharides It exerts significant level of antioxidant and antimicrobial activities against the pathogenic bacteria	Kulkarni et al. (1997), Rahi and Soni (2007), Sharma et al. (2013), Tewary and Muller (1992)
	Papad[a]	*Candida krusei* and *S. cerevisiae*	Carbohydrates (60 g/100 g), proteins (26 g/100 g), fat, minerals, calcium, iron, carotene and crude fibre	It is good appetiser and digestive food. It is claimed that it absorbs the fatty material from the mouth	Aidoo et al. (2006), Renu and Waghray (2016)

(continued)

Table 23.1 (continued)

Types of fermented foods	Name of the fermented food	Associated microbes	Nutritional composition	Health benefits	References
	Haria	Mould, yeast, lactic acid bacteria (*Lb. fermentum*) and *Bifidobacterium* sp.	pH (3.5), reducing sugar (mg/mL) (6.5), total sugar (mg/mL) (51), protein content (mg/mL) (0.63 maltooligosaccharides), pyranose sugar derivatives, vitamins, minerals and phenol-rich beverage	It is believed that it protects from gastrointestinal ailments like dysentery, diarrhoea, amebiasis, acidity and vomiting. It exerted significant level of antioxidant activity. It is effective against inflammation of body parts, diarrhoea and urinary problems, expelling worms and as a treatment of cholera	Ghosh et al. (2014, 2015a, b), Ray et al. (2016)
Plant-based fermented products	*Pickle*	Lactic acid bacteria	Carbohydrate, dietary fibres, minerals, vitamins, free reducing sugar (450.15 μg/g) and protein content (8.573 ± 1.01 μg/g) (unpublished data)	It exerts the health benefits associated with the lactic acid bacteria. It showed antimicrobial and antioxidant activities	Unpublished data
	Toddy	Yeast (*S. cerevisiae*), acetic acid bacteria, lactic acid bacteria, specifically *Leuc.* sp., *Lb.* sp. and *Zymomonas* sp.	pH (5.5), sucrose (12–15%), protein (0.23%), fat (0.02%), mineral, ascorbic acid 5.7 mg/100 ml) and alcohol (5–6%). There is very little reducing sugar, although glucose, fructose, maltose and raffinose are present	*Toddy* is believed to be good for the health particularly for eyesight and also serves as a sedative. It is also a mild laxative relieving constipation. It is prescribed as a tonic for those recovering from chickenpox	Karthikeyan et al. (2014), Sekar and Mariappan (2007), Steinkraus (1996)

	Mahua	Yeast (*S. cerevisiae*)	pH (3.8), alcohol (10–15%)	*Mahua* is used as an alcoholic drink. There is no such report about the health benefits of *Mahua*. However, the substrate *Mahua* flower exerts anthelmintic, antibacterial, analgesic and hepatoprotective effects	Chandra (2001), Katiyar et al. (2011), Umadevi et al. (2011), Verma et al. (2010)
	Karadi	Unknown	Unknown	It is used in the treatment of digestive problems, especially in constipation	Satish Kumar et al. (2013)
Dairy-based fermented products	*Dahi*	*Lb. alimentarius*, *Lb. bifermentans*, *Lb. paracasei*, *Lb. delbrueckii* subsp. *indicus*, *Lb. acidophilus*, *Lb. bulgaricus*, *Lb. cremoris*, *Lb. helveticus*, *Lb. fermentum*, *Lc. lactis*, *Streptococcus lactis*, *Streptococcus cremoris*, *Streptococcus thermophilus*, *Pd. acidilactici*, *Pd. pentosaceus*, *Weissella cibaria*, *Saccharomycopsis* sp. and *Candida* sp.	pH (4.2–4.3), free reducing sugar (3.2%), hydrosoluble protein (2.75%), free amino acids (1.10%), vitamins (thiamine, riboflavin, folic acid, niacin), proteins, essential amino acids, lactic acid (1.6% titratable acidity)	*Dahi* exerts antimicrobial effects against the pathogenic bacteria due to the high content of lactic acid and production of antimicrobial agents such as bacteriocin, antimicrobial peptides by the participating lactic acid bacteria. It is also effective in reducing cardiovascular diseases and tumours and increases the immunity	Adak et al. (2013), Bambha et al. (1973), Laxminarayana and Shankar (1980), Dave et al. (1992), Singh and Deodhar (1993), Sinha and Sinha (2000)

(continued)

Table 23.1 (continued)

Types of fermented foods	Name of the fermented food	Associated microbes	Nutritional composition	Health benefits	References
	Misti dahi	*Lb. acidophilus*, *Lc. lactis* subsp. *lactis*, *Lb. delbrueckii* subsp. *bulgaricus*, *Streptococcus salivarius* subsp. *thermophilus* and *S. cerevisiae*	pH (5.1), free reducing sugar (3.78%), hydrosoluble protein (1.25%), free amino acids (0.54%), lactic acid (0.96% titratable acidity)	It protects the individual from different gastrointestinal ailments. It exerts significant antimicrobial activities against pathogenic bacteria (*Clostridium perfringens*, *Escherichia coli*, *Shigella dysenteriae*, *Vibrio cholerae*, *Staphylococcus aureus* and *Salmonella typhi*). It also exhibits significant antioxidant activities	Adak et al. (2013), Ghosh and Rajorhia (1990), Gupta et al. (2000)
	Bandel cheese	*Lactic acid bacteria*	Not reported	Not reported	–
	Lassi[a]	*Lactic acid bacteria* and *dahi*-associated microbes	No such data is available for *lassi* prepared in West Bengal and Odisha. However, considering similar preparation process of *lassi* in other Indian states, the following might be the nutritional components. pH (3.83–4.34), lactic acid (0.67–1.0% titratable acidity), fat (2.10–4.24%), protein (2.78–3.6%), total sugar (12.31–14.30%), amino acids, peptides, vitamins, minerals	It helps in digestion. It is effective in gastrointestinal ailments such as diarrhoea, dysentery, etc.	Munde (2015), Padghan et al. (2015), Tamang et al. (2016)

	Chhena pora	*Not reported*	Protein, fat, vitamins, calcium	It is believed that the food strengthen the bones and good for digestion	–
Fish-based fermented products	Dry salted *Khainga*	Lactic acid bacteria and *Bacillus* sp.	Drying process increase the protein, lipids, vitamins and minerals	Not reported	Chukwu (2009), Thapa et al. (2007)
	Sukuti or shutki[a]	*Lb. lactis* subsp. *cremoris*, *Lb. lactis* subsp. *lactis*, *Lb. plantarum*, *Lb. mesenteroides*, *Enterococcus faecium*, *Enterococcus faecalis*, *Pd. pentosaceus*, *Candida chiropterorum*, *C. bombicola and Saccharomycopsis* spp.	Not reported	Not reported	Rapsang et al. (2011), Thapa (2016)

[a]No such data are available for these foods and beverages. However, considering the similar products in other states of India, it can be assumed that these products might contain similar microflora and nutrients and exhibit identical health benefits

of traditional fermented products can be categorized into (1) cereal- and pulse-based fermented products, (2) plant-based fermented products, (3) dairy-based fermented products, and (4) fish-based fermented products (Table 23.1).

23.3 Cereal- and Pulse-Based Fermented Products

Cereals are considered one of the staple foods in West Bengal and Odisha. Among the cereals, rice is the major producing cereal in West Bengal and Odisha. The people in this region take rice as a staple food because of its high calorific value (1460–1560 kJ) (FAO 1993) and availability. Rice provides a moderate amount of protein (6.3–7.1%) but devoid of lysine (FAO 1993). Another important ingredient in food fermentation is *pulse*. It is estimated that 442 thousand tonnes of *pulses* are produced in 2012 in West Bengal and Odisha (Inbasekar 2014). *Pulse* contains a high amount of protein, carbohydrate, dietary fibre, fat, and vitamins (Ofuya and Akhidue 2005). Hence, it is considered good nutritive food. Recent research showed that *pulses* are effective in cardiovascular diseases, cancer, hypertension and gastro-intestinal disorders (Hu 2003; Ray et al. 2015). Commonly used *pulses* in this region are *black gram*, *soybean*, *Bengal gram*, *red gram* and *green gram*. *Pulses* are locally known as '*dal*'.

In West Bengal and Odisha, the traditional fermented foods are generally prepared from most common types of available cereals such as rice and pulses mainly *black gram*. The popular fermented foods made up by cereals and *pulses* in this region are fermented rice cake (*pitha*), sour rice (*panta bhat*), fermented legume (*bari and papad*) and rice-based fermented beverage (*haria or handia*).

23.3.1 Fermented Rice Cake (*Pitha*)

Fermented rice cake is popularly known as *pitha* in West Bengal and Odisha. *Pitha* is made up from either fermented rice or rice-legume batter.

Flow sheet of traditional method of *Pitha* preparation in West Bengal and Odisha is presented in Fig. 23.1.

Culinary and mode of consumption For its preparation, the rice is first soaked in water for 8–12 h and air-dried (Fig. 23.1). Then the soaked rice is ground to make the fine flour by using a traditional grinder. The soaked legumes are pasted by traditional utensils. The rice flour and the pasted legume are then mixed with water to prepare the batter which is allowed to ferment for 4–5 h in summer and 10–12 h in winter. The fermented batter is fried over a pan. The food is taken at breakfast and lunch.

Socio-economy and ethnical or religious values During the Paush Sankranti (last day of ninth Bengali month called Paush, middle of December, winter season) or Raja Sankranti (summer), preparing and sharing of *pitha* are traditional culture.

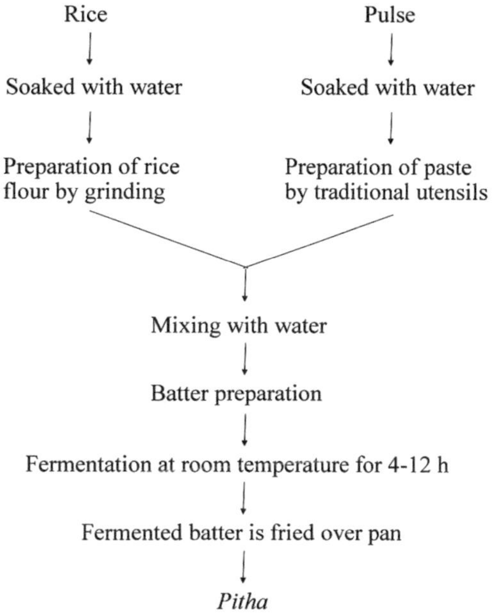

Fig. 23.1 Traditional method of *Pitha* preparation in West Bengal and Odisha

Microbiology Although these types of *pitha* are very popular in this region, surprisingly no such research efforts have been made to find out its nutritional and microbiological composition.

There are different kinds of *Pitha*: *Chakuli pitha*, *Enduri pitha*, *Munha pitha*, *Chhuchipatra pitha* and *Podo pitha* (Ray et al. 2016; Ray and Swain 2013).

23.3.2 *Chakuli Pitha*

Chakuli pitha, a round-shaped flattened pancake is similar to a popular South Indian food called *dosa*. The food is popular in West Bengal and Odisha.

Flow sheet of traditional method of *Chakuli pitha* preparation in West Bengal and Odisha is presented in Fig. 23.2.

Culinary and mode of consumption The principal ingredients are rice (*Oryza sativa*) and *black gram dal* (*Phaseolus mungo*). The batter is prepared by the coat-free *black gram dal* and rice and left for fermentation about 10–12 h (Ray et al. 2016). After fermentation, the batter is fried over a pan with a structure of round shape.

Socio-economy and ethnical or religious values The food is generally prepared during the Paush Sankranti or Raja Sankranti by the women in houses. After preparation, the food is usually offered to the God at first and then consumed by the other members.

Black gram dal (coat free)
↓
Soaked with water
↓
Mixing of rice flour with soaked dal
↓
Batter preparation
↓
Fermentation at room temperature for 10-12 h
↓
Fermented batter is shaped round and fried over pan
↓
Chakuli pitha

Fig. 23.2 Traditional method of *Chakuli pitha* preparation in West Bengal and Odisha

Microbiology Its microbiological composition is unknown so far. Considering its preparation process which is similar to the *dosa*, it is likely that lactic acid bacteria and yeast might play the major role in this fermentation.

Nutritional value Looking into its nutrient enrichment, it was suggested that the amount of vitamin B, vitamin C, zinc and iron are increased during fermentation. Moreover, fermentation helps to increase the bioavailability of the minerals by degrading the phytate, which is an antinutrient in cereals (Sahoo et al. 2017).

Optimization and commercialization This product is usually prepared in the household. Best of our knowledge, the food is not sold in the market.

23.3.3 *Enduri Pitha*

Enduri pitha is the flavoured cake prepared in West Bengal and Odisha.

Flow sheet of traditional method of *Enduri pitha* preparation in West Bengal and Odisha is presented in Fig. 23.3.

Culinary and mode of consumption It is prepared by rice and *black gram dal*. In its preparation, the fermented batter is kept and folded in turmeric (*Curcuma longa* L.) leaf followed by cooking over steam.

Socio-economy and ethnical or religious values Same as *Chakuli pitha.*

Microbiology *Lactobacillus* (*Lb.*) *fermentum* is one of the fermenting bacteria in this fermentation (Ray and Swain 2013).

Nutritional value It was reported that it helps in boosting the immune system (Roy et al. 2007) and acts as a protective agent in different seasonal infections (Ray et al. 2016). The health benefits probably come from the turmeric leaf and the participating lactic acid bacteria.

Optimization and commercialization Unknown.

Water soaked black gram dal

↓

Mixing of rice flour with soaked dal

↓

Batter preparation

↓

Fermentation at room temperature

↓

Fermented batter is kept and folded in turmeric (*Curcuma longa* L.) leaf

↓

Cooking over steam

↓

Enduri pitha

Fig. 23.3 Traditional method of *Enduri pitha* preparation in West Bengal and Odisha

23.3.4 *Munha Pitha*

Munha pitha is a spongy fermented cake which is very famous in West Bengal and Odisha.

Flow sheet of traditional method of *Munha pitha* preparation in West Bengal and Odisha is presented in Fig. 23.4.

Culinary and mode of consumption The principal ingredients are rice and *black gram dal* which are used for batter preparation (Ray and Swain 2013). Usually, a short fermentation time is required. After fermentation, the fermented batter is kept in cloth which is then placed in the mouth of an earthen pot containing half-filled water. Then it is boiled until the batter becomes spongy. Usually, this food is served with sugar and vegetable curry.

Socio-economy and ethnical or religious values Same as *Chakuli pitha.*

Microbiology The microbiology and nutritional values of this product are unexplored so far. Clearly, a detailed analysis is needed.

Nutritional value Unknown.

Optimization and commercialization Unknown.

23.3.5 *Chhuchipatra Pitha*

Chhuchipatra pitha is a square-shaped *pitha*. The food is nutritious, palatable and delicious (Ray and Swain 2013).

Flow sheet of traditional method of *Chhuchipatra pitha* preparation in West Bengal and Odisha is presented in Fig. 23.5.

Culinary and mode of consumption The batter is similar to the *Chakuli* batter. However, sometimes the curd (fermented milk) is added for better fermentation (Ray et al. 2016). Then the fermented batter is spread as a thin smear over the pan.

Fig. 23.4 Traditional method of *Munha pitha* preparation in West Bengal and Odisha

Batter preparation by mixing of rice flour and black gram dal

↓

Fermentation at room temperature for short time

↓

Keeping of the fermented batter in cloth

↓

Placing of the cloth in the mouth of an earthen pot containing half-filled water

↓

Boiling of the batter until it becomes spongy

↓

Munha pitha

Fig. 23.5 Traditional method of *Chhuchipatra pitha* preparation in West Bengal and Odisha

Batter preparation by mixing of rice flour and black gram dal

↓

Fermentation at room temperature for 10-12 h by mixing of curd

↓

Spreading of fermented batter as a thin smear over the pan

↓

Keeping of coconut, cheese, and sugar in the center of the pancake

↓

Folding in a square shape and frying more

↓

Chhuchipatra pitha

Then coconut, cheese and sugar are kept in the centre of the pancake followed by folding in a square shape and then allow frying more.

Socio-economy and ethnical or religious values It is generally prepared during Raja Sankranti (summer) festival in rural Odisha.

Microbiology As the curd (locally known as *dahi*) is added for fermentation, it is likely that the food might contain lactic acid bacteria. Again, no such studies have been conducted to explore its microbial contents.

Nutritional value Unknown.

Optimization and commercialization Unknown.

23.3.6 *Podo Pitha*

Podo pitha is a slow-cooked *pitha*.

Flow sheet of traditional method of *Podo pitha* preparation in West Bengal and Odisha is presented in Fig. 23.6.

Batter preparation by mixing of rice flour and black gram dal

↓

Fermentation at room temperature for 2-4 h

↓

Mixing of fermented batter with coconut, cashew nuts, and sugar

↓

Wrapping with the Banana or Sal leaves

↓

Baking slowly for 5 - 10 h in the traditional earthen oven

↓

Podo pitha

Fig. 23.6 Traditional method of *Podo pitha* preparation in West Bengal and Odisha

Culinary and mode of consumption Like the other *pitha*, rice and *black gram dal* are used for the batter preparation. The fermentation time is only 2–4 h. The fermented batter is mixed with coconut, cashew nuts and sugar (Ray and Swain 2013). The mixture is wrapped with the banana (*Musa paradisiaca* L) or sal (*Shorea robusta* C.F. Gaertn) leaves and baked slowly for 5–10 h in the traditional earthen oven which is covered by hot charcoal (Ray and Swain 2013). The outside of the product is usually slightly burnt, whereas the white spongy soft inside.

Microbiology Not reported so far.

Nutritional value It was reported that the food is energy rich because of the high carbohydrate content, sugar and fibres (Ray and Swain 2013).

Optimization and commercialization Unknown.

23.3.7 Sour Rice

Sour rice is popularly known as *panta bhat* in West Bengal and *pokhalo/pakhala* in Odisha. It is a naturally fermented rice product which is consumed during lunch and breakfast.

Flow sheet of traditional method of *panta bhat/pokhalo/pakhala* preparation in West Bengal and Odisha is presented in Fig. 23.7.

Culinary and mode of consumption In its preparation, the boiled rice is cooled down at room temperature followed by the addition of adequate water. Then it is kept in the room temperature for 10–12 h for fermentation. This is a natural fermentation; no starter is required. However, sometimes curd, salt and some vegetables are added for better fermentation and taste enhancement (Ray and Swain 2013). The sour rice is consumed with onion and cooked spicy vegetables.

Microbiology Lactic acid bacteria [especially *Lb. casei*, *Lb. bulgaricus*, *Pediococcus* (*Pd.*) *acidilactici*, *Streptococcus thermophilus*] and yeast [*Saccharomyces* (*S.*) sp.] actively participated in this fermentation (Ray et al. 2016).

Fig. 23.7 Traditional method of *panta bhat*/*pokhalo*/*pakhala* preparation in West Bengal and Odisha

Boiling of rice
↓
Cooling at room temperature
↓
Addition of water
↓
Fermentation at room temperature for 10-12 h
↓
Panta bhat / pokhalo / pakhala
↓
Serving with onion and spicy potatoes

Nutritional value After fermentation sour rice is enriched with vitamins (B complex and K) and minerals (iron, sodium, potassium, calcium, etc.) (Ray et al. 2016). It is an energy-rich food and helps to restore the gastrointestinal flora during intestinal ailments such as irritable bowel syndrome, duodenal ulcers, Crohn's disease, etc. (Choi et al. 2014; Ray and Swain 2013). Moreover, the food was also reported to prevent constipation (Ray et al. 2016).

Optimization and commercialization Unknown.

23.3.8 *Jalebi*

Jalebi is a popular Indian sweetened fermented product. It is usually sold in *mela* (social gathering) and sweet shops.

Flow sheet of traditional method of *Jalebi* preparation in West Bengal and Odisha is presented in Fig. 23.8.

Culinary and mode of consumption It is made from maida (wheat flour), *dahi* (curd) and water (Steinkraus 1996). The batter is fermented for 4–6 h in room temperature. For its preparation, the batter is squeezed through an embroidered hole (about 4 mm in diameter) in a thick and durable cotton cloth and dispensed in the hot oil (Tamang et al. 2016) in a manner as it gets a unique shape (round shaped with a network like arrangement; Fig. 23.9). The fried one is immersed in sugar syrup for 2–5 min.

Microbiology *Lb. fermentum*, *Lb. buchneri*, *Streptococcus lactis*, *Streptococcus faecalis* and *S. cerevisiae* were found in the fermented batter (Sekar and Mariappan 2007).

Nutritional value The pH in the fermented batter is around 3–4. Moreover, the content of amino nitrogen and free sugar were also decreased during fermentation (Steinkraus 1996).

Optimization and commercialization *Many* small cottage industries (sweet shops) now prepare this food. The usual price of the product is Rs. 150 per kg.

Fig. 23.8 Traditional method of *Jalebi* preparation in West Bengal and Odisha

Preparation of batter by mixing of maida (wheat flour), *dahi* (curd), and water

↓

Fermentation at room temperature for 4 - 6 h

↓

Keeping of batter in a cloth

↓

Batter is squeezed through an embroidered hole dispensed in the hot oil

↓

Immersion of fried one in sugar syrup for 2-5 min

↓

Jalebi

Fig. 23.9 *Jalebi*

23.3.9 *Bari*

Bari is a pulse-based traditional fermented food. Different types of pulses (*Urad dal*, *Chana dal*, *Matar dal*) are used in its preparation (Sha et al. 2013).

Flow sheet of traditional method of *Bari* preparation in West Bengal and Odisha is presented in Fig. 23.10.

Culinary and mode of consumption In its preparation, the pulses are soaked with water for 10–12 h followed by grinding to make the paste. Then some spices (chilli, coriander, cumin seeds, etc.) and salts are added to that paste. The dough is pressed to make an oval or round shape and kept in a cloth followed by sun drying for 5–10 days. Due to this drying process, moisture content is very low in *bari*. Hence, *bari* can be preserved in a container for 6 months to 1 year. *Bari* (Fig. 23.11) is consumed after cooking with vegetable curry or directly after frying.

Socio-economy and ethnical or religious values *Bari* is very popular in West Bengal and Odisha. Consumption of *bari* during occasion is a traditional culture (Sha et al. 2013).

Fig. 23.10 Traditional method of *Bari* preparation in West Bengal and Odisha

Soaking of the pulses (*Uraddal*, *Chanadal*, *Matardal*) in water
↓
Grinding of soaked pulses and preparation of paste
↓
Addition of spices (chili, coriander, cumin seeds, etc.) and salts
↓
Pressing of dough to make an oval or round shape
↓
Sun drying for 5 - 10 days
↓
Bari

Fig. 23.11 *Bari* prepared from *Matar dal*

Microbiology *B. subtilis*, *Candida famata*, *Candida curvata*, *Candida parapsilosis*, *Candida krusei*, *Candida vartiovaarai*, *Cryptococcus humicolus*, *Debaryomyces tamarii*, *Debaryomyces hansenii*, *Hansenula anomala*, *Geotrichum candidum*, *S. cerevisiae*, *Kluyveromyces marxianus*, *Enterococcus faecalis*, *Trichosporon beigelii, and Wingea robertsii* were isolated from *bari* prepared in another region (Rahi and Soni 2007).

Nutritional value It is a protein-rich food. It reduces flatulence-causing oligosaccharides (Kulkarni et al. 1997; Tewary and Muller 1992).

Optimization and Commercialization Bari is usually prepared in the household and small cottage industries. The product is available in the market. Due to an easily available substrate and short preparation time make the product inexpensive (approximate cost is Rs. 150 per kg).

23.3.10 *Goyna Bari*

Goyna bari is a decorated bari. It is designed like the traditional ornaments.

Fig. 23.12 Traditional method of *Goyna bari* preparation in West Bengal and Odisha

Soaking of the pulses (*Uraddal, Chanadal, Matardal*) in water

↓

Grinding of soaked pulses and preparation of paste

↓

Fermentation at room temperature for 1-2 h

↓

Keeping fermented batter in a cloth and squeezing through a hole into a dish containing poppy seed

↓

Sun drying for 5 - 10 days

↓

Goyna bari

Flow sheet of traditional method of *Goyna bari* preparation in West Bengal and Odisha is presented in Fig. 23.12.

Culinary and mode of consumption The preparation process of *Goyna bari* is illustrated in Fig. 23.12. The fermented batter is kept in a cloth and squeezed through a hole into a dish containing poppy seeds. The poppy seeds help to maintain the texture of the *Goyna bari* and to easily separate the *Goyna bari* form its base. Then the *Goyna bari* (Fig. 23.13) is dried under sunlight and kept in the airtight container. Surprisingly there are no such reports on its microbial and nutritional compositions.

Socio-economy and ethnical or religious values In West Bengal particularly in East Midnapore district, the shape of the *bari* is decorated like ornaments. This is an ancient food art of this region. Particularly the women are experts in making of this *Goyna* (ornaments) *bari.*

Microbiology Our group is now engaged to explore this *Goyna bari* scientifically. Initial results showed that yeast and lactic acid bacteria are the predominant microbes in this food (Unpublished data, Manuscript in preparation).

Nutritional value Moreover, *Goyna bari* is a nutrient-rich product and exerts a significant level of antioxidant and antimicrobial activities against the pathogenic bacteria (Unpublished data, Manuscript in preparation).

Optimization and commercialization The people in the West Midnapore district prepare it in the house and sell in the nearby states. The approximate price of the product is Rs. 300 per kg.

23.3.11 *Papad*

Papad is a pulse-based food. A circular sheet-like material is fried in oil and consumed as snacks.

Flow sheet of traditional method of *papad* preparation in West Bengal and Odisha is presented in Fig. 23.14.

Fig. 23.13 *Goyna bari*

Fig. 23.14 Traditional method of *papad* preparation in West Bengal and Odisha

Mixing of black gram, Bengal gram, and lentil, red gram or green gram with rice flour

↓

Addition of water, oil, and salt to make dough

↓

Fermentation for 1-6 h

↓

Rolling of dough to make a thin circular sheets

↓

Frying in oil

↓

Papad

Culinary and mode of consumption The raw ingredients are *black gram*, *Bengal gram* and lentil (*Lens culinaris*), *red gram* or *green gram*, a small quantity of peanut oil and common salt. In its preparation, flour of the above-mentioned pulses is mixed with rice flour and water and knead into dough with oil and salt (Shurpalekar 1986). Sometimes the spices are also added to enhance the taste and flavour. The dough is then allowed to ferment for 1–6 h. Then the dough is rolled into thin circular sheets and left for drying in room temperature (Fig. 23.15a). The dried sheet can be preserved for 4–6 months. The sheet is fried in oil and served as snacks (Fig. 23.15b). Usually, it is taken after a heavy meal and also as snacks.

Socio-economy and ethnical or religious values *Papad* consumption after a heavy meal is an ancient culture in West Bengal and Odisha. During any occasion, it is usually served during lunch or dinner. Sometimes it is also consumed as snacks.

Microbiology *Candida krusei* and *S. cerevisiae* were the principal microbes in this dough fermentation (Aidoo et al. 2006).

Nutritional value *Papad* contained a significant amount of carbohydrates, proteins, fat, minerals, calcium, iron, carotene and crude fibre (Renu and Waghray

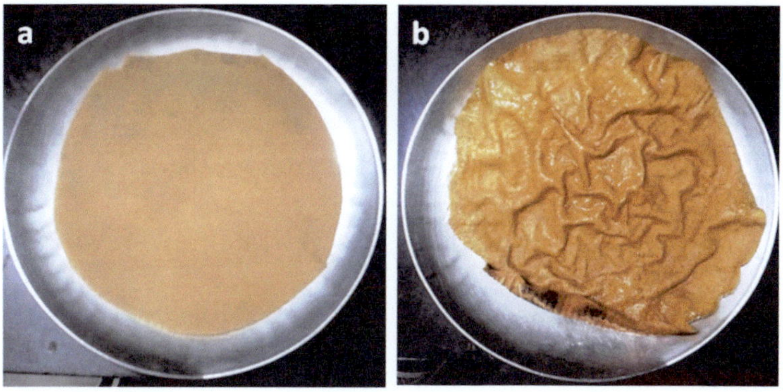

Fig. 23.15 Dried *papad* (**a**) and fried *papad* (**b**)

2016). It is a good appetiser and digestive food as it contains dietary fibre. It is suggested that it absorbs the fatty material from the mouth. Renu and Waghray (2016) claimed that the health benefits of the *papad* come from the used raw ingredients.

Optimization and commercialization As it is one of the regularly consumed foods in these states, the commercialization and the market value are increasing day by day. It is usually sold in a packet, and the cost is approximately Rs. 100–200 per kg.

23.4 Fermented Beverage

23.4.1 *Haria* or *Handia*

A popular ethnic rice-based alcoholic fermented beverage in this area is known as *Haria* (in West Bengal) or *handia* (in Odisha). This drink is very popular among the tribal people. The principal ingredient of this beverage is low-graded boiled rice which is mixed with the traditional starter called *Bakhar* (Ghosh et al. 2014). Six different types of plant (*Cissampelos pareira*, *Diospyros melanoxylon*, *Lygodium flexuosum*, *Orthosiphon rubicundus*, *Ruellia tuberosa* and *Terminalia alata*) parts and the rice dust are used in the preparation of *Bakhar* (Dhal et al. 2010; Ghosh et al. 2014). This starter is very unique as no old inoculum is added during its preparation. The endophytic microbes present in the plant materials are solely responsible for the rice fermentation.

Flow sheet of traditional method of *Haria* preparation in West Bengal and Odisha is presented in Fig. 23.16.

Culinary and mode of consumption In its preparation, the charred boiled rice is mixed with the *Bakhar* (Fig. 23.17a; about 2–3 g of starter was added to 200 g of boiled rice) and allowed to ferment within an earthen pot for 4–5 days (Ghosh et al. 2014). The fermented material (Fig. 23.17b) is diluted with drinking water and sieved by the cloth. The glutinous material is then consumed with spicy vegetables.

Fig. 23.16 Traditional method of *Haria* preparation in West Bengal and Odisha

Boiling of rice
↓
Mixing of traditional starter, *Bakhar* with the boiled rice
↓
Fermentation for 3-5 days at room temperature
↓
Diluted with the drinking water
↓
Sieving by fine cloth
↓
Haria
↓
Consumption with spicy vegetables

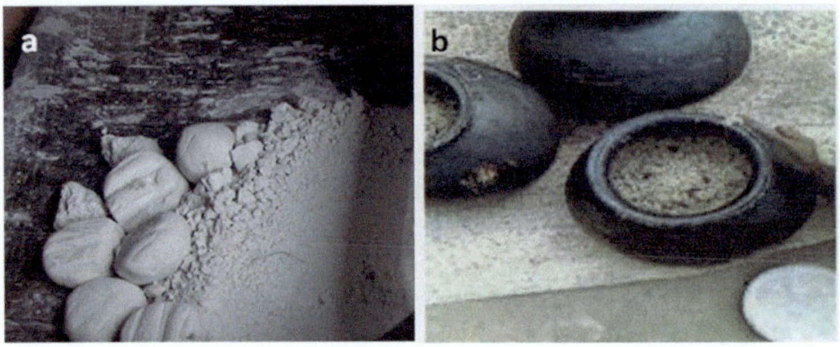

Fig. 23.17 Plant-based starter, *Bakhar* tablet and dust (**a**); rice-based fermented beverage, *Haria* (**b**)

Socio-economy and ethnical or religious values *Haria* is very popular among the tribal groups of West Bengal and Odisha. They regularly consume it as a staple food. During any occasion, sharing and consumption of *Haria* are traditional culture.

Microbiology Mould, yeast, lactic acid bacteria and *Bifidobacterium* sp. were the predominant microbes in *Haria* (Ghosh et al. 2015a). In the initial days of fermentation, mould and yeast grew well due to the aerobic environment and produced different amylolytic enzymes. The enzymes degraded the complex starch and produced the simple sugars which are then used by other bacteria such as lactic acid bacteria, *Bifidobacterium* sp., etc. for their growth (Ghosh et al. 2015a). The partial anaerobic condition inside the pot (due to consumption of oxygen by mould and yeast) also helped to grow lactic acid bacteria, *Bifidobacterium* sp.

Nutritional value *Haria* contained probiotic bacteria (*Lb. fermentum*), prebiotic component (maltooligosaccharides which stimulate the growth of probiotic bacteria, inhibit the growth of intestinal pathogens, and are nutritive for infant and aged persons), different pyranose derivatives

Cutting the tip of an un-opened peduncle of date tree, coconut palm, or Asian palmyra palm

↓

Collection of sap by placing of an earthen pot in the tree

↓

Fermentation for 3-5 days 6 - 8 h

↓

Toddy or *Tari*

Fig. 23.18 Traditional method of *Toddy* or *Tari* preparation in West Bengal and Odisha

(2,3,4,5-tetra-O-acetyl-1-deoxy-β-D-glucopyranose, β-D-mannopyranose pentaacetate, β-D-galactopyranose pentaacetate and 1,2,3,6–tetra-O-acetyl-4-O-formyl-D-glucopyranose which have significant antioxidant, immunostimulatory and antimutagenic activities), phenolics and flavonoids (showed antioxidant activities) (Ghosh et al. 2015a, b). Ray et al. (2016) suggested that *Haria* could protect from different gastrointestinal ailments like diarrhoea, dysentery, amebiasis and vomiting.

Optimization and commercialization Considering its popularity among the tribal groups, some small cottage industries now prepare and sell *Haria*. Due to the low cost of rice and *Bakhar*, the product price is only Rs. 10 per 250 mL.

23.4.2 *Toddy* or *Tari*

Toddy or Tari is an alcoholic drink mainly prepared in West Bengal and Odisha.

Flow sheet of traditional method of *Toddy or Tari* preparation in West Bengal and Odisha is presented in Fig. 23.18.

Culinary and mode of consumption The principal ingredients are the sap from date tree (*Phoenix sylvestris* L.), coconut palm (*Cocos nucifera* L.) or Asian palmyra palm (*Borassus flabellifer*) (Karthikeyan et al. 2014). The sap is collected from these trees by cutting the tip of an unopened peduncle. An earthen pot (containing little amount of old *Toddy or Tari*) is placed in the tree to collect the sap. It takes only 6–8 h to complete the fermentation. The fresh sap contains sucrose (12–15% by weight) and other monosaccharides which favour the growth of fermenting microflora. The shelf life of this drink is only 12–24 h. After fermentation, a white colour sweet material is consumed with some spicy vegetables.

Socio-economy and ethnical or religious values The product is consumed as a refreshing drink in this region.

Microbiology Acetic acid bacteria and lactic acid bacteria, specifically *Leuconostoc* (*Leuc.*) sp., *Lactobacillus* sp. and *Zymomonas* sp., were involved in the

Fig. 23.19 Traditional method of *Mahua* preparation in West Bengal and Odisha

Mixing of mahul flower with the starter, *Bakhar*

↓

Mixing of traditional starter, *Bakhar* with the boiled rice

↓

Fermentation for 5-10 days at room temperature

↓

Distilled by traditional process

↓

Mahua or *Mahuli*

early fermentation (Steinkraus 1996). Later *S. cerevisiae* produced the alcohol (Sekar and Mariappan 2007).

Nutritional value The alcohol percentage is only 4–6%. *Toddy* is believed to be good for the health particularly for eyesight and also used as a sedative (Sekar and Mariappan 2007). Moreover, it is a mild laxative relieving constipation (Sekar and Mariappan 2007). It is prescribed as a tonic for those recovering from diseases such as chickenpox (Steinkraus 1996).

Commercialization The shelf life of *Toddy or Tari* is very short (6–12 h). Hence, it is sold immediately after processing. The price of 250 mL of *Toddy or Tari* is Rs. 15–25.

23.4.3 *Mahua* or *Mahuli*

Mahua (West Bengal) or *Mahuli* (Odisha) is a distilled fermented beverage made from mahul flowers (Ray and Swain 2013).

Flow sheet of traditional method of *Mahua* preparation in West Bengal and Odisha is presented in Fig. 23.19.

Culinary and mode of consumption In its preparation, the mahul flower is washed and mixed with the starter *Bakhar* (plant-based starter) (Dhal et al. 2010) and an adequate amount of water. Generally, 10 kg of *Mahua* flowers is mixed with 300 g of *Bakhar* and 32 litres of water (Behera et al. 2016). In another method, 4 kg of *Mahua* flowers are mixed with 6 kg molasses, 300 g of *Bakhar* and 32 litres of water (Behera et al. 2016). After fermentation, the product is distilled by the traditional distillation process (Ray and Swain 2013).

Socio-economy and ethnical or religious values This product is popular among the tribal people of this region. During any types of occasions, consumption of *Mahua* (West Bengal) or *Mahuli* is traditional culture. It is consumed as a refreshing drink.

Microbiology Unknown.

Nutritional value The distilled product is diluted with water, and the alcohol percentage is 10–15% (Ray and Swain 2013).

Fig. 23.20 Traditional method of *Aamer achar* preparation in West Bengal and Odisha

Mangoes

↓

Washing, peeling, and slicing of mangoes

↓

Cooking the slices with salt and spices

↓

Putting the mixture in jar

↓

Keeping the jar in sun for 1-2 weeks
(shaking the jar at least twice a day)

↓

Storage at ambient temperature
(cool and dry place)

↓

Mango pickle (*Aamer Achar*)

Optimization and commercialization The product is usually bottled in glass or plastic. Since it is popular among the tribal community, it is sold only in the tribal population-rich location.

23.5 Plant-Based Fermented Products

23.5.1 *Achar/Chadnee*

Pickle (*achar*) is one of the oldest methods of food preservation. Pickle is usually made from chopped fruits (cucumber, olive, tomato, mango, lemon) and vegetables (cabbage, cauliflower, mustard vegetable). The name of the pickle is given based to their substrate such as mango pickle (*Aamer achar* in Bengali), lemon pickle (*lebur achar*), tomato pickle (*tomato chadnee*), etc. The most common *pickle*s in this region are mango or lemon *pickle*.

Flow sheet of traditional method of *Aamer achar* preparation in West Bengal and Odisha is presented in Fig. 23.20.

Culinary and mode of consumption The preparation process of mango pickle is shown in Fig. 23.20. At first, the chopped fruits and vegetables are lightly fried in oil. Then salt, chilli, turmeric and some other spices are added there. After that, the mixture is kept in a container and allowed to ferment for 1–2 weeks. *Pickle* is mainly two types: sour (Fig. 23.21a) and sweet (Fig. 23.21b). In a sweet pickle, a high amount of sugar is added.

Socio-economy and ethnical or religious values *Pickle* is popular among all age group people. At the end of a heavy meal, the consumption of *pickle* or *chadnee* is an ancient culture.

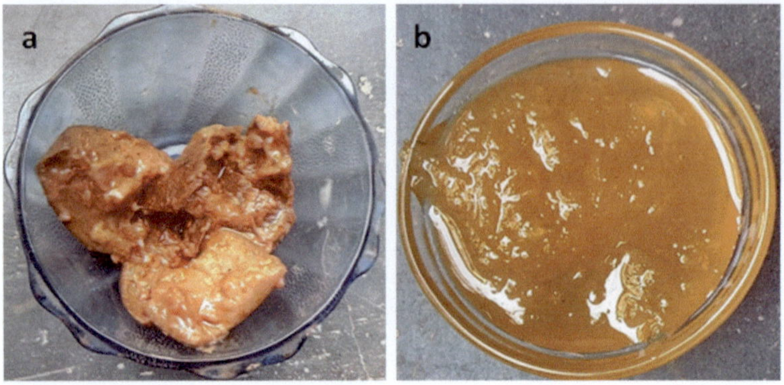

Fig. 23.21 Sour *mango pickle* (**a**) and sweet *pickle* (**b**)

Microbiology The naturally occurring lactic acid bacteria actively participate in this fermentation.

Nutritional value The bacteria produce lactic acid and as a result, the *pickle* becomes acidic (sour). *Pickle* is enriched with dietary fibres, minerals and vitamins. It exerts the health benefits associated with lactic acid bacteria.

Optimization and commercialization: Due to the low pH, high salt concentration and low moisture content, the *pickle* can be stored 6 months to 1 year. Nowadays, several small-scale industries have started producing and marketing *pickle* considering its popularity.

23.5.2 *Karadi*

Karadi is an ethnic fermented bamboo shoot (*Bambusa arundinacea* L.) product of Odisha (Ray and Swain 2013).

Flow sheet of traditional method of *Karadi* preparation in Odisha is presented in Fig. 23.22.

Culinary and mode of consumption The tip of the youngling bamboo is sliced into small pieces and dipped into water for 1 day for fermentation (Ray and Swain 2013). After fermentation, the product is cooked with the other vegetables. Sometimes the powdered fermented product is sun-dried and stored for 1 year (Panda and Padhy 2007).

Socio-economy and ethnical or religious values The product is used in the treatment of digestive problems, especially in constipation (Satish Kumar et al. 2013).

Microbiology Unknown.

Nutritional value Unknown.

Optimization and Commercialization Unknown.

Fig. 23.22 Traditional method of *Karadi* preparation in Odisha

Collection of the tip of youngling bamboos

↓

Slicing into pieces

↓

Fermentation for 1 day by dipping into water

↓

Karadi

23.6 Dairy-Based Fermented Products

Dairy-based fermented foods in West Bengal and Odisha are usually naturally fermented. The ethnic people use the black-slopping method where the previously fermented product is used as an inoculum to ferment the new batch (Josephsen and Jespersen 2004). The examples of some dairy-based products in this region are *dahi* (curd), *paneer*, *lassi* and *chhena poda*. These dairy-based foods and beverage are rich in proteins, carbohydrates, vitamins and minerals and exert different therapeutic activities (Sarkar 2008).

23.6.1 *Dahi*

Dahi is naturally fermented milk (lactic acid fermentation) consumed as a refreshing food (Tamang et al. 2016). The word *Dahi* comes from Sanskrit word "dadhi" (Yegna Narayan Aiyar 1953).

Flow sheet of traditional method of *Dahi*preparation in West Bengal and Odisha is presented in Fig. 23.23.

Culinary and mode of consumption In its traditional preparation, milk (from cow or buffalo) is boiled and cooled at room temperature. A small amount of old *dahi* (served as inoculum) is added on that and allowed to ferment for 6–12 h.

Socio-economy and ethnical or religious values *Dahi* (Fig. 23.24) is consumed on a regular basis by the people of this region. The people usually offer *Dahi* to god during worship. It is believed that *Dahi* is a holy product.

Microbiology A number of lactic acid bacteria (*Lb. alimentarius*, *Lb. bifermentans*, *Lb. paracasei*, *Lb. delbrueckii* subsp. *indicus*, *Lb. acidophilus*, *Lb. bulgaricus*, *Lb. cremoris*, *Lb. helveticus*, *Lb. fermentum*, *Lactococcus* (*Lc.*) *lactis*, *Streptococcus lactis*, *Streptococcus cremoris*, *Streptococcus thermophilus*, *Pd. acidilactici*, *Pd. pentosaceus*, *Weissella cibaria*) and yeast (*Saccharomycopsis* sp. and *Candida* sp.) were isolated from *Dahi* (Dewan and Tamang 2007; Patil et al. 2010; Ramakrishnan 1980).

Nutritional value *Dahi* is nutritious and easy to digest. The number of vitamins (thiamine, riboflavin, folic acid, niacin) (Atreja and Deodhar 1987; Bambha et al. 1973; Laxminarayana and Shankar 1980; Singh and Deodhar 1993), proteins, essential amino acids and lactic acid was significantly increased in *dahi* (Boghra 1988; Sarkar 2008). *Dahi* exerted the antimicrobial effects against the pathogenic

Fig. 23.23 Traditional method of *Dahi* preparation in West Bengal and Odisha

Milk (from cow or buffalo)
↓
Boiling of milk
↓
Cooling down at room temperature
↓
Addition of old *dahi* (served as inoculum)
↓
Fermentation for 6 - 12 h at room temperature
↓
Dahi

Fig. 23.24 *Dahi*

bacteria due to the high content of lactic acid and production of antimicrobial agents such as bacteriocin, antimicrobial peptides by the participating lactic acid bacteria (Dave et al. 1992; Srinivasan et al. 1995). It is also effective in reducing cardiovascular diseases and tumours and increasing immunity (Sinha and Sinha 2000).

Optimization and commercialization Due to its huge consumption, *Dahi* is produced in the home and sweet shops regularly. The sweet shop keeper uses earthen pot for preparation of this product. Each shop keeper has its own starter culture for *Dahi* production. Hence, the sensory properties are different from one shop to another. The cost of *Dahi* is Rs. 15–25 per 100 g.

23.6.2 *Misti Dahi*

Misti dahi is a sweetened fermented milk product prepared mainly in West Bengal. It is an indigenous product of West Bengal. It is mildly acidic, a sweetened product which appears in a light brown (like caramel) colour.

Fig. 23.25 Traditional method of *misti dahi* preparation in West Bengal

Milk (from cow or buffalo)

↓

Boiling of milk with sugar (or sugarcane)
for 2 - 3 h in low flame

↓

Cooling down at room temperature

↓

Addition of old *misti* *dahi* (served as inoculum)

↓

Fermentation for 6 - 12 h at room temperature

↓

Misti dahi

Flow sheet of traditional method of *misti dahi* preparation in West Bengal is presented in Fig. 23.25.

Culinary and mode of consumption Traditionally, the milk is boiled with sugar (sometimes sugarcane) for 2–3 h in low flame. The boiling helps to develop the unique red-brown colour and makes it slightly viscous. As like the *dahi* preparation, after cooling the milk, previously prepared old *misti dahi* is added and allow fermenting for 6–12 h in the earthen pot at room temperature.

Socio-economy and ethnical or religious values As like *dahi*, *misti dahi* (Fig. 23.26) is also offered to god. In many occasions in West Bengal, *misti dahi* is served after the lunch.

Microbiology *Lb. acidophilus*, *Lc. lactis* subsp. *lactis*, *Lb. delbrueckii* subsp. *bulgaricus*, *Streptococcus salivarius* subsp. *thermophilus* and *S. cerevisiae* were isolated from *misti dahi* (Ghosh and Rajorhia 1990; Gupta et al. 2000).

Nutritional value It is believed that the *misti dahi* is highly energy-rich food (due to the high sugar content), and it protects the individual from different gastrointestinal ailments.

Optimization and commercialization Usually *misti dahi* is prepared in the sweet shops. Nowadays, several companies started producing *misti dahi* commercially. The approximate cost of the product is Rs. 20–30 per 100 gram.

23.6.3 *Bandel Cheese*

Bandel cheese is an indigenous unripe, salted soft variety of *cheese*. The name is derived from the place Bandel located near Kolkata, West Bengal. The *Bandel cheese* is unique due to its dry, crumbly, smoky and aromatic flavour.

Flow sheet of traditional method of *Bandel cheese* preparation in West Bengal and Odisha is presented in Fig. 23.27.

Culinary and mode of consumption There are two varieties of *Bandel cheese* – plain and smoked. The raw ingredient is cow milk. After preparing the curds, the

Fig. 23.26 *Misti dahi* which appears in a light brown (like caramel) colour

whey is separated by the addition of lemon juice. Then it is moulded and drained in small containers. It is further smoked in the fire as required for smoked *cheese* preparation. The finished *Bandel cheese* is flat round shaped. The *cheese* is salted for long preservation. Just before the consumption, it is soaked in water overnight to remove the excess salt.

Socio-economy and ethnical or religious values This product is not the indigenous product of West Bengal and Odisha. This unique art of cheesemaking was probably introduced by the Portuguese during their colonization in the late sixteenth century.

Microbiology Although it is a popular cheese product in the Kolkata region, there are no scientific evidence about its microbial and nutritional compositions. As *dahi* is one of the important ingredients in *Bandel cheese*, it is likely that the lactic acid bacteria participate in the *dahi* fermentation may engage in the *Bandel cheese* fermentation. Clearly, a detailed study is needed to explore this traditional *cheese* product.

Optimization and Commercialization The product is mainly popular in the Bandel region of West Bengal. There are some shops in Bandel preparing and selling this unique kind of *cheese*.

23.6.4 *Lassi*

Lassi is a dairy-based low-fat beverage which is consumed in the summer as a refreshing drink.

Flow sheet of traditional method of *lassi* preparation in West Bengal and Odisha is presented in Fig. 23.28.

Culinary and mode of consumption It is prepared by mixing *dahi* with water, salt, sugar and rose water (Padghan et al. 2015). There are two main variants of *lassi*: *sweetened lassi and* salted *lassi. Sweetened lassi* (Fig. 23.29) *is prepared by mixing dahi with sugar syrup and flavour* (Shuwu et al. 2011). For *salted lassi*, an adequate amount of salt is added instead of sugar.

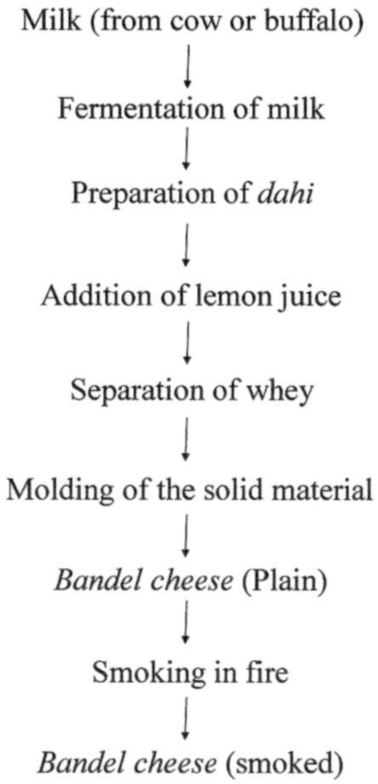

Fig. 23.27 Traditional method of *Bandel cheese* preparation in West Bengal and Odisha

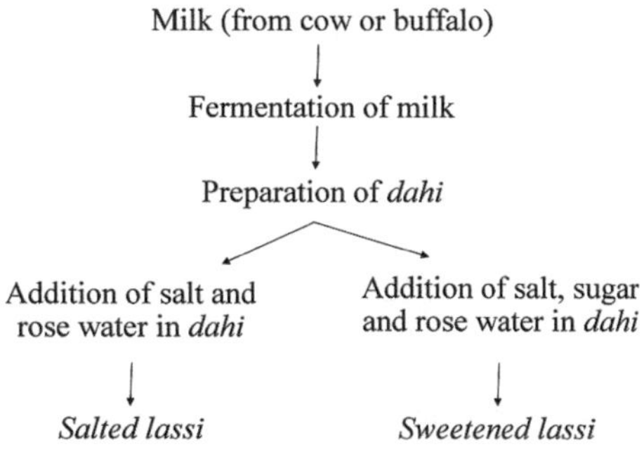

Fig. 23.28 Traditional method of *lassi* preparation in West Bengal and Odisha

Socio-economy and ethnical or religious values There is an ancient culture of preparing of *lassi* in this region. The people usually drink it after doing heavy work. There is a ritual to offer *lassi* to the guests who visit the host's home in the summer time.

Microbiology Lactic acid bacteria.

Nutritional value The health benefits of *lassi* mainly come from the bioactive components present in *dahi* (due to the action of lactic acid bacteria) and spices. It is a nutritive (amino acids, peptides, vitamins, minerals, etc.) and digestive drink (Padghan et al. 2015). It is effective in gastrointestinal ailments such as diarrhoea, dysentery, etc. (Padghan et al. 2015; Tamang et al. 2016). Nowadays, it is designated as a probiotic drink as the participating bacteria have been established as the probiotic bacteria (Patidar and Prajapati 1998).

Fig. 23.29 *Lassi*

Optimization and commercialization Many small shops prepare this drink during summer. Nowadays, some companies also started packaging *lassi* in a bottle and sell in the market. The cost of 500 mL of lassi is Rs. 30–50.

23.6.5 *Chhena Poda*

Chhena (*cheese*) *poda* (*baked*) is a quintessential burnt cottage cheese dessert of Odisha. It is originated in Nayagarh (a town in Odisha) in the first half of the twentieth century.

Flow sheet of traditional method of *Chhena poda* preparation in West Bengal and Odisha is presented in Fig. 23.30.

Culinary and mode of consumption In its preparation, the homemade fresh cheese is mixed with sugar, cashew nut and raisins and is wrapped by the leaves of the *sal* tree (http://www.uppercrustindia.com/oldsite/12crust/twelve/season8.htm). Then it is baked in a charcoal oven until it becomes brown in colour (Fig. 23.31). It has a shelf life of 5 days. The food is very popular due to its unique flavour which probably comes during baking.

Socio-economy and ethnical or religious values Unknown.

Microbiology Unknown.

Nutritional value The food contains a significant amount of protein, fat, calcium and vitamins. It is believed that the food strengthens the bones and good for digestion. However, the best of our knowledge, this food has been scientifically unexplored till date.

Optimization and commercialization Some sweet shops in West Bengal and Odisha produce and sell this product. The cost is Rs. 500 per kg.

Fig. 23.30 Traditional method of *Chhena poda* preparation in West Bengal and Odisha

Milk (from cow or buffalo)
↓
Boiling of milk in low flame for 1 h
↓
Addition of lime and stirring
↓
Curdling of the milk
↓
Separation of whey by using fine cloth
↓
Collection of the solid material– *Chenna*
↓
Washing with cold water for several times
↓
Addition of sugar, chopped dry fruits with Chenna and mixing it properly
↓
Baking the mixture for 30 - 40 min in an oven
↓
Chenna poda

Fig. 23.31 *Chhena poda*

23.7 Fish-Based Fermented Products

Fish is an important component of the human diet in West Bengal and Odisha. More than 90% of the population in this region consume fish. As fish is a perishable food, drying of fish can preserve it for a long time. In its preparation, salt is applied to the fishes (3–4 kg/100 kg of raw fish) to remove the water (Payra et al. 2016). It takes 7–10 days for complete drying. When the salt is added into fish, a certain degree of fermentation occurs (Panda et al. 2017) and that eventually increases the nutritive value and the sensory properties.

23.7.1 Dry Salted *Hilsa*

During the rainy season, a large amount of *hilsa* (*Tenualosa ilisha*) fish is caught. To preserve the excess *hilsa*, it is salted and dried.

Flow sheet of traditional method of dry salted *hilsa* preparation in West Bengal and Odisha is presented in Fig. 23.32.

Culinary and mode of consumption Salt is added to *hilsa* fish followed by fermentation in a brine solution for a period of 4–6 months (Ray and Swain 2013). Then the fishes are dried for 10–15 days. The dried fishes can be preserved for 1–2 years (Ray and Swain 2013). Dried *hilsa* appears in pink glossy colour.

Socio-economy and ethnical or religious values This product is very popular in this region and India due to its unique flavour, aroma and texture (Ray and Swain 2013).

Microbiology The microbial and nutritional quality of this food has been unexplored till now.

Nutritional value Unknown.

Optimization and commercialization Due to its long shelf life, the product is exported in other countries and sold in another part of India.

23.7.2 Dry Salted *Khainga*

Dry salted *Khainga* is a popular fish product in Odisha.

Flow sheet of traditional method of dry salted *Khainga* preparation in West Bengal and Odisha is presented in Fig. 23.33.

Culinary and mode of consumption The preparation procedure of dry salted *Khainga* (*Mugil cephalus*) fish is similar to dry salted *hilsa* (Ray and Swain 2013). The fermentation takes place after removing the water from the fish. Like dried *hilsa*, the fermentation time is 4–6 months and drying time is 10–15 days (Ray and Swain 2013). This fermented fish is popular in Odisha due to its unique flavour and taste.

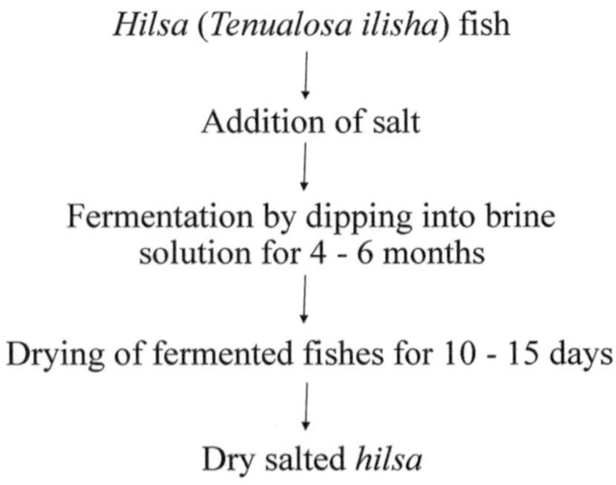

Fig. 23.32 Traditional method of dry salted *hilsa* preparation in West Bengal and Odisha

Fig. 23.33 Traditional method of dry salted *Khainga* preparation in West Bengal and Odisha

Khainga (*Mugil cephalus*) fish

↓

Addition of salt

↓

Fermentation by dipping into brine solution for 4 - 6 months

↓

Drying of fermented fishes for 10 - 15 days

↓

Dry salted *Khainga*

Socio-economy and ethnical or religious values Unknown.

Microbiology Considering the microorganisms involved in similar products (*karati*, *bordia* and *lashim*) in Assam (Thapa et al. 2007), it can be hypothesized that the lactic acid bacteria and *Bacillus* sp. might have significant impacts in this fermentation.

Nutritional value It was also reported that the drying process can increase the protein, lipids, vitamins and minerals (Chukwu 2009).

Optimization and commercialization Like dry salted *hilsa*, dry salted *Khainga* is also exported to other countries. The approximate cost of 1 kg of this product is Rs. 200.

23.7.3 *Sukuti* or *Shutki*

Sukuti or *Shutki* is a very popular sun-dried fish product in West Bengal.

Flow sheet of traditional method of *Shutki* preparation in West Bengal is presented in Fig. 23.34.

Culinary and mode of consumption Fishes are collected, washed, rubbed with salt and dried in the sun for 4–7 days (Thapa 2016). The product can be stored for 3–4 months (Thapa 2016). *Sukuti* is consumed as soup and curry.

Microbiology Bacteria (*Lb. lactis* subsp. *cremoris, Lb. lactis subsp. lactis, Lb. plantarum, Lb. mesenteroides, Enterococcus faecium, Enterococcus faecalis, Pd. pentosaceus) and yeasts (Candida chiropterorum, C. bombicola and Saccharomycopsis* spp.*)* were isolated from Sukuti prepared in other parts of India (Rapsang et al. 2011; Thapa 2016). Considering this, it can be assumed that similar bacteria might be involved in the Shutki fermentation in West Bengal and Odisha.

Nutritional value Not reported.

Optimization and commercialization *Sukuti* is exported to other countries. The approximate cost is Rs. 200–300 per kg.

Fig. 23.34 Traditional method of *Shutki* preparation in West Bengal

Collection and washing of fishes

↓

Rubbing with salt

↓

Drying of fishes in sun for 4 - 7 days

↓

Sukuti or *Shutki*

23.8 Conclusion

The large variety of traditional fermented foods in West Bengal and Odisha is due to the climatic variation and cultural diversity. These factors eventually influence in crop and microbes variety. Most of these fermented products are prepared by using the indigenous knowledge which people acquired from their ancestors. In most of the cases, they are completely unaware of the participating microbes and their effects in fermentation. During preparation of this manuscript, we were surprised to know that some of the fermented foods and beverages produced in this region have been scientifically unexplored or partially characterized. Moreover, due to urbanization and modernization, traditional knowledge is ignored. Hence, it is likely that very soon the indigenous traditional knowledge will be disappeared; modern civilization will be unaware of the hidden health benefits of these foods. Considering this, it is necessary to explore these foods scientifically (biochemical, microbiological and nutritive aspects) and protect them by legal tools like IPR (intellectual property rights). Since the people unknowingly use microbes and follow the old age methods, there is a demand for standardization of these foods by modern science and that may be helpful to increase the functionality of the ethnic foods.

References

Adak A, Parua S, Maity C, Ghosh K, Halder SK, Das Mohapatra PK, Mondal KC (2013) Potentialities of newly isolated *Bacillus subtilis* and *Lactobacillus* sp. for curd preparation and a comparative study of its physico-chemical parameters with other marketed curds. Indian J Exp Biol 51:910–918

Aidoo KE, Rob Nout M, Sarkar PK (2006) Occurrence and function of yeasts in Asian indigenous fermented foods. FEMS Yeast Res 6:30–39

Atreja S, Deodhar A (1987) Nutrient status during shrikhand making. J Food Sci Technol 24:266–267

Bambha P, Setty PS, Nambudripad V, Laxminarayana H (1973) Changes in vitamin B complex content of milk during the production of dahi (fermented milk). Indian J Anim Sci 43:210–215

Munde Balasaheb Bansi (2015) Quality evaluation of lassi sold in Kolhapur city. M.Sc. thesis, Mahatma Phule Krishi Vidyapeeth, Ahmednagar, p 73

Behera S, Ray R, Swain M, Mohanty R, Biswal AK (2016) Traditional and current knowledge on the utilization of Mahua (L.) flowers *Madhuca latifolia* by the Santhal tribe in Similipal biosphere reserve, Odisha, India. Ann Trop Res 38:94–104

Boghra V (1988) Role of physico-chemical status of major minerals and trace elements in selected indigenous milk products. Doctoral dissertation, National Dairy Research Institute

Chandra D (2001) Analgesic effect of aqueous and alcoholic extracts of *Madhuka Longifolia* (Koeing). Indian J Pharmacol 33:108–111

Choi J-S, Kim J-W, Cho H-R, Kim K-Y, Lee J-K, Ku S-K, Sohn JH (2014) Laxative effects of fermented rice extract (FRe) in normal rats. Toxicol Environ Heal Sci 6:155–163

Choudhury S (2013) Ancient Odisha was centre of Aryans. https://www.dailypioneer.com/2013/state-editions/ancient-odisha-was-centre-of-aryans.html. Accessed 14 Mar 2019

Chukwu O (2009) Influences of drying methods on nutritional properties of tilapia fish (*Oreochromis nilotieus*). World J Agric Sci 5:256–258

Dave R, Dave J, Sannabhadti S (1992) Antibacterial activity of buffalo milk dahi prepared using *S. thermophilus* strains as starter culture. Indian J Dairy Sci 45:607–607

Dewan S, Tamang JP (2007) Dominant lactic acid bacteria and their technological properties isolated from the Himalayan ethnic fermented milk products. Antonie Van Leeuwenhoek 92:343–352

Dhal N, Pattanaik C, Reddy CS (2010) Bakhar starch fermentation, a common tribal practice in Orissa. Indian J Tradit Knowl 9:279–271

FAO (1993) Rice in human nutrition. In: Juliano BO (ed) FAO food and nutrition series, No. 26. In collaboration with International Rice Research Institute (IRRI), Los Banos, Laguna (Philippines), pp 35–48

Ghosh J, Rajorhia G (1990) Selection of starter culture for production of indigenous fermented milk product (Misti dahi). Lait 70:147–154

Ghosh K, Maity C, Adak A, Halder SK, Jana A, Das A, Parua S, Das Mohapatra PK, Pati BR, Mondal KC (2014) Ethnic preparation of Haria, a rice-based fermented beverage, in the province of lateritic West Bengal, India. Ethnobot Res Appl 12:39–49

Ghosh K, Ray M, Adak A, Dey P, Halder SK, Das A, Jana A, Parua S, Das Mohapatra PK, Pati BR, Mondal KC (2015a) Microbial, saccharifying and antioxidant properties of an Indian rice based fermented beverage. Food Chem 168:196–202

Ghosh K, Ray M, Adak A, Halder SK, Das A, Jana A, Parua S, Vágvölgyi C, Das Mohapatra PK, Pati BR, Mondal KC (2015b) Role of probiotic *Lactobacillus fermentum* KKL1 in the preparation of a rice based fermented beverage. Bioresour Technol 188:161–168

Gode P (1943) Some notes on the history of Indian dietetics with special reference to the history of jalebi. New Indian Antiques 6:169–181

Goswami G, Baruah H, Boro RC, Barooah M (2016) Fermentation reduces anti-nutritional content and increases mineral availability in poita bhat. Asian J Chem 28:1929–1932

Gupta R, Bimlesh M, Joshi V, Prasad D (2000) Microbiological, chemical and ultrastructural characteristics of Mishti Doi (sweetened Dahi). J Food Sci Technol 37:54–57

Hu FB (2003) Plant-based foods and prevention of cardiovascular disease: an overview. Am J Clin Nutr 78:544S–551S

Inbasekar K (2014) Pulses production in India: challenges and strategies. Econ Aff 59:403

Josephsen J, Jespersen L (2004) Starter cultures and fermented products. In: Hui YH, Meunier-Goddik L, Hansen ÅS, Josephsen J, Nip WK, Stanfield PS, Toldrá F (eds) Handbook of food and beverage fermentation technology. CRC Press, Boca Raton, pp 23–49. ISBN: 9780203913550

Karthikeyan R, Suresh Kumar K, Singaravadivel K, Alagusundaram K (2014) Volatile elements of coconut toddy (*Cocos Nucifera*) by gas chromatography–mass spectrometry. J Chromatogr Sep Tech 5:2

Katiyar S, Tandon M, Chandekar A, Upamanyu N (2011) Pharmacognostic standardization, phytochemical investigation and the anthelmintic evaluation of the extract of *Madhuca indica* JF GMEL flowers. Pharmacology 3:892–903

Khan R (2017) The Tibeto-Burman and Austro-Asiatic ancestry of Bengalis. https://www.gnxp.com/WordPress/2017/10/07/the-tibeto-burman-and-austro-asiatic-ancestry-of-bengalis/. Accessed 14 Mar 2019

Kulkarni S, Manan J, Agarwal M, Shukla I (1997) Studies on physico-chemical composition, packaging and storage of blackgram and greengram wari prepared in Uttar Pradesh. J Food Sci Technol 34:119–122

Laxminarayana H, Shankar P (1980) Fermented milk in human nutrition. Indian Dairyman 32:121–129

Nair BM, Prajapati JB (2008) The history of fermented foods. In: Farnworth ER (ed) Handbook of fermented functional foods. CRC Press, Boca Raton, pp 17–42. ISBN: 9781420053265

Ofuya Z, Akhidue V (2005) The role of pulses in human nutrition: a review. J Appl Sci Environ Manag 9:99–104

Padghan P, Mann B, Sharma R, Kumar A (2015) Studies on bio-functional activity of traditional Lassi. Indian J Tradit Knowl 1:124–131

Panda T, Padhy RN (2007) Sustainable food habits of the hill-dwelling Kandha tribe in Kalahandi district of Orissa. Indian J Tradit Knowl 6:103–105

Panda SH, Ray RC, El Sheikha AF, Montet D, Worawattanamateekul W (2017) Fermented fish and fish products: an overview. In: Montet D, Ray RC (eds) Aquaculture microbiology and biotechnology. CRC Press, Boca Raton, pp 132–172. ISBN: 9781138113756

Patidar S, Prajapati J (1998) Standardisation and evaluation of lassi prepared using *Lactobacillus acidophilus* and *Streptococcus thermophilus*. J Food Sci Technol 35:428–431

Patil MM, Pal A, Anand T, Ramana K (2010) Isolation and characterization of lactic acid bacteria from curd and cucumber. Indian J Biotechnol 9:166–172

Payra P, Maity R, Maity S, Mandal B (2016) Production and marketing of dry fish through the traditional practices in West Bengal coast: problems and prospect. Int J Fish Aquat Stud 4:118–123

Rahi D, Soni S (2007) Applications and commercial uses of microorganisms. In: Soni SK (ed) Microbes: a source of energy in twenty-first century. Jai Bharat Printing Press, Delhi, pp 71–126. ISBN: 9788189422141

Ramakrishnan C (1980) Studies on Indian fermented foods. Baroda J Nutr 6:1–54

Rapsang GF, Kumar R, Joshi S (2011) Identification of *Lactobacillus pobuzihii* from tungtap: a traditionally fermented fish food, and analysis of its bacteriocinogenic potential. Afr J Biotechnol 10:12237–12243

Ray RC, Swain MR (2013) Indigenous fermented foods and beverages of Odisha, India: an overview. In: Joshi VK (ed) Indigenous fermented foods of South Asia. CRC Press, Boca Raton. ISBN: 9781439887837

Ray S, Raychaudhuri U, Chakraborty R (2015) Rice-, pulse-, barley-, and oat-based fermented food products. Cereal Foods World 60:218–223

Ray M, Ghosh K, Singh S, Mondal KC (2016) Folk to functional: an explorative overview of rice-based fermented foods and beverages in India. J Ethnic Foods 3:5–18

Renu R, Waghray K (2016) Development of papads: a traditional savoury with purslane, *Portulaca oleracea*, leaves. Health Scope 5:e26735

Roy A, Moktan B, Sarkar PK (2007) Traditional technology in preparing legume-based fermented foods of Orissa. Indian J Tradit Knowl 6:12–16

Sahoo S, Lenka C, Biswal G (2017) Knowledge and awareness about health benefits of indigenous fermented foods: a comprehensive study. Int J Food Sci Nutr 2:2

Sarkar S (2008) Innovations in Indian fermented milk products—a review. Food Biotechnol 22:78–97

Satish Kumar R, Kanmani P, Yuvaraj N, Paari K, Pattukumar V, Arul V (2013) Traditional Indian fermented foods: a rich source of lactic acid bacteria. Int J Food Sci Nutr 64:415–428

Sekar S, Mariappan S (2007) Usage of traditional fermented products by Indian rural folks and IPR. Indian J Tradit Knowl 6:111–120

Sha SP, Ghatani K, Tamang JP (2013) Dalbari, a traditional pulses based fermented food of west Bangal. Int J Agric Food Sci Technol 4:6–10

Sharma N, Handa S, Gupta A (2013) Comprehensive study of different traditional fermented foods/beverages of Himachal Pradesh to evaluate their nutrition impact on health and rich biodiversity of fermenting microorganisms. Int J Res Appl Nat Soc Sci 1:19–28

Shurpalekar S (1986) Papads. In: Reddy N, Pierson M, Salunkhe D (eds) Other legume-based fermented foods. CRC Press, Boca Raton, pp 191–217

Shuwu M, Ranganna B, Suresha K, Veena R (2011) Development of value added lassi using honey. Mysore J Agric Sci 45:757–763

Singh R, Deodhar A (1993) Quantitative assessment of total, bound and free forms of thiamine and riboflavin in cow milk and curd. Indian J Dairy Sci 46:525–525

Sinha P, Sinha R (2000) Importance of good quality dahi in food. Indian Dairyman 52:45–47

Srinivasan R, Sarkar S, Pramanik K, Kuila R, Misra A (1995) Isolation and characterization of *Lactococcus Species* producing bacteriocins. Indian J Dairy Sci 48:596–602

Steinkraus K (1996) Handbook of indigenous fermented food. CRC Press, Taylor & Francis Group, New York, p 792. ISBN: 0824793528

Tamang JP (2009) Himalayan fermented foods: microbiology, nutrition, and ethnic values. CRC Press, Taylor & Francis Group, New York, p 295. ISBN: 9781420093247

Tamang JP, Thapa N, Bhalla TC (2016) Ethnic fermented foods and beverages of India. In: Tamang JP (ed) Ethnic fermented foods and alcoholic beverages of Asia. Springer, New Delhi, pp 17–72. ISBN: 978-81-322-2800-4

Tewary H, Muller H (1992) The fate of some oligosaccharides during the preparation of wari, an Indian fermented food. Food Chem 43:107–111

Thapa N (2016) Ethnic fermented and preserved fish products of India and Nepal. J Ethnic Foods 3:69–77

Thapa N, Pal J, Tamang JP (2007) Microbiological profile of dried fish products of Assam. Indian J Fisheries 54:121–125

Tiwari S, Mahanta D (2007) Ethnological observations fermented food products of certain tribes of Arunachal Pradesh. Indian J Tradit Knowl 6:106–110

Umadevi M, Maheswari C, Jothi R, Paleti SK, Reddy YS, Narayanan RV (2011) Hepatoprotective activity of flowers of *Madhuca longifolia* (Koen.) Macbr. Against paracetamol-induced hepatotoxicity. Res J Pharm Technol 4:259–262

Verma KS, Neerja S, Rajshree S, Abha A (2010) Phytochemical screening and therapeutic profiling of *Madhuca indica* JF Gmel. Vegetos 23:109–115

Yegna Narayan Aiyar A (1953) Dairying in ancient India. Indian Dairyman 5:77